Periglacial Landscapes of Europe

Marc Oliva · Daniel Nývlt ·
José M. Fernández-Fernández
Editors

Periglacial Landscapes of Europe

Editors
Marc Oliva
University of Barcelona
Barcelona, Spain

José M. Fernández-Fernández
Department of Geography
Universidad Complutense de Madrid
Madrid, Spain

Daniel Nývlt
Department of Geography
Masaryk University
Brno, Czechia

ISBN 978-3-031-14897-2 ISBN 978-3-031-14895-8 (eBook)
https://doi.org/10.1007/978-3-031-14895-8

Cover illustration: "Relict rock glacier at Clòt Der Os (Tuc de Bacivèr, Central Pyrenees, Spain)", © Julia García-Oteyza

This Springer imprint is published by the registered company Springer Nature Switzerland AG
The registered company address is: Gewerbestrasse 11, 6330 Cham, Switzerland

Foreword: Periglaciation Past and Present

It is about time we renew our recognition that history matters. During the last half of the twentieth century, periglacial geomorphology evolved from a discipline integral to the general preoccupation of the earth sciences with the history of Earth to a process-based and process-monitoring study of environments undergoing seasonal freezing and thawing (Church 2017). The field left behind largely descriptive work to develop quantitative measurements, experimental techniques and analytical statements based upon physical principles (e.g., Lachenbruch 1962; Seppälä 1982; Mackay 1997). The time scales of interest for landform development in the periglacial realm range from a few hours, in the formation of needle ice (Outcalt 1970), to several centuries for the development of pingos (Mackay 1998). The field now includes physically based consideration of the active layer and permafrost.

For periglacial geomorphology, the transition from descriptive to quantitative research began about 1960. It is well illustrated by P.J. Williams's work on slope movement (1959) and J.R. Mackay's (1963) memoir on the Mackenzie Delta. These publications illustrate the application of *immanent*, i.e., universal, principles to understanding of the periglacial environment: in Williams's case it was the use of soil mechanics and for Mackay, hydraulics. These principles are inherently ahistorical. In many cases, especially over times scales of a few decades, immanent approaches are useful, as demonstrated by successful infrastructure engineering. Nevertheless, there are reported circumstances where environmental history has left a legacy that modifies infrastructure performance, perhaps catastrophically (e.g., Skempton et al. 1991). Simpson (1963) referred to contingent aspects of a landscape as comprising the unique circumstances, or *configuration*, that modify the choice of appropriate immanent principles to be applied to specific problems. He was thinking in terms of the time scales associated with geology, but the argument is just as sound when considered in the context of geomorphology. Configuration is inherently historical for it is the legacy of past events.

Configuration is a key element in understanding the periglacial environment, for ground-ice conditions are always a product of site history and they control the spatial variation of terrain responses to surface disturbance, including climate change (Mackay 1970; Kokelj et al. 2017; Burn et al. 2021). Similarly, soil texture principally governs the frost susceptibility of a soil, its capacity for frost heave (Burt and Williams 1976) and hence the development of hummocks and sorted circles (Mackay 1980; Hallet et al. 1988).

In North America and Russia, periglacial studies have been dominated by research on the permafrost environment, while in China, periglacial research on the Qinghai-Tibet Plateau has become very active in the last 25 years. In Europe, periglacial studies have been in progress for over a century and are the most catholic, ranging from investigations of active periglacial processes, including freeze-thaw shattering, solifluction, cryoturbation and permafrost creep in Scandinavia and the high mountains of the continent (e.g., Delaloye et al. 2010; Deprez et al. 2020), through sedimentary research on Quaternary stratigraphy (e.g., Murton and Kolstrup 2003), to the critical legacy of periglacial environments on *in situ* engineering behaviour of soils and on other elements of geomorphology (Hutchinson 1991; Ballantyne and Harris 1994). Europe is in some ways the periglacial continent. It most certainly is for those who carry what T.S. Eliot called the *historical sense*, that is an awareness of the presence of the past in what we see and encounter today.

At present, permafrost science and engineering are preoccupied with questions concerning thawing of the ground. The thawing period will extend over centuries for even thin permafrost (Burn 1998) and take millennia where ice rich ground is tens or hundreds of metres thick. Nevertheless, near-surface permafrost may thaw this century for there is little prospect of short-term control over climate change with the global atmospheric concentration of carbon dioxide already higher than at any point in the Quaternary (Burn et al. 2021). Periglacial environments may ultimately persist where the microclimate permits, especially at high elevations. Seen in this light, the periglacial record of Europe indicates the future of the world's permafrost environments. The treatment that the editors have brought together in this book may have been assembled primarily with a continental audience in mind, but it is relevant to the global community and has done us a great service.

C. R. Burn
President, International Permafrost
Association, 2020–2024
Carleton University
Ottawa, ON, Canada
Christopher.Burn@carleton.ca

References

Ballantyne CK, Harris C (1994). The periglaciation of Great Britain. Cambridge University Press, Cambridge. p 330

Burn CR (1998) The response (1958 to 1997) of permafrost and near-surface ground temperatures to forest fire, Takhini River valley, southern Yukon Territory. Canadian Journal of Earth Sciences 35(2):184–199

Burn CR, Lewkowicz AG, Wilson MA (2021) Long-term field measurements of climate induced thaw subsidence above ice wedges on hill slopes, western Arctic Canada. Permafrost and Periglacial Processes 32:261–276

Burn CR, Cooper M, Morison SR, Pronk T, Calder JH (2021) The CFES scientific statement on climate change—its impacts in Canada, and the critical role of Earth scientists in mitigation and adaptation. Geoscience Canada 48:59–72

Burt TP, Williams PJ (1976) Hydraulic conductivity in frozen soils. Earth Surface Processes 1:349–360

Church M (2017) Reconstructing periglacial geomorphology: The contribution of J. Ross Mackay. Permafrost and Periglacial Processes 28:517–522

Delaloye R, Lambiel C, Gärtner-Roer I (2010) Overview of rock glacier kinematyics research in the Swiss Alps. Geographica Helvetica 65:135–145

Deprez M, De Kock T, De Schutter G, Cnudde V (2020) A review on freeze-thaw action and weathering of rocks. Earth-Science Reviews 203:103143

Hallet B, Prestrud Anderson S, Stubbs CW, Gregory EC (1988) Surface soil displacements in sorted circles, western Spitzbergen. In: Permafrost Fifth International Conference, Proceedings, August 2–5, 1988. Trondheim, Tapir, v. 1:770–775

Hutchinson JN (1991) Periglacial and slope processes. In: Forster A, Culshaw MG, Cripps JC, Little JA, Moon CF (eds) Quaternary Engineering Geology. Geological Society, London, Engineering Geology Special Publications, 7:283–331

Kokelj SV, Lantz TC, Tunnicliffe J, Segal R, Lacelle D (2017) Climate driven thaw of permafrost preserved glacial landscapes, northwestern Canada. Geology 45:371–374

Lachenbruch AH (1962) Mechanics of thermal contraction cracks and ice-wedge polygons in permafrost. Special Paper 70. Boulder, CO: Geological Society of America.

Mackay JR (1963) The Mackenzie Delta, N.W.T. Geographical Branch Memoir 8. Ottawa, ON, Department of Mines and Technical Surveys.

Mackay JR (1970) Disturbances to the tundra and forest tundra environment of the western Arctic. Canadian Geotechnical Journal **7**:420–432

Mackay JR (1980) The origin of hummocks, western Arctic coast, Canada. Canadian Journal of Earth Sciences 17:996–1006

Mackay JR (1997) A full-scale field experiment (1978–1995) on the growth of permafrost by means of lake drainage, western Arctic coast: a discussion of the method and some results. Canadian Journal of Earth Sciences 34 17–33

Mackay JR (1998) Pingo growth and collapse, Tuktoyaktuk Peninsula area, western Arctic coast, Canada: a long-term field study. Géographie physique et Quaternaire 52:271–323

Murton JB, Kolstrup E (2003) Ice-wedge casts as indicators of palaeotemperatures: Precise proxy or wishful thinking? Progress in Physical Geography 27:155–170

Outcalt SI (1970) A study of time dependence during serial needle ice events. Archiv für Meteorologie, Geophysik und Bioklimatologie, Serie A 19:329–337

Seppälä M (1982) An experimental study of the formation of palsas. In: French HM (ed) The Roger J.E. Brown Memorial Volume, Proceedings of the Fourth Canadian Permafrost Conference, 2–6 March, Calgary, Alberta. Ottawa, ON, National Research Council of Canada. pp 36–42

Simpson GG (1963) Historical science. In: The Fabric of Geology. Edited by CC Albritton, Jr. Reading, MA, Addison-Wesley. pp 24–48

Skempton AW, Norbury D, Petley DJ, Spink TW (1991) Solifluction shears at Carsington, Derbyshire. In: Forster A, Culshaw MG, Cripps JC, Little JA, Moon CF (eds) Quaternary Engineering Geology, Geological Society, London, Engineering Geology Special Publications, 7:381–387

Williams PJ (1959) An investigation into processes occurring in solifluction. American Journal of Science 257:481–490

Preface

Joined to the vast Asian continent during the Uralian orogeny, Europe has been considered a peninsula of Asia, a western appendix. Its comparatively reduced size does not imply the simplicity of the territory but rather the opposite, manifested in many faces. This continent encompasses due to a long geological history a great heterogeneity of landscapes, from the eastern extensive plains and the Anatolian Plateau shaking hands with Asia to the peripheral mountains of the Balkan, Italian, Iberian and Scandinavian peninsulas. This continental wide spectrum of landscapes is complemented by the singular characteristics of several surrounding islands and archipelagos that extend beyond the boundaries of the continent to the North Atlantic, the Arctic and the southern Mediterranean Sea.

European landscapes are amongst the most difficult ones to be studied as Europe is in essence a humanised continent resulting from its long history and the superposition of different civilizations and ways to rationalise and use the environment. The compartmentation of the territory is enhanced by the numerous mountain ranges distributed across the continent, and is manifested in different cultures, languages and nations that have shaped the western and eastern cultures. Indeed, Europe has been often referred as the 'old continent', being not only the cradle of the western civilization, or the origin of the colonisers that extended its boundaries and influence the rest of the world, but also the origin of a great part of the Philosophy, rationalism, empiricism and the knowledge that feeds Science as a whole and many of the Earth Science disciplines such as Geography, Geology, Meteorology, Climatology, or Physics and Chemistry of the Earth, amongst many others. In fact, the origin of the periglacial research must be found in Europe, where the tradition on the study of the legacy of the mountain glaciations and cold processes in general is quite long. Indeed, this continent includes some of the most studied mountain areas (i.e. European Alps) on Earth.

However, within the realm of the cold region processes in Europe, significantly more attention has probably been paid to the glacial footprints and the reconstruction of the past glacial evolution as a mean to reconstruct former climate evolution. One could conclude that it is logical given to the obvious relationship between glaciers and climate, its visual and spectacular character and the vast areas that were covered by the Pleistocene ice sheets and mountain glaciers not only in Europe. However, glacial processes are telling us only part of the general picture of the cold climate areas: the glaciated areas and when there are (were) glaciers. In other words, what happens when glaciers disappear, or where the cold climate with insufficient precipitations does (did) not allow the formation of glaciers? That is a great opportunity to open the treasure chest of the periglacial landforms, sediments and landscapes; they may be not as spectacular as the glacial ones, but have also a great potential to study the climatic alterations in the past and at present-day focusing on major sentinels such as rock glaciers, palsas, pingos, solifluction lobes, patterned ground and many other landforms linked to different processes within the periglacial domain, affected by permafrost, or seasonal frost conditions. In addition, a better knowledge of these landforms and processes is crucial in a continent essentially humanised, where periglacial dynamics may cause serious damages to extensive urbanised areas, dense networks of communication and transport infrastructures and extensive industrial areas, especially in the Northern European and high-altitude regions.

Therefore, with this humble collective work that you have now in your hands we aim to collect and present the vast knowledge on the European periglacial landscapes and processes that has been accumulated during the last century. And to do it in a very didactic way, with lots of pictures, diagrams and schematic figures in order to reach an audience beyond the academia. We do not only intend to synthesise all the previous works, but to better understand their interrelationships and to uncover the current knowledge gaps and pose them as new challenges and opportunities to recover new hidden information on the past climate evolution and track the impact of the current climatic change. And last and not least, we aim to highlight the valuable potential of the periglacial landscapes and features, and the urgent need of protecting and preserving them for the next generations. As global change advances, this may be one of the last opportunities to raise awareness in society.

Barcelona, Spain — Marc Oliva
Brno, Czechia — Daniel Nývlt
Madrid, Spain — José M. Fernández-Fernández

Contents

Periglacial Landforms in Central and Eastern Europe

Periglacial Landforms in Northern Europe

Conclusions

Introduction

Marc Oliva, José M. Fernández-Fernández, and Daniel Nývlt

Periglacial research in Europe has progressed substantially over the last decades in parallel to a better comprehension of the past and present-day cold-climate geomorphological processes that have shaped the current landscape of this continent. Advances on glacial geomorphology have been complemented with a better characterisation of the spatial distribution of periglacial landforms and deposits in Europe as well as of their palaeoenvironmental and palaeoclimatic significance. In this sense, the progress on dating techniques have allowed to refine the chronology of the development of periglacial features, which is crucial to understand the time scales, climate conditions and geomorphological processes behind their formation.

Europe includes some of the best studied areas with regards to their periglacial evolution, which is demonstrated by the large amount of studies on periglacial processes and landforms from the Mediterranean basin to the polar regions, and from isolated Atlantic islands to the vast steppes of Eurasia. Thus, the objective of this book is to update the vast amount of information produced over the last few decades on the periglacial landscapes of Europe, synthesising the most recent advances on periglacial research across the European mountains and the lowlands once reshaped by cryogenic processes. The book pursues a double purpose: educational, including a large number of figures that complement the text and make easier the interpretation of the different chapters; and academic, being a reference for the future studies on the periglacial dynamics across the Old Continent.

M. Oliva (✉)
Universitat de Barcelona, Barcelona, Spain
e-mail: marcoliva@ub.edu

J. M. Fernández-Fernández
Universidad Complutense de Madrid, Madrid, Spain
e-mail: jmfernandez@edu.ulisboa.pt

D. Nývlt
Masaryk University, Brno, Czech Republic
e-mail: daniel.nyvlt@sci.muni.cz

M. Oliva et al. (eds.), *Periglacial Landscapes of Europe*,
https://doi.org/10.1007/978-3-031-14895-8_1

With that aim, this book includes an initial chapter where Filipa Naughton and co-authors present a reconstruction of the climatic evolution since the penultimate glacial cycle to the present day using a wide range of marine records from the Atlantic Ocean and Mediterranean Sea and ice core records from Greenland Ice Sheet; the climatic background is of crucial importance to understand the climatic phases that have been (un)favourable for periglacial activity in the past, and thus, to better frame the palimpsest of the current distribution of periglacial landforms and deposits.

The next part of the book comprises 12 chapters divided in three sections following major latitudinal regions: Southern, Central and Northern Europe. Periglacial evidence in each of these studied regions is examined by the leading scientists, who have long investigated these areas. The chapters focused on the different areas follow a similar structure; first, the main geographical and climatic characteristics of the study areas are analysed, together with the history of periglacial research in the area. Subsequently, a longer section (the core of each chapter) describes the distribution and types of periglacial phenomena existing in the region, trying to distinguish between the active ones and those inherited from past conditions. Finally, each chapter finishes with a conclusive section highlighting the unique characteristics of the region within the context of European periglacial landscapes.

The first section follows a W-E transect, from Iberia to Anatolia. Firstly, Marc Oliva and co-authors present the main periglacial features existing in the Iberian Peninsula, which are mostly concentrated in mountain ranges. Subsequently, Adriano Ribolini describes the processes and landforms of periglacial origin distributed across the Italian Peninsula, outside the Alps. Manja Žebre and Emil Gachev present the periglacial phenomena of the Balkans, outside the Carpathian Mountains, whereas Attila Çiner and Akif Sarikaya do the same for the Anatolian Peninsula. Finally, Mauro Guglielmin summarises the periglacial landforms existing in several Mediterranean islands, which are mostly inherited from past colder periods.

Periglacial phenomena in central Europe is examined in four chapters. Andreas Kellerer-Pirklbauer and co-authors provide a detailed analysis of active and relict periglacial features across the Alps, one of the regions where periglacial research has been more profusely developed. Later, Piotr Migoń and Jarosław Waroszewski examine the periglacial landscapes in the Variscan ranges, where the geomorphic and sedimentary records show evidence of past widespread periglacial environments. Finally, Zofia Rączkowska presents the impact that periglacial morphogenesis has had on the mountain landscapes of the Carpathians. And finally, Barbara Woronko and Maciej Dąbski examine the implications that periglacial remodelling had on the landscape of the North European Plain during Quaternary cold stages.

Three chapters comprise the section on Northern Europe. Firstly, Colin Ballantyne and Julian Murton describe the rich variety of landforms and deposits that periglaciation left in Great Britain and Ireland, including minor active features. John Matthews and Atle Nesje examine the periglacial landscapes of Scandinavia, where active and relict landforms are widespread and have been a reference for periglacial research in Europe. Finally, José M. Fernández-Fernández and co-authors summarise how periglacial processes have shaped in the past and continue to shape nowadays the very dynamic landscapes of Iceland.

The book concludes with a chapter prepared by the editors that seeks to summarise the impact of periglacial dynamics in European landscapes, from Late Pleistocene times to present-day, including how human activities have transformed periglacial landscapes.

In short, this book does neither constitute a compendium of study cases nor a handbook of periglacial dynamics, but it is designed as an interlaced connection of concepts and examples across Europe with a rich and well-illustrated material. This approach will be useful for academicians working in similar topics as well as for non-academicians to better understand the landform features existing in their regions. Indeed, the orographic complexity of Europe, with unevenly distributed mountain ranges, peninsulas, bays, islands, etc. combined with Quaternary climate variability and advances and retreats of glaciers and ice sheets has resulted in the variety and uniqueness of periglacial landscapes in a continent, where these landscapes should be preserved for future generations.

Regional Setting

Quaternary Climate Variability and Periglacial Dynamics

Filipa Naughton, Maria Fernanda Sánchez Goñi, and Samuel Toucanne

1 Climatic Framework of Europe During the Last Glacial Cycle

The Last Glacial Cycle (LGC, ~116–15 ka) encompasses the most recent glacial period of a series of 10 glacial-interglacial cycles that occurred during the last ca. 1 million years. The cycles from 700 ka onwards, identified from changes in the marine oxygen isotopic ratio of calcite shells ($\delta^{18}Oc$), are characterised by a dominant 100,000-year cyclicity originally triggered by changes in insolation (Shackleton and Opdyke 1973). During the last 800 ka, they are represented by 20 Marine Isotope Stages (MIS) from MIS 20 to MIS 1, where odd and even numbers broadly correspond to interglacial and glacial periods, respectively (Lisiecki and Raymo 2005; Railsback et al. 2015). After attaining a minimum of ice volume extent during Last Interglacial (MIS 5e), the LGC started at ~116 ka, the so-called last glacial inception (MIS 5e/5d transition). The ice volume progressively increased through MIS 5 and an additional decrease in summer insolation triggered the ice volume maxima, -80 m of relativc sea level, centred at 65 ka during MIS 4 (Waelbroeck et al. 2002) (Fig. 1). The subsequent increase in summer insolation produced a partial deglaciation giving way to the MIS

The original version of this chapter was revised: The missing letter 'a' in section heading has been corrected. The correction to this chapter is available at https://doi.org/10.1007/978-3-031-14895-8_17

F. Naughton (✉)
Portuguese Institute for Sea and Atmosphere (IPMA), Rua Alfredo Magalhaes Ramalho 6, 1495-006 Lisboa, Portugal
e-mail: filipa.naughton@gmail.com

Center of Marine Sciences (CCMAR), Algarve University, Campus de Gambelas, 8005-139 Faro, Portugal

M. F. Sánchez Goñi
Ecole Pratique Des Hautes Etudes (EPHE, PSL University), 33615 Pessac, France

UMR 5805, University of Bordeaux, EPOC, 33615 Pessac, France

S. Toucanne
Univ Brest, CNRS, Geo-Ocean, 29280 Plouzane, Ifremer, France

M. Oliva et al. (eds.), *Periglacial Landscapes of Europe*,
https://doi.org/10.1007/978-3-031-14895-8_2

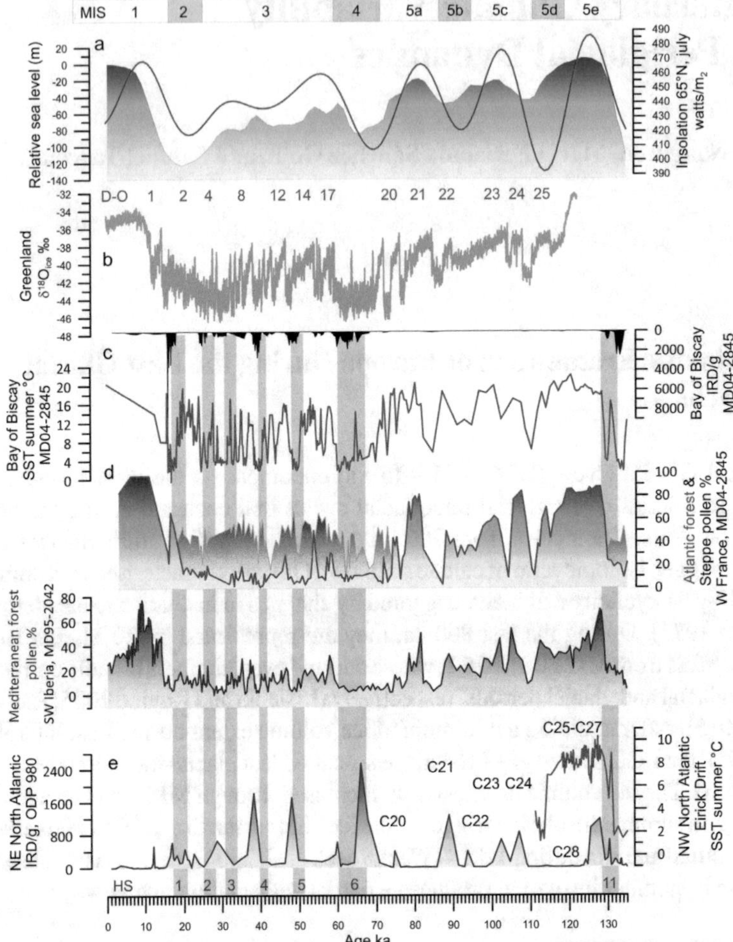

Fig. 1 Millennial scale palaeoenvironmental changes in Europe, Greenland and the North Atlantic during the last climatic cycle. a) July insolation at 65° N (red) (Berger and Loutre, 1991) and relative sea-level changes as an indicator for ice-volume changes (Waelbroeck et al. 2002), b) Greenland ice core δ^{18}O record (Rasmussen et al. 2014). Dansgaard-Oeschger (D-O) warming events (red), c) Foraminifera-based summer sea surface temperature (SST, red) and Ice Rafted Debris (IRD, black) records of core MD04-2845 from the Bay of Biscay (45°N) (Sánchez-Goñi et al. 2008), d) Atlantic temperate forest, steppe (orange, mainly Poaceae, Cyperaceae and Asteraceae) pollen percentage records from the Bay of Biscay (western France) and Mediterranean forest pollen percentage records from the SW Iberian margin (Sánchez-Goñi et al. 2008 and unpublished data), e) IRD concentration record from site ODP 980 (blue), NE North Atlantic. Grey intervals indicate the HS 6 to 1 and HS 11 and the cold events C20 to C24 (Bond and Lotti 1995; McManus et al. 1994). Summer SST from the NW North Atlantic indicating the cold events of the Last Interglacial C27 to C25 (Irvalı et al. 2016). MIS: Marine Isotope Stages. Cold events C19 and C18 are not detected in ODP site 980 (McManus et al. 1999) but recorded in the nearby site V29-191 (McManus et al. 1994). Cold event C28 follows HS 11 although the chronological uncertainties of both ODP 980 and the Erik Drift sites do not show it in the figure

3 intermediate ice volume period lasting from ~60 ka to 27 ka. The maximum global ice volume of the LGC was reached during MIS 2, 26–19.5 ka (Clark et al. 2009), and preceded the last deglaciation, composed of a series of climate shifts between 19.5 ka and 11.7 ka.

Geomorphological data and model simulations show that at ca. 115 ka ice sheets developed over Scandinavia and in Northern Eurasia, particularly during MIS 4 (73–60 ka), whilst large ice sheets close to Atlantic moisture sources, i.e. the Western European Ice Sheet (EIS), reached their maximum extent at the global Last Glacial Maximum (LGM, 26–19.5 ka) (Hughes et al. 2013; Batchelor et al. 2019; Palacios et al. 2022). Ice-sheets in Northern Eurasia were probably of similar size or even more extensive in MIS 5b (~85 ka) and MIS 5d (~115 ka) compared to MIS 4. Data and models also suggest that western Europe was more glaciated during the LGM than Eastern Europe and Northern Eurasia. Also, the partial marine-based nature of the EIS during the LGC seems to be marked by a higher susceptibility to rapid and frequent ice-sheet collapse (Batchelor et al. 2019). This high susceptibility is associated with the strong variability of the Greenland climate revealed by the analyses of water isotopes in deep ice cores from the centre of the Greenland Ice Sheet, known as the Dansgaard-Oeschger cycles (Dansgaard et al. 1984; Johnsen et al. 1992; Rasmussen et al. 2014) (Fig. 1). These cycles lasted 500–4000 years during MIS 3 and MIS 4, but were several millennia longer during MIS 5d-a (Rasmussen et al. 2014), supporting that climate was particularly variable during the middle part of the LGC (Bond et al. 1999). The abrupt transitions between the cold periods identified by low oxygen isotope ratios (δ^{18}Oice) to mild periods with high δ^{18}Oice are designed as Dansgaard-Oeschger (D-O) warming events. The LGC was punctuated by twenty-five D-O warming events in the atmosphere over Greenland, and were characterised by a high amplitude (7–16 °C) and rapid (within a few decades) warming event followed by a progressive decrease in temperature and a final abrupt cooling (Wolff et al. 2010). The D-O warming event and the progressive cooling phase form the Greenland Interstadial (GI), and the final cooling event leading to the cold phase form the Greenland Stadial (GS). The GI phases lasted between 100 and 2,600 years (Wolff et al. 2010; Rasmussen et al. 2014). A wide array of global climatic proxies such as CO_2 and CH_4 atmospheric concentrations from the air bubbles trapped in the Antarctic and Greenland ice cores have furthermore demonstrated the link between the initial observations in Greenland and the global expression of the D-O cycles (e.g. Brook et al. 1996). Marine sedimentary data also indicate that the D-O cycles were associated with 15–30 m of global sea-level increase during GI (Sierro et al. 2009).

Interestingly, the study of ice-rafted debris (IRD) found in North Atlantic deep-sea sedimentary sequences show that the LGC was also punctuated by six massive and repeated episodes of iceberg discharges, every ~7000 years, from the north-eastern Laurentide Ice Sheet and the Hudson Strait ice-stream. These Heinrich events (HE), as they are called, substantially cooled the surface of the North Atlantic (Heinrich 1988; Bond et al. 1993) (Fig. 1). A HE is defined as the event of the North American ice-rafted debris deposition resulting from the melting of the icebergs in the North Atlantic, preferentially between 45° N and 50° N, while the Heinrich Stadial

(HS) is the up to 3,000 years cold phase, in which occurred the HE (Sánchez Goñi and Harrison 2010). Moderate iceberg discharges from the Laurentide Ice Sheet, associated with the North Atlantic cold events C27 to C18 (Fig. 1), occurred during MIS 5 and MIS 4 (McManus et al. 1994; Shackleton et al. 2003; Oppo et al. 2006) but, paradoxically, those punctuating the MIS 5a/4 transition, C20 to C18 associated with GS 21 to GS 19, corresponded with warm sea surface temperatures (SST) in the mid-latitudes of the Western European margin (Sánchez Goñi et al. 2013). Weak iceberg discharges from the Icelandic, British-Irish, and Scandinavian Ice Sheets and with higher frequency than the HE also occurred between the HEs (Elliot et al. 2001) associated with the other GSs. Iceberg discharges have been simulated to last from 50–200 years (Roche et al. 2004) up to 1000–1500 years (Ziemen et al. 2019). Most of the HSs and non-HSs are related to cold sea surface conditions in the North Atlantic Ocean, and they are associated with decreases in the strength of the Atlantic Meridional Overturning Circulation (AMOC) (Henry et al. 2016; Lynch-Stieglitz, 2017). The AMOC is defined as the zonally integrated component of surface and deep currents in the Atlantic Ocean and it plays a crucial role in the Earth's climate by regulating the global transport of heat and freshwater. The tropical warm and saline surface waters flow northward densifying progressively and sink preferentially in the Norwegian Sea, between north of Iceland and south of Svalbard, and Irminger seas (Lozier et al. 2019). They release heat and humidity toward the atmosphere and form a southward flow of colder, deep waters that are part of the thermohaline circulation.

Iceberg discharges substantially slowdown the arrival of warm waters to Northern latitudes decreasing the deep-water formation and consequently cooling and drying the North Atlantic Ocean and the atmosphere (Ganopolski and Rahmstorf 2001). However, the fact that HEs occurred during a complex and longer period, i.e. the HS, and that lags the sea surface cooling (Barker et al. 2015) indicate that HE may be a consequence of AMOC weakening. This initial weakening, the origin of whose remains a subject of debate, may result from enhanced ice surface melting of the European and Pacific proximal ice settings (Boswell et al. 2019; Toucanne et al. 2022). The latter could cause, in turn, oceanic sub-surface sea warming, the ultimate destabilisation of the North Atlantic marine-terminating ice streams and the massive iceberg discharges corresponding to the HE (Álvarez-Solas and Ramstein 2011).

The expression of D-O cycles and HSs in European environments has been investigated by analysing speleothems, loess sequences and the pollen grains preserved in terrestrial and deep-sea cores, the latter allowing a direct comparison between D-O cycles and HSs. The general picture observed in Europe is a significant change in vegetation cover and soil characteristic between stadial and interstadial (Fletcher et al. 2010a). The stadials were characterised by much drier and colder conditions (steppe vegetation, loess development in continental Europe (Moine et al. 2017) and the interstadials were associated with forest expansion (Fig. 1). The Abrupt Climate change and Environmental Responses (ACER) pollen data synthesis (Sánchez Goñi et al. 2017) reveals the following features for the European vegetation during the succession of D-O events. Forest dominated Europe at latitudes lower than 40° N as well as north of 40° N during warmer and long interstadials (i.e., those associated with D-O events 12 and 14). In Western Europe pollen data reveal that the amplitude of Atlantic and Mediterranean forest expansions differs for any given D-O warming during the period encompassing MIS 4, 3 and 2 (73–14.7 ka). In the Western Mediterranean below 40° N, D-O 16–17 and D-O 8 and 7 were associated with strong expansion of forest cover, probably amplified by the high seasonality induced by the minima in precession at that time, contrasting with weak expansion of forest cover during D-O 14 and 12. The opposite pattern is revealed at the Atlantic sites, where the strongest forest development during D-O 12 and 14 was amplified by the maximum in obliquity affecting latitudes above 40° N (Sánchez Goñi et al. 2008). Some modelling simulations, in qualitative agreement with pollen data, indicate that the European vegetation responded quickly, by ~200 years, to the abrupt D-O warming events generated by the strengthening of the AMOC (Woillez et al. 2013). Associated with these vegetation changes, repeated inner-alpine ice fluctuations during D-O cycles of MIS 3 are also observed (Mayr et al. 2019; Martinez-Lamas et al. 2020).

As mentioned above, during the last interglacial-glacial transition (MIS 5a/4) the regional SST in the Western European margin remained relatively high, while those on the adjacent continent, identified by the replacement of temperate by boreal forest, were relatively cool and progressively decreasing. Superimposed on this long-term change, three cold-driven boreal forest regional expansions during GS 21 to 19 alternated with three temperate forest expansions during GI 21 to 19. In NW Iberia, cooling episodes were marked by the development of heathlands that alternated with oak forest increase during the warming events (Sánchez Goñi et al. 2013). During the short-term atmospheric cooling events SSTs, paradoxically, also increased or remained warm, at around 18 °C. Moderate North American iceberg discharges in the North Atlantic certainly deviated pinned the warm Gulf Stream and North Atlantic Current southwards to the European margin as described for example in Keffer et al. (1988) producing an increase in the land-sea thermal contrast and enhancing evaporation. Carried northward by the westerly winds, this moisture was the most likely source of the snowfall that formed the ice caps during MIS 4 (Sánchez Goñi et al. 2013). The cold-land warm-sea decoupling documented for the cooling events of the last glaciation in Western Europe from 42° N to 45° N is distinctive and promoted the regional development of boreal forest and heathlands. In contrast, the environments of Southern Iberia were dominated by semi-desert during the cold phases, which intervened between increases in Mediterranean woodland reflecting warm summers and humid winters.

Well-chronologically constrained deep-sea and terrestrial pollen records have shown furthermore a complex phase of sequences within the HSs and the contrasting regional impact of HEs on the European continent. Regarding the magnitude of the HE cooling, the NW Mediterranean borderlands (SE France and NE Iberia) experienced colder and drier conditions during HS 5 compared to HS 4 leading to a higher expansion of steppe vegetation. By contrast, during HS 4, the massive freshwater input in the North Atlantic may have led to the stratification of the Mediterranean water column and consequently limited upward mixing of cold water, resulting in regional atmospheric warming and wetting compared to HS 5 that allowed the maintenance of some temperate deciduous oak woodland stands (Sánchez Goñi et al. 2020). Similarly, HS 5 appears to be particularly cold and dry in NE Greece as shown by the dominance of steppe formations (Müller et al. 2011). In contrast, an area of relative ecological stability has been identified in SW Greece, where temperate tree populations survived throughout the LGC as the result of continued moisture availability and varied topography (Tzedakis et al. 2002). Regarding the complex structure of HSs, a wet to dry continental hydrological pattern has been identified in the Iberian Peninsula (Fletcher and Sánchez Goñi 2008; Naughton et al. 2009). In the north-Western Mediterranean region, a three-phase structure, wet-mild/dry-cold/wet-mild intervals, have been detected in the Gulf of Lyon and the adjacent landmasses (Sánchez Goñi et al. 2020). The multiphase structure of the HSs is also supported by both the Northern North Atlantic (Wary et al. 2018) and Greenland ice core records (Guillevic et al. 2014).

Despite the three last decades of research, the mechanisms at the origin of D-O cycles and HEs still remain elusive. A key role has been attributed to the variations of the AMOC as well as the Arctic sea ice. An important challenge in the modelling community is to be able to identify the original forcing, internal or external, of such a climatic variability and to make the link with the HSs (Landais et al. 2022). If the cause of the HSs remains enigmatic, the production of meltwater by mid-latitude proximal ice sheets from Europe and North America, could have triggered the initial AMOC weakening. This ice melting necessarily implies warming summer temperatures as cold conditions in the North Atlantic region and over Greenland are recorded at the same time. The mismatches between ice-core temperature oscillations in the polar regions and ice-margin fluctuations in mid-latitudes would, therefore, call for a paradigm shift in our understanding of past rapid climate changes (Toucanne et al. 2022).

2 Climate Changes During the Last Deglaciation

The last deglaciation was triggered by the increase of summer insolation at the Northern Hemisphere and consequent retreat of North American and Eurasian ice sheets, the increase in greenhouse gas concentrations and changes in other amplifying feedbacks that cause distinctive regional and global responses (Fig. 2) (e.g.; Clark et al. 2012; Shakun et al. 2015). Globally, the last deglaciation started at ~20–19 ka (ka: age in kiloanni) (e.g.; Denton et al. 2010; Carlson and Clark, 2012; Clark et al. 2012) and ended during the present day interglacial at ~6.8 ka (e.g.; Carlson et al. 2007, 2008). Multiple abrupt climate shifts were noticed in the North Atlantic, Greenland and Europe, superimposed on this long-term warming trend (Fig. 2) (e.g.; Mangerud et al. 1974; Dansgaard et al. 1982; Oeschger et al. 1984; Mix and Ruddiman, 1985; Alley et al. 1993; Alley and Clark, 1999). The most recognised abrupt climate episodes that punctuated the last deglaciation are the Heinrich Stadial 1 (HS 1, ~18–14.7 ka), the Bølling-Allerød interstadial (B-A, ~14.7–12.9 ka) and the Younger Dryas (YD, 12.9–11.7 ka) (Fig. 2).

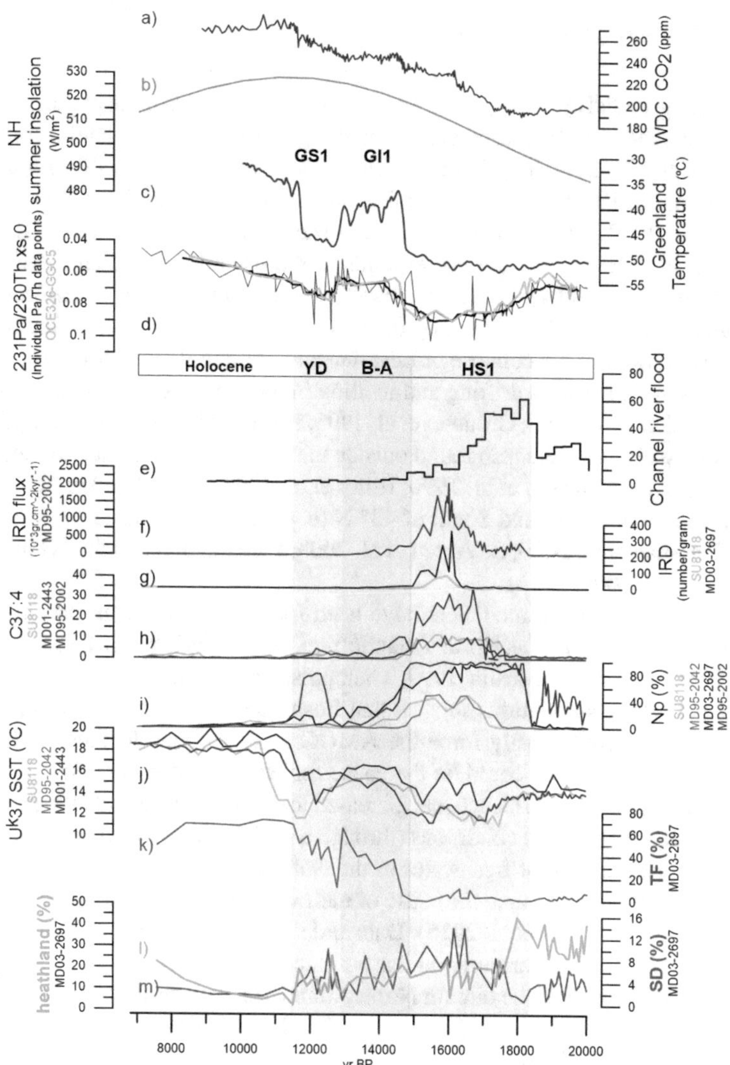

Fig. 2 a) Atmospheric CO_2 concentrations reconstructed from West Antarctic Ice Sheet Divide ice core (WDC) (Marcott et al. 2014); b) summer insolation at 65ºN (Berger and Loutre 1991); c) Greenland temperature (Buizert et al. 2014a, b); d) ex231Pa0/ex230Th0 from composite North Atlantic records (blue line) (black bold line: smoothed record) (Ng et al. 2018) and from the western North Atlantic (light blue) (OCE326-GGC5; McManus et al. 2004); e) Channel River Flood (number of flood events per 250 years) (Toucanne et al. 2015); f) IRD western French margin (Ménot et al. 2006) and g) in the western Iberian margin (Bard et al. 2000; Naughton et al. 2016); h) Meltwater discharges (Tetra-unsaturated alkenone $C_{37:4}$) from western Iberian and French margins (Bard et al. 2000; Ménot et al. 2006; Martrat et al. 2007); i) NP: planktic polar foraminifera *Neogloboquadrina pachyderma* abundances (%) from western Iberia and French margins (Salgueiro et al. 2014; Zaragosi et al. 2001); j) alkenone-derived SST records ($U^{k'}_{37}$ SST) from Iberian margin (Bard et al. 2000; Pailler and Bard, 2002; Martrat et al. 2007); k) TF: Temperate forest abundances (%) in NW Iberian margin record MD03-2697 (Naughton et al. 2016); l) Heathland; m) SD: Semi-desert plants. Light blue bands: HS 1 and YD; Light salmon band: B-A

2.1 Heinrich Stadial 1 (HS 1)

HS 1 is a long stadial episode (~18 to 14.7 ka) that includes two main phases: an early phase associated with increased surface melting of the Southern, land-terminating margins of the Laurentide (LIS) and Fennoscandian Ice Sheet (FIS) and a latest phase during which their marine-terminating ice-streams rapidly collapsed (e.g.; Barker et al. 2009, 2015; Sánchez Goñi and Harrison 2010; Toucanne et al. 2015; Boswell et al. 2019). The later corresponds to the so-called Heinrich event 1 (HE 1) (e.g.; Hemming, 2004). HE 1 was firstly identified in deep-sea cores located in the 'Ruddiman belt' (~43–53° N), by the anomalous presence of ice-rafted detritus (IRD), resulting from the collapse of Eurasian and Laurentide Ice Sheets and consequent debris-rich icebergs drifting and melting in the North Atlantic (e.g.; Heinrich 1988; Broecker et al. 1992; Grousset et al. 1993; Elliot et al. 1998; Hemming, 2004). Thin layers with IRD were also found outside the 'Ruddiman belt', at latitudes north of 53° N (e.g., Rasmussen et al. 1996; Elliot et al. 1998, 2001; Voelker et al. 1998; Van Kreveld et al. 2000) and South of 43° N (e.g.; Lebreiro et al. 1996; Bard et al. 2000; Chapman et al. 2000; de Abreu et al. 2003; Fletcher and Sánchez Goñi 2008; Naughton et al. 2009; 2016).

For long time climate modellers have been focused in exploring the impact of the melting icebergs in the North Atlantic region, i.e. during HE 1 (e.g.; Rahmstorf 1994; Seidov and Maslin 1999; Ganopolski and Rahmstorf 2001; Menviel et al. 2014). These simulations suggest that freshwater pulses in the North Atlantic and Nordic seas could easily force the AMOC to shut down, leading to the idea that iceberg calving (HE) could be the cause for Heinrich stadial (HS) conditions. However, others support evidence that increased iceberg calving could rather enhance and/or prolong cold stadial conditions (through a positive feedback on the AMOC in response to the addition of freshwater in the North Atlantic), suggesting that HE 1 was a consequence rather than the cause of meltwater pulses (McManus et al. 1999; Clark et al. 2007; Barker et al. 2015). Data and simulations further support the idea of a subsurface oceanic warming, under a reduced AMOC situation, as a possible trigger for the melting of the circum North Atlantic marine-terminating ice-streams and the subsequent HE 1 (Alvarez-Solas et al. 2010; 2013; Marcott et al. 2011; Bassis et al. 2017; He et al. 2020).

Most North Atlantic records show an extreme cooling (Fig. 2) contrasting with other data and climate simulations that indicate warm conditions at the onset of HS 1 (Fig. 2). In particular, SST profiles and the expansion of polar planktonic foraminifera *Neogloboquadrina pachyderma* reveal extreme cold conditions during the early phase of HS 1 in the North Atlantic and Mediterranean regions (e.g.; Bond and Lotti, 1995; Bard et al. 2000; Zaragosi et al. 2001; Hemming, 2004; Peck et al. 2006; Eynaud et al. 2007; 2009; Naughton et al. 2009; 2016; Penaud et al. 2009; Martrat et al. 2014). In contrast, Greenland ice core oxygen isotope $\delta^{18}O$ records do not show any cooling at the onset of HS 1 (He et al. 2021). Increased seasonality, with warm summers and severe cold winters have been proposed to explain this thermal contrasting signal (e.g.; Bromley et al. 2014a, b; Toucanne et al. 2015;

Boswell et al. 2019; Wittmeier et al. 2020; Fersi et al. 2021; He et al. 2021). During summer, warm atmospheric conditions would have favoured the surface melting of land-terminating margins and the resulted input of meltwater in the North Atlantic would have triggered the AMOC to almost shutdown (e.g.; Broecker 1994; McManus et al. 2004; Toucanne et al. 2015; Ng et al. 2018), favouring the expansion of sea ice during winter and leading to a cooling of the surface ocean (Denton et al. 2010).

Reduced temperate forest and a strong *Pinus* forest decline revealed by Iberian margin deep-sea pollen records testify the extreme cooling in Iberia at the onset of HS 1 (Fig. 2) (Roucoux et al. 2005; Fletcher and Sánchez Goñi 2008; Naughton et al. 2009; 2016). Similarly, speleothems from Southern France and Spain indicate atmospheric cold conditions in Western Europe during this early phase of HS 1 (Genty et al. 2006; Moreno et al. 2010). The cold signal might be likely the result of prolonged and extreme cold winters that result from the AMOC reduction and widespread winter sea-ice in the North Atlantic. This would lead to substantial changes in the atmospheric circulation patterns, as revealed by the complex spatial distribution of precipitation with contrasting wet or dry signals detected in the Iberian margin and speleothem records (e.g.; Genty et al. 2006; Naughton et al. 2009; 2016; Moreno et al. 2010; Pérez-Mejías et al. 2021).

The second phase of HS 1 (onset: ~16.2 ± 0.3 ka; Landais et al. 2018 and end: 14.7 ka Rasmussen et al. 2014), associated with the collapse of the LIS and EIS and marine terminating ice-streams, and consequent massive iceberg release in the North Atlantic, marks the initially defined HE 1 (e.g.; Hemming, 2004; Andrews and Voelker 2018). This phase is marked by large deposits of dolomite-rich IRD from LIS sources in the North Atlantic deep sea records, known as Heinrich layer 1 (e.g.; Bond et al. 1992; Broecker et al. 1992; Hemming 2004; Barker et al. 2009; Hodell et al. 2017; Andrews and Voelker 2018). The introduction of large quantities of meltwater, from the LIS and EIS marine-terminating ice-streams in the North Atlantic contribute to the sustained reduction of the AMOC and amplified North Atlantic cooling (Fig. 2) (Bond et al. 1992; Cortijo et al. 1997; McManus et al. 2004; Eynaud et al. 2007; Scourse et al. 2009; Ng et al. 2018). Cold winter SST persisted since the onset of HS 1, in the Iberian and French margins (Bard et al. 2000; Pailler and Bard, 2002; Naughton et al. 2009; 2016; Fersi et al. 2021). In Iberia, semi-desert plants expanded during the maximum deposition of IRD's and contracted towards the end of HS 1, suggesting a drying phase followed by moisture increase (Fig. 2) (e.g.; Fletcher and Sánchez Goñi 2008; Naughton et al. 2016). Several speleothem records from Northern Iberia and SW France show very cold and dry regional conditions during HE 1 (e.g.; Genty et al. 2006; Moreno et al. 2010).

2.2 *The Bølling-Allerød Interstadial (B-A)*

The B-A interstadial was the first warm and wet phase of the last deglaciation in the Northern Hemisphere that occurred after HS 1, i.e. between 14.7 ka and 12.9 ka (Fig. 2) (e.g.; Severinghaus and Brook, 1999; Hoek, 2009; Denton et al. 2010;

Rasmussen et al. 2014; Naughton et al. 2016). It was firstly noticed in two Danish records (the Bølling and the Allerød sites) based mainly on birch tree remains (Hartz and Milthers 1901; Iversen 1942, 1954). The presence of the alpine species *Dryas octopetala* in two distinct layers within the B-A revealed that this warm phase was however punctuated by the Older Dryas and the intra-Allerød abrupt cooling episodes (e.g.; Mangerud et al. 1974; Von Grafenstein et al. 1999; Brauer et al. 1999; Hoek, 2009). The B-A is marked by an abrupt temperature increase of about 4 to 10 ºC in the North Atlantic region, as revealed by SST estimates, the reduction of polar foraminifera *N. pachyderma* (Fig. 2) (e.g.; Bond and Lotti 1995; Bard et al. 2000; Cacho et al. 2001; Zaragosi et al. 2001; Pailler and Bard 2002; Martrat et al. 2007; 2014; Rodrigues et al. 2010; Salgueiro et al. 2014; Repschlager et al. 2015; Naughton et al. 2016) and the expansion of subtropical foraminiferal assemblages in the sea surface water masses of the mid-latitudes (e.g.; Chapman et al. 2000; de Abreu et al. 2003). An abrupt increase of 10 ºC, D-O warming event 1, followed by a gradual decrease in atmospheric temperatures are noticed in the Greenland ice cores (Buizert et al. 2014a, b). An increase of ~3 to 5 ºC in atmospheric summer temperatures is reconstructed for Northern and Southern Europe (Renssen and Isarin, 2001; Dormoy et al. 2009). Thus, the B-A is considered the terrestrial counterpart of the D-O 1 over Greenland (Fig. 2) (Rasmussen et al. 2014).

The B-A warming of the Northern Hemisphere was triggered by the increase of boreal summer insolation and CO_2 atmospheric concentrations since the last stages of HS 1, together with the strengthening of the AMOC (Fig. 2) (e.g.; Severinghaus and Brook 1999; Liu et al. 2009; Denton et al. 2010; Shakun et al. 2012; Zhang et al. 2017; Obase and Abe-Ouchi 2019). This warming favoured forest development in Europe (e.g.; Litt and Stebich, 1999; Hoek, 2009; Fletcher et al. 2010b; Naughton et al. 2016 and references therein), but its maximum expansion occurred latter during the Allerød, suggesting a deficit in moisture availability at the onset of the B-A (e.g.; Naughton et al. 2016 and references therein). This moisture availability deficit is also supported by Southern Europe speleothem records and pollen-based quantitative climate estimates from the Western Mediterranean region (Genty et al. 2006; Dormoy et al. 2009; Moreno et al. 2010). The displacement to the north of the polar front and jet stream (Eynaud et al. 2009; Naughton et al. 2016) with a more meandered configuration (Naughton et al. in press) would explain the deficit in precipitation at the onset of the Bølling in Europe (Naughton et al. 2016).

2.3 The Younger Dryas (YD)

The YD is the latest stadial event of the last deglaciation in the Northern Hemisphere, that occurred after the B-A and prior to the onset of the Holocene, i.e. between 12.9 ka and 11.7 ka, (Fig. 2) (e.g., Alley et al. 1993; Alley and Clark 1999; Tarasov and Peltier, 2005; Denton et al. 2010; Carlson, 2013). This stadial was firstly noticed in Scandinavian terrestrial records by the rapid development of the cold-tolerant *Dryas octopetala* (Iversen, 1954; Mangerud et al. 1974; Mangerud, 2021). The YD started at the same time as the Greenland Stadial-1 (GS-1) (e.g., Rasmussen et al. 2014; Naughton et al. 2019; Mangerud, 2021; Reinig et al. 2021), or was delayed about ~170/185 yr from the onset of GS-1 (e.g.; Brauer et al. 2008; Rach et al. 2014; Obreht et al. 2020) depending on the time response of certain climate indicators (e.g., Mangerud, 2021; Naughton et al. 2022).

Several hypotheses have been proposed to explain the causes of the YD. Some of them support that meltwater discharges from the LIS and/or FIS into the North Atlantic or Arctic Ocean (e.g.; Tarasov and Peltier 2005; Carlson et al. 2007; Murton et al. 2010; Muschitiello et al. 2015), caused by maxima of incoming solar radiance in the Northern Hemisphere, leading to substantial changes in oceanic and atmospheric circulation patterns (e.g.; Broecker 2003; McManus et al. 2004; Wünsch, 2006; Eisenman et al. 2009; Carlson, 2013; Ritz et al. 2013). Others support evidence of an extra-terrestrial (ET) impact event on or near the LIS that caused the above mentioned freshwater discharges (e.g., Firestone et al. 2007; Kennett et al. 2009; Sweatman 2021). Another proposal, based on numerical climate models, suggests that the YD results from combining processes such as AMOC slowdown, moderate negative radiative forcing and an altered atmospheric circulation (Renssen et al. 2015).

It is also known that both the AMOC slowdown and reduction of the North Atlantic Deep Water (NADW) formation, during the YD, have favoured the expansion of the winter sea ice and contributed for a cooling of 2 to 4 ºC and drying in the North Atlantic and over Europe (e.g.; Boyle and Keigwin 1987; Manabe and Stouffer 1997; Isarin et al. 1998; Rahmstorf 2002; McManus et al. 2004; Denton et al. 2005; Shakun and Carlson, 2010; Renssen et al. 2015; Ng et al. 2018), while of 5 to 9 ºC in Greenland ice records (Buizert et al. 2014a, b).

North Atlantic and European very high temporal resolution records have shown, however, that the impact of the YD climate was marked by a series of climate shifts with a complex hydroclimate spatial distribution in Europe in each phase (a, b, c and d) (Fig. 3) (e.g.; Magny and Begéot 2004; Brauer et al. 2008; Bakke et al. 2009; Lane et al. 2013; Rach et al. 2014; Baldini et al. 2015; Bartolomé et al. 2015; Naughton et al. 2019). They suggest that these climate shifts, with complex spatial moisture distribution across Europe, could not be explained by AMOC slowdown alone, but also by coupled changes in atmospheric circulation patterns. They particularly suggest that the combining changes of the AMOC strength and the North Atlantic sea-ice extent produced substantial latitudinal shifts of the polar jet stream and, therefore, changes in the position and strength of the westerlies across Western Europe that are in turn responsible for sea-land moisture transfer (e.g.; Renssen and Isarin 1997; Magny and Begéot 2004; Brauer et al. 2008; Bakke et al. 2009; Lane et al. 2013; Rach et al. 2014; Naughton et al. 2019; Rea et al. 2020).

The onset of the YD (~12.9–12.7 cal. ka BP; YDa) is marked by a progressive cooling and drying in the Eastern North Atlantic region (including the central and South-Western Europe) and over Greenland (Fig. 3) (e.g.; Alley et al. 1993; Brauer et al. 1999; Rasmussen et al. 2006; Bakke et al. 2009; Thornalley et al. 2011; Rach et al. 2014; Bartolomé et al. 2015; Muschitiello et al. 2015; Naughton et al. 2019). This cooling was triggered by the gradual slowdown of AMOC (Fig. 3) (McManus

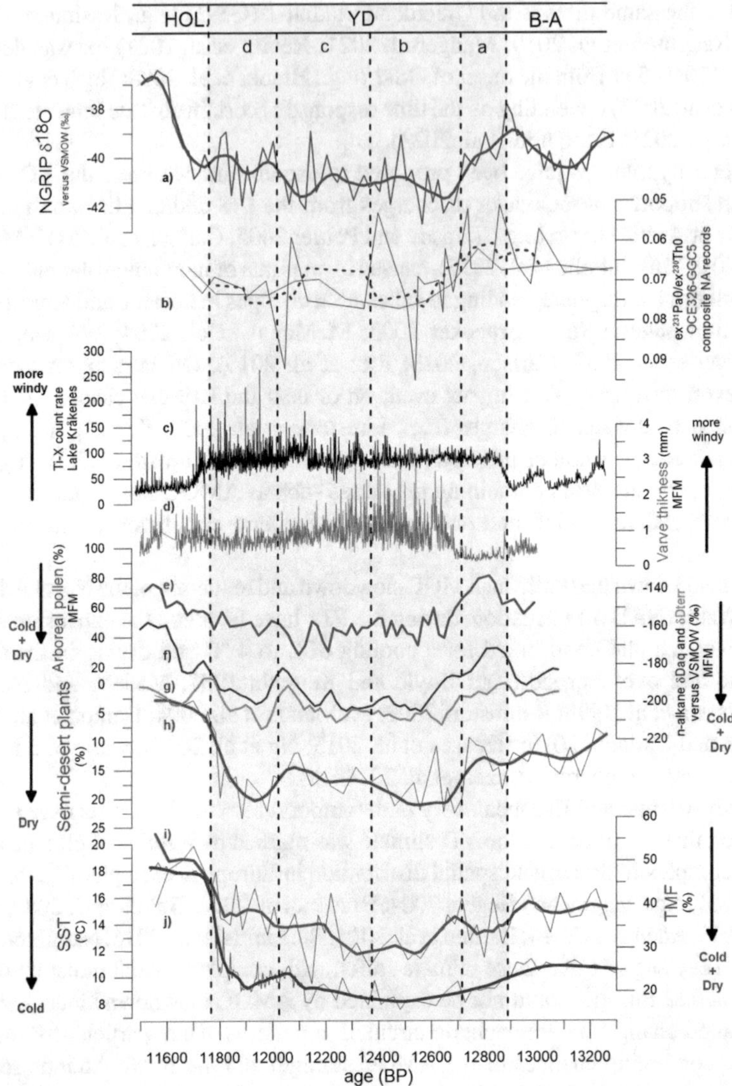

Fig. 3 a) North Grip (NGRIP) δ^{18}O versus VSMOW (in cal yr BP) (Rasmussen et al. 2006); b) ex231Pa0/ex230Th0 from core OCE326-GGC5 (red line) (McManus et al. 2004) and from a composite of North Atlantic records (thin grey line) (blue dashed line: smoothed record) (Ng et al. 2018); c) Ti count rate in Lake Kråkenes (Bakke et al. 2009); d) varve thickness and e) Arboreal pollen percentages at Lake Meerfelder Maar (MFM) (Brauer et al. 2008); f) δD values of the nC23 (aquatic plants, daq, blue line); g) δD values of the nC29 alkanes (higher terrestrial plants, dDterr, green line) at MFM (Rach et al. 2014); h) semi-desert plants percentages and i) Temperate Mediterranean forest (TMF) from western Iberian Peninsula (Naughton et al. 2019); j) SST from western Iberian margin (Naughton et al. 2019)

et al. 2004; Ng et al. 2018) and by the slight expansion of the sea ice in the North Atlantic (Cabedo-Sanz et al. 2013; Müller and Stein, 2014). The meridional thermal contrast in the North Atlantic was still reduced (Naughton et al. 2019) and the jet stream and the westerlies were relatively weakened in Western Europe (Fig. 3) (e.g. Rach et al. 2014; Bartolomé et al. 2015; Naughton et al. 2019).

The coldest phase of the YD (YDb) was detected in the Eastern North Atlantic, Western Europe and over Greenland during 12.7–12.4 cal. ka BP (Fig. 3) (e.g.; Brauer et al. 1999; Von Grafenstein et al. 1999; Rasmussen et al. 2006; Bakke et al. 2009; Dormoy et al. 2009; Fletcher et al. 2010b; Chabaud et al. 2014; Rach et al. 2014; Baldini et al. 2015; Bartolomé et al. 2015; Naughton et al. 2019). In contrast, the Western North Atlantic mid-latitudes were relatively warm (Carlson et al. 2008) making the meridional thermal contrast between the high- and mid- latitudes of the North Atlantic very high (Naughton et al. 2019). This thermal contrast was likely a response to the maximum slowdown of the AMOC (Fig. 3) (McManus et al. 2004; Ng et al. 2018) and consequent decrease in the northward heat transport, causing a relative warming in the Western North Atlantic mid-latitudes (Carlson et al. 2008) and very cold conditions in the North Atlantic high-latitudes and Eastern North Atlantic mid-latitudes (e.g.; Bakke et al. 2009; Naughton et al. 2019). This also favoured the decrease of the NADW formation and the southward extension of the sea ice (e.g.; Bakke et al. 2009; Thornalley et al. 2011; Cabedo-Sanz et al. 2013; Müller and Stein, 2014). As a consequence to the steep meridional thermal contrast, the polar jet stream strengthen and reached its southernmost position making the westerlies became stronger at the sea ice edge in Central Western Europe (Brauer et al. 2008; Bakke et al. 2009; Rach et al. 2014), which was demonstrated by climate simulations (Isarin et al. 1998). Although westerlies were not so intense at the mid-latitudes, the horizontal thermal contrast between the Western and Eastern North Atlantic would have favour some moisture delivery to Western Iberian Peninsula even if conditions in this region were mainly drier as a response to maxima of AMOC reduction (Naughton et al. 2019). This phase was followed by a progressive warming and increase in moisture conditions, between ~12.4 and 12.0 cal. ka BP (YDc), in South-Western and Central Europe and in the North Atlantic (Fig. 3) (e.g.; Bakke et al. 2009; Rach et al. 2014; Repschlager et al. 2015; Naughton et al. 2019). This warming was favoured by the slight AMOC re-invigorating (McManus et al. 2004; Ng et al. 2018), causing the sea ice and polar jet stream to retreat further north (e.g.; Thornalley et al. 2011; Cabedo-Sanz et al. 2013; Pearce et al. 2013; Müller and Stein, 2014; Gil et al. 2015), as revealed by the latitudinal switch of the westerlies from Central to Northern Europe (Fig. 3) (Bakke et al. 2009). Although the polar jet stream was located in Northern Europe, some moisture was delivered to Central and Southern Europe via the southern branch of the North Atlantic drift (Naughton et al. 2019).

Finally, the YD termination, or the YD-Holocene transition (YDd; 12.0–11.7 cal. ka BP) is marked by very unstable temperature and precipitation conditions in the North Atlantic and over Europe (Fig. 3) (Bakke et al. 2009; Rach et al. 2014; Naughton et al. 2019; Rea et al. 2020). These complex conditions seem likely to be the result of successive shifts in the mode of operation of the AMOC as part of its recovery at the onset of the Holocene (Naughton et al. 2019) and a more unsteady North Atlantic sea-ice cover in the North Atlantic and Nordic seas (Cabedo-Sanz et al. 2013; Pearce et al. 2013; Gil et al. 2015). This contributed to substantial changes in the strength and position of westerlies and a complex hydroclimatic scenario in the Central and Southern Europe (Bakke et al. 2009; Rach et al. 2014; Naughton et al. 2019; Rea et al. 2020).

2.4 The Holocene

The present-day interglacial, the Holocene (Gibbard et al. 2005), started at 11.7 cal. ka BP (Rasmussen et al. 2006; Steffensen et al. 2008; Walker et al. 2009) and was marked by an abrupt warming at the Younger Dryas (YD)-Holocene transition in a variety of records from Europe (e.g.; Mangerud et al. 1974; Bjorck et al. 1996; Litt et al. 2001; Moréllon et al. 2018 and references therein), North Atlantic mid-latitudes (e.g.; Naughton et al. 2007a; b; Fletcher et al. 2010b; Chabaud et al. 2014; Gomes et al. 2020) and in Greenland ice cores (Fig. 4) (Alley et al. 2003; Steffensen et al. 2008; Walker et al. 2009; Vinther et al. 2009). The Holocene interglacial started just after the peak maxima of Northern Hemisphere summer insolation (Fig. 4) (Berger and Loutre, 1991). Interestingly, the warmest conditions (also known as the Holocene Thermal Maximum: HTM) were attained, several millennia later than the peak maxima of Northern Hemisphere summer insolation, caused by the persistence of larger remaining ice sheets in the Northern Hemisphere (mainly LIS, but also Greenland Ice Sheet—GIS) until ~7 cal. ka BP (e.g.; Carlson et al. 2008; Renssen et al. 2009; Blaschek and Renssen, 2013; Bova et al. 2021). Data-model comparison shows that strong summer insolation during the Early Holocene favoured the disintegration of Northern Hemisphere ice sheets and consequent melting that cause a relative weak AMOC (Blaschek and Renssen 2013). This weak AMOC had a cooling effect in the Early Holocene, at least at a regional scale, compensating the relatively strong orbitally-forced boreal insolation. However, the timing of the HTM varied spatially as revealed by numerous terrestrial and marine proxy records and climate simulations (e.g.; Kaufman et al. 2004; 2020; Renssen et al. 2009; 2012). The HTM started earlier than 8 cal. ka BP in regions not strongly affected by the remnant of LIS, while after 8 cal. ka BP in other regions (Renssen et al. 2012). Furthermore, the warming was stronger in the polar regions than in the tropics, where the amplitude of warming was reduced during the HTM (Renssen et al. 2012). Some Southern Europe (e.g., Massif Central pollen-based temperature reconstructions, France) and North Atlantic mid-latitudes (Western Iberian margin SST) records show a progressive warming phase in the early part of the Holocene, between 11.7 ka and 10.5 cal. ka BP and that the HTM was detected earlier than 8 cal. ka BP, around 10.5 cal. ka BP (Rodrigues et al. 2010; Martin et al. 2020). SST in southern Iberian margin attain maximum values at around 10.5 ka BP, while pollen assemblages from the same records show maxima of forest development 1000 years later, i.e. by 9.5 cal. ka BP (e.g., Chabaud et al. 2014; Gomes et al. 2020). As forest, several other terrestrial climate indicators not

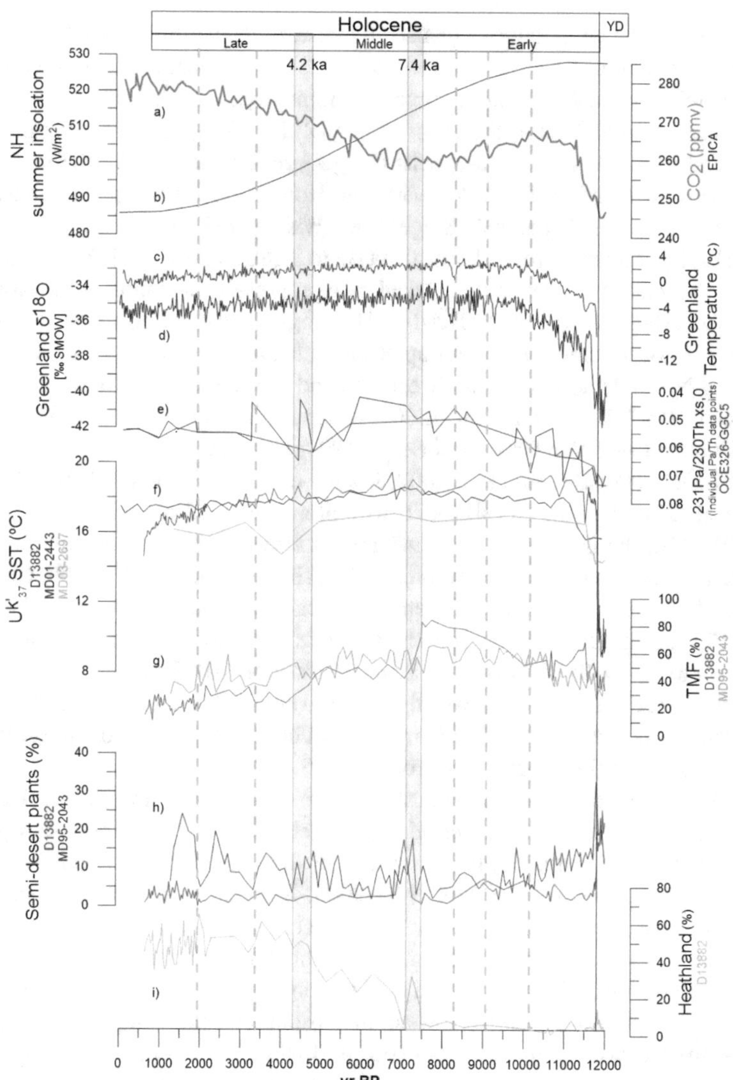

Fig. 4 a) atmospheric CO_2 concentrations reconstructed from EPICA DOME C ice core (EDC) (EPICA community members, 2004); b) summer insolation at 65ºN (Berger and Loutre, 1991); c) Greenland temperature (Vinther et al. 2009); d) Greenland ice core $\delta^{18}O$ record (Rasmussen et al. 2014); f) ex231Pa0/ex230Th0 from composite North Atlantic records (blue line) (black bold line: smoothed record) (Ng et al. 2018) and from the western North Atlantic (light blue) (OCE326-GGC5; McManus et al. 2004); e) alkenone-derived SST records ($U^{k'}_{37}$ SST) from Iberian margin (Martrat et al. 2007; Rodrigues et al. 2009; Naughton et al. 2016; Gomes et al. 2020); g) TMF: Temperate forest abundances (%); h) Semi-desert plants in i) Heathland Southwestern Iberian margin records (Rodrigues et al. 2009; Fletcher et al. 2013; Gomes et al. 2020); Light blue bands: millennial scale events that separates the Holocene Sub-series/Sub-epochs; Dashed light blue lines: other millennial scale events mentioned in the text

only dependent on temperature, but also on moisture availability, show that optimum continental conditions were attained by 9.5 ka BP, or later in the Iberian Peninsula (e.g.; Moréllon et al. 2018 and references therein; Gomes et al. 2020).

Despite the variations in the exact timing of the HTM, the Holocene has been subdivided in 3 Sub-series/Sub-epochs (Stage/Age) (Walker et al. 2019): an Early Holocene (Greenlandian Stage/Age), Middle Holocene (Northgrippian Stage/Age) and Late Holocene (Meghalayan Stage/Age) (Walker et al. 2019).

The Early Holocene, from 11.65 cal. ka BP to 8.186 ka BP (~11.7–8.2 cal. Ka BP) (Walker et al. 2019), is generally marked by a progressive warming in the Northern Hemisphere (Fig. 4) (Wanner et al. 2014). The progressive warming that characterises the Early Holocene ends with the abrupt 8.2 ka cold event (Walker et al. 2019) evident in several North Atlantic records, Greenland and Europe (e.g.; O'Brien et al. 1995; Alley et al. 1997; Bond et al. 1997, 2001; Klitgaard-Kristensen et al. 1998; Bianchi and McCave; 1999; Von Grafenstein et al. 1999; Nesje and Dahl, 2001; Tinner and Lotter, 2001, 2006; Baldini et al. 2002; Magny et al. 2004; Muscheler et al. 2004; Veski et al. 2004; Naughton et al. 2007b; Moréllon et al. 2018 and references therein). In the Western Iberian margin, the most abrupt forest contraction occurred later, at 7.4 cal. ka BP, with no subsequent forest recovering (Fletcher et al. 2010b; Chabaud et al. 2014; Gomes et al. 2020). This centennial scale event has been detected, between 7.5 and 7.0 cal. ka BP, in several other records from the Mediterranean region (Bar-Matthews et al. 1999; Desprat et al. 2013; Siani et al. 2013) as well as from other mid-latitude regions around the world (e.g., Hou et al. 2019 and references therein), supporting evidence that this abrupt event might had an important role in ending the HTM at least in regions located far from the LIS.

The Middle Holocene, from ~8.2 cal. ka BP to 4.2 cal. ka BP (Walker et al. 2019), was considered a relative warm period (Wanner et al. 2014). In Northern Europe, the HTM occurred at this time frame as revealed by data-model comparison and proxy-based reconstructions (e.g.; Davis et al. 2003; Renssen et al. 2009; Seppä et al. 2009). However, several SST North Atlantic records (e.g.; Rodrigues et al. 2009; Leduc et al. 2010) and the globally stacked temperature anomalies between 90º N and 30º N (Marcott et al. 2014) show a gradual cooling during this interval that follow the decrease in Northern Hemisphere summer insolation contrasting with the gradual increase in greenhouse concentrations (Fig. 4). In Southern Europe, vegetation cover and proxy-based reconstructions show a gradual cooling (e.g.; Dormoy et al. 2009; Chabaud et al. 2014; Gomes et al. 2020). However, vegetation changes in the Iberian Peninsula during this period are mainly controlled by precipitation, which in turn varied spatially (e.g., Jalut et al. 2009; Fletcher et al. 2013; Chabaud et al. 2014; Gomes et al. 2020).

The Middle Holocene ends with the well-known 4.2 ka event (e.g.; Mayewski et al. 2004; Renssen et al. 2022) or with the Alboran Cooling (AC) event 2 (Cacho et al. 2001), or the 5.4–4.5 ka event (Fletcher et al. 2013). The main impact of the 4.2 ka event was in the tropics instead of the mid-to-high latitudes of the Northern Hemisphere (e.g.; Bradley and Bakke, 2019). In Western Iberia this event is marked by an important forest contraction with no subsequent recovery (Fig. 4). (e.g.; Fletcher et al. 2013; Chabaud et al. 2014; Gomes et al. 2020).

The Late Holocene, covering the last 4.2 cal. ka BP (Walker et al. 2019) is marked by the continuing SST cooling in the North Atlantic (e.g.; Rodrigues et al. 2009; Leduc et al. 2010; Marcott et al. 2014) and increasing dryness over Europe (Fig. 4) (e.g.; Fletcher et al. 2013; Chabaud et al. 2014; Gomes et al. 2020).

Superimposed on these climatic sub-series, the North Atlantic records detect a millennial-scale variability associated with short and weak events of iceberg discharges with SST anomalies of ca. 2 °C, including both episodes mentioned above (Bond et al. 1997; 2001). Pollen records form the SW Iberian margin and the Mediterranean and Marmara Sea show an alternation of moderate contractions and expansions of the Southern European warm-temperate forest (Fig. 4) (Combourieu-Nebout et al. 2009; Valsecchi et al. 2012; Fletcher et al. 2013; Chabaud et al. 2014). Episodes of forest decline occurred at ca. 10.1, 9.2, 8.3, 7.4, 5.4–4.5 and 3.7–2.9 cal. ka BP, and between 1.9 cal. ka BP and 1.3 cal. ka BP. These episodes indicate repeated drier conditions in Southern Europe that can be tentatively correlated with increases in North Atlantic ice rafting, i.e. the Bond's events. Wavelet analysis of the arboreal pollen percentages record from the Alboran Sea (Western Mediterranean) confirms a ~900-yr periodicity prior to and during the Early Holocene and the dominance of a ~1750-yr periodicity after 6 cal. ka BP. The ~900-yr periodicity has counterparts in numerous North Atlantic and Northern Hemisphere palaeoclimate records. Comparisons between these climatic changes and those recorded in Morocco, Iceland, Norway and Israel suggest that the ~1750-yr Mid- to Late-Holocene oscillation reflects shifts between a prevailing strong and weak state of the westerlies, similar to the positive and negative modes of the North Atlantic Oscillation (Fletcher et al. 2013). The Mid- to Late-Holocene millennial oscillation in the westerlies appears closely coupled to North Atlantic surface ocean circulation dynamics, and may have been driven by an internal oscillation in deep water convection strength (Fletcher et al. 2013).

3 Climate Variability and Environmental Dynamics During the Quaternary

Superimposed to the long-term 100 ka glacial-interglacial cycles of the last 700 ka, characterised by the growth and demise of northern hemisphere ice sheets, millennial scale climate changes have been recorded in ice, marine and terrestrial records. During the LGC, this climatic variability, termed D-O cycles, was particularly large and frequent probably related to the instability of the partial marine-based nature of the LIS and EIS. At a Northern Hemispheric scale, these cycles were synchronous. Some of the D-O cycles were associated with episodes of large iceberg fragmentation of the North American ice caps, called the HS, and related to the complex reorganisation of oceanic and atmospheric circulation leading to distinct regional responses. Millennial-scale climate shifts also punctuated the orbitally induced last deglaciation, namely HS 1, B-A and YD, driven by changes in the Northern Hemisphere ice sheets retreat and consequent changes in both the ocean and atmospheric dynamics. Changes in glacial dynamics had also a great impact in climate evolution during the Holocene until the complete decay of the LIS at around 6.8 cal kyr BP.

References

Alley RB, Clark PU (1999) The deglaciation of the Northern hemisphere: a global perspective. Annu Rev Earth Planet Sci 27(1):149–182

Alley RB, Marotzke J, Nordhaus WD, Overpeck JT, Peteet DM, Pielke RA, Pierrehumbert RT Jr, Rhines PB, Stocker TF, Talley LD, Wallace JM (2003) Abrupt climate change. Science 299:2005–2010

Alley RB, Mayewski PA, Sowers T, Stuiver M, Taylor KC, Clark PU (1997) Holocene climatic instability: a prominent widespread event 8200 years ago. Geology 25:483–486

Alley RB, Meese DA, Shuman CA, Gow AJ, Taylor KC, Grootes PM, White JWC, Ram M, Waddington ED, Mayewski PA, Zielinski GA (1993) Abrupt increase in Greenland snow accumulation at the end of the Younger Dryas event. Nature 362:527–529

Alvarez-Solas J, Charbit S, Ritz C, Paillard D, Ramstein G, Dumas C (2010) Links between ocean temperature and iceberg discharge during Heinrich events. Nat Geosci 3(2):122–126

Alvarez-Solas, J., Ramstein, G., 2011. On the triggering mechanism of Heinrich events: Proceedings of the National Academy of Sciences108, no. 50, E1359-E1360.

Alvarez-Solas J, Robinson A, Montoya M, Ritz C (2013) Iceberg discharges of the last glacial period driven by oceanic circulation changes. Proc Natl Acad Sci 110(41):16350–16354

Andrews JT, Voelker AHL (2018) "Heinrich events" (& sediments): A history of terminology and recommendations for future usage. Quatern Sci Rev 187:31–40

Bakke J, Lie O, Heegaard E, Dokken T, Haug GH, Birks HH, Dulski P, Nilsen T (2009) Rapid oceanic and atmospheric changes during the Younger Dryas cold period. Nature Geosciences 2:202–205

Baldini JUL, McDermott F, Fairchild IJ (2002) Structure of the 8200-year cold event revealed by a speleothem trace element record. Science 296:2203–2206

Baldini LM, McDermott F, Baldini JUL, Arias P, Cueto M, Fairchild IJ, Hoffmann DL, Mattey DP, Müller W, Nita DC, Ontañón R, Garciá-Moncó C, Richards DA (2015) Regional temperature, atmospheric circulation, and sea-ice variability within the Younger Dryas Event constrained using a speleothem from northern Iberia. Proc Natl Acad Sci 419:101–110

Bard E, Rostek F, Turon J-L, Gendreau S (2000) Hydrological impact of Heinrich events in the subtropical northeast Atlantic. Science 289:1321–1324

Barker S, Chen J, Gong X, Jonkers L, Knorr G, Thornalley D (2015) Icebergs not the trigger for North Atlantic cold events. Nature 520(7547):333–336

Barker S, Diz P, Vautravers MJ, Pike J, Knorr G, Hall IR, Broecker WS (2009) Interhemispheric Atlantic seesaw response during the last deglaciation. Nature 457(7233):1097–1102

Bar-Matthews M, Ayalon A, Kaufman A, Wasserburg GJ (1999) The Eastern Mediterranean paleoclimate as a reflection of regional events: Soreq cave. Israel. Earth and Planetary Science Letters 166(1–2):85–95

Bartolomé M, Moreno A, Sancho C, Stoll HM, Cacho I, Spötl C, Belmonte A, Edwards RL, Cheng H, Hellstrom JC (2015) Hydrological change in Southern Europe responding to increasing North Atlantic overturning during Greenland Stadial 1. Proc Natl Acad Sci 112:6568–6572

Bassis JN, Petersen SV, Mac Cathles L (2017) Heinrich events triggered by ocean forcing and modulated by isostatic adjustment. Nature 542(7641):332–334

Batchelor CL, Margold M, Krapp M, Murton DK, Dalton AS, Gibbard PL, Stokes CR, Murton JB, Manica A (2019) The configuration of Northern Hemisphere ice sheets through the Quaternary. Nat Commun 10(1):3713

Berger A, Loutre MF (1991) Insolation values for the climate of the last 10 million years. Quatern Sci Rev 10:297–317

Bianchi GG, McCave IN (1999) Holocene periodicity in North Atlantic climate and deep-ocean flow South of Iceland. Nature 397:515–517

Bjorck S, Kromer B, Johnsen S, Bennike O, Hammarlund D, Lemdahl G, Possnert G, Rasmussen TL, Wohlfarth B, Hammer CU, Spurk M (1996) Synchronised terrestrial–atmospheric deglacial records around the North Atlantic. Science 274:1155–1160

Blaschek M, Renssen H (2013) The Holocene thermal maximum in the Nordic Seas: the impact of Greenland Ice Sheet melt and other forcings in a coupled atmosphere–sea-ice–ocean model. Climate of the past 9(4):1629–1643

Bond GC, Lotti R (1995) Iceberg discharges into the North Atlantic on millennial time scales during the last glaciation. Science 267(5200):1005–1010

Bond G, Broecker W, Johnsen S, McManus J, Labeyrie L, Jouzel J, Bonani G (1993) Correlations between climate records from North Atlantic sediments and Greenland ice. Nature 365:143–147

Bond G, Heinrich H, Broecker W, Labeyrie L, McManus J, Andrews J, Huon S, Jantschik R, Clasen S, Simet C, Tedesco K (1992) Evidence for massive discharges of icebergs into the North Atlantic ocean during the last glacial period. Nature 360(6401):245–249

Bond G, Kromer B, Beer J, Muscheler R, Evans M, Showers W, Hoffmann S, Lotti-Bond R, Hajdas I, Bonani G (2001) Persistent solar influence on North Atlantic climate during the Holocene. Science 294:2130–2136

Bond G, Showers W, Cheseby M, Lotti R, Almasi P, deMenocal P, Priore P, Cullen H, Hajdas I, Bonani G (1997) A pervasive millennial-scale cycle in North Atlantic Holocene and glacial climates. Science 278:1257–1266

Bond, G.C., Showers, W., Elliot, M., Lotti, R., Hajdas, I., Bonani, G., Johnson, S., 1999. The North Atlantic's 1–2 kyr climate rhythm: relation to Heinrich events, Dansgaard/Oeshger cycles and the Little Ice Age, Mechanisms of Global Climate Change at Millennial Time Scale, American Geophysical Union, 35–58.

Boswell SM, Toucanne S, Pitel-Roudaut M, Creyts TT, Eynaud F, Bayon G (2019) Enhanced surface melting of the Fennoscandian Ice Sheet during periods of North Atlantic cooling. Geology 47(7):664–668

Bova S, Rosenthal Y, Liu Z, Godad SP, Yan M (2021) Seasonal origin of the thermal maxima at the Holocene and the last interglacial. Nature 589(7843):548–553

Boyle, E.A., Keigwin, L.D., 1987. North Atlantic thermohaline circulation during the past 20,000 years linked to high-latitude surface temperature: Nature 330, 35–40.

Bradley, R.S., Bakke, J., 2019. Is there evidence for a 4.2 ka BP event in the Northern North Atlantic region?. Climate of the Past 15(5), 1665–1676.

Brauer A, Endres C, Gunter C, Litt T, Stebich M, Nengendank JFW (1999) High resolution sediment and vegetation responses to Younger Dryas climate change in varved lake sediments from Meerfelder Maar, Germany. Quaternary Science Reviwes 18:321–329

Brauer A, Haug GH, Dulski P, Sigman DM, Negendank JFW (2008) An abrupt wind shift in Western Europe at the onset of the Younger Dryas cold period. Nature Geosciences 1:520–523

Broecker W, Bond G, Klas M, Clark E, McManus J (1992) Origin of the Northern Atlantic's Heinrich events. Clim Dyn 6(3):265–273

Broecker WS (1994) Massive iceberg discharges as triggers for global climate change. Nature 372(6505):421–424

Broecker WS (2003) Does the trigger for abrupt climate change reside in the oceans or in the atmosphere? Science 300(5625):1519–1522

Bromley GR, Putnam AE, Rademaker KM, Lowell TV, Schaefer JM, Hall B et al (2014a) Younger Dryas deglaciation of Scotland driven by warming summers. Proc Natl Acad Sci 111(17):6215–6219

Bromley GR, Putnam AE, Rademaker KM, Lowell TV, Schaefer JM, Hall B, Winckler G, Birkel SD, Borns HW (2014b) Younger Dryas deglaciation of Scotland driven by warming summers. Proc Natl Acad Sci 111(17):6215–6219

Brook EJ, Sowers T, Orchardo J (1996) Rapid variations in atmospheric methane concentration during the past 110,000 years. Science 273:1087–1091

Buizert C, Gkinis V, Severinghaus JP, He F, Lecavalier BS, Kindler P et al (2014a) Greenland temperature response to climate forcing during the last deglaciation. Science 345(6201):1177–1180

Buizert C, Gkinis V, Severinghaus JP, He F, Lecavalier BS, Kindler P, Leuenberger M, Carlson AE, Vinther B, Masson-Delmotte V, White JW (2014b) Greenland temperature response to climate forcing during the last deglaciation. Science 345(6201):1177–1180

Cabedo-Sanz P, Belt ST, Knies J, Husum K (2013) Identification of contrasting seasonal sea ice conditions during the Younger Dryas. Quaternary Science Review 79:74–86

Cacho I, Grimalt JO, Canals M, Sbaffi L, Shackleton NJ, Schönfeld J, Zahn R (2001) Variability of the western Mediterranean Sea surface temperature during the last 25,000 years and its connection with the Northern Hemisphere climatic changes. Paleoceanography 16(1):40–52

Carlson AE (2013) The Younger Dryas Climate Event. Encyclopedia of Quaternary Science 3:126–134

Carlson, A.E., Clark, P.U., 2012. Ice sheet sources of sea level rise and freshwater discharge during the last deglaciation. Reviews of Geophysics 50(4).

Carlson AE, Clark PU, Haley BA, Klinkhammer GP, Simmons K, Brook EJ, Meissner KJ (2007) Geochemical proxies of North American freshwater routing during the Younger Dryas cold event. Proc Natl Acad Sci 104:6556–6561

Carlson AE, Oppo DW, Came RE, LeGrande AN, Keigwin LD, Curry WB (2008) Subtropical Atlantic salinity variability and Atlantic meridional circulation during the last deglaciation. Geology 36:991–994

Chabaud L, Sánchez Goñi MF, Desprat S, Rossignol L (2014) Land-sea climatic ~ variability in the eastern North Atlantic subtropical region over the last 14,200 years: atmospheric and oceanic processes at different timescales. The Holocene 24:787–797

Chapman MR, Shackleton NJ, Duplessy JC (2000) Sea surface temperature variability during the last glacial-interglacial cycle: assessing the magnitude and pattern of climate change in the North Atlantic. Palaeogeogr Palaeoclimatol Palaeoecol 157:1–25

Clark PU, Dyke AS, Shakun JD, Carlson AE, Clark J, Wohlfarth B, Mitrovica JX, Hostetler SW, McCabe AM (2009) The last glacial maximum. Science 325(5941):710–714

Clark PU, Hostetler SW, Pisias NG, Schmittner A, Meissner KJ (2007) Mechanisms for an~ 7-kyr climate and sea-level oscillation during marine isotope stage 3. Geophysical Monograph-American Geophysical Union 173:209

Clark PU, Shakun JD, Baker PA, Bartlein PJ, Brewer S, Brook E, Carlson AE, Cheng H, Kaufman DS, Liu Z, Marchitto TM (2012) Global climate evolution during the last deglaciation. Proc Natl Acad Sci 109(19):E1134–E1142

Combourieu Nebout N, Peyron O, Dormoy I, Desprat S, Beaudouin C, Kotthoff U, Marret F (2009) Rapid climatic variability in the west Mediterranean during the last 25 000 years from high resolution pollen data. Climate of the past 5(3):503–521

Cortijo E, Labeyrie L, Vidal L, Vautravers M, Chapman M, Duplessy JC, Elliot M, Arnold M, Turon JL, Auffret G (1997) Changes in sea surface hydrology associated with Heinrich event 4 in the North Atlantic Ocean between 40°N and 60°N. Earth Planet Sci Lett 146:29–45

Dansgaard W, Clausen HB, Gundestrup N, Hammer CU, Johnsen SF, Kristinsdottir PM, Reeh N (1982) A new Greenland deep ice core. Science 218(4579):1273–1277

Dansgaard W, Johnsen S, Clausen HB, Dahl-Jensen D, Gundestrup N, Hammer CU, Oeschger H (1984) North Atlantic climatic oscillations revealed by deep Greenland ice cores. In: Hansen JE, Takahashi T (eds) Climate processes and climate sensitivity: Washington. American Geophysical Union, pp 288–298

Davis BAS, Brewer S, Stevenson AC, Guiot J, Contributors D (2003) The temperature of Europe during the Holocene reconstructed from pollen data. Quatern Sci Rev 22:1701–1716

de Abreu L, Shackleton NJ, Schonfeld J, Hall M, Chapman M (2003) Millennial-scale oceanic climate variability off the W Iberian margin during the last two glacial periods. Mar Geol 196:1–20

Denton GH, Alley RB, Comer GC, Broecker WS (2005) The role of seasonality in abrupt climate change. Quatern Sci Rev 24(10–11):1159–1182

Denton GH, Anderson RF, Toggweiler JR, Edwards RL, Schaefer JM, Putnam AE (2010) The last glacial termination. Science 328(5986):1652–1656

Denton GH, Hughes TJ (1981) The Last Great Ice Sheets. Wiley Interscience, New York, p 484

Desprat S, Combourieu-Nebout N, Essallami L, Sicre MA, Dormoy I, Peyron O, Siani G, Bout Roumazeilles V, Turon JL (2013) Deglacial and Holocene vegetation and climatic changes in the southern Central Mediterranean from a direct land–sea correlation. Climate of the past 9(2):767–787

Dormoy I, Peyron O, Combourieu Nebout N, Goring S, Kotthoff U, Magny M, Pross J (2009) Terrestrial climate variability and seasonality changes in the Mediterranean region between 15 000 and 4000 years BP deduced from marine pollen records. Climate of the past 5:615–632

Eisenman I, Bitz CM, Tziperman E (2009) Rain driven by receding ice sheets as a cause of past climate change. Paleoceanography 24(4)

Elliot M, Labeyrie L, Bond G, Cortijo E, Turon J-L, Tisnerat N, Duplessy J-C (1998) Millennial-scale iceberg discharges in the Irminger Basin during the last glacial period: Relationship with the Heinrich events and environmental settings. Paleoceanography 13:433–446

Elliot M, Labeyrie L, Dokken T, Manthé S (2001) Coherent patterns of ice-rafted debris deposits in the Nordic regions during the last glacial (10–60 ka): Earth and Planetary Science Letters 194, 151–163

EPICA community members (2004) EPICA community members: 8 glacial cycles from an Antarctic ice core. Nature 429:623–628

Eynaud F, de Abreu L, Voelker A, Schönfeld J, Salgueiro E, Turon J.-L, Penaud A, Toucanne S, Naughton F, Sánchez-Goñi MF, Malaizé B, Cacho I (2009) Position of the Polar Front along the western Iberian margin during key cold episodes of the last 45 ka. Geochemistry, Geophysics, Geosystems Q07U05. https://doi.org/10.1029/2009GC002398. ISSN: 1525–2027.

Eynaud F, Zaragosi S, Scourse J, Mojtahid M, Bourillet JF, Hall IR, Penaud A, Locascio M, Reijonen A (2007) Deglacial laminated facies on the NW European continental margin: the hydrographic significance of British-Irish Ice Sheet deglaciation and Fleuve Manche paleoriver discharges. Geochem Geophys Geosyst 8. https://doi.org/10.1029/2006GC001496

Fersi W, Penaud A, Wary M, Toucanne S, Waelbroeck C, Rossignol L, Eynaud F (2021) Imprint of seasonality changes on fluvio-glacial dynamics across Heinrich Stadial 1 (NE Atlantic Ocean). Global Planet Change 204. https://doi.org/10.1016/j.gloplacha.2021.103552

Firestone RB, West A, Kennett JP, Becker L, Bunch TE, Revay ZS, Schultz PH, Belgya T, Kennett DJ, Erlandson JM, Dickenson OJ, Goodyear AC, Harris RS, Howard GA, Kloosterman JB, Lechler P, Mayewski PA, Montgomery J, Poreda R, Darrah T, Que Hee SS, Smith AR, Stich A, Topping W, Wittke JH, Wolbach WS (2007) Evidence for an extraterrestrial impact 12,900 years ago that contributed to the megafaunal extinctions and the Younger Dryas cooling. Proc Natl Acad Sci 104(41):16016–16021

Fletcher WJ, Sánchez Goñi MF, Allen JRM, Cheddadi R, Combourieu-Nebout N, Huntley B, Lawson I, Londeix L, Magri D, Margari V, Müller UC, Naughton F, Novenko E, Roucoux K, Tzedakis PC (2010a) Millennial-scale variability during the last glacial in vegetation records from Europe. Quatern Sci Rev 29(21–22):2839–2864

Fletcher WJ, Debret M, Goñi MFS (2013) Mid-Holocene emergence of a low-frequency millennial oscillation in western Mediterranean climate: Implications for past dynamics of the North Atlantic atmospheric westerlies. The Holocene 23(2):153–166

Fletcher WJ, Sánchez Goñi MF (2008) Orbital- and sub-orbital-scale climate impacts on vegetation of the western Mediterranean basin over the last 48,000 yr. Quatern Res 70:451–464

Fletcher WJ, Sánchez Goñi MF, Peyron O, Dormoy I (2010b) Abrupt climate changes of the last deglaciation detected in a western Mediterranean forest record. Climate of the past 6:245–264

Ganopolski A, Rahmstorf S (2001) Rapid Changes of glacial climate simulated in a coupled climate model. Nature 409:153–158

Genty D, Blamart D, Ghaleb B, Plagnes V, Causse C, Bakalowicz M, Zouari K, Chkir N, Hellstrom J, Wainer K, Bourges F (2006) Timing and dynamics of the last deglaciation from European and North African d13C stalagmite profiles d comparison with Chinese and South Hemisphere stalagmites. Quatern Sci Rev 25:2118–2142

Gibbard PL, Smith AG, Zalasiewicz J, Barry TL, Cantrill D, Coe AL, Cope JWC, Gale AS, Gregory FJ, Powell JH, Rawson PF, Stone P (2005) What status for the Quaternary? Boreas 34:1–6

Gil IM, Keigwin LD, Abrantes F (2015) The deglaciation over Laurentian Fan: History of diatoms, IRD, ice and fresh water. Quatern Sci Rev 129:57–67

Gomes SD, Fletcher WJ, Rodrigues T, Stone A, Abrantes F, Naughton F (2020) Time-transgressive Holocene maximum of temperate and Mediterranean forest development across the Iberian Peninsula reflects orbital forcing. Palaeogeogr Palaeoclimatol Palaeoecol 550:109739. https://doi.org/10.1016/j.palaeo.2020.109739

Grousset FE, Labeyrie L, Sinko JA, Cremer M, Bond G, Duprat J, Cortijo E, Huon S (1993) Patterns of ice-rafted detritus in the glacial North Atlantic (40–55°N). Paleoceanography 8:175–192

Guillevic M, Bazin L, Landais A, Stowasser C, Masson-Delmotte V, Blunier T, Eynaud F, Falourd S, Michel E, Minster B, Popp T, Prié F, Vinther BM (2014) Evidence for a three-phase sequence during Heinrich Stadial 4 using a multiproxy approach based on Greenland ice core records: Climate of the Past 10(6):2115–2133

Hartz N, Milthers V (1901) Det senglaciale Ler i Alleröd Teglvaerksgrav. Meddelelser Danmarks Geologisk Forening 8:31–59

He C, Liu Z, Otto-Bliesner BL, Brady EC, Zhu C, Tomas R, Buizert C., Severinghaus JP (2021) Abrupt Heinrich Stadial 1 cooling missing in Greenland oxygen isotopes. Science Adavances 7(25):eabh1007. https://doi.org/10.1126/sciadv.abh1007

He C, Liu Z, Zhu J, Zhang J, Gu S, Otto-Bliesner BL, Brady E, Zhu C, Jin Y, Sun J (2020) North Atlantic subsurface temperature response controlled by effective freshwater input in "Heinrich" events. Earth Planet Sci Lett 539:116247

Heinrich H (1988) Origin and consequences of cyclic ice rafting in the northeast Atlantic ocean during the past 130,000 years. Quatern Res 29:142–152

Hemming SR (2004) Heinrich events: massive late Pleistocene detritus layers of the North Atlantic and their global climate imprint. Reviews of Geophysics, 42.

Henry LG, McManus JF, Curry WB, Roberts NL, Piotrowski AM, Keigwin LD (2016) North Atlantic ocean circulation and abrupt climate change during the last glaciation: Science 353(6298):470–474

Hodell DA, Nicholl JA, Bontognali TR, Danino S, Dorador J, Dowdeswell JA, Einsle J, Kuhlmann H, Martrat B, Mleneck-Vautravers MJ, Rodríguez-Tovar FJ (2017) Anatomy of Heinrich Layer 1 and its role in the last deglaciation. Paleoceanography 32(3):284–303

Hoek ZW (2009) "Bølling-Allerød Interstadial". Encyclopedia of Paleoclimatology and Ancient Environments. Encyclopedia of Earth Sciences Series. Encyclopedia of Earth Sciences Series, 100–103. https://doi.org/10.1007/978-1-4020-4411-3_26. ISBN 978–1–4020–4551–6

Hou M, Wu W, Cohen DJ, Zhou Y, Zeng Z, Huang H, Zheng H, Ge Q (2019) Evidence for a widespread climatic anomaly at around 7.5–7.0 cal ka BP. Climate of the Past Discussions 1–50

Hughes PD, Gibbard PL, Ehlers J (2013) Timing of glaciation during the last glacial cycle: evaluating the concept of a global 'Last Glacial Maximum' (LGM): Earth-Science Reviews125, 171–198

Irvalı N, Ninnemann US, Kleiven HKF, Galaasen EV, Morley A, Rosenthal Y (2016) Evidence for regional cooling, frontal advances, and East Greenland Ice Sheet changes during the demise of the last interglacial. Quatern Sci Rev 150:184–199

Isarin RFB, Renssen H, Vandenberghe J (1998) The impact of the North Atlantic ocean on the Younger Dryas climate in north-western and central Europe. J Quat Sci 13:447–453

Iversen J (1942) En pollenanalytisk Tidsfaestelse af Ferskvandslagene ved Norre Lingby. Meddelelser Danmarks Geologisk Forening 10:130–151

Iversen J (1954) The late-glacial flora of Denmark and its relation to climate and soil. – Danm. Geol. Unders. Ser. II 80:87–119

Jalut G, Dedoubat JJ, Fontugne M, Otto T (2009) Holocene circumMediterranean vegetation changes: climate forcing and human impact. Quatern Int 200:4–18

Johnsen SJ, Clausen HB, Dansgaard W, Fuhrer K, Gundestrup N, Hammer CU, Iversen P, Jouzel J, Stauffer B, Steffensen JP (1992) Irregular glacial interstadials in a new Greenland ice core. Nature 359:311–313

Kaufman D, McKay N, Routson C, Erb M, Davis B, Heiri O, Jaccard S, Tierney J, Dätwyler C, Axford Y, Brussel T (2020) A global database of Holocene paleotemperature records. Scientific Data 7(1):1–34

Kaufman DS, Ager TA, Anderson NJ, Anderson PM, Andrews JT, Bartlein PJ, Brubaker LB, Coats LL, Cwynar LC, Duvall ML, Dyke AS (2004) Holocene thermal maximum in the western Arctic (0–180 W). Quatern Sci Rev 23(5–6):529–560

Keffer T, Martinson DG, Corliss BH (1988) The position of the Gulf Stream during Quaternary glaciations. Science 241(4864):440–442

Kennett DJ, Kennett JP, West A, Mercer C, Hee SSQ, Bement L, Bunch TE, Sellers M, Wolbach WS (2009) Nanodiamonds in the Younger Dryas boundary sediment layer. Science 323:94

Klitgaard-Kristensen D, Sejrup HP, Haflidason H, Johnsen S, Spurk M (1998) A regional 8200 cal. yr BP cooling event in northwest Europe, induced by final stages of the Laurentide ice-sheet deglaciation? Journal of Quaternary Science 13(2):165–169

Landais A, Capron E, Masson-Delmotte V, Toucanne S, Rhodes R, Popp T, Vinther B, Minster B, Prié F (2018) Ice core evidence for decoupling between midlatitude atmospheric water cycle and Greenland temperature during the last deglaciation. Climate of the Past 14(10): 1405–1415

Landais A, Sánchez Goñi MF, Toucanne S, Rodrigues T, Naughton F (2022) Abrupt climatic variability: Dansgaard-Oeschger events. In Palacios, D., Hugues, P. D., Garcia-Ruiz, J. M., and Andrés, N. (eds) European Glacial Landscapes, Elsevier

Lane CS, Brauer A, Blockley SPE, Dulski P (2013) Volcanic ash reveals time-transgressive abrupt climate change during the Younger Dryas. Geology 41:1251–1254

Lebreiro SM, Moreno JC, McCave IN, Weaver PPE (1996) Evidence for Heinrich layers off Portugal (Tore Seamount: 39°N, 12°W). Mar Geol 131:47–56

Leduc G, Schneider R, Kim JH, Lohmann G (2010) Holocene and Eemian sea surface temperature trends as revealed by alkenone and Mg/Ca paleothermometry. Quatern Sci Rev 29(7–8):989–1004

Lisiecki L, Raymo ME (2005) A Pliocene-Pleistocene stack of 57 globally distributed benthic δ 18 O records. Paleoceanography 20, PA1003.

Litt T, Brauer A, Goslar T, Merkt K, Balaga K, Muller H, Ralska-Jasiewiczowa M, Stebich M, Negendank JFW (2001) Correlation and synchronisation of lateglacial continental sequences in Northern central Europe based on annually laminated lacustrine sediments. Quatern Sci Rev 20:1233–1249

Litt T, Stebich M (1999) Bio- and chronostratigraphy of the Lateglacial in the Eifel region, Germany. Quatern Int 61:5–16

Liu Z, Otto-Bliesner B, He F, Brady E, Thomas R, Clark PU, Carlson AE, LynchStieglitz J, Curry W, Brook E, Erickson D, Jacob R, Kutzbach J, Chen J (2009) Transient climate simulation of last deglaciation with a new mechanism for Bølling-Allerød warming. Science 325:310–314

Ljungqvist FC (2011) The spatio-temporal pattern of the mid-Holocene thermal maximum. Geografie 116:91–110

Lozier MS, Li F, Bacon S, Bahr F, Bower AS, Cunningham SA, Jong MFd, Steur Ld, deYoung B, Fischer J, Gary SF, Greenan BJW, Holliday NP, Houk A, Houpert L, Inall ME, Johns WE, Johnson HL, Johnson C, Karstensen J, Koman G, Bras IAL, Lin X, Mackay N, Marshall DP, Mercier H, Oltmanns M, Pickart RS, Ramsey AL, Rayner D, Straneo F, Thierry V, Torres DJ, Williams RG, Wilson C, Yang J, Yashayaev I, Zhao J (2019) A sea change in our view of overturning in the subpolar North Atlantic. Science 363(6426):516–521

Lynch-Stieglitz J (2017) The Atlantic Meridional Overturning Circulation and abrupt climate change. Ann Rev Mar Sci 9:83–104

Magny M, Bégeot C (2004) Hydrological changes in the European midlatitudes associated with freshwater outbursts from Lake Agassiz during the Younger Dryas event and the early Holocene. Quatern Res 61(2):181–192

Manabe S, Stouffer RJ (1997) Coupled ocean-atmosphere model response to freshwater input: Comparison to Younger Dryas event. Paleoceanography 12:321–336

Mangerud J (2021) The discovery of the Younger Dryas, and comments on the current meaning and usage of the term. Boreas 50:1–5. https://doi.org/10.1111/bor.12481

Mangerud J, Andersen ST, Berglund BE, Donner JJ (1974) Quaternary stratigraphy of Norden, a proposal for terminology and classification. Boreas 3:109–126

Marcott SA, Bauska TK, Buizert C, Steig EJ, Rosen JL, Cuffey KM, Fudge TJ, Severinghaus JP, Ahn J, Kalk ML, McConnell JR (2014) Centennial-scale changes in the global carbon cycle during the last deglaciation. Nature 514:616–619

Marcott SA, Clark PU, Padman L, Klinkhammer GP, Springer SR, Liu Z, Otto-Bliesner BL, Carlson AE, Ungerer A, Padman J, He F (2011) Ice-shelf collapse from subsurface warming as a trigger for Heinrich events. Proc Natl Acad Sci 108(33):13415–13419

Martin C, Menot G, Thouveny N, Peyron O, Andrieu-Ponel V, Montade V, Davtian N, Reille M, Bard E (2020) Early Holocene Thermal Maximum recorded by branched tetraethers and pollen in Western Europe (Massif Central, France). Quatern Sci Rev 228:106–109

Martinez-Lamas R, Toucanne S, Debret M, Riboulot V, Deloffre J, Boissier A, Cheron S, Pitel M, Bayon G, Giosan L, Soulet G (2020) Linking Danube River activity to Alpine Ice-Sheet fluctuations during the last glacial (ca. 33–17 ka BP): Insights into the continental signature of Heinrich Stadials. Quaternary Science Reviews 229, 106136.

Martrat B, Grimalt JO, Shackleton NJ, de Abreu L, Hutterli MA, Stocker TF (2007) Four climate cycles of recurring deep and surface water destabilizations on the Iberian Margin. Science 317:502–507

Martrat B, Jimenez-Amat P, Zahn R, Grimalt JO (2014) Similarities and dissimilarities between the last two deglaciations and interglaciations in the North Atlantic region. Quatern Sci Rev 99:122–134

Mayewski PA, Rohling EE, Stager JC, Karlen W, Maasch, ´ KA, Meeker LD, Meyerson EA, Gasse F, van Kreveld S, Holmgren K, Lee-Thorp J, Rosqvist G, Rack F, Staubwasser M, Schneider RR, Steig EJ (2004) Holocene climate variability. Quatern Res 62:243–255

Mayr C, Stojakowits P, Lempe B, Blaauw M, Diersche V, Grohganz M, Correa ML, Ohlendorf C, Reimer P, Zolitschka B (2019) High-resolution geochemical record of environmental changes during MIS 3 from the Northern Alps (Nesseltalgraben, Germany). Quatern Sci Rev 218:122–136

McManus JF, Bond GC, Broecker WS, Johnsen S, Labeyrie L, Higgins S (1994) High-resolution climate records from the North Atlantic during the last interglacial. Nature 371:326–329

McManus JF, Francois R, Gherardi J-M, Keigwin LD, Brown-Leger S (2004) Collapse and rapid resumption of Atlantic meridional circulation linked to deglacial climate changes. Nature 428:834–837

McManus JF, Oppo DW, Cullen JL (1999) A 0.5-million-year record of millennial-scale climate variability in the North Atlantic. Science 283(5404), 971–975

Ménot G, Bard E, Rostek F, Weijers JW, Hopmans EC, Schouten S, Damsté JSS (2006) Early reactivation of European rivers during the last deglaciation. Science 313(5793):1623–1625

Menviel L, Timmermann A, Friedrich T, England MH (2014) Hindcasting the continuum of Dansgaard-Oeschger variability: mechanisms, patterns and timing. Climate of the past 10(1):63–77

Mix AC, Ruddiman WF (1985) Structure and timing of the last deglaciation: Oxygen-isotope evidence. Quatern Sci Rev 4(2):59–108

Moine O, Antoine P, Hatté C, Landais A, Mathieu J, Prud'homme C, Rousseau D.-D (2017) The impact of Last Glacial climate variability in west-European loess revealed by radiocarbon dating of fossil earthworm granules. Proceedings of the National Academy of Sciences 114(24):6209–6214

Morellon M, Aranbarri J, Moreno A, Gonzalez-Samperiz P, Valero-Garces BL (2018) Early Holocene humidity patterns in the Iberian Peninsula reconstructed from lake, pollen and speleothem records. Quatern Sci Rev 181:1–18

Moreno A, Stoll HM, Jimenez-Sánchez M, Cacho I, Valero-Garces B, Ito E, Edwards LR (2010) A speleothem record of rapid climatic shifts during last glacial period from Northern Iberian Peninsula. Global Planet Change 71:218–231. https://doi.org/10.1016/j.gloplacha.2009.10.002

Müller J, Stein R (2014) High-resolution record of late glacial and deglacial sea ice changes in Fram Strait corroborates ice-ocean interactions during abrupt climate shifts. Earth Planet Sci Lett 403:446–455

MÜLLER, Ulrich C et al. (2011) The role of climate in the spread of modern humans into Europe. Quaternary Science Reviews 30(3–4):273–279

Murton JB, Bateman MD, Dallimore SR, Teller JT, Yang Z (2010) Identification of Younger Dryas outburst flood path from Lake Agassiz to the Arctic Ocean. Nature 464:740–743

Muscheler R, Beer J, Vonmoos M (2004) Causes and timing of the 8200 yr BP event inferred from the comparison of the GRIP 10Be and the tree ring Δ14C record. Quatern Sci Rev 23:2101–2111

Muschitiello F, Pausata FSR, Watson JE, Smittenberg RH, Salih AAM, Brooks SJ, Whitehouse NJ, Karlatou-Charalampopoulou A, Wohlfarth B (2015) Fennoscandian freshwater control on Greenland hydroclimate shifts at the onset of the Younger Dryas. Nat Commun 6:8939

Naughton F, Bourillet J-F, Sánchez Goñi MF, Turon J-L, Jouanneau J-M (2007a) Long-term and millennial-scale climate variability in north-western France during the last 8 850 years. The Holocene 17:939–953

Naughton F, Sánchez Goñi MF, Desprat S, Turon J-L, Duprat J, Malaizé B, Joly C, Cortijo E, Drago T, Freitas MC (2007b) Present-day and past (last 25 000 years) marine pollen signal off western Iberia. Mar Micropaleontol 62:91–114

Naughton F, Sánchez Goñi MF, Kageyama M, Bard E, Cortijo E, Desprat S, Duprat J, Malaizé B, Joli C, Rostek F, Turon J-L (2009) Wet to dry climatic trend in north western Iberia within Heinrich events. Earth Planet Sci Lett 284:329–342

Naughton F, Sánchez Goñi MF, Landais A, Rodrigues T, Vazquez-Riveiros N, Toucanne S (in press) The Bølling-Allerød Interstadial. European Glacial Landscapes; Last Deglaciation; volume 2, Part II Climate changes during the Last Deglaciation in the Eastern North Atlantic region; Chapter 6

Naughton F, Sánchez Goñi MF, Rodrigues T, Salgueiro E, Costas S, Desprat S, Duprat J, Michel E, Rossignol L, Zaragosi S, Abrantes F (2016) Climate variability across the last deglaciation in NW Iberia and its margin. Quatern Int 414:9–22. https://doi.org/10.1016/j.quaint.2015.08.073

Naughton F, Costas S, Gomes SD, Rodrigues T, Desprat S, Bronk-Ramsey C, Salgueiro E, Sanchez Goñi MF, Renssen H, Trigo, R, Oliveira, Zoelker AHL, Abrantes F. 2019. Coupled ocean and atmospheric changes during the Younger Dryas in southwestern Europe. Quaternary science Reviews 212:108–120. https://doi.org/10.1016/j.quascirev 2019.03.033

Nesje A, Dahl SO (2001) The Greenland 8200 cal yr BP event detected in loss-on ignition profiles in Norwegian lacustrine sediment sequences. J Quat Sci 16:155–166

Ng, H.C., Robinson, L.F., McManus, J.F., Mohamed, K.J, Jacobel, A.W., Ivanovic, R.F., Gregoire, L.J., Chen, T., 2018. Coherent deglacial changes in western Atlantic Ocean circulation. Nature Communications 9(1). https://doi.org/10.1038/s41467-018-05312.

O'Brien SR, Mayewski PA, Meeker LD, Meese DA, Twickler MS, Whitlow SI (1995) Complexity of Holocene climate as reconstructed from a Greenland ice core. Science 270:1962–1964

Obase T, Abe-Ouchi A (2019) Abrupt Bølling-Allerød warming simulated under gradual forcing of the last deglaciation. Geophysical Research Letters 46(20):11397–405

Obreht I, Wörmer L, Brauer A, Wendt J, Alfken S, De Vleeschouwer D, Elvert M, Hinrichs KU (2020) An annually resolved record of Western European vegetation response to Younger Dryas cooling. Quatern Sci Rev 231:106198

Oeschger H, Beer J, Siegenthaler U, Stauffer B, Dansgaard W, Langway CC (1984) Late glacial climate history from ice cores. Climate Processes and Climate Sensitivity 29:299–306

Oppo DW, McManus JF, Cullen JL (2006) Evolution and demise of the Last Interglacial warmth in the subpolar North Atlantic. Quatern Sci Rev 25:3268–3277

PAGES 2k Consortium (2013) Continental-scale temperature variability during the last two millennia. Nat Geosci 6:339–346

Pailler D, Bard E (2002) High frequency palaeoceanographic changes during the past 140 000 yr recorded by the organic matter in sediments of the Iberian Margin. Palaeogeogr Palaeoclimatol Palaeoecol 181:431–452

Palacios D, Hughes PD, García-Ruiz JM, Andrés N (2022) European Glacial Landscapes. Elsevier, Maximum extent of glaciations

Pearce C, Seidenkrantz MS, Kuijpers A, Masse G, Reynisson NF, Kristiansen, SM (2013) Ocean lead at the termination of the Younger Dryas cold spell. Nature Communications 4

Peck VL, Hall IR, Zahn R, Elderfield H, Grousset F, Hemming SR, Scourse JD (2006) High resolution evidence for linkages between NW European ice sheet instability and Atlantic Meridional Overturning Circulation. Earth Planet Sci Lett 243(3–4):476–488

Penaud A, Eynaud F, Turon JL, Zaragosi S, Malaize B, Toucanne S, Bourillet J-F (2009) What forced the collapse of European ice sheets during the last two glacial periods (150 ka BP and 18 ka cal BP)? Palynological evidence. Palaeogeogr Palaeoclimatol Palaeoecol 281(1–2):66–78

Pérez-Mejías C, Moreno A, Bernal-Wormull J, Cacho I, Osácar MC, Lawrence E, Cheng H (2021) Oldest Dryas hydroclimate reorganization in the eastern Iberian Peninsula after the iceberg discharges of Heinrich Event 1. Quatern Res 101:67–83

Rach O, Brauer A, Wilkes H, Sachse D (2014) Delayed hydrological response to Greenland cooling at the onset of the Younger Dryas in western Europe. Nature Geosciences 7(1):109–112

Rahmstorf S (1994) Rapid climate transitions in a coupled ocean–atmosphere model. Nature 372(6501):82–85

Rahmstorf S (2002) Ocean circulation and climate during the past 120,000 years. Nature 419:207–214

Railsback LB, Gibbard PL, Head MJ, Voarintsoa NRG, Toucanne S (2015). An optimized scheme of lettered marine isotope substages for the last 1.0 million years, and the climatostratigraphic nature of isotope stages and substages. Quaternary Science Reviews 111(0):94–106

Rasmussen SO, Andersen KK, Svensson AM, Steffensen JP, Vinther BM, Clausen HB, Ruth U (2006). A new Greenland ice core chronology for the last glacial termination. Journal of Geophysical Research 111(D6), [D06102]. https://doi.org/10.1029/2005JD006079

Rasmussen SO, Bigler M, Blockley SP, Blunier T, Buchardt SL, Clausen HB, Cvijanovic I, Dahl-Jensen D, Johnsen SJ, Fischer H, Gkinis V, Guillevic M, Hoek WZ, Lowe JJ, Pedro JB, Popp T, Seierstad IK, Steffensen JP, Svens-son AM, Vallelonga P, Vinther BM, Walker MJC, Wheatley JJ, Winstrup M (2014) A stratigraphic framework for abrupt climatic changes during the Last Glacial period based on three synchronized Greenland ice-core records: refining and extending the INTIMATE event stratigraphy. Quatern Sci Rev 106:14–28

Rasmussen TL, Thomsen E, van Weering TCE, Labeyrie L (1996) Rapid changes in surface and deep water conditions at the Faeroe Islands Margin during the last 58 ka. Paleoceanography 11:757–771

Rea BR, Pellitero R, Spagnolo M, Hughes PD, Ivy-Ochs S, Renssen H, Ribolini A, Bakke J, Lukas S, Braithwaite RJ (2020) Atmospheric circulation over Europe during the Younger Dryas. Science Advances 6(50):eaba4844

Reinig F, Wacker L, Jöris O, Oppenheimer C, Guidobaldi G, Nievergel D, Adolphi F, Cherubini P, Engels S, Esper J, Land A, Lane C, Pfanz H, Remmele S, Sigl M, Sookdeo A, Büntgen U (2021) Precise date for the Laacher See eruption synchronizes the Younger Dryas. Nature 595:66–69

Renssen H (2022) Climate model experiments on the 4.2 ka event: The impact of tropical sea-surface temperature anomalies and desertification. The Holocene 32(5):378–389

Renssen H, Isarin RFB (1997) Surface temperature in NW Europe during the Younger Dryas: AGCM simulation compared with temperature reconstructions. Clim Dyn 14(1):33–44

Renssen H, Isarin RFB (2001) The two major warming phases of the last deglaciation at similar to 14.7 and similar to 11.5 ka cal BP in Europe: Climate reconstructions and AGCM experiments. Global Planet Change 30:117–153

Renssen H, Mairesse A, Goosse H, Mathiot P, Heiri O, Roche DM, Nisancioglu KH, Valdes PJ (2015) Multiple causes of the Younger Dryas cold period. Nature Geosciences 8:946–949

Renssen H, Seppä H, Crosta X, Goosse H, Roche DM (2012) Global characterization of the Holocene thermal maximum. Quatern Sci Rev 48:7–19

Renssen H, Seppä H, Heiri O, Roche DM, Goosse H, Fichefet T (2009) The spatial and temporal complexity of the Holocene thermal maximum. Nature Geosciences 2(6):411–414

Repschläger, J., Weinelt, M., Kinkel, H., Andersen, N., Garbe-Sch€onberg, D., Schwab, C., 2015. Response of the subtropical North Atlantic surface hydrography on deglacial and Holocene AMOC changes. Paleoceanography. https://doi.org/10.1002/2014PA002637.

Ritz SP, Stocker TF, Grimalt JO, Menviel L, Timmermann A (2013) Estimated strength of the Atlantic overturning circulation during the last deglaciation. Nature Geosciences 6(3):208–212. https://doi.org/10.1038/ngeo1723

Roche D, Paillard D, Cortijo E (2004) Constraints on the duration and freshwater release of Heinrich event 4 through isotope modelling. Nature 432(7015):379–382

Rodrigues T, Grimalt JO, Abrantes F, Naughton F, Jose-Abel Flores J-A (2010) The last glacial-interglacial transition (LGIT) in the eastern mid-latitudes of the North Atlantic: abrupt sea surface temperature change and sea level implications. Quatern Sci Rev 29:1853–1862

Rodrigues T, Grimalt JO, Abrantes FG, Flores JA, Lebreiro SM (2009) Holocene interdependences of changes in sea surface temperature, productivity, and fluvial inputs in the Iberian continental shelf (Tagus mud patch). Geochemistry, Geophysics, Geosystems 10(7).

Roucoux KH, de Abreu L, Shackleton NJ, Tzedakis PC (2005) The response of NW Iberian vegetation to North Atlantic climate oscillations during the last 65 kyr. Quatern Sci Rev 24(14–15):1637–1653

Salgueiro E, Naughton F, Voelker AHL, de Abreu L, Alberto A, Rossignol L, Duprat J, Magalhães VH, Vaqueiro S, Turon J-L, Abrantes F (2014) Past circulation along the western Iberian margin: a time slice vision from the Last Glacial to the Holocene. Quatern Sci Rev 106:316–329

Sánchez Goñi MF, Bard E, Landais A, Rossignol L, d'Errico F (2013) Air-sea temperature decoupling in western Europe during the last interglacial-glacial transition. Nat Geosci 6:837–841

Sánchez Goñi MF, Desprat S, Daniau A-L, Bassinot FC, Polanco-Martinez JM, Harrison SP, ACER, m., (2017) The ACER pollen and charcoal database: a global resource to document vegetation and fire response to abrupt climate changes during the last glacial period. Earth System Science Data 9:679–695

Sánchez-Goni MF, Fourcade T, Salonen S, Lesven J, Frigola J, Swingedouw D, Sierro FJ (2020) Muted cooling and drying of NW Mediterranean in response to the strongest last glacial North American ice surges. GSA Bull. https://doi.org/10.1130/B35736.1

Sánchez-Goni MF, Harrison SP (2010) Millennial-scale climate variability and vegetation changes during the Last Glacial: Concepts and terminology. Quatern Sci Rev 29(21–22):2823–2827

Sánchez-Goni MF, Landais A, Fletcher W, Naughton F, Desprat S, Duprat J (2008) Contrasting impacts of Dansgaard-Oeschger events over a western European latitudinal transect modulated by orbital parameters. Quatern Sci Rev 27: 1136–1151

Scourse JD, Haapaniemi AI, Colmenero-Hidalgo E, Peck VL, Hall IR, Austin WE, Knutz PC, Zahn R (2009) Growth, dynamics and deglaciation of the last British-Irish ice sheet: the deep-sea ice-rafted detritus record. Quatern Sci Rev 28(27–28):3066–3084

Seidov D, Maslin M (1999) North Atlantic deep water circulation collapse during Heinrich events. Geology 27(1):23–26

Seppä H, Bjune AE, Telford RJ, Birks HJB, Veski S (2009) Last nine-thousand years of temperature variability in Northern Europe. Climate of the Past 5:523–535

Severinghaus JP, Brook EJ (1999) Abrupt climate change at the end of the last glacial period inferred from trapped air in polar Ice. Science 286:930–934. https://doi.org/10.1126/science.286.5441.930

Shackleton NJ, Opdyke ND (1973) Oxygen isotope and paleomagnetic stratigraphy of Equatorial Pacific core V28–238: oxygen isotope temperatures and ice volumes. Quatern Res 3:39–55

Shackleton NJ, Sánchez Goñi MF, Pailler D, Lancelot Y (2003) Marine Isotope Substage 5e and the Eemian Interglacial. Global Planet Change 757:1–5

Shakun J, Clark P, He F, Marcott SA, Mix AC, Liu Z, Otto-Bliesner B, Schmittner A, Bard E (2012) Global warming preceded by increasing carbon dioxide concentrations during the last deglaciation. Nature 484:49–54. https://doi.org/10.1038/nature10915

Shakun JD, Carlson AE (2010) A global perspective on Last Glacial Maximum to Holocene climate change. Quatern Sci Rev 29:1801–1816

Shakun JD, Clark PU, He F, Lifton NA, Liu Z, Otto-Bliesner BL (2015) Regional and global forcing of glacier retreat during the last deglaciation. Nat Commun 6(1):1–7

Siani G, Magny M, Paterne M, Debret M, Fontugne M (2013) Paleohydrology reconstruction and Holocene climate variability in the South Adriatic Sea. Climate of the past 9(1):499–515

Sierro FJ, Andersen N, Bassetti MA, Berné S, Canals M, Curtis JH, Dennielou B, Flores JA, Frigola J, Gonzalez-Mora B, Grimalt JO, Hodell DA, Jouet G, Pérez-Folgado M, Schneider R (2009) Phase relationship between sea level and abrupt climate change. Quatern Sci Rev 28(25):2867–2881

Steffensen JP, Andersen KK, Bigler M, Clausen HB, Dahl-Jensen D, Fischer H, Goto-Azuma K, Hansson M, Johnsen SJ, Jouzel J, Masson-Delmotte V (2008) High-resolution Greenland ice core data show abrupt climate change happens in few years. Science 321(5889):680–684

Sweatman MB (2021) The Younger Dryas impact hypothesis: Review of the impact evidence. Earth-Science Reviews 218, 103677. https://doi.org/10.1016/j.earscirev.2021.103677

Tarasov L, Peltier WR (2005) Arctic freshwater forcing of the Younger Dryas cold reversal. Nature 435:662–665

Thornalley DJR, Elderfield H, McCave IN (2011) Reconstructing North Atlantic deglacial surface hydrography and its link to the Atlantic overturning circulation. Global Planet Change 79(3–4):163–175

Tinner W, Lotter AF (2001) Central European vegetation response to abrupt climate change at 8.2 ka. Geology 29(6), 551–554

Tinner W, Lotter AF (2006) Holocene expansions of Fagus silvatica and Abies alba in Central Europe: where are we after eight decades of debate? Quatern Sci Rev 25(5–6):526–549

Toucanne S, Naughton F, Rodrigues T, Vázquez-Riveiros N, Goñi MFS (2022) Abrupt (or millennial or suborbital) climatic variability: Heinrich events/stadials. In: European Glacial Landscapes. Elsevier, pp 181–187

Toucanne S, Soulet G, Freslon N, Jacinto RS, Dennielou B, Zaragosi S, Eynaud F, Bourillet JF, Bayon G (2015) Millennial-scale fluctuations of the European Ice Sheet at the end of the last glacial, and their potential impact on global climate. Quatern Sci Rev 123:113–133

Tzedakis PC, Lawson IT, Frogley MR, Hewitt GM, Preece RC (2002) Buffered tree population changes in a Quaternary refugium: evolutionary implications. Science 297:2044–2047

Valsecchi V, Sánchez Goni MF, Londeix L (2012) Vegetation dynamics in the Northeastern Mediterranean region during the past 23 000 yr: insight from a new pollen record from the Sea of Marmara (core MD01–2430). Climate os Past Discussions 8, 4183–4221. www.clim-past-discuss.net/8/4183/2012/, https://doi.org/10.5194/cpd-8-4183-2012.

van Kreveld S, Sarnthein M, Erlenkeuser H, Grootes P, Jung S, Nadeau MJ, Pflaumann U, Voelker A (2000) Potential links between surging ice sheets, circulation changes, and the Dansgaard-Oeschger cycles in the Irminger Sea, 60–18 ka. Paleoceanography 15:425–442

Veski S, Seppä H, Ojala AEK (2004) The cold event 8200 years ago recorded in annually laminated lake sediments in Eastern Europe. Geology 32:681–684

Vinther BM, Buchardt SL, Clausen HB, Dahl-Jensen D, Johnsen SJ, Fisher DA, Koerner RM, Raynaud D, Lipenkov V, Andersen KK, Blunier T (2009) Holocene thinning of the Greenland ice sheet. Nature 461(7262):385–388

Voelker AHL, Sarnthein M, Grootes PM, Erlenkeuser H, Laj C, Mazaud A, Nadeau M-J, Schleicher M (1998) Correlation of marine 14C ages from the Nordic seas with the GISP2 isotope record: Implications for 14C calibration beyond 25 ka BP. Radiocarbon 40:517–534

Von Grafenstein U, Erlenkeuser H, Brauer A, Jouzel J, Johnsen SJ (1999) A mid-European decadal isotope-climate record from 15,500 to 5000 years B.P. Science 284:1654–1657

Waelbroeck C, Labeyrie L, Michel E, Duplessy JC, McManus JF, Lambeck K, Balbon E, Labracherie M (2002) Sea-level and deep water temperature changes derived from benthic foraminifera isotopic records. Quatern Sci Rev 21:295–305

Walker M, Head MJ, Lowe J, Berkelhammer M, BjÖrck S, Cheng H, Cwynar LC, Fisher D, Gkinis V, Long A, Newnham R (2019) Subdividing the Holocene Series/Epoch: formalization of stages/ages and subseries/subepochs, and designation of GSSPs and auxiliary stratotypes. J Quat Sci 34(3):173–186

Walker M, Johnsen S, Rasmussen SO, Popp T, Steffensen JP, Gibbard P, Hoek W, Lowe J, Andrews J, Björck S, Cwynar LC (2009) Formal definition and dating of the GSSP (Global Stratotype Section and Point) for the base of the Holocene using the Greenland NGRIP ice core, and selected auxiliary records. Journal of Quaternary Science: Published for the Quaternary Research Association 24(1):3–17

Walker MJ, Berkelhammer M, Björck S, Cwynar LC, Fisher DA, Long AJ, Lowe JJ, Newnham RM, Rasmussen SO, Weiss H (2012) Formal subdivision of the Holocene Series/Epoch: a Discussion Paper by a Working Group of INTIMATE (Integration of ice-core, marine and terrestrial records) and the Subcommission on Quaternary Stratigraphy (International Commission on Stratigraphy). J Quat Sci 27(7):649–659

Wanner H, Mercolli L, Grosjean M, Ritz SP (2014) Holocene climate variability and change; a data-based review. J Geol Soc 172(2):254–263

Wanner H, Solomina O, Grosjean M, Ritz SP, Jetel M (2011) Structure and origin of Holocene cold events. Quatern Sci Rev 30:3109–3123

Wary M, Eynaud F, Kissel C, Londeix L, Rossignol L, Lapuyade J, Castéra M-H, Billy I (2018) Spatio-temporal dynamics of hydrographic reorganizations and iceberg discharges at the junction between the Northeast Atlantic and Norwegian Sea basins surrounding Heinrich event 4. Earth Planet Sci Lett 481:236–245

Wittmeier HE, Schaefer JM, Bakke J, Rupper S, Paasche Ø, Schwartz R, Finkel RC (2020) Late Glacial mountain glacier culmination in Arctic Norway prior to the Younger Dryas. Quatern Sci Rev 245:106461

Woillez MN, Kageyama M, Combourieu-Nebout N, Krinner G (2013) Simulating the vegetation response in western Europe to abrupt climate changes under glacial background conditions. Biogeosciences10(3):1561–1582

Wolff EW, Chappellaz J, Blunier T, Rasmussen SO, Svensson AC (2010) Millennial-scale variability during the last glacial: The ice core record. Quatern Sci Rev 29:2828–2838

Wünsch C (2006) Abrupt climate change: An alternative view. Quatern Res 65:191–203

Zaragosi S, Eynaud F, Pujol C, Auffret GA, Turon JL, Garlan T (2001) Initiation of the European deglaciation as recorded in the northwestern Bay of Biscay slope environments (Meriadzek Terrace and Trevelyan Escarpment): a multi-proxy approach. Earth Planet Sci Lett 188(3–4):493–507

Zhang X, Knorr G, Lohmann G, Barker S (2017) Abrupt North Atlantic circulation changes in response to gradual CO 2 forcing in a glacial climate state. Nat Geosci 10(7):518–523

Ziemen FA, Kapsch ML, Klockmann M, Mikolajewicz U (2019) Heinrich events show two-stage climate response in transient glacial simulations. Climate of the past 15(1):153–168

Periglacial Landforms Across Europe

Marc Oliva, José M. Fernández-Fernández, and Daniel Nývlt

The European continent extends over a vast surface of ~ 10 million km^2 from subtropical to polar latitudes. Most of the landmass corresponds to mainland Europe (including Great Britain), although a number of islands and archipelagos are also found surrounding the continent.

Mainland Europe extends from latitudes 36 to 71°N, and from longitudes 9°W to 66°E. The southern fringe is defined by the Mediterranean Sea and the northern edge by the Barents Sea. The western façade of the continent is washed by the waters of the Atlantic Ocean, whereas the eastern boundary is defined by the Ural Mountains (Fig. 1). In addition, there are several other islands and archipelagos surrounding Europe along its southern, western and northern fringes that have been affected by periglacial dynamics during the Quaternary or are still under cold-climate regimes with glacial and cryonival processes at almost sea level.

Despite the mid-low latitudes of Mediterranean islands (34–42°N), most of them encompass extensive mountainous terrain with peaks above 2,400 (i.e. Corsica, Crete) to 3,300 m (i.e. Sicily), and several others exceeding 1,400 m (e.g. Sardinia, Cyprus, Kefalonia, Mallorca, etc.). In the North Atlantic Ocean Basin at latitudes of 32–38°N, the volcanic islands of Açores and Madeira also include peaks ranging between 1,800 and 2,300 m that were affected by cold-climate processes during the Quaternary. Following the Mid-Atlantic Ridge, at higher latitudes (63–66°N), Iceland is home of a wide range of periglacial processes and landforms, both active and relict,

M. Oliva (✉)
Universitat de Barcelona, Barcelona, Spain
e-mail: marcoliva@ub.edu

J. M. Fernández-Fernández
Universidad Complutense de Madrid, Madrid, Spain
e-mail: jmfernandez@edu.ulisboa.pt

D. Nývlt
Masaryk University, Brno, Czech Republic
e-mail: daniel.nyvlt@sci.muni.cz

M. Oliva et al. (eds.), *Periglacial Landscapes of Europe*,
https://doi.org/10.1007/978-3-031-14895-8_3

Chapter numbers

3 Iberian Peninsula. 4 Italian Peninsula. 5 Balkans. 6 Anatolian Peninsula. 7 Mediterranean islands. 8 Alps. 9 Central European Variscan Ranges. 10 Carpathians. 11 North European Plain. 12 British Isles. 13 Scandinavia. 14 Iceland.

Fig. 1 Distribution of the regions examined in this book

those inherited from previous cold periods. Other islands and archipelagos above the Arctic Circle that have a high percentage of their landmass covered by glaciers—such as Svalbard (~60%) and Jan Mayen (~30%)—or others that were formerly glaciated—such as the Faroe Islands and Bear Island – include abundant periglacial features in their ice-free landscape.

The configuration of the relief of mainland Europe is characterized by the numerous peninsulas, capes, bays and high mountain ranges exceeding 2,500–3,000 m, which has favoured the existence of a wide spectrum of climate regimes. Most of these ranges are located in the central-southern part of the continent, and form part of the Alpine-Himalayan orogenic belt, including the Pyrenees, the Alps or the Carpathians. The central-northern and eastern parts of the continental landmass are mostly characterized by a relatively flat terrain, with extensive flat areas only interrupted by mid-low mountain ranges (<1,500 m), occasionally exceeding 2,000 m in the Tatra Mountains or the Scandinavian Alps.

To the purposes of this book, we have examined the periglacial landscapes of Europe dividing the continent in three major regions following a latitudinal zonation:

Southern Europe: The southern fringe of the continent includes the Mediterranean peninsulas as well as several islands. The first section focuses on the westernmost peninsula that constitutes the SW corner of Europe, Iberia (Chapter 4). Subsequently, the next chapters are centred on the Italian Peninsula outside the Alps (Chapter 5) and the Balkan Peninsula (Chapter 6)—without the Carpathians—that are studied in a single chapter. The Anatolian Peninsula, including the different volcanoes and massifs existing in this large landmass, is also included in this section (Chapter 7). Finally, we also include a chapter focusing on the Mediterranean islands (Chapter 8), which include moderately high mountains where periglacial landscapes are also present.

Central and Eastern Europe: The Central part of Europe encompasses both mountain areas with current periglacial domain, as well as lowlands with traces of the past widespread periglacial conditions. The highest European mountains, the Alps (Chapter 9) represent the mountain area, where periglacial landforms and processes are extensively studied for more than a century. The summit areas of the Variscan ranges of Central Europe (Chapter 10) are also affected by current periglacial dynamics, however most of these mountains bear traces of the past more extensive periglacial realm. The Carpathians (Chapter 11) located in Central and Eastern Europe represent highly distinctive mountain belt strongly affected by north-south and west-east gradient in periglacial processes and associated landforms. The only lowland area representing the classical area of periglacial research, where past periglacial dynamics was extensively studied since the World War II, is the North European Plain (Chapter 12).

Northern Europe: The north-western quadrant of Europe, to the west of the North European Plain encompasses the Scandinavian Peninsula as the largest of the continent, and some islands from 50° to 80° N. This region includes both some of the oldest and youngest volcanic areas of the European continent, which in turn is reflected in a wide landscape variety. The first section starts at the southern part of this region, and is focused in the British Islands, including Great Britain and Ireland (Chapter 13). The following chapter go to the east, examining the periglacial dynamics and landscapes of Scandinavia (Chapter 14) covering a wide range of over 15° of latitude. And finally, this region is closed with the mid-oceanic Europe, i.e. Iceland (Chapter 15), an active area where the cold-climate processes associated to its location close to the Arctic Polar Circle interact with those linked to the volcanic activity, producing a unique landscape.

Periglacial Landforms in Southern Europe

The Iberian Peninsula

Marc Oliva, Enrique Serrano, José M. Fernández-Fernández, David Palacios, Marcelo Fernandes, José M. García-Ruiz, Juan Ignacio López-Moreno, Augusto Pérez-Alberti, and Dermot Antoniades

1 Geographical Framework

The Iberian Peninsula covers 582,925 km^2 in the SW corner of Europe, between latitude 36° 01′ and 43° 47′ N and longitude 9° 30′ W and 3° 19′ E. The diversity of landscapes is conditioned by its rough orography, varied lithology, relatively high average altitude (~660 m) as well as its boundary location between two water masses (Atlantic Ocean vs. Mediterranean Sea), two climatic settings (subtropical high pressure belt vs mid-latitude westerlies) and two biomes (Europe vs. Africa) (Oliva et al. 2018a).

M. Oliva (✉)
Department of Geography, Universitat de Barcelona, Barcelona, Spain
e-mail: marcoliva@ub.edu

E. Serrano
Department of Geography, Universidad de Valladolid, Valladolid, Spain

J. M. Fernández-Fernández · M. Fernandes
Instituto de Geografia e do Ordenamento do Território (IGOT), Universidade de Lisboa, Lisbon, Portugal

D. Palacios
Department of Geography, Universidad Complutense de Madrid, Madrid, Spain

J. M. García-Ruiz · J. I. López-Moreno
Instituto Pirenaico de Ecología (IPE-CSIC), Campus de Aula Dei, Zaragoza, Spain

A. Pérez-Alberti
CRETUS Instituto. Departamento de Edafoloxía e Química Agrícola, Universidade de Santiago de Compostela, Santiago de Compostela, Spain

D. Antoniades
Department of Geography and Centre for Northern Studies, Université Laval, Quebec City, QC, Canada

M. Oliva et al. (eds.), *Periglacial Landscapes of Europe*,
https://doi.org/10.1007/978-3-031-14895-8_4

The relief of the Iberian Peninsula is characterized by the presence of mountain ranges in its periphery, with extensive flat and relatively high (700–800 m) areas—locally known as *Mesetas*—in the central sector, and narrow plains along the coastal margins (Fig. 1). The peninsula's coastlines are linear, with no major headlands or bays. There are six major mountain systems in Iberia that are generally aligned W-E, with summits exceeding 2000 m (the Cantabrian Mountains and the NW, Central and Iberian ranges), and in some cases 3000 m (the Pyrenees and the Betic Range).

As a consequence of the rough orography, continental and maritime air masses from distinct geographical origins affecting Iberia result in highly variable climates across the peninsula, with numerous microclimates even within individual mountain systems. Precipitation is mostly concentrated between October and May, driven by the low-pressure systems associated with the polar front and mid-latitude westerlies, whereas the influence of the Azores anticyclone usually brings dry conditions during the summer (Trigo et al. 2004). Precipitation in Iberia reaches its maximum at the peninsula's NW, decreasing towards the S and E, whereas mean annual air temperatures (MAAT) follow an opposite pattern. The elevation reinforces lapse rates, enhancing precipitation and decreasing temperatures. The highest rainfalls are

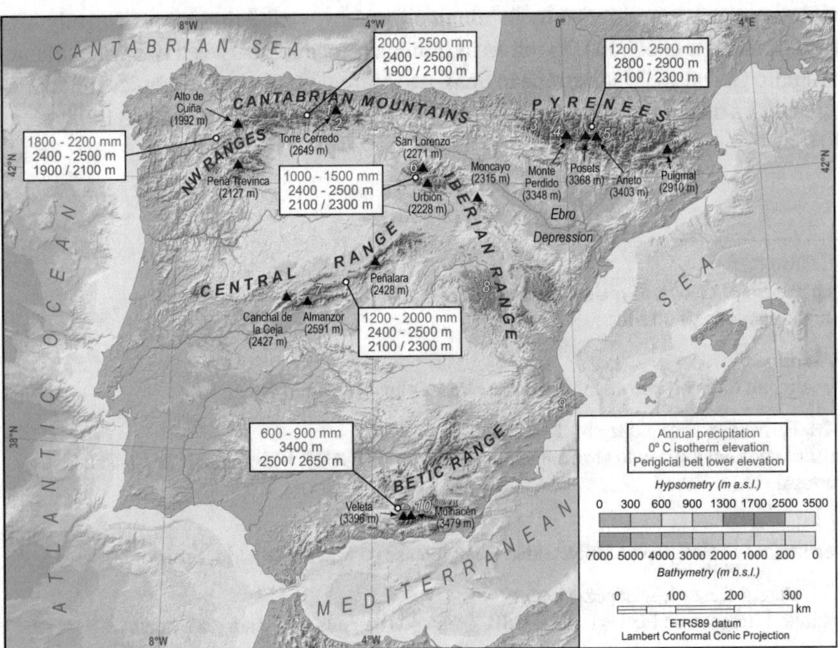

Placenames referred in the text

1 Picos de Europa: Jou Negro, Forcadona. 2 Espigüete. 3 Aragón Subordán. 4 Tendeñera, Telera, Tucarroya, Argualas, Cinca. 5 Maladeta, Besiberri, La Paúl. 6 Demanda. 7 Gredos. 8 Albarracín. 9 Aitana. 10 Sierra Nevada.

Fig. 1 Location of the main Iberian mountain ranges and local names mentioned in the text. The map also includes an estimation of the precipitation, elevation of the 0 °C isotherm and minimum altitude of the present-day periglacial belt recorded in NW mountain ranges influenced by the Atlantic Ocean, with annual means exceeding 2000–2500 mm, while the minima of 600–900 mm are registered in the semiarid Sierra Nevada massif. The elevation of the 0 °C isotherm increases from ~2400–2500 m in the Cantabrian Mountains to ~2950 m in the Pyrenees, and sits at summit level in the Sierra Nevada, at 3400 m (Oliva et al. 2016). A seasonal snowpack develops above 1500 m in most of the mountain areas, except in the Betic mountains where snow cover generally starts ~2000 m (Alonso-González et al. 2019).

Cold-climate geomorphological processes are currently limited to the highest parts of the main mountain ranges. Here, the landscape has been shaped by glacial dynamics during Pleistocene glacial cycles, but also by postglacial environmental processes including periglacial, slope and alluvial processes and shallow- and deep-seated landslides (Oliva et al. 2019). Depending on the combination of temperatures and moisture regimes, the glaciated environment during Quaternary cold stages was more or less extensive as reflected by fluctuations of the Equilibrium-Line Altitude (ELA). Together with climate, topography controlled the morphology of ice masses: whereas Alpine glaciers developed mostly in young mountains, ice caps formed in old ranges, particularly in the NW Iberian corner where extensive remnants of erosion surfaces are found (Oliva et al. 2019). In addition, the latitude as well as the influence of sea surface water temperatures (cool Atlantic Ocean vs warm Mediterranean Sea), determined the elevation of the ELA that generally increased towards the S and E (Pérez-Alberti et al. 2004).

Intense periglacial dynamics prevailed in ice-free landscapes above the glaciated environments, where permafrost was widespread. Evidence of enhanced cold conditions is also found in the lowlands of some Iberian areas, where seasonal frost features—inactive under present-day climates—developed during glacial periods (Oliva et al. 2016). Following Quaternary glaciations, when temperatures increased during interglacial periods, the periglacial belt migrated to higher elevations and cryonival processes reshaped the formerly glaciated domain. This is true of the current interglacial, the Holocene, when only small remnants of Quaternary glaciers remain in the highest massifs, which are mostly affected by periglacial processes.

However, the degree of preservation of glacial and periglacial features is constrained by local lithologies, which sometimes impedes the reconstruction of past environmental conditions. In the Iberian mountain ranges, there is a wide range of lithologies: whereas some ranges are composed of relatively homogeneous materials with mostly metamorphic and crystalline rocks (e.g. NW ranges, Central Range) or widespread thick Mesozoic carbonates (e.g. Cantabrian Mountains), others are made of a combination of Palaeozoic crystalline and metamorphic rocks and Mesozoic carbonates (e.g. Pyrenees, Iberian and Betic ranges).

2 The Development of Periglacial Research in the Iberian Peninsula

The interest in the landscape of the Iberian mountains drew the attention of numerous naturalists and scientists over the last two centuries. Due to the uniqueness of the ice masses in the Mediterranean mountain landscape, the first studies of cold-climate geomorphological processes focused on the glaciers that existed at the end of the Little Ice Age in the Pyrenees, the Cantabrian Mountains and the Sierra Nevada (Oliva et al. 2019). These studies often included sketches and paintings of these features and notes on their topographical characteristics, altitudes and dimensions (Boissier 1839; Bide 1893; Briet 1902). Subsequently, early scientists from Central Europe exported knowledge of glacial geomorphology developed in the Alps to the Iberian ranges, and wrote the first descriptions of geomorphic features left by glaciers during past Quaternary glaciations (Penck 1883; Obermaier 1916).

Over the following several decades the first direct references to periglacial dynamics shaping the landscape of the highest sectors of Iberian ranges appeared (García-Sainz 1935; Dresch 1937), although the tipping point for glacial and periglacial research was the organization of the International Union for Quaternary Research conference in Barcelona-Madrid in 1957 (Gómez-Ortiz and Vieira 2006). This meeting promoted networking and favoured the internationalisation of the few research teams working on cold-climate geomorphological processes in Spain and Portugal, which resulted in the completion of several PhD theses and reference works in the forthcoming decades centred on the glacial and periglacial geomorphology of certain Iberian massifs (Messerli 1965; Asensio-Amor and González-Martín 1974; Serrat 1977; García-Ruiz 1979; Pérez-Alberti 1979; Gómez-Ortiz 1980; Vilaplana 1983).

Until the 1980s and 1990s, periglacial studies were basically centred on the identification and description of periglacial phenomena through geomorphological mapping, as well as tentative dating of inactive landforms (González-Martín and Pellicer 1988; García-Ruiz et al. 1990; Chueca et al. 1994). Attempts at monitoring periglacial dynamics carried out in other mid-latitude mountain environments, such as the Alps, were progressively imported to Iberian mountains, where monitoring of present-day processes and soil thermal regimes in different periglacial features began (Chueca and Julián 1994; Gómez-Ortiz et al. 1999; Serrano et al. 2001, 2006; Ramos and Vieira 2003; García-Ruiz et al. 2004; Oliva et al. 2009, 2014a). Over the last decade, the dynamics and thermal monitoring of currently active landforms has been complemented with reconstructions of ancient environments based on the dating of periglacial landforms (Andrés et al. 2018; Oliva et al. 2011, 2009, 2014a; Palacios et al. 2016; Serrano and Agudo 2004).

3 Distribution and Chronology of Periglacial Landforms

Taking into account French's (2007) definition of the periglacial zone as those non-glacial areas where MAAT oscillates from −2 to +3 °C, the present-day periglacial environment in Iberia is confined to the highest mountains. In contrast to other mid- and high-latitude regions, periglacial dynamics solely prevail in uninhabited regions where no major infrastructure or settlements exist.

Periglacial activity is mainly driven by seasonal frost conditions, with active processes >1800 m in the NW mountains and >2500 m in the southern Sierra Nevada. In permafrost-free terrain, the ground >2000 m generally remains frozen during the cold season, although thermal regime dynamics are highly dependent on topography (slope, aspect) and snowpack conditions (thickness, evolution, extent). The length of the freezing season depends on the altitude and latitude of the mountain range. In Sierra Nevada, MAAT at 3400 m is 0.5 °C, and daily temperatures are negative almost half the year, with 108 days recording freezing-thawing cycles (2001–2016; Gómez-Ortiz et al. 2019). Ground temperatures in different periglacial landforms distributed between 2800 and 3300 m, however, oscillated between 3.4 and 0.5 °C, and the number of surface freezing–thawing cycles ranged from 18 to 66. In the Cantabrian Mountains, seasonal frost is widespread and mean annual ground temperatures were only slightly negative near ice patches (Pisabarro et al. 2015). At 2200 m in the Forcadona glacial cirque, temperature oscillated between 0.6 and 1.1 °C at 10 cm depth in a talus cone and a moraine; here, notable freezing-thawing cycles occurred only at the cirque wall, where ca. 47 cycles were recorded annually despite a higher annual average temperature of 3.1 °C (Ruiz-Fernández et al. 2017). In the Pyrenees, the 0 °C isotherm is located at 2945 m on the Monte Perdido massif (López-Moreno et al. 2019) and MAAT at 2000 m is ~5 °C. In this range alone, permafrost is currently found across extensive areas at elevations >2650 m on northern slopes and >2900 m on southern sides, with seasonal frost prevailing at lower elevations (Serrano et al. 2019a). Recently, direct measurements have confirmed the existence of rockwall permafrost on the north face of the Vignemale Peak (France) above 3000 m (Rico et al. 2021). Analogue conditions can be easily found in the Spanish Pyrenees, and frequent rockfalls in recent years at high elevations are suspected to be linked to permafrost degradation (Oliva et al. 2018b). Sporadic permafrost also exists at lower elevations of this range, as well as in the highest northern cirques of the Cantabrian Mountains and Sierra Nevada at >2150–2200 and >3000 m, respectively (Oliva et al. 2016), where loggers located in glacial cirques recorded mean temperatures of 0.1 and −1 °C at 0.5 and 1.5 m, respectively (Ruiz-Fernández et al. 2017; Gómez-Ortiz et al. 2019). Here, permafrost has aggraded since the end of the LIA on debris layers covering ice masses that were buried by rockfalls triggered by post-LIA paraglacial dynamics (Serrano et al. 2018b). However, permafrost and associated landforms (e.g. rock glaciers) are currently showing signs of accelerated degradation (Gómez-Ortiz et al. 2014, 2019). A particular case is underground permafrost detected in ice caves in the Pyrenees and Cantabrian Mountains at elevations below the periglacial belt down to ~2000 m. These caves, with perennial ice bodies accompanied by seasonal cryospeleothems, are indicative of endokarstic permafrost (Gómez-Lende 2016). They are inherited from past cold periods as their distribution does not show a clear linkage with MAAT (Serrano et al. 2018a).

Given that the intensity and spatial distribution of periglacial processes in Iberian mountains are determined by altitude and latitude, they follow the climate patterns described above. In addition, similar to other mountains of the Mediterranean region, winter precipitation is a key driver of periglacial dynamics due to its influence on the duration and thickness of snow cover, which in turn control soil moisture and ground thermal regime (Palacios et al. 2003; Zhang 2005). Climate also conditions the vegetation cover and the degree of soil development that, together with other local factors such as lithology and topography, are also crucial for the occurrence of periglacial processes in mountain environments (Vieira et al. 2003; Oliva et al. 2014b). In fact, topography, and specifically aspect, produces distinct intensities of periglacial dynamics within the same mountain ranges, as it controls the radiative balance on northern and southern slopes (Höllermann 1985). Most periglacial studies have shown that periglacial features in the Iberian mountains currently show a barely active to inactive pattern (Oliva et al. 2016). Landforms that show dynamic behaviour are mostly concentrated in the highest cirques and summit plateaus of the highest massifs, particularly those that are northerly-exposed (Serrano et al. 2006; Gómez-Ortiz et al. 2014, 2019; González-García et al. 2017).

Consequently, the widespread occurrence of inactive periglacial phenomena under current climate conditions suggests that they must have formed during colder past periods, driven by a more severe cold regime. The Iberian mountains therefore now encompass a wide range of periglacial phenomena resulting from both past and present climate conditions. The expansion of the periglacial belt during Quaternary glacial cycles was dependent on the intensity of cold conditions, and was confined to the highest altitudes during interglacials such as the Holocene. During glaciations, periglacial processes were active almost down to sea level in NW Iberia (Pérez-Alberti et al. 2004) as well as in the Spanish central *Meseta* at ca. 700 m (Serrano et al. 2010) and even in the Ebro Depression at 300 m a.s.l. (Rodríguez-Ochoa et al. 2019), particularly micro-morphological features caused by cryogenic processes, such as involutions and wedges, although they are not necessarily related to the presence of permafrost. Most landforms inherited from Pleistocene glacial phases as well as from the last deglaciation, such as rock glaciers, were associated with permafrost conditions (Andrés et al. 2018; García-Ruiz et al. 2016; Palacios et al. 2016). As temperatures increased during the Holocene, the majority of inactive periglacial landforms that developed during the last several millennia, such as solifluction features, formed under a seasonal frost regime (Oliva et al. 2009, 2011). In addition, during the second half of the current interglacial period, and especially between 5 and 4 ka, periglacial and nivation processes were reactivated in the subalpine belts of some mountain ranges (e.g. Pyrenees, Cantabrian Mountains) due to the impact of forest clearing by humans for grazing areas (Pérez-Díaz et al. 2015; Nieuwendam et al. 2016). In these areas, deforestation favoured a decrease in the altitude of the timberline, which enhanced erosion as well as solifluction and nival dynamics at elevations between 200 and 500 m lower than in undisturbed mountain areas (García-Ruiz et al. 2020).

3.1 Seasonally Frozen Ground Features

Seasonal frost action in the ground is highly dependent on climate conditions: cold intensity and duration, ground moisture regimes, freeze-thaw cycles, etc. In general, most Iberian mountains >1600–2000 m record several months per year with negative mean monthly temperatures to slightly positive at lower elevations. Under these circumstances, and particularly in non–vegetated and frost-susceptible soils, cryogenic microforms associated with frost heave and frost creep develop.

In mid-latitude mountain areas such as in Iberia, differential frost heave in areas with shallow and diurnal ground freezing can result in the formation of small-scale **patterned ground features** (Ballantyne 2018). Needle-ice (or pipkrake) growth is widespread in Iberian mountains at elevations down to 1500 m, particularly during the snow-melting period on moist and unvegetated fine-grained soils where ice needles can reach 4–5 cm length or more, particularly on regoliths with a high percentage of stones and a sandy-loamy mixture, causing the displacement of stones, creeping and micro-mudflows (Arnáez-Vadillo and García-Ruiz 2000). Repeated freeze-thaw cycles act as sorting mechanisms, leading to the development of small-scale sorted circles that evolve to sorted stripes as the gradient of slopes increases. Patterned ground is a common feature in the Pyrenees, both in seasonal frost and permafrost areas (Serrano et al. 2019a). The diameter of these patterned-ground features depends on microtopographical conditions, including the abundance of fine-grained sediments and clasts, together with prevailing climate conditions (Feuillet and Mercier 2012). Generally, active sorted circles include clasts ranging between 5 and 25 cm, although in some cases at elevations above 3000 m (i.e. Tucarroya or Posets) they can reach 150 cm (Serrano et al. 2020). In some cases, stone stripes can reach lengths of tens of meters. In the Cantabrian Mountains, patterned ground is mostly found in karstic depressions filled with till >2200 m. The diameter of the stone circles ranges from 40 to 150 cm, whereas stone stripes are metre-scale (Serrano et al. 2011). Present-day ground thermal regimes favour the formation of such features at elevations above 1800–2200 m in Atlantic-influenced ranges, above 2500 m in the Pyrenees and 2800–2900 m in the Sierra Nevada (Oliva et al. 2016). Inherited patterned ground structures, probably developed during the LIA, may be also found on many other Iberian massifs, particularly on gentle, wide divides, sometimes in areas affected by plant colonization such as the Moncayo and Demanda sierras, Iberian Range (Arnáez-Vadillo and García-Ruiz 2000). At even lower elevations, such as at 1500 m in Aitana massif (Marco et al. 2018) or 1700 m in the Demanda massif (Arnáez-Vadillo and García-Ruiz 2000), we found decimetric steps in gentle slopes defined by vegetation in their riser known as **garlands** (Table 1). Most of the massifs also include larger patterned ground landforms that are relict under present-day climate conditions. These metre-sized features are mostly distributed at relatively flat summit surfaces that were not severely glaciated during the Last Glacial Cycle and subsequent cold phases, developed under permafrost conditions. They also tend to form stone stripes, block streams or block slopes as the slope increases (Fig. 2). These inactive landforms are observed in the Pyrenees at 2500–2600 m, >1000 m (particularly >1900 m) in the Cantabrian Mountains and at 3200–3300 m Sierra Nevada (Oliva et al. 2016).

In poorly drained areas (flat and gently sloping) of the present-day periglacial environment, particularly inside glacial cirques and headwaters of U-shaped valleys, we found fields with tens or hundreds of vegetation mounds known as **earth hummocks**. They have been observed in the Pyrenees, NW Ranges, Central Range and Sierra Nevada. These features are separated by shallow troughs and are mostly composed of highly organic-rich layers, sometimes alternating with mineral units, and tend to be small semi-circular features, with a rise of 0.1–0.4 m and a diameter of 0.3–0.6 m.

3.2 *Rock Weathering and Derived Landforms*

Traditionally, weathering in cold-climate regions has been associated with the mechanical disintegration of rocks although recent research has shown increasing evidence of the impact of chemical and biotic weathering in the configuration of the

Table 1 Minimum elevation (m) of periglacial landforms in Iberian ranges (active landforms in blue, inactive in red)

Processes	Landforms	Pyrenees	Cantabrian mountains	NW ranges	Central range	Iberian range	Sierra Nevada	Other
Seasonal frost	Patterned ground	2200 2700	2000 1800–2200	1800–2000	2400	1950	2800 3200	Garlands, >1500–1700
	Earth hummocks	2100–2300	–	–	2200	–	2700	
Rock weathering	Blockfields	2200–2400 2600–2800	1000 2200	1600–2000	2000	1950	2800 3200	
	Periglacial karst	2000–3300	1700–2400	–	–	–	–	
Talus slopes	Talus cones	1100–1700 2100–2900	1100–1700 1900–2200	450–1400 1900–2100	1300–1800 2000–2500	800	1600 2500	
	Debris/mud flows	1100–1700 2100–2900	1800–2200	1900–2100	2200–2400	1000	2700	
	Rock glaciers	1750–2500 2500–2850	950–2480	1500–1950	2200	1740	2500–3200 (S) 3100 (N)	
	Protalus lobes	1750–2500 2500–2850	1650–2350	1500–1600	–	–	2400–3200 (S) 3100 (N)	
	Block streams	2200	1900	700–1300	–	1700	2500	
	Protalus ramparts	2300	1900	1500 1750	2300	1800	2650 2900	
Periglacial mass movement	Solifluction landforms	2000 2200–2400	1800–2200 2100–2300	1800–2000	1800–2200 2000–2400	1700	2500 2900	
	Landslides	1600–1800 2100	1450–1800 2100	1600	1800 2200	1650	2200 2600	
	Nivation features	1900–2400	1700–2000	1750	1900–2400	1200	2800	
	Cryoplanation surfaces	2400–2800		–	–	1800	2800 3200	
Other	Ice wedge features	–	1000	–	–	–	–	200–1000 (central Iberia)
	Frost mounds	2700–3050	2100–2200	–	–	–	–	–
	Ice caves	2000–3060	1945–2000	–	–	–	–	–

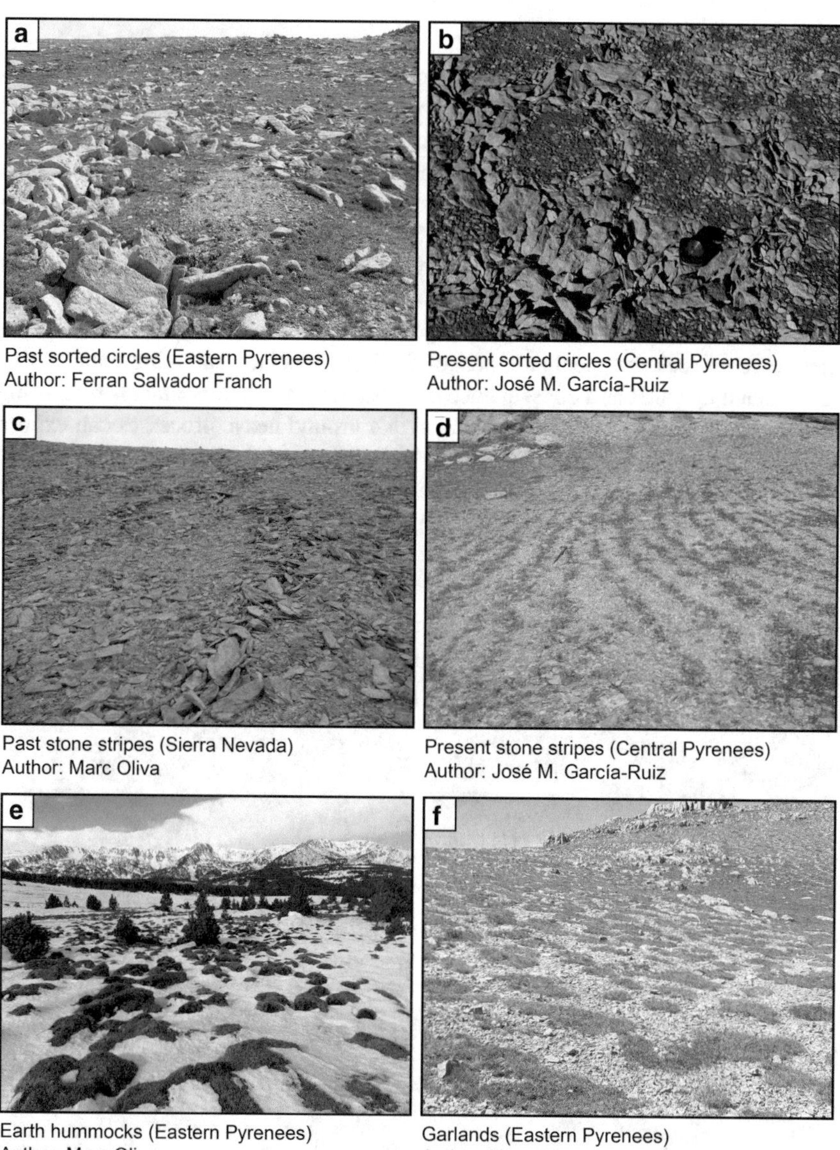

Fig. 2 Pictures of seasonally frozen ground features

mountain landscape (Thorn et al. 2011). Frost weathering is effective in all Iberian mountains, particularly in autumn and spring when freeze-thaw cycles are more frequent than in winter, when temperatures are predominantly negative.

The intense weathering of the bedrock of most summit surfaces of Iberian ranges results in the fracturing of rocks and the production of angular clasts that accumulate in thick debris mantles known as **blockfields**. Here, the absence of fine-grained sediments impedes the formation of patterned ground features. Instead, the weathered rock debris that covers the underlying bedrock forms an open work structure. Blockfields are present in all Iberian massifs, although they tend to be more abundant in lithologies more prone to generate boulder-sized debris. The highest peaks of most Palaeozoic mountains in W-NW Iberia include large open blockfields on their summits (Fig. 3), where elongated clasts are found in (semi)vertical positions (e.g. in the Sierra de la Demanda, Iberian Range; Arnáez-Vadillo and García-Ruiz 2000). These erected rock fragments are part of thick periglacial weathering mantles that were pushed upwards as a consequence of upheaving. Early experiments measuring this process in Iberian mountains revealed that ground heave processes can expose small centimetre-scale particles buried 5–10 below from the surface (Gómez-Ortiz 1980; Vieira et al. 2003). The large erected clasts across the modern summit surfaces must therefore have been exposed during previous colder periods with more intense periglacial dynamics than today.

Fig. 3 Pictures of rock weathering and derived landforms

Tors are landforms resulting from differential weathering of bedrock over thousands of years, and are found in most Iberian mountain ranges outside the environments glaciated during the last glaciations. They have been used as a geoindicator for establishing the lowest extent of Quaternary glaciers, as well as for identifying ice-free areas above the glacial domain (Cowton et al. 2009; Vieira et al. 2020). They are well preserved in gentle granitic areas, with dimensions between a few and several tens of meters.

Periglacial karst processes affect the limestone massifs, mainly of the Pyrenees and the Cantabrian Mountains, where chemical weathering is enhanced by cold-climate conditions. Together with frost-shattering, snow processes have reshaped the formerly glaciated terrain above 2000 m. Snow melting enhances karst dissolution, which promotes the widespread development of karren terrain (including microlandforms, i.e. rillenkarren and rinnenkarren, in vertical and horizontal surfaces) in high massifs of the Pyrenees (Monte Perdido, Tendeñera or Telera) and in the Cantabrian Mountains (Picos de Europa or Espigüete) (Fig. 3). Chemical and frost weathering generates dissolution furrows through rock disaggregation and cracking, which results in sharp edges. This process supplies small debris particles to the slopes surrounding glacio-karstic depressions that can be subsequently remobilized by slope dynamics, namely solifluction and debris flows (Serrano and González-Trueba 2004). In addition, near long-lasting snow patches small nivation hollows develop, together with small cone-shaped sinkholes that frequently contain (semi)permanent snow patches.

3.3 Talus Slopes and Associated Phenomena

On steep slopes, the weathering of the bedrock favours very active rockfall dynamics that result in the formation of debris accumulation at the foot of the rock ridges, which are subsequently reshaped by cascade surface processes generating a broad spectrum of periglacial features. Such debris accumulations are the main factor, together with frost shattering, for the development of regularized slopes, which occurs in many Iberian mountain areas, e.g. sierras of Albarracín and Demanda, Iberian Range (García-Ruiz et al. 1998).

Talus deposits are accumulations of loose, coarse and angular debris supplied by rockfalls. They are widespread in all mountain ranges and lithologies (Oliva et al. 2016). The weathering of bedrock favours frost shattering and the accumulation of debris at the foot of peaks and rocky ridges, and is enhanced by debris flows and avalanches (Fig. 4). These processes are very active inside glacial cirques near the summits of all massifs, particularly abundant in young Alpine ranges with sharp peaks and steep slopes, and sparser in old mountain ranges in the NW and W massifs where smoother slopes prevail. The monitoring of talus cones in the Cantabrian Mountains has shown evidence of very intense dynamics at 1900–2200 m, with morphological changes of 0.5–1 m year^{-1} at distal points and average accumulation rates of 1.6–26.2 mm year^{-1} (Serrano et al. 2019b). The development of the talus

Fig. 4 Pictures of talus slopes and associated phenomena

cones distributed in glacial cirques and high valleys began during phases of intense paraglacial dynamics once glacial ice had retreated, namely during Termination-1. Post-glacial warming favoured the upward shift of forests, which in some cases have partially covered talus slopes. In the highest northern cirques, intense post-LIA rock-fall activity generated talus cones covering the last remnants of Quaternary glaciers and permafrost with thick debris layers (Serrano et al. 2018b). The degradation of these ice-rich permafrost masses (**thermokarst**) is detected from the presence of subsidence and collapse features on the surface of the debris mantles (Gómez-Ortiz et al. 2014).

Talus cones accumulating at the foot of steep rock slopes and alluvial fans at the exit of mountain streams are reworked in parallel by other processes, such as rapid slope movements. Depending on the nature of the sediments, **debris flows** or **mud flows** triggered by intense rainfall or rapid snowmelt events redistribute the unconsolidated sediments. They usually form levées at both sides of the channel and a fan-shaped distal deposit. As in other mid-latitude mountain environments, their track is often followed in winter by **avalanches** that contribute to reshaping the debris slopes. These occur in all Iberian mountains (Palacios et al. 2003; Pedraza et al. 2004), although their magnitude and impact depends on topographical, lithological and climatic factors.

As in other mid-latitude mountains of the Mediterranean region, **relict rock glaciers** constitute the best geomorphological indicator of the occurrence of permafrost conditions in the Iberian mountains. They are concentrated in Iberian Mountains glaciated during the Pleistocene, except those distributed in more humid massifs and/or composed of limestones (Western Pyrenees, Basque Mountains, Eastern Cantabrian Mountains and western NW Ranges); the Central Range does not follow this pattern, as these features are almost absent despite the greater aridity and crystalline lithology. Rock glaciers are mostly located inside glacier cirques, but also on valley slopes in areas where glacial erosion was relatively weak during the Last Glacial Cycle (Palacios et al. 2017b; Andrés et al. 2018). These landforms are found at elevations of 950–2480 m in the Cantabrian Mountains (Gómez-Villar et al. 2011; Pellitero et al. 2011), 1500–1960 m in the NW ranges (Oliva et al. 2016), 1700–2200 m Iberian Range (García-Ruiz et al. 2020a), 1740–2850 m in the Central Pyrenees (Serrano and Agudo 2004; Fernandes et al. 2018), and 2500–3200 m in the Sierra Nevada (Palma et al. 2017). They are mostly exposed in northern slopes in the Central Pyrenees (Serrano and Agudo 2004) and Cantabrian Mountains (Pellitero et al. 2011), whereas in the Sierra Nevada these landforms mostly prevail on east-exposed slopes (Palma et al. 2017). In the Iberian mountains, rock glaciers have moderate dimensions between ~1 and 60 ha, with an average length of ~300 m and width of ~200 m, although some can reach larger sizes. Under favourable climate conditions, rock glaciers develop in any lithology: in the Central Pyrenees, they are distributed in diverse lithological settings including plutonic, metamorphic, calcareous and detrital bedrock (Serrano and Agudo 2004; Fernandes et al. 2018), sandstones and conglomerates in the Iberian Range (García-Ruiz et al. 2020a), quartzites and conglomerates in the Cantabrian Mountains (Redondo-Vega et al. 2004; Gómez-Villar et al. 2011; Pellitero et al. 2011) and mica schists in the Sierra Nevada (Gómez-Ortiz et al. 2013). Today, active rock glaciers only exist in the Central Pyrenees (Serrano and Agudo 2004) and Sierra Nevada (Gómez-Ortiz et al. 2014), revealing the current existence of mountain permafrost in these areas. Monitoring of active rock glaciers in the Pyrenees showed highly variable annual horizontal displacement rates of 17.7–40 cm year^{-1} in the Argualas rock glacier (1991–2000; Serrano et al. 2019a), 9–10.9 cm year^{-1} in the

Posets (2001–2011; Serrano et al. 2019a), 8.7–13.3 cm year^{-1} in the NW Besiberri (1993–2003; Chueca and Julián 2011), 12–13.8 cm year^{-1} in the Western Maladeta (2008–2012; González García et al. 2011; Serrano et al. 2019a), and 15–65 cm year^{-1} in La Paúl (2013–2018; Martínez-Fernández et al. 2019; Serrano et al. 2019a). In Sierra Nevada, an incipient rock glacier formed in the post-LIA deglaciated Veleta cirque showed higher vertical (3.97 m) than horizontal displacements (1.25 m) from 2001 to 2016 as a consequence of the degradation of underlying buried ice and permafrost (Gómez-Ortiz et al. 2019). Research into the timing of formation of these relict landforms has significantly advanced over the last decade through the application of cosmic-ray exposure dating. This method showed their final stabilization following periods of deglaciation with intense paraglacial slope readjustment, namely during the Bølling–Allerød interstadial (~15–14 ka) in the Cantabrian Mountains, SE Pyrenees and Iberian Range, whereas their activity continued even until the Mid-Early Holocene in the Central Pyrenees and Sierra Nevada (Palacios et al. 2017a).

Although less studied than rock glaciers in the Iberian mountains, **protalus lobes** are also good indicators of frozen ground creep. Up to ten active protalus lobes have been identified in the highest massifs of the Pyrenees (González-García et al. 2017) and only one in the Mulhacén cirque, Sierra Nevada (Serrano et al. 2018b). They are generally located at the distal section of talus slopes defined by a lobate morphology, in elevation belts ranging from 1670 to 2340 m in the Cantabrian Mountains (Pellitero et al. 2011), 2400–3200 m in Sierra Nevada (Palma et al. 2017) and 1740–2960 m in the Central Pyrenees (Fernandes et al. 2018; Serrano et al. 2019a). Protalus lobes have moderate dimensions between 0.1 and 7.5 ha, with an average length of ~60 m and width of ~180 m. These features predominantly occur on north-facing slopes, except in the Sierra Nevada where they also tend to be located in east-facing slopes (Palma et al. 2017). Protalus lobes are found in different lithological bedrock, such as in plutonic, metamorphic, calcareous and detrital areas in the Central Pyrenees (Serrano and Agudo 2004; Fernandes et al. 2018), quartzites and conglomerates in the Cantabrian Mountains (Pellitero et al. 2011), and mica schists in the Sierra Nevada (Palma et al. 2017). In Iberia, only one active protalus lobe has been monitored; this feature, located in the Maladeta massif at 2850–2960 m, yielded annual horizontal velocities of 3.8 and 10.8 cm year^{-1} (2008–2013; González-García et al. 2017). Most protalus lobes are currently inactive and must have developed during past periods, however nothing is known about their absolute age of formation.

The occurrence of **block streams** in Iberian mountains has been associated with the existence of permafrost (Oliva et al. 2016). These block streams occupy valley bottoms in extremely cold environments, when stone-banked lobes, garlands, large debris flows and gravity processes delivered large volumes of blocks and boulders from the neighbouring slopes to stream beds, which have almost no capacity for downstream transfer of coarse sediment. No information is available on the age of these landforms (Peña-Monné et al. 2010). They are inactive under present-day climate regimes, and the massive presence of lichens indicates that in all massifs they are inherited from colder phases. They developed during the Last Glacial Cycle both above the glaciated environment in glacier-free slopes as well as at lower elevations near the glacier margins (Oliva et al. 2016). They are best preserved in Paleozoic massifs above 1300 m in the NW ranges, 1700 m in the Iberian Range, 1900 m in the Cantabrian Mountains, 2200 m in the Pyrenees and 2500 m in the Sierra Nevada. Their length ranges from several tens of meters to 2.6 km in the case of the longest in Iberia located in the Albarracín massif (Peña-Monné et al. 2010), with a width from several meters to ca. 250 m (Fig. 4).

Protalus (or pronival) **ramparts** are cryonival deposits associated with non-glacial conditions. They are found at different altitudes in all Iberian ranges, usually showing arcuate morphologies and decametric dimensions that can reach lengths of ca. 100 m and ridges heights up to 1.5–2 m. They are still an active landform in the highest elevations, generally north-exposed, at the foot of long-lasting snow patches and at the distal parts of the slopes; in the Central Pyrenees, 60% of these landforms above 2300 m were active (Serrano and Agudo 2004). The relict protalus ramparts inherited from past periods distributed in many mountain ranges show distinct degrees of preservation that vary depending on topography and lithological setting. Some formed during the LIA, such as in Gredos massif at 2300 m, where lichenometric and geomorphological studies revealed that a protalus rampart located at 2300 m—at the foot of one of the latest long-lasting snow patches in central Iberia—was active during that cold phase and became relict by ca. 1870 AD (Sancho et al. 2001).

3.4 Periglacial Mass Movement and Slope Evolution

Soils in periglacial environments, both in seasonal frost and permafrost terrain, can move downslope even on very gentle slopes. Despite annual rates of periglacial mass movements on the order of mm- to cm-scale displacements, they affect extensive zones in Iberian mountains.

Together with talus cones, **solifluction landforms** are the most abundant periglacial phenomena in Iberian Mountains extending across slopes above 1800–2200 m in the northern ranges and 2500 m in Sierra Nevada (Oliva et al. 2016). A wide range of solifluction features is found in Iberian mountains (Fig. 5), such as those identified in Sierra Nevada: turf-banked lobes, stone-banked lobes, turf-banked terraces, stone-banked terraces, ploughing boulders and solifluction steps (Matsuoka 2001; Oliva et al. 2009). Here, a morphometric analysis of these features revealed a wide range of morphologies with diverse degrees of vegetation cover and abundance of gravels on their surface. Moreover, they tend to be longer than wider, although most of the landforms range from 2 to 6 m, with a riser lower than 0.6 m; 90% of the landforms occurred on slopes between 5 and 15° (Oliva et al. 2009). Solifluction processes are more intense in areas with shallow soils and unconsolidated sediments, particularly where weathering of the parent materials results in fine-grained particles (e.g. lutites or schists). Monitoring studies of present-day solifluction processes in Iberia, however, have indicated the very weak activity of these landforms, mostly driven by ground thermal regimes and snow cover (García-Ruiz et al. 2021). In the Pyrenees,

Solifluction lobe (Sierra Nevada)
Author: Marc Oliva

Ploughing boulder (Central Pyrenees)
Author: José M. García-Ruiz

Stratified scree (Central Pyrenees)
Author: José M. García-Ruiz

Snowbeds and solifluction lobe (Sierra Nevada)
Author: Marc Oliva

Frost mounds (Central Pyrenees)
Author: Enrique Serrano

Ice caves (Cantabrian Mountains)
Author: Enrique Serrano

Fig. 5 Pictures of periglacial mass movement and other features

between 2050 and 2260 m, movement rates ranged from 2.6 to 3.6 cm year^{-1} (1990–1992; Chueca and Julián 1994). In the Picos de Europa, at 2400 m, the displacement of stone-banked lobes was quantified at 0.2–1.8 cm year^{-1} (Brosche 1994), whereas solifluction terraces and ploughing boulders yielded lower values of <0.5 cm year^{-1} (Sanjosé et al. 2016). These values were much lower in the Sierra Nevada, where half of landforms did not report any movement and the others yielded displacements from 0.2 to 0.9 cm year^{-1} (2005–2011; Oliva et al. 2014c). In this range, an accurate paleoenvironmental reconstruction was carried out based on solifluction records that showed an alternation between organic-rich layers and mineral deposits suggesting that: (i) periods with enhanced solifluction activity occurred 5–4, 3.6–3.4, 3–2.8, 2.5–2.3, 1.8–1.6, 0.85–0.7 and 0.5–0.15 ka cal BP, and (ii) solifluction processes shifted upslope to higher elevations since 4.2 ka in response to more arid conditions (Oliva et al. 2011).

Some irregularly distributed periglacial deposits are still of uncertain origin. This is true of **stratified screes** (also referred to as *grèzes litées* and *éboulis ordonnés*), which extend widely throughout most Iberian mountains. Unfortunately, there are few detailed studies of such deposits because their analysis requires the existence of sections (opened by road construction, river undermining and gravel mining) that expose their internal structure. Regardless, our understanding has advanced in the last two decades, including through direct or indirect dating. Their structure shows alternating 10- to 30-cm thick matrix- and clast-supported beds, including variable clast sizes. Stratified screes appear across a wide elevation range, from ~ 300 m to those near the highest divides of the Pyrenees, where they are currently active. They are thought to represent the lower limit of the periglacial belt, and as such occupy shady canyons at lower elevations (<1000 m), such as in the Iberian Range (Peña Monné et al. 2018), Central Pyrenees (García-Ruiz et al. 2001), Cantabrian Mountains (Ruiz-Fernández et al. 2019) and northwest Iberia (Pérez-Alberti et al. 2009). Stratified scree development is attributed to the confluence of several processes, including not only frost shattering and gravity as in most of mountain screes, but also stone-banked solifluction lobes, frost-coated clast flows, debris flows and nivo-aeolian transport accompanied by pipkrake (needle-ice) activity (García-Ruiz et al. 2001), whose superposition results in alternating beds with contrasting structures. Ages obtained to date show that most stratified screes developed during cold stages of Termination-1. In the Cinca Valley, Central Pyrenees, the age of pollen in their lower beds is 22.8 ka cal BP, suggesting deposition started during the Last Glacial Maximum (LGM) and finished at the Early Holocene at 10–9 U/Th ka BP (García-Ruiz et al. 2001). A stratified scree in the Central Ebro Depression was aged 17.1 ka cal BP (Valero-Garcés et al. 2004), whereas Holocene ages for stratified screes were reported in Picos de Europa (Ruiz-Fernández et al. 2019) and the Sierra de Albarracín (Peña Monné et al. 2018). Here, old stratified screes have been also found, with a minimum age of 43.5 ka cal BP, similar to northwest Iberia (Viana-Soto and Pérez-Alberti 2019).

Iberian mountains also preserve evidence of rapid mass movements affecting more localized portions of the slopes within the present-day periglacial belt or in ancient periglacial environments. Currently, **landslides** occur mostly in spring during the snowmelt period as well as after intense/persistent rainfall events. Likewise, geomorphic evidence of past rapid downslope movement is found on many massifs, particularly in bedrock that efficiently produces fine-grained sediments, such as valleys in the Central and Eastern Pyrenees, where several types of slope failures occurred during the paraglacial readjustment of slopes following deglaciation (Jarman et al.

2014; Fernandes et al. 2020). Some landslides occurred in permafrost terrain, and thus likely represent **active-layer failures**.

Periglacial slopes are also affected by **nivation** processes that can alter hillside morphology, particularly near perennial or long-lasting snowpatches. These are found in topographically sheltered areas (nivation hollows) inside the glacial cirques or in lee-windward slopes of the highest massifs where summer temperatures are cold enough to prevent their complete melting, namely the Pyrenees, Cantabrian Mountains, Central Range, and Sierra Nevada. Snow creep and snowmelt runoff generate nival stone pavements and snow polished bedrock. In addition, persistent snowpatches provide sustained water supply downslope that can favour small-scale sediment redistribution and enhance solifluction activity (Oliva et al. 2014d). Field experiments carried out at elevations of 2200–2500 m in Gredos and Peñalara massifs, Central Range, have shown the strong influence that snow duration exerts on the distribution of vegetation in the periglacial belt (Palacios and García 1997; Muñoz et al. 2007). In mountains under a continental Mediterranean climate with a long dry summer season, such as the Sierra Nevada, groundwater supply comes mostly from these long-lived snowfields that enhance erosion, favour small-scale sediment redistribution and enhance solifluction activity (Oliva et al. 2014d). Nivation hollows continue to expand today, particularly in heavily weathered bedrock, and may coalesce to form larger niches (Palacios et al. 2004). It is only during periods of scarce snowfall, such as over the last two decades in central Iberian massifs, that niche growth can be less effective due to the rapid colonisation by vegetation (Andrés and Palacios 2010).

Bedrock surfaces often separated by rock scarps in flat or low-gradient slopes have been traditionally defined as **cryoplanation** terraces. In Iberian mountains, horizontal summit surfaces with small rocky outcrops protruding the debris mantles have been traditionally assigned to cryoplanation processes (e.g. Gómez-Ortiz 1980). However, their development and age of formation has not yet been studied in detail. The clasts resulting from the bedrock destruction in these relatively flat surfaces may be reworked by frost heave forming patterned ground features.

3.5 Others

In the Iberian Peninsula, there are also other less abundant periglacial phenomena, some of them indicative of past permafrost conditions. This is the case of **ice wedge morphologies** that have been observed in fine-grained sediments of Pleistocene fluvial terraces of central Iberia at elevations of 700–1000 m (Serrano et al. 2010).

Occasionally, **frost mounds** have been observed in the Pyrenees (i.e. Maladeta, La Paúl, Posets and Tucarroya massifs, at 2550–3050 m) and Picos de Europa (i.e. Jou Negro cirque, at 2100 m) linked to the occurrence of permafrost conditions. These meter-sized features are mostly distributed within LIA glaciated environments, frequently on the debris cover burying relict ice-patches, as well as on ice-cored moraines (Serrano et al. 2018b). In poorly-sorted sediments, water coming from melting snow and ice patches flowing across the sediments, and between the ice and sediments, favours water refreezing and ice segregation that enhance the formation of frost mounds (Serrano et al. 2019a).

Ice caves exist in the Pyrenees and in the Cantabrian Mountains, even at elevations between 500 and 800 m below the regional 0 °C isotherm. They are indicators of endokarstic permafrost and, to date, >170 ice caves have been identified in these two areas (Serrano et al. 2018a). They show different endoclimatic regimes resulting from distinct outer MAAT and snow conditions. Whereas in the Pyrenees they are distributed within a wide elevation belt of 2000–3060 m, in the Cantabrian Mountains they appear at lower elevations of 1945–2000 m. As a whole, the estimated mass balances of Iberian ice caves show a negative trend (Belmonte-Ribas et al. 2014; Gómez-Lende 2016; Serrano et al. 2018a). The dating of organic fragments preserved in the ice suggest that the onset of ice accumulation in some caves in the Pyrenees started at 6.1 ka (Sancho et al. 2018), whereas in others as well as in the Picos de Europa occurred prior and during the LIA (Serrano et al. 2018a).

4 The Significance of Periglacial Dynamics in the Iberian Peninsula Within the European Context

The scientific understanding of the geographical distribution and morphogenic evolution of the periglacial landforms existing across the Iberian Peninsula has significantly improved over the last several decades. The impact of past and current periglacial processes in the landscape of the high mountain ranges—also including other low and mid-altitude regions—has been widely examined by a number of researchers, particularly in the 1970s–1990s and with less intensity since 2005. Significant efforts have focused on the present-day monitoring of the movement and ground thermal regimes of present-day periglacial landforms, particularly rock glaciers. Substantial advances were also made in the reconstruction of cold environments in the Iberian Peninsula using new chronological techniques that allowed a better understanding of glacial oscillations during the Last Glacial Cycle and subsequent deglaciation, although this progress was not widely applied in periglacial studies. To date, the developmental chronology of inactive periglacial phenomena in the Iberian Peninsula is still vague, with only a few surface exposure dates for relict rock glaciers, since most periglacial landforms are extremely fragile or small, and are therefore not well preserved. Moreover, they occur at altitudes where the absence of organic carbon hinders radiocarbon dating, or they are composed of a mixture of clasts and fine particles that frequently make cosmogenic exposure dating inapplicable. In any case, most studies conclude that cold environments linked to Termination-1 are responsible for periglacial landforms whose development requires long-term cold conditions (e.g. rock glaciers, deep-seated landslides related to deglaciation, block streams), whereas the LIA was a critical period for the development of landforms that benefit from short-term cold conditions (e.g. solifluction lobes, protalus ramparts, patterned ground), and for the reactivation of processes inherited since the Younger Dryas and located in the most suitable topographic areas, close to the main divides.

Quaternary climate changes, together with the complex Iberian orography and the existence of multiple climate regimes, has favoured the occurrence of a wide variety of periglacial landforms across different elevation belts (Fig. 6). This may have resulted in different spatio-temporal patterns for the formation, development and stabilization of inactive features. Today, in the first decades of the twenty-first century, active periglacial processes prevail only in the highest mountains little impacted by human activities. Here, low temperatures and abundant winter snowfalls control cryonival dynamics during the snow-free season. Together with climate, lithological and topographical settings influence the intensity of periglacial dynamics. As a result, periglacial features today show a weakly active to inactive pattern driven mostly by seasonal frost above between 1800 (NW ranges) and 2500 m (Sierra Nevada). Extensive permafrost areas only exist in the Pyrenees above 2650–2900 m, whereas isolated permafrost environments are also distributed in the highest northern cirques in the Cantabrian Mountains and Sierra Nevada.

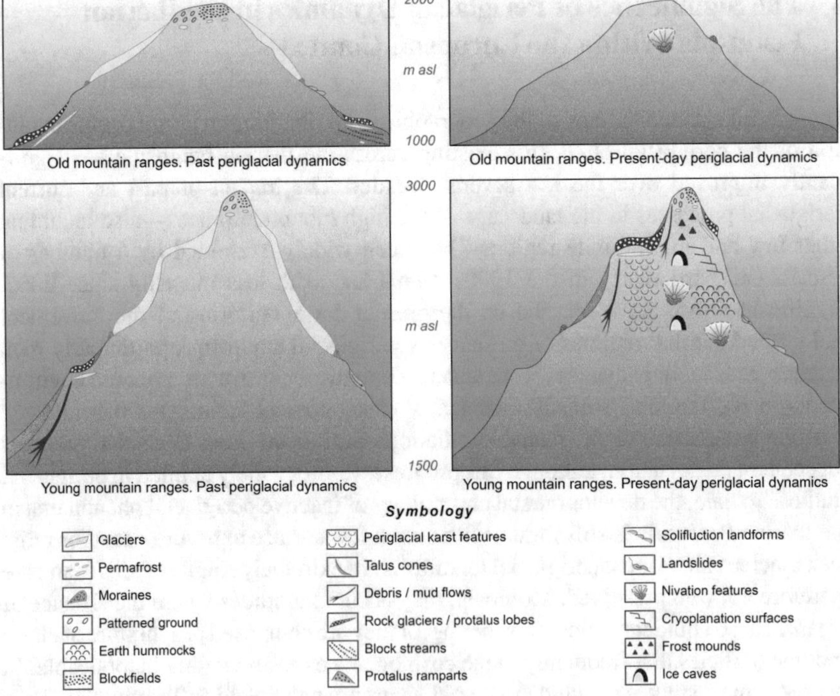

Fig. 6 Conceptual sketch with the development of periglacial landforms during the Last Glacial Cycle and the present-day in two different Iberian mountain ranges, such as the Atlantic-influenced mountain ranges and the Pyrenees

As such, most periglacial features developed during past colder periods, such as during the LGM, when intense periglacial processes developed across the non-glaciated mountain summits as well as below the glaciated terrain ca. 1000 m below the present-day periglacial environment. Indeed, the intense cold prevailing in Iberia favoured the occurrence of periglacial conditions in the central Meseta and coastal areas in the NW corner. Warming that started in the Northern Hemisphere by 19–20 ka favoured an accelerated glacial retreat, and periglacial dynamics occupied recently deglaciated environments, accompanied by the establishment of periglacial activity at higher elevations. After the millennial-scale glacial advances during Termination-1, most of the rock glaciers existing in the Iberian mountains formed under the intense paraglacial readjustment that reshaped the glacial cirques and valley heads. These features, which are the best geoindicators of permafrost conditions in the Mediterranean mountains, are best preserved in metamorphic and granitic areas, such as in the NW ranges and Pyrenees. With the temperature increase at the onset of the Holocene, periglacial dynamics have only remained active at the highest elevations. Depending on temperature and moisture regimes, the periglacial belt has shifted in altitude by hundreds of meters, expanded down-valleys during colder phases, or shifted upwards during warmer periods. The most recent elevation change of the periglacial belt occurred from the end of the LIA until the present, when periglacial processes show only weak to moderate activity in the highest areas of the highest ranges.

Despite all these advances in our comprehension of the periglacial environment in the Iberian mountains, there are still scientific gaps that will need to be addressed in the next decades. Knowledge for specific time periods and certain mountain ranges is still limited. Most periglacial research has been conducted in the Pyrenees and the Sierra Nevada, but the chronology of formation of periglacial features, as well as their geographical distribution in other mountains, such as in the Cantabrian Mountains, is still in development. Indeed, their dynamics during certain periods, i.e. the Holocene, remain poorly investigated. Together with a better knowledge of ancient periglacial environments, researchers should carry on monitoring periglacial dynamics, namely (i) the kinematic control of certain landforms (e.g. rock glaciers); and (ii) ground thermal regimes, particularly in areas where permafrost degradation may affect the stability of rock walls (i.e. the highest massifs in the Pyrenees).

The relevance of periglaciation in the Iberian Peninsula thus rests on the diversity of periglacial phenomena across the landscape, and the fact that they are the southernmost periglacial environments in Europe. A better understanding of the geomorphological response of the last remnants of permafrost, as well as seasonal frost phenomena, to rapid warming may be useful for anticipating the impact that projected future climate scenarios may have on high mountain ecosystems in southern Europe.

Acknowledgements Marc Oliva is supported by the Ramón y Cajal Program (RYC-2015-17597) and the Research Group ANTALP (Antarctic, Arctic, Alpine Environments; 2017-SGR-1102) funded by the Government of Catalonia through the AGAUR agency. This chapter complements the research topics examined in the PALEOGREEN project (CTM2017-87976-P) and the NUNANTAR project (02/SAICT/2017—32002) of the Fundação para a Ciência e a Tecnologia, Portugal.

References

Alonso-González E, López-Moreno JI, Navarro-Serrano F et al (2019) Snow climatology for the mountains in the Iberian Peninsula using satellite imagery and simulations with dynamically downscaled reanalysis data. Int J Climatol n/a.

Andrés N, Gómez-Ortiz A, Fernández-Fernández JM et al (2018) Timing of deglaciation and rock glacier origin in the southeastern Pyrenees: a review and new data. Boreas 47:1050–1071

Andrés N, Palacios D (2010) Cobertura nival y distribución de la temperatura en el suelo en las cumbres de la Sierra de Guadarrama. Cuad Investig Geográfica 36:7–36

Arnáez-Vadillo J, García-Ruiz JM (2000) El periglaciarismo en el Sistema Ibérico noroccidental. In: Peña-Monné JL, Sánchez-Fabre M, Lozano MV (eds) Procesos y formas periglaciares en la montaña mediterránea. Instituto de Estudios Turolenses, Teruel, pp 113–126

Asensio-Amor I, González-Martín JA (1974) Formas de crioturbación en altos niveles cuaternarios del valle del Jarama. Estud Geográficos 35:579–591

Ballantyne CK (2018) Periglacial geomorphology. Wiley Blackwell

Belmonte-Ribas Á, Sancho C, Moreno A et al (2014) Present-day environmental dynamics in ice cave a294, central pyrenees, spain. Geogr Fis e Din Quat 37:131–140

Bide J (1893) Deuxième excursión dans la Sierra Nevada. Annuaire du Club Alpine Français

Boissier CE (1839) Viaje botánico al sur de España durante el año 1837. Granada, Fundación

Briet L (1902) Le glacé de Mont-Perdú. Rev Club Alp Fr secc Sud Ouest 183–186

Brosche KU (1994) Ergebnisse von Abtragungsmessungen an periglazialen Solifluktions schutt decken in vier Hochgebirgen der Ibersichen Halbinsel (Picos de Europa, Peña Prieta, Sierra de Urbión und Sierra Nevada). Eiszeittalter Gegenwart 44:28–55

Chueca J, Gómez-Ortiz A, Lampre F, Peña JL (1994) El periglaciarismo heredado y actual de la cordillera pirenaica y del sistema costero catalán. In: Gómez Ortíz A, Simón M, Salvador F (eds) Periglaciarismo en la Península Ibérica, Canarias y Baleares. Sociedad Española de Geomorfología, pp 93–117

Chueca J, Julián A (2011) Besiberris glacigenic rock glacier (Central Pyrenees, Spain): Mapping surface horizontal and vertical movement (1993–2003). Cuad Investig Geogr 37:7–24

Chueca J, Julián A (1994) Cuantificación de movimientos en masa lentos en medios de montaña: Pirineo Central. Iturralde 173–196

Cowton T, Hughes PD, Gibbard PL (2009) Palaeoglaciation of Parque Natural Lago de Sanabria, northwest Spain. Geomorphology 108:282–291

Dresch J (1937) De la Sierra Nevada au Grand Atlas, formes glaciaires et de nivation. Mélanges Géographie d'orientalisme Offer à EF Gautiers 194–212

Fernandes M, Oliva M, Vieira G (2020) Paraglacial slope failures in the Aran valley (Central Pyrenees). Quat Int. https://doi.org/10.1016/j.quaint.2020.07.045

Fernandes M, Palma P, Lopes L et al (2018) Spatial distribution and morphometry of permafrost-related landforms in the Central Pyrenees and associated paleoclimatic implications. Quat Int 470:96–108

Feuillet T, Mercier D (2012) Post-little ice age patterned ground development on two pyrenean proglacial areas: From deglaciation to periglaciation. Geogr Ann Ser A Phys Geogr 94:363–376

French HM (2007) The periglacial environment. Wiley

García-Ruiz JM (1979) El glaciarismo cuaternario en la Sierra de la Demanda (Logroño- Burgos, España). Cuad Investig Geogr e Hist 5:325

García-Ruiz JM, Alvera B, Del Barrio G, Puigdefabregas J (1990) Geomorphic processes above timberline in the Spanish Pyrenees. Mt Res Dev 10:201–214

García-Ruiz JM, Arnáez J, Sanjuán Y et al (2021) Landscape changes and land degradation in the subalpine belt of the Central Spanish Pyrenees. J Arid Environ 186:104396

García-Ruiz JM, Chueca A, Julián J (2004) Los movimientos en masa del Alto Gállego. Geogr Física Aragón Asp Gen y temáticos

García-Ruiz JM, Ortigosa L, Pellicer F, Arnáez-Vadillo J (1998) Geomorfología glaciar del Sistema Ibérico. In: Gómez-Ortiz A, Pérez-Alberti A (eds) Las huellas glaciares de las montañas españolas. Universidad de Santiago de Compostela, Santiago de Compostela, pp 347–381

García-Ruiz JM, Palacios D, Fernández-Fernández JM et al (2020a) Glacial stages in the Peña Negra valley, Iberian Range, northern Iberian Peninsula: Assessing the importance of the glacial record in small cirques in a marginal mountain area. Geomorphology 362:107195

García-Ruiz JM, Palacios D, González-Sampériz P et al (2016) Mountain glacier evolution in the Iberian Peninsula during the Younger Dryas. Quat Sci Rev 138:16–30

García-Ruiz JM, Tomás-Faci G, Diarte-Blasco P et al (2020) Transhumance and long-term deforestation in the subalpine belt of the central Spanish Pyrenees: An interdisciplinary approach. Catena 195:104744. https://doi.org/10.1016/j.catena.2020.104744

García-Ruiz JM, Valero-Garcés B, González-Sampériz P et al (2001) Stratified scree in the Central Spanish Pyrenees: Palaeoenvironmental implications. Permafr Periglac Process 12:233–242

García-Sainz L (1935) Morfología glaciar y periglaciar de la región de la Noguera (cuenca Cinca-Segre). Boletín la Soc Geográfica Nac 54:64–110

Gómez-Lende M (2016) Cuevas heladas en el Parque Nacional Picos de Europa. Fronteras subterráneas del hielo en el Macizo Central. OAPN, Ministerio de Agricultura, Alimentación y Medio Ambiente

Gómez-Ortiz A (1980) Estudio geomorfológico del Pirineo catalán: morfogénesis glacial y periglacial de los altos niveles y vertientes merdionales de los macizos de Calmquerdós, Tossa Plana de Llés y Port Negre (Cerdanya-Alt Urgell). PhD Thesis. University of Barcelona

Gómez-Ortiz A, Oliva M, Salvador-Franch F et al (2019) Monitoring permafrost and periglacial processes in Sierra Nevada (Spain) from 2001 to 2016. Permafr Periglac Process. https://doi.org/10.1002/PPP.2002

Gómez-Ortiz A, Oliva M, Salvador-Franch F et al (2014) Degradation of buried ice and permafrost in the Veleta cirque (Sierra Nevada, Spain) from 2006 to 2013 as a response to recent climate trends. Solid Earth 5:979–993

Gómez-Ortiz A, Palacios D, Palade B et al (2013) La evolución glaciar de Sierra Nevada y la formación de glaciares rocosos. Bol la Asoc Geogr Esp 139–162

Gómez-Ortiz A, Palacios D, Ramos M et al (1999) Degradación de permafrost en Sierra Nevada y repercusiones geomorfológicas: el caso del Corral del Veleta. Resultados preliminares. Boletín la Asoc. Geógrafos Españoles 7–21

Gómez-Ortiz A, Vieira G (2006) La investigación en geomorfología periglaciar en España y Portugal. Finisterra XLIV:119–137

Gómez-Villar A, González-Gutiérrez RB, Redondo-Vega JM, Santos-González J (2011) Distribución de glaciares rocosos relictos en la cordillera cantábrica. Cuad Investig Geogr 37:49–80

González-García M, Serrano E, Sanjosé JJ, González-Trueba JJ (2017) Surface dynamic of a protalus lobe in the temperate high mountain. Western Maladeta, Pyrenees. Catena 149:689–700. https://doi.org/10.1016/j.catena.2016.08.011

González García M, Serrano E, Sanjosé Blasco JJ, González Trueba JJ (2011) Dinámica superficial y estado actual del glaciar rocoso de la Madaleta Occidental (Pirineos). Cuad Investig Geográfica 37:81

González-Martín JA, Pellicer F (1988) Rasgos generales del periglaciarismo de la Península Ibérica: dominio continental de las tierras del interior. Cuad Investig Geogr 14:23–80

Höllermann P (1985) The periglacial belt of mid-latitude mountains from a geoecological point of view. Erdkunde 39:259–270

Jarman D, Calvet M, Corominas J et al (2014) Large-scale rock slope failures in the eastern Pyrenees: identifying a sparse but significant population in paraglacial and parafluvial contexts. Geogr Ann Ser A Phys Geogr 96:357–391

López-Moreno JI, Alonso-González E, Monserrat O et al (2019) Ground-based remote-sensing techniques for diagnosis of the current state and recent evolution of the Monte Perdido Glacier, Spanish Pyrenees. J Glaciol 65:85–100

Marco JA, Giménez P, Padilla A (2018) Procesos de clima frío en el extremo NE peninsular de la Cordillera. Cuaternario y Geomorfol 32:39–56

Martínez-Fernández A, Serrano E, Sanjosé JJ et al (2019) Geomatic methods applied to the change study of the La Paúl rock glacier, Spanish pyrenees. Int Arch Photogramm Remote Sens Spat Inf Sci—ISPRS Arch 42:1771–1775

Matsuoka N (2001) Solifluction rates, processes and landforms: a global review

Messerli B (1965) Beiträge zur Geomorphologie der Sierra Nevada (Andalusien). Juris Verlag, Zürich

Muñoz J, García A, Andrés N, Palacios D (2007) La vegetación del Ventisquero de la Condesa (Sierra de Guadarrama, Madrid) y sus condicionantes termo-nivales. Bol la Asoc Geogr Esp 29–52

Nieuwendam A, Ruiz-Fernández J, Oliva M et al (2016) Postglacial landscape changes and cryogenic processes in the Picos de Europa (Northern Spain) reconstructed from geomorphological mapping and microstructures on quartz grains. Permafr Periglac Process 27:96–108

Obermaier H (1916) Los glaciares cuaternarios de Sierra Nevada. Trab Del Mus Nac Ciencias Nat 17:1–68

Oliva M, Gómez Ortiz A, Palacios D et al (2014a) Environmental evolution in Sierra Nevada (South Spain) since the Last Glaciation, based on multi-proxy records. Quat Int 353:195–209

Oliva M, Gómez Ortiz A, Salvador F et al (2014b) Long-term soil temperature dynamics in the Sierra Nevada, Spain. Geoderma 235–236:170–181

Oliva M, Ortiz AG, Estremera DP et al (2014c) El cuaternario en el macizo de Sierra Nevada: Evolución paleoambiental y paisaje a partir de la interpretación de registros naturales y documentos de época. Scr Nov 18:1–16

Oliva M, Ortiz AG, Franch FS, Catarineu MS (2014d) Present-day solifluction processes in the semi-arid range of Sierra Nevada (Spain). Arctic, Antarct Alp Res 46:365–370

Oliva M, Palacios D, Fernández-Fernández JM et al (2019) Late Quaternary glacial phases in the Iberian Peninsula. Earth-Science Rev 192:564–600

Oliva M, Ruiz-Fernández J, Barriendos M et al (2018a) The little ice age in Iberian mountains. Earth-Science Rev 177:175–208

Oliva M, Schulte L, Gómez-Ortiz A (2009) Morphometry and Late Holocene activity of solifluction landforms in the Sierra Nevada, southern Spain. Permafr Periglac Process 20:369–382

Oliva M, Schulte L, Ortiz AG (2011) The role of aridification in constraining the elevation range of Holocene solifluction processes and associated landforms in the periglacial belt of the Sierra Nevada (southern Spain). Earth Surf Process Landforms 36:1279–1291

Oliva M, Serrano E, Gómez-Ortiz A et al (2016) Spatial and temporal variability of periglaciation of the Iberian Peninsula. Quat Sci Rev 137:176–199

Oliva M, Žebre M, Guglielmin M et al (2018b) Permafrost conditions in the Mediterranean region since the Last Glaciation. Earth-Science Rev 185:397–436

Palacios D, Andrés N, Luengo E (2004) Tipología y evolución de los nichos de nivación en la Sierra de Guadarrama, España. Boletín la Real Soc Española Hist Nat Sección geológica 99:141–158

Palacios D, de Andrés N, Gómez-Ortiz A, García-Ruiz JM (2017a) Evidence of glacial activity during the Oldest Dryas in the mountains of Spain. Geol Soc Spec Publ 433:87–110

Palacios D, de Andrés N, Luengo E (2003) Distribution and effectiveness of nivation in Mediterranean mountains: Peñalara (Spain). Geomorphology 54:157–178

Palacios D, García-Ruiz JM, Andrés N et al (2017b) Deglaciation in the central Pyrenees during the Pleistocene–Holocene transition: Timing and geomorphological significance. Quat Sci Rev 162:111–127

Palacios D, García M (1997) The influence of geomorphologic heritage on present nival erosion: Peñalara, Spain. Geogr Ann Ser A, Phys Geogr 79:25–40

Palacios D, Gómez-Ortiz A, Andrés N et al (2016) Timing and new geomorphologic evidence of the last deglaciation stages in Sierra Nevada (southern Spain). Quat Sci Rev 150:110–129

Palma P, Oliva M, García-Hernández C et al (2017) Spatial characterization of glacial and periglacial landforms in the highlands of Sierra Nevada (Spain). Sci Total Environ 584–585:1256–1267

Pedraza J, Martín J, Sanz M, Carrasco R (2004) El Macizo de Peñalara (Sistema Central Español). Morfoestructura y modelado. Boletín la Real Soc Española Hist Nat Sección Geológica 99:185–196

Pellitero R, Serrano E, González Trueba JJ (2011) Glaciares rocosos del sector central de la Montaña Cantábrica : indicadores paleoambientales. Cuad Investig Geográfica 37:119

Peña-Monné JL, Lozano MV, Sánchez-Fabre M et al (2010) Las acumulaciones de clima frío de la Sierra de Albarracín. In: Peña-Monné JL, Sánchez-Fabre M, Lozano MV (eds) Las formas del relieve de la Sierra de Albarracín. Centro de Estudios de la Comunidad de Albarracín (CECAL), pp 163–188

Peña Monné JL, Pérez Alberti A, Sampietro Vattuone MM et al (2018) The Holocene stratified screes from Sierra de Albarracín (Iberian Ranges, Spain) and their paleoenvironmental significance. Holocene 28:478–491

Penck A (1883) La période glaciaire dans les Pyrénées. etin la Société d'Histoire Nat Toulouse 19:200

Pérez-Alberti A (1979) Nuevas observaciones sobre glaciarismo y periglaciarismo en el NW de la Península Ibérica: la Galicia sudoriental. Acta geológica hispánica 14:441–444

Pérez-Alberti A, López-Bedoya J, Cunha P (2009) Sedimentological Analysis of Cold-Climate, Stratified Slope Deposits of Galicia, Nw Iberia. 27th IAS Meet Sedimentol 2009 636

Pérez-Alberti A, Valcárcel M, Blanco R (2004) Pleistocene glaciation in Spain. In: Ehlers J, Gibbard PL (ed) Quaternary Glaciations- Extent and Chronology: Part I: Europe (2). Elsevier, London, pp 389–394

Pérez-Díaz S, López-Sáez JA, Galop D (2015) Vegetation dynamics and human activity in the Western Pyrenean Region during the Holocene. Quat Int 364:65–77

Pisabarro A, Serrano E, González-Trueba JJ (2015) Régimen térmico de suelos del macizo central de Picos de Europa (España). Pirineos 170:e010

Ramos M, Vieira G (2003) Active Layer and permafrost monitoring in Livingston Island, Antarctic. First results from 2000 to 2001. Proc 8th Int Conf Permaforst 2001:929–933

Redondo-Vega JM, Gómez-Villar A, González-Gutierrez RB (2004) Localización y caracterización morfométrica de los glaciares rocosos relictos de la Sierra de Gistredo (Montaña Cantábrica, León). Cuad Investig Geográfica 30:35

Rico I, Magnin F, López-Moreno JI et al (2021) First evidence of rock wall permafrost in the Pyrenees (Southwestern Europe, 42°N). Permafr Periglac Process

Rodríguez-Ochoa R, Olarieta JR, Santana A, Castañeda C, Calle M, Rhodes E, Bartolomé M, Peña-Monné JL, Sancho C (2019) Relict periglacial soils on Quaternary terraces in the Central Ebro Basin (NE Spain). Permafr Periglac Process 30:364–373. https://doi.org/10.1002/ppp.2005

Ruiz-Fernández J, García-Hernández C, Fernández S del C (2019) Holocene stratified scree of praón (Picos de Europa, cantabrian mountains) = Los derrubios estratificados holocenos de praón (Picos de Europa, montañas cantábricas). Cad do Lab Xeol Laxe 41:23–46

Ruiz-Fernández J, Oliva M, Hrbáček F et al (2017) Soil temperatures in an Atlantic high mountain environment: the Forcadona buried ice patch (Picos de Europa, NW Spain). Catena 149:637–647

Sancho C, Belmonte Á, Bartolomé M et al (2018) Middle-to-late Holocene palaeoenvironmental reconstruction from the A294 ice-cave record (Central Pyrenees, northern Spain). Earth Planet Sci Lett 484:135–144

Sancho LG, Palacios D, De Marcos J, Valladares F (2001) Geomorphological significance of lichen colonization in a present snow hollow: Hoya del cuchillar de las navajas, sierra de gredos (spain). Catena 43:323–340

Sanjosé JJ de, Serrano E, Gómez-Lende M (2016) Análisis Geomático De Bloques Aradores Y Lóbulos En Los Puertos De Áliva (Picos De Europa, Cordillera Cantábrica). Polígonos Rev Geogr 28:123

Serrano E, Agudo C (2004) Rock glaciers and deglaciation on Aragonese Pyrenean high mountain (Spain). Bol R Soc Esp Hist Nat 99:159–172

Serrano E, Agudo C, Delaloyé R, González-Trueba JJ (2001) Permafrost distribution in the Posets massif, Central Pyrenees. Nor J Geogr 55:245–252

Serrano E, Gómez-Lende M, Belmonte Á et al (2018a) Ice Caves in Spain. Elsevier Inc.

Serrano E, González-Trueba JJ (2004) Morfodinámica Periglaciar En El Grupo Peñavieja. Cuaternario y Geomorfol 73–88

Serrano E, González-Trueba JJ, Sanjosé JJ, Del Río LM (2011) Ice patch origin, evolution and dynamics in a temperate high mountain environment: The Jou Negro, Picos de Europa (NW Spain). Geogr Ann Ser A Phys Geogr 93:57–70

Serrano E, Lende MG, Ignacio J et al (2019a) Periglacial environments and frozen ground in the central Pyrenean high mountain area: ground thermal regime and distribution of landforms and processes. Permafr Periglac Process 30:292–309

Serrano E, López-Moreno JI, Gómez-Lende M et al (2020) Frozen ground and periglacial processes relationship in temperate high mountains: a case study at Monte Perdido-Tucarroya area (The Pyrenees, Spain). J Mt Sci 17:1013–1031

Serrano E, Oliva M, González-García M et al (2018b) Post-little ice age paraglacial processes and landforms in the high Iberian mountains: a review. Land Degrad Dev 29:4186–4208

Serrano E, Pellitero R, Otero M (2010) Huellas pleistocenas de frío intenso en la Cuenca del Duero: cuñas de arena relictas en las terrazas del Pisuerga. In: Úbeda X, Vericat D, Batalla RJ (eds) Avances de la Geomorfología en España, 2008–2010b. Sociedad Española de Geomorfología, Universitat de Barcelona, pp 417–420

Serrano E, San José JJ, Agudo C (2006) Rock glacier dynamics in a marginal periglacial high mountain environment: flow, movement (1991–2000) and structure of the Argualas rock glacier, the Pyrenees. Geomorphology 74:285–296

Serrano E, Sanjosé JJ, Gómez-Gutiérrez Á, Gómez-Lende M (2019b) Surface movement and cascade processes on debris cones in temperate high mountain (Picos de Europa, northern Spain). Sci Total Environ 649:1323–1337

Serrat D (1977) L 'estudi geomorfològic del Pirineu oriental (Puigmal i Costabona). PhD Thesis. University of Barcelona

Thorn CE, Darmody RG, Dixon JC (2011) Rethinking weathering and pedogenesis in alpine periglacial regions: Some Scandinavian evidence. Geol Soc Spec Publ 354:183–193

Trigo RM, Pozo-Vázquez D, Osborn TJ et al (2004) North Atlantic oscillation influence on precipitation, river flow and water resources in the Iberian Peninsula. Int J Climatol 24:925–944

Valero-Garcés BL, González-Sampériz P, Navas A et al (2004) Paleohydrological fluctuations and steppe vegetation during the last glacial maximum in the central Ebro valley (NE Spain). Quat Int 122:43–55

Viana-Soto A, Pérez-Alberti A (2019) Periglacial deposits as indicators of paleotemperatures. A case study in the Iberian Peninsula: The mountains of Galicia. Permafr Periglac Process 30:374–388

Vieira G, de Castro E, Gomes H et al (2020) The Estrela Geopark—from planation surfaces to glacial erosion. Springer International Publishing

Vieira G, Mora C, Ramos M, Benett M (2003) Ground temperature regimes and geomorphological implications in a Mediterranean mountain (Serra da Estrela, Portugal). Geomorphology 52:57–72

Vilaplana JM (1983) Estudi del glacialisme de les valls de la Valira d'Ordino i d'Arinsal. PhD Thesis. University of Barcelona

Zhang T (2005) Influence of the seasonal snow cover on the ground thermal regime: an overview. Rev Geophys 43

The Italian Peninsula

Adriano Ribolini and Carlo Giraudi

1 Geographical Framework

The Italian Peninsula encompasses the territory that extends in a NW-SE direction from the southern edge of the Alps to the Sicily Island. The Mediterranean Sea, which surrounds all sides of the peninsula except the northern one, takes local names depending on the stretch of the coast (Fig. 1). The Italian Peninsula is dominated by the Apennine mountain range, which geographically begins at 44°30′ N (Cadibona Pass) and ends at around 39°30′ N. The Apennines constitutes the geographical and geological axis of the peninsula, giving it a slightly arched shape. The elevations of the main peaks are very variable, but many of them exceed 2,000 m a.s.l. in various sectors of the range. In the northern Apennines (44°30′–42°50′ N) the highest peaks vary between 2,054 and 2,165 m, with Mt Cimone being the highest elevation (2,165 m). In the Central Apennines (42°50′–41°30′ N) many mountains far exceed 2,400 m, with the Gran Sasso as the highest peak (2,915 m). In the Southern Apennines (41°30′–40°00′ N), the mountain range rarely exceeds 2,000 m, being the Pollino the most elevated massif (2,266 m).

During the Miocene-Pliocene, the collision between the western continental margin of the Adria plate and the Sardinian-Corsican block led to the geological structuring of the Apennine chain (Bosellini 2017). In these regards, the southernmost part of the peninsula (Calabrian Arc) does not geologically belong to the Apennines, being formed by a different tectonic mechanism. However, the mountainous massifs at the southern limit of the peninsula (e.g.; the Sila Massif) are geographically included in the southern Apennines. Flysch sequences, limestones, and marls constitute the prevailing rocks in the Northern Apennines, together with ophiolites

A. Ribolini (✉)
Dipartimento di Scienze della Terra, Università di Pisa, Pisa, Italy
e-mail: adriano.ribolini@unipi.it

C. Giraudi
ENEA, Centro Ricerche Saluggia, Saluggia, Italy

M. Oliva et al. (eds.), *Periglacial Landscapes of Europe*,
https://doi.org/10.1007/978-3-031-14895-8_5

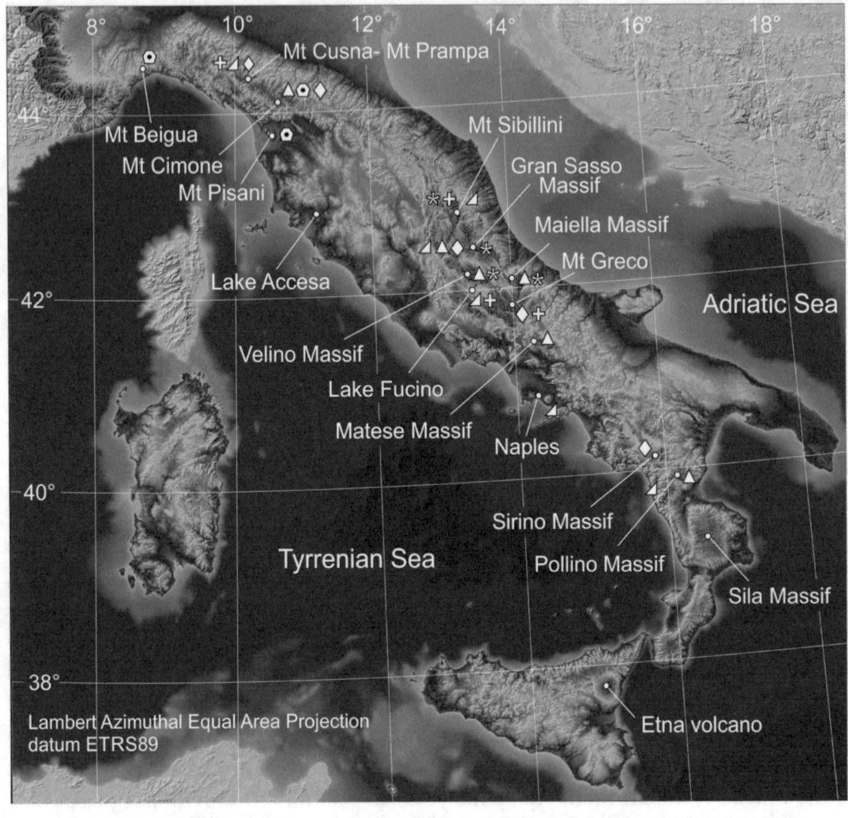

Fig. 1 The Italian Peninsula and key locations described in the main text. Digital terrain model from Tarquini et al. (2007)

(peridotites, gabbros, serpentinites) outcropping in the northernmost part (APAT 2004; Bosellini 2017). Extensive outcrops of carbonate rocks dominate the central Apennines (APAT 2004; Bosellini 2017). The bedrock of the southern Apennines is essentially made of limestones and dolomites (Matese, Sirino and Pollino massifs), while mica-schists, gneiss and granites constitute the southernmost part of the chain (Sila and Aspromonte massifs) (APAT 2004).

The orographic characteristics of the Apennine chain (i.e. peaks exceeding 2,000 m at a short distance from the coast) and its latitudinal development (i.e. t 1,400 km in the NW-SE direction) determine highly different climatic settings. Furthermore, the orientation of the axis of the chain (NW-SE) with respect to the general pathway of mid-latitude low pressure systems of Atlantic origin causes an orographic barrier effect, with consequent disparity in the amount of precipitation between the western (wetter) and eastern (drier) flanks of the chain. In general, the climate of the Apennines can be considered cool temperate (Cf in the Köppen-Geiger classification) (Beck et al. 2018), with mean annual air temperatures (MAAT) ranging from 6 to 9.9 °C, a mean air temperature of the coldest month (MTCO) oscillating between 0 and −3 °C, and a mean air temperature of the warmest month (MTWO) between 15 and 19.9 °C. In the higher sectors of the range, however, the MAAT varies between 3 and 5.9 °C, the MTCO is lower than −3 °C, and the MTWO varies between 10 and 14.9 °C (Fratianni and Acquaotta 2017). These characteristics correspond to a cold temperate (Dw) climate (Beck et al. 2018).

The average annual rainfall along the Apennine range varies from 1,000 to 1,500 mm a^{-1}. However, numerous mountain sectors are affected by average precipitation between 1,500 and 2,000 mm a^{-1} and even more (Fratianni and Acquaotta 2017). Among the latter, the western slopes of the chain constitute an orographic barrier that concentrates the influence of the westerly air masses (e.g. the Apuan Alps, with about 2,500 mm a^{-1} of precipitation in correspondence of the main watershed). Precipitation exhibits a maximum at the end of autumn, a main minimum in summer, and a relative minimum at the end of winter (Fratianni and Acquaotta 2017).

The regional average (annual) 0 °C isotherm is at 2,400–2,500 m (Oliva et al. 2018), i.e. above most of the slopes of the Apennines with the exception of some limited parts of the central sector of the chain. However, the lowering of the altitude of the 0 °C isotherm during the cold Pleistocene and Holocene periods favoured periglacial processes and the formation of permafrost in many areas of the Apennines. Most of the main resulting landforms are no longer active today, and the permafrost is considered to have disappeared. However, some periglacial processes could still be active in the higher part of the slopes.

Likewise glacial chronology (Frezzotti and Narcisi 1989; Giraudi and Frezzotti 1997, Giraudi 1999, 2015), also the periglacial landforms benefitted from dated volcanic tephra and aeolian layers. Indeed, stratigraphic exposures and hand-corings across the Apennines revealed that these layers can be found interbedded in periglacial deposits, as well as in sediment infilling closed-depressions opened on their surface or formed immediately upvalley. In this sense, the following tephra and aeolian layers are used as stratigraphic markers:

- The Biancavilla-Montalto tephra, produced by eruptions of the Etna Volcano in the 19–17 cal ka BP interval (Albert et al. 2013);
- The Neapolitan Yellow Tuff (NYT), a tephra generated by an eruption of the Campi Flegrei caldera and ^{40}Ar/^{39}Ar dated at 14.9 ± 0.4 cal ka BP (Deino et al. 2004);
- A quartz-rich loess, most likely sourced from the Libyan desert (Frezzotti and Giraudi 1989; Giraudi et al. 2013). A peat layer covering the loess in the Central Apennine was dated at 15.7 ± 0.8 cal ka BP (Frezzotti and Giraudi 1989). Thereby, the end of the aeolian sedimentation was estimated at about 16 cal ka BP (Giraudi 2015).

These stratigraphic markers, together with some ^{14}C dates, made it possible to frame the phases of post-LGM glacial retreat in the Apennines. The data collectively indicate that the glacial retreat was punctuated by stadial of readvance/standstill occurring between ~18 and 16 cal ka BP (Oldest Dryas), one just after the deposition of the NYT (~15 cal ka BP, tentatively attributable to the Older Dryas) and one younger than 14–13 cal ka BP (Younger Dryas) (Giraudi 2015). The glacial readvances/standstills post-LGM and before the Younger Dryas correspond to the Fontari Stade of the Apennine glacial chronology (Giraudi and Frezzotti 1997).

2 The Development of Periglacial Research in the Italian Peninsula

Alike the Italian Alps, studies on the periglacial environment in the Italian Peninsula are not numerous, and only in recent decades were based on modern investigation techniques capable of providing information beyond simple notes about landforms existence. The reasons for this low development of research are found in the low number of periglacial landforms and, in many cases, in their non-univocal climatic significance. Therefore, for a long time the periglacial geomorphology was marginal to the glacial research, where a quite continuous progress of knowledge was observed, even animated by scientific disputes (see Ribolini et al. 2022a for a synthesis).

The review by Nangeroni (1952) on periglacial phenomena in Italy marks an important point as it puts a series of features, many of them about Quaternary stratigraphy, under the same umbrella, i.e. processes and landforms promoted by non-glacial cold conditions. Among the papers listed in this review, it is worth mentioning those of Rovereto (1939) and Conti (1940) on the periglacial landforms of the Ligurian Apennines; Segre (1947) and Demangeot (1951) on solifluction lobes and sorted soils in the Central Apennines; and Tricart and Cailleaux (1953) on the morphological and stratigraphic evidence of the cold Quaternary periods.

It is then necessary to wait until the 60s and 70s for new reviews on periglacial geomorphology (Gentileschi 1967a, b; Kelletat 1969; Hollermann 1977; Damangeot 1978). These studies refer to multi-annual and regional research activities, and have the merit of underlining several aspects of periglacial landforms, highlighting their palaeoclimatic relevance and also comparing the results with other geographical areas in the Mediterranean context (Kelletat 1969). It became even more evident that also in the mountains of the Italian Peninsula there were, albeit less numerous and smaller, periglacial landforms similar to those of the Alps, such as patterned grounds, solifluction lobes, ploughing blocks, nivation hollows and block streams.

In the 80–90s, research on the periglacial environment saw a new impetus particularly concerning talus slope landforms (for references see Boenzi 1980; Dramis 1983; Coltorti and Dramis 1988). This is also the moment when some periglacial features began to be included in schemes of the Pleistocene and Holocene environmental evolution principally based on glacial evidence (Giraudi 1996; Giraudi and

Frezzotti 1997). It is worth noting that the first paper that specifically described a rock glacier dates to those years (Giraudi 1988).

The most modern attempts to correlate periglacial landforms with other climate proxies (such as lacustrine/alluvial stratigraphic sequence, phases of aeolian sedimentation, lake-level oscillations, insolation trend, Greenland isotopic chronologies) appear only in the twenty-first century (Giraudi et al. 2001; Giraudi 2005).

3 Distribution and Chronology of Periglacial Landforms

The complex orography of the Italian Peninsula can partly explain the heterogeneous distribution of the periglacial landforms. Indeed, their highest concentration is found in the central Apennines, where not only the highest elevations are reached, but it is also the geographic sector where the relief remains at altitudes above 1000 m for larger extensions. Here, both minor (e.g. patterned grounds, solifluction/solifluction lobes, ploughing blocks) and large-scale landforms (e.g. stratified screes, rock glaciers, protalus rampart and block streams) are present. In the southern Apennines there are almost exclusively rock glaciers and protalus ramparts at the highest altitudes, and stratified screes at altitudes close to sea level. In the northern Apennines, there are only a few rock glaciers and protalus ramparts, stratified screes, block streams and few minor landforms (e.g. solifluction lobes) (Oliva et al. 2018 and references therein).

The elevation of the periglacial landforms is very variable across the peninsula, from almost the sea level (stratified scree in the Sorrento Gulf, near Naple; Brancaccio 1968) up to more than 2,400 m (patterned ground and rock glaciers in the Gran Sasso and Maiella massif) (Dramis and Kotarba 1994; Giraudi 2001). The currently active landforms, or active during the Late Holocene, are exclusive to the central Apennines, while in the remaining sectors of the range landforms that were active in the Late Pleistocene are well-preserved, i.e. those formed during the Last Glacial Maximum (LGM) or the cold periods that characterised the deglaciation (i.e. Lateglacial, Termination I) (Oliva et al. 2018).

3.1 Seasonally Frozen Ground Features

The presence of **patterned ground** features in the central Apennines was first reported by Sacco (1908), and his descriptions are compatible with features such as **sorted stripes**. In the following years, similar features are described in the report by Segre (1947) on Mt Velino (central Apennines). Finally, patterned grounds (**sorted-circles, sorted stripes, ploughing blocks**) are documented in the archive of the crionival landforms of the Gran Sasso and Maiella massifs of Gentileschi (1967a, b) (Fig. 2,a, b). In other sectors of the central Apennines (Sibillini Mountains), decimetric slope-steps bordered by a vegetation rim are consistent with **garland** features

Fig. 2 Seasonal frozen-ground features and rock weathering landforms from various sectors of the Apennine mountains. Patterned ground: **a** small-scale sorted circle; **b** sorted stripes (Maiella Massif, Central Apennines); **c** decimetric slope-steps bordered by a vegetation rim, i.e. garlands (Mts Sibillini, Central Apennines); **d** earth hummocks in the Campo Felice area (Mt Velino Massif); **e** block stream (open-work structure with evidence of erected blocks) (foreground) and tor (background) (Mt Beigua, Norther Apennines); **f** small ice-cored frost mound in the recently deglaciated area of the Calderone Glacier (Gran Sasso Massif, Central Apennines) (photo taken by C. Giraudi in 2003)

(Fig. 2c). However, the first modern and detailed analysis of these features was only published at the beginning of the twenty-first century (Giraudi 2001): at about 2,400 m in the Gran Sasso Massif (Campo Pericoli area), pattern ground features composed of parallel alignments of coarse stones and intervening stripes of fine-grained sediments (i.e. sorted stripes) were investigated. The few centimetres thick surface layer affected by cryoturbation processes covers an organic-rich soil layer that yielded radiocarbon ages of 3.7–3.4 cal ka BP and 3–2.8 cal ka BP. The morphostratigraphic considerations and the ^{14}C ages therefore indicate that the sorted stripes formed over the last 2,000 years, therefore potentially during the Little Ice Age when the nearby Calderone glacier experienced a marked expansion (Giraudi 2001; Hughes et al. 2006). The sorted stripes found in the Conca degli Invalidi area (Gran Sasso Massif) are also of Late Holocene age, as the detrital layer in which they formed covers a soil dated at 1.6–1.4 cal ka BP (Giraudi 2005).

In some flat areas of the Central Apennines there are fields of few dm-high mounds composed of silty-clayey layers, covered by herbaceous vegetation and known as **earth hummocks** (Fig. 2d).

3.2 Rock Weathering (Cold-Induced) Landforms

In the sector of the Apennine facing the Ligurian sea, some **block fields** were observed in the area of Mt Beigua (1,287 m). Here, the rock weathering of the ophiolitic bedrock (serpentinites and metabasalts) led not only to the formation of block fields characterised by an open-work texture with angular-tabular blocks but also of **tors** in the slope uppermost parts (Firpo et al. 2006) (Fig. 2e).

Block fields are also reported in results of geomorphological surveys in the northern (Alpe delle Tre Potenze and Mt Giovo areas) (Castaldini et al. 1998, 2009) and central Apennines (Berti et al. 2006), but detailed studies are missing yet.

3.3 Talus Slope Landforms

In addition to block fields, the Mt Beigua area (Northern Apennine) is characterised also by the existence of several **block streams** that, given their widespread and dimensions, attracted the attention since the beginning of the twentieth century (Sacco 1934; Conti 1940) (Fig. 2e). However, only recently these were correctly interpreted thanks to a modern research approach (Firpo et al. 2006). Morphological investigations revealed the existence of flow structures (i.e. transverse and longitudinal ridges) on the surface of the block streams. The analysis of the detrital surface texture showed an open-work fabric, composed of blocks with maximum diameters in dm, frequently imbricated and showing a frequent vertical dipping. Moreover, the pattern of the block long-axis orientation shows preferential directions (Firpo et al. 2006).

Block streams are also documented in Mt Pisani, a sub-Apennine mountainous relief that reaches maximum elevations of about 1,000 m (Pappalardo and Putzolu 1995). Here, surface texture analysis evidenced deposits composed of heterometric angular blocks forming an open-work structure. Surface undulations are doubtfully interpreted as flow structures.

Stratified screes are surely among the most studied landforms in the Italian Peninsula and are widespread at various altitudes starting from sea level up to more than 2,000 m (Fig. 3a). In the 1980–1990s, a series of systematic studies began in various sectors of the Apennine chain (Bernini et al. 1978; Castiglioni et al. 1979; Boenzi 1980; Gruppo Ricerca Geomorfologia-CNR 1982; Dramis 1983; Coltorti and Dramis 1988; Frezzotti and Giraudi 1992; Coltorti et al. 1983; Coltorti and Dramis 1995; Miccadei et al. 1998). The greatest concentration of stratified screes is observed in the central and partly southern Apennines, associated to the prevalent outcrops of carbonate rocks. Here, in addition to the morphological observations, the various studies documented the textural and structural characteristics of the debris accumulations, confirming the role of frost action in the mechanism of sediments production and of frost-promoted ground creep in depositional processes (Coltorti et al. 1983; Coltorti and Dramis 1988; Frezzotti and Giraudi 1992; Scarciglia et al. 2003; Robustelli and Scarciglia 2006).

There are other talus-slope landforms whose formative processes are not exclusive to the periglacial environment, but whose increased frequency/intensity is indicative of an environment characterised by persistent snow cover/ground ice favouring solifluction phenomena, as well as bedrock cryo-weathering leading to debris accumulations. Interestingly, some of these landforms show morphostratigraphic relationships with soils typical of milder climate condition. This is the case of the numerous evidence of **scree deposits (non-stratified) and solifluction deposits** that covered organic-rich soils (Fig. 3b). All these landforms and deposits (not exclusive to the periglacial environment), together with the others present on the talus slopes (e.g. patterned grounds, stratified screes, solifluction), collectively concur in characterising the active processes in the periglacial climatic phases that affected the Italian Peninsula (Giraudi 2005).

In these regards, the observations on the scree deposits undertaken in some sectors of the Central Apennines (Aremogna plain, Fucino plain, Rieti plain, and Gran Sasso massif) are crucial. Indeed, in addition to the stratigraphic data, these deposits returned the evidence of morphostratigraphic relationships with buried palaeosoils, tephra layers and lake deposits (Frezzotti and Giraudi 1989; Lorenzoni et al. 1992; Giraudi 1995, 1996), which made it possible to reconstruct the Pleistocene and Holocene phases of formation of these sedimentary bodies. The radiocarbon ages indicate that Pleistocene construction phases occurred: (i) after about 21–10 ka BP (maximum age), (ii) during two Late Glacial periods framed between the deposition of the Biancavilla-Montalto tephra (19–17 cal ka BP) and the NYT (14.9 ± 0.4 cal ka BP), and (iii) just before 12.9–12.4 cal ka BP (Giraudi 1995, 1996).

In the Holocene, the oldest date of a soil covered by scree debris was 2.2–1.9 cal ka BP (Campo Pericoli, Gran Sasso Massif). The numerous other chronological data of this type in the Central Apennines show the clustering of debris production phases

Stratified-scree deposit (Mts Sibillini, Central Apennines)
A. Ribolini

Scree deposit (Mt Sirente, Central Apennines)
A. Ribolini

Solifluction lobe (Velino Massif, Central Apennines)
C. Giraudi

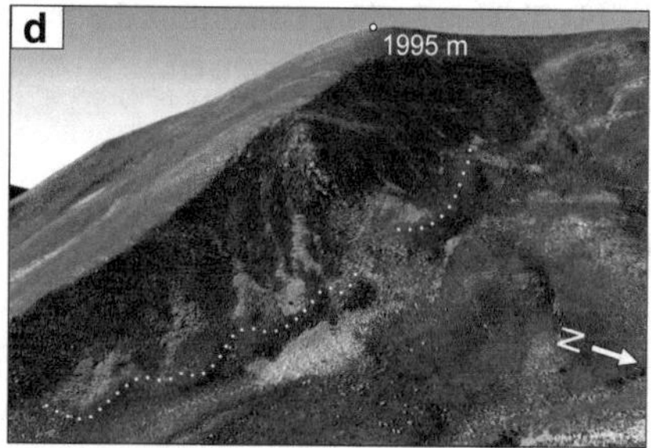

Protalus rampart (Sirino Massif, Southern Apennines)
Google Earth

Fig. 3 Talus slope landforms from various sectors of the Apennines. **a** Stratified-scree deposit ("Lame Rosse" site, Mts Sibillini, central Apennines; **b** scree-deposit; **c** solifluction lobe (Velino Massif, central Apennines); **d** oblique satellite image (source Google Earth®, acquisition date 2017) of a protalus rampart at the base of a north-faced slope of the Sirino Massif (Southern Apennines)

between about 1.5 and 1.0 cal ka BP (Giraudi 2005). This chronological dataset indicates that the Late Holocene was particularly characterised by an alternation of cold (periglacial dynamics, increase of slope debris supply) and milder (pedogenesis) climate phases.

Solifluction lobes are reported in various explanatory notes associated with geomorphological maps, regional geomorphological descriptions, and inventories of periglacial landforms of specific Apennine sectors (Gentileschi 1967a, b; Chelli and Tellini 2006 and references therein). Furthermore, solifluction processes were recognised as possibly responsible for depositional features in stratigraphic sections (Giraudi 1995; Chelli and Tellini 2006). However, their morphostratigraphic relationships with other landforms were rarely examined in detail, making relationships with palaeoenvironmental conditions uncertain. The observations in the central Apennines (Gran Sasso massif, Mts Sibillini and Mt Greco) are an exception (Giraudi 2005). Here, the accumulations of solifluction deposits composed of gravel layers supported by a silty matrix led to the formation of a few decimetres-thick lobes and terraces (Fig. 3c). The solifluction accumulation in some cases overlain an organic silty soil, the dating of which returned maximum ages (*post quem*) for their formation. The chronological data in the Gran Sasso massif (920–670 cal year BP), Sibillini massif (1,070–940 cal year BP) and Mt Greco (1,400–1,270 cal year BP) collectively indicate active solifluction dynamics during the Late Holocene (Giraudi 2005).

Protalus ramparts are reported in geomorphological and geological sketch maps of areas of the Northern Apennines (e.g.; Castaldini et al. 1998; Mariani et al. 2018), although without a detailed characterisation. The information about these landforms is more documented in the central Apennines (Giraudi 1988; Savelli et al. 1995; Jaurand 1998). The most complete analyses are those from the Matese and Sirino massifs, both in the southern Apennines (Giraudi 1998a,b,1999) (Fig. 3d). Here, the sediments infilling closed-depressions backside a protalus rampart ridge revealed the existence of a layer rich in volcanic minerals, corresponding to the pedogenesis of the NYT. This finding constrains the protalus rampart formation to an age older than about 15 ka (Giraudi 1998a, b).

3.4 Rock Glaciers

After the first modern observations by Giraudi (1988), it became evident that **rock glaciers** were largely diffused across the Apennines, being present in all sectors of the mountain chain. The research undertaken during the following years was summarised in Giraudi (2002), but to date no real inventories of Italian Peninsula rock glaciers are available (Dramis et al. 2003).

All the geomorphological observations agree that the rock glaciers of the Apennines are relict landforms, with the sole possible exception of a rock glacier on the Maiella massif (Central Apennines), where the geomorphological characteristics (steepness of the perimeter scarps, general turgid shape) indicate that the presence of permafrost (or its recent disappearance) cannot be excluded (Dramis and Kotarba 1992; Bisci et al. 2003).

Most of the Apennine rock glaciers were formed in areas glacierised during the LGM, particularly in north-facing slopes. In the northern Apennines they are not very widespread and poorly studied, briefly in explanatory notes of general geomorphological surveys (GRG-CNR 1982; Carton and Panizza 1988; Chelli and Tellini 2002; Castaldini et al. 2009) and in a few papers, however, not specifically dedicated to these landforms (e.g. Tommaselli and Agostini 1990). The central Apennines host most of the rock glaciers of the Italian Peninsula, developed from ca. 1,570 m up to 2,550 m. Permafrost creeping involved glacial deposits in most cases, but rock glaciers originating from slope deposits are also common. The largest landforms are present in the Maiella and Velino massifs, where they reach maximum lengths of ca. 1 km (Dramis and Kotarba 1994; Giraudi 2002). In the southern Apennines only one rock glacier has been identified on the northern slope of Mt Pollino (Giraudi 1998, 2002).

In the central and southern Apennines, ages of formation of rock glaciers (or better the maximum age of their deactivation) were proposed based on their morphostratigraphic relationships with LGM/Late Glacial moraines. Furthermore, the analyses of the sediments (up to 3–4 m thick sequences of aeolian sediments, tephra layers, palaeosoils and lacustrine deposits) preserved in depressions formed on the surface or behind rock glacier ridges helped to constrain the timing of the dynamic deactivation of the rock glacier, not excluding that permafrost may have remained for a longer time. The morphostratigraphic relationships, chronologically-supported, constitute the backbone of the proposed phases of formation of the Apennine rock glaciers (Giraudi 2002). These phases of formation are based on data coming largely from the central Apennines and to a limited extent from the southern Apennines Furthermore, rock glaciers present on north-facing slopes are the majority. Therefore, the rock glaciers of the Northern Apennines and the few ones on the South-facing slopes may not fit with the proposed subdivision.

The rock glaciers whose fronts are at altitudes between 1500 and 1600 m involved in the permafrost creeping the first post-LGM retreat moraines, remobilising sediments during the paraglacial phase (onset of the Fontari Stade), which in the Apennines can be dated to about 19–18 ka (Giraudi and Frezzotti 1997; Giraudi 2015; Ribolini et al. 2022b). Moreover, the 16 ka aeolian deposit was found in the deposits infilling depressions on the surface of these rock glaciers, predating their formation to the 18–16 ka interval (Giraudi 2002).

Numerous rock glaciers with front elevation varying between 1,700 and 1,800 m show morphostratigraphic relationships with post-LGM stadial moraines, indicating a formation occurred at the same time of their deposition or shortly after them (Fig. 4) (Giraudi 2002). The sediments infilling surface's depressions showed evidence of the existence of the NYT (14.9 $\pm$ 0.4 cal ka BP) and of a soil produced by its reworking. The NYT and the absence of the 16 ka aeolian deposit constrain the age of this rock glacier generation between 16 and 15 ka BP (Giraudi 2002).

Rock glaciers with fronts at about 1,900 m formed within glacial cirques and their source material is almost always a glaciogenic deposit (Fig. 4). The rock glaciers are not covered by the NYT, while a lacustrine deposit within a surface depression returned a radiocarbon age of 9.5–9.4 cal ka BP (Giraudi and Frezzotti 1997). These data collectively point to the Younger Dryas as a possible age of formation, as the intra Bølling-Allerød cold phases were too short-lived and not cold enough to favour the development of these features (Ribolini et al. 2022c).

The rock glaciers with fronts at ca. 2,000 m formed from glaciogenic deposits within glacial cirques (Fig. 4). The deposits covering their surfaces returned the evidence of a palaeosoil, radiocarbon dated at 3-5–3.3 cal ka BP (Giraudi 2002). These rock glaciers thereby developed between the Younger Dryas and about 3 ka BP, presumably during the cold episodes of the Early-Mid Holocene.

There are few rock glaciers with fronts above 2,300 m, all located within N-NE facing glacial cirques and dominated by steep walls. A palaeosoil interbedded in deposits covering a rock glacier returned a radiocarbon age of 760–660 cal year BP. This last generation of rock glaciers therefore should have formed between about 3.3 and 0.7 cal ka BP (Giraudi 2002). The rock glacier at the base of the NE slope of Mt Amaro in the Maiella massif (elevation of the front about 2,550 m) belongs to this group. The possible current existence of permafrost is still under discussion (Bisci et al. 2003; Dramis and Kotarba 1994), but if confirmed this landform would mark the southern limit of discontinuous permafrost in the Italian Peninsula (Fig. 5).

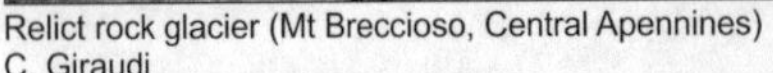

Relict rock glacier (Gran Sasso Massif, Central Apennines)
C. Giraudi

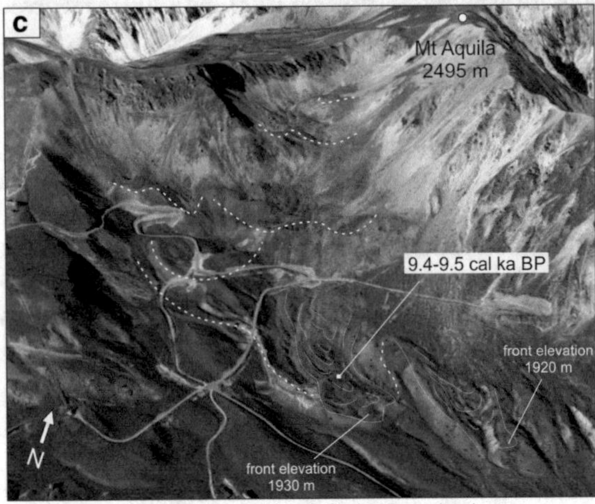

Relict rock glaciers (Gran Sasso Massif, Central Apennines)
Google Earth

Fig. 4 Rock glaciers. **a** Relict rock glacier from central Apennine (Mt Breccioso), front elevation about 1,750 m; **b** relict rock glacier from central Apennine (Gan Sasso Massif), front elevation about 2,270 m; **c** oblique satellite image (source Google Earth®, acquisition date 2017) of rock glaciers in the Campo Imperatore area from Central Apennines (Gran Sasso Massif), front elevations about 1,920 m. These latter involved Lateglacial moraines (Fontari Stade, Oldest Dryas) in the permafrost creeping. Continuous thin white lines: main rock glacier ridges; discontinuous thick white lines: moraine ridges (from Ribolini et al. 2022b, modified). The age of lacustrine deposit infilling a depression on the rock glacier surface is reported (Giraudi and Frezzotti 1997; Giraudi 2002)

3.5 The Contribution of Periglacial Landforms in Reconstructing the Palaeoenvironmental Evolution of the Italian Peninsula

The morphostratigraphic relationships with moraines, tephras, lake deposits, and palaeosoils make it possible to frame the periglacial landforms formation within a palimpsest of cold events, which occurred since the Late Pleistocene. The chronological constraints show how these events match with those recognised in other terrestrial and marine climate proxies of the Mediterranean region, as well as with

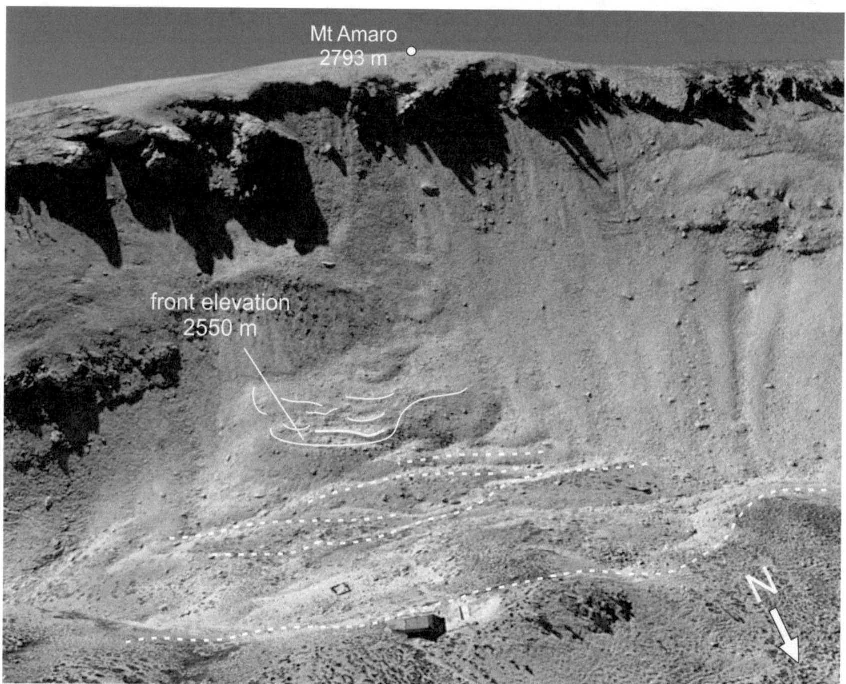

Fig. 5 Oblique satellite image (source Google Earth, acquisition date 2017) of a rock glacier on the NE slope of Mt Amaro in the Maiella massif (central Apennines). Continuous white lines: main rock glacier ridges; discontinuous white lines: Lateglacial moraine ridges (according to Dramis and Kotarba 1994)

those inferred from Greenland isotopic chronologies. Given the available dataset, palaeoclimatic inferences are only related to the medium-high altitude environments of the Apennines.

The development of rock glaciers during the post-LGM period suggest a general environmental aridity, or at least, climate conditions characterised by a low precipitation regime. To retrieve palaeoclimatic information in these regards, Giraudi and Frezzotti (1997) combined the Equilibrium-Line Altitude (ELA) derived from post-LGM recessional moraines with the MAAT inferred from the discontinuous permafrost limits represented by the rock glaciers fronts. The results of the calculation depicted an environment where precipitations (snowfall) were similar or lower than today during the early cold phases (Oldest Dryas, cold events of Bølling-Allerød), and strongly lower than at present during the Younger Dryas (up to 600–800 mm a^{-1} lower). This arid condition matches well with the low-stands reconstructed for some Apennine major lakes during those phases (e.g. Lake Fucino, Lake Accesa), as well as with the disappearance of the minor/ephemeral ones (e.g. Lake Campo Felice) (Giraudi 1989, 1998b, 2002, 2017) (Fig. 6).

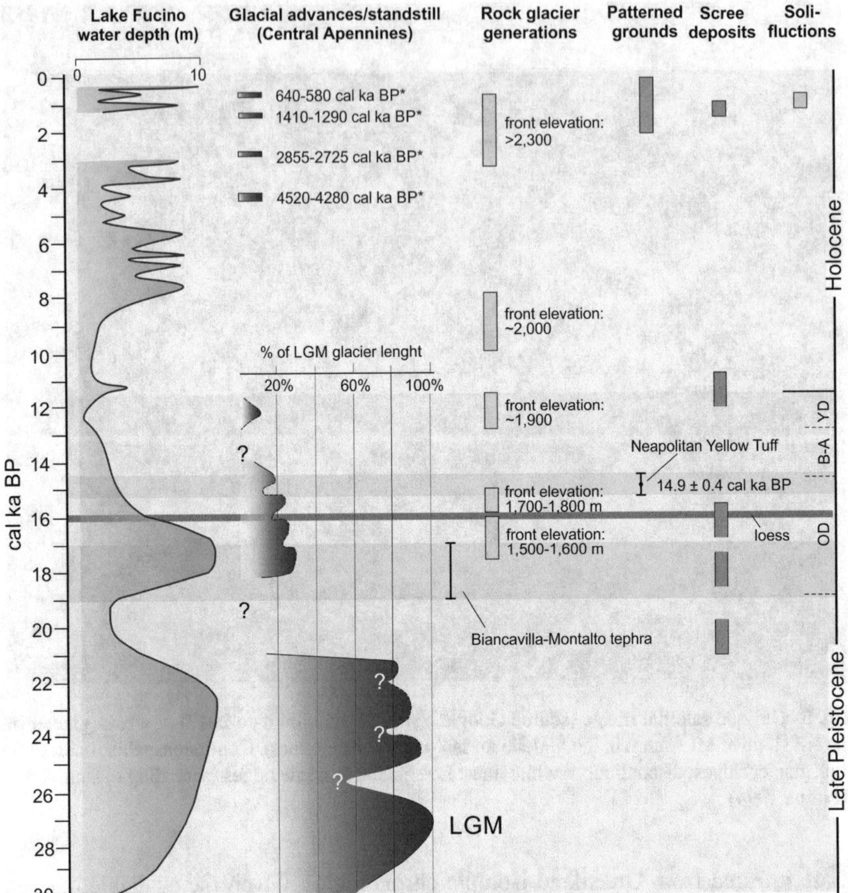

Fig. 6 Chronological scheme of periglacial landforms occurrence in the Italian Peninsula, along with water-depth variations of Lake Fucino, and glacial advances/standstills. All the data are from Central Apennines. Grey bands: dated tephra layers (Deino et al. 2004; Albert et al. 2013). Brown band: aeolian quartz-rich sand layer (loess), found in several sites of the Apennines (Giraudi 2015). YD: Younger Dryas; B-A: Bølling-Allerød; OD: Oldest Dryas (*Data sources* Lake Fucino water depth, Giraudi 1989, 1998, 2002, 2017; glacial advances, Giraudi and Frezzotti 1997; Giraudi 2011, 2015; Ribolini et al. 2022b; rock glaciers: Giraudi 2002; patterned ground, scree deposits and solifluctions, Giraudi 2001). *ages of Calderone Glacier advances (Giraudi 2011))

An alternation of cold phases with intense frost shattering processes is evidenced by (stratified) scree deposits that overlap soils typically formed under milder and wetter conditions. Also in this case, after a first phase about after 21 ka, the other scree deposits fall into the cold periods occurred during the post-LGM deglaciation (Giraudi 2005) (Fig. 6).

The palaeosoils existing in sediments covering rock glaciers with fronts higher than 2,000 m predate their formation to the Early Holocene, remarking how also the cold phases of this period (e.g. Pre-boreal, 8.2 ka event) were characterised by arid condition at higher altitudes. In this case, it is unclear the correlation with low lake levels due to the low temporal resolution of the palaeolimnological data (Giraudi 2002), as it happens for Late Holocene rock glaciers with fronts higher than 2,300 m. However, these surely formed before the period in which Lake Fucino experienced its maximum level, i.e. during the Little Ice Age (Giraudi 2002) (Fig. 6).

The chronology of the aggradation phases of (stratified) scree deposits and of solifluction phenomena collectively indicate that the Apennine environment experienced cold phases during the Late Holocene (<3 ka), with an increase of bedrock frost-shattering and frozen ground creep alternating with period of biostasis with development of soils (Giraudi 2005; Giraudi et al. 2011) (Fig. 6). By broadening the picture of the observations, it is possible to note that during the Holocene phases of aggradation of the scree deposits and the occurrence of solifluction lobes, the Apennine mountains also recorded phases of alluvial sedimentation (Giraudi 2005) favoured by the availability of frostshattered slope debris poorly protected by soil and vegetation (Coltorti and Dramis 1988) with a consequent increase in the solid load of the rivers (Giraudi 2014).

4 The Significance of Periglacial Dynamics in the Italian Peninsula Within the European Context

Despite the southern latitudes, the relatively modest mountain elevations and the close proximity to the Mediterranean Sea of all inland areas, the Italian Peninsula exhibits a fair number of periglacial landforms. However, they always remained in the background of research, especially that dedicated to the Quaternary glaciers of the Apennine chain. Only recently some landforms were included in reconstructions of paleoenvironmental evolution. This represents a missed opportunity because, despite essentially relics, several periglacial landforms are well preserved, and in some cases are comparable with those of mountain ranges where they are widely studied (e.g. the Alps).

The analysis of dated tephras and loess deposits gave a great impetus to the understanding of the climatic-environmental contribution of periglacial processes, because it made possible to frame their occurrence in the palimpsest of cold events detected by other environmental records (Fig. 6).

This chronological framework allows for two main considerations. The timing of the periglacial processes confirms that also the mountain environments of the Italian Peninsula were sensitive to cold climatic events recorded in terrestrial/marine climate proxies and in Greenland isotopic stratigraphy (e.g. Oldest Dryas, Older Dryas, Younger Dryas). The correlations with other Mediterranean contexts are possible, with the aim of comparing synchronism and intensity of periglacial processes.

The correlation is complicated because the periglacial dynamics of the Italian Peninsula are chronologically constrained by minimum/maximum ages of formation/stabilization (*ante* and *post quem* ages). However, taking these limitations into account, correlations with regions at comparable latitudes deserve to be attempted. In the Iberian Peninsula, the stabilization of most of the rock glaciers occurred at about 15–14 ka (Palacios et al. 2017). Therefore it is chronologically consistent with the dynamic deactivation of the rock glaciers of the central Apennines (fronts at 1,700–1,800 m), whose surfaces are covered by the NYT (about 15 cal ka BP) (Fig. 6). The formation of the stratified screes in the Iberian Peninsula began after about 23 cal ka BP, it experienced construction phases during the Termination I (at about 17 cal ka BP) and in Early Holocene (10–9 H/Th ka BP) (Ruiz-Fernández et al. 2019; García-Ruiz et al. 2001; Valero-Garcés et al. 2004). This chronological succession fits relatively well with the time intervals of deposition of scree deposits in the Apennines, where after a first post-LGM phase (<21 cal ka BP), depositions between 10 and 15 cal ka BP and just before 12 cal ka BP are documented (Fig. 6).

The comparison with the Balkan Peninsula is complicated by the limited availability of chronological constraints, and also by climatic conditions different from the Italian Peninsula (in the past as now), as this region is subjected also to an atmospheric circulation sourced in the Scandinavia and Siberia which could have impacted on local periglacial dynamics. Similarly to the Apennines, it is believed that the main Balkan rock glaciers (Mts Pindus and Parnassos sectors) formed in post-LGM cold phases (Termination I) (references) and that in the Early Holocene they became relict, with the exception of some landforms above 2,450 m in the massifs of Rila (country) and Pirin, (Onaca et al. 2020). Therefore, in the Balkans as in the Apennines, the existence of permafrost is doubtful but cannot be excluded.

Although the puzzle is still incomplete and uncertain, a general examination of the data on the periglacial landforms of the Italian Peninsula underlines how they represent a good opportunity to implement the paleoenvironmental reconstructions of the Termination I and the Holocene with evidence of (non-glacial) cold climate-driven processes. In these regards, investigations to detect the existence of permafrost ice and thermal monitoring of ground surfaces/bedrock to investigate the rates of freeze/thaw cycles and amplitudes could be a significant step forward, which must be associated with the increase of the geomorphological and chronological datasets of all periglacial landforms.

Acknowledgements The project "The termination I. The environmental and palaeoclimatic variations occurred during the 25–11 ka period" (leader A. Ribolini) supported this work (Progetti di Ricerca di Ateneo PRA, 2020–2021, University of Pisa).

References

Albert PG, Tomlinson EL, Lane CS, Wulf S, Smith VC, Coltelli M, Keller J, Lo Castro D, Manning CJ, Müller W, Menzies MA (2013) Late glacial explosive activity on Mount Etna: implications for proximal–distal tephra correlations and the synchronisation of Mediterranean archives. J Volcanol Geotherm Res 265:9–26

APAT (2004). Carta Geologica d'Italia. Scala 1:250 000. Selca, Firenze

Beck HE, Zimmermann NE, McVicar TR, Vergopolan N, Berg A, Wood EF (2018) Present and future Köppen-Geiger climate classification maps at 1-Km resolution. Sci Data 5 (180214).

Bernini M, Carton A, Castaldini D, Cremaschi M (1978) Segnalazione di un deposito di versante di tipo Grèzes liteès a sud di M. Prampa (alto Appennino reggiano). Gruppo Studio del Quaternario Padano 4:153–172

Berti C, Miccadei E, Piacentini T (2006) La Marsica (Abruzzo meridionale). In: Chelli A, D'Aquila P, Firpo M, Ginesu S, Guglielmin M, Pappalardo M, Pecci M, Piacentini T, Queirolo C, Robustelli G, Scarciglia F, Sias S, Tellini C (eds) Testimoni di una montagna scomparsa. Contributo alle metodologie d'indagine delle forme periglaciali relitte. Problematiche e applicazioni in differenti ambienti morfodinamici. Collana «Quaderni della Montagna». vol 8. Bonomia University Press, Bologna, pp 70–73

Bisci C, Dramis F, Fazzini M, Guglielmin M (2003) Climatic conditions and sporadic permafrost in the Maiella Massif (Central Apennines, Italy). Geogr Fis Din Quat 26(1):3–13

Boenzi F (1980) Some evidence of Quaternary cold periods in Southern Italy: data and reflections. Riv Geogr Fisica Dinamica Quaternaria 3:16–20

Bosellini A (2017). Outline of the geology of Italy. In: Soldati M, Marchetti M (eds) Landscapes and landforms of Italy. Springer, pp 21–27

Brancaccio L (1968) Genesi e caratteri delle forme costiere nella Penisola Sorrentina. vol 96. Bollettino della Società Geologica Italiana, pp 169–180

Carton A, Panizza M (1988) Il paesaggio fisico dell'Alto Appennino emiliano. Grafis Ed., Bologna, 182 p

Castaldini D, Caredio F, Puccinelli A (1998) Geomorfologia delle valli del Rio delle Pozze e del Torrente Motte (Abetone, Appennino Tosco-Emiliano). Geogr Fis Din Quat 21:177–204

Castaldini D, Valdati J, Ilies DC (2009) Geomorphological and Geoturistic Maps of the Upper Tagliole Valley (Modena Apennines, Norther Italy). Memorie Descrittive Carta Geologica d'Italia LXXXVII:29–38

Castiglioni GB, Girardi A, Sauro U, Tessari F (1979) Grézes Litées e falde detritiche stratificate di origine crionivale. Rivista di Geografia Fisica e Dinamica Quaternaria 2:64–82

Chelli A, Tellini C (2002) Geomorphological features of the Bratica Valley (Northern Apennines, Italy). Geogr Fis Din Quat 25(1):45–60

Chelli A, Tellini C (2006) L'Appennino. In: Chelli A., D'Aquila P, Firpo M, Ginesu S, Guglielmin M, Pappalardo M, Pecci M, Piacentini T, Queirolo C, Robustelli G, Scarciglia F, Sias S, Tellini C (eds.) Testimoni di una montagna scomparsa. Contributo alle metodologie d'indagine delle forme periglaciali relitte. Problematiche e applicazioni in differenti ambienti morfodinamici. Collana «Quaderni della Montagna». vol 8. Bonomia University Press, Bologna, pp 70–73

Coltorti M, Dramis F, Pambianchi G (1983) Stratified slope-waste deposits in the Esino River basin, Umbria-Marche Apennines, central Italy. Polarforschung 53(2):59–66

Coltorti M, Dramis F (1988) The significance of stratified slope-waste deposits in the Quaternary of Umbria-Marche Apennines, central Italy. Zeitschrift fur Geomorphologie, Supplementband 71:59–70

Coltorti M, Dramis F (1995) The chronology of Upper Pleistocene stratified slope-waste deposits in central Italy. Permafrost Periglac Process 6(3):235–242

Conti S (1940) La nivazione e la morfologia periglaciale nell'Appennino ligure occidentale. Bollettino della Società Geologica Italiana 59(1):69–94

Damangeot J (1951) Observation sur le sols en gradins de l'Apennin Central. Rev Géomorphol Dynam 3:110–119

Demangeot J (1978) L'étage périglaciaire au Gran Sasso (2912 m), Italie centrale, in: Colloque sur le périglaciaire d'altitude du domaine méditerranéen et abords, Strasbourg-Université Louis Pasteur, 12–14 mai 1977. Association géographique d'Alsace, pp 209–220

Deino A, Orsi G, de Vita S, Piochi M (2004) The age of The Neapolitan Yellow Tuff caldera-forming eruption (Campi Flegrei caldera-Italy) assessed by 40Ar/39Ar dating method. J Volcanol Geoth Res 133:157–170

Dramis F (1983) Morfogenesi di versante nel Pleistocene superiore in Italia: i depositi detritici stratificati. Geogr Fis Din Quat 6:180–182

Dramis F, Kotarba A (1992) Southern limit of relict rock glaciers, Central Apennines, Italy. Permafr Periglac Process 3:257–260

Dramis F, Kotarba A (1994) Geomorphological evidences of high-mountain permafrost in central Apennnines. Geogr Fis Din Quat 17:29–39

Dramis F, Giraudi C, Guglielmin M (2003) Rock glacier distribution and paleoclimate in Italy. In: Phillips M, Springman SM, Arenson LU (eds) Permafrost, vol 1. Swets & Zelinger, Kisse, pp 199–204

Firpo M, Guglielmin M, Queirolo C (2006) Relict blockfields in the Ligurian Alps (Mount Beigua, Italy). Permafrost Periglac Process 17:71–78

Fratianni S, Acquaotta F (2017) The climate of Italy. In: Soldati M, Marchetti M (eds) Landscapes and landforms of Italy. Springer, pp 29–38

Frezzotti M, Giraudi C (1989) Evoluzione geologica tardo-pleistocenica ed olocenica del Piano di Aremogna (Roccaraso - Abruzzo): implicazioni climatiche e tettoniche. Memorie della Società Geologica Italiana 42:5–19

Frezzotti M, Narcisi B (1989) Identificazione di un andosuolo, possibile livello guida per la cronostratigrafia olocenica dell'Appennino Centrale. Memorie Società Geologica Italiana 42:351–358

Frezzotti M, Giraudi C (1992) Evoluzione geologica Tardo-Pleistocenica ed olocenica del conoide complesso di Valle Majelama (massiccio del Velino, Abruzzo) – Il Quaternario 5(1):33–50.

García-Ruiz JM, Valero-Garcés B, González-Sampériz P, Lorente A, Martí-Bono C, Beguería S, Edwards L et al (2001) Stratified scree in the Central Spanish Pyrenees: Palaeoenvironmental implications. Permafr Periglac Process 12:233–242

Gentileschi ML (1967a) Forme crionivali sul Gran Sasso d'Italia. Bolletino della Società geografica Italiana 9, 8(1–3):34–61

Gentileschi ML (1967b) Forme crionivali sulla Maiella. Bolletino della Società geografica Italiana 9, 8(7–9):325–350

Giraudi C (1988) Segnalazione di scarpate di faglia post-glaciali nel Massiccio del Gran Sasso (Abruzzo): implicazioni tettoniche, rapporti tra tettonica recente e morfologia, paleosismicità. Memorie della Società Geologica Italiana 41:627–635

Giraudi C (1989) Lake levels and climate for the last 30, 000 years in the Fucino area (Abruzzo-Central Italy)—a review. Palaeogeogr Palaeoclimatol Palaeoecol 70:249–260

Giraudi C (1995) I detriti di versante ai margini della Piana del Fucino (Italia centrale): significato paleoclimatico ed impatto antropico. Il Quat, Ital J Quat Sci 8:203–210

Giraudi C (1996) The effect of the Younger Dryas and of Heinrich Events on the climatic and environmental evolution of Central Italy. Il Quat, Ital J Quat Sci 9:533–540

Giraudi C (1998a) The Late Pleistocene deglaciation on Mt Sirino and Mt Pollino (Basilicata, Calabria - Southern Italy). Il Quat, Ital J Quat Sci 11:247–254

Giraudi C (1998b) Late Pleistocene and Holocene Lake Level Variations in Fucino Lake (Abruzzo - Central Italy) Inferred from Geological, Archaeological and Historical Data. Palaoklimaforschung- Palaeoclimate Research 25:1–18

Giraudi C (1999) Datazione e correlazione di depositi glaciali con l'uso di tephra e loess: il caso del Matese (Campania-Molise). Il Quat, Ital J Quat Sci 12:11–15

Giraudi C (2001) Segnalazione di strutture a strisce parallele (sorted stripes) su detrito del Gran Sasso d'Italia. Il Quat, Ital J Quat Sci 14:5–8

Giraudi C (2002) I rock glaciers tardo-pleistocenici ed olocenici dell'Appennino - Età, distribuzione, significato paleoclimatico. Il Quat, Ital J Quat Sci 15:45–52

Giraudi C (2005) Middle to Late Holocene glacial variations, periglacial processes and alluvial sedimentation on the higher Apennine massifs (Italy). Quatern Res 64:176–184

Giraudi C (2011) Middle Pleistocene to Holocene glaciations in the Italian Apennines. In: Ehlers J, Gibbard PL, Hughes PD (eds) Quaternary glaciations—extent and chronology. A closer look. Developments in Quaternary Science, Elsevier, vol 15, pp 211–219

Giraudi C (2014) Coarse sediments in Northern Apennine peat bogs and lakes: New data for the record of Holocene alluvial phases in peninsular Italy. The Holocene 24 (8): 932–943

Giraudi C (2015) The Upper Pleistocene deglaciation on the Apennines (Peninsular Italy). Cuadernos de Invesrtigación Geográfica 41(2):337–358

Giraudi C (2017) Climate evolution and forcing during the last 40 ka from the oscillations in Apennine glaciers and high mountain lakes, Italy. J Quat Sci 32:1085–1098

Giraudi C, Frezzotti M (1997) Late Pleistocene glacial events in the Central Apennines, Italy. Quat Res 48(3):280–290

Giraudi C, Magny M, Zanchetta G, Drysdale RN (2011) The Holocene climatic evolution of Mediterranean Italy: A review of the continental geological data. Holocene 21(1):105–115

Giraudi C, Zanchetta G, Sulpizio R (2013) A Late-Pleistocene phase of Saharan dust deposition in then high Apennine mountains (Italy). AlpE MediterrEan Quat 26:111–122

Gruppo Ricerca Geomorfologia-CNR (1982) Geomorfologia del territorio di Febbio tra il M. Cusna ed il F. Secchia (Appennino emiliano). Geogr Fis E Din Quat 5:285–360

Hollermann P (1977) Die periglaziale Höhenstufe der Gebiete in einem West-Ost_Profil von Nordiberian zum Kaukasus. Dritte Folge 31:2

Hughes PD, Woodward JC, Gibbard PL (2006) Quaternary glacial history of the mediterranean mountains. Prog Phys Geogr 30(3):334–364

Jaurand E (1998) Glaciers Disparù de l'Appenin: Geomorphologie et Paleoenvironement Glaciaires de l'Italie Peninsulaire. Monographie de l'Universite de la Sorbonne, Paris

Kelletat D (1969) Verbreitung und Vergesellschaftung rezenter Periglazialerscbeinungen im Apennin. Gottinger Geogr Abh 48:114.

Lorenzoni P, Raglione M, Brunamonte F, Michetti AM, Pennacchioni M (1992) Stratigrafia dei depositi di versante tardo-quaternari del Bacino di Rieti: la sezione de "La Casetta." Studi Geologici Camerti 1:145–153

Mariani GS, Cremaschi M, Zerboni A, Zuccoli L, Trombino L (2018) Geomorphology of the Mt. Cusna Ridge (Northern Apennines, Italy): evolution of a Holocene landscape. J Maps 14(2):392–401. https://doi.org/10.1080/17445647.2018.1480976

Miccadei E, Barberi R, Cavinato GP (1998) La geologia quaternaria della conca di Sulmona (Abruzzo, Italia centrale). Geol Romana 34(1):59–86

Nangeroni G (1952) I fenomeni di morfologia periglaciale in Italia. Rivista Geografica Italiana 59:1–15

Onaca A, Ardelean F, Ardelean A, Magori B, Sîrbu F, Voiculescu M, Gachev E (2020) Assessment of permafrost conditions in the highest mountains of the Balkan Peninsula. CATENA,185,104288, https://doi.org/10.1016/j.catena.2019.104288.

Oliva M, Žebre M, Guglielmin M, Hughes PD, Çiner A, Vieira G, Bodin X, Andrés N, Colucci RR, García-Hernández C, Mora C, Nofre J, Palacios D, Pérez-Alberti A, Ribolini A, Ruiz-Fernández J, Sarıkaya MA, Serrano E, Urdea P, Valcárcel M, Woodward JC, Yıldırım C (2018). Permafrost conditions in the Mediterranean region since the Last Glaciation. Earth Sci Rev 185:397–486. https://doi.org/10.1016/j.earscirev.2018.06.018.

Palacios D, de Andrés N, Gómez-Ortiz A, García-Ruiz JM (2017) Evidence of glacial activity during the Oldest Dryas in the mountains of Spain. Geol Soc Spec Publ 433:87–110

Pappalardo M, Putzolu PP (1995) Le "sassaie" dei Monti Pisani, "colate di pietre" tipiche dell'ambiente periglaciale. Atto Società Toscana di Scienze Naturali, Memorie, Seria A 101:323–342

Ribolini A, Spagnolo M, Giraudi C (2022a) The Italian Peninsula. In: Palacios D, Hughes P, García-Ruiz J, Nuria A (eds) European glacial landscapes: last deglaciation, vol 1. Amsterdam, Elsevier, https://www.sciencedirect.com/science/article/pii/B9780128234983000017

Ribolini A, Spagnolo M, Giraudi C (2022b) The Italian mountains: glacial landforms during main deglaciation (18.9–14.6 ka). In: Palacios D, Hughes P, García-Ruiz J, Nuria A (eds) European glacial landscapes: last deglaciation, Vol. 2. Amsterdam, Elsevier.

Ribolini A, Spagnolo M, Giraudi C (2022c) The Italian mountains: glacial landforms from the Bølling-Allerød Interstadial (14.6–12.9 ka). In: Palacios D, Hughes P, García-Ruiz J, Nuria A (eds) European glacial landscapes: last deglaciation, Vol. 2. Amsterdam, Elsevier

Robustelli G, Scarciglia F (2006) Mormanno e Praia a Mare (Calabria settentrionale). In: Chelli A, D'Aquila P, Firpo M, Ginesu S, Guglielmin M, Pappalardo M, Pecci M, Piacentini T, Queirolo C, Robustelli G, Scarciglia F, Sias S, Tellini C (eds) Testimoni di una montagna scomparsa. Contributo alle metodologie d'indagine delle forme periglaciali relitte. Problematiche e applicazioni in differenti ambienti morfodinamici. Collana Quaderni della Montagna, vol 8. Bonomia University Press, Bologna, pp 81–92

Rovereto G (1939) Liguria geologica. Memorie della Società geologica Italiana 2:743

Ruiz-Fernández J, García-Hernández C, Fernández S del C (2019) Holocene stratified scree of praón (Picos de Europa, cantabrian mountains). Los derrubios estratificados holocenos de praón (Picos de Europa, montañas cantábricas). Cadernos do Lab. Xeolóxico de Laxe 41:23–46

Sacco F (1908) Glacialismo ed erosione nella Majella. Atti della Società Italiana di Scienze Naturali 47(4):269.

Sacco F (1934) Il glacialismo nel gruppo di Voltri. Atti della reale Accademia delle Scienze di Torino 70:96–105

Savelli D, Nesci O, Basile M (1995) Evidenze di un apparato glaciale pleistocenica sul Massiccio del Catria (Appennino Marchigiano). Geogr Fis Din Quat 18:331–335

Scarciglia F, Terribile F, Colombo C, Cinque A (2003) Late Quaternary climatic changes in Northern Cilento (Southern Italy): an integrated geomorphological and paleopedological study. Quat Int 106–107:141–158

Segre AG (1947) Suoli e strutture da nivazione nell'Appennino Centrale. L'Universo 27:805–814

Tarquini S, Isola I, Favalli M, Mazzarini F, Bisson M, Pareschi MT, Boschi E (2007) TINITALY/01: a new triangular irregular network of Italy. Ann Geophys

Tommaselli M, Agostini N (1990) Vegetation patterns and dynamics on a rock glacier in the Norther Apennines. Pireneos 136:33–46

Tricart J, Cailleaux A (1953) Action du froid Quaternaire en Italie péninsulaire. Actes IV Congrès INQUA, Roma-Pisa I:136–142

Valero-Garcés BL, González-Sampériz P, Navas A, Machín J, Delgado-Huertas A, Peña-Monné JL, Sancho-Marcén CS, Stevenson T, Davis B (2004). Paleohydrological fluctuations and steppe vegetation during the last glacial maximum in the central Ebro valley (NE Spain). Quat Int 122:43–55

The Balkans (Without Carpathians)

Manja Žebre and Emil M. Gachev

1 Geographical Framework

The Balkan Peninsula lies in south-east Europe and is bounded by the Adriatic, Ionian, Aegean and Black Seas. The boundaries on the mainland, however, are less well defined. Here, the northern boundary is defined by the Danube River and its delta to the east and the Sava, Krka, Vipava and Soča rivers to the west. The eastern boundary is defined at the Bosporus Strait. To the west and south, the offshore islands of Croatia and Greece lie within the boundaries of the Balkan Peninsula. According to this delimitation, the Balkan Peninsula covers a total area of ~508,000 km^2.

The Balkan Peninsula is very mountainous, forming a mosaic of four main mountain systems (i.e., Dinaric Mountains, Balkan Mountains/Stara Planina, Pindus Mountains and Rila-Rhodope Massif) separated by internal tectonic depressions and large valleys (Oliva et al. 2018). In addition, there is also a number of high mountains surrounded by the main Balkan mountain systems between the Drin and Struma rivers, known by the common names of West Macedonian Zone (Šar Planina, Korab, Pelister, Galičica, Strogovo, Jablanica, Bistra, Suva gora; after Arsovski 1960), Pelagonian Massif (Jakupica, Nidze; after Milevski 2015) and Dardanian Massif (Osogovska Planina, Belasica; after Alexiev 2002). These mountain ranges are predominantly located in the country of North Macedonia and are therefore hereafter referred to as "the North Macedonia Mountains". The summits of all the main mountain ranges, which are generally oriented towards NW–SE, exceed 2000 m a.s.l.. The highest summit is Musala (2925 m) in the Rila-Rhodope Massif. In general, there is a contrast in topography between the eastern and western Balkan Peninsula. In the

The original version of this chapter was revised: Belated correction has been updated. The correction to this chapter is available at https://doi.org/10.1007/978-3-031-14895-8_17

M. Žebre (✉)
Geological Survey of Slovenia, Ljubljana, Slovenia
e-mail: manja.zebre@geo-zs.si

E. M. Gachev
South-West University "Neofit Rilski", Blagoevgrad, Bulgaria

Geodesy and Geography, National Institute of Geophysics, BAS, Sofia, Bulgaria

M. Oliva et al. (eds.), *Periglacial Landscapes of Europe*,
https://doi.org/10.1007/978-3-031-14895-8_6

east, the slopes are rather gentle, while in the west the Dinaric Mts. and Pindus Mts. rise steeply from the coast. Most of the lowlands of the Balkan Peninsula cover the northern part, which includes the southernmost extension of the Pannonian Basin and plains between the Carpathians in the north and Stara Planina in the south. There are also extensive lowlands along the Maritsa River (Thracian Plain) between the Rila-Rhodope Massif and the Black Sea (Reed et al. 2004) (Fig. 1).

The Balkan Peninsula has a very diverse geology. The Dinaric Mts., for example, are dominated by very thick Mesozoic carbonates. Mesozoic carbonates in combination with ophiolitic rocks are also the dominant lithologies in the Pindus Mts. The Rila-Rhodope Massif is composed mainly of Precambrian and Palaeozoic crystalline rocks, whereas mountains within the Balkan Range (Stara Planina, Srednogorie) mainly consist of Palaeozoic and Mesozoic sedimentary and metamorphic rocks, volcanic rocks and granitic intrusions (European Geological Data Infrastructure).

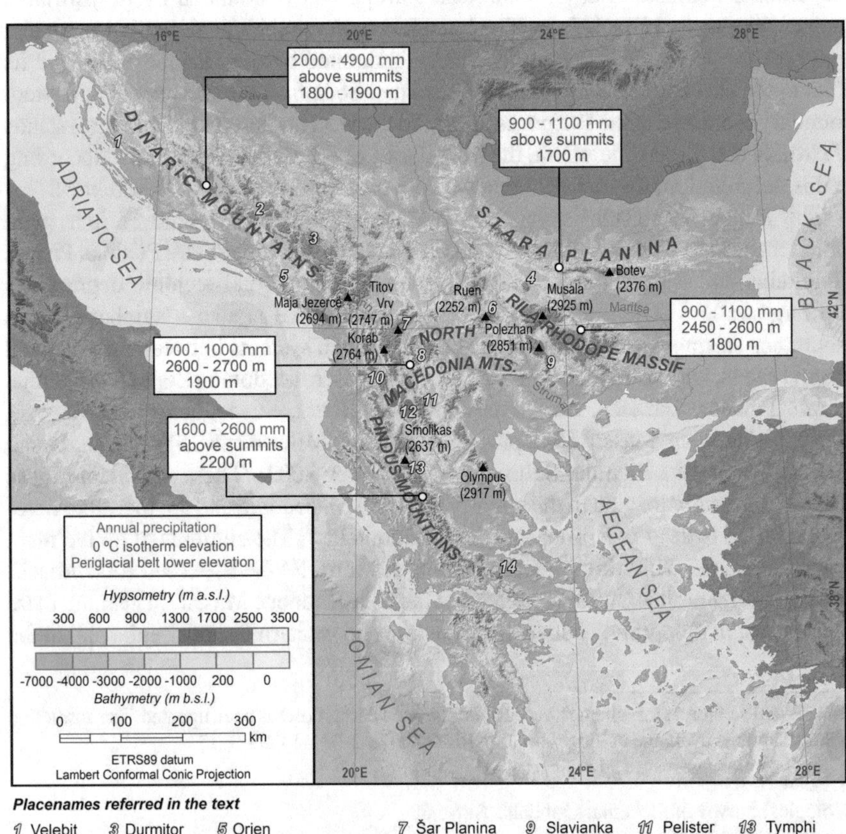

Fig. 1 Location of the main mountain systems in the Balkan Peninsula and local names mentioned in the text. The map also includes an estimation of the precipitation, altitude of the 0 °C isotherm and minimum altitude of the present-day periglacial belt

Other mountains on the central Balkan Peninsula are composed of various lithologies, such as marble (Jakupica, Suva Gora), limestone (Bistra, Jablanica, Galičica), shale (Šar Planina), granite (Pelister), or schist (Osogovska Planina) (see Milevski 2015).

According to Köppen–Geiger climate classification, the Balkan Peninsula is divided into cold climate without dry season (Dfb, Dfc) in the mountainous areas, temperate climate without dry season (Cfa, Cfb) in the north and north-east, and temperate climate with dry summer (Csa, Csb) in the south and south-west. Some areas in Greece have arid, steppe and cold climate (BSk) (Kottek et al. 2006; Rubel et al. 2017). Winters in the Balkan Peninsula are characterized by strong temperature fluctuations. Generally, westerly winds transport mild and moist Mediterranean air towards the Balkan Peninsula, but occasionally the large-scale atmospheric circulation changes, resulting in colder conditions with northerly flow from Scandinavia or north-easterly flow from Siberia (Anagnostopoulou et al. 2017). On the contrary, heat waves usually occur when there is a ridge of high pressure over Europe. This system causes the inflow of continental air masses from the north-east or east. Another source of air masses that cause heat waves is advection from the south (Tomczyk 2016). The mean annual precipitation (MAP) over the Balkan Peninsula is highly variable and is influenced by the presence of mountain ranges, with large differences between west and east, and between lowlands and mountains (Furlan 1977). The south-west slopes of the Pindus Mts. and Dinaric Mts. generally have MAP above 2000 mm (Furlan 1977). The latter is also the area with the highest annual precipitation in Europe, where the average over the period 1931–1960 was 4926 mm, with 8063 mm measured in 1937 at the meteorological station Crkvice (940 m) on Mt. Orjen (Dinaric Mts.; Magaš 2002). Due to the precipitation shadow, the mountain basins to the east receive much lower MAP (500–700 mm). The same is true for the highest parts of the Rila-Rhodope Massif and Stara Planina, which receive less than 1100 mm MAP (Marinova et al. 2017). The elevation of the 0 °C isotherm is above the highest summits in most areas apart from the Rila-Rhodope Massif, where it is at ~2450–2600 m (Grunewald 2014; Onaca et al. 2020). Here, the only present-day permafrost was modelled in the form of isolated patches (Obu et al. 2019). The mean annual air temperature (MAAT) on the highest summits of most of the other mountain ranges ranges between 0 and 5 °C, according to WorldClim 1.4 data for 1960–1990 (Hijmans et al. 2005).

Therefore, the current climatic conditions favour the occurrence of cold-climate geomorphological processes only in a few mountain ranges in the Balkan Peninsula. Instead, the mountains of the Balkan Peninsula were characterised by the presence of small ice fields and valley glaciers during the Quaternary cold stages (Hughes et al. 2006b, 2010, 2011; Kuhlemann et al. 2009, 2013; Žebre and Stepišnik 2016; Çiner et al. 2019; Žebre et al. 2019, 2021; Ruszkiczay-Rüdiger et al. 2020; Sarıkaya et al. 2020), having their Equilibrium-Line Altitudes (ELAs) even as low as ~1250 m in the west with a strong west-east increasing gradient (Hughes et al. 2010; Žebre et al. 2019). Periglacial processes took place in the ice-free areas above the glaciated environments (Oliva et al. 2018), but also in the lowlands, especially in karst depressions, due to the favourable microclimate. Periglacial processes were generally more widespread in the continental, more arid eastern part of the Balkan Peninsula than in the wetter western part, which is nowadays reflected in a more extensive relict periglacial environment in the east compared to the west. Current glacial environment with small glaciers (0.005–0.05 km^2) is present only in the Pirin, Prokletije and Durmitor Mts. (Gachev et al. 2016).

2 The Development of Periglacial Research in the Balkan Peninsula

The first studies on cold-climate geomorphological processes in the Balkan Peninsula published in the late nineteenth and early twentieth centuries focused on glacial environments (Cvijić 1897, 1899; Penck 1900; Grund 1902; Janković 1903; Annaheim 1939), although some information on periglacial landforms can be also found in these pioneering studies. Among the first publications in which periglacial landforms were a special focus were the works of Glovnia (1958, 1962b, 1968) in the mid-twentieth century. In these studies, the presence of various periglacial features in the Rila Mt., such as rock glaciers, block fields, patterned ground, solifluction lobes and ice needles, was mentioned for the first time. In the same period, Popov (1962) published a detailed geomorphological study of Golemia Kazan glacio-karstic cirque in the Pirin Mt., which included descriptions and mapping of periglacial karst features in marble. These early periglacial studies also mentioned the occurrence of block streams on the Pelister Mt. (Stojadinoviќ 1962) and a rock glacier on the Jakupica Mt. (Manakoviќ 1962). Studies focused mainly on mapping and inventory of periglacial landforms without quantitative analyses were published at the end of the twentieth century for the Rila Mt. (Velchev 1999), Pirin Mt. (Velchev and Kenderova 1994), Prokletije Mts. (Palmentola et al. 1995) and Jablanica Mt. (Kolčakovski 1999).

Since the 2010s, periglacial research in the Balkan Peninsula has become more extensive and specialised. The first rock glacier inventories based on satellite imagery were assembled for the Rila Mt. (Gikov and Dimitrov 2010) and the Pirin Mt. (Dimitrov and Gikov 2011) and later verified in the field (Dimitrov and Velchev 2012; Gachev et al. 2017; Gachev 2018). While several other studies documented rock glaciers in various mountain ranges of the Dinaric Mts. (Milivojević et al. 2008; Gachev (2019), Pindus Mts. (Hughes et al. 2003; Palmentola and Stamatopoulos 2006; Allard et al. 2020) and North Macedonia Mts. (Kuhlemann et al. 2009; Milevski 2015), the most complete rock glacier inventory of the entire Balkan Peninsula was published recently (Magori et al. 2020). The only comprehensive study in which thermal and geophysical measurements were performed to reveal the activity status of rock glaciers were conducted in the Rila and Pirin Mts. (Onaca et al. 2020). Other periglacial features have been studied in the context of research on debris flows (Kenderova and Vassilev 2002; Kenderova et al. 2013, 2014; Baltakova et al. 2018), frost weathering and creep processes (Gachev 2009a, b; Kenderova et al. 2015, 2018), and environmental conditions for the occurrence of periglacial processes in the Rila-Rhodope Massif (Rachev et al. 2017). Apart from comprehensive periglacial studies in the Rila-Rhodope Massif and recently some ice cave studies in the Dinaric Mts. (e.g., Paar et al. 2013; Buzjak et al. 2018; Kern et al. 2018; Mihevc 2018), the other mountains of the Balkan Peninsula have received considerably less attention in periglacial research.

3 Distribution and Chronology of Periglacial Landforms

Only the highest mountains of the Balkan Peninsula currently experience periglacial conditions, where a range of cold, non-glacial processes takes place. According to French's (2007) definition of the periglacial zone (i.e., non-glacial areas where MAAT oscillate between −2 °C and +3 °C), the periglacial environment in the Balkan Peninsula is limited to areas above ~1700–2200 m (considering WorldClim 1.4 data for 1960–1990; Hijmans et al. 2005). However, a definition of the periglacial zone is not a clear-cut and there are many other areas where (micro)climatic conditions are favourable for periglacial processes to take place, such as karst depressions. As for permafrost conditions, only isolated patches above ~2400 m still exist in the Rila and Pirin Mts. (Dobinski 2005; Obu et al. 2019; Onaca et al. 2020). According to Brown et al. (2001) and Gruber (2012), isolated patches of permafrost may also occur in the North Macedonia Mts. (Korab, Šar Planina, Jakupica), Dinaric Mts. (Prokletije, Durmitor, Volujak, Maglić), Stara Planina and Mt. Olympus. A special case is a sporadic underground permafrost in ice caves that exists down to ~650 m in the Dinaric Mts. (Mihevc 2018). Relict periglacial features in the Balkan Peninsula occur above 1400 m (Kanev 1990; Magori et al. 2020), but they can be sporadically present even down to 1000 m (Kanev 1990; Perica et al. 2010).

The degree of periglacial processes in modifying the mid-latitude landscapes depends largely upon the lithology and persistence of periglacial conditions (Ballantyne and Harris 1994; Murton 2021). Certain lithologies, such as chalk, slate and schist are more susceptible to frost action than others, for example dense crystalline rocks and some sedimentary rocks (French 2018; Murton 2021). There may exist a link between lithology and degree of periglaciation in the Balkan Peninsula. For example, periglacial features are less common (or simply less studied) in the western Balkan Peninsula, where carbonate sedimentary rocks prevail, whereas metamorphic rocks build some of the highest parts of the eastern Balkan Peninsula, where landscape modified by periglacial processes is widespread. A major part of periglacial landscape in the Balkan Peninsula has developed within the pre-existing glacial landscape. Here, both timing and duration of postglacial periglaciation can help understanding the extent and degree of periglacial modification. While the duration of postglacial periglaciation in the Balkan Peninsula before the Last Glacial Maximum (LGM) is unknown, the periglacial conditions after the LGM are believed to last longer in the Pindus and Parnassus Mts. than in other mountains of the Balkan Peninsula, resulting in larger rock glaciers there (Magori et al. 2020). Moreover, persistence of periglacial conditions in mid-latitude mountains not only depends on the altitude of the area, but also on the seasonal snow cover, which has an effect on the ground thermal regime (Zhang 2005). Snow cover can significantly reduce the seasonal freezing depth and can be responsible for the absence of permafrost in certain locations (Zhang 2005). Long-lasting and thick snow cover in the mountains of the western Balkan Peninsula might have been responsible for more limited periglacial modification of the landscape than in the eastern part.

3.1 *Seasonally Frozen Ground Features*

Traces of **patterned ground** were recorded on the summit surface of the eastern and central Rila Mt. (Glovnia 1962a, 1968). Typical, especially for the summit cols made of schists and other metamorphic rocks, is the formation of barren surfaces with no grass cover where only loose material is exposed (Fig. 2a). These are a result of both weathering and wind activity. On windy summit ridges made of crystalline rocks (granite, sienite) insolation weathering is strong, and winds blow out the loose weathering products. In result, sculptures of rounded rocks on sandy grounds were formed on numerous locations in the Pirin (Spano pole plateau, Fig. 2b) and Rila Mts. (the cols Razdela and Bialata prast; Glovnia 1958). Patterned ground was also observed on Mt. Tymphi (Hagedorn 1969; Hughes 2004). Sorted circles were described in numerous locations in the south-west Rila Mt. (Glovnia 1958) where metamorphic rocks prevail and disaggregate into tiles. **Stone pavements** are also well spread in the alpine zone of the Rila Mt., also on granites, but especially on metamorphic rocks. These features usually form below cirque walls, where snow patches retain for the longest. The main forming agent for them is the periodic freezing and thawing of meltwater from snow.

There are also several caves, where the cave morphology enhances periglacial dynamics, resulting in patterned ground (Zupan Hajna 2007; Obu et al. 2018; Blatnik et al. 2020). Active patterned ground formations have been described from many cave entrances in the Dinaric Mts. of Slovenia (Zupan Hajna 2007; Mihevc 2009; Obu et al. 2018), which are often exposed to seasonal freezing (Blatnik et al. 2020). In two Slovenian caves, seasonal vertical displacements due to freezing and thawing were found to be 5–15 cm, and in extreme cases up to 30 cm (Mihevc 2009).

3.2 *Rock Weathering and Derived Landforms*

Blockfields (also known as "stone seas") have formed on ridge tops and open slopes, where slope inclination is low enough to prevent debris from rolling or sliding down to the base of the slope. Blockfields form in crystalline rocks that tend to fragment into coarse debris due to frost action. Such landforms have been reported from several locations in the Rila Mt. above 2100 m, Stara Planina (Midzor's and Vezhen's northern slopes; Glovnia 1958, 1962a; Velchev et al. 2000) and Vitosha

Weathered summit surface (Rila Mt) Author: Emil Gachev

Rounded granite blocks on the summit surface (Pirin Mt.) Author: Emil Gachev

Blockfield in andesite (Vitosha Mt.) Author: Emil Gachev

Stone stripes (Rila Mt.) Author: Emil Gachev

"Zlatnite mostove" stone river (Vitosha Mt.) Author: Emil Gachev

"The Sphynx" tor (Rila Mt.) Author: Emil Gachev

Fig. 2 Landforms, generated by rock weathering under periglacial conditions

Mt. (Jordanov 1977; Stoyanov 2014). However, the most well-known stone sea in the Balkan Peninsula is located on the northern slope of the Vitosha Mt. This 1 km^2 size stone sea with an average slope angle of 25° is made of angular andesite clasts of various size (boulders to pebbles) and has been most probably formed by in situ disintegration of staircase series of rock escarpments (Fig. 2c).

Block streams or **stone rivers** are common for areas of the Balkan Peninsula that are built of granite and other coarse-grained intrusive rocks, such as the Rila-Rhodope Massif, Stara Planina, Osogovska planina, Vitosha Mt. and Pelister Mt. (Glovnia 1958; Jordanov 1977; Kanev 1988; Kolčakovski 1994; Velchev et al. 1994; Stoyanov 2014; Ribolini et al. 2018). They are less common in hard brittle, fine-grained and karstifiable rocks, because they either form soil cover relatively quickly due to mechanical weathering, or dissolve rather than disintegrate. As a result, these landforms are not widespread in the Dinaric and Pindus Mts., although they have been identified on the steep slopes of some karst depressions, at the bottom of the funnel-like depressions, and on the slopes of highest peaks of the Velebit Mt. (Dinaric Mts.; Perica et al. 2010). Block streams are commonly found on high altitude open slopes above 2000 m, which were not severely affected by glaciation. They are normally composed of in situ weathered coarse material, where finer fractions are periodically removed, such as areas of long-lasting seasonal snow cover. Conditions for block stream formation are optimal in the Musala area (Rila Mt.) above 2600 m, within steadier slopes, where block streams often join laterally and resemble blockfields (Fig. 2d). Block streams are also spread in other areas of the Rila and Pirin Mts., especially on granite. Here, soil cover is often folded before their front downslope, which indicate their present slow motion. Block streams in Stara Planina are found at lower altitudes (down to 1300–1500 m), they are much smaller in size, more sporadic, and, at least in the forest zone, are considered to be of a relict character (Petrov 2008). On the northern slope of the Vitosha Mt., numerous relict block fields of andesitic debris are spread down to even 1000–1100 m. Stone rivers are elongated block streams that usually run along stream and river channels. They are distinct elements of the landscape of the Vitosha Mt. and around Vezhen peak in Stara Planina. The stone rivers of the Vitosha Mt. (Fig. 2e) are formed in various intrusive rocks, mainly sienite and monzonite, which have been simultaneously subjected to both insolation and frost weathering. Surface runoff serves as main agent for the removal of fine-grained material, which ensures the prolonged and ongoing weathering. Here, the stone rivers are spread between 1250 and 1900 m and appear almost on all aspects (Jordanov 1977; Stoyanov and Gachev 2012; Stoyanov 2014).

Tors (also known as "rock castles") are categorized as products of periglacial activity by a number of authors (e.g., Glovnia 1971). Their presence on the highest peaks indicates intensive periglacial processes in the ice-free areas above the local ELA. They are typical for the smooth summit surfaces built of magmatic rocks. Tors are common features on the summit surfaces of the Rila Mt. and on the plateaux of the Vitosha Mt., where they occupy the highest parts of the smooth watershed ridges (Glovnia 1962b; Jordanov 1977; Velchev et al. 2000). They are also found in the Pirin Mt. and in the central Stara Planina. Some of them are popular among tourists and bear their own names, such as Sfinksa (the Sphynx, Fig. 2f) in the Rila Mt., Kafadyklidy (the Head) in Stara Planina and Ushite (the Ears) in the Vitosha Mt.

In periglacial zone, where both frost weathering and dissolution have an effect, special conditions develop for the evolution of **karst features**. Here, periglacial processes play an important role in the development of new and re-shaping of previously existing karst closed depressions (Ćalić 2011) as well as influencing other karst features (Fig. 3a,b). Karst depressions are complex features, which, in periglacial zone, are often partially filled with talus or solifluction lobes. Small size solution dolines and snow kettles, i.e., vertical or almost vertical features with flat floor, are believed to be the most common type of karst depressions in the periglacial belt (Veress 2017). They mostly occur above the tree line, where the environment is suitable for snow accumulation and there is sediment on the floor of depressions promoting vertical water movement (Veress 2017). A great part of the Balkan Peninsula is characterized by a well-developed karst landscape. This is especially true for the Dinaric and Pindus Mts., where karst is developed on limestone and dolostone. Here, the high-elevated karst plateaux were under the influence of glacial, but also periglacial processes. The latter are still taking effect on some of the highest summits and within karst depressions due to temperature inversion (Perica et al. 2010). Periglacial karst developed in marble, such as the one in the Rila-Rhodope Massif (Pirin, Slavianka, Falakro) and Jakupica Mt. in North Macedonia, is of specific character. Marble is a metamorphic rock with unique combination of carbonate chemical composition and crystalline mineral structure. On one hand, marble is water soluble and highly prone to karstification, which determines the lack of surface runoff and minor role of fluvial processes on the surface. On the other hand, crystalline structure, and the high pressure at which the rock cooled, determines greater rate of fragmentation of the rock on the surface due to weathering processes. As a result, marbles influenced by frost weathering tend to produce larger debris size material compared to limestones, resulting in the formation of a boulder size scree accumulations, "Ice cream ball" blocks (Gachev 2017a), weaker development of linear karren, and increased rates of frost weathering fragmentation (Fig. 3c–f). In the Pirin Mt., the active periglacial karst processes in marble occur above 2100 m (Grunewald and Scheithauer 2008, 2011; Mitkov 2020).

3.3 *Talus Slopes and Associated Phenomena*

The presence of **rock glaciers** in the Balkan Peninsula was documented for the highest sections of the Rila-Rhodope Massif (Glovnia 1968; Gikov and Dimitrov 2010; Dimitrov and Gikov 2011; Dimitrov and Velchev 2012; Gachev et al. 2017; Gachev 2020a; Onaca et al. 2020), Pindus Mts. (Hughes et al. 2003; Palmentola and Stamatopoulos 2006), Prokletije Mts. (Palmentola et al. 1995; Milivojević et al. 2008), and North Macedonia Mts. (Jakupica, Jablanica, Šar Planina and Korab Mts.; Manakoviḱ 1962; Kolčakovski 1999; Kuhlemann et al. 2009; Milevski 2015). A total of 224 rock glaciers ranging from 1412 to 2674 m and accounting for ~17 km^2 have been identified in the Balkan Peninsula as part of a regional rock glacier inventory (Magori et al. 2020). The mean elevation of their termini is 2180 m, being lowest in the Pindus Mts. and Parnassus (1850 m) and highest in the Rila-Rhodope Massif (2340 m). They are unevenly distributed across the mountains of the Balkan Peninsula. 73% of all identified rock glaciers are concentrated in the Rila, Pirin and Prokletije Mts. (Magori et al. 2020; Fig. 4a–c). In mountain ranges where granites

Periglacial karst (Dinaric Mountains) Author: Manja Žebre

Mechanically weathered surface in flysch (Durmitor Mt.) Author: Emil Gachev

"Ice cream balls" scree accumulation in marble (Pirin Mt.) Author: Emil Gachev

Boulder size screes and suffosion doline (Pirin Mt.) Author: Emil Gachev

Slopes subjected to strong mechanical weathering (Pirin Mt.) Author: Emil Gachev

Weathered summit surfaces and doline (Slavyanka Mt.) Author: Emil Gachev

Fig. 3 Periglacial karst features: in limestone (a, b) and marble (c–f)

occur, such as Pirin Mt., the density of rock glaciers is significantly higher. The number of rock glaciers is also high in the eastern Prokletije Mts., where magmatic and silicate metamorphic rocks prevail (gabbro, serpentinite, quartzite; Fig. 4e). Conversely, the dominance of limestone in the Dinaric Mts. and to some extent in the Pindus Mts., accompanied with higher precipitation amounts has unfavourable effect on rock glaciers' development there (Magori et al. 2020). The few rock glaciers that have been observed in limestone and marble show that such landforms do not occur in carbonate environments unless there are special geological conditions, such as the presence of non-carbonate layers in the bedrock below the debris accumulation, which prevent water from escaping downwards. For example, rock glaciers in limestone (such as those in Mt. Tymphi and Korab Mts., Fig. 4d) are formed in areas that are in contact with flysch, containing non-carbonate sediments. The only two rock glaciers in the marble part of the Northern Pirin are found in marginal cirques near the geological contact between carbonates and silicates, where debris originates from a marble area, but has accumulated on a silicate bedrock (Gachev 2020a; Fig. 4f). Talus rock glaciers prevail over debris rock glaciers (Magori et al. 2020), suggesting that the main source of the material originates from a talus slope and not an ice-cored moraine. The rock glacier formation period is assumed to coincide with the global LGM in the Pindus Mts. (Allard et al. 2020) and Parnassus (Magori et al. 2020), whereas rock glaciers elsewhere in the Balkan Peninsula likely formed later during the Late Glacial (Milivojević et al. 2008; Kuhlemann et al. 2009, 2013; Ruszkiczay-Rüdiger et al. 2020). These are assumptions based on moraine ages, ELAs and/or mean elevation front of rock glaciers, whereas direct numerical dating of rock glaciers has not been carried out in this region yet. Morphological studies and Schmidt hammer relative dating carried out on Musala rock glacier (Rila Mt.) provided indications for a compound morphology and at least two-stage formation: an earlier formation of a lower section (debris rock glacier), and a later of an upper section (talus rock glacier; Gachev 2018). Rock glacier activity in the Balkan Peninsula likely ceased in the early Holocene because of increasingly warm temperatures (Magori et al. 2020) and most rock glaciers are nowadays considered relict. However, results from some recent studies revealed that rock glaciers including permafrost are present in the Rila and Pirin Mts. above 2450 m (Onaca et al. 2020).

Single linear ridges exhibiting arcuate morphologies that form close to the foot of a steep slope (<70 m) are regarded as **protalus (or pronival) ramparts** (Matthews et al. 2017). These cryonival features are widespread in the Rila and Pirin Mts. Here, 143 protalus/pronival ramparts are present between 2080–2700 m (Magori et al. 2017). They usually occupy higher parts of cirque slopes, as is the case for the Golemia Kazan and Banski suhodol cirques, whereas some are formed at fronts of existing ice bodies (so called protalus-moraines, Popov 1962, 1964; Grunewald and Scheithauer 2008, 2011). Protalus ramparts have also been mapped on Mt. Pelister (Ribolini et al. 2018) and dated on the Jablanica Mt. in North Macedonia (Ruszkiczay-Rüdiger et al. 2020). Here, the two cosmogenic ages (13.1 ± 0.5 ka and 13.2 ± 0.4 ka) suggest that they co-existed with small cirque glaciers during ~13 ka. To date these features have not been systematically mapped elsewhere in the Balkan Peninsula.

Talus deposits are widespread in periglacial areas as mechanical weathering, especially frost action, accelerates the regular supply of debris to talus cones. Controlling factors for their formation are the presence of exposed bedrock, steep slopes, and availability and configuration of tectonic cracks and disruptions. Best conditions are provided on the steep slopes of relict glacial cirques, in mountains built in magmatic and metamorphic rocks such as in the Rila Mt, and Pirin Mt. (Glovnia

Rock glacier in granite (Rila Mt.)
Author: Emil Gachev

Rock glacier in granite (Pirin Mt.)
Author: Emil Gachev

Rock glacier in gneiss (Rila Mt.)
Author: Emil Gachev

Rock glacier in limestone (Korab Mt.)
Author: Emil Gachev

Rock glacier in quartzite (Prokletije Mts.)
Author: Emil Gachev

Rock glacier in marble (Pirin Mts.)
Author: Emil Gachev

Fig. 4 Examples of rock glaciers in the mountains of the Balkan Peninsula

1958, 1962b, 1968; Popov 1962; Kanev 1990; Velchev 1999). Talus slopes related to periglacial environment in carbonate rocks have been described from some parts of the Dinaric Mts. (Mt. Prenj; Lepirica 2008), North Macedonia Mts. (Mt. Galičica; Ribolini et al. 2011) and Pindus Mts. (Mt. Tymphi; Hughes et al. 2006a). Talus deposits are accumulated at the foot of rock walls, while fans are formed at lower ends of gullies. The debris size is determined by a rock type. Limestone and dolomite will normally produce smaller (sometimes even pebble) size debris (Gachev 2021). Larger debris size are talus deposits in marble. These are found in the northern Pirin Mt. within Pleistocene glacio-karstic cirques above 2400 m, where frost action in marble becomes highly pronounced (Gachev 2017a, 2020a; Mitkov 2020). Even coarser talus deposits are produced in magmatic intrusive rocks (e.g., granite) that can be found in the Rila Mt. and part of the Pirin Mt. Presently, above 2400 m the rockfall activity is often still existent, whereas talus cones at lower elevations are partially or completely covered by vegetation and thus essentially considered relict features. Here, talus modification by other processes, such as gullying, debris flows and snow avalanching, is taking place (Velchev and Kenderova 1994; Velchev 1999).

Debris flows occur in all the mountains of the Balkan Peninsula (e.g., Kenderova et al. 2013; Baltakova et al. 2018), although they are more characteristic for the steep marble slopes of the Pirin Mt. and Mt. Slavianka, silicate rocks in the Rila Mt. and granitic rocks in Stara Planina. Debris flows on the south-western slopes of the Pirin Mt. are 1.5–2.5 km long and descend more than 1000 m down to Vlahina River, where they enter granitic terrain. Debris flows are widespread also on the northern slopes, where they are a lot shorter (max. 300–500 m), apart from a tree-like debris flow system on the northern slope of Cherna mogila (2688 m) where the main, 4.2 km long channel runs along the dry floor of the Banski suhodol glacial valley and reaches down to 1300 m (Mitkov 2020). Although much shorter and lower in number, debris flows in marble are also formed above 2000 m on the south-east slopes of the Gotsev peak (2212 m) in Mt. Slavianka. In the Rila Mt., the most prominent features of this kind are developed on the southern slopes, where a series of torrential gullies, so called "Zlite potoci" (The Evil Streams), have horizontal lengths in the order of 1.5–2 km, and some descend more than 1400 m down to the valley of Rilska River. Debris flows in other parts of the Rila Mt. are widespread above 2400–2500 m (Glovnia 1959, 1971; Velchev 1999). Much shorter (250–500 m) debris flows are locally spread on Stara Planina above 2000 m. In the Dardanian Massif, the largest debris flows are also formed in marble, where they descent deep into the forest belt down to 700–800 m (Gachev 2021). In winter, debris flow tracks serve as avalanche tracks. Avalanches are very common for the Rila and Pirin Mts. (Mironski and Peev 1951; Stoyanov 2011). They have been studied in detail on the western slope of Todorka Peak (Pirin Mt.), which is located near the ski-tracks of Bansko resort (Panayotov 2011). Avalanches often occur in the highest massifs of the Dinaric Mts. (Prokletije, Durmitor, Komovi, Bjelasnica), on Šar Planina and Korab Mt., and on the northern slopes of Stara Planina (Mironski and Peev 1951). There are only two avalanche-prone locations on the Vitosha Mt., but due to its proximity to Sofia, avalanches here take casualties almost every year.

3.4 Periglacial Mass Movement and Slope Evolution

Solifluction landforms require developed soil cover and seasonally frozen ground below. They are observed mostly in areas made of silicate rocks that tend to accumulate water. Another factor is the relatively high altitude, which provides appropriate thermal conditions. Solifluction lobes, terraces and ploughing blocks are observed mostly in the Rila Mt. and in the high central part of the northern Pirin Mt. above 2150–2200 m (Glovnia 1962b, 1968; Kanev 1988; Velchev and Kenderova 1994), but also in the North Macedonia Mts. (Kolčakovski 1994, 1999) and the Dinaric Mts. (Lepirica 2008; Annys et al. 2014). Solifluction acts mostly in spring (and autumn) during periods of diurnal temperature fluctuations across 0 °C. The softened surface soil layer slides over a rigid bedrock or above a temporarily frozen weathering mantle, and the pace of the sliding depends on the active layer thickness and the degree of water saturation. In this case, the melting snow cover is not a direct factor, but rather an indirect one (water supply). Solifluction lobes in the Rila Mt. develop as a result of the ground's seasonal thermal and water state variations, and appear on gentle slopes, up to 6° gradients (Glovnia 1958). Ice wedges have a particular role in the process, as they lift up some frozen fragments of the regolith and scatter the weathering cover. The diurnal peak of solifluction intensity is in the late morning hours. **Ploughing blocks** are typical for the silicate parts of the Rila and Pirin Mts. above 2000–2100 m. Some blocks in the south-western Rila Mt., which show fresh traces of sliding, have dimensions of up to 6 m (Glovnia 1958). Usually, a grassy ridge is formed in front of a ploughing block, while a trench is formed behind them. In some cases, however, when the block is within a field of downsliding weathered material, the block, which is submerged in the regolith, can move slower than the finer fractions, in which case microforms are reversed: the trench before the front, and a covering regolith on the upper (back) side on the block. Ploughing blocks were also found on Galičica Mt. (Kolčakovski 1994, 1999) and Osogovska Planina (Milevski 2015). In the latter, their movement was quantified to 2–3 cm yr^{-1}. Solifluction terraces are staircase strips of scattered grass cover parallel to the slope. When meltwater enters the soil layer, they concentrate on locations where soil is thinner, and subsequently, when they freeze, they exert more pressure in these locations. Finally, this causes scattering of grass cover and downsliding of its strips (Kolčakovski 1997). The presence of solifluction terraces was documented in the central Stara Planina (Glovnia 1964), Osogovska Planina (Velchev et al. 1994), Pirin and Rila Mts. (Glovnia 1958), and Galičica Mt. (Kolčakovski 1994, 1999). Solifluction landforms are less typical for karstic areas, unless some special conditions are present (e.g., presence of impermeable layers, such as flysch). Among the exceptions are solifluction terraces, formation of which is controlled by infiltrating meltwater from snow that freeze in the course of the diurnal cycle.

The highest mountains of the Balkan Peninsula lie at least 200–300 m below the present climatic snow line. However, there are numerous locations, which provide conditions for annual preservation of snow and its gradual compaction into firn and even ice. These conditions are met with a combination of local topoclimatic factors, such as altitudes above 2000 m, carbonate bedrock, strong shading in deep relict north to north-east facing glacial cirques, considerable winter precipitation and great accumulation of avalanche and windblown snow. The lower altitude limit of such permanent firn/ice features is 2400 m in the eastern Balkan Peninsula, whereas it drops to 1800–2000 m in the western part mainly due to the increase in winter precipitation. The lowest permanent **ice patch** is found in the Prokletije Mt. (Dinaric Mts.) at 1600–1800 m. Such firn-ice bodies normally cover floors of glacio-karstic depressions and are remnants of former small glaciers that existed during the Little Ice Age (LIA; Wilkinson 2011; Gachev 2017b, 2020b; see also Fig. 5a,b). In the autumn of 2017, after the greatest recession of snow patches in the cirques around Mt. Jezerce (Prokletije Mts.), permafrost layers were observed at some locations (Gachev 2020b). However, there are also small glaciers (glacierets) with sizes in the range of 0.005–0.05 km^2 that still survive in the recent climate. They have been recorded in the Prokletije, Durmitor and Pirin Mts. between 1900 and 2600 m (Gachev et al. 2016; Gachev 2017b).

Associated with the presence of snow patches is a set of geomorphic processes called **nivation**, which may produce nivation hollows. These are found in topographically sheltered areas, such as karst depressions, glacial cirques, or shady lee-windward slopes in Stara Planina, Osogovska Planina, Vitosha (Georgiev 1991; Velchev et al. 1994, 2000; Stoyanov and Gachev 2012), Galičica (Milevski 2015; Gromig et al. 2018) and Dinaric Mts. (Lepirica 2008; see Fig. 5c). Nivation hollows in karst areas will often have a form of steep- to vertical-sided dolines that trap and conserve snow, which accelerates corrosion (Ford and Williams 2007). They are called snow kettles (or schachtdolinen or kotlič, Fig. 5d; Kunaver 1983) and are typically found in the Dinaric Mts. Some of the nivation hollows at highest elevations develop into embryonic cirques. A seasonal cryolithozone commonly forms along the margins of long-lasting snow patches (Glovnia 1958, 1968). When these thaw in spring and early summer, abundant meltwaters soak the ground and provoke downward sliding of the regolith. In result, solifluction lobes (Fig. 5e) and local landslides are observed. These can be found in the Rila and Pirin Mts. (Glovnia 1959; Kanev 1990).

3.5 *Ice Caves*

A great part of the Balkan Peninsula is composed of carbonate rocks, such as limestone and dolomite, which makes this area rich in caves. Some caves host perennial accumulations of ice and they are called **ice caves** (Fig. 5f). This ground ice is usually older than 2 years, which is why ice caves are commonly considered as sporadic permafrost phenomena (e.g., Holmlund et al. 2005; Luetscher et al. 2005, 2013; Kern et al. 2011; Hausmann and Behm 2011), although they often exist at altitudes with an outside MAAT well above 0 °C (Holmlund et al. 2005; Stoffel et al. 2009; Obleitner and Spötl 2011). Existence of permanent frozen materials in caves is conditioned by a combination of favourable cave morphology, local topography and

Fig. 5 Periglacial slope evolution and long-lasting snow and ice related landforms

climatic conditions (Colucci et al. 2016; Perșoiu and Lauritzen 2018). Majority of caves containing permanent frozen materials in the Balkan Peninsula, the so-called cryo-caves (sensu Colucci et al. 2016), occur above 1000 m (Barović et al. 2018; Buzjak et al. 2018; Mihevc 2018; Nešić and Ćalić 2018; Temovski 2018). The lowest reported ice cave entrance in the Balkan Peninsula is Laniško brezno in the Dinaric

Mts. of Slovenia at 645 m (Mihevc 2018). In the cave register of Slovenia, 551 caves with permanent ice have been reported so far and many more, where seasonal snow or ice occur (Mihevc 2018). There are 321 known cryo-caves in the Dinaric Mts. of Croatia, among which more than 70% are located on the Velebit Mt. (Buzjak et al. 2016, 2018). More than 80 caves with permanent or seasonal frozen materials have been reported so far from Bosnia and Herzegovina (Dinaric Mts.; Mulaomerović et al. 2006; Barović et al. 2018), 25 from North Macedonia (Jakupica and Krčin Mts.; Temovski 2018), 27 from Serbia (Carpatho-Balkanides and Dinaric Mts.; Nešić and Ćalić 2018), and over 16 from Montenegro (Durmitor Mt. and Prokletije Mts. in the Dinaric Mts.; Barović et al. 2018). Ice caves exist also in other Balkan Peninsula countries (Greece, Bulgaria, Albania), but they have not been reported systematically. Few data are available on the age of the cave ice deposits in the Balkan Peninsula. The study from three Velebit ice caves demonstrated that the oldest ice dated so far is likely to be older than 3500 years (Kern et al. 2018), while ice from other caves is ~1000 and ~500 years old (Paar et al. 2013; Kern et al. 2018).

4 The Significance of Periglacial Dynamics in the Balkans within the European Context

Among cold-climate geomorphological processes, those of glacial origin have always been the focus of research in the mountains of the Balkan Peninsula, whereas periglacial landscapes have been much less studied. In the last decade, the glacial chronology for the Last Glacial Cycle in the Balkan Peninsula has improved significantly, which may also contribute to a better understanding of the chronology of inactive periglacial landforms. However, there has also been some progress in periglacial research in recent decades, especially on rock glaciers, although still without absolute age dating.

Most periglacial research has been conducted in the Rila-Rhodope Massif, where periglacial landforms seem to be most widespread and diverse. Here, a variety of periglacial landforms (e.g., patterned ground, blockfields, block streams, solifluction landforms) have been detected and the largest number of rock glaciers in the entire Balkan Peninsula have been identified. Periglacial landscape also appears to be extensive in the North Macedonia Mts., though less well studied. Here, ploughing blocks, block streams, rock glaciers and protalus ramparts, among others, have been studied in detail. On the other hand, ice caves and periglacial karst are the focus of periglacial studies in the Dinaric Mts. and elsewhere in carbonate-dominated mountains. The presence of a great diversity of periglacial landforms across different mountain ranges of the Balkan Peninsula is due to multiple climate regimes with large differences in precipitation between the west and the east, the latter being much drier. Not only the different climate, but also the diverse geology plays an important role in the observed great diversity of periglacial landforms. The silicate geological background, which is predominant in the mountains of the eastern Balkan Peninsula,

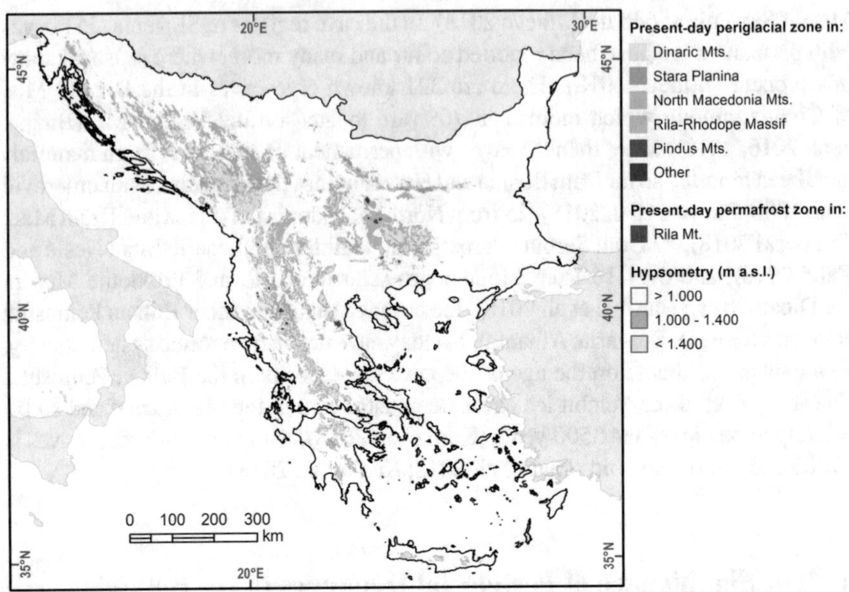

Fig. 6 Present-day periglacial zone (as defined by French [2007]) in the mountain ranges of the Balkan Peninsula extracted from WorldClim 1.4 data for 1960–1990 (Hijmans et al. 2005), and present-day permafrost zone after Obu et al. (2019). The map also shows hypsometry above 1000–1400 m, which is likely to present the minimum altitude of the past periglacial belt

has favoured the formation of abundant periglacial landforms, such as blockfields and block streams of various size, patterned ground features and rock glaciers. On the other hand, the wide distribution of carbonate rocks in the south and west of the peninsula has determined the development of periglacial karst. Specifically for rock glaciers as one of the most distinctive periglacial landforms, observations in the mountains of the Balkan Peninsula show that they mostly occur in silicate igneous and metamorphic rocks, while their formation in carbonate rocks is much rarer and requires special geological conditions, such as the presence of layers with low water permeability beneath the debris accumulation.

Periglacial landforms can reach elevations as low as 1000 m in the Balkan Peninsula, but those of active type are found only above 1700–2200 m (Fig. 6). Most rock glaciers in the mountains of the Balkan Peninsula are nowadays considered relict. However, rock glaciers including permafrost appear above 2450–2600 m in the Rila-Rhodope Massif. Above this altitude, permafrost may still be existent in the Rila-Rhodope Massif, but only in the form of isolated patches. The rest of the mountains are likely permafrost-free. The only exception is underground permafrost observed in ice caves, which can be found in the Dinaric Mts. even down to ~650 m. Permafrost was much more widespread in the past. Considering the mean elevation of the rock glaciers front across the Balkan Peninsula, the lower limit of former discontinuous permafrost was 2180 m, and across different mountains it was in the range of 1850–2340 m. Therefore, most periglacial landforms must have formed during colder periods in the past. Although there are almost no data for the Balkan Peninsula to determine the absolute age of periglacial landforms, it is assumed that at least the rock glaciers and protalus ramparts were formed during the global LGM or Late Glacial. For some of them, a multistage and even polygenetic origin has been recognized. Their activity probably ceased in the early Holocene due to temperature rise. Most periglacial landforms (e.g., rock glaciers, protalus ramparts) developed in areas previously modified by glaciers, but some also formed on glacier-free mountain tops (e.g., tors, patterned ground) or even below the glaciated landscape. Periglacial conditions during cold stages were typical of most mountain slopes, and in many areas extended down to their feet. Traces of periglacial activity are still recognisable as relict landforms in mid- and low mountain belts and exert an impact on present geomorphic processes and landform diversity.

Acknowledgements Manja Žebre is supported by the Slovenian Research Agency, research core funding No. P1-0419. Emil Gachev is supported by the South-west University "Neofit Rilski" (Bulgaria), through the research resources of the "Air and Waters" Laboratory of the Faculty of Mathematics and Natural Sciences, and the funding of the University research projects: RP-A 13/17, RP-A 4/18 and RP-A 3/20.

References

Alexiev G (2002) Geomorphological regionalization. In: Geography of Bulgaria. Bulgarian Academy of Sciences, Sofia, pp 104–105

Allard JL, Hughes PD, Woodward JC et al (2020) Late Pleistocene glaciers in Greece: a new 36Cl chronology. Quat Sci Rev 245:106528. https://doi.org/10.1016/j.quascirev.2020.106528

Anagnostopoulou C, Tolika K, Lazoglou G, Maheras P (2017) The exceptionally cold January of 2017 over the Balkan Peninsula: a climatological and synoptic analysis. Atmosphere 8:252. https://doi.org/10.3390/atmos8120252

Annaheim H (1939) Die Eiszeit im Rila Gebirge. Petermanns Geogr Mitteilungen 2:41–49

Annys K, Frankl A, Spalević V et al (2014) Geomorphology of the Durmitor Mountains and surrounding plateau Jezerska Površ (Montenegro). J Maps 10:600–611. https://doi.org/10.1080/17445647.2014.909338

Arsovski M (1960) Some characteristics of the tectonic features of the central part of Pelagonian horst anticlinorium and its relation to the Vardar Zone. Pap Geol Surv SRM 7:76–94

Ballantyne CK, Harris C (1994) The periglaciation of Great Britain. Cambridge University Press, Cambridge

Baltakova A, Nikolova V, Kenderova R, Hristova N (2018) Analysis of debris flows by application of GIS and remote sensing: case study of western foothills of the Pirin Mountains (Bulgaria). In: Chernomorets SS, Gavardashvili GV (eds) 5th International Conference. Debris Flows: disasters, risk, forecast, protection. Publishing House "Universal," Tbilisi, Georgia, pp 22–32

Barović G, Kicińska D, Mandić M, Mulaomerović J (2018) Ice caves in Montenegro and Bosnia and Herzegovina. In: Perşoiu A, Lauritzen S-E (eds) Ice Caves. Elsevier

Blatnik M, Culver DC, Gabrovšek F et al (2020) Significant findings from karst sediments research. In: Knez M, Otoničar B, Petrič M et al. (eds) Karstology in the classical karst. Springer

Brown J, Ferrians OJ, Heginbottom JA, Melnikov ES (2001) Circum-arctic map of permafrost and ground ice conditions. Digital media

Buzjak N, Bočić N, Paar D et al (2018) Ice caves in Croatia. In: Perşoiu A, Lauritzen S-E (eds) Ice Caves. Elsevier

Buzjak N, Bočić N, Paar D, Dubovečak V (2016) Geographical distribution and ice cave types in Croatia. In: 7th International Workshop on Ice Caves, Postojna, 2016. pp 37–38

Ćalić J (2011) Karstic uvala revisited: Toward a redefinition of the term. Geomorphology 134:32–42. https://doi.org/10.1016/J.GEOMORPH.2011.06.029

Çiner A, Stepišnik U, Sarıkaya MA et al (2019) Last Glacial Maximum and Younger Dryas piedmont glaciations in Blidinje, the Dinaric Mountains (Bosnia and Herzegovina): insights from 36Cl cosmogenic dating. Mediterr Geosci Rev 1:25–43. https://doi.org/10.1007/s42990-019-0003-4

Colucci RR, Fontana D, Forte E et al (2016) Response of ice caves to weather extremes in the southeastern Alps, Europe. Geomorphology 261:1–11. https://doi.org/10.1016/j.geomorph.2016.02.017

Cvijić J (1897) Traces of former glaciers on Rila. In: Glas Srpske Kraljevske Akademije. Beograd, pp 1–405

Cvijić J (1899) Glacijalne i morfološke studije o planinama Bosne, Hercegovine i Crne Gore. Državna štamparija Kraljevine Srbije, Beograd

Dimitrov P, Gikov A (2011) Relict rock glaciers identification and mapping in Pirin mountain using aerial and satellite images. In: Space, Ecology, Safety. pp 256–263

Dimitrov P, Velchev A (2012) The relict rock glaciers as a morphological feature in Rila Mountain's alpine zone. Annu J Sofia Univ, Fac Geol Geogr, Book 2 Geogr 103:97–112

Dobinski W (2005) Permafrost of the Carpathian and Balkan Mountains, Eastern and Southeastern Europe. Permafr Periglac Process 16:395–398

European Geological Data Infrastructure Geological Map. http://www.europe-geology.eu/onshore-geology/geological-map/. Accessed 2 Nov 2021

Ford D, Williams PD (2007) Karst hydrogeology and geomorphology. Wiley, Chichester

French HM (2007) The periglacial environment, 3rd edn. Wiley, Chichester, UK

French HM (2018) The periglacial environment, 4th edn. Wiley, Chichester, UK

Furlan D (1977) The climate of Southeast Europe. In: Wallen CC (ed) Climates of Central and Southern Europe. World Survey of Climatology Volume 6. Elsevier, Amsterdam, pp 185–235

Gachev E (2009a) Research of some geomorphic processes in Golemia Kazan cirque, Pirin Mountains. Probl Geogr 2:96–104

Gachev E (2009b) Indicators for modern and recent climate change in the highest mountain areas of Bulgaria. Landf Anal 10:33–38

Gachev E (2017a) High mountain relief in marble in Pirin Mountains, Bulgaria: structure, specifics and evolution. Revista de Geomorfologie 19:118–135. https://doi.org/10.21094/rg.2017.012

Gachev E (2017b) The unknown Southernmost Glaciers of Europe. In: Godone D (ed) Glaciers evolution in a changing world. InTech, Rijeka, pp 77–102.

Gachev E (2018) Characteristics and evolution of the rock glacier near Musala Peak (Rila Mountains, Bulgaria). Studia Geomorphologica Carpatho-Balcanica 51–52:25–42

Gachev E (2020a) Rock glaciers in mixed lithology: a case study from Northern Pirin. Revista de Geomorfologie 22:61–72

Gachev E (2020b) Small glaciers in the Dinaric Mountains after eight years of observation: on the verge of extinction? Acta Geogr Slov 60:191–211. https://doi.org/10.3986/AGS.8092

Gachev E (2021) Periglacial landforms and the geological controlling factors: examples from the highest mountains of the Balkan Peninsula. J Bulg Geogr Soc 44:39–47. https://doi.org/10.3897/jbgs.e68982

Gachev E, Magori B, Sirbu FS et al (2017) Field observations of rock glaciers in the mountains of Bulgaria. Probl Geogr 3:55–66

Gachev E, Stoyanov K, Gikov A (2016) Small glaciers on the Balkan Peninsula: state and changes in the last several years. Quatern Int 415:33–54. https://doi.org/10.1016/J.QUAINT.2015.10.042

Georgiev M (1991) Physical Geography of Bulgaria. Sofia University Press

Gikov A, Dimitrov P (2010) Identification and mapping of the relict rock glaciers in the Rila Mountain using aerial and satellite images. In: Space, ecology, safety. Sofia, Bulgaria, pp 252–259

Glovnia M (1958) Geomorphological researches in the South-west part of Rila. Annu J Sofia Univ, FGG, Book 3 Geogr 51:1–101

Glovnia M (1959) About periglacial relief in Bulgaria. J Bulg Geogr Soc 2:15–23

Glovnia M (1962a) Studies of glacial morphosculpture in the Eastern part of Rila Mountains. Annu J Sofia Univ, FBGG, Book 3 Geogr 55:1–50

Glovnia M (1962b) A contribution towards studying of the periglacial morphosculpture in the Rila mountains. Announcements of Bulgarian Geographic Society 3:47–55

Glovnia M (1964) On the issue about the glacial and periglacial relief in the massif of Botev, Central Stara planina. J Bulg Geogr Soc 4:147–155

Glovnia M (1968) Studies of glacial morphosculpture in the southern part of Central Rila. Annu J Sofia Univ, FBGG, Book 2 Geogr 61:37–68

Glovnia M (1971) Comparative geomorphological studies of periglacial morphosculpture of Southern Carpathians and Rila Mountains. Annu J Sofia Univ, FBGG, Book 2 Geogr 64:27–43

Gromig R, Mechernich S, Ribolini A et al (2018) Evidence for a Younger Dryas deglaciation in the Galicica Mountains (FYROM) from cosmogenic 36Cl. Quatern Int 464:352–363. https://doi.org/10.1016/j.quaint.2017.07.013

Gruber S (2012) Derivation and analysis of a high-resolution estimate of global permafrost zonation. Cryosphere 6:221–233. https://doi.org/10.5194/tc-6-221-2012

Grund A (1902) Neue Eiszeitspuren aus Bosnien und der Hercegovina. Globus 81:149–150

Grunewald K (2014) Climate change and related management issues in the mountains of South-eastern Europe—the Pirin National Park in Bulgaria. eco.mont (J Prot Mt Areas Res) 5. https://doi.org/10.1553/eco.mont-5-1s49

Grunewald K, Scheithauer J (2008) Klima-und Landscaftgeschichte Sudosteuropas. Rekonstruktion anhand von Geoarchiven im Piringebirge (Bulgarien). Rhombos, Berlin

Grunewald K, Scheithauer J (2011) Landscape development and climate change in southwest Bulgaria (Pirin Mountains). Springer Netherlands, Dordrecht

Hagedorn J (1969) Beiträge zur Quartärmorphologie griechischer Hochgebirge. Göttinger Geogr Abh 50

Hausmann H, Behm M (2011) Imaging the structure of cave ice by ground-penetrating radar. Cryosphere 5:329–340. https://doi.org/10.5194/tc-5-329-2011

Hijmans RJ, Cameron SE, Parra JL et al (2005) Very high resolution interpolated climate surfaces for global land areas. Int J Climatol 25:1965–1978. https://doi.org/10.1002/joc.1276

Holmlund P, Onac BP, Hansson M et al (2005) Assessing the palaeoclimate potential of cave glaciers: the example of the scărişoara ice cave (Romania). Geogr Ann Ser B 87:193–201. https://doi.org/10.1111/j.0435-3676.2005.00252.x

Hughes PD (2004) Quaternary glaciation in the Pindus Mountains, Northwest Greece

Hughes PD, Gibbard PL, Woodward JC (2003) Relict rock glaciers as indicators of Mediterranean palaeoclimate during the Last Glacial Maximum (Late Würmian) in northwest Greece. J Quat Sci 18:431–440. https://doi.org/10.1002/jqs.764

Hughes PD, Gibbard PL, Woodward JC (2006a) Middle Pleistocene glacier behaviour in the Mediterranean: sedimentological evidence from the Pindus Mountains, Greece. J Geol Soc 163:857–867. https://doi.org/10.1144/0016-76492005-131

Hughes PD, Woodward JC, Gibbard PL et al (2006b) The glacial history of the Pindus mountains, Greece. J Geol 114:413–434. https://doi.org/10.1086/504177

Hughes PD, Woodward JC, van Calsteren PC et al (2010) Pleistocene ice caps on the coastal mountains of the Adriatic Sea. Quat Sci Rev 29:3690–3708. http://dx.doi.org/10.1016/j.quascirev.2010.06.032

Hughes PD, Woodward JC, van Calsteren PC, Thomas LE (2011) The glacial history of the Dinaric Alps, Montenegro. Quatern Sci Rev 30:3393–3412. https://doi.org/10.1016/j.quascirev.2011.08.016

Janković P (1903) Glacial traces in Pirin. In: Glas Srpske Kraljevske Akademije

Jordanov L (1977) Vitosha mountain. Zemizdat, Sofia

Kanev D (1988) To the secrets of relief in Bulgaria. Narodna prosveta, Sofia

Kanev D (1990) Geomorphology of Bulgaria. Sofia University Press

Kenderova R, Vassilev I (2002) Climate changes in Bulgaria—cause of debris flow activation. Geography Issues 1

Kenderova R, Baltakova A, Ratchev G (2013) Debris Flows in the Middle Struma Valley, Southwest Bulgaria. In: Loczy D (ed) Geomorphological impacts of extreme weather. Springer Netherlands, Dordrecht

Kenderova R, Baltackova A, Rachev G (2014) Debris flows in the Middle Struma valley. Annu J Sofia Univ, FGG, Book 2 Geogr 106:13–40

Kenderova R, Baltakova A, Krenchev D, Rachev G (2018) Creep process in the Pirin Mountains. Comptes rendus de l'Academie bulgare des Sciences 71:236–242

Kenderova R, Rachev G, Baltakova A et al (2015) Variations in soil surface temperature in the Pirin high mountain area (Bulgaria) and their relation to slope processes activity (preliminary results). Comptes rendus de l'Academie bulgare des Sciences 68:1027–1035

Kern Z, Bočić N, Sipos G (2018) Radiocarbon-Dated Vegetal Remains from the Cave Ice Deposits of Velebit Mountain, Croatia. Radiocarbon 60:1391–1402. https://doi.org/10.1017/RDC.2018.108

Kern Z, Széles E, Horvatinčić N et al (2011) Glaciochemical investigations of the ice deposit of Vukušić Ice Cave, Velebit Mountain, Croatia. Cryosphere 5:485–494. https://doi.org/10.5194/tc-5-485-2011

Kolčakovski D (1994) High-mountain areas of the mountains Jablanitsa, Galichitsa and Pelister. Ecol Environ Prot 1:43–51

Kolčakovski D (1997) Preliminary knowledge about the phenomena of girland soils (girland forms) of the high mountains in the Republic of Macedonia. Maced J Ecol Environ 5:41–48

Kolčakovski D (1999) Glacial and periglacial relief on the mountain Jablanica. Annu Inst Geogr, FNSM, Skopje 33–34:15–38

Kottek M, Grieser J, Beck C et al (2006) World Map of the Köppen-Geiger climate classification updated. Meteorol Z 15:259–263. https://doi.org/10.1127/0941-2948/2006/0130

Kuhlemann J, Gachev E, Gikov A et al (2013) Glaciation in the Rila Mountains (Bulgaria) during the Last Glacial Maximum. Quatern Int 293:51–62. https://doi.org/10.1016/J.QUAINT.2012.06.027

Kuhlemann J, Milivojević M, Krumrei I, Kubik PW (2009) Last glaciation of the Šara Range (Balkan peninsula): increasing dryness from the LGM to the Holocene. Austrian J Earth Sci 102:146–158

Kunaver J (1983) Geomorfološki razvoj Kaninskega pogorja s posebnim ozirom na glaciokraške pojave = Geomorphology of the Kanin Mountains with special regard to the glaciokarst. Geografski zbornik = Acta geographica 22:197–346

Lepirica A (2008) Geomorphological characteristics of the Massif Prenj. Acta Carsologica 37:307–329. https://doi.org/10.3986/ac.v37i2-3.154

Luetscher M, Borreguero M, Moseley GE et al (2013) Alpine permafrost thawing during the Medieval Warm Period identified from cryogenic cave carbonates. Cryosphere 7:1073–1081. https://doi.org/10.5194/tc-7-1073-2013

Luetscher M, Jeannin P-Y, Haeberli W (2005) Ice caves as an indicator of winter climate evolution: a case study from the Jura Mountains. Holocene 15:982–993. https://doi.org/10.1191/0959683605hl872ra

Magaš D (2002) Natural-geographic characteristics of the Boka Kotorska area as the basis of development. Geoadria 7:51–81

Magori B, Onaca A, Gachev E, Urdea P (2017) The geomorphological characteristics of rock glaciers and protalus ramparts in the Rila and Pirin Mountains. In: 9th International Conference on Geomorphology. New Delhi, pp 356–356

Magori B, Urdea P, Onaca A, Ardelean F (2020) Distribution and characteristics of rock glaciers in the Balkan Peninsula. Geogr Ann Ser B 102:354–375. https://doi.org/10.1080/04353676.2020.1809905

Manakoviќ D (1962) Nivation processes and landforms on the mountain Jakupica. Annal of FNSM 13:47–57

Marinova T, Malcheva K, Bocheva L, Trifonova L (2017) Climate profile of Bulgaria in the period 1988–2016 and brief climatic assessment of 2017. Bulg J Meteorol Hydrol 22:2–15

Matthews JA, Wilson P, Mourne RW (2017) Landform transitions from pronival ramparts to moraines and rock glaciers: a case study from the Smørbotn cirque, Romsdalsalpane, Southern Norway. Geogr Ann Ser B 99:15–37. https://doi.org/10.1080/04353676.2016.1256582

Mihevc A (2009) Cryoturbation of the sediments at the cave entrances: case studies from Skednena jama, Potočka zijalka and Bestažovca Cave. In: Steguweit L (ed) Hugo Obermaier Society for Quaternary Research and archaeology of the Stone Age. 51st Annual Meeting in Ljubljana. Ljubljana, pp 26–27

Mihevc A (2018) Ice Caves in Slovenia. In: Perşoiu A, Lauritzen S-E (eds) Ice Caves. Elsevier

Milevski I (2015) General geomorphological characteristics of the Republic of Macedonia. Geogr Rev 48:5–25

Milivojević M, Menković L, Ćalić J (2008) Pleistocene glacial relief of the central part of Mt. Prokletije (Albanian Alps). Quatern Int 190:112–122. https://doi.org/10.1016/j.quaint.2008.04.006

Mironski N, Peev H (1951) Snow avalanches. Fizkultura, Sofia

Mitkov I (2020) Evolution and present state of glacial and cryogenic relief in the marble parts of the Pirin Mountains

Mulaomerović J, Zahirović D, Handžić E (2006) Katastar speleoloških objekata Bosne i Hercegovine

Murton JB (2021) What and where are periglacial landscapes? Permafr Periglac Process 32:186–212. https://doi.org/10.1002/ppp.2102

Nešić D, Ćalić J (2018) Ice Caves in Serbia. In: Perşoiu A, Lauritzen S-E (eds) Ice Caves. Elsevier

Obleitner F, Spötl C (2011) The mass and energy balance of ice within the Eisriesenwelt cave, Austria. Cryosphere 5:245–257. https://doi.org/10.5194/tc-5-245-2011

Obu J, Košutnik J, Overduin PP et al (2018) Sorted patterned ground in a karst cave, Ledenica pod Hrušico, Slovenia. Permafr Periglac Process 29:121–130. https://doi.org/10.1002/ppp.1970

Obu J, Westermann S, Bartsch A et al (2019) Northern Hemisphere permafrost map based on TTOP modelling for 2000–2016 at 1 km2 scale. Earth-Sci Rev 193:299–316. https://doi.org/10.1016/j.earscirev.2019.04.023

Oliva M, Žebre M, Guglielmin M et al (2018) Permafrost conditions in the Mediterranean region since the Last Glaciation. Earth Sci Rev 185:397–436. https://doi.org/10.1016/j.earscirev.2018.06.018

Onaca A, Ardelean F, Ardelean A et al (2020) Assessment of permafrost conditions in the highest mountains of the Balkan Peninsula. CATENA 185:104288. https://doi.org/10.1016/j.catena.2019.104288

Paar D, Buzjak N, Sironić A, Horvatinčić N (2013) Paleoklimatske Arhive Dubokih Jama Velebita. In: 3. znanstveni skup geologija kvartara u Hrvatskoj. Zagreb, Croatia, pp 39–40

Palmentola G, Baboci K, Gruda GJ, Zito G (1995) A note on rock glaciers in the Albanian Alps. Permafr Periglac Process 6:251–257

Palmentola G, Stamatopoulos L (2006) Preliminary data about sporadic permafrost on Peristeri and Tzoumerka massifs (Pindos chain, Northwestern Greece). Revi sta de geomorfologie 8:17–23

Panayotov M (2011) Avalanches on the northwestern slope of peak Todorka (Pirin Mts, SW Bulgaria) and their influence on forests. Phytologia Balcanica 17:237–246

Penck A (1900) Die Eiszeit auf der Balkanhalbinsel. Globus 78:133–178

Perica D, Lončar N, Lozić S (2010) The influence of nivation and cryofraction on periglacial relief formation on Velebit Mt. (Croatia). Geol Croat 63:271–282. https://doi.org/10.4154/GC.2010.22

Perşoiu A, Lauritzen S-E (2018) Introduction. In: Perşoiu A, Lauritzen S-E (eds) Ice Caves. Elsevier

Petrov G (2008) Periglacial relief in the area of Chumerna peak. Probl Geogr 3:117–125

Popov V (1962) Morphology of the cirque Golemia Kazan—Pirin Mountains. Announc Inst Geogr VI:86–99

Popov V (1964) Observations over the snow patch in the cirque Golemia Kazan, Pirin Mountains. Announc Inst Geogr 8:198–207

Rachev G, Kenderova R, Nikolova N et al (2017) Meteorological, hydrological and geomorphological studies in Begovitsa river watershed—results from 2012–2015 period. Annu J Sofia Univ, FGG, Book 2 Geogr 109:17–33

Reed JM, Kryštufek B, Eastwood WJ (2004) The physical geography of the Balkans and nomenclature of place names. In: Griffiths HI, Kryštufek B, Reed JM (eds) Balkan biodiversity: pattern and process in the European hotspot. Springer Netherlands, Dordrecht, pp 9–22

Ribolini A, Bini M, Isola I et al (2018) An Oldest Dryas glacier expansion on Mount Pelister (Former Yugoslavian Republic of Macedonia) according to 10Be cosmogenic dating. J Geol Soc 175:100–110. https://doi.org/10.1144/jgs2017-038

Ribolini A, Isola I, Zanchetta G et al (2011) Glacial features on the Galicica Mountains, Macedonia: Preliminary report. Geogr Fis Din Quat 34:247–255. https://doi.org/10.4461/GFDQ.2011.34.22

Rubel F, Brugger K, Haslinger K, Auer I (2017) The climate of the European Alps: Shift of very high resolution Köppen-Geiger climate zones 1800–2100. Meteorol Z 26:115–125. https://doi.org/10.1127/metz/2016/0816

Ruszkiczay-Rüdiger Z, Kern Z, Temovski M et al (2020) Last deglaciation in the central Balkan Peninsula: geochronological evidence from the Jablanica Mt. (North Macedonia). Geomorphology 351:106985. https://doi.org/10.1016/j.geomorph.2019.106985

Sarıkaya MA, Stepišnik U, Žebre M et al (2020) Last glacial maximum deglaciation of the Southern Velebit Mt. (Croatia): insights from cosmogenic 36Cl dating of Rujanska Kosa. Mediterr Geosci Rev 2:53–64. https://doi.org/10.1007/s42990-020-00030-9

Stoffel M, Luetscher M, Bollschweiler M, Schlatter F (2009) Evidence of NAO control on subsurface ice accumulation in a 1200 yr old cave-ice sequence, St. Livres ice cave, Switzerland. Quat Res 72:16–26. https://doi.org/10.1016/j.yqres.2009.03.002

Stojadinoviќ Č (1962) Stone rivers and screes on Pelister. Geogr Rev 1:45–51

Stoyanov K (2011) Evaluation of the avalanche danger in Northwest Rila Mountain. In: Zhelezov G (ed) Sustainable development in mountain regions. Springer Netherlands, Dordrecht, pp 95–102

Stoyanov K (2014) Relief of Vitosha. In: Popov A (ed) The natural wealth of Vitosha Nature Park

Stoyanov K, Gachev E (2012) Recent landform evolution in Bulgaria. In: Lóczy D, Stankoviansky M, Kotarba A (eds) Recent landform evolution. Springer Netherlands, Dordrecht

Temovski M (2018) Ice Caves in FYR of Macedonia. In: Perşoiu A, Lauritzen S-E (eds) Ice Caves. Elsevier

Tomczyk A (2016) Impact of atmospheric circulation on the occurrence of heat waves in southeastern Europe. Idojaras 120:395–414

Velchev A (1999) Glacial and cryogenic relief in some parts of Musala watershed in Rila Mountain. Annu J Sofia Univ, FGG, Book 2 Geogr 89:7–21

Velchev A, Kenderova R (1994) Some views about the Pleistocene and Holocene evolution of the valley of Mozgovitsa river. Annu J Sofia Univ, FGG, Book 2 Geogr 85:29–42

Velchev A, Konteva M, Penin R, Todorov N (2000) The landscape features of Midzhour (Chiprovska planina). Annu J Sofia Univ, FGG, Book 2 Geogr 90:111–131

Velchev A, Todorov N, Kostadinov K (1994) Development and contemporary state of the subalpine landscapes at the Osogovo Mountain. Annu J Sofia Univ, FGG, Book 2 Geogr 85:181–199

Veress M (2017) Solution DOLINE development on GLACIOKARST in alpine and Dinaric areas. Earth Sci Rev 173:31–48. https://doi.org/10.1016/J.EARSCIREV.2017.08.006

Wilkinson R (2011) A multi-proxy study of Late Holocene environmental change in the Prokletije Mountains, Montenegro and Albania

Žebre M, Sarıkaya MA, Stepišnik U et al (2019) First 36Cl cosmogenic moraine geochronology of the Dinaric mountain karst: Velež and Crvanj Mountains of Bosnia and Herzegovina. Quat Sci Rev 208:54–75. https://doi.org/10.1016/j.quascirev.2019.02.002

Žebre M, Sarıkaya MA, Stepišnik U et al (2021) An early glacial maximum during the last glacial cycle on the northern Velebit Mt. (Croatia). Geomorphology 392:107918. https://doi.org/10.1016/j.geomorph.2021.107918

Žebre M, Stepišnik U (2016) Glaciokarst geomorphology of the Northern Dinaric Alps: Snežnik (Slovenia) and Gorski Kotar (Croatia). J Maps 12:873–881. https://doi.org/10.1080/17445647.2015.1095133
Zhang T (2005) Influence of the seasonal snow cover on the ground thermal regime: an overview. Rev Geophys 43:RG4002. https://doi.org/10.1029/2004RG000157
Zupan Hajna N (2007) Barka depression, a denuded shaft in the area of Sneznik Mountain, Southwest Slovenia. J Cave Karst Stud 69:266–274

The Anatolian Peninsula

Attila Çiner and Mehmet Akif Sarıkaya

1 Geographical Framework

Turkey is a mountainous country covering 783,000 km^2 in the eastern Mediterranean. The Anatolian Peninsula (also known as Asia Minor) constitutes the main part of Turkey, bordering several seas; the Mediterranean Sea to the south, the Aegean Sea to the west, the inland Marmara Sea to the northwest and the Black Sea to the north. It forms an intercontinental bridge between Europe and Asia, at latitudes 36–42 °N and longitudes 26–45 °E. Situated between one of the world's most tectonically active regions, the Anatolian Peninsula constitutes a relatively small orogenic plateau (mean altitude 1132 m a.s.l.), fringed by narrow coastal plains that have profoundly affected the geologic, geomorphic, and climatic evolution of the eastern Mediterranean (Çiner et al. 2013). Mountain ranges surround it in the north (the eastern Black Sea Mountains, also known as the Pontides), and in the south (the Taurus Mountains) with elevations surpassing 3000 m in places containing a few peaks exceeding 4000 m (Kuzucuoğlu 2019a; Kuzucuoğlu et al. 2019) (Fig. 1). Because of progressively rising elevations from west towards east, the eastern part of the peninsula comprises high mountainous landscapes, including Mount Ağrı (5137 m, also known as Mt. Ararat), the country's highest peak.

The Anatolian Peninsula's geology is quite complex. The Anatolian plate has been extruding toward the west with respect to Eurasia since the Miocene as a result of extension in the Aegean and the Arabia-Eurasia collision (Şengör and Yılmaz

A. Çiner (✉) · M. A. Sarıkaya
Eurasia Institute of Earth Sciences, Istanbul Technical University, Istanbul, Turkey
e-mail: cinert@itu.edu.tr

M. A. Sarıkaya
e-mail: masarikaya@itu.edu.tr

M. A. Sarıkaya
Department of Geography, Universitat de Barcelona, Montalegre 6–8, 3r floor, 08001 Barcelona, Spain

M. Oliva et al. (eds.), *Periglacial Landscapes of Europe*,
https://doi.org/10.1007/978-3-031-14895-8_7

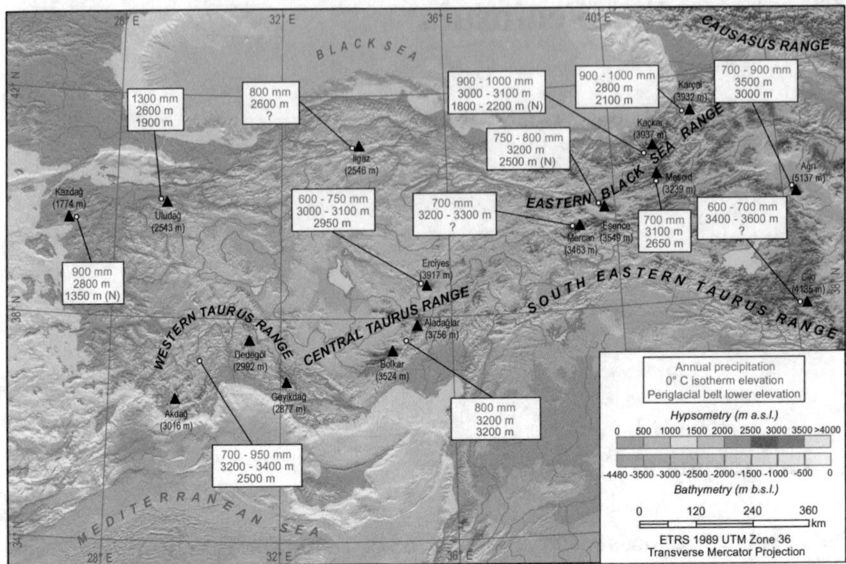

Fig. 1 Locations of the Anatolian mountains mentioned in the text. The map shows an estimation of the mean annual precipitation and elevation of the 0 °C isotherm on the mountaintops mentioned above. Precipitation and temperatures data for the period 1960–1990 were obtained from Hijmans et al. (2005), spatial resolution 30 arc-seconds, modified after WorldClim 1.4 (www.worldclim.org). Minimum altitudes of the present-day periglacial belt were also shown in the white boxes

1981; McKenzie 2020; Şengör and Yazıcı 2020). The eastern Black Sea Mountains comprise quartz-bearing lithologies of primarily plutonic and volcanic rocks. The volcanoes dispersed in the central parts, especially in the Cappadocia region, are mostly andesitic to rhyolitic in composition with Late Miocene ignimbrites and numerous Quaternary basaltic cinder cones and maars (Aydar et al. 2012; Kuzucuoğlu 2019b; Mouralis et al. 2019). On the other hand, the Taurus Mountains along the Mediterranean coast are composed of platform carbonate rocks with some ophiolitic thrust sheets (Monod 1977; Özgül 1984).

The northern and southern mountain ranges that run parallel to the coasts and the overall topographic rise eastwards create an extreme climatic variability in Turkey. The Aegean and Mediterranean coastal regions in the west and south face the North Atlantic Oscillation (NAO) influenced atmospheric circulation (Mediterranean cyclonic tracks), with annual precipitation varying from 600 to 1300 mm. The northern Black Sea region receives humidity from the Black Sea and is also influenced by the Siberian high-pressure system (Türkeş and Erlat 2005). Because of the orographic effect related to the steep topography, the eastern coast of the Black Sea collects the highest amount of annual precipitation in the peninsula, reaching more than 2000 mm. However, values decrease substantially at mountaintops as most rainfall occurs in the coastal areas. This is especially true for the Black Sea region mountaintops, where the rainfall values ca. 1000 mm, are less than half of those recorded at the coastline (Fig. 1). It is also proposed that the precipitation on the peninsula is sensitive to the variations of the surface temperature of the surrounding seas (Bozkurt and Sen 2011). Towards the interior, in central Anatolia, the rain shadow of the surrounding highlands creates humidity reduction, and precipitation decrease with increasing elevation, averaging about 400 mm yr^{-1}.

Regional temperature differences in Anatolia are significant, demonstrating the continentality effect of the climate. Mean annual air temperature (MAAT) varies from <4 °C on the highest mountains of the eastern Anatolian plateau to >20 °C in the southwestern coastal regions, and ca. 10 °C at central Turkey (Taştan and Yılmaz 2015). The northeast Black Sea and Taurus mountains have MAAT of 11 °C and 13 °C, respectively. The corresponding ranges for mean January and July temperatures are <2 °C to <3 °C, and >22 °C to >24 °C, respectively. These climatic data reveal a similar range even though they are within different latitude ranges from northeast to southwest Turkey.

Several Anatolian mountains show traces of past glacial activities, where cirques, sporadic glaciers, glacial lakes, U-shaped valleys containing diverse types of moraines, and periglacial features are widespread (Palgrave 1872; Penther 1905; Louis 1944; Kurter 1991; Çiner 2004; Sarıkaya et al. 2011; Çalışkan et al. 2012; Bayrakdar et al. 2015, 2020a; Doğukan et al. 2015; Yavaşlı et al. 2015; Satır 2016; Sarıkaya and Çiner 2019; Oliva et al. 2018, 2020; Altınay et al. 2020; Azzoni et al. 2020, 2022). Towards the interior, high volcanoes such as Ağrı (5137 m), Süphan (4058 m) and Erciyes (3917 m) also contain active glaciers and periglacial features (Penther 1905; Erinç 1951; Messerli 1964; Kurter 1991; Sarıkaya et al. 2009; Sarıkaya 2012; Sarıkaya and Tekeli 2014). However, a recent study by Kesici (2021) indicates that the glacier filling the Mt. Süphan caldera (ca. 3000 × 400 m) at the end of the nineteenth century had completely disappeared at the beginning of the twenty-first century.

A recent work by Evans et al. (2021) in the southwestern Taurus Mountains reported a detailed inventory of 85 cirques. Their results indicate that the cirques are comparable to other cirques elsewhere but narrower and often less developed. Their lengths are somewhat higher than their widths, which average slightly more than 500 m with axial gradients ca. 22°. In another study in the Teke Peninsula that covers part of the western Taurus Range, Çılğın and Bayrakdar (2020) concluded that the shadow effect and the height of the mountain primarily controlled cirque floor altitudes, but also precipitation played a significant role.

Another recent study using satellite images and topography maps reported 660 cirque lakes in 28 different mountains (Öztürk et al. 2021). Of these, 77% are located in the Eastern Black Sea Mountains. Although several glaciated peaks and cirques are found in the Taurus Mountains Range, glacial lakes are relatively few and sometimes totally absent because of karst prone lithologies. The authors calculated the average altitude of all cirque lakes in Anatolia to be 2882 m, which corresponds to the Late Pleistocene permanent snow altitudes. Nearly all cirque lakes (99.5%) were classified as small lakes with an average cirque lake area of 13.3 m^2.

A comprehensive analysis of all Cosmic-Ray Exposure (CRE) dated moraine data on the Mediterranean gives insight into the region's glacial history and proposes a model of rapid mountain glacier meltdown and enhanced valley-side debris supply during Heinrich Stadials (Allard et al. 2021). A review paper by Sarıkaya and Çiner (2015) and recent book chapters by Akçar (2022a, b, c) provide accounts of CRE-based chronologies in Turkey. Moraines in the western and central Taurus Mountains, predominantly in Mt. Sandıras (2295 m) (Sarıkaya et al. 2008), Mt. Akdağ (3016 m) (Sarıkaya et al. 2014), Mt. Dedegöl (2992 m) (Zahno et al. 2009; Çılğın 2015; Köse et al. 2019; Candaş et al. 2020), Mt. Geyikdağ (2877 m) (Çiner et al. 2015, 2017; Sarıkaya et al. 2017), Mt. Bolkar (3524 m) (Çiner and Sarıkaya 2017), and Mt. Aladağlar (3756 m) (Zreda et al. 2011) yield ^{36}Cl CRE ages ranging between ca. 50 to 5 ka.

Last Glacial Cycle (MIS 5d–3) moraines that pre-date the global Last Glacial Maximum (LGM) are only reported from three sites in Turkey (Akçar 2022b). Moraines representing the maximum ice extent at Mt. Uludağ yielded CRE ages of ca. 30 ka and ca. 50 ka (Akçar et al. 2014). On Mt. Akdağ (the western Taurus Range), Sarıkaya et al. (2014) reported moraines dated between ca. 30 and 40 ka that were less advanced than the global LGM moraines. Along the eastern Black Sea Mountains, on Mt. Kaçkar, the maximum ice extent dates back to ca. 56 ka (Reber et al. 2014). Although the a/synchronicity of the LGM in the Mediterranean region is still debated (e.g., Clark et al. 2009; Hughes and Woodward 2008; Hughes et al, 2013), a consensus exists on the rather synchronous LGM extent (ca. 21 ka; MIS 2) observed on the Taurus Mountains (e.g., Sarıkaya and Çiner 2017), the eastern Black Sea Mountains (Akçar et al. 2007, 2008), the Lesser Caucasus Mountains (Dede et al. 2017) and some other mountains in Anatolia such as Mt. Erciyes (Sarıkaya et al. 2009) and Mt. Uludağ (Zahno et al. 2010; Akçar et al. 2014, 2015; Akçar 2022c).

Following the cold conditions of the LGM, rapid deglaciation occurred in the mountains of Anatolia (e.g., Sarıkaya and Çiner 2017). Two cold spells, namely the Older Dryas and a comparatively shorter Younger Dryas (YD), allowed glaciers to readvance (Çiner et al. 2015; Sarıkaya and Çiner 2017). Even though possible Holocene glacial deposits are reported from the southeastern Taurus Mountains (e.g., Mt. Cilo, 4135 m) (İzbıkak 1951; Erinç 1953) and major volcanoes such as Mt. Ararat (Sarıkaya 2012; Azzoni et al. 2017), ^{36}Cl CRE ages exist only from Mt. Erciyes in central Anatolia. Sarıkaya et al. (2009) reported Early Holocene (ca. 9 ka) and Late Holocene (ca. 4 ka) glacial advances from the Erciyes Volcano. During the Little Ice Age (LIA), glaciers were more widespread in Turkey and the Mediterranean mountains in general (Hughes et al. 2006, 2020; Hughes 2014; Oliva et al. 2020). For example, the uppermost moraines in Mt. Uludağ were ascribed to LIA by Erinç (1952). In the same mountain, Zahno et al. (2010) ^{10}Be CRE dated a boulder from the innermost moraine with insignificant cosmogenic nuclide concentration, likely indicating an LIA expansion, in accordance with Birman (1968), who also cautiously suggested a very young age for this moraine. A small debris-covered glacier in Mt. Aladağlar (Altın 2006; Gürgen et al. 2010) is undoubtedly a relic LIA feature. Although numerous small glaciers exist in the eastern Black Sea Mountains, no LIA moraines were described from Mt. Kaçkar (Doğu et al. 1993; Akçar et al. 2007, 2008; Bayrakdar and Özdemir 2010; Reber et al. 2014).

Apart from the Mt. Ararat ice cap (Blumental 1958; Sarıkaya 2012; Azzoni et al. 2017, 2020) and Mt. Cilo Uludoruk glacier (<3 km long) (Azzoni et al. 2022), only tiny glaciers exist today in Anatolia (Kurter 1991; Çiner 2004; Akçar and Schlüchter 2005; Sarıkaya and Tekeli 2014).

2 The Development of Periglacial Research in the Anatolian Peninsula

Contrary to the glacial research, periglacial studies attracted little attention, and an inventory of periglacial landforms in Turkey is missing. Some very limited investigations conducted by foreign pioneers (e.g., Klaer 1962, 1965; Messerli 1967; Birman 1968) contain descriptive accounts of periglacial activity, mainly related to rock glaciers, in the Turkish mountains. For instance, rock glaciers are reported from Mt. Bolkar (Blumenthal 1956) and Mt. Aladağ (Blumenthal 1952). Blumenthal (1947) interpreted similar formations as debris deposited by floods generated by exceptionally heavy rains in the Taurus Range. In his work at Mt. Erciyes, Klaer (1962, pp. 57–65) attributed the probable age of the main rock glacier to post-Wisconsin, referring to LIA or Neoglacial times. Birman (1968), in his comprehensive survey entitled "Glacial reconnaissance in Turkey", also mentioned the presence of rock glaciers in numerous mountains.

Besides rock glaciers, foreign scientists also described periglacial landforms encountered in the mountainous regions. Although most accounts are related to glacial landforms, a few gave details of the periglacial deposits. One of them is related to Mt. Karagöl near Giresun town along the Black Sea coast to the north, where Planhol and Bilgin (1964, pp. 500–504) described stone circles and talus deposits in detail on the northern slopes of Mt. Karagöl starting from 1800 m. On the contrary, the southern slopes were primarily devoid of periglacial landforms.

Later, native scientists also depicted several periglacial landforms (e.g., Erinç 1955, 1957; Erinç et al. 1961; Türkeş and Öztürk 2008). However, these works lack quantitative age data and are often published in Turkish and local journals. Recently, a few young Turkish researchers (e.g., Öztürk 2012; Dede et al. 2021) started to show interest in this understudied subject.

3 Periglacial Landforms, Processes and Deposits

During the LGM and the last glacial period in general, periglacial processes below the LGM snowline, approximately between 2300–2500 m in the eastern Black Sea Mountains (Erinç 1952; Messerli 1967; Çiner 2004; Sarıkaya et al. 2011) and 2400–2600 m in the Taurus Mountains were active, and seasonal freezing and thawing chiefly controlled the development of periglacial landforms. It appears that the size of glacier expansions and retreats probably mimicked the periglacial landforms developments in these zones.

It is estimated that Turkey's ca. 20,250 km^2 (3% of land surfaces) is covered by permafrost (Çalışkan et al. 2013). Gorbunov (1978) tentatively estimated the lower limit of permafrost occurrence in Turkey at 3800 to 4000 m. However, there is no information on its depth and active layer thickness underpinning the importance of future studies. Below, we provide an overview of the periglacial landforms distribution and relative chronology since the last glacial cycle based on literature data and field observations.

3.1 Rock Glaciers

Rock glaciers are by far the most described periglacial landforms in Turkey. Using satellite image interpretation, Gorbunov (2012) recognised ca. 600 rock glaciers in Turkey, of which ca. 200 were active. According to the study, published only as abstract, active and inactive forms are located between 3400 and 2800 m, with some fossil forms going down to 2200–2300 m. The rock glaciers were up to 1200–1300 m long, the largest being 2000 m in length. Unfortunately, the absence of mapping and a proper publication related to this work undermines and limits its credibility.

On the other hand, Ünal (2013) identified and mapped 55 rock glaciers in the Black Sea region, southeastern Anatolia and Mt. Erciyes. Such study revealed that the rock glacier situated on the volcano's eastern side between 3200 and 2900 m was active and unvegetated (Ünal and Sarıkaya 2013) (Fig. 2a). It covered an area of ca. 0.84 km^2 in 2012, measured from ASTER (Advanced Spaceborne and Thermal Emission and Reflection Radiometer) satellite image analysis. This rock glacier was already known from earlier works (e.g., Erinç 1951; Klaer 1962; Birman 1968). Birman (1968) described the rock glacier extending from the lower third of the cirque headwall to 0.5–1 km downslope. He interpreted the rock glacier as a stable one because of the presence of lichens and some soil development on its surface. Nevertheless, he also pointed out the presence of clearly less vegetation cover than the moraines below and the relatively recent activity implied by very steep sides, well-developed crescentic ridges, and a very hummocky toe zone.

Following the cold conditions of the LGM, the rock glaciers' development on the higher parts of the mountains and the formation of outwash fans and plains in lower elevations increased at the onset of Holocene due to global climate amelioration. This was predominantly observed in the eastern Black Sea Mountains (e.g., Akçar et al. 2007), and Mt. Mercan (also known as Mt. Munzur; 3463 m) (e.g., Bilgin 1972; Çılğın 2013). Nevertheless, Yeşilyurt and Doğan (2010) point out the misinterpretation of several debris-covered glaciers as rock glaciers on Mt. Munzur. For example, in the upper Başyayla Valley in the eastern Black Sea Mountains, numerous active rock glaciers were described (Çalışkan 2016; Gürgen and Yeşilyurt 2012; Reber et al. 2014). In Kavron Valley, Akçar and Schlüchter (2005) attributed the destruction of the LIA moraines to the intense activity of the rock glaciers.

Dede et al. (2015) described five rock glaciers, each facing east-northeast and developed between 2730 to 3070 m, in Mt. Karçal (3932 m) of the Lesser Caucasus

Fig. 2 (**a**) An active rock glacier (RG) in Üçker Valley of Erciyes Volcano (geologist for scale at the centre-left); (**b**) An active rock glacier (RG) and protalus lobes (PL) in Mt. Karçal (for ^{36}Cl CRE ages see Dede et al. 2017); (**c**) Rock glacier (RG) inside a moraine loop (M) in Mt. Geyikdağ (for ^{36}Cl CRE ages see Çiner et al. 2017); (**d**) Hacer Valley (Mt. Aladağlar) view from the Yedigöller Plateau. Talus deposits (PL) and moraines (M in the forest area) are well-developed (for ^{36}Cl CRE ages of moraines see Zreda et al. 2011); (**e**) Glaciokarst in Çimi Valley of Mt. Geyikdağ (background mountains) (for ^{36}Cl CRE ages of moraines (M) see Sarıkaya et al. 2017); (**f**) A mid-Holocene rock avalanche (RA) (for ^{36}Cl CRE ages see Hashemi et al. 2022) sourced from the cirque (c) area. Rock avalanche overlies the LGM hummocky moraines (HM) and lateral moraines (LM) (for ^{36}Cl CRE ages see Çiner et al. 2015), in the Namaras Valley of Mt. Geyikdağ, central Taurus Range

Mountains. Two inactive and three active rock glaciers are described from five valleys, covering a total area of 0.78 km^2 (Fig. 2b). A relict rock glacier (Karçal rock glacier; terminus at 2400 m) was ^{36}Cl CRE dated at 15.7 ± 1.3 ka (Dede et al. 2017), which constitutes the first-ever numerical age account of a rock glacier in Turkey.

The expansion of rock glaciers was somewhat limited in the Taurus Mountains because of the well-developed karst that restricted surface flow. Mt. Dedegöl and

Mt. Barla constitute the only locations where rock glaciers were reported from the western Taurus Range (Delanoy and Maire 1983; Altınay et al., 2022). Even though some are inactive, others situated between 2500 and 2800 m are active and contain metre size calcareous blocks. Most of the rock glaciers described in Mt. Geyikdağ (central Taurus Range) by Arpat and Özgül (1972) are, in fact, not rock glaciers but hummocky moraines (Çiner et al. 1999, 2015). A talus-derived and fresh-looking rock glacier in the same mountain was ^{36}Cl CRE dated to the onset of Holocene (ca. 12 ka) and constitutes the only available exposure age from the region (Çiner et al. 2017) (Fig. 2c). However, the stratigraphic position of the rock glacier and a close-by moraine age prove that the rock glacier boulders contain excessive nuclide inheritance implying a true age that could be as young as a few ka.

Hacer Valley in Mt. Aladağlar is one of the deepest valleys in the central Taurus Range, where several successive YD to Holocene moraines (Zreda et al. 2011) and well-developed talus deposits (Fig. 2d) are preserved (Köse et al. 2021). Another rock glacier from a cirque area in Mt. Karanfil in the central Taurus Range was recently ^{36}Cl CRE dated and yielded an age (ca. 10 ka) that pinpoints the onset of Holocene (Köse et al., 2022). On the southeastern Taurus Range, rock glaciers also developed in front of the quickly thawing glaciers of Mt. İhtiyar Şahap (also known as Mt. Kavuşşahap; 3650 m) (Doğu 2019). Together with several moraine groups, active and fossil rock glaciers were carefully mapped in the Narlıca Valley of Mt. Kavuşşahap by Yeşilyurt et al. (2018), although their ages are still unknown.

3.2 Glaciokarst

Glaciokarst is a karstic landscape glaciated primarily during the last glacial period. Most of the Taurus Mountains are composed of platform carbonates prone to glaciokarst (Nazik et al. 2019). Indeed, a study carried out in Mt. Geyikdağ in the central Taurus Range indicates the presence of numerous karstic depressions (dolines and uvalas), mainly in altitudes 1600–2400 m, some covered with moraines (Şimşek et al. 2019) (Fig. 2e). In some cases, the karstification continued under the moraine covered doline, creating collapse dolines. In the same mountain, aeolian transport of fine-grained outwash sediments filling dolines was also reported (Şimşek et al. 2019). In their study of highly-fractured and jointed doline systems attaining a maximum of 187 doline/km^2 in the central Taurus Range, Öztürk et al. (2018) indicated that the densest doline zone corresponds to the alpine and periglacial area above the treeline.

3.3 Mass Movements

Post-LGM deglaciation induced rock-slope instabilities that produced numerous landslides and rockfalls in Turkish mountainous areas. Several colluvial fans composed of angular rock fragments also developed at the base of rocky slopes.

Small to large talus cones are present in almost all post-glacial terrains, especially where valley glaciers once prevailed, and constitute an undeniable part of any glacial geomorphological map made in Turkey. A recent study carried out in Emli Valley of Mt. Aladağlar by Utlu et al. (2020a) conluded that larger talus deposits (mainly talus cones) developed in areas where the elevation decreases regularly with homogeneous slope conditions. Most fresh-looking talus cones are found in cirque areas and up valley sections, but their monitoring has not been part of any study yet. Besides, regional inventory of talus deposits and formation age is also entirely lacking. For instance, in a survey on Mt. Ararat's southern slopes, Avcı (2007) argued that during the past glacial/periglacial climates, mass movements generated debris on the slopes and blocky colluvium in the valley floor. Especially, the debris released as a result of the rapid melting of the glaciers on Mt. Ararat (Sarıkaya 2012) and fast snow melting events induce large debris flows in this area during the summer periods. Episodic debris flows in the east-facing sections of the mountain can develop into catastrophic events with high frequency and magnitude due to rapid glacial melting.

Another example of a devastating debris flow occurred at Senirkent village on July 13, 1995, in the northern parts of Mt. Barla (Tunusluoglu et al. 2007, 2008), one of the glacierised mountains (Altınay et al., 2022) of central Anatolia. The Senirkent debris flow, which resulted in the deaths of 74 people, has a 1300 m elevation difference in source and depositional area at a very short distance of 3 km. As observed in various congeneric high mountain landscapes in Europe, the material exposed by glacial retreat and high mechanical weathering has made this area more susceptible to debris flows occurrence.

A comprehensive study carried out by Görüm (2018) on the northern part of the Anatolian Plateau revealed that 1290 large landslides (>1 km^2) occurred mainly in a terrain with a hillslope relief >1 km. The results also indicate that besides the hillslope relief and steepness, the effects of lithology type and tectonics are more important than climate on controlling the hillslope erosion. Another study that examines the spatiotemporal variations of fatal landslides in Turkey points out the increasing anthropogenic effect on their formation (Görüm and Fidan 2021). The only reported snowmelt triggered landslide without any other precursor is reported from central Anatolia (Gokceoglu et al. 2005). To the date, there are no numerical chronologies available on landslides in Turkey.

On Mt. Karagöl (3107 m), along the eastern Black Sea Range, recent solifluction, frost creep, and mass movements such as rockfalls, talus, talus creeps, rock avalanches, and rock flows were reported by Turoğlu (2009), but they need to be dated. A recent study by Utlu et al. (2020b) investigated the rockfall characteristics of the Kazıklıali Gorge (0.25 km^2) on Mt. Aladağlar using unmanned air vehicle (UAV) technology and rockfall modelling. The authors created an orthophoto and 3 cm resolution digital surface model using UAV images to define the rockfall properties. Later, fallen blocks larger than 0.5 m in diameter were digitized as polygons in GIS to produce density maps and calculate frequency distributions. The results indicate a total of 10,348 fallen rocks in Kazıklıali Gorge.

On the other hand, regional distribution and timing of rockfalls and rock avalanches associated with post-glacial warming are better known, although more age data are needed to come to a clear conclusion. For instance, Bayrakdar et al. (2020b) ^{36}Cl CRE dated 18 limestone boulders from the Akdağ rock avalanche from the western Taurus Range, the largest (3×10^8 m^3) known bedrock landslide deposits in Turkey. Apparently, the main collapse occurred in Early Holocene 8.3 ± 1.4 ka, followed by secondary failures. Another glaciokarstic rock avalanche originated in the northern flank of Mt. Geyikdağ in the central Taurus Range, covering 0.430 km^2 with an estimated average thickness of 10 m indicate two successive events ^{36}Cl CRE dated to Mid to Late Holocene (Hashemi et al., 2022) (Fig. 2f).

3.4 Patterned Ground

Patterned grounds encompass distinct and often symmetrical forms that develop due to freezing and thawing of the ground in periglacial regions. They vary in size and shape and can be sorted, i.e. made of fine and coarse sediments, or unsorted, showing different sizes and heights (Ballantyne 2018). On quasi horizontal grounds, they occur as isolated circles to polygonal nets or closely spaced mounds (Hallet 2013). On the other hand, they usually appear as stripes aligned downslope on gentle slopes. Several sorted and unsorted patterned grounds were observed in mountainous regions in Turkey.

Oliva et al. (2018) mapped the distribution of permafrost-related features since the last glacial cycle in the Anatolian Peninsula (their Fig. 1). A time-independent map of the periglacial areas, primarily associated with glaciated regions, was recently presented by Dede et al. (2021) (their Fig. 1). Stone circles, stone clusters, block currents, and garlands were described from Mt. Ilgaz (2546 m), a small mountain in northwest Anatolia (Kızılkaya et al. 2019; Dede et al. 2020, 2021). Thúfurs, which are perennial hummocks, up to 50 cm high and 1–2 m wide, and congeliturbation deposits (cold-climate variety of solifluction) that form by the breakup of soil caused by the pressure exerted by freezing water in pores, were also reported at ca. 2070–2400 m. The authors calculated the ground temperatures during their development to vary between 2.5 °C and 4 °C. On the other hand, while block streams and cryoplanation surfaces were interpreted as inactive, stone circles up to 50 cm in diameter and garlands were classified as active (Erinç et al. 1961).

In the westernmost part of the Anatolian Peninsula, garland soils at the summit and northeast facing slopes of Mt. Kazdağ (ca. 1700 m) were reported by Bilgin (1960). Besides, the author also encountered relict block streams composed of granite and schist at 1350 m facing northeast.

Mt. Uludağ (2543 m) in western Turkey is probably one of the best-described mountains related to periglacial landforms, where Erinç (1949, 1957) recognised two distinct periglacial levels. The first one is found between 1900–2300 m and is characterised by garland soils developed on slopes up to 40° steep (Fig. 3A). The second level lies over 2300 m and is chiefly represented by stone accumulations, up to 0.5 m in diameter, stone stripes and stone circles preserved on flat surfaces (Fig. 3B) (Öztürk 2012). Türkeş and Öztürk (2008, 2011) pointed out that while several alpine plant species are active in the garland development in Mt. Uludağ, only two types of *Festuca sp.* are efficient in the circle formation.

On the Yedigöller plateau of Mt. Aladağlar, in central Taurus Range, stone stripes are also reported at 3200 m (Bayarı et al. 2003, 2019). In the same range, on Mt. Geyikdağ, Çiner (2003) described stone rings composed of 1–10 cm angular limestone pebbles scattered around red soil. Garlands and polygonal soils located at ca. 2500 m near Yedigöller Lake on Mt. Esence (also known as Mt. Keşiş; 3549 m) (Akkan and Tuncel 1993), and stone rings at 2650 m on Mt. Mescid (3239 m) are also reported (Atalay 1983).

On Mt. Karagöl (3107 m), along the eastern Black Sea Range, stone circles (0.25–1 m in diameter) and ovoid depressions (1–1.5 m in diameter) are reported at ca. 1800 and ca. 1900 m respectively (de Planhol and Bilgin 1964). In the close by Mt. Karadağ (also known as Mt. Gavurdağ; 3331 m), periglacial features, such as nivation hollows, talus cones and rock falls, were also described (Bilgin 1969; Gürgen 2001). On Elevit and Hacıvanak glacial valleys of Mt. Göller (3328 m), Çiçek et al. (2006) reported the presence of garlands, stone circles and solifluction terraces mainly on the south-facing slopes.

3.5 *Frost Weathering*

Frost weathering is a collective term used to mechanical weathering implemented by the stress caused by the freezing and thawing of water. The term is used as an umbrella term for processes such as frost shattering, frost wedging and cryofracturing on timescales ranging from minutes to years. Although it is mainly a product of cold climates, it may occur in any environment where water is available at sub-freezing temperatures (between −3 and −8 °C) (Hales and Roering 2007). In Anatolia, although frost weathered in situ parent rocks are observed, frost weathered, fractured and dislocated moraine blocks are more common, especially in granitic or volcanic rocks (Fig. 3C). Tors created by the erosion and weathering of granitic rocks are also reported nearby the Deniz Lake on Mt. Kaçkar (Bayrakdar and Özdemir 2010).

3.6 *Ice Wedges and Ice Caves*

An ice wedge is a crack in the ground formed by ice and can measure a few metres in length and several metres deep. The only known cause of a fossil ice-wedge pseudomorph is reported from the southeast of Acıgöl village at ca. 1400 m in Cappadocia (Mouralis 2003). The ice wedge pseudomorph is ca. 1 m deep and >0.5 m wide at its upper part and is developed in red alluvial fine-grained sediments (Fig. 3D). Grey coloured, >1 m thick volcanic surge deposits fill the ice-wedge, which is covered by basaltic scoria dated at 32 ka.

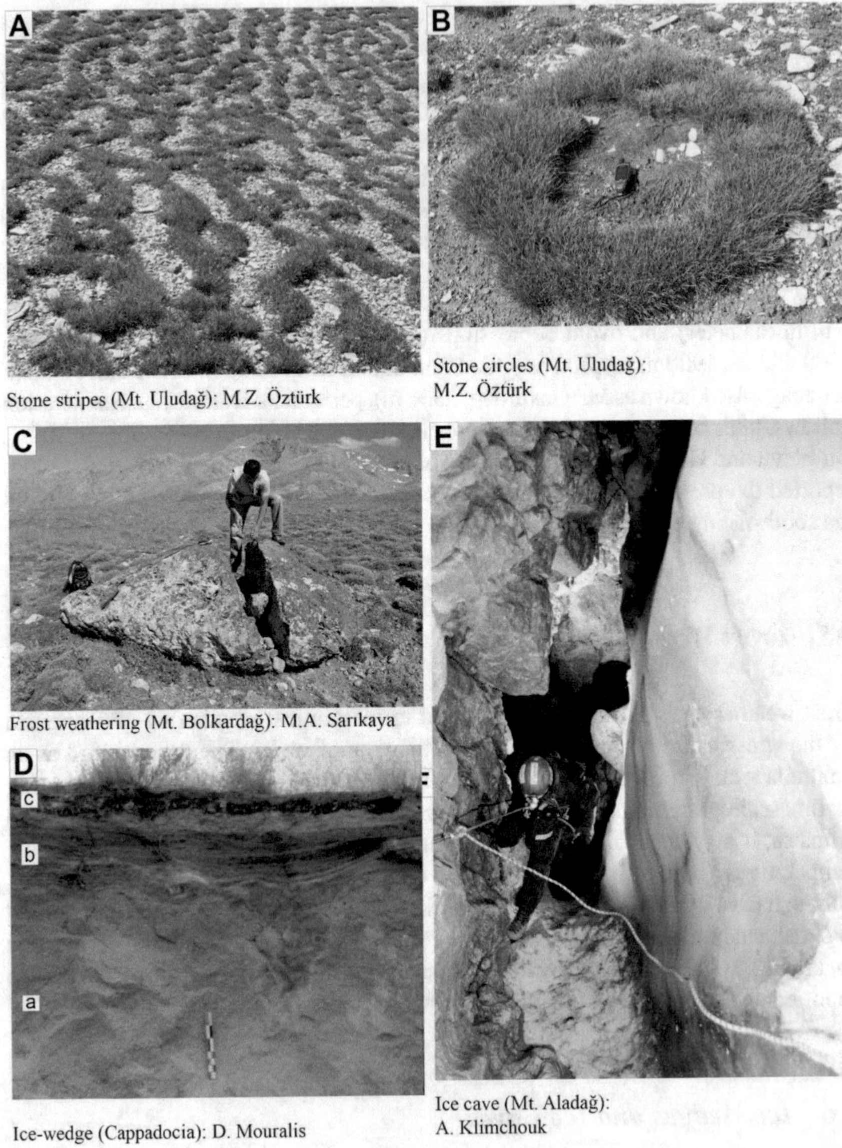

Fig. 3 Patterned ground examples from Mt. Uludağ; (**A**) Stone stripes and (**B**) stone circles (garlands) developed on the slopes and horizontal surfaces, respectively. Two types of *Festuca* sp. play a significant role in developing garlands in the soil. (**C**) Frost weathered and cracked gabbro boulder from a moraine in Mt. Bolkardağ (for ^{36}Cl CRE ages see Çiner et al. 2017). (**D**) A fossil ice-wedge from the southeast of Acıgöl village at *ca.* 1400 m in Cappadocia. The scale is 0.5 m long. (a) red alluvial fine-grained sediments, (b) volcanic surge deposits filling the ice-wedge, which is itself covered by (c) basaltic scoria dated to 32 ka. (**E**) Ice in the so-called Pyrr's Victory cave at Mt. Aladağlar of central Taurus Range. A 130 m deep shaft is located at 3380 m. The shaft contains a 90-m continuous column of ice and firn snow and an ice cone of blurred ice-rich in limestone clasts at the bottom area

Permanent ice in caves is also considered a form of permafrost (Zorn et al. 2020). Although not surveyed regularly, a karstic ice cave at ca. 3000 m containing columnar ice of ca. 130 m long along a shaft was described at the Aladağlar of central Taurus Range (Bayarı et al. 2003, 2019; Klimchouk et al. 2006). The ice is probably inherited from the LIA (Fig. 3E) and shows signs of shrinking. Up to date, there are no other reported ice caves in Turkey.

4 The Significance of Periglacial Dynamics in the Anatolian Peninsula Within the European Context

Considerable progress has been made in reconstructing cold palaeoenvironments in the Anatolian Peninsula thanks to the increasing use of CRE and other dating methods. Therefore, Anatolia's Late Pleistocene palaeoglacial deposits are currently well described and makeup one of the best-dated accounts of its kind (Hughes and Woodward 2017). Indeed, LGM temperature drops between 8 and 11 °C (Sarıkaya et al. 2008) gave rise to the advance of glaciers and the development of periglacial landforms at lower elevations and on the non-glaciated mountain summits. The warming that followed LGM gave rise to an enhanced glacial meltdown, which is now well established in the Anatolian Peninsula (Sarıkaya and Çiner 2015). The onset of the Holocene marked the increasing development of periglacial landforms, especially rock glaciers.

However, contrary to glacial deposits, the research on periglacial landforms is still in its beginning in Turkey. Although some works have focused on the rock glaciers and a few descriptive periglacial landforms, current monitoring of the movement of present-day periglacial landforms is lacking. Therefore, upcoming work should emphasise the inventories of all periglacial futures and accurate descriptions of landforms. Numerical ages are also crucially missing, with a few exceptions from rock glaciers. In a world where climatic oscillations become increasingly essential to be forecasted, understanding the timing and formation conditions of the periglacial landscapes and deposits is of prime importance.

Acknowledgements This review chapter was possible mainly thanks to The Scientific and Technological Research Council of Turkey (TÜBİTAK) funding. Financial support to our work on glacial landscapes and palaeoclimate studies since 2000 via TÜBİTAK projects (107Y069, 110Y300, 112Y139 and 114Y139), TÜBİTAK—NSF (National Science Foundation—USA) Co-Project No: 101Y002, and TÜBİTAK—SLOVENIA Research Agency (ARRS) Co-Project No: 118Y052 are greatly acknowledged. During the last 20 years, numerous colleagues and students contributed to the fieldwork and laboratory analyses, and we are grateful for their help and companionship in the field. We also appreciate helpful comments and suggestions by the book editors, which increased the quality of the manuscript.

References

Akçar N (2022a) The Anatolian Peninsula. In: Palacios et al. (eds) European glacial landscapes: maximum extent of glaciations. Elsevier, pp 149–157. https://doi.org/10.1016/B978-0-12-823498-3.00018-2

Akçar N (2022b) The Anatolian mountains: glacial landforms prior to the Last Glacial Maximum. In: Palacios et al. (eds) European glacial landscapes: maximum extent of glaciations. Elsevier, pp 333–337. https://doi.org/10.1016/B978-0-12-823498-3.00027-3

Akçar N (2022c) The Anatolian Mountains: glacial landforms from the Last Glacial Maximum. In: Palacios et al. (eds) European glacial landscapes: maximum extent of glaciations. Elsevier, pp 497–504. https://doi.org/10.1016/B978-0-12-823498-3.00016-9

Akçar N, Schlüchter CA (2005) Paleoglaciations in Anatolia: a schematic review and first results. Eiszeit Gegenw 55:102–121

Akçar N, Yavuz V, Ivy-Ochs S, Kubik PW, Vardar M, Schlüchter CA (2007) Paleoglacial records from Kavron Valley, NE Turkey: Field and cosmogenic exposure dating evidence. Quatern Int 164–165:170–183

Akçar N, Yavuz V, Ivy-Ochs S, Kubik PW, Vardar M, Schlüchter CA (2008) A Case for a down wasting Mountain Glacier during the Termination-I, Verçenik Valley NE Turkey. J Quat Sci 23:273–285

Akçar N, Yavuz V, Ivy-Ochs S, Reber R, Kubik PW, Zahno CA, Schlüchter CA (2014) Glacier response to the change in atmospheric circulation in the eastern Mediterranean during the Last Glacial Maximum. Quat Geochronol 19:27–41

Akçar N, Yavuz V, Yeşilyurt S, Ivy-Ochs S, Reber R, Bayrakdar CA, Kubik PW, Zahno CA, Schlunegger F, Schlüchter CA (2015) Synchronous last glacial maximum across the Anatolian Peninsula. Geological Society, London, Special Publications, 433. https://doi.org/10.1144/SP433.7

Akkan E, Tuncel M (1993) Esence (Kesiş) Dağlarında Buzul Şekilleri, A.Ü. Coğrafyası Araştırma ve Uygulama Merkezi Dergisi 2:225–240

Allard J, Hughes P, Woodward JCA (2021) Heinrich Stadial aridity forced Mediterranean-wide glacier retreat in the last cold stage. Nat Geosci 14:197–205. https://doi.org/10.1038/s41561-021-00703-6

Altın T (2006) Aladağlar ve Bolkar dağları üzerinde görülen periglasiyal jeomorfolojik şekiller. Türk Coğrafya Dergisi 46:105–122

Altınay O, Sarıkaya MA, Çiner A (2020) Late-glacial to Holocene glaciers in the Turkish Mountains. Mediterr Geosci Rev 2(1):119–133. https://doi.org/10.1007/s42990-020-00024-7

Altınay O, Sarıkaya MA, Çiner A, Žebre M, Stepišnik U, Yıldırım CA (2022) Glacial chronology and polygenetic-polycyclic evolution of Mt. Barla, Western Taurus, Turkey. Geomorphology. 416, Doi:10.1016/j.geomorph.2022.108424

Arpat E, Özgül N (1972) Orta Toroslar'da Geyik dağı yöresinde kaya buzulları. Bull Miner Res Explor, Ankara 78:30–35

Atalay İ (1983) Mescid dağının glasyal morfolojisi. Ege Coğrafya Dergisi 2(1):31–48

Avcı M (2007) Noah's Ark: its relationship to the Telçeker earthflow, Mount Ararat, Eastern Turkey. Bull Eng Geol Environ 66:377–380

Aydar E, Schmitt AK, Çubukçu HE, Akın L, Ersoy O, Şen E, Duncan RA, Atıcı G (2012) Correlation of ignimbrites in the central Anatolian volcanic province using zircon and plagioclase ages and zircon compositions. J Volcanol Geotherm Res 213–214:83–97

Azzoni RS, Bollati IM, Pelfini M, Sarıkaya MA, Zerboni A (2022) Geomorphology of a recently deglaciated high mountain area in Eastern Anatolia (Turkey). J Maps DOI: 10.1080/17445647.2022.2035269

Azzoni RS, Sarıkaya MA, Fugazza D (2020) Turkish glacier inventory and classification from high-resolution satellite data. Mediterr Geosci Rev 2:153–162. https://doi.org/10.1007/s42990-020-00029-2

Azzoni RS, Zerboni A, Pelfini M, Garzonio CAA, Cioni R, Meraldi E, Smiraglia CA, Diolaiuti GA (2017) Geomorphology of Mount Ararat/Ağri Daği (Ağri Daği Milli Parki, Eastern Anatolia, Turkey). J Maps 13(2):182–190. https://doi.org/10.1080/17445647.2017.1279084

Ballantyne CAK (2018) Periglacial Geomorphology. Wiley Blackwell

Bayari S, Klimchouk A, Sarıkaya M, Nazik L (2019) Aladağlar Mountain range: a landscape-shaped by the Interplay of Glacial, Karstic, and Fluvial Erosion. In: Kuzucuoğlu CA, Çiner A, Kazancı N (eds) Landforms and landscapes of Turkey. "World Geomorphological Landscapes" Series, Migon P (ed). Springer Nature Pub., pp 423–435

Bayarı S, Zreda M, Çiner A, Nazik L, Törk K, Özyurt N, Klimchouk A, Sarıkaya MA (2003) The extent of Pleistocene ice cap, glacial deposits and glaciokarst in the Aladağlar Massif: Central Taurids Range, Southern Turkey. The XVI INQUA Congress, Reno, Nevada USA, Geological Society America Abstracts, pp 144–145

Bayrakdar CA, Özdemir H (2010) Kaçkar Dağı'nda bakı faktörünün glasiyal ve periglasiyal topografya gelişimi üzerindeki etkisi. Türk Coğrafya Dergisi 54:1–13

Bayrakdar CA, Çılgın Z, Döker MF, Canpolat E (2015) Evidence of an active glacier in the Munzur Mountains, eastern Turkey. Turkish J Earth Sci 24:56–71. https://doi.org/10.3906/yer-1403-7

Bayrakdar CA, Çılğın Z, Keserci F (2020a) Traces of late quaternary glaciations and paleoclimatic interpretation of Mount Akdağ (Alanya, 2451 m), Southwest Turkey. Mediterr Geosci Rev 2:135–151. https://doi.org/10.1007/s42990-020-00026-5

Bayrakdar CA, Gorum T, Çılğın Z, Vockenhuber CA, Ivy-Ochs S, Akçar N (2020b) Chronology and geomorphological activity of the Akdag rock avalanche (SW Turkey). Front Earth Sci 8:295

Bilgin T (1960) Kaz Dağı ve üzerindeki periglasyal şekiller hakkında. Türk Coğrafya Dergisi 20:114–123. https://dergipark.org.tr/tr/pub/tcd/issue/21265/228296

Bilgin T (1969) Gavurdağ Kütlesinde Glasiyal ve Periglasiyal Topografya Şekilleri., İst. Üniv. Coğ. Enst. Yay. No:58, İstanbul

Bilgin T (1972) Munzur Dağları doğu kısmının glasiyal ve periglasiyal morfolojisi, İstanbul Üniv. Yayınları, 1757, 69, 85p

Birman JH (1968) Glacial reconnaissance in Turkey. Geol Soc Am Bull 79:1009–1026

Blumenthal MM (1947) Das palaeozoische Fenster von Belemedik und sein mesozoisher Kalkrahmen (Cilicischer Taurus). M.T.A. Publ, no. D3, Ankara

Blumenthal MM (1952) Das taurische Hochgebirge des Aladağ, neuere Forschungen zu seiner Geographie, Stratigraphie und Tektonik. M.T.A. Publ., no. D6, Ankara

Blumenthal MM (1956) Yüksek Bolkardağ'ın kuzey kenar bölgelerinin ve batı uzantılarının jeolojisi. M.T.A. Enst, Yayınl. Ankara, 156 p

Blumenthal MM (1958) From Mount Ağrı to Kaçkar Mountains. Die Alpen 34:125–137

Bozkurt D, Şen OL (2011) Precipitation in the Anatolian Peninsula: sensitivity to increased SSTs in the surrounding seas. Clim Dyn 36:711–726. https://doi.org/10.1007/s00382-009-0651-3

Çalışkan O (2016) Dört farklı soğuk ortam topoğrafyası tek bir buzullaşma alanı üzerinde gözlenebilir mi? Karçal Dağları örneği (Doğu Karadeniz Bölümü/Türkiye). Marmara Coğrafya Dergisi 33:368–389

Çalışkan O, Çalışkan A, Abacı A, Topgümüş CA, Demirkesen E, Dikenoğlu G, Demir H, Ayantaş T, Balcı T (2013) Türkiye buzküresinin jeoistatistik modellerle belirlenmesi. Coğrafyacılar Derneği Yıllık Kongresi Bildiriler Kitabı, Fatih Üniversitesi, İstanbul, pp 554–562

Çalışkan O, Gürgen G, Yılmaz E, Yeşilyurt S (2012) Bolkar Dağları kuzeydoğusunun glasyal morfolojisi ve döküntüyle örtülü buzulları. Uluslararası İnsan Bilimleri Dergisi 9(1):890–911

Candaş A, Sarıkaya MA, Köse O, Şen ÖL, Çiner A (2020) Modelling Last Glacial Maximum ice cap with Parallel Ice Sheet Model to infer palaeoclimate in south-west Turkey. J Quat Sci 935–950. https://doi.org/10.1002/jqs.3239

Çiçek İ, Gürgen G, Tunçel H, Doğu AF (2006) Doğu Karadeniz dağları'nın glasyal morfolojisi. The symposium of first International studies of geography (The Mountainous Areas of Caucasus and Anatolia on Pleistocene and today), 9–13 June 2003, pp 28–42

Çiner A (2003) Sedimentary facies analysis and depositional environments of the Late Quaternary moraines in Geyikdağ (Central Taurus Mountains). Geol Bull Turkey 46(1):35–54 (in Turkish)

Çiner A (2004) Turkish glaciers and glacial deposits. In: Ehlers J, Gibbard PL (eds) Quaternary glaciations: extent and chronology, Part I: Europe. Elsevier Publishers, Amsterdam, pp 419–429

Çiner A, Sarıkaya MA (2017) Cosmogenic 36Cl Geochronology of late Quaternary glaciers on the Bolkar Mountains, south central Turkey. In: Hughes P, Woodward J (eds) Quaternary Glaciation in the Mediterranean Region. Geological Society of London Special Publication, 433, pp 271–287. https://doi.org/10.1144/SP433.3

Çiner A, Deynoux M, Çörekçioğlu E (1999) Hummocky moraines in the Namaras and Susam valleys, Central Taurids, SW Turkey. Quat Sci Rev 18(4–5):659–669

Çiner A, Sarıkaya MA, Yıldırım CA (2015) Late Pleistocene piedmont glaciations in the Eastern Mediterranean; insights from cosmogenic 36Cl dating of hummocky moraines in southern Turkey. Quatern Sci Rev 116:44–56. https://doi.org/10.1016/j.quascirev.2015.03.017

Çiner A, Sarıkaya MA, Yıldırım CA (2017) Misleading old age on a young landform? The dilemma of cosmogenic inheritance in surface exposure dating: moraines vs. rock glaciers. Quat Geochronol 42:76–88. https://doi.org/10.1016/j.quageo.2017.07.003

Çiner A, Strecker MR, Bertotti G (2013) Late Cenozoic Evolution of the Central Anatolia Plateau: preface, Turkish. J Earth Sci 22:i–ii

Çılğın Z (2013) Ovacık Ovası (Tunceli) ve Munzur Dağlarının Güneybatı Aklanının Jeomorfolojisinde Buzullaşmaların Etkisi. Kilis Üniversitesi Sosyal Bilimler Dergisi 3(6):103–121

Çılğın Z (2015) Dedegöl Dağı Kuvaterner buzullaşmaları. Türk Coğrafya Dergisi 64:19–37

Çılğın Z, Bayrakdar CA (2020) Teke Yarımadası'ndaki (Güneybatı Anadolu) glasiyal sirklerin morfometrik özellikleri. Türk Coğrafya Dergisi 74:107–121. https://doi.org/10.17211/tcd.729978

Clark PU, Dyke AS, Shakun JD, Carlson AE, Clark J, Wohlfarth B, Mitrovica JX, Hostetler SW, McCabe AM (2009) The last glacial maximum. Science 325:710–714

Dede V, Çiçek İ, Sarıkaya MA, Çiner A, Uncu L (2017) First cosmogenic geochronology from the Lesser Caucasus: Late Pleistocene glaciation and rock glacier development in the Karçal Valley, NE Turkey. Quatern Sci Rev 164:54–67. https://doi.org/10.1016/j.quascirev.2017.03.025

Dede V, Çiçek İ, Uncu L (2015) Formations of rock glacier in Karçal Mountains, (in Turkish). Bull Earth Sci Appl Res Cent Hacet Univ 36(2):61–80

Dede V, Dengiz O, Demirağ Tİ, Türkeş M, Gökçe CA, Serin S (2020) Ilgaz Dağları Periglasyal Şekillerinde Oluşmuş Toprakların Fizikokimyasal Özellikleri ile Bazı Erozyon Duyarlılık Parametreleri Arasındaki İlişkilerin Belirlenmesi. Coğrafi Bilimler Dergisi 18:99–123. https://doi.org/10.33688/aucbd.689755

Dede V, Dengiz O, Zorlu BŞ, Zorlu K (2021) Ilgaz Dağları'nda yükseltiye bağlı sıcaklık değişiminin periglasyal şekillerdeki toprak özellikleri üzerine etkisi. Türk Coğrafya Dergisi (78):23–32. https://doi.org/10.17211/tcd.1002568

Delannoy JJ, Maire R (1983) Le Massif de Dedegöl dag (Taurus Occidental, Turquie). Recherches de géomorphologie glaciaire et karstique. Bulletin de l'Association de Géographie Française 491:43–53

de Planhol X, Bilgin T (1964) Glaciaire et périglaciaire quaternaires et actuels dans le massif du Karagöl (Chaines pontiques, Turquie). Revue de Géographie Alpine 52(3):497–512

Doğu AF, Somuncu M, Çiçek İ, Tuncel H, Gürgen G (1993) Kaçkar Dağı'nda buzul şekilleri, yaylalar ve turizm. Ankara Üniversitesi Türkiye Coğrafyası Araştırma ve Uygulama Merkezi Dergisi 157–183

Doğu (2019) Pleistocene Glacier Heritage and Present-Day Glaciers in the Southeastern Taurus (İhtiyar Şahap Mountains). In Kuzucuoğlu CA, Çiner A, Kazancı N (eds.), *Landforms and landscapes of Turkey*. "World Geomorphological Landscapes" Series, Migon P (ed). Springer Nature Pub. 413–422.

Doğukan DY, Tucker CAJ, Melocik KA (2015) Change in the glacier extent in Turkey during the Landsat Era. Remote Sens Environ 163:32–41

Erinç S (1949) Research on glacial morphology of Mount Uludag. Rev Geogr Inst Univ Istanbul 11–12:79–94

Erinç S (1951) The glacier of Erciyes in Pleistocene and Post-glacial epochs. Rev Geogr Inst Univ Istanbul 1:82–90

Erinç S (1952) Glacial evidences of the climatic variations in Turkey. Geogr Ann 34(1–2):89–98
Erinç S (1953) From Lake Van to Mount Cilo. Rev Geogr Inst Univ Istanbul 2:84–106
Erinç S (1955) Periglacial features on the Mount Honaz (SW-Anatolia). Rev Geogr Inst Univ Istanbul 2:185–187
Erinç S (1957) About the Uludag periglacial. Rev Geogr Inst Univ Istanbul 8:91–94
Erinç S, Bilgin T, Bener M (1961) Ilgaz Üzerinde Periglasyal Şekiller. İstanbul Üniversitesi Coğrafya Enstitüsü Dergisi 12:151–160
Evans IS, Çılğın Z, Bayrakdar CA, Canpolat CA (2021) The form, distribution and palaeoclimatic implications of cirques in southwest Turkey (Western Taurus). Geomorphology. https://doi.org/10.1016/j.geomorph.2021.107885
Gokceoglu CA, Sonmez H, Nefeslioglu HA, Duman TY, Can T (2005) The 17 March 2005 Kuzulu landslide (Sivas, Turkey) and landslide-susceptibility map of its near vicinity. Eng Geol 81(1):65–83. https://doi.org/10.1016/j.enggeo.2005.07.011
Gorbunov AP (1978) Permafrost Investigations in High-Mountain Regions. Arct Alp Res 10(2):283–294
Gorbunov AP (2012) Rock glaciers, kurums, glaciers and permafrost in the mountains of Turkey, Geographical Review. Earth Cyrosphere XVI(2):3–8 (in Russian)
Görüm T (2018) Tectonic, topographic and rock-type influences on large landslides at the northern margin of the Anatolian Plateau. Landslides 16(2):333–346
Görüm T, Fidan S (2021) Spatiotemporal variations of fatal landslides in Turkey. Landslides 18(5):1691–1705. https://doi.org/10.1007/s10346-020-01580-7
Gürgen G (2001) Karadağ (Gümüşhane) çevresinin glasyal morfolojisi ve turizm potansiyeli. AÜ Türkiye Coğrafyası Dergisi 8:109–132
Gürgen G, Yeşilyurt S (2012) Karçal Dağı Buzulları (Artvin). Coğrafi Bilimler Dergisi 10(1):91–104
Gürgen G, Çalışkan O, Yılmaz E, Yeşilyurt S (2010) Yedigöller platosu ve Emli vadisinde (Aladağlar) döküntü örtülü buzullar. E-J New World Sci Acad 5:98–116
Hales TC, Roering J (2007) Climatic controls on frost cracking and implications for the evolution of bedrock landscapes. J Geophys Res: Earth Surf 112(F2). https://doi.org/10.1029/2006JF000616
Hallet B (2013) Stone circles: form and soil kinematics. Phil Trans R Soc A 371:20120357. https://doi.org/10.1098/rsta.2012.0357
Hashemi K, Sarıkaya A, Görüm T, Wilcken KM, Çiner A, Žebre M, Stepišnik U, Yıldırım CA (2022) The Namaras rock avalanche; a mid-to-late Holocene paraglacial activity in the Central Taurus Mountains, SW Turkey. Geomorphology 408, https://doi.org/10.1016/ j.geomorph.2022.108261
Hughes PD (2014) Little Ice Age glaciers on the Mediterranean mountains. Mediterranée 112:63–79
Hughes PD, Woodward JCA (2008) Timing of glaciation in the Mediterranean mountains during the Last Cold Stage. J Quat Sci 23:575–588
Hughes PD, Woodward J (2017) Quaternary glaciation in the Mediterranean mountains: a new synthesis. In: Hughes P, Woodward J (eds) Quaternary Glaciation in the Mediterranean Region. Geological Society of London Special Publication, 433. https://doi.org/10.1144/SP433.14
Hughes PD, Gibbard PL, Ehlers J (2013) Timing of glaciation during the last glacial cycle: evaluating the concept of a global 'Last Glacial Maximum' (LGM). Earth Sci Rev 125:171–198
Hughes PD, Woodward JCA, Gibbard PL (2006) Late Pleistocene glaciers and climate in the Mediterranean region. Glob Planet Change 46:83–98
Hughes P, Fletcher W, Bell B, Braithwaite RJ, Cornelissen HL, Fink D, Rhoujjati A et al (2020) Late Pleistocene glaciers to present-day snowpatches: a review and research recommendations for the Marrakech High Atlas. Mediterr Geosci Rev 2:163–184. https://doi.org/10.1007/s42990-020-00027-4
İzbırak R (1951) Cilo Dağı ve Hakkari ile Van Gölü çevrelerinde coğrafya araştırmaları, Ankara. Üniversitesi Dil Tarih Coğrafya Fakültesi Yayınları 67(4):149p
Kesici Ö (2021) Küresel ısınma açısından Süphan dağı buzullarındaki değişmeler. Doğu Coğrafya Dergisi 26(46):159–176

Kızılkaya R, Dede V, Dengiz O, Ay A (2019) Ilgaz Dağları'nda farklı periglasyal şekiller üzerinde oluşmuş topraklara ait özelliklerin dehidrogenaz enzim aktivitesine etkisi. Toprak Bilimi ve Bitki Besleme Dergisi 7:121–127

Klaer W (1962) Untersuchungen zur klimagenetischen Geomorphologie in den Hochgebirgen Vorderasiens: Heidelberger Geog. Arb., no. 11, 135 p

Klaer W (1965) Geomorphologische Unterschungen in den Randgebirgen des Van-See (Ostanatolien). Zeitshrift für Geomorphologie 9(3):346–356

Klimchouk A, Bayarı S, Nazik L, Törk K (2006) Glacial destruction of cave systems in high mountains, with a special reference tot he Aladağlar massif, central Taurus, Turkey. Acta Carsologica 35(2):111–121

Köse O, Sarıkaya MA, Çiner A, Candaş A (2019) Late Quaternary glaciations and cosmogenic ^{36}Cl geochronology of Mount Dedegöl, south-west Turkey. J Quat Sci 34(1):51–63. https://doi.org/10.1002/jqs.3080

Köse O, Sarıkaya MA, Çiner A, Yıldırım CA (2021) Glacial geomorphology of the Aladağlar, Central Taurus Mountains, Turkey. J Maps 17(2):101–113. https://doi.org/10.1080/17445647.2021.1883137

Köse O, Sarıkaya MA, Çiner A, Candaş A, Wilcken K (2022) Cosmogenic 36Cl chronology and modelling of paleoglaciers on Mt Karanfil, Central Taurus Range, Turkey. Quaternary Science Reviews 291, Doi:10.1016/j.quascirev.2022.107656

Kurter A (1991) Glaciers of Middle East and Africa—Glaciers of Turkey, In: Williams RS, Ferrigno JG (eds) Satellite image atlas of the World. USGS Professional Paper, 1386-G-1, pp 1–30

Kuzucuoğlu CA (2019a) The physical geography of Turkey: an outline. In: Landforms and landscapes of Turkey. "World Geomorphological Landscapes" Series, Migon P (ed). Springer Nature Pub., pp 3–5. https://doi.org/10.1007/978-3-030-03515-0-2

Kuzucuoğlu CA (2019b) Geology and geomorphology of the Cappadocia volcanic Province, Turkey. Mediterr Geosci Rev 1:163–166. https://doi.org/10.1007/s42990-019-00015-3

Kuzucuoğlu CA, Çiner A, Kazancı N (2019) The geomorphological regions of Turkey. In: Kuzucuoğlu CA, Çiner A, Kazancı N (eds) Landforms and landscapes of Turkey. "World Geomorphological Landscapes" Series, Migon P (ed). Springer Nature Pub., pp 41–178. https://doi.org/10.1007/978-3-030-03515-0-4

Louis HL (1944) Evidence for Pleistocene glaciation in Anatolia (in German). Geol Rundsch 34(7–8):447–481

McKenzie D (2020) The structure of the lithosphere and upper mantle beneath the Eastern Mediterranean and Middle East. Mediterr Geosci Rev 2:311–326. https://doi.org/10.1007/s42990-020-00038-1

Messerli B (1964) Der Gletscher am Erciyas Dagh und das Problem der rezenten Schneegrenze im anatolischen und mediterranen Raum. Geographica Helvetica 19(1):19–34

Messerli B (1967) Die eiszeitliche und die gegenwartige Vergletscherung in Mittelmeerraum. Geographica Helvetica 22:105–228

Monod O (1977) Recherches géologiques dans le Taurus occidental au sud de Beyşehir (Turquie). PhD Thesis, Université Paris Sud, Orsay [unpublished]

Mouralis D (2003) Les complexes volcaniques quaternaires de Cappadoce (Göllüdag et Acigöl) - Turquie : évolutions morphodynamiques et implications environnementales (Thèse doctorat). Université Paris 12, Paris, 303 p

Mouralis D, Pastre JF, Kuzucuoğlu CA, Türkecan A, Guillou H (2019) Tephrostratigraphy and chronology of the Quaternary Göllüdağ and Acıgöl volcanic complexes (Central Anatolia, Turkey). Mediterr Geosci Rev 1:179–202. https://doi.org/10.1007/s42990-019-00010-8

Nazik L, Poyraz M, Karabıyıkoğlu M (2019) Karstic landscapes and landforms in Turkey. In: Catherine Kuzucuoğlu A, Kazancı ÇN (eds) Landscapes and landforms of Turkey. Springer, Cham, pp 181–196. https://doi.org/10.1007/978-3-030-03515-0_5

Oliva M, Sarıkaya MA, Hughes P (2020) Holocene and earlier glaciations in the Mediterranean Mountains. Mediterr Geosci Rev 2:1–4. https://doi.org/10.1007/s42990-020-00025-6

Oliva M, Žebre M, Guglielmin MM, Hughes P, Çiner A, Vieria G, Bodin X, Andrés N, Colucci RR, García-Hernández CA, Mora CA, Nofre J, Palacios D, Pérez-Alberti A, Ribolini A, Ruiz-Fernández J, Sarıkaya MA, Serrano E, Urdea P, Valcárcel M, Woodward J, Yıldırım CA (2018) Permafrost conditions in the Mediterranean basin since the Last Glaciation. Earth Sci Rev 185:397–436. https://doi.org/10.1016/j.earscirev.2018.06.018

Özgül N (1984) Geology of the central Taurides. In: Proceedings of the International Symposium on Geology of the Taurus Belt, pp 26–29

Öztürk MZ (2012) Uludağ'daki periglasiyal süreçlerin, periglasiyal yerşekillerinin ve bunları denetleyen etmenlerin incelenmesi, Nilüfer Akkılıç Kütüphanesi Yayınları, No: 10, 161 p

Öztürk MZ, Şimşek M, Şener MF, Utlu M (2018) GIS based analysis of doline density on Taurus Mountains, Turkey. Environ Earth Sci 77:536

Öztürk MZ, Şimşek M, Utlu M (2021) Anadolu'nun sirk gölleri. Türk Coğrafya Dergisi 78:49–60. https://doi.org/10.17211/tcd.998089

Palgrave WG (1872) Vestiges of the glacial period in northeastern Anatolia. Nature 5:444–445

Penther A (1905) Eine Reise in das Gebiet des Erdschias-Dagh (Kleinasien), 1902. Abhandlungen der k. k. Geography Gesellschaft in Wien, 6, 1

Reber R, Akçar A, Yesilyurt S, Yavuz V, Tikhomirov D, Kubik PW, Schlüchter CA (2014) Glacier advances in northeastern Turkey before and during the global Last Glacial Maximum. Quatern Sci Rev 101:177–192

Sarıkaya MA (2012) Recession of the ice cap on Mount Ağrı (Ararat), Turkey, from 1976 to 2011 and its climatic significance. J Asian Earth Sci 46:190–194

Sarıkaya MA, Çiner A (2015) Late Pleistocene glaciations and paleoclimate of Turkey. Bull Miner Res Explor (MTA) 151:107–127

Sarıkaya MA, Çiner A (2017) The late Quaternary glaciation in the Eastern Mediterranean. In: Hughes P, Woodward J (eds) Quaternary Glaciation in the Mediterranean Mountains. Geological Society of London Special Publication, 433, pp 289–305. https://doi.org/10.1144/SP433.4

Sarıkaya MA, Çiner A (2019) Ice in paradise: glacial heritage landscapes of Anatolia. In: Kuzucuoğlu CA, Çiner A, Kazancı N (eds) Landforms and landscapes of Turkey. "World Geomorphological Landscapes" Series, Migon P (ed). Springer Nature Pub., pp 397–411. https://doi.org/10.1007/978-3-030-03515-0-20

Sarıkaya MA, Tekeli AE (2014) Satellite inventory of glaciers in Turkey, In: Kargel JS, Leonard GJ, Bishop MP, Kääb A, Raup B (eds) Global Land Ice Measurements from Space. Praxis-Springer (Publisher), Berlin Heidelberg, 876 pp. ISBN: 978-3-540-79817-0, 465-480

Sarıkaya MA, Çiner A, Haybat H, Zreda M (2014) An early advance of glaciers on Mount Akdağ, SW Turkey, before the global Last Glacial Maximum; insights from cosmogenic nuclides and glacier modeling. Quatern Sci Rev 88:96–109. https://doi.org/10.1016/j.quascirev.2014.01.016

Sarıkaya MA, Çiner A, Yıldırım CA (2017) Cosmogenic ^{36}Cl glacial chronologies of the Late Quaternary glaciers on Mount Geyikdağ in the Eastern Mediterranean. Quat Geochronol 39:189–204. https://doi.org/10.1016/j.quageo.2017.03.003

Sarıkaya MA, Çiner A, Zreda M (2011) Quaternary Glaciations of Turkey. In: Ehlers J, Gibbard PL, Hughes PD (eds) Quaternary Glaciations- extent and chronology; a closer look. Elsevier Publications, Developments in Quaternary Science, Vol. 15, Amsterdam, The Netherlands, pp 393–403. https://doi.org/10.1016/B978-0-444-53447-7.00030-1

Sarıkaya MA, Zreda M, Çiner A (2009) Glaciations and paleoclimate of Mount Erciyes, central Turkey, since the Last Glacial Maximum, inferred from 36Cl cosmogenic dating and glacier modeling. Quatern Sci Rev 28(23–24):2326–2341

Sarıkaya MA, Zreda M, Çiner A, Zweck CA (2008) Cold and wet Last Glacial Maximum on Mount Sandıras, SW Turkey, inferred from cosmogenic dating and glacier modeling. Quatern Sci Rev 27(7–8):769–780

Satır O (2016) Comparing the satellite image transformation techniques for detecting and monitoring the continuous snow cover and glacier in Cilo mountain chain Turkey. Ecol Ind 69:261–268

Şengör AMCA, Yazıcı M (2020) The aetiology of the neotectonic evolution of Turkey. Med Geosc Rev 2:327–339. https://doi.org/10.1007/s42990-020-00039

Şengör AMCA, Yilmaz Y (1981) Tethyan evolution of Turkey: a plate tectonic approach. Tectonophysics 75:181–241
Şimşek M, Mutlu M, Poyraz M, Öztürk MZ (2019) Surface karst geomorphology of the Mt. Geyik and relationship between glacial and karst geomorphologies in the mount. Aegean Geogr J 28(2):97–110
Taştan F, Yılmaz FH (2015) Türki̇ye meteorolojik parametreleri̇ni̇n i̇stati̇sti̇ksel anali̇zi̇ (1970–2015). Orman ve Su i̇işleri̇ Bakanliği Meteoroloji Genel Müdürlüğü Yayınları. Ankara
Tunusluoglu MCA, Gokceoglu CA, Nefeslioglu HA, Sonmez H (2008) Extraction of potential debris source areas by logistic regression technique: a case study from Barla, Besparmak and Kapi mountains (NW Taurids, Turkey). Environ Geol 54(1):9–22
Tunusluoglu MCA, Gokceoglu CA, Sonmez H, Nefeslioglu HA (2007) An artificial neural network application to produce debris source areas of Barla, Besparmak, and Kapi Mountains (NW Taurids, Turkey). Nat Hazard 7(5):557–570
Türkeş M, Erlat E (2005) Climatological responses of winter precipitation in Turkey to variability of the North Atlantic Oscillation during the period 1930–2001. Theor Appl Climatol 81:45–69
Türkeş M, Öztürk MZ (2008) Uludağ periglasiyal jeomorfolojisi. Ulusal jeomorfoloji Sempozyumu, pp 387–395
Türkeş M, Öztürk MZ (2011) Uludağ'da Girland ve Çember Oluşumları (Garland and Circle Formations on Uludağ). Coğrafi Bilimler Dergisi / Turkish J Geogr Sci 9(2):239–257. https://doi.org/10.1501/Cogbil_0000000127
Turoğlu H (2009) Aksu Deresi Havzası (Giresun) Periglasiyal Sahasında Kütle Hareketleri. Türk Coğrafya Dergisi 52:41–54
Ünal A (2013) Geographic distribution of Rock Glaciers in Turkey: the case study of Erciyes Rock Glacier. MSca. Thesis, Fatih University, 56 p
Ünal A, Sarıkaya MA (2013) Erciyes kaya buzulunun uzaktan algılama yöntemiyle incelenmesi (2001–2012). Coğrafyacılar Derneği Yıllık Kongresi Bildiriler Kitabı 19–21 Haziran 2013, Fatih Üniversitesi, İstanbul, pp 1–5
Utlu M, Öztürk MZ, Şimşek M (2020a) Emli vadisindeki (Aladağlar) talus depolarının kantitatif analizlere gore incelenmesi. In: Birinci S, Kaymaz ÇK, Kızılkan Y (eds) Coğrafi perspektifle dağ ve dağlık alanlar, pp 51–72
Utlu M, Öztürk MZ, Şimşek M (2020b) Rockfall analysis based on UAV technology in Kazıklıali Gorge, Aladağlar (Taurus Mountains, Turkey). Int J Environ Geoinformatics 7(3):239–251
Yavaşlı DD, Tucker CAJ, Melocik KA (2015) Change in the glacier extent in Turkey during the Landsat Era. Remote Sens Environ 163:32–41
Yeşilyurt S, Doğan U (2010) Munzur Dağları'nın buzul jeomorfolojisi: CBS ve uzaktan algılama yöntemleri ile bir değerlendirme, TÜCAUM VI. Ulusal Coğrafya Sempozyumu, Ankara, pp 287–288
Yeşilyurt S, Doğan U, Akçar N (2018) Traces of late Quaternary glaciations in the Narlıca Valley, Kavuşşahap Mountains. Türk Coğrafya Dergisi 70:99–108. https://doi.org/19.17211/tcd.415232
Zahno CA, Akçar N, Yavuz V, Kubik PW, Schlüchter CA (2009) Surface exposure dating of Late Pleistocene glaciations at the Dedegöl Mountains (Lake Beyşehir, SW Turkey). J Quat Sci 24:1016–1028
Zahno CA, Akçar N, Yavuz V, Kubik PW, Schlüchter CA (2010) Chronology of Late Pleistocene glacier variations at the Uludağ Mountain, NW Turkey. Quatern Sci Rev 29:1173–1187
Zorn M, Komac B, Carey A, Hrvatin M, Ciglič R, Lyons B (2020) The disappearing cyrosphere in the Southeastern Alps: introduction to special issue. Acta Geogr Slov 60–2(2020):109–124
Zreda M, Çiner A, Sarıkaya MA, Zweck CA, Bayarı S (2011) Remarkably extensive Early Holocene glaciation in Aladağlar Central Turkey. Geology 39(11):1051–1054. https://doi.org/10.1130/G32097.1

The Mediterranean Islands

Mauro Guglielmin

1 Geographical Framework

The Mediterranean Sea includes several archipelagos and some major islands, five of which exceed 8000 km^2 (Sicily, Sardinia, Cyprus, Corsica and Crete in order of extension, Fig. 1). The main islands are very mountainous, with elevations exceeding 1800 m a.s.l. (hereafter as a.s.l. unless otherwise is stated) in Sicily (Mt. Etna, 3350 m), Corsica (Mt. Cinto, 2706 m), Crete (Lefka Ori, 2452 m), Cyprus (Mt. Tróodos, 1953 m), Sardinia (Mt. Gennargentu, 1834 m).

Lithologies are highly variable, with abundant limestones, granites and also volcanic rocks in the case of the highest elevation of the Mediterranean Sea, the Mt. Etna. Tectonically, Sardinia and Corsica are mainly characterized by the large Hercynic granite batolite while the other islands generally show a much more complicated tectonic arrangement related to the hercynian or alpine phases. Climate is generally mild and quite dry, with the altitude of the 0 °C mean annual air temperature (MAAT) isotherm increasing from N to S and from W to E. MAAT at Mt. Cinto is very close to the 0 °C (0.3 °C, at 2700 m), while it is ca. 400 m higher at Mt. Etna (3180 m). Precipitation values range between <500 mm (i,e. inner areas of Sicily and Sardinia) and >1500 mm on top of Mt. Etna (French and Guglielmin 2021). Glacial evidence is reported in Corsica, on Mt. Etna in Sicily, and in the two main mountains areas of Crete: Mt. Lefka Ori (2453 m) and Mt. Idi (2456 m) although only the former has been dated at 18 ka cal BP (Kuhlemann et al. 2008).

Considering modern climate, Haeberli et al. (1993) proposed the possible existence of permafrost at the summit surfaces of both Mt. Etna and Mt. Cinto. According to French (2017), the zonification of periglacial activity (it may occur below the + 3 °C MAAT isotherm), the periglacial processes may theoretically prevail at elevations ranging from 2160 m in the East to 2750 m on Mt. Etna in Sicily to the West.

M. Guglielmin (✉)
Department of Theoretical and Applied Sciences, Insubria University, Varese, Italy
e-mail: mauro.guglielmin@uninsubria.it

M. Oliva et al. (eds.), *Periglacial Landscapes of Europe*,
https://doi.org/10.1007/978-3-031-14895-8_8

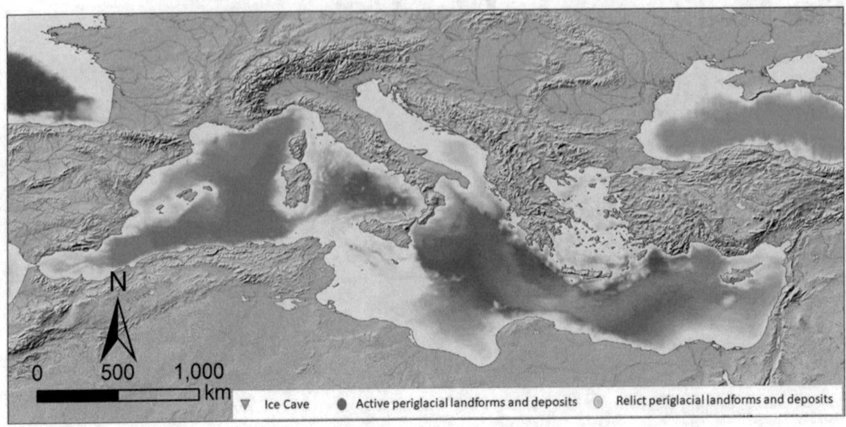

Fig. 1 Distribution of the ice caves as well as active and inactive periglacial landforms and deposits discussed in this chapter

Therefore, under the current climatic conditions, we can conclude that only a few massifs in Corsica (Mt. Cinto, Mt. Rotondo and Mt. Resinoso) and Sicily (Etna) are located within the current periglacial zone.

2 The Development of Periglacial Research in the Mediterranean Islands

While past glacial activity in Mediterranean islands has received some attention (e.g. Hughes et al. 2006, Hughes and Woodward 2017), periglacial processes have been almost neglected. Only recently, a synthesis on permafrost conditions in the Mediterranean region provided some information about past and present periglacial dynamics in the main Mediterranean islands (Oliva et al. 2018).

During the second half of the 20th century, periglacial research in the Mediterranean islands has demonstrated the existence of periglacial deposits of Pleistocene age in many regions, although the interpretation of some features still remains unclear. The occurrence of periglacial features such as stratified slope deposits, block streams, block slopes and debris flows has been reported in many islands, namely the Balearic Islands (Rosselló, 1977), Sardinia (Ginesu, 1990; Ginesu and Sias 2006), Sicily (Hugonie 1982; Agnesi et al. 1993) and Elba (Bortolotti et al., 2015), and patterned ground features have been described by Poser (1957) in Crete.

Ice caves—associated with permafrost conditions well below the lower periglacial belt in southern Europe (Colucci and Guglielmin, 2019)—were already described in the 1990s on Mt. Etna (Marino 1992; Hughes and Woodward 2009; Scoto et al. 2016), and more recently on Mt. Lefka Ori in Crete (Pennos et al., 2018). Moreover, granitic weathering features are also very common in both periglacial and cryotic environments, with a typical cavernous weathering called tafoni (Tuckett, 1884). Such features are included within the azonal landforms in French's reference handbook "The Periglacial Environment" (French 2017), and that is the reason why these features are presented here. That book included several examples from Corsica and Sardinia, which were some of the first areas of the world where such landforms were studied (i.e. Tuckett 1884; Brandmeier et al. 2011; French and Guglielmin 2021).

3 Distribution and Chronology of Periglacial Landforms

3.1 Active Periglacial Landforms and Deposits

On Mt. Etna, above 2900 m there is the only permafrost—frozen ground occurrence, which consists in icy layers formed within nivo-aeolian deposits (Oliva et al. 2018). Periglacial conditions can be found in Corsica three different massifs—from N to S: Mt. Cinto, Mt. Rotondo and Mt. Resinoso—present active slope deposits and debris flows generally above 2100 m. Patterned ground is still probably active and occur above 2230 m and sorted stripes have been observed in Mt. Rotondo at 2300 m, on the southern slope, close to the summit. Conversely, on the northern slope there is an incipient rock glacier, whose front rests at 2175 m. On the northern slope of the Mt. Cinto, a probably active rock glacier has its front at 2335 m with widespread solifluction lobes at ca. 2400 m. Southward on the northern slope of the Mt. Resinoso, solifluction lobes developed at ca. 2200–2300 m. According to the existing literature, no research on these features has been carried out until now and therefore these observations are original but based only on visual inspection of Google Earth imagery and thus needs further confirmation through field observation and geophysical investigations (Fig. 3a).

At present, there are still some ice caves in the north-western slope of Mt. Etna volcano at 2030–2040 m, the most famous one being the "*Grotto del Gelo*" (Cave of Frost). The cave is a 120-m-long and 7-m wide lava tube that can be accessed through a 10 × 5 m wide entrance at its upper end (Santagata et al. 2017). The lava tube formed during the historic long-lasting eruption of 1614–1624 CE. Despite its geological setting and latitude, after about 20 years from the last phases of the eruption—coinciding with the Minimum Maunder, 1645–1715 CE—subterranean freezing inside the cave started to occur. Beside the development of seasonal ice formations (seasonal lake ice, ice stalactites, stalagmites and columns generally located close to the entrance), Marino (1992) stated that some ice was forming also up to 1981 CE. Subsequently, it underwent a strong shrinkage probably due to the strong and rapid air warming but also to the 1981 eruption, whose vent broke out at a few tens of metres from the cave, modifying the inner temperature. Despite its likely origin during the Little Ice Age (LIA), some perennial ground ice is still present in the deepest zone with an ice extension of ca. 240 m^2, and volume lower than 270 m^3 (Scoto et al., 2016).

In addition, over 100 ice caves have been also reported in Crete on Mt. Lefka Ori (Pennos et al., 2018). Among these, the biggest is Skud (Speleo 2017), located on the northern slope of the mountain at 1915 m, with a 46-m-deep vertical pit that hosts a 15-m-thick ice deposit at its bottom (Pennos et al., 2018). Although the ice body is fully shielded from solar radiation, it is reasonable to think that ice is not forming under the recent climatic conditions, thus being relict, likely formed during the LIA. The Skud cave represents the southernmost European ice cave, at 35° N (Pennos et al., 2018).

3.2 *Relict Periglacial Landforms and Deposits*

In the Balearic Islands, there is evidence of periglacial activity during the last glacial cycle, with scree deposits located at elevations above 1200 m in the Tramuntana limestone massif, Mallorca (Rosselló, 1977). No evidence of permafrost conditions has been detected at this altitude on this island (Fig. 2).

In Sardinia, there are numerous landforms associated with past periglacial, or even cryotic conditions mainly in the northern and central part of the island. Stratified slope deposits were found in the "costiera Turritana" northward to Sassari at very low altitudes (below 500 m; Federici et al. 1987), while several basaltic block fields and block streams were described by Ginesu (1990) on the northern slopes of the Pranu Mannu plateau at altitudes of ca. 600 m, in the central part of the island

Fig. 2 Examples of scree deposits located at elevations above 1200 m in the limestone Tramuntana massif (photos kindly provided by Francesc X. Roig Munar)

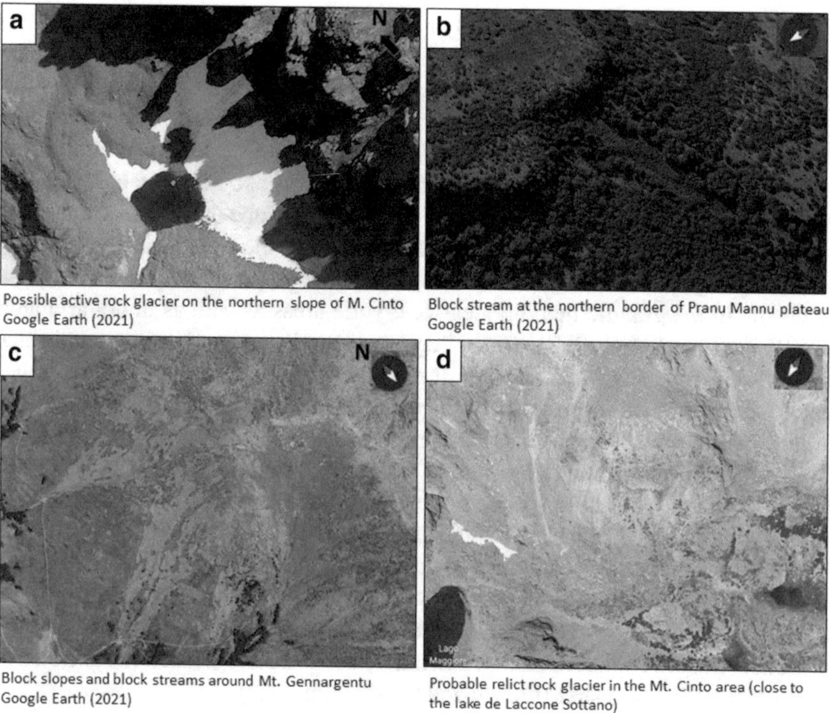

Fig. 3 Examples of Google Earth views of different periglacial landforms: (**a**) possible active rock glacier on the northern slope of M. Cinto; (**b**) block streams located along the northern border of the of Pranu Mannu plateau; (**c**) block slopes and block streams around Mt. Gennargentu; and (**d**) probable relict rock glacier located in the Mt. Cinto area (close to the lake de Laccone Sottano)

(Fig. 3b). Ginesu and Sias (2006) suggested also a periglacial origin for other block accumulations on different lithologies in the higher relief of Sardinia (Gennargentu, Limbara, Perdasdefogu) at elevations exceeding 600 m. A couple of these are block slopes that originate block streams on N-NW slopes of the Mt. Gennargentu, reaching an altitude of 1320 m (Fig. 3c). More recently, Ginesu et al. (2014) interpreted some block deposits found on Asinara Island also as block streams. In this case, block streams are composed by granitic subrounded blocks located close to the coast and even below the present sea level (4 m deep) in Cala Arena.

In Corsica, inactive debris flows and slope deposits occur at several places, but they are almost completely unstudied. On the southern slope of the Mt. Cinto, a quite large (ca. 120 × 400 m) debris accumulation closes the lake "de Laccone Sottano". Such accumulation shows some flow structures that may be associated with a relict rock glacier whose front reaches 2170 m (Fig. 3d). Kuhlemann et al. (2009) recorded block fields and tors above 2200 m, interpreted as possible periglacial landforms.

Stratified slope deposits (*èboulis ordonnès*) of Late Glacial age have been reported in various locations of Sicily (Agnesi, 2004), and in some places like the Aegadi islands and the Palermo Mounts overlying beach deposits of Upper Pleistocene age (Agnesi et al. 1993; Hugonie 1982).

Periglacial landforms have also been detected in Crete, where Poser (1957) found patterned ground phenomena and solifluction lobes above 1800 m. The author highlighted that the lower limit of periglacial activity in this island was represented by boulder pavements extending down to elevations of 800 m. Cemented stratified scree of unknown age are also present on the southern slope of Mt. Idi (Guglielmin, unpublished), probably related to the coldest phases of the last glacial cycle.

In the periglacial environment, cavernous weathering phenomena are widespread especially in the granitic outcrop areas of cryotic areas as well as in hot arid areas (French, and Guglielmin 2021) where they are generally active (i.e. Guglielmin et al. 2005; Strini et al. 2008). Therefore, the widespread distribution of these (relict) features in the islands suggests the occurrence of cryotic or periglacial environments during the Pleistocene or azonal phenomena as suggested recently by French and Guglielmin (2021). Among cavernous weathering, tafoni are the most impressive features, characterised by a typical hollow with concave inner walls, overhanging visors and a gently sloping debris-covered floor (Turkington and Phillips 2004). They are widespread especially in Corsica (where the name "Tafone" was created) and in Sardinia, where Variscan granites crop out regardless the distance from the current shoreline. These cavernous weathering landforms are mainly related to the salt action, but are enhanced also by thermal stress and biological action and also by the cryofracturing of quartz (French and Guglielmin 2002).

Active tafoni occur almost everywhere close to the Sardinia and Corsica coasts, where granites crop out normally at elevations lower than 35 m (Fig. 4a). They appear more rounded and eroded at the lowest elevations, being directly washed by the waves during the storms (French and Guglielmin 2021; Fig. 4b). It is remarkable that in several places at low elevations (ca. 25 m) tafoni appeared buried by aeolianites of Pleistocene age (Coltorti et al., 2010, Fig. 4c). Some tafoni were also found at or below the current sea level (Fig. 4d). The occurrence of Tafoni under the present sea level suggest a probable triggering of these features during the Last Glacial Maximum in periglacial conditions related to an arid and cold environment. According to French and Guglielmin (2021), tafoni are still active in Sardinia under the current climate conditions below an altitude of 6 m, with a likely maximum growth rate of ca. 26 mm ka^{-1}, lower than the 40 mm ka^{-1} that Brandmeier et al. (2011) calculated from sites at higher elevation (150 m) on the neighbouring Corsica Island. In such island, Brandmeier et al. (2011) made an extensive analysis of tafoni distributed almost everywhere from the coast up to almost 2000 m and found no statistical correlation between altitude, distance to the sea and size of the weathering form. However, they did find a decrease of the weathering rate between the coast to the inland massifs and with the increase of the elevation and the exposure time calculated by means of cosmogenic nuclides. On the contrary, in Sardinia there is a general decrease of their activity from the coast to towards the interior of the island (French and Guglielmin 2021).

In Sardinia, relict or at least inactive tafoni are quite widespread in almost all areas where granites outcrop. In particular, it is remarkable the example described

Fig. 4 Examples of active tafoni located along the shoreline at Capo Testa in the northern Sardinia; (**a**) Tafone located at ca 35 m and still active although a thick weathering rind on the top suggests a quite old age of formation; (**b**) tafone at low elevation (<4 m) appear eroded and "fresh"; (**c**) tafone sealed by Pleistocene aeolianites; and (**d**) partially submerged tafone during low tide (in **c** and **d**, H.M. French stands as scale)

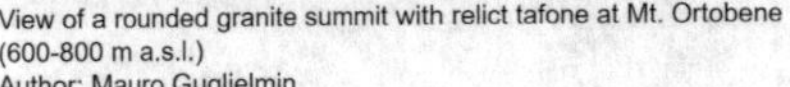

Fig. 5 Examples of relict tafoni at Mt. Ortobene at 600–800 m: (**a**) view of a rounded granite summit with tafone; (**b**) detail of several tafoni that are visible within the vegetation cover

by French and Guglielmin (2021), located on the Mount Ortobene massif (40°32 N, 9°33 E) at altitudes ranging from 600 to 800 m (see Fig. 5a). Here the bedrock consists of a monzogranite of Carboniferous age with large K-feldspars fractured along two main regional directions (N–S and E–W). The weathering hollows show various sizes, from decimetrical to several metres long, but they are on average deeper and larger than along the coast. Moreover, they are partially covered by shrub-forest vegetation, lichens and mosses, and the weathering rind is seen quite thick (>1 cm) although rock varnish is absent (Fig. 5b). Tafoni are also very common in the Elba Island and some Aegean Islands (Hejl, 2005).

4 The Significance of Periglacial Dynamics in The Mediterranean Mountains Within the European Context

Despite their reduced extent, the occurrence of (relict) periglacial landforms—including some potential permafrost indicators—have a remarkable paleoclimatic significance suggesting the prevailing climatic conditions during cold-climate periods during the Late Pleistocene in the Mediterranean islands. However, periglacial phenomena in these islands have been little addressed to date. Some features, such as the dating of the tafone could provide insights on the spatial patterns of the Pleistocene palaeoclimate in southern Europe, as they have been defined as indicators of the past aridity even in maritime locations. In addition, research on the temperature and moisture conditions in the ice caves existing in Crete and Sicily, together with the timing of ice accumulation, can provide new knowledge on the impact that the periglacial and cryotic conditions prevailing during the LIA had in the southern limits of the periglacial belt in Europe.

Acknowledgements I would like to thank Dr. Francesc X. Roig Munar for providing the pictures of slope deposits in the Balearic Islands and Dr. Stefano Ponti to reassemble the figures of this manuscript.

References

Agnesi V (2004) Sicily in the last one million years. Bocconea 17:23–33

Agnesi V, Macaluso T, Orrù P, Ulzega A (1993) Paleogeografia dell'arcipelago delle Egadi (Sicilia) nel Pleistocene superiore/Olocene. NatSic 17(1–2):3–22

Bortolotti V, Pandeli E, Principi G (2015) Geological map of Elba Island 1:25,000 Scale, Dream, Italia.

Brandmeier JM, Kuhlemann J, Krumrei I, Kappler A, Kubik PW (2011) New challenges for tafoni research. A new approach to understand processes and weathering rates. Earth Surf Process Landf 36:839–852

Coltorti M, Melis E, Patta D (2010) Geomorphology, stratigraphy and facies analysis of some late Pleistocene and Holocene key deposits along the coast of Sardinia (Italy). Quat Int 222:16–30

Colucci RR, Guglielmin M (2019) Climate change and rapid ice melt: Suggestions from abrupt permafrost degradation and ice melting in an alpine ice cave. Prog Phys Geogr 43(4):561–573

Federici PR, Ginesu S, Oggiano G (1987) Genesi ed evoluzione della pianura costiera turritana (Sardegna settentrionale). Geogr Fis Dinam Quat 10:103–121

French HM (2017) The periglacial environment, 4th edn. John Wiley and Sons, Chichester, UK

French HM, Guglielmin M (2002) Observations on Granite Weathering Phenomena, Mount Keinath, Northern Victoria Land, Antarctica. Permafrost Periglac Process 13:231–236

French H, Guglielmin M (2021) Tafoni weathering is an azonal process: examples from Antarctica. Sardinia and Australia. Geomorphology 375:107556

Ginesu S (1990) Periglacial deposits in Sardinia: the blockstreams near Pranu Mannu. Geogr Fis Din Quat 13(2):179–181

Ginesu S, Sias S (2006). La Sardegna. In: Chelli A, D'Aquila P, Firpo M, Ginesu S, Guglielmin M, Pappalardo M, Pecci M, Piacentini T, Queirolo C, Robustelli G, Scarciglia F, Sias S, Tellini C (eds) Testimoni di una montagna scomparsa. Contributo alle metodologie d'indagine delle forme periglaciali relitte. Problematiche e applicazioni in differenti ambienti morfodinamici. Collana «Quaderni della Montagna», vol 8. Bologna: Bonomia University Press, p 73–80

Ginesu S, Carboni D, Congiatu PP (2014) Relict forms of disappeared mountain. The periglacial deposits in Asinara Island-Sardinia (Italy). J Environ Tour Anal 2(1):19–32

Guglielmin M, Cannone N, Srini A, Lewkowicz AG (2005) Biotic and abiotic processes on granite weathering landforms in a cryotic environment, Northern Victoria Land. Antarctica Permafr Periglac Process 16:69–85

Haeberli W, Guodong C, Gorbunov AP, Harris SA (1993) Mountain permafrost and climatic change. Permafrost Periglac Process 4:165–174

Hejl E (2005) A pictorial study of tafoni development from the 2nd millennium BC. Geomorphology 64:87–95

Hughes PD, Woodward JC (2009) Glacial and periglacial environments. In: Woodward JC (ed) The physical geography of the Mediterranean. Oxford University Press, Oxford, pp 353–383

Hughes PD, Woodward J (2017) Quaternary glaciation in the Mediterranean mountains: a new synthesis. In: Hughes P, Woodward J (eds) Quaternary glaciation in the Mediterranean mountains, geological society of London Special Publication, 433, p 1–23

Hughes PD, Woodward JC, Gibbard PL (2006) Quaternary glacial history of the Mediterranean mountains. Prog Phys Geogr 30(3):334–364

Hugonie G (1982) Mouvementes tectoniques et variation de la morphogenese au Quaternaire en Sicilie septentrionale. Revue De Geologie Dynamique Et Geographie Physique 23:3–14

Kuhlemann J, Rohling EJ, Krumrei I, Kubik P, Ivy-Ochs S, Kucera M (2008) Regional synthesis of Mediterranean atmospheric circulation during the last glacial maximum. Science 321:1338–1340

Kuhlemann J, Milivojević M, Krumrei I, Kubik PW (2009) Last glaciation of the Šara Range (Balkan Peninsula): Increasing dryness from the LGM to the Holocene. Austrian J Earth Sci 102:146–158

Marino A (1992) Nota preliminaire sul fenomeno glaciologico della Grotta del Gelo (Mt. Etna). Geogr Fis Din Quat 15:127–132

Oliva M, Žebre M,Guglielmin M. et al. (2018) Permafrost conditions in the Mediterranean region since the last Glaciation. Earth-Science Reviews 185:397–436

Pennos C, Styllas M, Sotiriadis Y, Vaxevanopoulos M (2018) Ice caves in Greece. In: Perşoiu A, Lauritzen SE (eds) Ice caves 2018. Elsevier, p 385–397

Poser J (1957) Klimamorphologische Probleme auf Kreta. Z Geomorphol 2:113–142

Rosselló V (1977) Screes periglaciares en la montaña mallorquina. V Coloquio de Geógrafos Españoles. AGE-Universidad de Granada, Granada, p 83–92

Santagata T, Vattano M, Sauro F, Spitaleri G, Giudice G, Corrado Bongiorno C (2017) Inside The Glaciers Project: Laser scanning of the Grotta Del Gelo (Mt. Etna, Italy). 17th International Congress of Speleology. Sydney.

Scoto F, Giudice G, Randazzo, L, Maggi, V (2016) Air circulation model and topographic survey of the "grotto del gelo", Mt. Etna Italy. In: Mihevc A, Zupan Hajna N, Gostinčar P (eds) Proceedings of the 7th International Workshop on Ice Caves, p. 71–73

Strini A, Guglielmin M, Hall K (2008) Tafoni development in a cryotic environment: an example from Northern Victoria Land, Antarctica. Earth Surf Process Landf 33:1502–1519

Tuckett F (1884) Notes on Corsica. Alp J 11:449–459

Turkington A, Phillips JD (2004) Cavernous weathering, dynamical instability and selforganization. Earth Surface Process and Landforms 29:665–675

Periglacial Landforms in Central and Eastern Europe

European Alps

Andreas Kellerer-Pirklbauer, Isabelle Gärtner-Roer, Xavier Bodin, and Luca Paro

1 Geographical Framework

1.1 Background and Delimitation of the Alps

Describing the periglacial landscapes of a given area requires a sound delimitation of it. Such a delimitation is essential in this contribution as we describe, interpret, quantify, and discuss different types of spatial data related to the periglacial realm in the European Alps (for simplicity hereafter termed as "Alps"). Delimiting the Alps in a spatial sense is, however, not straightforward. Possibilities for delimitation are manifold, either using physiographical (e.g., topography, elevation, geology) or economical-political criteria based on administrative boundaries. Commonly, a mixed delineation approach is used based on both criteria groups (Price et al. 2004; Veit 2002; Bätzing 2015). In this contribution we use the delimitation by the Alpine Convention (2020) accepted by all eight-member countries covering an area of 190,900 km^2, with a north–south expansion of 5° (43°29' to 48°21'N), a west–east

A. Kellerer-Pirklbauer (✉)
Institute of Geography and Regional Science, Cascade—The Mountain Processes and Mountain Hazards Group, University of Graz, Graz, Austria
e-mail: andreas.kellerer@uni-graz.at

I. Gärtner-Roer
Department of Geography, University of Zurich, Zurich, Switzerland
e-mail: isabelle.roer@geo.uzh.ch

X. Bodin
Laboratoire EDYTEM, CNRS, Université Savoie Mont Blanc, Chambéry, France
e-mail: xavier.bodin@univ-smb.fr

L. Paro
Department of Natural and Environmental Risks—Unit of Monitoring and Geological Studies, Regional Agency for Environmental Protection (ARPA) of Piemonte, Turin, Italy
e-mail: luca.paro@arpa.piemonte.it

M. Oliva et al. (eds.), *Periglacial Landscapes of Europe*,
https://doi.org/10.1007/978-3-031-14895-8_9

one of 11.5° (4°55' to 16°36'E), extending for some 1200 km, and with a maximum width of 280 km (Fig. 1A). Regarding neighbouring mountain systems, at Colle di Cadibona the Alps are linked to the Apennines, via the Vienna Basin with the Carpathians, and by the approximate line from Maribor via Ljubljana and Postojna to Gorizia with the Dinarides (Fig. 1A). We further subdivided the Alps into the two "classical" parts, Western Alps (42.2% of the total area) and Eastern Alps (57.8%) as well as into five major sectors Southern-Western Alps (hereafter abbreviated as SWA; 20.3%), Northern-Western Alps (NWA; 21.9%), Central-Eastern Alps (CEA; 20.2%), Northern-Eastern Alps (NEA; 18.6%), and Southern-Eastern Alps (SEA; 19.0%) following the classification schemes by Grassler (1984) and Marazzi (2005).

1.2 General Geographic Setting of the Alps

The Alps are the most prominent and extensive mountain range system entirely located in Europe. The highest mountain is Mont Blanc or Monte Bianco reaching 4807.8 m a.s.l. (measured by the *Ordre des géomètres-experts de Haute-Savoie* in September 2021) situated basically on the border of France and Italy. Some 130 peaks of the Alps exceed an elevation of 4000 m. The mountain range reaches sea level at the Mediterranean Sea in Eastern-most Italy (Monfalcone; Fig. 1A) as well as along the French, Italian and Monegasque coast in the west. From a hypsometric point of view, 29 km^2 (0.01%) of the entire Alps exceed 4000 m, whereas 2,325 km^2 (1.2%) and 33,230 km^2 (17.4%) are in the range of <4000–3000 and <3000–2000 m, respectively. The hypsometric curves for the five major sectors of the Alps (Fig. 1B) reveal that the highest percentage of areas exceeding 2000 m is found in the CEA (30.9%) followed by the NWA (29.1%). In contrast, only 3.9% of the NEA exceed this elevation. On average for the entire Alps, 18.7% or 35,584 km^2 exceed an elevation of 2000 m.

The Alps emerged by plate collision during the Alpine orogenesis, a mountain-building period that began about 65 million years ago. Between 135 and 15 million years ago, the tectonic nappes of the Alps formed in relation with a subduction zone until the Penninic ocean was completely overridden by the north-directed Adriatic plate. The major present-day tectonic units of the Alps (South Alpine, Eastern Alpine, Penninic, Helvetic) reflect this evolution. Each unit is characterised by Mesozoic-to-Neogene-aged cover sediments such as limestone or dolomite and a prealpine crystalline basement (Fitzsimons and Veit 2001; Veit 2002). From a geological point of view, the Alps are, therefore, rather young. Today's surface relief, as we see it in the Alps, can largely be attributed to deformation, thrusting, and erosion processes that emerged particularly during the last millions of years.

Despite the complex geological and tectonic structure in the Alps, a relatively clear zonal arrangement of the rock zones and lithology emerges on a north–south profile—especially in the Eastern Alps. This lithological pattern was already recognised by Johann Wolfgang von Goethe on a trip to Italy in 1786 (Wolff 1986). The symmetry, well visible in the Eastern Alps, consists in a crystalline central zone widely lacking

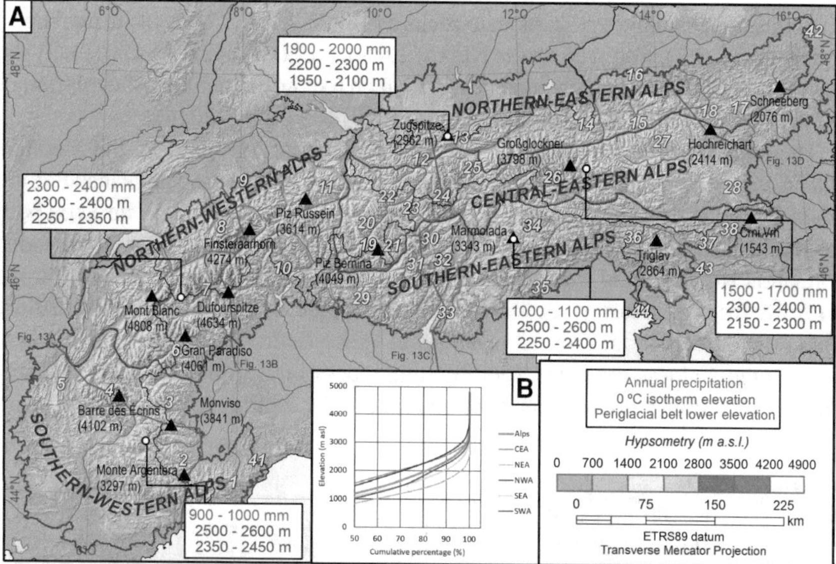

Placenames referred in the text

SWA: 1 Ligurian Alps. 2 Maritime Alps: Argentera Massif, Nice Prealps. 3 Cottian Alps: Queyras Massif. 4 Dauphiné Alps: Taillefer Massif, Écrins Massif, Combeynot Massif. 5 Dauphiné Prealps. NWA: 6 Graian Alps: Mont Blanc Massif, Lanzo Massif, Vanoise Massif, Tarentaise Massif. 7 Valais Alps/Pennine Alps. 8 Bernese Alps. 9 Bernese Prealps. 10 Lepontine Alps. 11 Glarner Alps. NEA: 12 Lechtal Alps. 13 Wetterstein Mountains. 14 Berchtesgaden Alps: Steinernes Meer. 15 Dachstein Massif. 16 Upper Austrian Prealps. 17 Hochschwab Mountains. 18 Ennstal Alps. CEA: 19 Bernina Alps. 20 Albula Alps. 21 Livigno Alps 22 Silvretta. 23 Sesvenna Alps. 24 Ötztal Alps. 25 Stubai Alps. 26 Hohe Tauern Range: Ankogel Mountains, Glockner Mountains, Schober Mountains, Großvenediger Mountains, Goldberg Mountains. 27 Niedere Tauern Range: Seckauer Tauern, Schladminger Tauern. 28 Lavantal Alps: Packalpe, Gleinalpe, Koralpe. SEA: 29 Bergamo Alps. 30 Ortles-Cevedale Group. 31 Adamello-Presanella Alps. 32 Brenta Mountains. 33 Garda Mountains. 34 Dolomites. 35 Venetian Prealps. 36 Julian Alps. 37 Kamnik Alps/Steiner Alps. 38 Karawanks and Pohorje.
Sites: 41 Colle di Cadibona. 42 Vienna Basin. 43 Ljubljana. 44 Monfalcone

Fig. 1 Overview of the European Alps. (A) Topography and delineation of the European Alps with its five major sectors Southern-Western Alps (SWA), Northern-Western Alps (NWA), Central-Eastern Alps (CEA), Northern-Eastern Alps (NEA), and Southern-Eastern Alps (SEA). Local names mentioned in the text are indicated); (B) hypsographic curves of the entire Alps and its five sectors. The map also depicts climatic means based on the climate normal 1981–2010 (annual precipitation and elevation of the 0 °C isotherm derived from air temperature using a 0.6 °C/100 m lapse rate) and the minimum elevation of the present-day periglacial belt for five selected locations (one per major sector): Sonnblick/CEA at 3109 m (source ZAMG), Zugspitze/NEA at 2960 m (German Weather Service), Pian Fedaia, Marmolada/SEA at 2063 m (Climatrentino), Col du Grand St-Bernard/NWA at 2472 m (MeteoSwiss) and, Saint-Paul-sur-Ubaye/SWA at 1908 m. (Météo France). The minimum altitude of the present-day periglacial belt is based on the potential treeline (see text). Data sources: ESA 2021 (topography), Alpine Convention (perimeter of Alps)

carbonate rocks (CEA), and north (NEA) and south (SEA) adjoining sedimentary zones with the widespread existence of carbonate rocks. Such rocks are favourable for cold-climate karst landforms, which are substantially different to non-karstic environments. Figure 2A depicts the distribution of continuous and discontinuous carbonate rocks in the Alps based on Goldscheider et al. (2004). As shown by this generalised map of karstifiable rocks, it can be stated that some 52.4% of the Alps consist of carbonate rocks, either continuously (41.2%) or discontinuously (11.2%).

The highest percentages of carbonate rocks are in the SEA (77.6% of its total area) and SWA (68.8%), followed by the NEA (61.4%), the NWA (42.8%), and finally the CEA (14.4%). For a more detailed description on tectonic structure and evolution of the Alps refer to, e.g., Schuster and Stüwe (2010) or Pfiffner (2014).

The Alps separate the marine west-coast climates of Europe from the Mediterranean areas of France, Italy, and the Balkan region related to the arch shape of the mountain range. In addition, the Alps create their own unique climate based on both the regional differences in elevation and relief, and the position of the mountains in relation to the frontal systems and thus mid-latitude cyclones that cross Europe generally from west to east. Prominent climatic features of the Alps include distinct climatic gradients in all three dimensions of space (latitude, longitude, and elevation). Based on these gradients, four different types of transitions are relevant: planetary (south-north: from Mediterranean climate in the south to the temperate climate

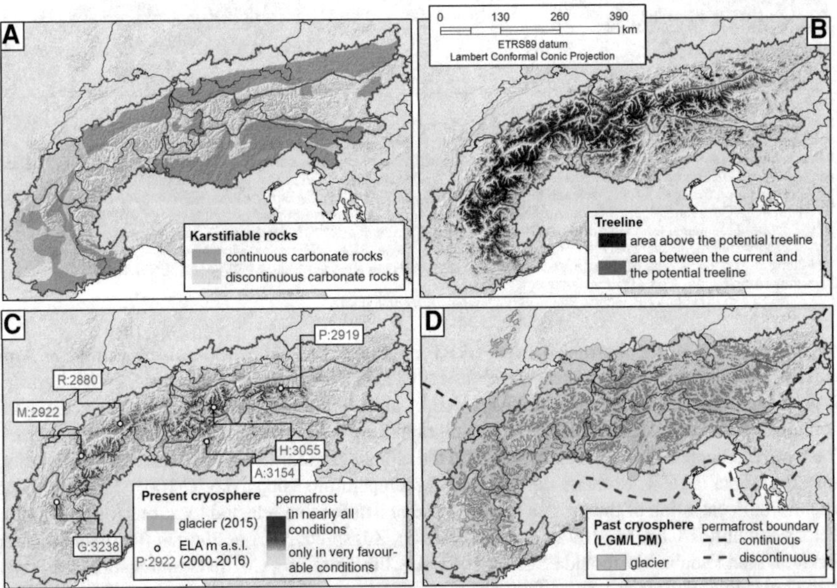

Fig. 2 Thematic maps of the Alps: (A) karstifiable rocks; (B) areas above the potential treeline as a proxy for the periglacial zone; (C) extent of glaciers (stage 2015), modelled permafrost at present, and the mean equilibrium line altitude (ELA) in the period 2000–2016 of selected glaciers: A = Adamello Glacier, G = Glacier de la Girose, H = Hintereisferner, M = Mer de Glace, P = Pasterze, R = Rhone; (D) extent of glaciers and the estimated southern boundaries of continuous and discontinuous permafrost in the Alps and its forelands during the Last Glacial Maximum (LGM) or Last Permafrost Maximum (LPM). Data sources (in addition to Fig. 1): Goldscheider 2004 (karstifiable rocks); Pecher et al. 2011 (potential treeline); Boeckli et al. 2012 (permafrost present); Paul et al. 2020 (glaciers present); Davaze et al. 2020 (ELA for selected glaciers); Ehlers and Gibbard 2004; Ehlers et al. 2011 (LGM glaciation); Van Vliet-Lanoë et al. 2004; Vandenberghe et al. 2014; Andrieux et al. 2016; Lehmkuhl et al. 2020 (permafrost boundaries during the LGM/LPM—25–17 ka BP)

in the north; greatest impact of the Alps on the near-surface air layer), west–east (from humid-oceanic climate to dry-continental at the eastern margin of the Alps), peripheral-central (humid-cool margin of the Alps versus dry-warm inner alpine climate), and hypsometric (causing a distinct altitudinal zonation) (Veit 2002).

Higher elevated areas are more exposed to radiation and experience, on average, lower temperatures and in general higher precipitation rates (at least at the Alpine margin). Furthermore, in the southern part of the Alps, the higher precipitation rates are in the prealpine regions. The north Piedmont and central-east Friuli Venezia Giulia are the rainiest regions of Italy. In detail, mean annual precipitation ranges from 500 mm in the Aosta plain and inner Alpine valleys to 3000 mm in some prealpine regions with the highest precipitation in summer and the lowest one in winter in the central and eastern part of the Italian Alps whereas two maxima in spring and autumn and a decrease of precipitation in summer and winter in the western Italian Alps (Crespi et al. 2017). The adiabatic lapse rate is higher in summer (0.7 °C/100 m) compared to winter (0.4 °C/100 m) where it might be even negative due to air temperature inversions (Schüepp et al. 1978). Mean annual air temperatures (MAAT) for different alpine sites (see Fig. 1 for location and data source) during the period 1981–2010 are 4.3 °C at 1908 m (Saint-Paul-sur-Ubaye/SWA), 3.1 °C at 2063 m (Pian Fedaia, Marmolada/SEA), −0.6 °C at 2472 m (Col du Grand St-Bernard/NWA), −4.3 °C at 2960 m (Zugspitze/NEA), and −5.1 °C at 3109 m (Sonnblick/CEA), respectively.

Freeze–thaw cycles have a substantial impact on physical weathering. The highest number of freeze–thaw cycles of air temperature in the Alps is, however, not located in the periglacial zone above the treeline but substantially lower at between 1400 (more humid Alpine margin) and 1700 m (rather dry central Alps) (Veit 2002). These numbers are valid for air temperature, but, in contrast, altitudinal trends of freeze–thaw cycles using ground temperature data are less clear because fewer time series are available for analysis and more complex interactions occur between different elements of local ground conditions, i.e., vegetation offset, nival offset, and thermal offset (French 2007; Kellerer-Pirklbauer 2019; Rist et al. 2020). The number of ice days (i.e., days with maximum temperature <0 °C) increase above the treeline characterising the climate in the periglacial belt. Both, the number of freeze–thaw cycles and the duration and strength of a freezing period are essential for near-surface weathering by freeze–thaw action if the volumetric-expansion model as well as the segregation-ice model are considered (e.g., Coutard and Francou 1989; Kellerer-Pirklbauer 2017; Draebing and Krautblatter 2019). For more details on the climate of the Alps refer to, e.g., Schär et al. (1998) or Veit (2002).

1.3 The Periglacial Belt in the Alps

The periglacial altitudinal zone in the Alps is characterised by periglacial processes and frost-action above the potential treeline but below the snowline (for discussion see, e.g., Höllermann 1985 and references therein, Veit 2002). Veit (2002) argued that a clear distinction must be made between periglacial landforms and periglacial processes as many of the solifluction-features just above the present-day treeline have been formed during cooler periods in the past and are inactive or even relict today. In contrast, periglacial landforms, which are active today and evolve further under current periglacial conditions, are located at higher elevations. However, the fact that even in present-day periglacial environments relict periglacial landforms are abundant and that process rates are often unclear complicates the delimitation of the lower limit of active periglacial landforms (Höllermann 1985). Morawetz (1971a) postulated that the lower limit of the periglacial belt in the Alps was at least 1500 m lowered during the coldest periods of the Pleistocene.

Solifluction landforms are commonly subdivided in terms of the presence or absence of vegetation at the riser and at the tread of the lobe or terrace. This absence or presence is used to differentiate between turf-banked and stone-banked features. Turf-banked forms comprise those with a complete vegetation cover (also termed as "bound" solifluction) and those that only have vegetation at the riser ("semi-bound") (Benedict 1970; Höllermann 1985; Matsuoka 2001). Höllermann (1985) argued that the present-day periglacial belt is divided into two sub-belts with indistinct boundaries. A lower sub-belt of turf-banked solifluction begins at the lower limit of active solifluction about 50–200 m above the present treeline with a vertical expansion of some 400–500 m. This sub-belt is characterised by the absence of active sorting processes. At higher elevations, an upper sub-belt of stone-banked solifluction and patterned ground is commonly developed marked by the lower limit of active patterned ground formation (Höllermann 1985). This is in line with the ecological differentiation of the periglacial belt in the Alps consisting of the alpine belt (ranging from the treeline to upper limit of the alpine meadow) and the sub-nival belt (above the upper limit of alpine meadow and below permanent snow and ice). According to Veit (2002), the upper boundary of the alpine belt is at 2400–2500 m at the margin of the Alps and at 2700–3000 m at its centre. Höllermann (1983) pointed out, nevertheless, that the distribution pattern of solifluction landforms of bound, semi-bound, and unbound types as well as small-patterned ground and nivation features are closely associated within the narrow periglacial belt and that the altitudinal zonation appears rather indistinct due to local effects. Höllermann (1985) concluded furthermore that the mean July temperature at the level of active solifluction comes close to 10 °C, thus, is at the upper level of potential tree growth. According to Lehmkuhl (2008), in the Western Alps and at a regional scale, the present lower-limit of turf-banked solifluction is at 2350 m and for stone-banked solifluction at 2750 m, respectively. The same two limit-values are 2200 and 2500 m, respectively, for the Eastern Alps.

Veit (2002) pointed out that in crystalline mountain regions, a belt of alpine grassland is typically found above the treeline followed by a belt dominated by gelifraction at higher elevations. In contrast, in the altitudinal gradient of alpine ecosystems in calcareous mountain regions, a krummholz belt follows the treeline, which itself is followed directly by a gelifraction belt. Alpine grassland is widely lacking in such regions apart from areas affected by human clearance. To sum up, for simplicity and because long-term monitoring data of periglacial landform evolution (apart from rock glaciers; cf. below) are widely lacking, we follow the suggestion by Veit (2002) and use the appearance of periglacial landforms above the treeline as the main criteria for the periglacial belt in the Alps. Therefore, the treeline might be used as a proxy of the lower limit of a presently active periglacial zone.

Pecher et al. (2011) performed a study about the potential and current treeline in the Alps. The latter is lowered related to land-use practises (alpine farming). The authors concluded that the mean elevation of the potential treeline at a regional scale is 2000 m at the margin and 2200 to 2350 m in central parts of the Alps. They further pointed out that the current treeline is on average about 350–400 m below the potential treeline in the central Alps whereas the difference is much smaller on the Alpine margin. Figure 2B depicts the distribution of areas above the potential treeline as well as the area between the current and the potential treeline. Results show that 12.0% of the entire Alps are above the potential treeline and further 15.4% are between the potential and the present treeline summing up to ca. 52,000 km^2. The highest percentage above the potential treeline is in the NWA (23.1%) followed by the CEA (19.1%). Less than 8% of the three areas of SWA, NEA, and SEA are below the treeline. The highest percentage of areas between the potential and current treeline are at the NWA (20.7%) and the CEA (20.1%) followed by the other three major sectors with values between 13.3% (SWA) and 10.0% (NEA) of the entire mountain groups or major sectors.

1.4 *The Cryosphere of the Alps at Present and During the LGM*

The upper limit of the periglacial belt is related to perennial snow and glacier ice, thus, to the climatic snow line and the lower extent of the alpine glaciers, respectively. The most recent alpine-wide glacier inventory dates to 2015–2017 and is based on Sentinel-2 data. Paul et al. (2020) derived a total glacier-covered area of 1806 $\pm$ 60 km^2 and considered 4395 glaciers larger than 0.01 km^2. Thus, only 0.95% of the entire Alps were covered by glaciers around 2015–2017. Figure 2C depicts this present distribution of glaciers in the Alps. Our spatial analysis of the data published by Paul et al. (2020) shows that by far the highest number of glaciers ($n = 2163$) covering an area of 1129 km^2 (62.5% of the total) are found in the NWA. The second largest area covered by glaciers is in the CEA with 498 km^2 aggregated by 1609 glaciers. Third comes the SEA with 308 glaciers covering 106 km^2. About 66 km^2 of the SWA are covered by 282 glaciers whereas only 6 km^2 (0.3% of the entire Alpine glaciated area) are glaciated in the NEA by 33 glaciers, which exceed the 0.01 km^2-threshold. The values presented in this paragraph imply that the upper periglacial belt is some extent controlled by glaciers in the two sectors NWA and CEA where they cover in total 1627 km^2. In contrast, due to the minor importance of glaciers in the other three sectors SWA, NEA, and SEA, the upper limit of the periglacial belt is mainly controlled by topography reaching its upper limit in the summit regions.

Perennially frozen (permafrost) areas, on the other hand, are substantially more widespread in the Alps than glaciers, and less regionally variable, according to the regional permafrost model APMOD (Alpine-wide Permafrost MODel) used to calculate the Alpine Permafrost Index Map (APIM) (Boeckli et al. 2012). The permafrost index is mainly based on documented permafrost evidence (Cremonese et al. 2011) and provides an indication of possible permafrost extent in the Alps. Boeckli et al. (2012) point out that the index, which ranges from 0 to 1, represents an indicator of the probability for permafrost occurrence, the spatial percentage of permafrost per cell, the thickness of the permafrost body for current climatic conditions, and/or a proxy of the mean annual ground temperature. The relative area of permafrost occurrence in relation to the total area of the Alps is 3.3% (6,220 km^2) when considering an index $\geq$0.5. Thus, although the modelled permafrost extent in the Alps is also very small in spatial extent, the area presumably influenced by permafrost at present is nevertheless 3.4 times larger than the glaciated areas.

During the climax of the Last Glacial Maximum (LGM) some 26–19 ka BP (Monegato et al. 2007), most of the Alps were covered by an interconnected Alpine glacier system or by local glaciers (Fig. 2D). In periglacial research, the term Last Permafrost Maximum (LPM) is sometimes used for the period of maximum permafrost extent occurring in the period of about 25–17 ka BP (Vandenberghe et al. 2014). Large piedmont glacier systems existed in the northern Alpine foreland whereas in the southern margin of the Alps, only large glacier tongues reached the Alpine margin where the build-up distinct terminal moraines as for instance at the southern end of Lake Garda (Ehlers et al. 2011). About 55% of the Alps as defined by the Alpine Convention were covered by glaciers during the LGM. The highest percentage of ice coverage was achieved in the NWA with 74.4% followed by the CEA and NEA where glaciers covered about 62% in both cases. Furthermore, in the SEA 54.5% of the total area were covered by LGM glaciers whereas this value is only 20.6% of the total area of the sector SWA. Considering the concurrent permafrost extent during the LGM or LPM, as proposed by Van Vliet-Lanoë et al. (2004), Vandenberghe et al. (2014) and Andrieux et al. (2016), it can be concluded that most of the unglaciated mountains in the CEA, NEA and NWA were influenced by continuous permafrost (Fig. 2D). In the two sectors SEA and SWA we can assume continuous permafrost conditions in the higher-elevated mountain areas and either continuous or sporadic permafrost (particularly in the SWA) in the lower-elevated ones.

In conclusion, continuous permafrost was a major factor in shaping the landscape in most of the unglaciated mountain areas in the Alps during the LGM or LPM, respectively. Furthermore, even though permafrost is of limited extent in the Alps today, strong seasonal or even diurnal frost plays a major role in the periglacial belt of the Alps reshaping the alpine and subalpine landscape at present. Finally, during the Alpine Late Glacial and the early Holocene periods, when the large glacier systems receded and revealed vast mountainous areas, the periglacial belt was able to extent and "occupy" previously glaciated land causing substantial landscape modification with periglacial landforms still visible today.

2 The Development of Periglacial Research in the European Alps

Periglacial research in the Alps has a long tradition exceeding 100 years. A first publication about rock glaciers and related block streams in the Alps was published in 1919 by André Chaix (Chaix 1919) studying three rock glaciers in the Swiss National Park. A few years later, Hermann (1925) published observations results of rock glaciers in a side valley of the Aosta Valley (Valsavarenche) in Italy, and Finsterwalder described and measured creep velocities of the Innere Ölgrube rock glacier, Austria (Finsterwalder 1928). Around the same time, early studies were published describing periglacial processes and landforms in the Austrian Alps (Kinzl 1928; Sölch 1928) or by Allix (1923) and Douvillé (1917) in the French Alps. As the research activity in the present and past periglacial belt of the Alps was manifold over time, space, and topics, we are not able to provide an encompassing insight into the periglacial research history of the Alps. Therefore, we give in this section a brief overview of the historic development of periglacial research in the European Alps by country highlighting some achievements.

2.1 Periglacial Research History in Austria

Research on periglacial landforms in the Austrian Alps was initiated in the 1920s. The main landform of interest during that time were rock glaciers focusing on the description and velocity of the landforms. Finsterwalder discovered during a mapping project in the Ötztal Alps in 1922 two rock glaciers (Innere Ölgrube, Krummgampenspitze). In 1923 and 1924 he carried out two terrestrial photogrammetric surveys to measure a velocity profile across the Innere Ölgrube rock glacier yielding a maximum velocity of 50 cm a^{-1} (Finsterwalder 1928; Kaufmann 2012). This was the first time that rock glacier displacement was measured in Austria and the entire Alps. In 1938, 1939, and 1953 measurements were repeated at this rock glacier by another pioneer of rock glacier monitoring, the Austrian Wolfgang Pillewitzer (Fig. 3A) (Pillewizer 1938, 1957).

Gerhold (1970) studied the glacial geology in the western Ötztal Alps working on morphology and classification of rock glaciers. The spatio-temporal distribution of relict rock glaciers in the Niedere Tauern Range was considered by Nagl (1976). Kerschner (1983, 1985) described relict rock glaciers and used their distribution to provide palaeoclimatic interpretations. The distribution of rock glaciers in the Austrian Alps in relation to climatic and biogeographic parameters was elaborated by Höllermann (1983). De Jong and Kwadijk (1998) discussed the origin and age of several rock glaciers in western-most Austria. In the course of history, several rock glacier inventories have been elaborated in Austria (Lieb 1996, 1998; Kellerer-Pirklbauer et al. 2012a, b; Krainer and Ribis 2012). The most-recent inventory lists

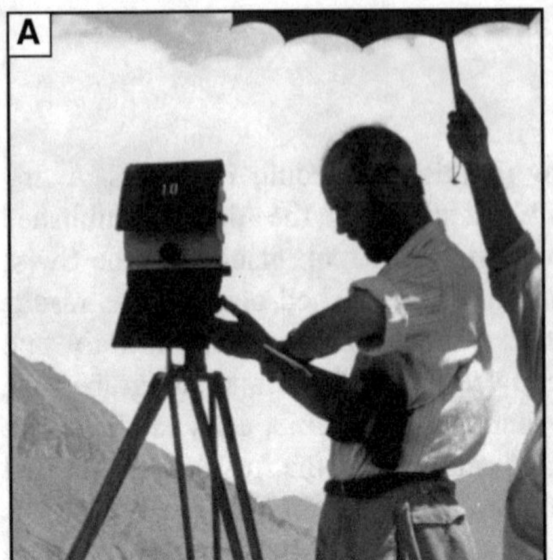

Wolfgang Pillewizer in 1954 carrying out a terrestrial photogrammetric survey in Asia
Source: Estate of Wolfgang Pillewizer

Maintenance of a rock temperature sensor in the 1980's at Roc noir de Combeynot, Écrins Massif, Dauphiné Alps (SWA), France
Source: Bernard Francou

First deformation measurements of a block stream (white line) in the Western Alps in 1961 at Punta d'Aprile, Lanzo Massif, Graian Alps (NWA), Italy - Source: Carlo F. Capello

First drilling on a rock glacier in 1987 at Murtèl/Corvatsch, Bernina Alps (CEA), Switzerland
Source: Daniel Von der Mühll

Fig. 3 Selected field impressions of important achievements of early periglacial research in the European Alps: (A) Wolfgang Pillewizer, a pioneer of rock glacier monitoring, carrying out a terrestrial photogrammetric survey with a phototheodolite during the German-Austrian Karakoram Expedition in 1954. A team member shadowed with an umbrella the lens of the camera; (B) the maintenance of an analogic device that records rock temperature in the North face of the Roc noir de Combeynot (SWA, 3000 m a.s.l.) is an engaged challenge and important for alpine rock weathering studies; (C) a first attempt of deformation measurements of a block stream in 1961 by Carlo Felice Capello at Punta d'Aprile (NWA) by painting a white line on several boulders perpendicular to the flow line (photo taken from Capello 1963); (D) first drilling on a rock glacier in 1987 at Murtèl/Corvatsch (CEA), initiating long-term rock glacier deformation and temperature data

5769 rock glaciers and protalus ramparts for Austria, 3460 of them are considered as relict whereas 2309 as intact (Wagner et al. 2020).

Research on alpine karst landscapes in the Austrian Alps was initiated in the mid nineteenth century. Simony (1847) discussed the formation of caves and exokarst landforms (i.e., karst features, which developed at the surface; Ford and Williams 2007) in the alpine limestone plateaus of the Northern Calcareous Alps focusing on the Dachstein Massif in the NEA. Bock (1913) was the first to transfer the research findings elaborated in the classic karst area of the Dinarides to the Alpine karst areas of Austria. Lehmann (1927) studied the zonal distribution of small-scale dissolution forms (karren) and dolines in alpine karst areas. In this regard, karren are defined as furrows or clefts caused primarily by solution of limestone surfaces. In the subsequent decades, surface landforms of Alpine karst were the main object of studies by different university institutes in Austria and the Vienna Speleological Institute (Bauer and Zötl 1972). In the recent past, a previously unknown karren type was detected in the periglacial belt of Austria named "hummocky karren" defined as randomly distributed, dispersed assemblage of small (in the range of mm to cm) hummocks and depressions in between (Plan et al. 2012).

Relict periglacial landforms at the eastern margin of the Alps (such as debris-mantled rectilinear slopes or hummock meadows—"Buckelwiesen" in Austria) and in the Alpine foreland (cryoturbation, ice-wedge pseudomorphs, valley asymmetry) have been a main research interest of Morawetz (1952, 1968). Morawetz (1971a) described debris-mantled rectilinear slopes from various mountain groups in the Eastern Alps and pointed out that such slopes are characterised by block fields, block strips, block patches or ploughing boulders indicating solifluction-related processes. Solifluction landforms and processes have been studied in Austria since the 1960s (Höllermann 1967; Stingl 1969) with most detailed studies in the 1990s (Jaesche 1999; Jaesche et al. 2003). After the turn of the millennium, however, solifluction research in Austria decreased substantially (Kellerer-Pirklbauer 2018).

In contrast to solifluction studies, research about rock glaciers and permafrost increased considerably since the 1990s in Austria (e.g., Lieb and Schopper 1991). Rock glaciers and glaciers in the Schober Mountains, Hohe Tauern Range, were analysed by Buchenauer (1990) formulating palaeoclimatic interpretations, which were recently revised by Reitner et al. (2016). Schmöller and Fruhwirth (1996) investigated the internal structure of Dösen rock glacier in the Ankogel Mountains, Hohe Tauern Range, using geophysical methods. At the same rock glacier, geomorphometric and movement monitoring using geodetic and photogrammetric methods was started in 1995 (Kaufmann 1996; Kellerer-Pirklbauer et al. 2017). Comparable monitoring was started in 1997 at Weissenkar rock glacier and in 1998 at Hinteres Langtalkar rock glacier, both in the Schober Mountains (Kaufmann and Ladstädter 2003; Avian et al. 2005; Kaufmann et al. 2006). Relict rock glaciers located in Eastern Austria (Niedere Tauern Range) were investigated by Untersweg and Schwendt (1996) focussing on geomorphological and hydrogeological aspects. Ventilated talus slopes at low elevations were studied by Wakonigg (2001) who explained undercooling and formation of permafrost by the chimney effect. Controlling factors of the microclimate in blocky surface layers of two rock glaciers was studied by Wagner et al. (2019) highlighting the importance of the seasonal snow cover pattern in addition to topography and microclimatic variability.

Terrestrial (TLS) and airborne laser scanning (ALS) as well as synthetic aperture radar (SAR) interferometry were used for the first time during the first decade of the present millennium to monitor rock glaciers and periglacial permafrost areas (Bauer et al. 2003; Kenyi and Kaufmann 2003; Avian et al. 2009). The relevance of the lithology on the spatial distribution of rock glaciers was studied by Kellerer-Pirklbauer (2007) whereas the relationship between glacier recession and formation of permafrost-related landforms was analysed by Kellerer-Pirklbauer and Kaufmann (2018). Regarding the age of periglacial landforms, absolute and relative dating of rock glaciers in Austria have been initiated about 10 years ago. Two relict rock glaciers in the Stubai Alps were dated for instance by cosmic ray exposure (CRE) dating (or termed as terrestrial cosmogenic-nuclide dating/TCND) using cosmogenic ^{10}Be (Ivy-Ochs et al. 2009). Further to the east, the Schmidt-hammer exposure-age dating (SHD) method was applied at several rock glaciers in the Hohe and Niedere Tauern Ranges (Kellerer-Pirklbauer 2008; Rode and Kellerer-Pirklbauer 2012). In recent years, TCND was applied there as well (Reitner et al. 2016; Steinemann et al. 2020).

In 2012, a special issue on permafrost and periglacial landforms in Austria and South Tyrol, Italy, was published (Krainer et al. 2012a) with papers related to: (a) modelling of mountain permafrost distribution and the distribution of rock glaciers and permafrost in the past and today (Avian and Kellerer-Pirklbauer 2012; Krainer and Ribis 2012; Kellerer-Pirklbauer et al. 2012a, b; Schrott et al. 2012), (b) internal structure and dynamics of rock glaciers (Hausmann et al. 2012; Krainer et al. 2012b), (c) permafrost monitoring (Kaufmann 2012; Hartmeyer et al. 2012; Schöner et al. 2012), and (d) climate change and natural hazards (Kellerer-Pirklbauer and Kaufmann 2012; Kern et al. 2012).

2.2 *Periglacial Research History in France*

During the early 20st century, some French geologists and geographers were enthusiastic about the observations made during exploratory expeditions in cold regions such as the Svalbard archipelago and questioned the role of frost in the processes driving the landforms like patterned ground (Douvillé 1917). Being among the first in the Alps, Allix (1923) described similar surface morphologies in the French Alps near Grenoble and even "rock glaciers" near the Mont-de-Lans Glacier, and debated the respective roles of snow, frost, and glacier in shaping the landscapes. Similarly, Gignoux (1931) and Demangeot (1941) focused on the sorting processes responsible of the polygons observed in the Col de la Leysse vicinity, in the upper Isère catchment, and pointed out the importance of the freeze–thaw cycles and the ice segregation. These two authors also mentioned the possible presence of a frozen layer in the ground, which they called with the Swedish word "tjäle" (literally translated "ground frost"), explaining the different patterned grounds observed in various regions of the French Alps (Tarentaise Valley and Upper Ubaye catchment).

Bringing a physicist view on the "rock glacier problem", the French glaciologist Louis Lliboutry suggested a combination of glacial origin of the ice and periglacial processes for explaining the morphology of the landforms he observed in the semi-arid Chilean Andes (Lliboutry 1955). Among other explanations, he insisted on the long-term sieving of fine material through what he called "mollisol" (meaning here a thin debris layer), which explains, according to him, the coarse-debris aspect of the surface of "old" rock glaciers (Lliboutry 1961). Mostly following Lliboutry's interpretation of processes in cold glacial-periglacial environments, authors like Julian (1966), Schweizer (1968) and Guiter (1972) tried to explain the past and present Alpine periglacial landforms, for instance attributing the origin of rock glaciers to either a receding glacier or to a "glacier d'éboulis" (literally translated "scree glacier") as proposed by Lliboutry (1961).

A few years after, combining field observation and in situ measurements of various parameters, Michèle Evin and Bernard Francou (Fig. 3B) started intensive research on mountain periglacial slope dynamics (Francou 1982; Evin and de Beaulieu 1985). This interest was concomitant to the efforts similarly done in other Alpine regions (especially in Switzerland and Austria) and to the experimental research undertaken by the Centre de Géomorphologie de Caen (Lautridou and Ozouf 1982) in tight collaboration with Albert Pissart and his colleagues in Belgium (Pissart 1987; Coutard 2019) and with French geoarcheologists at INRAP (Institut National de Recherches Archéologiques Préventives). This led to significant advances in the understanding of the effects of periglacial processes on mountain slopes. Among other noticeable facts, Francou and Reynaud (1992) started in 1980 a geodetic survey of the surface displacement of the Laurichard rock glacier, which is still active and constitutes among the longest time series of permafrost creeping globally (Thibert and Bodin 2022). Evin and Fabre (1990) carried out numerous vertical electrical resistivity surveys to better constrain the internal structure of rock glaciers and the distribution of mountain permafrost in the Southern French Alps. Another pioneering work of Francou focused on the thermal regime of frozen rock faces, relating the intensity and frequency of freeze–thaw cycles to the production of debris by frost shattering (Coutard and Francou 1989).

In the more recent years, the interest on Alpine periglacial landforms and dynamics increased in France and strong research groups established in Paris (Laboratoire PRODIG), Grenoble (Institut de Géographie Alpine), and Chambéry (Laboratoire EDYTEM). In 2018, they organised the 5th European Conference on Permafrost, in Chamonix, which gathered more than 500 participants. Finally, it is worth noting that the French community of periglacial specialists started in 1974 to annually meet within the working group "Régionalisation du périglaciaire" (a section of the "Comité National Français de Géographie"), later renamed into "Association Française du Périglaciaire". Focusing on both circum-polar and Alpine environments and fed with observational, experimental, and modelling approaches. All the research works of this active group is accessible through a dedicated webpage (https://afdpblog.wordpress.com/publications/). In parallel, since 2010, a permafrost monitoring network is federating the observational activities of the French research group, focusing on the standard parameters of the ECV (Essential Climate Variable) Permafrost, with focus on rock glaciers and rock walls (https://permafrance.osug.fr). These two typical environments of mountain permafrost, and the climatic and environmental controls on the associated geomorphological processes (rockfall activity, rock glacier dynamics) are at present the main research focuses of the French teams (Bodin et al. 2016; Magnin et al. 2017; Ravanel et al. 2017).

2.3 Periglacial Research History in Germany

Many German scientists—and geomorphologists in particular—contributed significantly to research on periglacial processes and landforms in the past. As the present occurrence of permafrost is limited to the highest mountain peaks in Germany (e.g., the Zugspitze, Wetterstein Mountains, NEA, with 2962 m, Boeckli et al. 2012) and as also the occurrence of current periglacial phenomena is limited in spatial extent as well as activity, most of the researchers in Germany performed their studies in arctic and subarctic environments, as well as in other mountain environments on Earth (e.g., Finsterwalder 1928). Especially after World War II, large attention has been devoted to alpine periglacial environments in all parts of the world (Karte 1982). These scientists mainly focused on the landform significance as climate indicators and the reconstruction of past climate conditions (e.g., Troll 1944; Karte 1983; Höllermann 1985; Barsch 1993; Lehmkuhl 2016). Dietrich Barsch—maybe the father of rock glacier research in Europe—intensively investigated rock glaciers in various parts of the world and contributed a lot to the process understanding of these peculiar landforms, widely accepted as geomorphological indicators for the occurrence of permafrost (Barsch 1996). He approached the phenomenon with a large variety of methods. Not only geomorphological, but also photogrammetric, geophysical, and hydrological analyses were applied by him to improve the understanding of landform development and the analysis of current dynamics.

As large areas in northern and southern Germany were influenced by glacial and periglacial processes during the LGM, another research focus was on palaeo-periglacial processes, such as solifluction or cryoturbation. The related landforms were described, classified and dated enabling the reconstruction of past climate conditions and environments in large detail (e.g., Raab et al. 2007).

More recently, an extensive geophysical and geotechnical monitoring was initiated at high-mountain peaks in Germany, e.g., at the Zugspitze, where permafrost still exists in some areas. These areas are highly affected by rising temperatures and show indications for permafrost warming and thawing. In reaction to current and long-term changes, related thermo-cryogenic processes can induce changes in slope stability and trigger gravitational processes such as rockfalls or landslides (Krautblatter et al. 2010; 2012).

2.4 Periglacial Research History in Italy

The alpine periglacial environment and processes are studied in Italy since about 100 years. The first studies were carried out by geographers with a naturalistic approach. As in many parts of Europe, these first researchers started their studies on the geomorphology of the Italian alpine cold environment some decades after studies, which focused exclusively on glaciers. Indeed, many studies were influenced by some foreign geographical schools, particularly by Austrian (Eduard Richter) and German (Eduard Brückner, Albrecht Penck) pioneers in glaciology and glacial geomorphology. In the second half of the nineteenth century, the study of glaciers was an important field of research, almost an obligatory step for most of the Italian geographers. The development of this issue benefited from the existence and creation of several national institutions: the Italian Alpine Club (C.A.I.), which greatly supported this research, the International Commission for the Study of Glaciers and the Italian Glaciological Committee, both created just before World War I (Scaramellini and Bonardi 2001).

Since the 1920s, the Italian researchers' attention turned to rock glaciers, the most characteristic and evident forms of the Alpine periglacial environment, firstly called in Italian language "colate di pietre" (literally translated "flow of stones") or "pietraie semoventi" ("self-propelled stones"). The first authors who dealt with these phenomena were Hermann (1925) in Valsavarenche (secondary valley of the Aosta Valley, NWA) and Capello. Capello (1947) published a brief note related to the rock glaciers in the Western Alps and subsequently two detailed articles concerning rock glaciers of the Gran Paradiso region (Capello 1959) and the Southern Graian Alps (Capello 1963). Capello (1963) described in detail the extensive debris bodies that largely cover the slopes of the Lanzo Massif (near Turin, Graian Alps, NWA) subsequently interpreted as block field and block stream (Fioraso and Spagnolo 2009). He established the first system for measuring the displacement of these debris bodies using lines drawn with white paint on the blocks (Fig. 3C). Unfortunately, no data of these measurements has been found. Periglacial-relevant descriptions related to the Valtellina (Nangeroni 1929) and cartographic representations concerning the Adamello-Presanella Alps, the Venetian Prealps, and the Dolomites in the SEA (Castiglioni 1961; 1974; Carta Geologica d'Italia 1977) were elaborated.

Different methodologies in periglacial research have been applied particularly since the late 1980s, including the geomorphological survey and mapping of landforms and active processes, the survey of vegetation species and associations indicating the presence or absence of permafrost, geophysical measurements, and the interpretation of remote sensing data. Rock glaciers got the most attention, especially regarding their morphology (Smiraglia 1992; Ribolini 2003; Guglielmin et al. 2004; Colucci et al. 2019) and distribution (Guglielmin and Smiraglia 1997; Ribolini 2001; Baroni et al. 2004; Seppi et al. 2012). Some specific aspects of this periglacial phenomenon, such as the mechanism of movement and the study of inner structures through geophysical prospecting, have been addressed since the 1990s by Smiraglia

(1990), Resnati and Smiraglia (1990), Fabre and Ribolini (2003), Ribolini et al. (2010), and Colombo et al. (2020).

In addition to the rock glaciers, some works have also been devoted to the other periglacial landforms in the Italian Alps, starting from general studies (e.g., Albertini et al. 1955; Nangeroni 1962; Capello 1969) to specific ones: loess deposits (Nangeroni 1959; Forno 1979; Cremaschi et al. 2015; Costantini et al. 2018; D'Amico et al. 2021), stratified slope deposits (Castiglioni et al. 1979; Pappalardo 1999; Pappalardo and Spagnolo 1999), block fields and block streams (Firpo et al. 2006; Fioraso and Spagnolo 2009; Paro 2011), protalus and pronival ramparts (Scotti et al. 2013; Colucci et al. 2016a), and earth hummocks (Pintaldi et al. 2016). At the same time, several publications have been issued on the permafrost distribution (Guglielmin et al. 1994; Guglielmin 1997; Ribolini and Fabre 2006; Carturan et al. 2016) and its relationship with climate change (Belloni et al. 1993; Dramis et al. 2003; Guglielmin 2003; Seppi et al. 2014; Colombo et al. 2016).

Since the end of 1990s, new European projects on permafrost monitoring were accomplished, involving Italian researchers and institutions as partners or lead partners, and monitoring networks have been established in the Alps. Following these transnational projects, more than 10 boreholes for permafrost monitoring have been established in all the Italian Alps (Harris et al. 2001; Mair et al. 2011). The deepest borehole (235 m) in the European mountain permafrost is located close to Stelvio Pass in the Central Italian Alps (border CEA/SEA) and monitored since 2010 (Guglielmin et al. 2018). Permafrost and water resources relationship was the focus of cross-border European projects between Italy and Austria (Mair et al. 2015) and between Italy and Switzerland (Carturan et al. 2016, Colombo et al. 2019, 2020, Rogora et al. 2020). Finally, the most recent line of research concerning the high elevated karst environments of the Italian Alps involves researchers with different backgrounds engaged in the study of ice caves (Maggi et al. 2018; Colucci and Guglielmin 2019; Vigna and Paro 2019).

2.5 *Periglacial Research History in Slovenia*

Periglacial landforms in Slovenia have been given less scientific attention in contrast to early studies on glacial landforms caused by (almost exclusively) past glaciation (e.g., Lucerna 1906), since the former is not particularly evident in the high mountains of the country (Natek 2007). In addition, as most of the high mountains of Slovenia consist of karstifiable rocks (mainly Julian and Kamnik Alps), research on periglacial landforms focused particularly on high-elevated karst features in these two areas, as well as in the Karawanks (Gams 1965; Herak 1972; Veress and Zentai 2004; Telbisz et al. 2019; Tóth and Veress 2019; Veress et al. 2019), and substantially less on non-karstic periglacial landforms such as the ones in the Pohorje Mountains further to the east (Gams 1959; Šifrer 1983; Natek 1993, 2007).

Melik (1935) discovered that frost weathering was a very effective geomorphic process in non-glaciated areas of the Slovenian Alps during the Pleistocene. He furthermore quantified an enhanced transport of scree down the slopes during the Pleistocene but did not use the term periglacial, as this term was not well-known during that time. However, the same author used in the second edition of his Slovenia monograph (Melik 1963)—as mentioned by Natek (2007)—the term periglacial thereby describing slopes in Slovenia, which were affected by low temperatures and frequent transitions of temperatures below the freezing point causing intensive sedimentation of unglaciated valleys during the Pleistocene.

Furthermore, rock glaciers and related landforms such as pronival and protalus ramparts have been studied and inventoried rather recently in the Slovenian Alps (Colucci et al. 2016a, b; Bocalli et al. 2019). Obu et al. (2018) studied sorted patterned ground in a karst cave at the south-eastern margin of the Alps. Finally, Stepišnik (2020) analysed and classified karst poljes in Slovenia and revealed that a high-elevated polje located southeast of the highest mountain of the country (Mt. Triglav, 2864 m) with an area of 0.05 km^2 is the only proglacial polje in the Slovenian Alps.

2.6 *Periglacial Research History in Switzerland*

Switzerland is a country whose image is closely related to geomorphology and especially high-alpine landscapes, as the Alps make up 60% of Switzerland's land area (Reynard et al. 2021). The Alpine relief and related climate conditions favor the existence and evolution of glacial and periglacial processes. While the Quaternary glaciations influenced almost the whole of Switzerland and left behind a variety of landforms (Schlüchter et al. 2021), todays glacial and periglacial processes are relevant above an altitude of about 1800 m. Here, valley glaciers exist and periglacial landforms, linked to the occurrence of freeze–thaw cycles and/or the occurrence of permafrost, dominate in many regions.

Since the nineteenth century, mountaineers, climbers, and painters visited the Alps to experience the spectacular landscape. While exploring the mountains, people provided first descriptions of landforms and related processes. In the context of discussions on the Ice Age theory, Louis Agassiz started his influential scientific investigations on Unteraarglacier in the Bernese Alps in 1840 (Agassiz et al. 1847). Later, also other cryospheric landforms were described and over time the analyses developed towards more process-oriented and quantitative studies.

The history of rock glacier research in Europe had its origin in the Eastern Swiss Alps, in the—at that time—newly founded Swiss National Park (Livigno Alps, Sesvenna Alps, Ortles Alps of the Ortles-Cevedale Group, mainly CEA). In 1918, Emile Chaix was the first to pay attention to the peculiar landforms he observed. He started a geomorphological monitoring, which was continued by André Chaix. By measuring stone lines on the rock glaciers of Val Sassa and Val da l'Acqua, André Chaix delivered the first evidence of the movement of rock glaciers and compared his observations to landforms in California, USA (Chaix 1923, 1943). The measurements were continued for several decades (Eugster 1973). Soon, rock glaciers were also investigated and mapped using photogrammetry (e.g., Messerli and Zurbuchen

1968) and geophysics (e.g. Barsch 1969). Another important chapter in rock glacier science began in 1987 with the first deep borehole ever drilled through a rock glacier (e.g., Von der Mühll and Haeberli 1990) (Fig. 3D). This was performed on the Murtèl-Corvatsch rock glacier (Upper Engadin, Bernina Alps, CEA) where Dietrich Barsch had already started measurements in the 1970s (Barsch and Hell 1975). Several boreholes followed on rock glaciers in Switzerland and allowed for a more detailed analysis of permafrost temperatures, landform composition (ice content, sediment characteristics), deformation at depth (e.g., Arenson et al. 2002) and movements at the surface (e.g., Kääb 1997). With the increased availability of remote sensing data such as space-borne radar interferometry, the quantification of rock glacier dynamics was enabled on a larger spatial scale (Strozzi et al. 2004).

Another characteristic geomorphological process of the periglacial belt is solifluction. Different processes related to solifluction (cf. Matsuoka 2001) such as repeated freeze–thaw processes in the subsurface lead to the formation of distinctive tongue-shaped lobes, which move downslope with velocities of cm per year (Veit 2002). Furrer (1954) was the first describing solifluction landforms and involved processes in detail, again in the Swiss National Park. His measurements were continued in a systematic monitoring program until today (Gamper 1984; Rist et al. 2013) and enabled the identification of influencing factors, such as air temperature and snow depth.

Nowadays, several periglacial landforms in Switzerland are monitored continuously by different in-situ and remote sensing methods as part of the Permafrost Monitoring Network Switzerland (PERMOS). PERMOS was founded in 2000 and involves long-term and systematic measurements of permafrost temperature, active layer thickness, deformation, and other characteristics, such as ice content (e.g., PERMOS 2021).

3 Distribution and Chronology of Periglacial Landforms

Periglacial landforms are considered to be formed by specific weathering conditions. Weathering in such environments very much depends on frost action, which is a collective term used to describe a number of processes strongly linked to the alternation of freezing and thawing of material (French 2007). Frost action through physical and chemical weathering but also cryoturbation has an important effect on the near-surface layer of alpine surfaces causing in-situ or autochthonous periglacial landforms. Characteristics of landforms derived from weathering in the periglacial zone of the European Alps are reported in Sect. 3.1, landforms related to cryoturbation caused by seasonally frozen material in flat to gently sloping areas are treated in Sect. 3.2, and landforms occurring on slopes, which are related to frost weathering, cryoturbation, and different forms of frost creep are of concern in Sect. 3.3. Finally, Sect. 3.4 deals with other periglacial-related landforms such as nivation forms, thermokarst features, and ice caves.

3.1 *Rock Weathering and Derived Landforms*

3.1.1 Context and Acting Processes

Two general models for weathering based on frost action are distinguished. First, the volumetric-expansion or hydro-fracturing model when local water expands by 9% in volume due to freezing (Matsuoka 1990, 2008) and second, the segregation-ice model where water potentially migrates within freezing ground supplying humidity for the progressive growth of ice lenses (Hallet et al. 1991). For both models of frost weathering, knowledge about the thermal regime and rock moisture are crucial. Such field studies have a long tradition in the periglacial belt of the Alps. Numerous measurements in the European Alps and supportive modelling approaches showed that rock temperatures in alpine environments are strongly influenced by slope inclination, slope aspect, local topo-climatic conditions, seasonal snow cover, and thermal properties of the rock (Coutard and Francou 1989; Matsuoka 1990, 2008; Gruber et al. 2004a, 2004b; Noetzli et al. 2007; Noetzli and Gruber 2009; Haberkorn et al. 2017; Kellerer-Pirklbauer 2017). Rock temperature data with an hourly or even minutes logging interval were considered already several decades ago as useful to assess the intensity of near-surface physical weathering and hence of potential rock shattering in the Alps (Coutard and Francou 1989; Matsuoka 2008; Hasler et al. 2012). Studies in the Alps showed furthermore that short time series are not representative in showing the average or range of potential weathering at a given site (Hasler et al. 2011; Kellerer-Pirklbauer 2017).

Coutard and Francou (1989) carried out a pioneer study in rock wall weathering focusing on two opposite-looking rock walls (southwest- and north-facing) in the French Alps (Fig. 3B). Their southwest-facing slope showed many superficial freeze–thaw-cycles (FTC) with short durations of freezing events compared to the north-facing slope with few superficial FTC but longer durations of freezing events. Field studies in the Swiss Alps revealed that during warming a rock slope expands whereas during cooling contraction occurs related to thermomechanical forcing (Hasler et al. 2012; Draebing et al. 2017). Such field studies showed furthermore that thermally induced stress propagates several tens of metres into the bedrock to depths where only small annual temperature variations exist (Hasler et al. 2012). In addition to that, data gathered by crackmeter in the Swiss Alps in combination with modelling and laboratory tests were used to quantify thermally and ice-induced rock and fracture kinematics considering the release of rockfalls and effects of climate change (Draebing et al. 2017; Draebing 2021). Studies in the Austrian Alps showed that predicted warmer atmospheric conditions in the future (Gobiet et al. 2014) will lead to shorter seasonal snow covers, to less severe freezing events, and to a shorter period within the frost-cracking window at alpine rock walls (Kellerer-Pirklbauer 2017). Permafrost and the thermal characteristics in steep alpine rock walls of Aiguille du Midi, French Alps, was studied by Magnin et al. (2015) using surface sensors and 10 m deep boreholes. These authors found out that thick snow has a warming effect on shaded areas, reduce active layer refreezing in winter, and delay its thawing in summer. In addition, Magnin et al. (2015) demonstrated that the high albedo of snow leads to cooler conditions at the rock surface in rock exposed to the direct solar radiation.

Rock moisture data are essential for estimating weathering rates (Matsuoka and Murton 2008). Studies in the Alps related to rock moisture conditions in the periglacial belt were for instance carried out by Sass (2005) who pointed out that rock moisture levels in rock walls are lower in winter compared to summer. Furthermore, he revealed that northerly exposed rock walls in the Alps are generally moister in comparison to rock faces oriented in a southerly direction. Earlier laboratory studies (Fagerlund 1979) and later field studies in the Alps (Matsuoka et al. 1997) showed that rocks are little damaged by frost shattering if they are below a critical degree of saturation. Draebing et al. (2017) carried out indirect measurements of cryosuction in alpine rock walls by using piezometers for monitoring crack water pressure. Based on various research activities in the Alps, Krautblatter et al. (2012) discussed research perspectives on alpine rock walls and related periglacial processes and landforms identifying four major topics: rock temperature measurement and modelling, remote sensing of rock walls, process understanding of rock mass instability, and flow propagation models of detached rock/ice masses.

Savi et al. (2015) studied the efficiency of frost-cracking processes in small catchment in the eastern Italian Alps applying a heat-flow model to analyse the variations of the frost-cracking intensity, which could have controlled the sediment production in the basin. The model results, combined with field observations, mapping, and quantitative geomorphic analyses, revealed that frost-cracking processes have had a primary role in the production of sediment where the intensity of sediment supply has been dictated and limited by the combined effect of temperature variations and conditions of bedrock preservation.

Rock weathering in alpine areas with carbonate rocks might be different compared to climatically similar areas located in metamorphic, igneous, or other sedimentary rock types that are not karstifiable. Rock weathering in carbonate areas is influenced by karstification processes such as carbonate solution, subsurface drainage, and distinct surface karst features collectively termed as exokarst (Ford and Williams 2007). In this regard, a clear distinction of the three related terms alpine karst, nival karst, and glaciokarst is of relevance for weathering studies in the Alps. Alpine karst is found in high-elevated karst areas in mountain environments above the treeline, nival karst is formed by karstification in glacial and periglacial environments under the influence of snow and its seasonal melt contributes forming for instance nivation hollows, and glaciokarst is a karst landscape, which was glaciated during the colder periods of the Pleistocene and displays still today major landforms of glacial influence (Ford and Williams 2007; Veress et al. 2019).

Pure alpine karst landforms, which evolved in the periglacial belt are difficult to define. Cold-climate weathering effects such as frost-shattering can totally mask the karstic effect of dissolution. Ford and Williams (2007) pointed out that strongly competing geomorphological processes such as frost shattering need to be weak to permit the rather slowly developing solution landforms to become dominant. Such competitive processes cause an altitudinal zonation of exokarst features in the Alps, starting with a doline (or sinkhole) zone at the treeline, followed by a higher karren zone (i.e., furrows or clefts caused by solution), and finally the highest zone where frost shattering (considered as "scherbenkarst"; cf. Fig. 5C) is predominant (Bauer 1962; Ford and Williams 2007). Based on studies in the Austrian Alps, Bauer and Zötl (1972) revealed that at elevations above 2200 m frost weathering becomes increasingly important and prevents the development of significant karren landforms.

3.1.2 Periglacial Landforms Derived from Weathering in Non-carbonate Rocks

Rock breakdown in alpine environments produces commonly angular, frost shattered detritus. Such mountain-top detritus is widespread in the Alps forming autochthonous diamicts or block fields, depending on the lithology. Ballantyne (1998) pointed out that on well-jointed rocks, mountain-top detritus has developed within a few millennia of exposure to periglacial weathering, whereas on massive lithologies, it has formed over much longer timescales. The latter was for instance possible in the Alps at flat or concave nunataks above the glacial maximum ice systems or in marginal areas of the Alps where glaciation was only limited. Figure 4A depicts typical fine-grained autochthonous mountain top detritus derived from calcareous mica-shist forming a diamict at an elevation of 2930 m. Figure 4B shows a well-developed coarse-grained autochthonous mountain top detritus derived from orthogneiss at an elevation of 2416 m. At the depicted site (Hochreichart, Seckauer Tauern, CEA), autochthonous block fields characterise the summit area (partly with sorting; cf. Fig. 7C and D) and coarse-grained solifluction-related landforms dominate morphologically the adjacent slopes. This site experienced only marginal glaciation during the colder periods of the Pleistocene (Kellerer-Pirklbauer 2019). At inclined hillsides of the summit area, block fields started to move downward forming block streams (cf. Fig. 11D) or different types of solifluction features (cf. Fig. 11C) leading to the formation of rectilinear or convex-concave debris-mantled slopes (cf. Fig. 11F) (Höllermann 1967, 1985; French 2007). In very steep mountain summits, mountain-top detritus is replaced by fractured summits and ridges (Fig. 4C) where loose debris gets transported mainly by gravity to the slope deposits below.

A further in situ landform caused (or at least reshaped) by periglacial weathering are tors although many hypotheses have been proposed to explain their origin. Tors are considered as free-standing rock outcrops that rise abruptly from the surrounding smooth and gentle slopes of a rounded summit area (Goudie 2004). Such tors are considered as summit tors (Ballantyne 2018). However, tors might also form on slopes as hillslope tors (French 2007), or valley-side tors (Ballantyne 2018). Tors are for instance widespread in the Eastern Alps in areas, which were not glaciated during the Pleistocene (Fig. 4E). There, tors are commonly linked to folk tales and were frequently considered as devil rocks ("Teufelstein" in German) or witch rocks ("Hexenstein") (Morawetz 1971b). Studies in the Eastern Alps revealed that tors are often linked to more resistant rock types such as quartzite, pegmatite, eclogite, and platy gneiss (Kieslinger 1927).

Fine-grained mountain-top detritus - active (Glockner Mountains, Hohe Tauern Range, CEA) Author: Andreas Kellerer-Pirklbauer

Coarse-grained mountain-top detritus/block field - relict (Seckauer Tauern, Niedere Tauern Range, CEA) - Author: Andreas Kellerer-Pirklbauer

Fractured summit - transitonal or relict (Ankogel Mountains, Hohe Tauern Range, CEA) Author: Andreas Kellerer-Pirklbauer

Rock slope disintegration and autochthonous block field formation - relict (Lanzo Massif, Graian Alps, NWA) - Author: Luca Paro

Tor consisting of crystalline rock - relict (Packalpe, Lavantal Alps, CEA) Author: Andreas Kellerer-Pirklbauer

Periglacial trimline separating a periglacial block slope from a cirque with a younger rock glacier - relict (Seckauer Tauern, Niedere Tauern Range, CEA) - Author: Andreas Kellerer-Pirklbauer

Fig. 4 Pictures of weathering features and related lanforms in non-carbonate rocks

Rinnenkarren (Steinernes Meer, Berchtesgaden Alps, NEA) Author: Andreas Kellerer-Pirklbauer

Kluftkarren or grikes (Steinernes Meer, Berchtesgaden Alps, NEA) Author: Andreas Kellerer-Pirklbauer

Frost-shattered ground or scherbenkarst, sheeps for scale (Steinernes Meer, Berchtesgaden Alps, NEA) - Author: Andreas Kellerer-Pirklbauer

Tor consisting of layered carbonate rock (tor at ca. 1850 m a.s.l, Monte Baldo, Garda Mountains, SEA) - Author: Andreas Kellerer-Pirklbauer

Dolines with late-lying snow (E - Hochschwab Mountains, F - Berchtesgaden Alps, both NEA) Author: Andreas Kellerer-Pirklbauer

A field of suffosion dolines where a thin layer of sediments covers bedrock (Hochschwab Mountains, NEA) - Author: Andreas Kellerer-Pirklbauer

Fig. 5 Pictures of weathering features and related lanforms in carbonate rocks

One example of periglacial weathering and related in situ landforms are described here from the Pohorje Mountains in the SEA. The Pohorje Mountains are low in elevation (maximum 1543 m at Črni Vrh) and are built of igneous and metamorphic rocks. During the cold periods of the Pleistocene, the Pohorje Mountains were not glaciated but subject to periglacial processes forming related landforms still visible today. Gams (1959) concluded that the snowline during the Last Glacial Maximum was at about 1300 m on eastern Pohorje and raised towards western Pohorje. He further argued that periglacial processes were efficient during that time based on thick layers of debris on the slopes and the extensive fluvio-periglacial sediments at the base of the Pohorje Mountains. Šifrer (1983) studied the fluvio-periglacial accumulation of sediments, especially in the upper Dravinja River basin in south-eastern part of the Pohorje Mountains. He explained the large quantities of debris in the upper reaches of valleys and the extensive existence of alluvial fans deposited by the Pohorje streams because of intensive frost weathering and fluvial transport during annual spring thawing periods. Natek (1993, 2007) described and morphometrically analysed 59 nivation hollows in the highest and central part of the Pohorje Mountains. Natek (2007) found in the same area evidence of periglacially-formed relict block fields (or block meer) located in a small but distinct cirque (Jezerc cirque) and a neighbouring—presumably nivation—hollow. Natek (2007) argued that denudation and other processes have destroyed or at least blurred smaller periglacial landforms such as patterned ground, stone stripes or solifluction features, while larger periglacial and glacial landforms such as the nivation hollows and cirques have been preserved until today. This described reshaping—and blurring—of relict periglacial landforms is presumably valid for many other present-day forested areas of the Alps.

Another in situ periglacial landform frequently found in the Alps are periglacial trimlines (Ballantyne 2018). Figure 4F depicts a periglacial trimline at a cirque NE of the summit of Hochreichart, CEA, Austria. The trimline marks the boundary between a lower zone of glacially modified bedrock and the higher-elevated palaeo-surface with sloping block fields. This line represents the maximum altitude of the local glacier system during the different glaciation periods during the entire Pleistocene. The block field area remained always above the local glacier and was exposed to intensive periglacial activity during the cooler periods of the Pleistocene. The rock glacier in the cirque is much younger and was formed in the Late Glacial and Early Holocene periods (Kellerer-Pirklbauer 2019).

3.1.3 Periglacial Landforms Derived from Weathering in Carbonate Rocks

Periglacial landforms derived from weathering of carbonate rocks are—as one example—very well developed in the Julian Alps (max. elevation Mt. Triglav 2864 m) and the Kamnik Alps (Grintovec 2558 m), both SEA. The karst in the two mountain groups is predominantly of high-mountain character developed in Triassic limestones and exposed to high annual precipitation rates in the range of 1600 to 3000 mm (Gams 1965). Herak (1972) found out that the alpine character in this area is expressed by karren, grooves, small sinkholes with vertical walls, and dolines above the forest line, which is there between 1500 and 1900 m. Potholes are especially numerous because of the primarily vertical water flow owing to the massive thickness of the carbonate rocks with up to 2000 m from the summit regions to the valleys. In contrast, caves—and thus endokarst features, which developed underground—are rare in this area (Herak 1972). Tóth and Veress (2019) described karren surfaces in the proglacial area of a small glacier at Mt. Triglav examining initial karstification. They concluded that rillenkarren appear at first, followed by kamenitzas (i.e., solution pans, Ford and Williams 2007) and rinnenkarren. Very early forms in the same proglacial area at Mt. Triglav are furthermore kluftkarren (or grikes), which are major joint- or fault-guided solutional clefts (Veress and Zentai 2004). Telbisz et al. (2019) pointed out that there are two well-developed glaciokarst areas in the Julian Alps, which are the Seven Lakes Valley in the Eastern Julian Alps with Mt. Triglav and the Canin plateau (max. elevation 2587 m) in the Western Julian Alps at the border to (and partly stretching into) Italy.

Stepišnik (2020) studied and classified karst poljes in Slovenia. Poljes are defined as basins or valleys in karstifiable rocks whose walls are relatively steep but whose bottom is flat (Ford and Williams 2007). The rather flat bottom of one of Stepišnik's four polje types is formed through glaciofluvial sedimentation. Stepišnik termed such poljes a "proglacial type" and concluded that these poljes are seldom and small in relation to other types. In fact, the high-elevated Velo polje located southeast of Mt. Triglav, Julian Alps, with an area of 0.05 km^2 is the only proglacial polje in the entire Slovenian Alps and with an infill of glaciofluvial sediments (Stepišnik 2020).

Various examples of weathering and related lanforms in carbonate rocks are depicted in Fig. 5. Figure 5A shows typical rinnenkarren at an elevation of 2100 m in the NEA. Rinnenkarren are also common at much lower elevations at sites lacking vegetation, thus such features are not pure alpine karst features, but are nevertheless widely distributed in the Alps. Karren often developed on the surface of glacially eroded limestone and thus are of Late Glacial to Holocene age (Plan 2016). Figure 5B shows kluftkarren or grikes at about 2000 m, also in the NEA. Grikes are major joint- or fault-guided solutional clefts, which are normally 1–10 m in length (Ford and Williams 2007). Such grikes often develop at the surface of limestone layers, which are sometimes terraced and thus have a stepped longitudinal profile termed as staircase limestone pavements ("Schichtreppenkarst"). Such staircase limestone pavements exist widespread in the NEA (Bögli 1980; Plan 2016). As for the rinnenkarren, kluftkarren also occur at much lower elevation and are thus not pure periglacial karst landforms. In Fig. 5C a karst landscape termed as frost-shattered ground ("scherbenkarst") is shown were freeze–thaw action caused the formation of frost-shattered clasts. Frost-shattered ground in karst areas is frequent at higher elevations where periglacial weathering is crucial in relation to carbonate solution.

Figure 5D shows an example of a hillslope tor in carbonate rocks at an elevation of about 1850 m at the valley head of Val del Parol, Monte Baldo (SEA). Dissolution of the horizontally layered carbonate rock acted stronger along the weaker bedding planes forming a sandwich or pancake like structure. Dolines are surface depressions caused by different processes, mainly by dissolution and underground collapse, which are not typical features of alpine karst areas (Ford and Williams 2007). However, dolines in the periglacial belt are nevertheless strongly influenced by the thermal (e.g., number and nature of freeze–thaw-cycles) and nival (e.g., long-lasting snow cover in the dolines) conditions. Figure 5E depicts a snow-covered solution doline where we do not see the bottom of the doline due to the late-lying snow patch (date 08.08.2015). Figure 5F shows a steep-sided doline favouring snow conservation with a remaining snow patch in late summer (date 27.08.2020). The latter doline type traps and conserves snow, becomes thus a site of accelerated corrosion, and is termed "schachtdoline", "schneedoline" or "kotlic" (Ford and Williams 2007). Finally, Fig. 5E shows a field of suffosion dolines, where fine-grained glacial and periglacial material was piped into epikarst cavities (Ford and Williams 2007; Plan 2016).

3.2 *Periglacial Sorting and Derived Landforms in Flat Areas*

3.2.1 Context and Acting Processes

The Alps, like all orogens recently formed, are characterised by a complex morphology, in which the tectonic forces that form the relief are continually opposed by the exogenous processes that erode and reshape it. In this context, flat or low-angle areas are quite rare and generally with limited extension.

In general, the slope evolution is mainly linked to erosional processes, while depositional processes have a secondary role. The four main factors that affect the morphology of the slope are the following: (a) lithology and structural characteristics, (b) climatic conditions, (c) type and intensity of the processes, and (d) duration of their action (time factor). The relationships between slope morphology and geomorphic factors are not so simple and there is no simple method to define all the processes that contributed to the formation of the slopes, nor to identify which factors determine their profile (Ritter et al. 2002).

In the Alps, the litho-structural control on the slope morphology is commonly evident. The climatic conditions, on the other hand, affect the type and intensity of the morphological processes. In cold climate at high elevations (and at lower elevations, which were affected by past cold climate conditions) there is growing recognition of strong periglacial control on bedrock erosion in mountain landscapes, including the shaping of low-relief surfaces along the slopes. The process of promoting low-angled slopes and level-bedrock surfaces is referred to as "cryoplanation" (French 2007 and references therein). Cryoplanation landforms on slopes might be terraces, in summit areas flats (Czudek 1995; French 2007). Hitherto, the hypothesis that frost action was crucial to the assumed Late Cenozoic rise in erosion rates remains nevertheless compelling and untested. Landscape evolution model may incorporate two key periglacial processes—regolith production via frost cracking and sediment transport via frost creep—which together are linked to variations in temperature and the evolving thickness of sediment cover (Egholm et al. 2015).

Although cryoplanation terraces display slope angles comparable to those of cryopediments (from 1° to 12°, French 2007), they are generally of smaller size, not exceeding a few hundred metres in length. Cryoplanation terraces display various planform shapes (e.g., elongated, crescent-shaped), but with a rather constant profile, with a gently sloping surface backed by a steep riser or a scarp that marks the upper end of the terrace, very similar to fluvial terrace sequences (Rixhon and Demoulin 2013). Cryoplanation terraces may be carved into every rock type. However, they are best developed on fine-grained and closely jointed rocks susceptible to blocklike disintegration, such as basalt, diorite, gabbro, granite, and quartzite. The first order (but nonexclusive) controls on the formation of cryoplanation terraces seem indeed to be the underlying bedrock structure and the slope aspect (Czudek 1995). As for the probable link between cryoplanation terraces and the presence of permafrost, it was (Demek 1969; Reger and Péwé 1976) and still is a debated issue (Matthews et al. 2019).

Egholm et al. (2015) stated that the highest summit flats in mid- to high-latitude mountains may have formed via frost action prior to the Quaternary. Deep cooling in the Quaternary accelerated mechanical weathering globally by significantly expanding the area subject to frost activity. Robl et al. (2015) observed that in steady state orogens, topographic gradients are expected to increase with elevation whereas the Alps feature a transition from increasing to decreasing slope with elevation. Beginning at an elevation of about 1500 to 2000 m, most slopes in the Alps tend to become less steep with increased elevation. Therefore, surfaces with a gentle slope in the alpine sectors are almost polygenetic (originating by different types of processes) and polycyclic (generated in different times). Nearly horizontal areas or smoothly inclined slopes in the Alps are formed by glacial and periglacial processes, fluvial erosion/deposition, aeolian processes, landslides and gravitational mass movements, ground collapsing and karst processes, differential erosion due to lithological contrast or linked to geological structure exhumation, and so on.

The complexity of flat surfaces regarding textural characteristics, or snow cover distribution and duration condition the evolution of periglacial processes and ground freezing also at the large scale. Some periglacial environments are underlain by permafrost, whereas others experience seasonal, intermittent or shorter periods of freezing. Their ground thermal regime is influenced by a dynamic buffer layer of snow, vegetation, organic material and water that thermally modulates the interactions between the atmospheric and ground climate (French 2007). Three groups of processes, structures and deposits characterise periglacial environments. (a) Ground cooling and freezing processes comprise thermal contraction, volumetric expansion of freezing liquid water, ice segregation, and syngenetic permafrost growth. (b) Ground warming and thawing processes consist of thermal expansion, thaw consolidation and thermal erosion. (c) Recurrent freeze–thaw processes involve cryoturbation and solifluction. These processes produce a variety of distinctive periglacial and permafrost structures and deposits (Murton 2021).

3.2.2 Needle Ice: Formation and Geomorphic Relevance

The most important periglacial processes relate to the formation and thawing of ground ice, which denotes all types of ice contained in freezing and frozen ground, irrespective of the form of occurrence, or origin of the ice (French 2007). Needle ice growth is one of the more widespread and easily visible, but less studied, climate related processes caused by nocturnal cooling to just below 0 °C (Matsuoka 2001). Needle ice is the usual term applied to the accumulation of slender, bristle-like ice crystals (needles) practically at, or immediately beneath, the surface of the ground (Washburn 1979). An alternative term for it is pipekrake ice (Van Vliet-Lanoë 2014). Among the freezing and thawing processes, the needle ice growth is surely the more widespread in the alpine environments. Although they are easily visible especially during freeze–thaw periods in bare ground of wherever nocturnal freezing occurs, only few studies were carried out on their formation, evolution and effects on the soil structure, landscape, vegetation, or ecosystems (French 2007).

The occurrence of needle ice depends on the interaction of soil and micro-climatic variables such as soil texture, soil moisture and temperature, soil cooling rate and the freezing duration. In a study carried out at 2670 m close to the Stelvio Pass (Italian Central Alps, border CEA/SEA) Ponti et al. (2018) observed that needle ice can develop with a relatively low ground water content (13.2%), at a relatively high minimum ground temperature (−0.3 °C) and with a low cooling rate (< 1.8 °C h^{-1}). Moreover, they observed that needle ice can form below a thin snow cover (<25 mm) that can enhance the sensible heat flow from the ground to the atmosphere and, therefore, promote the cooling of the near surface ground. Statistically, the minimum air temperature results in the leading factor for the needle ice growth (Ponti et al. 2018). In this case the total frost heave seems to be related to the abundance of fine material.

Regarding geomorphological impact, the primarily diurnal formation and subsequent melting of needle ice causes a process called needle ice creep, which is one of the four types of solifluction (Matsuoka 2001). Needle ice creep is a special kind of diurnal frost creep and occurs when fine-grained surface debris is lifted by ice needles perpendicular to the surface and falls during thawing following gravity causing a downward movement of material. Furthermore, needle ice formation acts as a disruptive agent, causing the disintegration of non-cohesive sediments and soil (Fig. 6A).

3.2.3 Sorting and Frost Patterned Ground Formation

Seasonal frost may develop from a few centimetres to some metres depending on the region, its altitude, its latitude, the grain size of the ground and the climate (French 2007). Usually, it reaches a deeper horizon with rising altitude or latitude. A consequence of seasonal ground freezing in both permafrost and non-permafrost areas is the tilting and upfreezing of stones (clasts) and other objects towards the ground surface (Fig. 6B). This widespread phenomenon is sometimes termed frost jacking or vertical frost-sorting and affects not only periglacial environments, but all areas where winter or diurnal ground freezing occurs. In periglacial environments it is an important component in the formation of patterned ground and in the formation of periglacial weathering mantles or block fields (Ballantyne 2018).

Patterned ground is formed by differential frost heave and may develop as single forms, self-organised nets, merging as coalescing forms, or as crack-guided nets

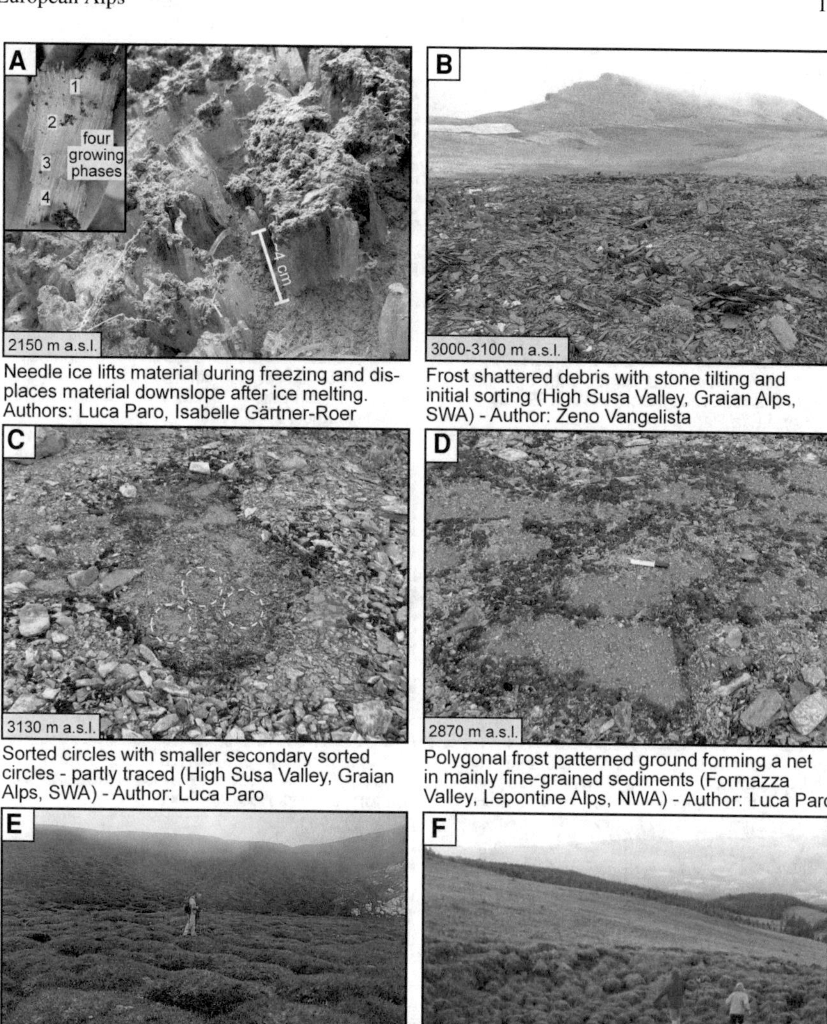

Fig. 6 Pictures of seasonally frozen ground features: present-day active examples. Inset picture in (A) shows well-developed needle ice with four distinct growing phases (hand for scale)

Fig. 7 Pictures of seasonally frozen ground features: present-day inactive or relict examples. The vegetation pattern at the depicted locations might either indicate inactivity (A–D) or more favourable growing conditions for plants (e.g., humidity) along the polygon borders (F)

(thermal or desiccation) (Van Vliet-Lanoë 2014). The affected rock material might be sorted or unsorted. The main geometric forms of patterned ground include circles, polygons, and stripes at slopes, and in case of interacting circles, a net might develop (Figs. 6C, D and 7A–F) (French 2007). As highlighted in Sect. 2, a lot of the research on periglacial forms and processes in the Alps has focused on rock glaciers and permafrost creep processes and distribution. Rather few recent works were dedicated to other active periglacial evidence, such as frost patterned ground, hummocks, or block fields. For example, none of such information is reported in the inventory of permafrost evidence for the Alps (Cremonese et al. 2011). The lesser scientific production concerning these characteristic morphological features of the periglacial environment is certainly due to their lesser presence, obviousness, and distribution in the different mountain ranges of the Alps, but also to a difficulty in the recognition linked to their pure size. Many of these small features are not readily discernible with remote (e.g., aerial photographs) observation; field surveys remain the best approach for their detection. Patterned ground in mountainous areas has, nevertheless, a high palaeogeographic significance as it is associated with cold environments and frequently with past permafrost conditions (Figs. 7A–D). Their sizes/diameter vary from decametric to decimetric. However, wind-blown summit areas with low snow cover allow for the activity of sorted circles, earth and peat hummocks, and some non-sorted stripes.

Sorting processes in the formation of patterned ground generally need no stronger frost action than solifluction or frost creep but remain confined to more specifically geo-ecological conditions with bare, fine-grained, and moist substrate (Fig. 6C). The distribution of active patterned ground is therefore largely dependent on local geoecological factors (Höllermann 1985). Many patterned ground features (i.e., sorted polygons, nets and/or stripes) in the Alps are overgrown by vegetation and seem to be inactive or relict (Figs. 7A–F). In some cases, only the inner secondary sorted circles are active whereas the outer circles are relict (Fig. 6C) (Warburton 1990). The plant cover varies considerably in relation to ground patterning originated by periglacial processes, especially frost heave, frost creep, gelifluction and ice segregation, giving rise to a mosaic of microhabitats sharply differing from each other as regards physical properties and microclimate (Fig. 7F). The distributional patterns of vascular plants, lichens, and bryophytes (i.e., liverworts, hornworts and mosses) are primarily affected by complex responses to substrate texture, soil moisture content and substrate disturbance (Cannone et al. 2004). Openings in the vegetation cover as potential sites for the development of patterned ground may result from recent deglaciation in the glacier forelands, edaphic aridity of a permeable substrate, long lasting snow patches, deflation, overgrazing and trampling, mining activities, and mountain tourism (Höllermann 1985). Some patterned ground features occur on very high elevated, exposed, often windswept, snow-free plateaus, benches and slopes, whereas some features are located at alpine lake margins at locations that may be submerged during the spring melt period (Millar and Westfall 2008).

More details regarding frost patterned ground distribution and description in the Alps are listed as follows without claiming to provide a complete and exhaustive list and in a (primarily) chronological order.

Allix (1923) provided an early description of patterned ground in the French Alps near Grenoble (SWA). The first somewhat extensive study is that of Kinzl (1928), who made numerous observations, between 2200 and 2700 m in various regions in Austria (Silvretta, Ötztal Alps, Hohe Tauern Range including Großvenediger Mountains, Ankogel Mountains, all CEA). Gignoux (1931) summarised first observations and early studies in the Swiss Alps (Selbsanft Massif, Glarus Alps, NWA, at 2600 m; Engadine, CEA, at 2600–3000 m), in the French Alps (Meije Massif, Oisans, Dauphiné Alps, SWA, at 2800–3400 m, SWA), and the ones of Kinzl (1928). Furrer (1955) discussed comprehensively different types of frost patterned ground features in the Alps. Giorelli and Pietracaprina (Veyret 1956) referred on polygonal soils in the Bormio region (Central Italian Alps, CEA). Jennings (1960) described an unusual occurrence of stone polygons in the French Alps. Marnezy (1977) studied in detail sorted stripes in the Vanoise Massif (French Graian Alps, NWA). Pissart (1977) and CNRS-Caen and Institut de Géographie Aix-en-Provence (1980) treated patterned ground landforms (circles, polygons, sorted and non-sorted stripes, steps) near Chambeyron in the Southern French Alps (Cottian Alps, SWA), at elevations above 2500 m.

The Italian research team of Gruppo Nazionale Geografia Fisica e Geomorfologia-CNR (1986) mapped and described stone pavements, patterned grounds, sorted stripes, hummocks and steps in the Peio Valley (Ortles-Cevedale Group, SEA). In the Ortles-Cevedale Group, the distributional patterns of plant species were analysed within periglacial microforms belonging to the collective groups of patterned grounds and sorted stripes by Gerdol and Smiraglia (1990). In this area, three altitudinal zones were previously identified by Recami (1967): the lower zone (2250–2550 m) with old and inactive landforms (earth hummocks, steps); the middle zone (2550–2800 m), with the richest and most varied mature landforms (patterned grounds); and the upper zone (up to ca. 3100 m), with young and active landforms (frost boils, patterned stripes). Zückert (1996) cited the presence of patterned ground and hummocks in the plateau of the Southern Hochschwab Mountains (Northern Limestone Alps, Styria, Austria, CEA).

Matsuoka et al. (2003a, b) revealed that small-scale sorted stripes and circles dominate over the alpine regions in the Upper Engadin, Swiss Alps (NWA), and many of these features are spaced at less than 30 cm intervals by coarse-grained margins shallower than 5 cm. Blum (2006) referred on frost-induced structural soils in the Tennengebirge, Salzburg (NEA). Marvánek (2010) reports about patterned ground in the Krumgampen Valley, Ötztal Alps, Austria (CEA). Filippa et al. (2011) reported stone circles in Gran Paradiso National Park (Western Italian Alps, NWA), at an elevation of 3028 m, on a gentle slope plateau, exposed to wind. Colombo et al. (2013) and D'Amico et al. (2019) reported the presence of some periglacial forms (such as patterned ground, frost boils, polygonal soils) in the basin of Sabbione Lake (Lepontine Alps, NWA, Fig. 6D) and in south-western Italian Alps (SWA), respectively.

Van Vliet-Lanoë (2014) gave a comprehensive overview regarding patterned ground and climate change in arctic and alpine regions presenting pictures of pattered ground in shales and dolomites at La Mortice, 3500 m in Haute Ubaye (Southern

French Alps, SWA). D'Amico et al. (2015) and Mania et al. (2016) referred about flat sites dominated by sorted or non-sorted circles, while at gentle slopes large sorted stripes prevail in the Graian Alps in north-western Italy (Piemonte and Valle d'Aosta regions, NWA). Eleven sorted stripes of coarse limestone debris had developed on 1 m thick silt-rich sediment, underlain by limestone bedrock in the Slovenian Ledenica pod Hrušico karst cave, 45 km SW of Ljubljana, SEA (Obu et al. 2018). Polygonal grounds forms have recently been discovered at an altitude of 1664 m, at the Col du Noyer in the French Dévoluy Mountains (Dauphiné Alps, France, SWA) (Pech et al. 2021). Serra et al. (2021) analysed a silty deposit covered by patterned ground at a Swiss mountain pass to get a better understanding of the Late Glacial-Holocene transition (Sanetsch Pass, Bernese Alps, NWA).

Finally, Höllermann (1985) summarised that in the Central Alps the lower limit of active patterned ground coincides with the upper limit of the alpine meadow vegetation, some hundred metres above the current treeline. A dense turf cover impedes sorting processes, and a closed vegetation cover limits the development of active patterned ground more than any cold climatic threshold. Nevertheless, small-scale patterned ground can also form below the treeline when vegetation-free areas exist.

3.2.4 Vegetated Mounds: Earth Hummocks

Earth or turf hummocks are a type of patterned ground comprising vegetated mounds with a diameter of 0.5–2 m, a bulge of 30–50 cm, and separated by a network of shallow troughs (Höllermann 1967; Keuschnig et al. 2007). Usually, they occur as hummock fields on flat or gently sloping terrain or in depression. Earth hummocks are regarded as a type of non-sorted circles and are commonly developed in silt rich soil containing few or no stones (French 2007). Commonly, such sites are characterised by humid soil conditions favouring frost sorting processes. A polygenetic origin is most likely and not all earth hummocks are of the same origin (French 2007). Other names for this landform are "þúfur" or "thúfur" in Iceland, "pounus" in Finnland, "buttes gazonnées" in France, "Bülten" in Germany or "Buckel" in Austria, "cuscinetti erbosi" in Italy (Troll 1944; Morawetz 1952; Nangeroni 1959; French 2007). However, confusion might arise particularly with the usage of Buckel and Bülten as such forms are commonly linked to whole suite of forming theories, which are unrelated to periglacial processes such as forest clearing, windthrow, karst solution and deglaciation (Ebers 1959). On gentle slopes, hummocks are sometimes elongated downslope. Commonly, earth hummocks in the Alps occur in areas of seasonal frost rather than permafrost at elevations above the treeline, where the soil is composed of fine-grained material at rather humid sites (Figs. 6E and F).

Studies on earth hummocks started in Austria in the 1920s (Sölch 1922) describing earth hummocks in the Stubalm mountains (CEA). The above-mentioned term "Buckel" or "Buckelwiese" (literally translated "hummock meadow") was introduced by Morawetz (1952) for such landforms in the Lavanttal Alps (SEA). Subsequently, these landforms were described in several other areas in Austria as for instance in the Seetaler Alpen, CEA (Eisenhut 1963). In a study of earth hummocks at

Peischlachtörl, Schober Mountains (CEA), Austria, Keuschnig et al. (2007) showed that differences exist in the vegetation between the top of the hummocks and the depressions between the hummocks. Earth hummocks are common on the flat to gently sloping alpine pastures as for instance near the Schönwies hut, Ötztal Alps, CEA (Kerschner et al. 2014). Tomaselli et al. (2018) mapped and studied three hummock areas in north-eastern Dolomites and in western Venetian Prealps (SE Italian Alps, SEA).

Hummock meadows are reported from several sites in the Northern Calcareous Alps (NEA) as for instance in Germany at Berchtesgaden (Königsee, Bischofswiesen), at Mittenwald (Krün, Elmau, Leutasch), in the Bayrischzell Valley, near Pfronten and at Oberjoch in the Allgäu; and in Austria at the Tannheimer Valley, or at the Achen Lake (Ebers 1959). Earth hummocks are also reported from the Western Alps. After first observations in the 1950s in Piemonte region (valleys of Ossola and lower Dora Baltea in NWA, and of Varaita and Maira in SWA, Veyret 1956), and in the 1960s in Ortles-Cevedale Group and Sole Valley (Southern Rhaetian Alps, SEA, Recami 1967), Paro (2011) reported about one site of earth hummocks in a frontal depression of a relict block stream in Lanzo Massif (Graian Alps, NWA) at 1313 m. Pintaldi et al. (2016) analysed earth hummocks in the LTER (Long Term Ecological Research) monitoring site of Tellinod, 2118 m, in the Aosta Valley (NWA). CNRS-Caen and Institut de Géographie Aix-en-Provence (1980) and Van Vliet-Lanoë (2014) described in detail silty earth hummocks at the Valonnet, at 2500 m, in a permafrost-free environment in the Southern French Alps (SWA).

3.2.5 Other Periglacial Processes and Landforms in Flat Areas

With regards to sedimentary successions in flat areas, diagnostic criteria for identifying "periglacial" or "cold-climate" river deposits and forms are not simple, and periglacial conditions during (syngenetic) or after (epigenetic) deposition must be deduced from sedimentary structures. These are periglacial deformation structures, for example, ice-wedge pseudomorphs, or cryoturbations potentially occurring at fluvial sediments which build up river terraces (French 2007). Ice-wedge pseudomorphs appear near the surface, which can abruptly narrow with increasing depth (Furrer and Fitze 1970). The associated cryoturbations are distinct deformations near the edges of the ice-wedge pseudomorphs or soil movements due to frost action. Ice-wedges pseudomorphs and cryoturbation in the Alps are hardly reported in literature (i.e., along the Wechsel road—B54—at the eastern margin of the Alps; Morawetz 1968), although they are well known in the northern Alpine foreland reaching several metres in depth (e.g., Bachmann 1966; Furrer and Fitze 1970; Van Husen and Reitner 2001). In contrast, only poorly developed cryoturbation and ice-wedge pseudomorphs have been reported from Pleistocene gravel and sand deposits in the eastern Alpine foreland around Graz (Morawetz 1968).

Relict large-scale polygonal patterns at the surface have been studied in detail by Pech et al. (2021) at the Col du Noyer, Massif du Dévoluy, Dauphiné Prealps (SWA) at an elevation of ca. 1660 m. Their orthophoto interpretation analysis resulted in an inventory of 342 polygons. Some polygons are very large exceeding 80 m^2 and more than 5 m in diameter. These dimensions exceed polygons currently active in the Alps. For instance, the diameters of the polygonal patterns in the Upper Ubaye and Mortice valleys (Southern French Alps, SWA) measure between 0.8 and 1.5 m in diameter (CNRS-Caen and Institut de Géographie Aix-en-Provence 1980). Pech et al. (2021) conclude that the relict large-scale polygonal patterns at Col du Noyer is a rare conservation of the past conditions for the formation of polygonal grounds in the Alps, i.e., suitable conditions for thermal-contraction cracking (French 2007). Active polygonal patterned grounds are found at altitudes above 2500 m, at least in several sites in the Hautes-Alpes, SWA (France), for example on the Plateau de Bure, in the Dévoluy, on the Mortice ridge, east of Vars, or at the foot of the Tête de Vautisse in the south of the Écrins (or Pelvoux) Massif (Pech et al. 2021). However, polygonal patterns are rare in the Alps (D'Amico et al. 2019).

Compact and dense layers with strong platy/lenticular structural aggregation, ice-wedge pseudomorphs, and large-scale cryoturbations were described by D'Amico et al. (2019) on stable surfaces between 600 and 1600 m. in the Upper Tanaro valley, Ligurian Alps, Italy (SWA), characterised by strongly developed relict periglacial surface morphologies such as large-scale patterned ground, block fields/block streams, or solifluction sheets. Weak signs of cryoturbation are reported in a well-developed palaeosol covered under the active periglacial stone cover at the Stolenberg Plateau, at 3030 m at the foot of the Monte Rosa Massif, Pennine Alps (NWA) (Pintaldi et al. 2021). Cryoturbation as folds and deformations in till deposits of main arc-shaped terminal moraines at Ivrea, NWA (Capello 1963), Varese, NWA, and Lago di Garda, SEA (Nangeroni 1959) in the Po Plain, and possible cryoturbations on the loess cover at Bagaggera (310 m), south of Lake Como, SEA (Oliva et al. 2018), have also been detected.

3.3 *Slope Processes and Related Landforms*

3.3.1 Context and Acting Processes

French (2007) pointed out that our understanding of slopes in cold-climate environments was (during that time) and partly still is limited. Despite substantial advances in the recent decades, the link between slope form and process often remains unclear. Many periglacial landscapes are presumed to be in disequilibrium with current climatic conditions or are still strongly morphologically influenced from past climatic conditions. One example of a landform caused by periglacial and possibly other (previously acting) processes are tors, where different hypotheses have been proposed to explain their origin (Goudie 2004; Ballantyne 2018). Furthermore, French (2007) pointed out that there is no slope landform or slope assemblage, which might be regarded as pure periglacial in nature, as discussed in Sect. 3.2. As a consequence of long-living morphologies on slopes, present periglacial processes are reshaping significantly older palaeo-slopes and—vice versa—palaeo-periglacial landforms, which were formed during the Late Pleistocene get currently reshaped by non-periglacial processes. In this section we elaborate, first, on rock walls and talus slopes, focus subsequently on rock glaciers and related landforms, discuss solifluction and derived landforms, and complete with debris-mantled slopes.

3.3.2 Rock Walls and Talus Slopes

Weathering on rock walls has been discussed in Sect. 3.1. In this section we focus on the geomorphic periglacial landsystem "rock wall and talus slope". Whereby, periglacial landsystems are assemblages of characteristic periglacial landforms and connected sediments. Rock wall and talus slope-landsystems are a very common component of the Alpine environment (Messenzehl et al. 2017). They shape the upper parts of the landscape with two typical and rather simple landforms: (a) a vertical or near-vertical rock outcrop or rock face, typically a few tens to several hundred metres high, exposed to mechanical and bio-chemical weathering actions; (b) a talus slope composed of gravitationally-accumulated rock debris located at the foot of the rock wall, laterally sorted and often also impacted and modified by other processes such as avalanches, debris flows, creep or solifluction. This model relates to the free-face model used by French (2007) with a rock wall, a talus slope and finally a gentler footslope. In the Alps, such geomorphic system was first investigated by Francou (1982) in the French Alps, with morphological, statistical, climatic, and physico-empirical approaches.

Active alpine rock wall and talus slopes including talus cones are typically found above ca. 1000–1500 m, where frost shattering starts to be significantly efficient (Figs. 8A and B), and become rare above ca. 3200–3400 m, where glaciers are predominant and actively transport debris downslope. It is commonly admitted that the accumulations of debris presently stored in Alpine valleys started to accumulate on the top of bedrock or older sediments once the LGM glaciers retreated. Locally, talus slopes can be found down to very low altitude, for instance in Austria, where Stiegler et al. (2014) studied the presence of permafrost at a low-elevated, undercooled talus slope in the Niedere Tauern Range (CEA), Austria at ca. 1000 m. Independently from the activity state of the feeding rock wall, micro-climatic conditions specific to the coarse material allow undercooling of the subsurface and limit the development of soil and vegetation over such terrain (Wakonigg 1996, 2001; Delaloye et al. 2003).

Geological parameters, such as lithology and structural characteristics (joint density, porosity), strongly control the development of both rock wall and talus slope, but a topographical predisposition caused by past glacio-geomorphological activity, such as glacial cirque, can also be important. Investigating more than 200 talus slopes in the Turtmann Valley (Valais—or Pennine—Alps, NWA), Messenzehl et al. (2017) showed that the spatial variability of rock fall activity is the product of complex interplay between geological settings, glacio-geomorphological history and present-day topo-climatic factors. Those latter especially influence the effects of freeze–thaw cycles on debris production (Sass 2005; Matsuoka 2008; Messenzehl

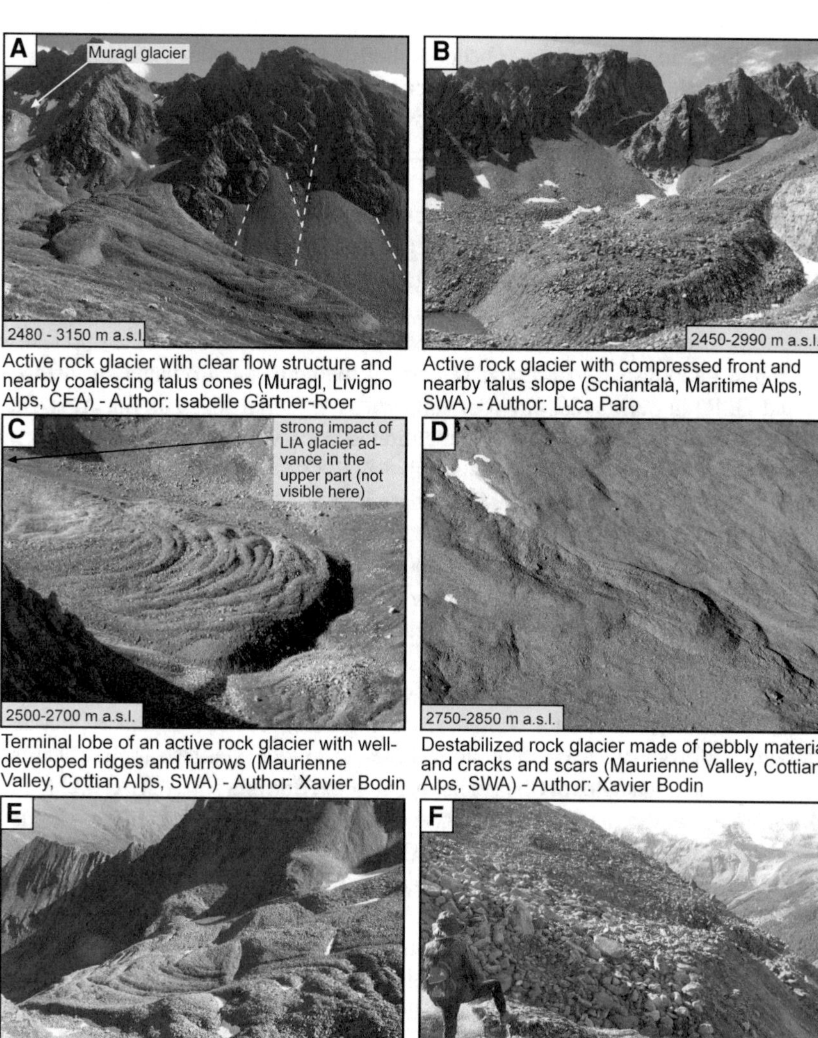

Fig. 8 Pictures of talus slopes, talus cones, and present-day active and/or destabilised rock glaciers

and Dikau 2017; Kellerer-Pirklbauer 2017), whereas the role of permafrost has also been pointed out, especially in a context of increasing temperature, but mostly in order to understand rock falls of large amplitude and low frequency, as exemplified by Ravanel et al. (2017) in the Mont Blanc massif.

Talus slopes in the Alps have been studied frequently using different types of geophysical techniques. Some studies focus on the internal structure and thickness of such sediment deposits (Hoffmann and Schrott 2003; Sass 2008), whereas other studies aimed to quantify permafrost existence or thickness of permafrost lenses (Stiegler et al. 2014; Kellerer-Pirklbauer 2019). Otto and Sass (2006) compared three geophysical methods for talus slope investigations in the Hungerli Valley, Turtmann, Valais Alps (NWA) using ground penetrating radar (GPR), electrical resistivity tomography (ERT), and seismic refraction (SR). Their combined geophysical results produced a detailed image of the studied talus slopes and adjacent landforms. Hauck and Kneisel (2008) elaborated a comprehensive book about the application of different geophysical techniques (electrical methods, electromagnetic methods, SR, and GPR) in periglacial environments with a regional focus on the Alps. Sass (2008), for instance, presented in the mentioned book GPR data used to determine thickness and inner structure of a talus slopes in the Lechtal Alps (NEA) and Stubai Alps (CEA). Schrott and Sass (2008) discussed the advances and limitations of three different types of field geophysics in geomorphology (GPR, ERT, SR) exemplified by case studies; one was a talus slope in the Stubai Alps (CEA). Geophysical results indicate that the uppermost part of a talus body is commonly characterised by a surface-parallel structure possibly related to a succession of stacked finer and coarser layers (Sass 2008).

Hauck et al. (2008) developed a model, which determines the volumetric fractions of rock/soil, water, air, and ice contents by using ERT and SR data and applied it at different landforms, one was a talus slope. Later, this model was also applied at alpine rock glacier, as for instance at Chastelets rock glacier, Bernina Alps (CEA) (Schneider et al. 2013). However, the first seismic measurements on a rock glacier in the Alps date back to 1968 on the Macun I rock glacier (Lower Engadin, Sesvenna Alps, CEA) in Switzerland, followed by the Murtèl-Corvatsch rock glacier in 1971 (Upper Engadin, Bernina Alps, CEA) (Barsch 1969, 1996). First geoelectric measurements on Alpine rock glaciers were carried a few years later in 1972 by Barsch and Schneider, although never published (Barsch 1996). The combination of several methods to quantify the inner structure of rock glaciers has also been applied in other rock glaciers in the Alps as for instance in the Ötztal Alps, Austria (CEA) by Hausmann et al. (2012), or in the Turtmann Valley, Valais Alps (NWA) by Merz et al. (2016) also using helicopter-borne ground-penetrating radar (relevant rock glacier indicated in Fig. 9F).

3.3.3 Rock Glaciers and Related Landforms

Rock glaciers as well as ice-cored moraines are considered as morphological indicators for the occurrence of permafrost in high mountain environments (e.g., Haeberli 1985; Barsch 1996; Humlum 2000). In accordance with the IPA Action Group "Rock glacier inventories and kinematics" (RGIK 2020), rock glaciers as landforms are defined here by their characteristic morphology related to long-term cohesive creep of ice-rich material. In details, active rock glaciers display convex and elongated shape with over-steepened fronts and lateral talus as well as a typical microtopography

Protalus rampart - relict (Brandstätter cirque, Seckauer Tauern, Niedere Tauern Range, CEA) Author: Andreas Kellerer-Pirklbauer

Rock glacier - relict (Hölltal [german "hell valley"], Seckauer Tauern, Niedere Tauern Range, CEA) Author: Andreas Kellerer-Pirklbauer

Rock glacier - pseudo-relict (Reichart cirque, Seckauer Tauern, Niedere Tauern Range, CEA) Author: Andreas Kellerer-Pirklbauer

Terminus of a relict rock glacier with complex ridge-and-furrow morphology (Combeynot Massif, Dauphiné Alps, SWA) - Author: Xavier Bodin

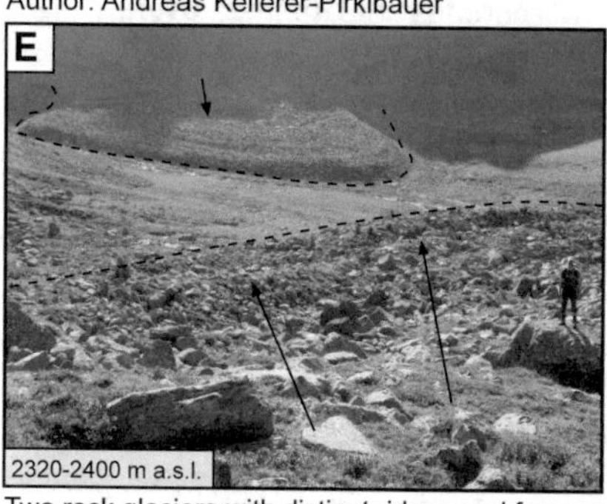

Two rock glaciers with distinct ridges and furrows - relict (Schober Mountains, Hohe Tauern Range, CEA) - Author: Andreas Kellerer-Pirklbauer

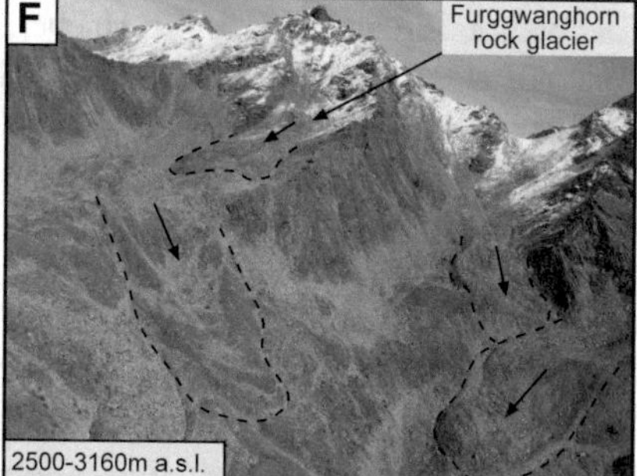

Rock glaciers of different generations, Hungerli hanging valley - relict to active (Turtmann, Valais Alps, NWA) - Author: Isabelle Gärtner-Roer

Fig. 9 Pictures of talus slopes and associated landforms: present-day relict, pseudo-relict, and intact (i.e., active or inactive) examples of furrows and ridges that reflect the internal process of deformation. Figure 8A–F depicts six different active rock glaciers in the Alps, ranging from the Maritime Alps and Cottian Alps (both SWA) in the southwest, to the Hohe Tauern Range (CEA) in the east. It is generally accepted that the morphological characteristics of rock glaciers are the product of high (excess) ice contents, as well as a complex flow behaviour controlled by negative (or near-zero) subsurface temperatures, water circulation and highly anisotropic material properties (Wahrhaftig and Cox 1959; Haeberli 1985; Haeberli et al. 1998; Merz et al. 2016; Cicoira et al. 2019).

Rock glaciers occur in different types of lithologies, which break down to particles of at least gravel size (Ikeda and Matsuoka 2006) and if other conditions are favourable (Barsch 1996). However, coarse-grained sediments favour cooler near-surface temperatures and hence rock glacier existence (Wagner et al. 2019). In the central Swiss Alps rock glacier occur in all kinds of crystalline rocks (Barsch 1996). The formation of rock glaciers in the south-western Alps are favoured by massive rocks such as granite, gneiss, limestone, and sandstone (Evin 1987). Kellerer-Pirklbauer (2007) studied the lithology and the distribution of 332 rock glaciers in central Austria ranking lithologies from more to less suitable for rock glacier formation: granite and granitic gneiss (orthogneiss), paragneiss, amphibolite, quartzite, mica schist, and carbonates (including marble). The stratigraphy of active rock glaciers is typically described by three layers. The uppermost metres are often characterised by large boulders allowing a high porosity and voids filled with air, snow or water. However, fine-grained ("pebbly"; Ikeda and Matsuoka 2006) rock glaciers are also found (Fig. 8D), bearing a large proportion of fine material and few coarse blocks, and show a similar but smoothed morphology and thinner overall thickness (Wahrhaftig and Cox 1959; Haeberli 1985; Kääb and Weber 2004; Ikeda and Matsuoka 2006; Frehner et al. 2015). This shallow layer is covering an ice-rich permafrost layer with 50–70% of ice and about 30–50% of debris (Barsch 1996; Arenson et al. 2002). The lowermost layer is often characterised by large boulders, that were deposited at the front and overridden by the advance of the landform (Swift et al. 2015). Borehole measurements showed that a large part of the deformation occurs in a thin shear horizon at a typical depth of 20–30 m (Arenson et al. 2002; Springman et al. 2012).

In relation to their ice content, flow behaviour and morphological characteristics, rock glaciers are distinguished into different activity types: active, inactive, pseudo-relict, relict and destabilised (Barsch 1996; Roer et al. 2008; Kellerer-Pirklbauer 2019). To distinguish between active, inactive, relict and destabilised ones, multi-temporal surface data of the rock glacier of interest derived from remote sensing or geodetic measurements are required (Kaufmann et al. 2021). The fifth type pseudo-relict cannot be distinguished from relict ones solely by surface observations or remote sensing data, instead, subsurface information about permafrost presence based on direct or indirect methods is needed (Kellerer-Pirklbauer 2019). Commonly, several landforms of different activity can be superimposed on each other (Fig. 9F), forming a sequence of different landform generations (Swift et al. 2015). Figure 9 depicts six examples of relict and pseudo-relict rock glaciers including a protalus rampart (Figs. 9A–E) as well as a more complex assemblage of active to relict rock glaciers in the Hungerli Valley, Valais Alps (NWA).

A comparison of morphologies and thermal conditions of active, inactive and relict rock glaciers is given by Ikeda and Matsuoka (2002). An overview of the morphology of destabilised rock glaciers, based on an inventory in the French Alps, is provided in Marcer et al. (2019). In a strict sense, the activity of a rock glacier is determined based on the knowledge on the movement of the landform. However, due to the difficulties to get kinematic data particularly in the past, many earlier studies assessed the activity by morphological characteristics (Wahrhaftig and Cox 1959; Haeberli 1985; Barsch 1996; Burger et al. 1999; Ikeda and Matsuoka 2002; Nyenhuis et al. 2005). Nowadays, an increasing number of remote sensing data help to detect and monitor rock glaciers even in remote areas (Strozzi et al. 2020; Kääb et al. 2021).

The typical morphology of an active rock glacier is a steep terminal front (> 35°) with loose boulders and without vegetation at the front and in many cases also at the surface (Figs. 8A–E). The front of advancing rock glacier might reach topographic positions where they might cause potential threat to lower elevated infrastructure or settlements (Fig. 8F). Inactive rock glaciers, which still contain widespread permafrost but do not move (due to climatic or topographic reasons, Barsch 1996; Ikeda and Matsuoka 2002), have a gentler front with stable boulders and partial or full vegetation. Related to the widespread occurrence of permafrost ice, active and inactive rock glaciers are grouped as "intact" rock glaciers (Haeberli 1985). Pseudo-relict rock glaciers appear to be visually relict with a subsided surface due to the almost complete thawing of the ice component, but still contain patches of permafrost in the upper part whereas the lower part might be partly covered by vegetation (Fig. 9C). This type is only detectable by ground temperature measurements (borehole), or geophysical measurements. The rock glacier depicted in Fig. 9C (Reichartkar rock glacier; "Kar" German word for "cirque") might be regarded as the *locus typicus* for pseudo-relict rock glaciers (Kellerer-Pirklbauer 2019). Relict landforms, finally, show a subsided surface due to the complete disappearance of the ice, with the surface morphology still visible but the front having lower slope angle and sometimes a dense vegetation (Roer 2007). Relict and pseudo-relict rock glaciers are important indicators for the former permafrost distribution and therefore they are considered for palaeoclimatic reconstructions (Kerschner 1985; Konrad et al. 1999; Frauenfelder et al. 2001). For several rock glaciers in the Alps their age was assessed by absolute and relative age-determination methods indicating typical ages of several thousand years (Frauenfelder and Kääb 2000; Haeberli et al. 2003; Kellerer-Pirklbauer 2008; Fuchs et al. 2013; Krainer et al. 2014; Reitner et al. 2016). Many nowadays relict rock glaciers were formed between Heinrich Stadial 1 and the Younger Dryas and stabilised during the early Holocene (Charton et al. 2021) or are even substantially older developing shortly after or even simultaneously with the retreat of pre-Bølling glaciers and stabilised around 15.7 ka (Steinemann et al. 2020).

The more recently investigated morphology of destabilised rock glaciers is typically characterised by surface "disturbances" (as compared to non-destabilised rock glaciers) such as cracks, crevasses and scarps (Roer et al. 2008; Bodin et al. 2016;

Scotti et al. 2017; Marcer et al. 2021). These features (e.g., visible in Figs. 8D and E) depict clearly distinctive mechanical processes compared to those of "normal" rock glaciers, probably induced by higher deformation related to near-zero ground temperature, significant water content, and topography (Ikeda et al. 2008; Roer et al. 2008; Cicoira et al. 2021).

Active rock glaciers typically creep downslope with velocities of several decimetres up to some metres per year (Roer 2007; Delaloye et al. 2010; Kaufmann et al. 2021). This creep behaviour results from internal deformation of the frozen debris and shearing in a narrow horizon at typical depths of around 20 m (Arenson et al. 2002; Cicoira et al. 2021). Reliable information on the kinematics of rock glaciers provides insights into the evolution of ice-rich permafrost on mountain slopes and facilitates the analysis of its dynamics (Roer et al. 2005a; Haeberli et al. 2006; Kääb et al. 2007; Bodin et al. 2009; Kellerer-Pirklbauer et al. 2018). Different terrestrial as well as remote sensing techniques are applied to quantify the kinematics of permafrost creep, i.e., horizontal surface velocities and vertical surface changes, respectively (Kääb et al. 1997, 2021; Lambiel and Delaloye 2004; Strozzi et al. 2004, 2020; Kääb 2005; Roer et al. 2005b; Haeberli et al. 2006; Roer 2007; Kaufmann et al. 2021). In 1923 and 1924 Finsterwalder carried out two terrestrial photogrammetric surveys to measure a velocity profile across the Innere Ölgrube rock glacier, Ötztal Alps, Austria (CEA), quantifying a maximum flow velocity of 50 cm a^{-1} (Finsterwalder 1928; Kaufmann 2012). This was the first time that rock glacier displacement was measured in the entire European Alps. Pillewizer (1957) began in 1938 to quantify flow velocities at the Hochebenkar rock glacier, also Ötztal Alps (CEA), which is still ongoing. This rock glacier has the worldwide longest record of flow velocities and has been measured since 1938 by terrestrial photogrammetry (Fig. 3A), since 1951 by terrestrial geodetic methods, and since 2008 by differential GPS (Vietoris 1958, 1972; Schneider and Schneider 2001; Kaufmann 2012; Hartl et al. 2016).

Measurements over the last decades indicated that many landforms accelerated, related to rising permafrost temperatures and changing water contents—in the Alps and in many parts of the world (Kääb et al. 2007; Ikeda et al. 2008; Delaloye et al. 2008, 2010; Eriksen et al. 2018; Kellerer-Pirklbauer et al. 2018; Cusicanqui et al. 2021). In addition, seasonal and short-term changes in horizontal velocities appear to be regionally synchronised, possibly reflecting a short-term response to changing ground-surface temperatures and changing hydrology (Kääb et al. 2007; Delaloye et al. 2010; Kenner et al. 2017; Cicoira et al. 2019). In the context of permafrost degradation, some rock glaciers show acceleration with extraordinary horizontal velocities of up to 100 m a^{-1} (Delaloye et al. 2010; Eriksen et al. 2018). A mass-balance approach of rock glacier evolution has been proposed by Kellerer-Pirklbauer and Kaufmann (2018) as well as Cusicanqui et al. (2021) over multi-decadal periods. The latter study suggests a complex influence of snow and debris accumulation as well as surface thawing and basal conditions. Kaufmann et al. (2021) revealed that despite maximum rock glacier creep velocities exceeding 4 m a^{-1} (2013–2020) and long-term photogrammetric data (1954–2018), no significant volumetric loss or negative mass balance was revealed for the Leibnitzkopf rock glacier (Schober

Mountains, Hohe Tauern Range, CEA) over a 64-year period. This implies that even ice-poor rock glaciers might be highly active.

Relict and pseudo-relict rock glaciers are of particular interest in marginal permafrost areas of the Alps as for instance Slovenia with only 3.6 km^2 with a permafrost index $\geq$0.5 (Boeckli et al. 2012). Colucci et al. (2016a, b) worked on rock glaciers, protalus ramparts and pronival ramparts in the north-eastern-most region of Italy (Friuli Venezia Giulia region, SEA) and Slovenia inventorying 53 rock glaciers and 66 protalus and pronival ramparts. Colucci et al. (2016a, b) mapped 11 rock glaciers and more than 30 protalus or pronival ramparts in the Slovenian part of the study area (mainly Julian Alps, Karawanks, and Kamnik Alps). Boccali et al. (2019) studied the geometry and estimated the palaeo-ice content of the 52 previously inventoried rock glaciers of Colucci et al. (2016a, b). One rock glacier in Slovenia was excluded by Boccali et al. (2019) from the previous inventory after field verification. Based on both inventories, 6 rock glaciers exist in the Julian Alps and two each in the Karawanks and Kamnik Alps summing up to 10 Slovenian rock glaciers. This is in a substantial contrast to Austria, where the most-recent inventory lists 5,769 rock glaciers and protalus ramparts for most of the alpine areas of the country (Wagner et al. 2020).

3.3.4 Solifluction and Derived Landforms

Solifluction is generally considered as the most widespread form of slope movement in periglacial environments (French 2007). In the Alps, large areas are dominated by diurnal freezing processes and seasonal freeze–thaw cycles, affecting depths of decimetres to metres (Haeberli and Gruber 2008). Here, thermal conditions and the temporary occurrence of ground ice are the most effective drivers for cracking, heaving, settling or cryoturbation processes and the development of related landforms (Swift et al. 2015). Where mixtures of fine and coarse clasts occur, sorting processes lead to frost patterned ground and at inclined slopes to sorted stripes (Fig. 10A, B, F), which might be even diagonal related to local topo-climatic conditions (Fritz 1976). Andersson (1906) described the solifluction process originally as "slow flowing from higher to lower ground of masses of waste saturated with water". As this definition is also valid for non-periglacial climates, often the term "gelifluction" was also used commonly in the past, implying the influence of seasonally or permanently frozen ground (French 2007; Jaesche 1999). Nowadays, however, solifluction is considered in a broader sense and combines the four processes of needle-ice creep (cf. Fig. 6A), frost creep, gelifluction (considered as the movement of saturated soil associated with ground thawing during spring), and plug-like flow (Matsuoka 2001).

Repeated freeze–thaw processes and gelifluction in the subsurface lead to the formation of characteristic tongue-shaped earth lobes with a typical width of several metres and a length of up to hundred metres. Matsuoka (2001) concluded that the riser height of a lobe might be indicative of the maximum depth of movement and prevailing freeze–thaw type. Smaller forms are often grouped, entire slopes

Detail of diagonal stone stripes (or diagonal garlands) - active (Hochschwab Mountains, NEA)
Author: Andreas Kellerer-Pirklbauer

Overview of diagonal stone stripes (diagonal garlands) - active (Hochschwab Mountains, NEA)
Author: Andreas Kellerer-Pirklbauer

Turf- and stone-banked solifluction lobes (Fallbichl, Goldberg Mountains, Hohe Tauern Range, CEA) - Author: Andreas Kellerer-Pirklbauer

Turf-banked solifluction lobes and block channels (Hinteres Langtalkar, Hohe Tauern Range, CEA)
Author: Andreas Kellerer-Pirklbauer

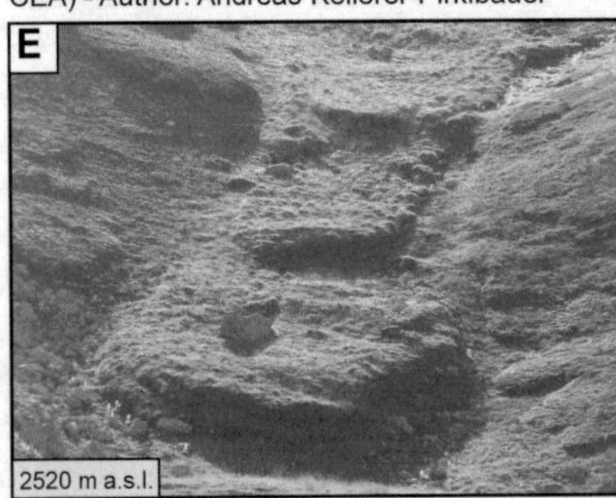

Three distinct generations of solifluction lobes near Sommeiller pass (High Susa Valley, Graian Alps, NWA) - Author: Luca Paro

Frost heave monitoring on sorted stripes using dilatometers - active (Valletta, Albula Alps, NEA)
Author: Andreas Kellerer-Pirklbauer

Fig. 10 Pictures of periglacial mass movement on slopes and other related landforms: present-day active examples

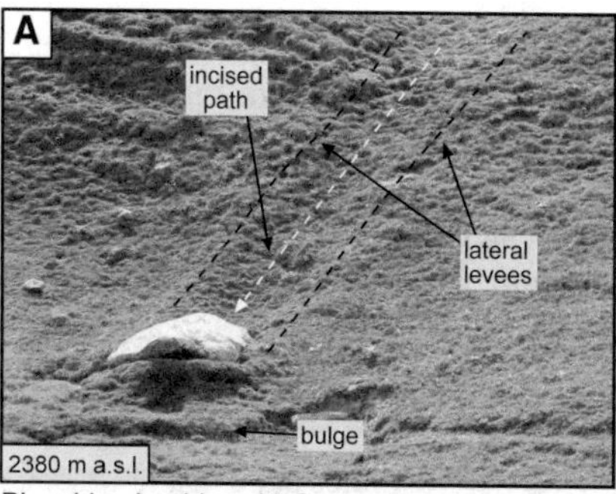

Ploughing boulder with former transport path, levees and bulge - relict (Gardetta plateau, Cottian Alps, SWA) - Author: Luca Paro

Group of ploughing boulders with a summit tor as source area - relict (Koralpe, Lavantaler Alps, CEA) - Author: Andreas Kellerer-Pirklbauer

Solifluction sheet - relict (Seckauer Tauern, Niedere Tauern Range, CEA)
Author: Andreas Kellerer-Pirklbauer

Block stream - relict (Seckauer Tauern, Niedere Tauern Range, CEA)
Author: Andreas Kellerer-Pirklbauer

Block stream - relict (Lanzo Massif, Graian Alps NWA)
Author: Luca Paro

Rectilinear and convexo-concavo debris-mantled slopes (Seckauer Tauern, Niedere Tauern Range, CEA) - Author: Andreas Kellerer-Pirklbauer

Fig. 11 Pictures of periglacial mass movement on slopes and other related landforms: present-day relict examples might be characterised by coalescing solifluction landforms (Fig. 10C), or individual solifluction lobes might show several generations (pulses) of lobes sitting on each other (Fig. 10E). Monitoring of solifluction might include surface velocities, vertical velocity profiles, depths of movement, and volumetric velocities, often on an annual basis (Matsuoka 2001). In the Alps, the surface velocity of solifluction lobes is in the range of 1–20 cm a^{-1} (Matsuoka 2001; Jaesche et al. 2002; Veit 2002; Kellerer-Pirklbauer 2018). Solifluction studies in the Alps suggest that solifluction intensity is strongly driven by the annual depth of soil freezing, by needle-ice creep, or by gelifluction (thus soil moisture) depending on the site. Furthermore, longer zero curtain periods, longer seasonal ground thawing periods, later start of the seasonal snow cover, more freeze–thaw cycles, and cooler early summer temperatures seem to promote solifluction (Kellerer-Pirklbauer 2018). The distribution of solifluction lobes has been documented in the Alps by, among others, Furrer (1954, 1965), Höllermann (1967), or Stingl (1969). Already in earlier times, they have been assigned as indicators for the periglacial altitudinal belt (Büdel 1937; Troll 1944).

In France, earliest measurements related to solifluction started in 1947 with the long-term monitoring of painted stone lines on the slopes of Chambeyron Group (Cottian Alps, SWA) and the detailed description of morphometry, granulometry, soil moisture and vegetation (Michaud 1950). These data were used to investigate climatic controls without meteorological data (Pissart 1993). In Switzerland, more precisely in the Swiss National Park (Livigno Alps, Sesvenna Alps, Ortles Alps of the Ortles-Cevedale Group, mainly CEA), Furrer (1954) was the first observing solifluction lobes and related processes in detail. His measurements were continued in a systematic monitoring programme until today and enabled the identification of influence factors, such as air temperature and snow depth (Gamper 1984; Rist et al. 2013). Another investigation site is the Blauberg at Furkapass in the Central Swiss Alps (Lepontine Alps, NWA) led by H. Veit and A. Rist (University of Bern, Switzerland). Here, solifluction occurs between 2380 and 2700 m with seasonal frost on the lower slope and permafrost conditions on the upper slope. Beside the full range of meteorological measurements, focus is given on horizontal velocities and vertical changes of the lobes on various temporal scales.

In Austria, solifluction landforms and processes have been studied by Höllermann (1967), Stingl (1969), Stocker (1973, 1984), Höfner (1993), or Klinge and Lehmkuhl (1998). A spatial focus in solifluction monitoring was an area close to the Glorer mountain refuge (Schober Mountains, Hohe Tauern Range, CEA), where a monitoring plot was initiated in 1985 (Veit and Höfner 1993; Veit et al. 1995) and extended in 1994 (Jaesche 1999; Jaesche et al. 2003). Solifluction monitoring activities at this site and generally in Austria, however, decreased substantially after the turn of the millennium. One of the most recent studies about solifluction in Austria was about five solifluction lobes in the Hohe Tauern Range, CEA (two sites are shown in Figs. 10C and D), which were monitored during a four-year period (Kellerer-Pirklbauer 2018). Annual rates of surface velocity, vertical velocity profiles, depths of movement, and volumetric velocities were quantified using near-surface markers and painted lines and related to air temperature, soil texture, and ground temperature-derived parameters. Results indicate mean surface velocity rates of 3.5–6.1 cm a^{-1}

(near-surface markers) and 6.2–8.9 cm a^{-1} (painted lines), respectively, indicating a high relevance of needle-ice creep at the studied landform. Lehmkuhl (1989) also discussed the solifluction lobes shown in Fig. 10C (Fig. 8 in his work) and pointed out that such turf-banked solifluction lobes could also termed as earth flows.

Ploughing boulders (or ploughing blocks) are commonly individual boulders that move downslope faster than their surrounding material by processes related to different types of solifluction, particularly to annual freeze–thaw cycles (Goudie 2004; Feuillet 2014). Ploughing boulders are some of the most widespread periglacial features occurring on moderately inclined slopes in all periglacial environments in the Alps (Höllermann 1967; Morawetz 1971a; Fritz 1976). They occur particularly in the same altitudinal belt as solifluction lobes and active forms move typically with a few centimetres per year. Source areas of boulder might be allochthonous (e.g., glacial erratic boulder) or autochthonous (rock outcrop further upslope). The rate of movement directly relates to the slope gradient (Ballantyne 2001). By moving downhill, the boulder pushes a bulge of fine material in front of it, which also shows itself as an upslope tail with lateral levees or ridges on both sides, indicating the former displacement path of the boulder (Fig. 11A). A depression is typically found directly behind the boulder. Even after inactivation, ploughing boulders with their morphological expressions of former movement remain to be distinct small-scale landforms in the landscape. Such examples are widespread in for instance the Lavantal Alps, Austria, CEA (Fig. 11B), where the most-recent main period of periglacial activity was during the Late Glacial period (Fig. 11B).

In contrast to ploughing boulders there are "braking boulders". Braking boulders (or "Bremsblock" in German, "blocchi frenanti" in Italian) are individual boulders that move downslope slower (or not at all) than their surrounding finer material by processes related to seasonal frost. Thus, in such cases only the fine material is displaced downslope, whereas the coarse material-component is basically stable (Höllermann 1967). In the brake boulder-case, a depression might be found directly in front of the boulder due to the rather fast removal of finer material (Höllermann 1967). Braking boulders were hardly considered in periglacial studies in the Alps. One of the few exceptions is a detailed geomorphological study by Klinge and Lehmkuhl (1998) working on periglacial landforms in the Piff Cirque, Hohe Tauern Range, Austria (CEA).

Further prominent landforms on solifluction-affected slopes in the Alps at present or in the past are, for instance, solifluction sheets, block streams or block channels. Solifluction sheets are large-scale and extensive landforms existing on slopes and which developed from mass-wasting of glacial or periglacial sediments mainly by solifluction (Fig. 11C). Block streams are covers of coarse debris that have accumulated by mass movement on slopes (Fig. 11D) or valley floors (Fig. 11E). Block streams—as well as block fields—occur on a wide range of lithologies but are particularly common on well-jointed metamorphic and igneous rocks that have weathered to produce abundant blocks and only limited amounts of fine material. Below the openwork surface layer, block streams usually contain an infill of fine sediments (Goudie 2004). The presence of extensive relict block streams and block fields is also reported from the Graian and Pennine—or Valais—Alps, NWA. There, the largest area covered by such features is located at the Ultramafic Lanzo Complex (Lanzo Massif, Graian Alps, NWA) where 25 km^2 (18% of the total Lanzo area) at an elevation of between 1000 and 1650 m are covered by block streams and block fields (Fig. 4D). These landforms could be related to permafrost conditions during the LGM (Fioraso and Spagnolo 2009; Paro 2011). Other block streams and block fields are found at the Alpine margin surrounding the Piedmont between Sangone Valley (50 km S Turin) and Oropa (150 km NE of Turin) (Paro 2011). Finally, block channels (German "Blockrinnen" or "Steinstreifen") are elongated strips of debris in the direction of the slope, often—but not always—following a gently incised trench on the slope (Flohr 1935). Such hardly studied features—at least in the Alps—are depicted in Fig. 10D.

3.3.5 Debris-Mantled Slopes

Debris-mantled slopes are widespread landforms in the present and past periglacial belt of the Alps (e.g., Richter 1900; Spreitzer 1960; Höllermann 1967; Morawetz 1971a; Fritz 1976). However, the link between slope form and process of periglacial slopes is still unclear (French 2007). Many of the debris-mantled slopes were significantly formed during the Pleistocene with minor reshaping during the Holocene. Such slopes might be convex, rectilinear (uniform), or concave without obvious bedrock outcrops, with hillslope tors, or with significant rock walls overlooking the slope (French 2007).

Convex slope elements generally indicate erosional processes, which increase with slope length. Soil creep and weathering has been identified as the dominating process combination for convex slopes (Goudie 2004). The uppermost part of the convex slope close to the drainage divide is also commonly affected by deflation processes related to an absence of a protective winter snow cover and strong winds at the crest (Fig. 11F). Rectilinear slopes generally point to spatially homogeneous erosion or denudation conditions, often related to solifluction processes forming widespread sheets, block streams, block strips, or similar landforms. Such slopes might retreat parallel causing a rectilinear—or uniform—shape. Therefore, in the German literature the term "Glatthang" (meaning straight slope) is often used (Spreitzer 1957). Morawetz (1971a) recognised 50 years ago that the debris covering rectilinear slopes mostly does not originate from rock walls overlooking the respective slope but develop in situ at the slope itself. This observation is related to an assumed balance between debris supply and debris removal at such slopes (French 2007). Another mechanism for the formation of rectilinear slopes might be through rock wall retreat and the formation of a rectilinear slope at the angle of rest of the accumulating talus downslope thus through gravitative processes (Goudie 2004). Such types of rectilinear slopes are sometimes termed as Richter denudation slope, named after Eduard Richter, who described such slopes in the Alps more than 120 years ago (Richter 1900). Finally, concave parts of the lower part of hillslopes are explained by sediment accumulation (Goudie 2004). However, as hillslopes in the Alps are complex phenomena with an evolutionary history over long timescales exceeding the entire Pleistocene, many interactions of processes and components can occur implying that different processes might lead to the formation of debris-mantled slopes.

Debris-mantled slopes in different areas in the eastern part of the CEA (Hohe Tauern Range, Niedere Tauern Range, Ötztal Alps, and Lavantal Alps) appear preferentially in southern, southwestern and western exposure, at altitudes of 2000–2500 m and at slope angles of 28–30° (Spreitzer 1960; Morawetz 1971a). All these slopes have frequently in common either a fairly closed grass or debris cover depending on elevation. The orientation pattern of such slopes at the eastern margin of the Alps is often related to the former redistribution of snow, causing the formation of glaciers on the lee side and of debris-mantled slopes on the stoss side (Richter 1900; Spreitzer 1960).

3.4 Other Periglacial-Related Landforms

3.4.1 Nivation-Related Landforms

Nivation is a morphogenetic term introduced more than 120 years ago by Matthes (1900) to describe and explain the processes associated with late-lying seasonal or perennial snow patches and landforms derived from them such as nivation benches or terraces and nivation hollows (Goudie 2004). Spatial absence or presence of nivation is related to topography and wind-driven redistribution of snow accumulating in suitable topo-climatic sites where a self-reinforcing mechanism (snow facilitates weathering and erosion→deepening of hollow→increased accumulation of snow→increased erosion) acts. A precise definition of nivation is, however, not straightforward as it is a collective noun identifying a set of geomorphic processes, comprised of several elements and of unknown relative importance (Thorn and Hall 1980).

Nivation hollows, as one example, were traditionally considered as a precursor of cirques, where processes acting under and around snow patches widened initial hollows until glaciers could become established. Matthes (1900) identified in this regard a landform continuum from nivation hollows to cirques (see Goudie 2004). Nevertheless, detailed field measurements related to nivation were not undertaken until the 1970s (Thorn 1988) even though the term was much used in periglacial geomorphology (Goudie 2004). Nivation hollows as precursors for cirques in areas, which were only marginally glaciated during the LGM have been reported exemplarily for the Pohorje Mountains in Slovenia (Natek 1993, 2007). Nivation hollows as present-day relict forms are nevertheless widespread in the Alps. Figure 12A shows one example from the CEA from a gentle-sloping area widely covered by relict block fields, solifluction forms, but with one nivation hollow with a distinct headwall thus forming an initial cirque.

Nivation studies in the Alps were never a particular focus of research. Maull (1958), who worked a lot in the Austrian Alps, described in the second edition of

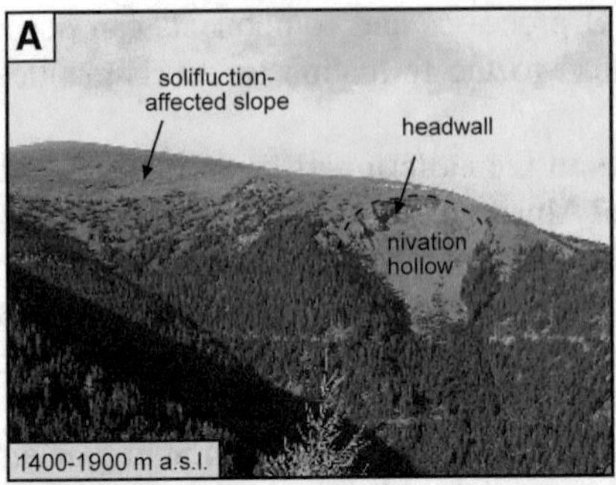

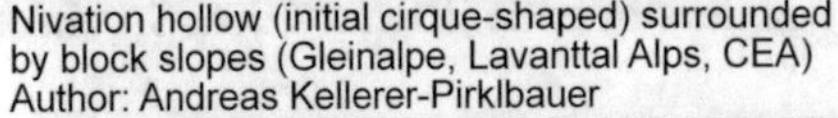
Nivation hollow (initial cirque-shaped) surrounded by block slopes (Gleinalpe, Lavanttal Alps, CEA) Author: Andreas Kellerer-Pirklbauer

Suffosion depressions on a ski slope in ice-rich permafrost (Val Thorens, Vanoise Massif, Graian Alps, NWA) - Author: Pierre-Allain Duvillard

Interaction of glacial and periglacial landforms at Gruben Valley (Valais Alps, NWA). In former times, Gruben glacier co-vered the upper part of the rock glacier, where several thermokarst lakes later developed. - Source: aerial image 2017 by Swissimage, Swisstopo

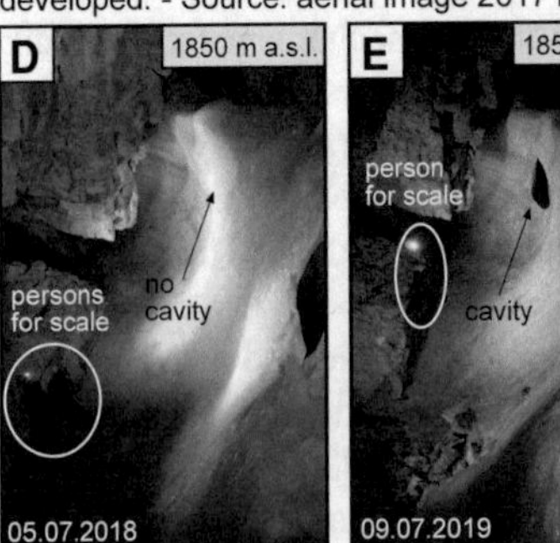

Ice changes at the Rem del Ghiaccio ice cave during a one year period (Ligurian Alps, SWA) Author: Bartolomeo Vigna

Massive ice stalagmites, Parzivaldom, Dachstein Rieseneishöhle ice cave (Dachstein Massif, NEA) - Author: Christian Bauer

Fig. 12 Pictures of further periglacial landforms: nivation hollow (relict), thermokarst features (active), and ice caves as specific form of periglacial landforms his handbook of geomorphology snow-erosion troughs or nivation hollows, thereby equating the term snow-erosion with nivation. He considered frequent freeze–thaw cycles, the erosional effect of melt water, and the downward-acting mechanical pressure effect of moving snow as the main causes for the formation of nivation hollows. He also described briefly the formation of pronival ramparts at the foot of nivation hollows. However, detailed descriptions and quantitative nivation data are missing is his book. Lehmkuhl (1989) worked on nival landforms in the Alps focusing on eight sites distributed over three regions: (a) the Hohe Tauern Range, Austria (two sites: Glockner Mountains; Ankogel Mountains, both CEA), (b) the Swiss Alps (four sites: Dente du Midi, Chablais Alps, NWA; Grand St-Bernard, NWA; Gemmi, Bernese Alps, NWA; Furka-Grimsel, Lepontine Alps and Uri Alps, NWA), and (c) the French Alps (two sites: Pelvoux—or Ecrins—Massif, Dauphiné Alps, SWA; Queyras, Cottian Alps, SWA). In his work he pointed out that a nival altitudinal belt is defined by snow-erosion including frost dynamical processes and that frost weathering and frost shattering have a major effect on reshaping initial nivation forms such as meltwater channels, nivation hollows, or nivation funnels. However, Lehmkuhl (1989) admitted that differencing acting processes in the nival and periglacial belts is not feasible. According to his work, the lower nival limit is located at the eight sites between 2300 m (Dente du Midi) and 2950 m (Queyras), the upper limit of it at between 2800 m (Glockner Mountains, Ankogel Mountains, Furka, Gemmi, Dente du Midi) and 3200 m (Queyras), and finally the lower limit of the periglacial belt between 2200 m (Glockner Mountains, Ankogel Mountains, Furka, Gemmi, Dente du Midi) and 2400 m (Queyras).

Zückert (1996), working in the Hochschwab area, NEA, for instance used the term “Alpine nivation-doline landscape”, which is characterised by the frequent occurrence of large nivation dolines breaking up the high-elevated plateaus into remnants of palaeo-surfaces and gentle summits and forming large gaps in the grass cover. Such nivation dolines are, according to Zückert, strongly influenced by nivation processes causing a doline asymmetry. Nivation dolines are the largest dolines of the periglacial belt of the Hochschwab area. Besides their considerable dimensions (depths of up to 80 m and diameters of up to 250 m), their asymmetrical morphology is striking: the mostly steeper, S- to E-exposed slopes and the bottom of the dolines consist of bare rock, N- to W-exposed slopes, on the other hand, have a largely closed soil and grass cover. This asymmetry results can be explained by uneven distribution of snow: The leeward boundary slopes and the bottom of the doline receive the greatest snow, are therefore exposed to the most intensive erosive action and are more rapidly deepened than the surrounding area. This also prevents the formation of a vegetation cover. This asymmetry of nivation hollows in the NEA was also described form other plateau-shaped areas by Haserodt (1965) or Fritz (1976). In present-day periglacial karst research, however, the term nivation doline is not much used anymore as the dominant factors for doline formation are considered to be either solution (which might be seen as a part of nivation), collapse or suffosion (Ford and Williams 2007; Plan 2016).

3.4.2 Massive Ground Ice Melting and Thermokarst Forms

Thermokarst is characterising landforms related to the degradation of ice-rich permafrost in soils and sediments, building depressions resembling those of karstic processes (French 2007). The related landforms (mounds, sink holes, slumps or lake basins) are caused by disturbances of the thermal equilibrium due to climate, geomorphic or vegetation changes (Lewkowicz and Way 2019). While thermokarst landforms are widespread and distinct in lowland arctic permafrost regions (big ponds in relation to the melt of massive ice bodies), they are only rarely described from mountain regions. In the Swiss Alps, for instance, thermokarst has been described at the Gruben site in the Saas Valley (Valais Alps, NWA), where glacial and periglacial processes interact directly (Kääb et al. 1997; Gärtner-Roer et al. 2021). Since the early twentieth century, the complex evolution of glaciers and permafrost in the Gruben cirque has led to the development of an interconnected system of small and partly dangerous—in terms of potential outburst floods—lakes (Haeberli et al. 2001). Figure 12C shows some spatial relationship and interaction of glacial and periglacial landforms at the Gruben site. Gruben glacier, including its debris-covered tongue, can be seen in the lower part of the aerial image. Gruben rock glacier is situated in the centre of the image. In former times, Gruben glacier covered the upper part of the rock glacier. In this connection zone, several thermokarst lakes developed after glacier recession. Nowadays, most of them are drained, but their past outlines can still be seen.

A rather similar complex of glacial and periglacial landforms is found in the Southern French Alps, on the upper Chauvet valley, whose dynamics of water filling and sudden outburst damaged several times, during the twentieth and twenty-first centuries, a road downvalley, at the confluence between the Chauvet stream and the Ubaye river (Southern French Alps, SWA). Ongoing studies (Cusicanqui et al. in progress) suggest that a thermokarst lake exists on ice-rich terrain, in a depression controlled by the bedrock morphology. The specific glacio-geomorphological dynamics of debris-covered and rock glaciers units that spatially co-exist and support and surround the thermokarst lake, are probably impacting the hydrology of the lake outflow. Another well investigated series of thermokarst lake in a glacial/periglacial context is found in the Vanoise massif, near the Col des Vés in Tignes ski resort, where it severely alters the stability of a chairlift pylon. There, as shown by Duvillard et al. (submitted) several thermokarst lakes formed during the last 70 years, and their drainage seems to be related to karstic conduits in the limestone that supports the glacial and periglacial landforms.

A comparable example of thermokarst has been found in Italy. Guglielmin et al. (2021) documented the development of four thermokarst depressions on an anthropogenic surface (ski slope) at high elevation close to the Stelvio Pass (Italian Central Alps, border CEA/SEA) at 2763 m. These landforms (thaw pits and thaw slides according to Jorgenson 2003) have elliptical morphologies with length of ca. 20–48 m, width 12.5–38 m and estimated depth of 3–4 m. Guglielmin et al. (2021) noted that the construction of the ski slope in 1987–1991 dramatically changed the surface characteristics in terms of grain size, slope, and vegetation. Indeed, the area containing the thermokarst depressions correspond to the frontal part of a former rock glacier, which was characterised by a coarse blocky surface with several troughs and ridges and free of vegetation. These natural characteristics were more favourable to the permafrost conditions than those of the anthropogenic surface of the ski slope (gravel with compacted sand, smooth surface, and scattered vegetation). During the excavation of ski slope, massive ice was found and melting of the remaining ice was also favoured by the surface changes. The depressions increased in number, size, and depth with time, and developed during a phase of permafrost degradation that started after 2002. They formed without any lake possibly due to the coarse grain size and high hydraulic conductivity. At the same time vegetation changed rapidly indicating important ecological changes (Guglielmin et al. 2021).

Similar features, though much less spatially extended, are the sink holes or depressions that form in the context of the melting of ground ice coupled with the internal suffosion of fine materials. Such small landforms are for instance commonly visible on terrain impacted by anthropic actions, such as ski slopes, especially when superficial removal of material leads to a subsequent active layer reforming and to localised collapse of the terrain. Figure 12B depicts an example of suffosion depressions, which were formed in the Val Thorens, Vanoise Massif, Graian Alps (SWA) at an elevation of 2750 m related to the construction of a ski slope (Duvillard et al. 2019).

An example of ground subsidence related to the melting of buried massive ice has been reported from the Schober Mountains (Hohe Tauern Range, CEA) by Kellerer-Pirklbauer and Kaufmann (2018). They studied two neighbouring cirques: one with a rock glacier and one with a glacier remnant. At the latter cirque, ground ice melting caused notable subsidence of a LIA terminal moraine. In the former cirque, a distinct rooting zone depression evolved during the last decade causing even a backward-movement of the rock glacier debris (i.e., in-valley creep) into the depression. In some years, an ephemeral lake was observed in the depression. Furthermore, massive debris-covered ice exposures have been reported (up to 20 m long and <2 m high) at the margin of the depression with a clear layered structure indicating sedimentary ice, common for the internal structure of active rock glaciers (Haeberli et al. 2006). Multi-temporal elevation change-analyses revealed continuous ground surface lowering at the rooting zone of the rock glacier since 1954 indicating long-term—and in this case natural—thermokarst effects at the studied site.

3.4.3 Ice Caves

Ice caves might be regarded as a specific form of permafrost, where the mean annual air temperature is—at least for some parts of the cave—below 0 °C (Spötl and Pavuza 2016). Considering this thought further, ice caves with their ice formations might be seen as one element of the periglacial domain. Research on ice in subsurface cavities has a long tradition in the Alps. In a recently published book dedicated to ice caves (Perşoiu and Lauritzen 2018), national ice cave-overviews were presented for Switzerland (Luetscher and Jeannin 2018), Austria (Spötl et al. 2018), Italy (Maggi et al. 2018), Germany (Meyer and Pflitsch 2018), and Slovenia (Mihevc 2018).

In Italy, as a first example, the most recent line of research concerning the study of the high elevated karstic environments of the Italian Alps involves researchers with different backgrounds engaged in the study of ice caves. The interest in ice caves dates back several centuries. One of the most discussed ice caves was the "Ghiacciaia del Moncodeno" (Bergamo Alps, SEA) used as a private ice reserve by a wealthy family. By the end of the fifteenth century, Leonardo da Vinci made some observations and took note in his personal diary after visiting probably the same ice cave, while he was a commissioner of nearby Valsassina (Brevini 2004). Such ice caves nowadays are undergoing drastic transformations due to atmospheric warming causing a strong and rather quick reduction of cave-ice masses (Maggi et al. 2018; Colucci and Guglielmin 2019; Vigna and Paro 2019). Ice caves in the Italian Alps are distributed along the entire karstic area, mainly in the CEA (Lombardy, Veneto, Trentino-Alto Adige-Südtirol, and Friuli Venezia Giulia regions) but also in the SWA (Maritime and Ligurian Alps in Piemonte region), with probably more than 1600 existing cryo caves, having within them permanent (multiyear) masses of ice, firn or permanent snow; some 150 of them with ice deposits (Maggi et al. 2018).

In Lombardy, as an example, 149 cryo caves are listed (Ferrario and Tognini 2016), which occur in a very large altitudinal distribution from between 1340 and 2840 m (both CEA and SEA). However, only 12 of them are considered as ice caves containing well known ice deposits and two of them were studied in the last 10 years. In the Trentino Alto Adige and Veneto regions information about ice caves is very scarce despite widespread karst regions. Detailed information regarding ice cave exists only for the Brenta Mountains (or Brenta Dolomites), Southern Raethian Alps (SEA), where a continuous survey since the 1980s shows a general trend of the reducing of cave ice mass (Borsato et al. 2006). In the Friuli Venezia Giulia region, ice caves are widespread in high elevated karstic environments. Recently, 1111 caves containing evidence of perennial snow or ice out of a total of 3068 known caves located above 1000 m have been identified by Colucci et al. (2016b) in the area. The highest ice cave percentage is reported in the Canin area, a high-altitude karstic massif on the Julian Alps (SEA). Here, 79 cryo-caves occur, with a mean altitude of 1928 m (Colucci et al. 2016b). In the Piedmont region, there are tens of caves with cryogenic deposits, but it is still not clear if some of them present permanent ice nowadays. Starting in 2016, Politecnico di Torino and Arpa Piemonte established a thermometric monitoring network in four ice caves with permanent ice (three in Ligurian Alps and one in Maritime Alps). "Rem del Ghiaccio" ice cave (Ligurian Alps, SWA) presents the greatest accumulation of ice with the most evident ice melting phase in the second part of the 2010s as also shown in Figs. 12D, E (Vigna and Paro 2019).

In Austria, as a second example, a first attempt to explore a former ice cave date back to 1592 (Spötl and Pavuza 2016). The long research tradition in Austria is also related to the high number of ice-bearing caves relative to the size of the country: 1200 Austrian caves are classified as ice caves containing perennial ice, firn, or snow (Spötl et al. 2018). The Northern Calcareous Alps (NEA) show a high density of such caves with some of the largest of their kind on Earth, i.e., Eisriesenwelt, Dachstein-Rieseneishöhle (Spötl et al. 2018). Figure 12F depicts exemplarily massive ice formations with impressive ice stalagmites inside the Parzivaldom at the Dachstein Rieseneishöhle (NEA), an extensive ice cave in the Austrian Alps. The best studied ice caves in Austria are the Eisriesenwelt, the Schönberg-Höhlensystem, the Dachstein-Mammuthöhle, and the Hundsalm Eis- und Tropfsteinhöhle (Spötl et al. 2018). Mass loss and gain of perennial cave ice depends on several environmental parameters including the cave geometry, the general climate setting, and the lack or presence of winter snow cover providing water supply to the cave during spring snowmelt (Spötl et al. 2018). Besides seasonal changes related to ablation and accumulation of various ice forms, multi-annual to long-term (0.1 to >1 ka) changes have also been reported from different Alpine ice caves. For example, Buchroithner and Gaisecker (2020) used a terrestrial laser scanner to monitor ice surface changes in the Eisriesenwelt cave between 2017 and 2020. Spötl and Pavuza (2016) presented data on the gradual decline of ice volume since the 1990s in the Dachstein-Mammut cave, regarded as typical for Alpine ice caves. May et al. (2011) drilled an ice core through a 7.1 m-thick ice body in the Eisriesenwelt cave revealing substantial changes in the ice formation process during the last several thousand years. Finally, long-term studies showed that no cave ice in Austria seems to be from the first part of the Holocene (Achleitner 1995; Herrmann et al. 2010; Spötl et al. 2014).

4 The Significance of Periglacial Dynamics in the Alps Within the European Context

The mountain range "European Alps" appears as one big system or geomorphic entity but shows many different landscapes. From the more karstic and glacier-free regions in the east to the almost Mediterranean regions in the south-west, passing through the "polar environment" of the high peaks, clear differences appear in climate, vegetation, geology, and thus geomorphic landforms. This circumstance is well visible in cross profiles through different parts of the Alps (Fig. 13). While relict periglacial landforms are distributed basically all over the Alps related either to glacier-free conditions and hence periglacial weathering during the LGM (Fig. 2D), or to periglacial landscape modification of deglaciated areas in the Late Glacial and Holocene periods. Active periglacial landforms concentrate mainly on the central parts of the Alps, where relief and related climate conditions favour frost-related processes and landforms. Mapping and analysing the distribution and activity of periglacial features in all kinds of past and present periglacial environments allows for the reconstruction of periglacial conditions after the LGM and mirrors the climate and glacial history.

Figure 13A depicts a 305 km long cross profile starting in Colombe in the Isère Valley outside the LGM glacier, crosses the Taillefer, Ecrins, and Queyras Massifs and Monviso to the settlement of Saluzzo in the Piemonte, makes a 90° bend to the south towards Cuneo, where it crosses the Maritime Alps all the distance to the Mediterranean coast at Nice. The lower limit of present periglacial landform reshaping is between 1900 m in the very west of the profile to almost 2500 m in

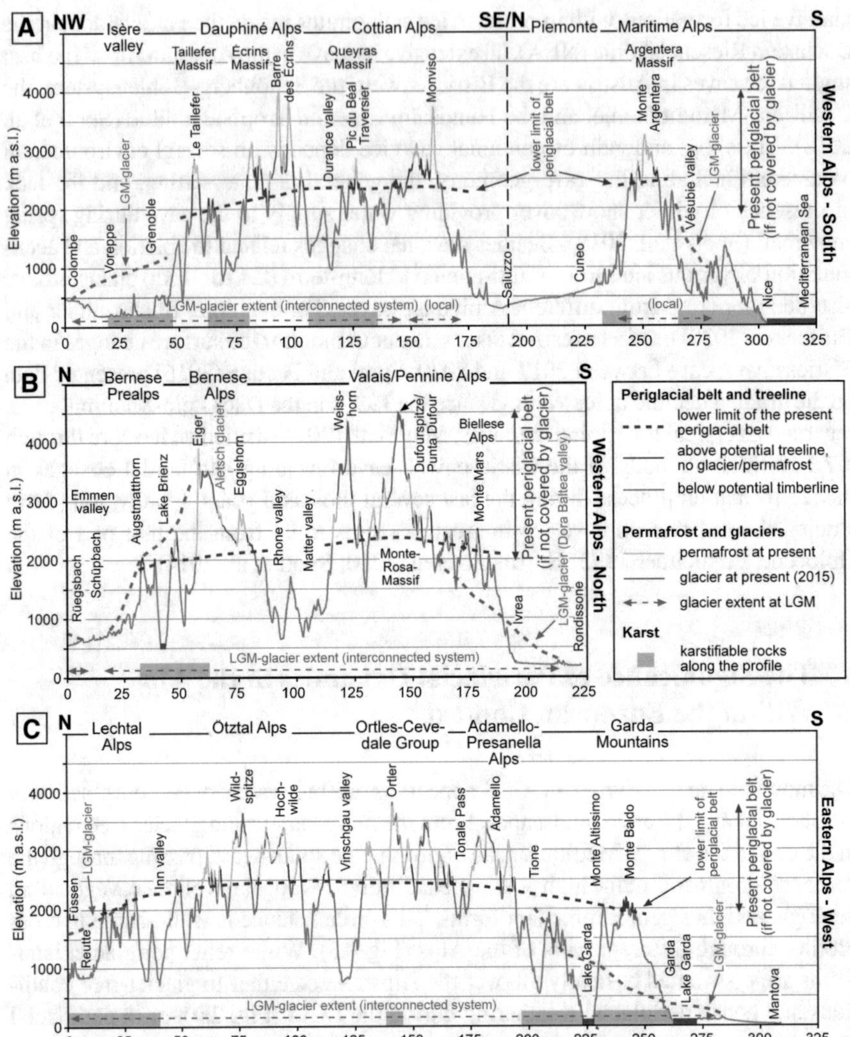

Fig. 13 Four selected cross profiles of the Western and Eastern Alps depicting the periglacial and glacial conditions at present in relation to karstifiable rocks and LGM-glaciation: (A) Western Alps—South: from Colombe via Saluzzo to Nice (305 km); (B) Western Alps—North: from Rüegsbach to Rondissone (223 km); (C) Eastern Alps—West: from Füssen to Mantova (309 km); (D) Eastern Alps—East: from Kirchdorf to Graz (139 km); (E) overview map with delineation of the present periglacial and glacial domains in the Alps based on our review. The course of the four profiles in relation to the five major sectors of the Alps is additionally shown

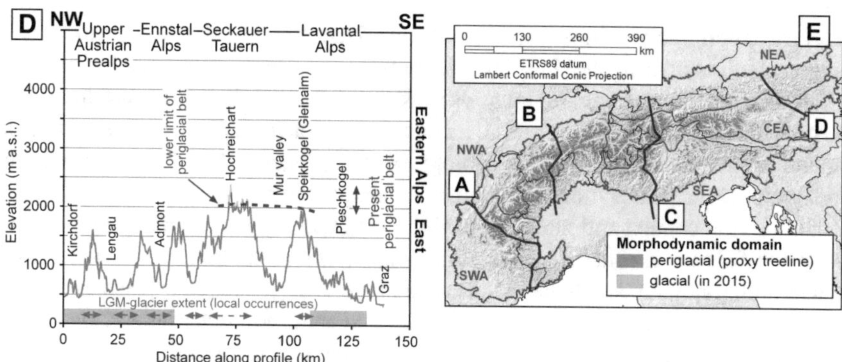

Fig. 13 (continued)

the Maritime Alps. The profile shown in Fig. 13B starts in the Swiss Prealps at Rüegsbach, which was glaciated during the LGM from a lobe coming from the NW, head to the south via the Bernese Prealps, Bernese Alps to the Aletsch glacier, the largest glacier in the Alps (ca. 75 km^2), crosses the Rhone valley and the Valais or Pennine Alps via the highest summit of Switzerland (Dufourspitze), and decents down via the Biellese Alps to the former Piemont glacier in the Dora Baltea basin around Ivrea ending at the proglacial LGM-area at Rondissone. The lowest lower limit of the present periglacial belt is about 1900 m in the Bernese Prealps, whereas the highest one is in the Valais Alps at about 2350 m. The third profile depicted in Fig. 13C shows a 309 km long section from Füssen in Germany via the Lechtal Alps and the Inn valley to the Öztal Alps, heads further via the Vintschgau valley into the Ortles-Chevedale Group, where it reaches its highest position at Ortler (3905 m). From there, the profile descents to the Tonale Pass, crosses the Adamello-Presanella Alps and reaches its last mountain group along the profile east of Lake Garda at Monte Baldo. From there, the profile descends via the LGM-glacier terminal moraine at Lake Garda to Mantova in the Po basin. The course of this profile is similar to the one presented by Höllermann (1967, 1983). Profile Nr. 3 shows a similar bulge of the lower limit of the periglacial belt in the central part of the Alps compared to the margin with about 2500 m at the former, whereas only 2000 m at the latter. The final profile presents the periglacial belt for a cross profile in the very east of the Alps starting in Kirchdorf in the Upper Austrian Prealps, crossing the Ennstal Alps, which consist of carbonate rocks, continue further to the eastern-most unit of the Niedere Tauern Range, the Seckauer Tauern, where the maximum elevation along this profile is reached with 2414 m (Hochreichart). The profile continues further to the SE reaching after 139 km the city of Graz. Along this profile, only the mountain summits are influenced by noteworthy present periglacial conditions. The four profiles can, thus, be seen as representative of the entire Alpine range with highest present periglacial process rates in the Northern-Western Alps and the Central-Eastern Alps.

Over the last 150 years, many European scientists worked in the Alps and contributed to a dense picture of geomorphological studies and the understanding of glacial and periglacial processes and landforms. Therefore, the Alps are the mountain range, which is investigated best and where most long-term analyses of periglacial landforms, such as solifluction lobes or rock glaciers, exist. But open questions still exist. In former decades, the scientific focus was more on the holistic understanding of mountain environments, including geology, climate, vegetation and geomorphology. Today, studies are much more focussed on single processes and landforms and, thus, sometimes the context and interaction within the landscape is forgotten. Our understanding of slopes in cold-climate environments is still partly limited (French 2007). In addition, many recent studies focus on climate-change related aspects and got a lot of attention, e.g., to rock glacier dynamics and changes. Nevertheless, also other periglacial landforms and processes are still of interest and deserve attention and continuous monitoring. In addition, it is valuable to use the early studies (again) and include them in the current analysis to generate a broader picture of periglacial morphodynamics during the last century.

Many collaborative projects have been conducted so far to improve knowledge about the periglacial environment. In particular, since the late 1990s, new European projects on permafrost monitoring have been financed, involving researchers and institutions from all over the continent, and in particular the Alps. Some most important transnational projects were "PACE—Permafrost and climate change in Europe" (1997–2001) (Harris et al. 2001), "PERMAdataROC—Elaboration d'une base de données et expérimentation de méthodes de mesure des mouvements gravitaires et des régimes thermiques des parois rocheuses à permafrost en haute montagne" (2006–2008) (Deline et al. 2008), "PermaNET—Permafrost long-term monitoring network"(2008–2011) (Mair et al. 2011), or "PrévRiskHauteMontagne—Exemplary actions of resilience of cross-border communities in the face of the natural risks of the high mountain environment" (2016–2017) (Ravanel et al. 2018). In addition to monitoring permafrost as an effect of climate change, these projects provided advanced tools for assessing related risks in mountain areas. A further important focus recently highlighted within European projects is the relationship between permafrost and water resources (e.g. "Permaqua", 2011–2015, Mair et al. 2015; "Reservaqua", 2019–2023), a growing potential problem in the continent's most populated mountain area. These European-founded projects bring together researchers and policy makers and other stakeholders for the integration of science and governance to face the future great challenges of sustainable development in the Alps.

These projects are worth to continue or reactivate. Furthermore, a closer cross-border cooperation of all bordering states of the Alps could help to compile an updated picture of the Alpine periglacial geomorphology. In this context, long-term studies (monitoring) are of great value. They provide the data basis for change analyses, in parallel to climate monitoring. Such data series should be carried out at various representative locations in the Alps to improve process understanding. This requires also improved collaborations with other disciplines such as geology, glaciology, geophysics, climatology or geochemistry. This enhanced understanding of ongoing changes in very valuable for applied studies (such as natural hazard) as well as decision making processes, as the Alpine landscape has a specific value.

The Alps are densely populated and "used" for hydropower production, recreation, etc., therefore, ongoing changes in the periglacial belt may influence nature, society, and infrastructure and by this may influence the hazard and risk situation in many regions. Knowing the natural processes is of great advantage when it comes to impact assessments. We understand that despite long traditions and comprehensive experiences in observing, describing, mapping, measuring, and modelling periglacial landforms in the Alps, future research is needed, which will help to better understand the impact of ongoing climate change on modifications and the upward shift of periglacial processes and active landforms that steadily reshape this impressive mountain chain in Central Europe.

References

Achleitner A (1995) Zum Alter des Höhleneises in der Eisgruben-Eishöhle im Sarstein (Oberösterreich). Die Höhle 46:1–5

Agassiz L, Guyot AH, Desor É (1847) Système glaciaire ou recherches sur les glaciers, leur mécanisme, leur ancienne extension et le rôle qu'ils ont joué dans l'histoire de la terre. Masson, Paris, p 598

Albertini R, Amedeo R, Capello CF, Dona F, Giacomini V, Giorcelli V (1955) Studi sui fenomeni crionivali (periglaciali partim) delle Alpi italiane, Fondaz. per i problemi montani dell'Arco alpino, Milano Frenshing, 11, Parma, p 148

Allix A (1923) Nivation et sols polygonaux dans les Alpes françaises. La Géographie XXXIX:431–438

Alpine Convention (2020) Perimeter of the Alpine Convention. https://www.atlas.alpconv.org/layers/geonode:Alpine_Convention_Perimeter_2018_v2

Andersson IG (1906) Solifluction, a component of subaerial denudation. J Geol 14:91–112

Andrieux E, Bertran P, Saito K (2016) Spatial analysis of the French Pleistocene permafrost by a GIS database: French Pleistocene permafrost database. Permafr Periglac Process 27:17–30. https://doi.org/10.1002/ppp.1856

Arenson LU, Hoelzle M, Springman S (2002) Borehole deformation measurements and internal structure of some rock glaciers in Switzerland. Permafr Periglac Process 13:117–135. https://doi.org/10.1002/ppp.414

Avian M, Kellerer-Pirklbauer A (2012) Modelling of potential permafrost distribution during the Younger Dryas, the Little Ice Age and at present in the Reisseck Mountains, Hohe Tauern Range. Austria. Austrian J Earth Sci 105(2):140–153

Avian M, Kaufmann V, Lieb GK (2005) Recent and Holocene dynamics of a rock glacier system: The example of Langtalkar (Central Alps, Austria). Norsk Geogr Tidsskr 59:149–156

Avian M, Kellerer-Pirklbauer A, Bauer A (2009) LiDAR for monitoring mass movements in permafrost environments at the cirque Hinteres Langtal, Austria, between 2000 and 2008. Nat Hazard Earth Sys 9:1087–1094

Bachmann F (1966) Fossile Strukturböden und Eiskeile auf jungpleistocänen Schotterflächen im nordostschweizerischen Mittelland. Doctoral dissertation, University of Zurich

Ballantyne CK (1998) Age and significance of mountain-top detritus. Permafr Periglac Process 9:327–345

Ballantyne CK (2001) Measurement and theory of ploughing boulder movement. Permafr Periglac Process 12:267–288. https://doi.org/10.1002/ppp.389

Ballantyne CK (2018) Periglacial Geomorphology. Wiley-Blackwell, ISBN: 978-1-405-10006-9; p 472

Baroni C, Carton A, Seppi R (2004) Distribution and behaviour of rock glaciers in the Adamello-Presanella Massif (Italian Alps). Permafr Periglac Process 15:243–259. https://doi.org/10.1002/ppp.497

Barsch D (1969) Studien und Messungen an Blockgletschern in Macun, Unterengadin. Z Geomorphol Supp 8:11–30

Barsch D (1993) Periglacial geomorphology in the 21st century. Geomorphology 7:141–163

Barsch D (1996) Rockglaciers: indicators for the present and former geoecology on high mountain environments. Springer, Berlin, Germany, p 331

Barsch D, Hell G (1975) Photogrammetrische Bewegungsmessungen am Blockgletscher Murtèl I, Oberengadin, Schweizer Alpen. Z Gletscherk Glazialgeol 11:111–142

Bätzing W (2015) Die Alpen: Geschichte und Zukunft einer europäischen Kulturlandschaft. C.H.Beck, Munich, p 484

Bauer F (1962) Nacheiszeitliche Karstformen in der österreichischen Kalkalpen. In: Proceedings of the 2nd International Congress of Speleology, Bari, pp 299–328

Bauer A, Paar G, Kaufmann V (2003) Terrestrial laser scanning for rock glacier monitoring. In: Proceedings of the 8th International Conference on Permafrost, Zurich, Switzerland, pp 55–60

Bauer F, Zötl J (1972) Chapter 7—Karst of Austria. In: Herak M, Stringfield VT (eds) Karst: Important Karst Regions of the Northern Hemisphere. Elsevier, Amsterdam, London, New York, pp 225–265

Belloni S, Carton A, Dramis F, Smiraglia C (1993) Distribution of permafrost, glaciers and rock glaciers in the Italian mountains and correlations with climate: An attempt to synthesize. In: Proceedings of the 6th International Conference on Permafrost, Beijing, China, pp 36–41

Benedict JB (1970) Downslope soil movement in a Colorado alpine region: Rates, processes, and climate significance. Arctic Alpine Res 2:165–226

Blum W (2006) Frost-induced structural soils in the Tennengebirge. Salzburg, Mitt Oster Bodenku G 73:133–143

Boccali C, Žebre M, Colucci RR (2019) Geometry and paleo-ice content of rock glaciers in the southeastern Alps (NE Italy—NW Slovenia). J Maps 15(2):346–355. https://doi.org/10.1080/17445647.2019.1595753

Bock H (1913) Charakter Des Mittelsteirischen Karstes. Mitt. Höhlenkunde 6(4):5–19

Bodin X, Krysiecki JM, Schoeneich P, Le Roux O, Lorier L, Echelard T, Peyron M, Walpersdorf A (2016) The 2006 Collapse of the Bérard Rock Glacier (Southern French Alps). Permafr Periglac Process 28:209–223. https://doi.org/10.1002/ppp.1887

Bodin X, Thibert E, Fabre D, Ribolini A, Schoeneich P, Francou B, Reynaud L, Fort M (2009) Two decades of responses (1986–2006) to climate by the Laurichard Rock Glacier, French Alps. Permafr Periglac Process 20:331–344. https://doi.org/10.1002/ppp.665

Boeckli L, Brenning A, Gruber S, Noetzli J (2012) Permafrost distribution in the European Alps: calculation and evaluation of an index map and summary statistics. Cryosphere 6(4):807–820. https://doi.org/10.5194/tc-6-807-2012

Bögli A (1980) Karst hydrology and physical speleology. Springer, Berlin, p 284

Borsato A, Miorandi R, Flora O (2006) I depositi di ghiaccio ipogei della Grotta dello Specchio e del Castelletto di Mezzo (Dolomiti di Brenta, Trentino): morfologia, età ed evoluzione recente. Studi Trent Sci Nat Acta Geol 81:53–74

Brevini F (2004) Rocce. Mondadori, Milano, 252 p

Buchenauer HW (1990) Gletscher-und Blockgletschergeschichte der westlichen Schobergruppe (Osttirol). Marburger Geogr Schriften 117:1–376

Buchroithner M, Gaisecker T (2020) Ice surface changes in Eisriesenwelt (Salzburg, Austria) based on LIDAR measurements between 2017 and 2020. Die Höhle 71:62–70

Büdel J (1937) Eiszeitliche und rezente Verwitterung and Abtragung im ehemals nicht vereisten Teil Mitteleuropas. Petermann Mitt Ergaenz 229:1–71

Burger KC, Degenhardt JJ Jr, Giardino JR (1999) Engineering geomorphology of rock glaciers. Geomorphology 31:93–132

Cannone N, Guglielmin M, Smiraglia C (1995) Relazioni tra forme periglaciali e caratteri della vegetazione di alta quota nell'area del Livignasco (Alta Valtellina). Rivista Geografica Italiana 102:91–111

Cannone N, Piccinelli S (2021) Changes of rock glacier vegetation in 25 years of climate warming in the Italian Alps. CATENA 206(2):105562. https://doi.org/10.1016/j.catena.2021.105562

Cannone N, Guglielmin M, Gerdol R (2004) Relationships between vegetation patterns and periglacial landforms in northwestern Svalbard. Polar Bio 27:562–571

Capello CF (1947) Le pietraie semoventi (rock glaciers) delle Alpi Occidentali. Natura 38:17–23

Capello CF (1959) Prime ricerche sulle "pietraie semoventi" del settore montuoso del Gran Paradiso. Riv Mens Club Alp Italiano 78(294–300):371–376

Capello CF (1963) Le morfologie crionivali (periglaciali) nelle Alpi Graie meridionali italiane. Pubblicazioni dell'Istituto di Geografia Alpina, Volume 3/Studi sulle morfologie crionivali 1, 124 p

Capello CF (1969) Le morfologie tardo-glaciali e crionivali della Val Maira (Piemonte), Pubbl. Ist. Geogr. Alp. 13, S.P.E., Torino, 79 p

Carta Geologica d'Italia (1977) Explanatory Notes of sheet 28 "La Marmolada" at scale 1:50.000. Geological Survey of Italy, Florence.

Carturan L, Zuecco G, Seppi R, Zanoner T, Borga M, Carton A, Dalla Fontana G (2016) Catchment-scale permafrost mapping using spring water characteristics. Permafr Periglac Process 27:253–270. https://doi.org/10.1002/ppp.1875

Castiglioni GB (1961) I depositi morenici del gruppo Adamello-Presanella con particolare riguardo agli stadi glaciali postwurmiani. Mem Ist Geol Min Univ Padova 23:1–131

Castiglioni GB (1974) Importanza dei processi periglaciali nel Pleistocene per l'evoluzione del rilievo nelle Prealpi Venete. Natura e Montagna 21:15–17

Castiglioni GB, Girardi A, Sauro U, Tessari F (1979) Grèzes litées e falde detritiche stratificate di origine crionivale. Geogr Fis Din Quat 2:64–82

Chaix A (1919) Coulées de blocs (rock-glaciers, rock-streams) dans le parc national suisse de la Basse-Engadine. Cr Séances Soc Phys Hist Nat Genève 36:12–15

Chaix A (1923) Les coulées de blocs du Parc National Suisse d'Engadine (Note préliminaire). Le Globe 62:1–35

Chaix A (1943) Les coulées de blocs du Parc National Suisse: nouvelles mesures et comparaison avec les "rock stream" de la Sierra Nevada de Californie. Le Globe 82:121–128

Charton J, Verfaillie D, Jomelli V, Francou B (2021) Early Holocene rock glacier stabilisation at col du Lautaret (French Alps): Palaeoclimatic implications. Geomorphology 394:107962. https://doi.org/10.1016/j.geomorph.2021.107962

Cicoira A, Beutel J, Faillettaz J, Vieli A (2019) Water controls the seasonal rhythm of rock glacier flow. Earth Planet Sc Lett 528:115844. https://doi.org/10.1016/j.epsl.2019.115844

Cicoira A, Marcer M, Gärtner-Roer I, Bodin X, Arenson LU, Vieli A (2021) A general theory of rock glacier creep based on in situ and remote sensing observations. Permafr Periglac Process 32:139–153. https://doi.org/10.1002/ppp.2090

CNRS-Caen and Institut de Géographie Aix-en-Provence (1980) Observations sur quelques formes et processus périglaciaires dans le Massif du Chambeyron (Alpes de Hautes-Provence), Revue de géogr alpine 68(4)349–382.https://doi.org/10.3406/rga.1980.221

Colombo N, Fratianni S, Giaccone E, Paro L (2013) Relationships among atmosphere-cryosphere-biosphere in a transitional glacial catchment (Sabbione Lake, northwestern Italian Alps). In: Proceedings of the 20th International Snow Science Workshop, Grenoble-Chamonix Mont-Blanc, pp 1201–1207

Colombo N, Giaccone E, Paro L, Buffa G, Fratianni S (2016) The recent transition from glacial to periglacial environment in a high-altitude alpine basin (Sabbione Basin, North-Western Italian Alps). Preliminary outcomes from a multidisciplinary approach. Geogr Fis Din Quat 39:21–36. https://doi.org/10.4461/GFDQ.2016.39.3

Colombo N, Salerno F, Martin M, Malandrino M, Giardino M, Serra E, Godone D, Said-Pullicino D, Fratianni S, Paro L, Tartari G, Freppaz M (2019) Influence of permafrost, rock and ice glaciers on chemistry of high-elevation ponds (NW Italian Alps). Sci Total Environ 685:886–901. https://doi.org/10.1016/j.scitotenv.2019.06.233

Colombo N, Ferronato C, Vittori Antisari L, Marziali L, Salerno F, Fratianni S, D'Amico ME, Ribolini A, Godone D, Sartini S, Paro L, Morra di Cella U, Freppaz M (2020) A rock-glacier-pond system (NW Italian Alps): Soil and sediment properties, geochemistry, and trace-metal bioavailability. CATENA 194:104700. https://doi.org/10.1016/j.catena.2020.104700

Colucci RR, Boccali C, Žebre M, Guglielmin M (2016a) Rock glaciers, protalus ramparts and pronival ramparts in the south-eastern Alps. Geomorphology 269:112–121. https://doi.org/10.1016/j.geomorph.2016.06.039

Colucci RR, Fontana D, Forte E, Potleca M, Guglielmin M (2016b) Response of ice caves to weather extremes in the Southeastern Alps, Europe. Geomorphology 261:1–11. https://doi.org/10.1016/j.geomorph.2016.02.017

Colucci RR, Forte E, Žebre M, Maset E, Zanettini C, Guglielmin M (2019) Is that a relict rock glacier? Geomorphology 330:177–189. https://doi.org/10.1016/j.geomorph.2019.02.002

Colucci RR, Guglielmin M (2019) Climate change and rapid ice melt: Suggestions from abrupt permafrost degradation and ice melting in an alpine ice cave. Prog Phys Geog 43(4):561–573. https://doi.org/10.1177/0309133319846056

Costantini EAC, Carnicelli S, Sauer D, Priori S, Andreetta A, Kadereit A, Lorenzetti R (2018) Loess in Italy: Genesis, characteristics and occurrence. CATENA 168:14–33. https://doi.org/10.1016/j.catena.2018.02.002

Coutard JP, Francou B (1989) Rock temperature measurements in two alpine environments: implications for frost shattering. Arctic Alpine Res 21:399–416

Coutard JP (2019) Retour sur 50 années de contributions du professeur Albert Pissart à la géomorphologie périglaciaire. Environ. Périglaciaires, Bulletin de l'Association Française du Périglaciaire 22(23):8–18

Cremaschi M, Zerboni A, Nicosia C, Negrino F, Rodnight H, Spötl C (2015) Age, soil-forming processes, and archaeology of the loess deposits at the Apennine margin of the Po plain (northern Italy): New insights from the Ghiardo area. Quat Int 376:173–188. https://doi.org/10.1016/j.quaint.2014.07.044

Cremonese E, Gruber S, Phillips M, Pogliotti P, Boeckli L, Noetzli J, Suter C, Bodin X, Crepaz A, Kellerer-Pirklbauer A, Lang K, Letey S, Mair V, Morra di Cella U, Ravanel L, Scapozza C, Seppi R, Zischg A (2011) Brief Communication: An inventory of permafrost evidence for the European Alps. Cryosphere 5:651–657. https://doi.org/10.5194/tc-5-651-2011

Crespi A, Brunetti M, Lentini G, Maugeri M (2017) 1961–1990 high-resolution monthly precipitation climatologies for Italy. Int J Climatol 38:878–895. https://doi.org/10.1002/joc.5217

Cusicanqui D, Rabatel A, Vincent C, Bodin X, Thibert E, Francou B (2021) Interpretation of volume and flux changes of the Laurichard Rock Glacier between 1952 and 2019, French Alps J Geophys Res-Earth 126(9):e2021JF006161. https://doi.org/10.1029/2021JF006161

Czudek T (1995) Cryoplanation terraces—a brief review and some remarks. Geogr Ann 77A(1–2):95–105

D'Amico ME, Casati E, Andreucci S, Martini M, Panzeri L, Sechi D, Abu El Khair D, Previtali F (2021) New dates of a Northern Italian loess deposit (Monte Orfano, Southern pre-Alps, Brescia). Journal Soils Sediments 21:832–841. https://doi.org/10.1007/s11368-020-02860-4

D'Amico ME, Gorra R, Freppaz M (2015) Small-scale variability of soil properties and soil–vegetation relationships in patterned ground on different lithologies (NW Italian Alps). CATENA 135:47–58. https://doi.org/10.1016/j.catena.2015.07.005

D'Amico ME, Pintaldi E, Catonia M, Freppaz M, Bonifacio E (2019) Pleistocene periglacial imprinting on polygenetic soils and paleosols in the SW Italian Alps. CATENA 174:269–284. https://doi.org/10.1016/j.catena.2018.11.019

Davaze L, Rabatel A, Dufour A, Hugonnet R, Arnaud Y (2020) Region-wide annual glacier surface mass balance for the European Alps from 2000 to 2016. Front Earth Sci 8:149. https://doi.org/10.3389/feart.2020.00149

De Jong MGG, Kwadijk JK (1998) Fossil rock glaciers in central Vorarlberg, Austria. Arctic Alpine Res 20:86–96
Delaloye R, Reynard E, Lambiel C, Marescot L, Monnet R (2003) Thermal anomaly in a cold scree slope (Creux Du Van, Switzerland) In: Proceedings of the 8th International Conference on Permafrost, Zurich, Switzerland, pp 175–180
Delaloye R, Perruchoud E, Avian M, Kaufmann V, Bodin X, Hausmann H, Ikeda A, Kääb A, Kellerer-Pirklbauer A, Krainer K, Lambiel C, Mihajlovic D, Staub B, Roer I, Thibert E (2008) Recent interannual variations of rock glacier creep in the European Alps. In: Proceedings of the 9th International Conference on Permafrost, Fairbanks, Alaska, pp 343–348
Delaloye R, Lambiel C, Gärtner-Roer I (2010) Overview of rock glacier kinematics research in the Swiss Alps. Seasonal rhythm, interannual variations and trends over several decades. Geogr Helvetica 65(2):135–145
Deline P, Jaillet S, Rabatel A, Ravanel L (2008) Ground-based LiDAR data on permafrost-related rock fall activity in the Mont-Blanc massif. In: Proceedings of the 9th International Conference on Permafrost, Fairbanks, Alaska, pp 349–354
Deline P, Gruber S, Amann F, Bodin X, Delaloye R, Faillettaz J, Fischer L, Geertsema M, Giardino M, Hasler A, Kirkbride M, Krautblatter M, Magnin F, McColl S, Ravanel L, Schoeneich P, Weber S (2021) Ice loss from glaciers and permafrost and related slope instability in high-mountain regions. In: Snow and ice-related hazards, risks, and disasters, Elsevier, pp 501–540
Demangeot J (1941) Contribution à l'étude de quelques formes de nivation. Rev Géographie Alp 29:337–352. https://doi.org/10.3406/rga.1941.4310
Demek J (1969) Cryogene processes and the development of cryoplanation terraces. Biul Peryglac 18:115–125
Douvillé R (1917) Sols polygonaux ou réticulés. La Géographie XXXI:241–251
Draebing D (2021) Identification of rock and fracture kinematics in high alpine rockwalls under the influence of altitude. Earth Surf Dynam 9:1–18. https://doi.org/10.5194/esurf-9-1-2021
Draebing D, Krautblatter M (2019) The efficacy of frost weathering processes in alpine rockwalls. Geophys Res Lett 46:6516–6524. https://doi.org/10.1029/2019GL081981
Draebing D, Haberkorn A, Krautblatter M, Kenner R, Philipps M (2017) Thermal and mechanical responses resulting from spatial and temporal snow cover variability in permafrost rock slopes, Steintaelli, Swiss Alps. Permafr Periglac Process 28:140–157. https://doi.org/10.1002/ppp.1921
Dramis F, Giraudi C, Guglielmin M (2003) Rock glacier distribution and paleoclimate in Italy. In: Proceedings of the 8th International Conference on Permafrost, Zurich, Switzerland, pp 199–204
Duvillard PA, Ravanel L, Marcer M, Schoeneich P (2019) Recent evolution of damage to infrastructure on permafrost in the French Alps. Reg Envion Change 19:1281–1293. https://doi.org/10.1007/s10113-019-01465-z
Ebers E (1959) Die Buckelwiesen: nicht Eiszeitalter, sondern Gegenwart. Eiszeitalter und Gegenwart 10(1):9 https://doi.org/10.23689/fidgeo-1479
Egholm DL, Andersen JL, Knudsen MF, Jansen JD, Nielsen SB (2015) The periglacial engine of mountain erosion—part 2: Modelling large-scale landscape evolution. Earth Surf Dynam 3:463–482. https://doi.org/10.5194/esurf-3-463-2015
Eisenhut M (1963) Über einige Beobachtungen an den Buckelalmen der Seetaleralpen. Mitt Naturwiss Ver Steiermark 93:17–21
Eugster H (1973) Bericht über die Untersuchungen des Blockstroms in der Val Sassa im Schweizerischen Nationalpark von 1917–1971. Ergebnisse Der Wissenschaftlichen Untersuchungen Im Schweizerischen Nationalpark 11:368–384
Ebohon B, Schrott L (2008) Modelling mountain permafrost distribution. A new permafrost map of Austria. In: Proceedings of the 9th International Conference on Permafrost, Fairbanks, Alaska, pp 397–402
Ehlers J, Gibbard PL (2004) Quaternary glaciations-extent and chronology: Part I: Europe. Elsevier, Amsterdam
Ehlers J, Gibbard PL, Hughes PD (2011) Quaternary glaciations-extent and chronology: a closer look. Elsevier, Amsterdam

Eriksen HØ, Rouyet L, Laukness TR, Berthling I, Isaksen K, Hindberg H, Larsen Y, Corner DG (2018) Recent acceleration of a rock glacier complex, Ádjet, Norway, documented by 62 years of remote sensing observations. Geophys Res Lett 45:8314–8323. https://doi.org/10.1029/2018GL077605

ESA (2021) EU-DEM v1.1—Copernicus land monitoring service. https://land.copernicus.eu/imagery-in-situ/eu-dem/eu-dem-v1.1/view

Evin M, de Beaulieu JL (1985) Nouvelles données sur l'âge de la mise en place et les phases d'activité du glacier rocheux de Marinet I (Haute-Ubaye, Alpes du Sud françaises). Méditerranée 56:21–30. https://doi.org/10.3406/medit.1985.2330

Evin M (1987) Lithologly and fracturing control of rock glaciers in southwestern Alps of France and Italy. Rock glaciers: a review of the knowledge base. Allen & Unwin, London, pp 83–106

Evin M, Fabre D (1990) The distribution of permafrost in rock glaciers of the southern Alps (France). Geomorphology 3:57–71. https://doi.org/10.1016/0169-555X(90)90032-L

Fabre D, Ribolini A (2003) The lower discontinuous permafrost boundary in the Argentera Massif (Maritime Alps, Italy): insight from rock glacier geoelectrical soundings. In: Proceedings of the 8th International Conference on Permafrost—Extended Abstracts, Zurich, Switzerland, pp 33–34

Fagerlund G (1979) Studies of the destruction mechanism at freezing of porous materials. Found Fr Etud Nord Actes Doc 6:73–79

Ferrario A, Tognini A (Eds) (2016) Catasto Speleologico Lombardo (Progetto Tu.Pa.Ca.). Federazione Speleologica Lombarda, 448 p

Feuillet T (2014) Ploughing Boulder. In: Encyclopedia of Planetary Landforms. Springer, New York, NY. https://doi.org/10.1007/978-1-4614-9213-9_272

Filippa G, Freppaz M, Zanini E (2011) Pedogenetic processes beneath high altitude sorted circles in the Italian Western Alps. Geophysical Research Abstracts 13:EGU2011–6421

Fink MH (1989) Ein Permafrostboden in den Kalkvoralpen bei Puchenstuben (Niederösterreich). Die Höhle 40:95–98

Finsterwalder S (1928) Begleitworte zur Karte des Gepatschferners. Z Gletscherkunde 16:20–41

Fioraso G, Spagnolo G (2009) I block stream del massiccio peridotitico di Lanzo (Alpi Nord-occidentali). Il Quaternario—Ital J Quat Sci 22(1):3–22

Firpo M, Guglielmin M, Queirolo C (2006) Relict blockfields in the Ligurian Alps (Mount Beigua, Italy). Permafr Periglac Process 17:71–78

Fitzsimons SJ, Veit H (2001) Geology and geomorphology of the European Alps and the Southern Alps of New Zealand. Mt Res Dev 21(4):340–349. https://doi.org/10.1659/0276-4741(2001)021[0340:GAGOTE]2.0.CO;2

Flohr EF (1935) Betrachtungen über die Bahnen der Schneeschmelzwässer. Ein Beitrag zum Problem der Blockrinnen (Steinstreifen). Zeitschr d Gesellschaft f Erdkunde zu Berlin, pp 353–369

Ford DC, Williams P (2007) Karst hydrogeology and geomorphology. Wiley, Chichester, p 562.https://doi.org/10.1002/9781118684986

Forno MG (1979) Il "loess" della Collina di Torino: revisione della sua distribuzione e della sua interpretazione genetica e cronologica. Geogr Fis Din Quat 2:105–124

Francou B (1982) Chutes de pierres et éboulisation dans les parois de l'étage périglaciaire. Rev. Géographie Alp. 70:279–300. https://doi.org/10.3406/rga.1982.2508

Francou B, Reynaud L (1992) 10 year surficial velocities on a rock glacier (Laurichard, French Alps). Permafr Periglac Process 3:209–213. https://doi.org/10.1002/ppp.3430030306

Frauenfelder R, Kääb A (2000) Towards a palaeoclimatic model of rock-glacier formation in the Swiss Alps. A Glaciol 31:281–286

Frauenfelder R, Haeberli W, Hoelzle M, Maisch M (2001) Using relict rockglaciers in GIS-based modelling to reconstruct Younger Dryas permafrost distribution patterns in the Err-Julier area, Swiss Alps. Norsk Geogr Tidsskr 55:195–202

Frehner M, Ling AHM, Gärtner-Roer I (2015) Furrow-and-ridge morphology on rockglaciers explained by gravity-driven buckle folding: a case study from the Murtèl rockglacier (Switzerland). Permafr Periglac Process 26(1):57–66. https://doi.org/10.1002/ppp.1831

Fritz P (1976) Gesteinsbedingte Standorts- und Formendifferenzierung rezenter Periglazialerscheinungen in den Ostalpen. Mitt Osterr Geogr G 118(2):237–273

French H (2007) The periglacial environment, 3rd edn. Wiley, Chichester, UK, p 458

Fuchs MC, Böhlert R, Krbetschek M, Preusser F, Egli M (2013) Exploring the potential of luminescence methods for dating Alpine rock glaciers. Quat Geochronol 18:17–33. https://doi.org/10.1016/j.quageo.2013.07.001

Furrer G (1954) Solifluktionsformen im Schweizerischen Nationalpark. Untersuchung und Interpretation auf morphologischer Grundlage. Ergebnisse Der Wissenschaftlichen Untersuchungen Im Schweizerischen Nationalpark IV 29:200–275

Furrer G (1955) Die Strukturbodenformen Der Alpen. Geogr Helv 10:193–213

Furrer G (1965) Die subnivale Höhenstufe und ihre Untergrenze in den Bündner und Walliser Alpen. Geogr Helvetica 20(4):185–192

Furrer G, Fitze P (1970) Beitrag zum Permafrostproblem in den Alpen. Vierteljahrsschrift Der Naturforschenden Gesellschaft in Zurich 115(3):353–368

Gärtner-Roer I, Brunner N, Delaloye R, Haeberli W, Kääb A, Thee P (2021) Glacier-permafrost relations in a high-mountain environment: 5 decades of kinematic monitoring at the Gruben site, Swiss Alps. The Cryosphere Discuss.https://doi.org/10.5194/tc-2021-208, in review

Gamper MW (1984) Controls and rates of movement of solifluction lobes in the eastern Swiss Alps. In: Proceedings of the 4th International Conference on Permafrost, Fairbanks, Alaska, pp 328–333

Gams I (1959) Pohorsko Podravje. Razvoj kulturne pokrajine. Dela 5, 231 p

Gams I (1965) Speleological characteristics of the Slovene karst. Nase Jame 7:41–50

Gerdol R, Smiraglia C (1990) Correlation between vegetation pattern and microtopography in periglacial areas of the Central Alps. Pirineos 135:13–28

Gerhold N (1970) Blockgletscher Im Ötztal. Tiroler Heimatblätter 10(11):107–114

Gignoux M (1931) Les sols polygonaux dans les Alpes et la genèse des sols polaires. Ann Géographie 40:610–619. https://doi.org/10.3406/geo.1931.11102

Gobiet A, Kotlarski S, Beniston M, Heinrich G, Rajczak J, Stoffel M (2014) 21st century climate change in the European Alps—a review. Sci Total Environ 493:1138–1151. https://doi.org/10.1016/j.scitotenv.2013.07.050

Goldscheider N, Chen Z, Auler AS, Bakalowicz M, Broda S, Drew D, Hartmann J, Jiang G, Goudie A (eds) (2004) Encyclopedia of geomorphology. Routledge, London, 1200 p https://doi.org/10.4324/9780203381137

Goudie AS (2004) Encyclopedia of Geomorphology. Routledge, ISBN: 0–415–27298–X; p 1156

Grassler F (1984) Alpenvereinseinteilung der Ostalpen (AVE). Berg 108:215–224

Gruber S, Hoelzle M, Haeberli W (2004a) Rock-wall temperatures in the Alps: modelling their topographic distribution and regional differences. Permafr Periglac Process 15:299–307. https://doi.org/10.1002/ppp.501

Gruber S, Hoelzle M, Haeberli W (2004b) Permafrost thaw and destabilization of Alpine rock walls in the hot summer of 2003. Geophys Res Lett 31:L13504. https://doi.org/10.1029/2004GL020051

Gruppo Nazionale Geografia Fisica e Geomorfologia-CNR (1986) Ricerche geomorfologiche nell'alta Val di Peio (Gruppo del Cevedale). Geogr Fis Din Quat 9:137–191

Guglielmin M (1997) Il permafrost alpino. Concetti, morfologia e metodi di individuazione. Quaderni di Geodinamica alpina e Quaternaria, CNR 5:117

Guglielmin M (2003) Observation on permafrost ground thermal regimes from Antarctica and the Italian Alps, and their relevance to global climate change. Global Planet Change 40:159–167

Guglielmin M, Lozej A, Tellini C (1994) Permafrost distribution and rock glaciers in the Livigno Area (Northern Italy). Permafr Periglac Process 5:1–12

Guglielmin M, Smiraglia C (eds) (1997) The rock glacier inventory of the Italian Alps. Archivio Comitato Glaciologico Italiano 3, 103 p

Guglielmin M, Camusso M, Polesello S, Valsecchi S (2004) An old relict glacier body preserved in permafrost environment: the Foscagno rock glacier ice core (Upper Valtellina, Italian Central Alps). Arct Antarct Alp Res 36:108–116

Guglielmin M, Donatelli M, Semplice M, Serra Capizzano S (2018) Ground surface temperature reconstruction for the last 500 years obtained from permafrost temperatures observed in the Share Stelvio Borehole, Italian Alps. Clim past 14:709–724

Guglielmin M, Ponti S, Forte E, Cannone N (2021) Recent thermokarst evolution in the Italian Central Alps. Permafr Periglac Process 32(2):537–537. https://doi.org/10.1002/ppp.2099Citations

Guiter V (1972) Une forme montagnarde: le rock-glacier. Rev Géographie Alp 60:467–487

Haberkorn A, Wever N, Hoelzle M, Phillips M, Kenner R, Bavay M, Lehning M (2017) Distributed snow and rock temperature modelling in steep rock walls using Alpine3D. Cryosphere 11:585–607. https://doi.org/10.5194/tc-11-585-2017

Haeberli W (2007) Formbildung durch periglaziale Prozesse. Geographie—Physische Geographie und Humangeographie. Spektrum, Heidelberg, pp 289–295

Haeberli W, Patzelt G (1982) Permafrostkartierung im Gebiet der Hochebenkar-Blockgletscher, Obergurgl, Ötztaler Alpen. Z Gletscherk Glazialgeol 18:127–150

Haeberli W, Gruber S (2008) Recent challenges for permafrost in steep and cold terrain: an Alpine perspective. In: Proceedings of the 9th International Conference on Permafrost, Fairbanks, Alaska, pp 597–605

Haeberli W (1979) Holocene push-moraines in Alpine permafrost. Geogr Annaler 61A(1–2):43–48

Haeberli W (1985) Creep of mountain permafrost. Mitteilungen der Versuchsanstalt für Wasserbau, Hydrologie und Glaziologie der ETH Zurich, p 77.

Haeberli W, Brandovà D, Burga C, Egli M, Frauenfelder R, Kääb A, Maisch M (2003) Methods for absolute and relative age dating of rock-glacier surfaces in alpine permafrost. In: Proceedings of the 8th International Conference on Permafrost, Zurich, Switzerland, pp 343–348

Haeberli W, Hoelzle M., Kääb A, Keller F, Vonder Mühll D, Wagner S (1998) Ten years after drilling through the permafrost of the active rock glacier Murtèl, eastern Swiss Alps: answered questions and new perspectives. In: Proceedings of the 7th International Conference on Permafrost, Yellowknife, pp 403–410

Haeberli W, Hallet B, Arenson L, Elconin R, Humlum O, Kääb A, Kaufmann V, Ladanyi B, Matsuoka N, Springman S, Vonder Mühll D (2006) Permafrost creep and rock glacier dynamics. Permafr Periglac Process 17:189–214

Haeberli W, Kääb A, Vonder Mühll D, Teysseire P (2001) Prevention of debris flows from outbursts of periglacial lakes at Gruben, Valais, Swiss Alps. J Glaciol 47:111–122

Hallet B, Walder J, Stubbs CW (1991) Weathering by segregation ice growth in microcracks at sustained sub-zero temperatures: verification from an experimental study using acoustic emissions. Permafr Periglac Process 2:283–300

Harris C, Haeberli W, Vonder Mühll D, King L (2001) Permafrost monitoring in the high mountains of Europe: the PACE project in its global context. Permafr Periglac Process 12:3–11

Hartl L, Fischer A, Stocker-Waldhuber M, Abermann J (2016) Recent speed-up of an Alpine rock glacier: an updated chronology of the kinematics of Outer Hochebenkar rock glacier based on geodetic measurements. Geogr Annaler 98A(2):129–141. https://doi.org/10.1111/geoa.12127

Hartmeyer I, Keuschnig M, Delleske R, Krautblatter M, Lang A, Schrott L, Prasicek G, Otto JC (2020) A 6-year lidar survey reveals enhanced Rockwall retreat and modified rockfall magnitudes/frequencies in deglaciating cirques. Earth Surf Dynam 8:753–768. https://doi.org/10.5194/esurf-8-753-2020

Haserodt K (1965) Untersuchungen zur Höhen- und Altersgliederung der Karstformen in den Nördlichen Kalkalpen. Münchner Geogr Hefte 27:1–114

Hasler A, Gruber S, Haeberli W (2011) Temperature variability and offset in steep alpine rock and ice faces. Cryosphere 5:977–988. https://doi.org/10.5194/tc-5-977-2011

Hasler A, Gruber S, Beutel J (2012) Kinematics of steep bedrock permafrost. J Geophys Res 117:F01016. https://doi.org/10.1029/2011JF001981

Herak M (1972) Chapter 3—Karst of Yugoslavia. Karst: Important Karst regions of the Northern Hemisphere. Elsevier, Amsterdam, London, New York, pp 25–83

Hartmeyer I, Keuschnig M, Schrott L (2012) A scale-oriented approach for the long-term monitoring of ground thermal conditions in permafrost-affected rock faces, Kitzsteinhorn, Hohe Tauern Range Austria. Austrian J Earth Sci 105(2):128–139

Hauck C, Kneisel C (2008) Applied geophysics in periglacial environments. Cambridge University Press, Cambridge, UK.https://doi.org/10.1017/CBO9780511535628

Hauck C, Bach M, Hilbich C (2008) A four-phase model to quantify subsurface ice and water content in permafrost regions based on geophysical data sets. In: Proceedings of the 9th International Conference on Permafrost, Fairbanks, Alaska, pp 675–680

Hausmann H, Krainer K, Brückl E, Ullrich C (2012) Internal structure, ice content and dynamics of Ölgrube and Kaiserberg rock glaciers (Ötztal Alps, Austria) determined from geophysical surveys. Austrian J Earth Sci 105(2):12–31

Hermann F (1925) I rockglaciers della Valsavarenche. Natura 16:139–142

Herrmann E, Pucher E, Nicolussi K (2010) Das Schneeloch auf der Hinteralm (Schneealpe, Steiermark): Speläomorphologie, Eisveränderung, Paläozoologie und Dendrochronologie. Die Höhle 61:57–72

Hoffmann T, Schrott L (2003) Determining sediment thickness of talus slopes and valley fill deposits using seismic refraction—a comparison of 2D interpretation tools. Z Geomorphol Supp 132:71–87

Höfner T (1993) Fluvialer sedimenttransfer in der periglazialen Höhenstufe der Zentralalpen, südliche Hohe Tauern, Ostirol—Bestandsaufnahme und Versuch einer Rekonstruktion der mittel-bis jungholozänen Dynamik. Bamberger Geogr Schriften 13:1–121

Höllermann P (1967) Zur Verbreitung rezenter periglazialer Kleinformen in den Pyrenäen und Ostalpen. Göttinger Geogr Abhandlungen 40:1–198

Höllermann P (1983) Blockgletscher als Mesoformen der Periglazialstufe: Studien aus europäischen und nordamerikanischen Hochgebirgen. Bonner Geogr Abhandlungen 67:1–73

Höllermann P (1985) The periglacial belt of mid-latitude mountains from a geoecological point of view. Erdkunde 39(4):259–270

Humlum O (2000) The geomorphic significance of rock glaciers: estimates of rock glacier debris volumes and headwall recession rates in West Greenland. Geomorphology 35:41–67

Ikeda A, Matsuoka N (2002) Degradation of talus-derived rock glaciers in the Upper Engadin, Swiss Alps. Permafr Periglac Process 13:145–161

Ikeda A, Matsuoka N (2006) Pebbly versus bouldery rock glaciers: morphology, structure and processes. Geomorphology 73(3–4):279–296. https://doi.org/10.1016/j.geomorph.2005.07.015

Ikeda A, Matsuoka N, Kääb A (2008) Fast deformation of perennially frozen debris in a warm rock glacier in the Swiss Alps: An effect of liquid water. J Geophys Res-Earth 113:F01021. https://doi.org/10.1029/2007JF000859

Ivy-Ochs S, Kerschner H, Maisch M, Christl M, Kubik PW, Schlüchter C (2009) Latest Pleistocene and Holocene glacier variations in the European Alps. Quaternary Sci Rev 28:2137–2149. https://doi.org/10.1016/j.quascirev.2009.03.009

Jaesche P (1999) Bodenfrost und Solifluktionsdynamik in einem alpinen Periglazialgebiet (Hohe Tauern, Osttirol). Bayreuther Geow Arbeiten 20:1–152

Jaesche P, Huwe B, Stingl H, Veit H (2002) Temporal variability of Alpine solifluction: a modelling approach. Geogr Helvetica 57:157–169

Jaesche P, Veit H, Huwe B (2003) Snow cover and soil moisture controls on solifluction in an area of seasonal frost, Eastern Alps. Permafr Periglac Process 14:399–410

Jennings JN (1960) On an unusual occurrence of stone polygons in the French Alp. Biul Peryglac 7:169–173

Jorgenson MT (2003) Thermokarst terrains. Treatise Geomorphol 8:313–324

Julian M (1966) Les montagnes du Haut Var. Esquisse Morphologique. Méditerranée 7:185–206. https://doi.org/10.3406/medit.1966.1197

Kääb A (1997) Oberflächenkinematik ausgewählter Blockgletscher des Oberengadins. Beiträge aus der Gebirgs-Geomorphologie. Mitteilung Der VAW-ETH Zürich 158:121–140

Kääb A (2005) Remote sensing of mountain glaciers and permafrost creep. Schriftenreihe Physische Geogr 48:1–264

Kääb A, Weber M (2004) Development of transverse ridges on rock glaciers: field measurements and laboratory experiments. Permafr Periglac Process 15:379–391

Kääb A, Frauenfelder R, Roer I (2007) On the response of Rockglacier creep to surface temperature increase. Glob Planet Change 56:172–187

Kääb A, Haeberli W, Gudmundsson GH (1997) Analysing the creep of mountain permafrost using high precision aerial photogrammetry: 25 years of monitoring Gruben rock glacier. Swiss Alps. Permafr Periglac Process 8(4):409–426

Kääb A, Strozzi T, Bolch T, Caduff R, Trefall H, Stoffel M, Kokarev A (2021) Inventory and changes of rock glacier creep speeds in Ile Alatau and Kungöy Ala-Too, northern Tien Shan, since the 1950s. Cryosphere 15:927–949. https://doi.org/10.5194/tc-15-927-2021

Karte J (1982) Development and present status of German periglacial research in the polar and subpolar regions. Polar Geogr 6:1–24 https://www.jstor.org/stable/41143141

Karte J (1983) Periglacial phenomena and their significance as climatic and edaphic indicators. GeoJournal 7/4/Polar Research:329–340 https://www.jstor.org/stable/41143141

Kaufmann V (1996) Der Dösener Blockgletscher—Studienkarten und Bewegungsmessungen. Arb Inst Geogr Univ Graz 33:141–162

Kaufmann V (2012) The evolution of rock glacier monitoring using terrestrial photogrammetry: the example of Äußeres Hochebenkar rock glacier (Austria). Austrian J Earth Sci 105(2):63–77

Kaufmann V, Ladstädter R (2003) Quantitative analysis of rock glacier creep by means of digital photogrammetry using multi-temporal aerial photographs: two case studies in the Austrian Alps. In: Proceedings of the 8th International Conference on Permafrost, Zurich, Switzerland, pp 525–530

Kaufmann V, Ladstädter R, Lieb GK (2006) Quantitative assessment of the creep process of Weissenkar Rock Glacier (Central Alps, Austria). In: Proceedings of the 8th International Symposium on High Mountain Remote Sensing Cartography (HMRSC-VIII), Kathmandu, La Paz, Bolivia, pp 77–86

Kaufmann V, Kellerer-Pirklbauer A, Seier G (2021) Conventional and UAV-based aerial surveys for long-term monitoring (1954–2020) of a highly active rock glacier in Austria. Front. Remote Sens. 2:732744. https://doi.org/10.3389/frsen.2021.732744

Kellerer-Pirklbauer A (2007) Lithology and the distribution of rock glaciers: Niedere Tauern Range, Styria. Austria. Z Geomorphol 51(2):17–38

Kellerer-Pirklbauer A (2008) The Schmidt-hammer as a Relative Age Dating Tool for Rock Glacier Surfaces: Examples from Northern and Central Europe. In: Proceedings of the 9th International Conference on Permafrost, Fairbanks, Alaska, pp 913–918

Kellerer-Pirklbauer A (2017) Potential weathering by freeze-thaw action in alpine rocks in the European Alps during a nine-year monitoring period. Geomorphology 296:113–131. https://doi.org/10.1016/j.geomorph.2017.08.020

Kellerer-Pirklbauer A (2018) Solifluction rates and environmental controls at local and regional scales in central Austria. Norsk Geogr Tidsskr 72:37–56. https://doi.org/10.1080/00291951.2017.1399164

Kellerer-Pirklbauer A (2019) Long-term monitoring of sporadic permafrost at the eastern margin of the European Alps (Hochreichart, Seckauer Tauern range, Austria). Permafr Periglac Process 30(4):260–277. https://doi.org/10.1002/ppp.2021

Kellerer-Pirklbauer A, Kaufmann V (2012) About the relationship between rock glacier velocity and climate parameters in central Austria. Austrian J Earth Sci 105(2):94–112

Kellerer-Pirklbauer A, Kaufmann V (2018) Deglaciation and its impact on permafrost and rock glacier evolution: New insight from two adjacent cirques in Austria. Sci Total Environ 621:1397–1414. https://doi.org/10.1016/j.scitotenv.2017.10.087

Kellerer-Pirklbauer A, Rieckh M (2016) Monitoring nourishment processes in the rooting zone of an active rock glacier in an alpine environment. Z Geomorphol Supp 60(3):99–121. https://doi.org/10.1127/zfg_suppl/2016/00245

Kellerer-Pirklbauer A, Lieb GK, Avian M, Carrivick J (2012a) Climate change and rock fall events in high mountain areas: numerous and extensive rock falls in 2007 at Mittlerer Burgstall, Central Austria. Geogr Ann A 94:59–78. https://doi.org/10.1111/j.1468-0459.2011.00449.x

Kellerer-Pirklbauer A, Lieb GK, Kleinferchner H (2012b) A new rock glacier inventory of the Eastern European Alps. Austrian J Earth Sci 105(2):78–93

Kellerer-Pirklbauer A, Lieb GK, Kaufmann V (2017) The Dösen Rock Glacier in Central Austria: a key site for multidisciplinary long-term rock glacier monitoring in the Eastern Alps. Austrian J Earth Sci 110(2). https://doi.org/10.17738/ajes.2017.0013

Kellerer-Pirklbauer A, Delaloye R, Lambiel C, Gärtner-Roer I, Kaufmann V, Scapozza C, Krainer K, Staub B, Thibert E, Bodin X, Fischer A, Hartl L, Morra di Cella U, Mair V, Marcer M, Schoeneich P (2018) Interannual variability of rock glacier flow velocities in the European Alps. In: 5th European Conference on Permafrost—Book of Abstracts, Chamonix, France, pp 396–397

Kenner R, Phillips M, Beutel J, Hiller M, Limpach P, Pointner E, Volken M (2017) Factors controlling velocity variations at short-term, seasonal and multiyear time scales, Ritigraben Rock Glacier, Western Swiss Alps. Permafr Periglac Process 28(4):675–684. https://doi.org/10.1002/ppp.1953

Kenyi LM, Kaufmann V (2003) Measuring rock glacier surface deformation using SAR interferometry. In: Proceedings of the 8th International Conference on Permafrost, Zurich, Switzerland, pp 537–541

Kern K, Lieb GK, Seier G, Kellerer-Pirklbauer A (2012) Modelling geomorphological hazards to assess the vulnerability of alpine infrastructure: The example of the Großglockner-Pasterze area. Austria. Austrian J Earth Sci 105(2):113–127

Kerschner H (1983) Late glacial paleotemperatures and paleoprecipitation as derived from permafrost; glacier relationships in the Tyrolean Alps, Austria. In: Proceedings of the 4th International Conference on Permafrost, Fairbanks, Alaska, pp 589–594

Kerschner H (1985) Quantitative paleoclimatic inferences from lateglacial snowline, timberline and rock glacier data, Tyrolean Alps, Austria. Z Gletscherk Glazialgeol 21:363–369

Kerschner H, Krainer K, Spötl C (2014) DEUQUA excursions: From the foreland to the Central Alps. Field trips to selected sites of Quaternary research in the Tyrolean and Bavarian Alps, Excursion guide of the field trips of the DEUQUA Congress in Innsbruck, Austria.https://doi.org/10.3285/g.00011

Keuschnig C, Krainer K, Erschbamer B (2007) Bodenstruktur, Temperaturen und Vegetation von Bültenböden am Peischlachtörl (nördliche Schobergruppe, Nationalpark Hohe Tauern, Österreich). Tuexenia 27:343–361

Kieslinger A (1927) Geologie und Petrographie der Koralpe III. Die Steinöfen des Koralpengebietes. Sitzungsber Akad Wiss Wien, math.-naturwiss Klasse. Abt I 136:79–94

Kinzl H (1928) Beobachtungen uber Strukturböden in den Ostalpen. Petermanns Mitt 74:261–265

Klinge M, Lehmkuhl F (1998) Zur Differenzierung des periglazialen Formenschatzes nach ökologischen Standortfaktoren im Piffkar (Hohe Tauern). Wissenschaftliche Mitteilungen Aus Dem Nationalpark Hohe Tauern 4:207–223

Konrad SK, Humphrey NF, Steig EJ, Clark DH, Potter N, Pfeffer WT (1999) Rock glacier dynamics and paleoclimatic implications. Geology 27(12):1131–1134

Krainer K, Bressan D, Dietre B, Haas JN, Hajdas I, Lang K, Mair V, Nickus U, Reidl D, Thies H, Tonidandel D (2014) A 10,300-year-old permafrost core from the active rock glacier Lazaun, southern Ötztal Alps (South Tyrol, northern Italy). Quaternary Res 83(2):24–335. https://doi.org/10.1016/j.yqres.2014.12.005

Krainer K, Ribis M (2012) A rock glacier inventory of the Tyrolean Alps (Austria). Austrian J Earth Sci 105(2):32–47

Krainer K, Kellerer-Pirklbauer A, Kaufmann V, Lieb GK, Schrott L, Hausmann H (2012a) Permafrost research in Austria: history and recent advances. Austrian J Earth Sci 105(2):2–11

Krainer K, Mussner L, Behm M, Hausmann H (2012b) Multi-disciplinary investigation of an active rock glacier in the Sella Group (Dolomites; Northern Italy). Austrian J Earth Sci 105(2):48–62

Krautblatter M, Huggel C, Deline P, Hasler A (2012) Research perspectives on unstable high-alpine bedrock permafrost: measurement, modelling and process understanding. Permafr Periglac Process 23(1):80–88. https://doi.org/10.1002/ppp.740

Krautblatter M, Verleysdonk S, Flores-Orozco A, Kemna A (2010) Temperature-calibrated imaging of seasonal changes in permafrost rock walls by quantitative electrical resistivity tomography (Zugspitze, German/Austrian Alps). J Geophys Res-Earth 115:F2. https://doi.org/10.1029/2008JF001209

Kunz J, Kneisel C (2020) Glacier-permafrost interaction at a thrust moraine complex in the glacier forefield Muragl. Swiss Alps. Geosciences 10:205. https://doi.org/10.3390/geosciences10060205

Lambiel C (2004) Delaloye R (2004) Contribution of real-time kinematic GPS in the study of creeping mountain permafrost: examples from the Western Swiss Alps. Permafr Periglac Process 15(3):229–241. https://doi.org/10.1002/ppp.496

Lautridou JP, Ozouf JC (1982) Experimental frost shattering: 15 years of research at the Centre de Géomorphologie du CNRS. Prog Phys Geog Earth Env 6:215–232. https://doi.org/10.1177/030913338200600202

Lehmann O (1927) Das Tote Gebirge als Hochkarst. Mitt Geol Ges Wien 70:201–242

Lehmkuhl F (1989) Geomorphologische Höhenstufen in den Alpen unter besonderer Berücksichtigung des nivalen Formenschatzes. Göttinger Geogr Abhandlungen 88:1–113

Lehmkuhl F (2008) The kind and distribution of mid-latitude periglacial features and alpine permafrost in Eurasia. In: Proceedings of the 9th International Conference on Permafrost, Fairbanks, Alaska, pp 1031–1036

Lehmkuhl F (2016) Modern and past periglacial features in Central Asia and their implication for paleoclimate reconstructions. Prog Phys Geog Earth Env 40(3):369–391. https://doi.org/10.1177/0309133315615778

Lehmkuhl F, Nett J, Pötter S, Schulte P, Sprafke T, Jary Z, Antoine P, Wacha L, Wolf D, Zerboni A, Hošek J, Marković S, Obreht I, Sümegi P, Veres D, Zeeden C, Boemke B, Schaubert V, Viehweger J, Hambach U (2020) Geodata of continuous and discontinuous permafrost during the last glacial maximum in Europe. CRC806-Database. https://doi.org/10.5880/SFB806.61

Lewkowicz AG, Way RG (2019) Extremes of summer climate trigger thousands of thermokarst landslides in a High Arctic environment. Nat Commun 10:1329. https://doi.org/10.1038/s41467-019-09314-7

Lieb GK (1996) Permafrost und Blockgletscher in den östlichen österreichischen Alpen. Arb Inst Geogr Univ Graz 33:9–125

Lieb GK (1998) High-mountain permafrost in the Austrian Alps (Europe). In: Proceedings of the 7th International Conference on Permafrost, Yellowknife, Canada, pp 663–668

Lieb GK, Schopper A (1991) Zur Verbreitung von Permafrost am Dachstein (Nördliche Kalkalpen, Steiermark). Mitt Naturwiss Ver Steiermark 121:149–163

Lliboutry L (1955) Origine et évolution des glaciers rocheux. Comptes Rendus Académie Sci 240:1913–1915

Lliboutry L (1961) Phénomènes cryonivaux dans les Andes de Santiago (Chili). Biul Peryglac 10:209–224

Lucerna R (1906) Gletscherspuren in den Steiner Alpen. Geographisches Jahresbericht Aus Österreich 4:9–74

Luetscher M, Jeannin PY (2018) Ice caves in Switzerland. In: Ice caves. Elsevier, Amsterdam, pp 221–235

Maggi V, Colucci RR, Scoto F, Giudice G, Randazzo L (2018) Ice caves in Italy. In: Ice caves, Elsevier, Amsterdam, pp 399–423

Magnin F, Deline P, Ravanel L, Noetzli J, Pogliotti P (2015) Thermal characteristics of permafrost in the steep alpine rock walls of the Aiguille du Midi (Mont Blanc Massif, 3842 m a.s.l). Cryosphere 9:109–121. https://doi.org/10.5194/tc-9-109-2015

Magnin F, Josnin JY, Ravanel L, Pergaud J, Pohl B, Deline P (2017) Modelling rock wall permafrost degradation in the Mont Blanc massif from the LIA to the end of the 21st century. Cryosphere 11:1813–1834. https://doi.org/10.5194/tc-11-1813-2017

Mair V, Zischg A, Lang K, Tonidandel D, Krainer K, Kellerer-Pirklbauer A, Deline P, Schoeneich P, Cremonese E, Pogliotti P, Gruber S, Böckli L (2011) PermaNET—Permafrost Long-term Monitoring Network. Synthesis report, INTERPRAEVENT Journal series 1/3, Klagenfurt.

Mair V, Lang K, Tonidandel D, Thaler B, Alber R, Lösch B, Tait D, Nickus U, Krainer K, Thies H, Himsperger M, Sapelza A, Tolotti M (2015) Progetto Permaqua. Permafrost e il suo effetto sul bilancio idrico e sull'ecologia delle acque di alta montagna. Ufficio Geologia e Prove Materiali, Provincia Autonoma di Bolzano.

Mania I, D'Amico ME, Freppaz M, Gorra R (2016) Driving factors of soil microbial ecology in alpine, mid-latitude patterned grounds (NW Italian Alps). Biol Fertil Soils 52(8):1135–1148

Marazzi S (2005) Atlante orografico delle Alpi. SOIUSA—Suddivisione orografica internazionale unificata del Sistema Alpino, Quaderni di cultura alpina 82–83, Priuli & Verlucca editori

Marcer M, Cicoira A, Cusicanqui D, Bodin X, Echelard T, Obregon R, Schoeneich P (2021) Rock glaciers throughout the French Alps accelerated and destabilised since 1990 as air temperatures increased. Commun Earth Environ 2:81. https://doi.org/10.1038/s43247-021-00150-6

Marcer M, Serrano C, Brenning A, Bodin X, Goetz J, Schoeneich P (2019) Evaluating the destabilization susceptibility of active rock glaciers in the French Alps. Cryosphere 13:141–155. https://doi.org/10.5194/tc-13-141-2019

Marnezy A (1977) Aspects du modèle périglaciaire dans le Vallon de la Rocheure (Massif de la Vanoise). Revue de Géographie Alpine 65(4):365–384

Marvánek O (2010) Periglacial features in the Krumgampen Valley, Ötztal Alps Austria. Moravian Geographical Reports 18(2):1–56

Matsuoka N, Abe M, Ijiri M (2003a) Differential frost heave and sorted patterned ground: Field measurements and a laboratory experiment. Geomorphology 52(1–2):73–85. https://doi.org/10.1016/S0169-555X(02)00249-0

Matsuoka N, Abe M, Ijiri M (2003b) Differential frost heave and sorted patterned ground: field measurements and a laboratory experiment. Geomorphology 52:73–85

Matsuoka N (1990) The rate of bedrock weathering by frost action: field measurements and a predictive model. Earth Surf Proc Land 15:73–90

Matsuoka N (2001) Solifluction rates, processes and landforms: A global review. Earth Sci Rev 55:107–134

Matsuoka N (2008) Frost weathering and rockwall erosion in the southeastern Swiss Alps: long-term (1994–2006) observations. Geomorphology 99:353–368. https://doi.org/10.1016/j.geomorph.2007.11.013

Matsuoka N, Murton J (2008) Frost weathering: recent advances and future directions. Permafr Periglac Process 19:195–210. https://doi.org/10.1002/ppp.620

Matsuoka N, Hirakawa K, Watanabe T, Moriwaki K (1997) Monitoring of periglacial slope processes in the Swiss Alps: the first two years of frost shattering, heave and creep. Permafr Periglac Process 8:155–177

Matthes FE (1900) Glacial sculpture of the Bighorn Mountains, Wyoming, United States Geological Survey, 21st Annual Report 1899–1900:167–190

Matthews JA, Wilson P, Winkler S, Mourne RW, Hill JL, Owen G, Hiemstra JF, Hallang H, Geary AP (2019) Age and development of active cryoplanation terraces in the alpine permafrost zone at Svartkampan, Jotunheimen, southern Norway. Quaternary Res 92(3):641–664. https://doi.org/10.1017/qua.2019.41

Maull O (1958) Handbuch der geomorphologie, 2nd edn. Deuticke, Vienna, p 574

May B, Spötl C, Wagenbach D, Dublyansky Y, Liebl J (2011) First investigations of an ice core from Eisriesenwelt cave (Austria). Cryosphere 5:81–93. https://doi.org/10.5194/tc-5-81-2011

Melik A (1935) Slovenija—Geografski opis [Slovenia—Geographical description], 1st volume, Ljubljana

Melik A (1963) Slovenija—Geografski opis [Slovenia. Geographical description], 1st general part, 2nd. revised edition. Ljubljana.

Merz K, Maurer H, Rabenstein L, Buchli T, Springman SM, Zweifel M (2016) Multidisciplinary geophysical investigations over an alpine rock glacier. Geophysics 81/1:WA147–WA157 https://doi.org/10.1190/GEO2015-0157.1

Messenzehl K, Dikau R (2017) Structural and thermal controls of rockfall frequency and magnitude within rockwall-talus systems (Swiss Alps). Earth Surf Proc Land 42(13):1963–1981. https://doi.org/10.1002/esp.4155

Messenzehl K, Meyer H, Otto JC, Hoffmann T, Dikau R (2017) Regional-scale controls on the spatial activity of rockfalls (Turtmann Valley, Swiss Alps)—A multivariate modeling approach. Geomorphology 287:29–45. https://doi.org/10.1016/j.geomorph.2016.01.008

Messerli B, Zurbuchen M (1968) Blockgletscher im Weissmies und Aletsch und ihre photogrammetrische Kartierung. Die Alpen 3:1–13

Meyer C, Pflitsch A (2018) Ice caves in Germany. In: Ice caves, Elsevier, Amsterdam, pp 371–384

Michaud (1950) Emploi des marques dans l'etude des mouvements du sol. Rev Geom Dyn l(4):180–189.

Mihevc A (2018) Ice caves in Slovenia. In: Ice caves, Elsevier, Amsterdam, pp 691–703

Millar C, Westfall RD (2008) Rock glaciers and related periglacial landforms in the Sierra Nevada, CA, USA; inventory, distribution and climatic relationships. Quaternary Int 188:90–104

Monegato G, Ravazzi C, Donegana M, Pini R, Calderoni G, Wick L (2007) Evidence of a two-fold glacial advance during the last glacial maximum in the Tagliamento end moraine system (Eastern Alps). Quaternary Res 68:284–302

Moosdorf N, Stevanovic Z, Veni G (2020) Global distribution of carbonate rocks and karst water resources. Hydrogeol J 28:1661–1677. https://doi.org/10.1007/s10040-020-02139-5

Morawetz S (1952) Periglaziale Erscheinungen auf der Koralpe. Mitt Osterr Geogr G 94:252–257

Morawetz S (1968) Zur Frage der periglazialen Erscheinungen im Gebiet zwischen Graz und Hartberg. Mitt Naturwiss Ver Steiermark 98:61–68

Morawetz S (1971a) Fragen der Hangentwicklung. Mitt Naturwiss Ver Steiermark 101:73–95

Morawetz S (1971b) Zur Geomorphologie des Steirischen Randgebirges. Mitt Naturwiss Ver Steiermark 100:84–104

Murton J. (2021) Periglacial processes and deposits, Encyclopaedia of Geology, 2nd edition, Elsevier, pp. 857–875

Nagl H (1976) Die Raum-Zeit-Verteilung der Blockgletscher in den Niederen Tauern und die eiszeitliche Vergletscherung der Seckauer Tauern. Mitt Naturwiss Ver Steiermark 106:95–118

Nangeroni G (1929) Grotte e laghi subglaciali, colate e mari di pietre. Natura 20:152–161

Nangeroni G (1959) I fenomeni periglaciali in Italia. Atti Acc Roveretana Agiati 6(B):43–62

Nangeroni G (1962) Les phénomènes périglaciaires en Italie. Biuletyn Peryglac 11:57–64

Natek N (1993) Geomorfološka karta 1:100,000 list Celje in analiza reliefa sekcije [Geomorphological map 1: 100,000 sheet Celje and section relief analysis]. Doctoral dissertation, University of Ljubljana

Natek N (2007) Periglacial landforms in the Pohorje Mountains. Dela 27:247–263

Noetzli J, Gruber S (2009) Transient thermal effects in alpine permafrost. Cryosphere 3:85–99. https://doi.org/10.5194/tc-3-85-2009

Noetzli J, Gruber S, Kohl T, Salzmann N, Haeberli W (2007) Three-dimensional distribution and evolution of permafrost temperatures in idealized high-mountain topography. J Geophys Res Earth Surf 112/F2 F02S13 https://doi.org/10.1029/2006JF000545

Nyenhuis M, Hoelzle M, Dikau R (2005) Rock glacier mapping and permafrost distribution modelling in the Turtmanntal, Valais Switzerland. Z Geomorphol 49(3):275–292

Obu J, Košutnik J, Overduin PP, Boike J, Blatnik M, Zwieback S, Gostinčar P, Mihevc A (2018) Sorted patterned ground in a karst cave, Ledenica pod Hrušico, Slovenia. Permafr Periglac Process 29:121–130. https://doi.org/10.1002/ppp.1970

Oliva M, Žebre M, Guglielmin M, Hughes PD, Çiner A, Vieira G, Bodin X, Andrés N, Colucci RR, García-Hernández C, Mora C, Nofre J, Palacios D, Pérez-Alberti A, Ribolini A, Ruiz-Fernández

J, Sarikaya MA, Serrano E, Urdea P, Valcárcel M, Woodward JC, Yildirim C (2018) Permafrost conditions in the Mediterranean region since the Last Glaciation. Earth Sci Rev 185:397–436. https://doi.org/10.1016/j.earscirev.2018.06.018

Otto JC, Sass O (2006) Comparing geophysical methods for talus slope investigations in the Turtmann valley (Swiss Alps). Geomorphology 76:257–272

Pappalardo M (1999) Observations on stratified slope deposits, Gesso Valley, Italian Maritime Alps. Permafr Periglac Process 10:107–111

Pappalardo M, Spagnolo M (1999) A peculiar stratified slope deposit in the Val Grande di Palanfré (Southern Maritime Alps). Bollettino Accademia Delle Scienze Di Torino, Atti Classe Scienze Fisiche 1333(1):1–11

Paro L (2011) Relationship between cryotic processes and block streams evolution in the Lanzo Ultrabasic Complex (western Alps, Italy). Doctoral dissertation, University of Torino

Paul F, Rastner P, Azzoni RS, Diolaiuti G, Fugazza D, Le Bris R, Nemec J, Rabatel A, Ramusovic M, Schwaizer G, Smiraglia C (2020) Glacier shrinkage in the Alps continues unabated as revealed by a new glacier inventory from Sentinel-2. Earth Syst Sci Data 12:1805–1821. https://doi.org/10.5194/essd-12-1805-2020

Pech P, Ajinca M, Abdulhak S, Hustache E, Simon L, Talon B (2021) The geoecological evaluation of the heritage interest of polygonal soils inherited in alpine mountains. the example of the Col du Noyer (Massif du Dévoluy, Hautes Alpes, France). Journal Alp Res 109(4). https://doi.org/10.4000/rga.8780

Pecher C, Tasser E, Tappeiner U (2011) Definition of the potential treeline in the European Alps and its benefit for sustainability monitoring. Ecol Indic 11:438–447

PERMOS (2021) Swiss Permafrost Bulletin 2019/2020. https://doi.org/10.13093/permos-bull-2021

Perşoiu A, Lauritzen SE (eds) (2018) Ice caves. Elsevier, Amsterdam. https://doi.org/10.1016/B978-0-12-811739-2.09990-3

Pfiffner OA (2014) Geology of the Alps. Wiley, 368 p.

Pillewizer W (1938) Photogrammetrische Gletscheruntersuchungen im Sommer 1938. Z Gesell f Erdkunde 1938(9/19):367–372

Pillewizer W (1957) Untersuchungen an Blockströmen der Ötztaler Alpen. Geomorph Abhandl Geograph Inst FU Berlin (Otto-Maull-Festschrift) 5:37–50

Pintaldi E, D'Amico ME, Siniscalco C, Cremonese E, Celi L, Filippa G, Prati M, Freppaz M (2016) Hummocks affect soil properties and soil-vegetation relationships in a subalpine grassland (North-Western Italian Alps). CATENA 145:214–226. https://doi.org/10.1016/j.catena.2016.06.014

Pintaldi E, D'Amico ME, Colombo N, Colombero C, Sambuelli L, De Regibus C, Franco D, Perotti L, Paro L, Freppaz M (2021) Hidden soils and their carbon stocks at high-elevation in the European Alps (North-West Italy). CATENA 198:105044. https://doi.org/10.1016/j.catena.2020.105044

Pissart A (1977) Apparition et évolution des sols structuraux périglaciaires de haute montagne. Expériences de terrain au Chambeyron (Alpes, France). Abhandl Akad Wissensch Gött Math-Phy Kla 31:142–156

Pissart A (1987) Weichselian periglacial structures and their environmental significance: Belgium, the Netherlands and northern France. In: Boardman J (ed) Periglacial processes and landforms in Britain and Ireland. Cambridge University Press, Cambridge, pp 77–85

Pissart A (1993) Understanding the controls on solifluction movements in different environments: a methodology and its application in the French Alps. Palaeoclimate Research 11:209–215

Plan L (2016) Oberflächenkarstformen. In: Höhlen und Karst in Österreich, Oberösterreichisches Landesmuseum, Linz, pp 23–34.

Plan L, Renetzeder C, Pavuza R, Körner W (2012) A new karren feature: hummocky karren. Intern J Speleology 41(1):75–81. https://doi.org/10.5083/1827-806X.41.1.8

Ponti S, Cannone N, Guglielmin M (2018) Needle ice formation, induced frost heave, and frost creep: A case study through photogrammetry at Stelvio Pass (Italian Central Alps). CATENA 164:62–70. https://doi.org/10.1016/j.catena.2018.01.009

Price MF, Lysenko I, Gloersen E (2004) Delineating Europe's mountains. Revue Geograph Alpine 92(2):75–86

Raab T, Leopold M, Völkel J (2007) Character, age, and ecological significance of Pleistocene periglacial slope deposits in Germany. Phys Geogr 28(6):451–473. https://doi.org/10.2747/0272-3646.28.6.451

Ravanel L, Magnin F, Deline P (2017) Impacts of the 2003 and 2015 summer heatwaves on permafrost-affected rock-walls in the Mont Blanc Massif. Sci Total Environ 609:132–143. https://doi.org/10.1016/j.scitotenv.2017.07.055

Ravanel L, Troilo F, Pogliotti P, Paro L, Morra di Cella U, Duvillard PA, Motta E (2018) Risques naturels émergents en haute montagne. Actions conduites dans le cadre du projet ALCOTRA PrévRisk Haute Montagne—Apports pour les décideurs et les pratiquants. Rapport de synthèse du WP3 du projet ALCOTRA PrévRisk Haute Montagne, 28 p

Recami E (1967) Fenomeni crionivali in Val di Sole. Natura Alpina 18(2):60–67

Reger RD, Péwé TL (1976) Cryoplanation terraces; indicators of a permafrost environment. Quaternary Res 6:99–109

Reitner JM, Ivy-Ochs S, Drescher-Schneider R, Hajdas I, Linner M (2016) Reconsidering the current stratigraphy of the Alpine Lateglacial: Implications of the sedimentary and morphological record of the Lienz area (Tyrol/Austria). E&G Quaternary Sci Journal 65(2):113–144. https://doi.org/10.3285/eg.65.2.02

Resnati C, Smiraglia C (1990) Determinazione della struttura interna del rock glacier di Val Pisella (Alta Valtellina) attraverso sondaggi elettrici verticali. Risultati e problemi. Geogr Fis Din Quat 13:171–177

Reynard E, Häuselmann P, Jeannin PY, Scapozza C (2021): Geomorphological landscapes in Switzerland. In: Landscapes and Landforms of Switzerland. World Geomorphological Landscapes, Springer, pp 71–80

RGIK (2020) Towards standard guidelines for inventorying rock glaciers: baseline concepts (v. 4.0). IPA Action Group rock glacier inventories and kinematics

Ribolini A (2001) Active and fossil rock glaciers in the Argentera Massif (Maritime Alps): surface ground temperatures and paleoclimatic significance. Z Gletscherk Glazialgeol 37:125–140

Ribolini A (2003) An active rock glacier in the southernmost permafrost environment of the Alps (Argentera Massif, Italy): four years of surface ground temperature monitoring. In: Proceedings of the 8th International Conference on Permafrost—Extended Abstracts, Zurich, Switzerland, pp 133–134

Ribolini A, Fabre D (2006) Permafrost existence in the rock glacier of the Argentera Massif, Maritime Alps. Italy. Permafr Periglac Process 17(1):49–63

Ribolini A, Guglielmin M, Fabre D, Bodin X, Marchisio M, Sartini S, Spagnolo M, Schoeneich P (2010) The internal structure of rock glaciers and recently deglaciated slopes as revealed by geoelectrical tomography: insights on permafrost and recent glacial evolution in the Central and Western Alps (Italy-France). Quaternary Sci Rev 29:507–521

Richter E (1900) Geomorphologische Untersuchungen in den Hochalpen. Petermanns Mitteilungen:132—Ergänzungsband 29:1–103

Rist A, Keller F, Schmid C, Gerber C, Vogel D, Bozzini C, Wunderle S, Veit H (2013) Langsam, aber stetig. Die Solifluktionsloben am Munt Chavagl. In: Atlas des Schweizerischen Nationalparks. Die ersten 100 Jahre. Nationalpark-Forschung Schweiz 99/1, Bern, Haupt

Rist A, Roth L, Veit H (2020) Elevational ground/air thermal gradients in the Swiss inner Alpine Valais. Arct Antarct Alp Res 52(1):341–360. https://doi.org/10.1080/15230430.2020.1742022

Ritter DF, Kochel RC, Miller JR (2002) Process geomorphology, 2nd edn, Waveland Press Inc, 560 p

Rixhon G, Demoulin A (2013) Evolution of slopes in a cold climate. Treatise on Geomorphology 8:392–415

Robl J, Prasicek G, Hergarten S, Stüwe K (2015) Alpine topography in the light of tectonic uplift and glaciation. Global Planet Change 127:34–49. https://doi.org/10.1016/j.gloplacha.2015.01.008

Rode M, Kellerer-Pirklbauer A (2012) Schmidt-hammer exposure-age dating (SHD) of rock glaciers in the Schöderkogel-Eisenhut area, Schladminger Tauern Range Austria. The Holocene 22(7):761–771. https://doi.org/10.1177/0959683611430410

Roer I (2007) Rockglacier kinematics in a high mountain geosystem. Bonner Geogr Abhandlungen 117:217 p

Roer I, Kääb A, Dikau R (2005a) Rockglacier acceleration in the Turtmann valley (Swiss Alps): Probable controls. Norsk Geogr Tidsskr 59:157–163

Roer I, Kääb A, Dikau R (2005b) Rockglacier kinematics derived from small-scale aerial photography and digital airborne pushbroom imagery. Z Geomorphol 49(1):73–87

Roer I, Haeberli W, Avian M, Kaufmann V, Delaloye R, Lambiel C, Kääb A (2008) Observations and considerations on destabilizing active rock glaciers in the European Alps. In: Proceedings of the 9th International Conference on Permafrost, Fairbanks, Alaska, pp 1505–1510

Rogora M, Somaschini L, Marchetto A, Mosello R, Tartari GA, Paro L (2020) Decadal trends in water chemistry of Alpine lakes in calcareous catchments driven by climate change. Sci Total Environ 708:1–14. https://doi.org/10.1016/j.scitotenv.2019.135180

Rolshoven M (1982) Alpines Permafrostmilieu in der Lasörlinggruppe/Nördliche Deferegger Alpen (Osttirol). Polarforschung 52(1–2):55–64

Ruszkiczay-Rüdiger Z, Kern Z (2016) Permafrost or seasonal frost? A review of paleoclimate proxies of the last glacial cycle in the East Central European lowlands. Quaternary Int 415:241–252. https://doi.org/10.1016/j.quaint.2015.07.027

Šifrer M (1983) Kvartarni Razvoj Škofjeloškega Hribovja. Geografski Zbornik 22:139–196

Sass O (2005) Rock moisture measurements: techniques, results, and implications for weathering. Earth Surf Proc Land 30:359–374

Sass O (2008) The use of GPR in determining talus thickness and talus structure. In: Applied Geophysics in Periglacial Environments, Cambridge University Press, Cambridge, UK, pp 165–171. https://doi.org/10.1017/CBO9780511535628

Savi S, Delunel R, Schlunegger F (2015) Efficiency of frost-cracking processes through space and time: An example from the eastern Italian Alps. Geomorphology 232:248–260. https://doi.org/10.1016/j.geomorph.2015.01.009

Scaramellini G, Bonardi L (2001) La géographie italienne et les Alpes de la fin du XIXe siècle à la Seconde Guerre mondiale. Revue De Géographie Alpine 89(4):133–158. https://doi.org/10.3406/rga.2001.3062

Schär C, Davies TD, Frei C, Wanner H, Widmann M, Wild M, Davis HC (1998) Current alpine climate. Views from the Alps: regional perspectives on climate change. MIT Press, Boston, pp 21–72

Schlüchter C, Akcar N, Ivy-Ochs S (2021) The Quaternary period in Switzerland. In: Landscapes and Landforms of Switzerland. World Geomorphological Landscapes. Springer, pp 47–69

Schmöller R, Fruhwirth RK (1996) Komplexgeophysikalische Untersuchung auf dem Dösener Blockgletscher (Hohe Tauern, Österreich). Arb Inst Geogr Univ Graz 33:165–190

Schneider B, Schneider H (2001) Zur 60jährigen Messreihe der kurzfristigen Geschwindigkeitsschwankungen am Blockgletscher im Äusseren Hochebenkar. Ötztaler Alpen. Tirol. Z Gletscherk Glazialgeol 37:1–33

Scotti R, Brardinoni F, Alberti S, Frattini P, Crosta GB (2013) A regional inventory of rock glaciers and protalus ramparts in the central Italian Alps. Geomorphology 186:136–149

Scotti R, Crosta GB, Villa A (2017) Destabilisation of creeping permafrost: the Plator Rock Glacier case study (Central Italian Alps). Permafr Periglac Process 28:224–236. https://doi.org/10.1002/ppp.1917

Schneider S, Daengeli S, Hauck C, Hoelzle M (2013) A spatial and temporal analysis of different periglacial materials by using geoelectrical, seismic and borehole temperature data at Murtèl-Corvatsch, Upper Engadin, Swiss Alps. Geogr Helvetica 68:265–280. https://doi.org/10.5194/gh-68-265-2013

Schöner W, Boeckli L, Hausmann H, Otto JC, Reisenhofer S (2012) Spatial patterns of permafrost at Hoher Sonnblick (Austrian Alps)—extensive field-measurements and modelling approaches. Austrian J Earth Sci 105(2):154–168

Schrott L, Sass O (2008) Application of field geophysics in geomorphology: advances and limitations exemplified by case studies. Geomorphology 93:55–73

Schrott L, Otto JC, Keller F (2012) Modelling alpine permafrost distribution in the Hohe Tauern region, Austria. Austrian J Earth Sci 105(2):169–183

Schüepp M, Bouët M, Bider M, Urfer C (1978) Regionale Klimabeschreibungen (1. Teil). Beiheft Annalen Schweiz, Met Anstalt, Zürich

Schuster R, Stüwe K (2010) Die Geologie der Alpen im Zeitraffer. Mitt Naturwiss Ver Steiermark 140:5–21

Schweizer G (1968) Le tardiglaciaire et le niveau des neiges permanentes dans les hautes montagnes des Alpes-Maritimes. L'exemple du bassin supérieur de la Tinée. Méditerranée 9:23–40. https://doi.org/10.3406/medit.1968.1263

Seppi R, Carton A, Zumiani M, Dall'Amico M, Zampedri G, Rigon R (2012) Inventory, distribution and topographic features of rock glaciers in the southern region of the Eastern Italian Alps (Trentino). Geogr Fis Din Quat 35:185–197

Seppi R, Zanoner T, Carton A, Bondesan A, Francese R, Carturan L, Zumiani M, Giorgi M, Ninfo A (2014) Current transition from glacial to periglacial processes in the Dolomites (South-Eastern Alps). Geomorphology 228:1–86. https://doi.org/10.1016/j.geomorph.2014.08.025

Serra E, Valla P, Gribenski N, Guedes Magrani F, Carcaillet J, Delaloye R, Grobéty B, Braillard L (2021) Geomorphic response to the Lateglacial-Holocene transition in high Alpine regions (Sanetsch Pass, Swiss Alps). Boreas 50(1):242–261. https://doi.org/10.1111/bor.12480

Simony F (1847) Kalkhöhlenbildung. Berichte über die Mittheilungen von Freunden der Naturwissenschaften in Wien (Ed. Wilhelm Haidinger) 1:55–59

Smiraglia C (1990) Misure di velocità superficiale al rock glacier orientale di Val Pisella (Gruppo del Cevedale, Alta Valtellina). Geogr Fis Din Quat 2:41–44

Smiraglia C (1992) Observations on the rock glaciers of Monte Emilius (Valle d'Aosta, Italy). Permafr Periglac Process 3:163–168

Sölch J (1922) Karbildungen in Der Stubalpe. Z Gletscherkunde 12:20–38

Sölch J (1928) Die Landformung der Steiermark (Grundzüge einer Morphologie). Verlag des naturwiss Ver Steiermark, Graz, p 221

Spötl C, Pavuza R (2016) Eishöhlen und Höhleneis. In: Höhlen und Karst in Österreich, Oberösterreichisches Landesmuseum, Linz, pp 139–154

Spötl C, Wimmer M, Pavuza R, Plan L (2018) Ice caves in Austria. In: Ice caves. Elsevier, Amsterdam, pp 237–262

Spreitzer H (1957) Zur Geographie des Kilikischen Ala Dag im Taurus. Festschr z Hundertjahrfeier der Geogr Ges Wien, pp 414–459

Spreitzer H (1960) Hangformung und Asymmetrie der Bergrücken in den Alpen und im Taurus. Z GeomorphoL Supp 1:211–236

Springman SM, Arenson LU, Yamamoto Y, Maurer H, Kos A, Buchli T, Derungs G (2012) Multidisciplinary investigations on three rock glaciers in the Swiss Alps: legacies and future perspectives. Geogr Annaler 94A(2):215–243

Steinemann O, Reitner JM, Ivy-Ochs S, Christl M, Synal HA (2020) Tracking rockglacier evolution in the Eastern Alps from the Lateglacial to the early Holocene. Quatern Sci Rev 241:106424. https://doi.org/10.1016/j.quascirev.2020.106424

Stepišnik U (2020) Kraška polja v Sloveniji / Karst poljes in Slovenia. Dela 53:23–43. https://doi.org/10.4321/dela.53.23-43

Stiegler C, Rode M, Sass O, Otto JC (2014) An undercooled scree slope detected by geophysical investigations in sporadic permafrost below 1000 m asl Central Austria. Permafr Periglac Process 25(3):194–207. https://doi.org/10.1002/ppp.1813

Stingl H (1969) Ein periglazialmorphologisches Nord-Süd-Profil durch die Ostalpen. Göttinger Geogr Abhandlungen 49:1–115

Stocker E (1973) Bewegungsmessungen und Studien an Schrägterrassen an einem Hangausschnitt in der Kreuzeckgruppe (Kärnten). Arb Inst Geogr Univ Salzburg 3:193–203

Stocker E (1984) Ergebnisse elfjähriger Messungen der Bodenbewegung in der alpinen Stufe der Kreuzeckgruppe (Kärnten). Wiener Geogr Schriften 59(60):27–35

Strozzi T, Kääb A, Frauenfelder R (2004) Detecting and quantifying mountain permafrost creep from in situ inventory, space-borne radar interferometry and airborne digital photogrammetry. Int J Remote Sens 25:2919–2931

Strozzi T, Caduff R, Jones N, Barboux C, Delaloye R, Bodin X, Kääb A, Mätzler E, Schrott L (2020) Monitoring rock glacier kinematics with satellite synthetic aperature radar. Remote Sensing 12(3):559. https://doi.org/10.3390/rs12030559

Swift DA, Cook S, Heckmann T, Moore J, Gärtner-Roer I, Korup O (2015) Ice and snow as land-forming agents. In: Snow and ice-related hazards, risks, and disasters. Elsevier, pp 167–199

Tampucci D, Gobbi M, Marano G, Boracchi P, Boffa G, Ballarin F, Pantini P, Seppi R, Compostella C, Caccianiga M (2017) Ecology of active rock glaciers and surrounding landforms: climate, soil, plants and arthropods. Boreas 46:185–198. https://doi.org/10.1111/bor.12219

Telbisz T, Tóth G, Ruban DA, Gutak JM (2019) Notable glaciokarsts of the World. In: Glaciokarsts, Springer Geography. Springer, Cham, pp 373–485. https://doi.org/10.1007/978-3-319-97292-3

Thibert E, Bodin X (2022) Changes in surface velocities over four decades on the Laurichard rock glacier (French Alps). Permafr Periglac Process 33:323–325. https://doi.org/10.1002/ppp.2159

Thorn CE (1988) Nivation: a geomorphic chimera. In: Advances in periglacial geo-morphology. Wiley, Chichester, pp 3–31

Thorn CE, Hall K (1980) Nivation: an arctic-alpine comparison and reappraisal. J Glaciol 25(91):109–124

Tomaselli M, Gualmini M, Petraglia A, Pontin A, Carbognani M, Gerdol R (2018) Three mires in the south-eastern Alps (northern Italy). J Maps 14(2):303–311. https://doi.org/10.1080/17445647.2018.1461692

Tóth G, Veress M (2019) Case studies on glaciokarst. In: Glaciokarsts, Springer Geography. Springer, Cham, pp 3353–272 https://doi.org/10.1007/978-3-319-97292-3

Troll C (1944) Strukturböden, Solifluction und Frostklimate der Erde. Geolog Rundschau 34:545–694

Untersweg T, Schwendt A (1996) Blockgletscher und Quellen in den Niederen Tauern. Mitt Österr Geolog Gesellsch 87:47–55

Vandenberghe J, French HM, Gorbunov A, Marchenko S, Velichko AA, Jin H, Cui Z, Zhang T, Wan X (2014) The Last Permafrost Maximum (LPM) map of the Northern Hemisphere: permafrost extent and mean annual air temperatures, 25–17 ka BP. Boreas 43:652–666. https://doi.org/10.1111/bor.1207

Van Husen D, Reitner JM (2001) An Outline of the Quaternary Stratigraphy of Austria. E&G Quaternary Sci Journal 60(2–3):366–387. https://doi.org/10.3285/eg.60.2-3.09

Van Vliet-Lanoë B (2014) Patterned ground and climate change. In Permafrost: distribution, composition and impacts on infrastructure and ecosystems, Nova Science Publishers, Inc, Results of the IPEV CRYOCLIM 2004 program, chapter 2, pp 67–106

Van Vliet-Lanoë B, Magyari Á, Meilliez F (2004) Distinguishing between tectonic and periglacial deformations of Quaternary continental deposits in Europe. Global Planet Change 43:103–127

Veit H, Höfner T (1993) Permafrost, gelifluction and fluvial transfer in the alpine/subnival ecotone, Central Alps, Austria: Present, past and future. Z Geomorphol Supp 92:71–84

Veit H (2002) Die Alpen—Geoökologie und Landschaftsentwicklung. Verlag Eugen Ulmer, Stuttgart

Veit H, Stingl H, Emmerich KH, John B (1995) Zeitliche und räumliche Variabilität solifluidaler Prozesse und ihre Ursachen: Eine Zwischenbilanz nach acht Jahren Solifluktionsmessungen (1985–1993) an der Meßstation Glorer Hütte, Hohe Tauern, Österreich. Z Geomorphol Supp 99:107–122

Veress M, Zentai Z (2004) Karros lejtőfejlődés a Triglav északi előterében (Karren slope development in the northern foreground of Triglav). Karsztfejlődés 9:177–196

Veress M, Telbisz T, Tóth G, Lóczy D, Ruban DA, Gutak JM (2019) Glaciokarsts, Springer Geography. Springer, Cham, 616 p. https://doi.org/10.1007/978-3-319-97292-3

Veyret P. (1956) Studi sui fenomeni crionivali (periglaciali partim) nette Alpi Italiane (fondazione per i pioblemi montani dell'Arco alpino). Revue de géographie alpine 44(3):148

Vietoris L (1958) Der Blockgletscher des äußeren Hochebenkares. Gurgler Berichte 1:41–45

Vietoris L (1972) Über die Blockgletscher des Äußeren Hochebenkars. Z Gletscherk Glazialgeol 8:169–188

Vigna B, Paro L (2019) Ghiacciai ipogei e grotte con depositi di ghiaccio e neve. In: Ultimi ghiacci, clima e ghiacciai nelle Alpi Marittime. Ed Soc Meteo Subalp, Moncalieri, Memorie dell'Atmosfera 11:307–313

Von der Mühll D, Haeberli W (1990) Thermal characteristics of the permafrost within an active rock glacier (Murtèl/Corvatsch, Grisons, Swiss Alps). J Glaciol 36:151–158

Wagner T, Pauritsch M, Mayaud C, Kellerer-Pirklbauer A, Thalheim F, Winkler G (2019) Controlling factors of microclimate in blocky surface layers of two nearby relict rock glaciers (Niedere Tauern Range, Austria). Geogr Ann A 10(4):310–333. https://doi.org/10.1080/04353676.2019.167095

Wagner T, Pleschberger R, Kainz S, Ribis M, Kellerer-Pirklbauer A, Krainer K, Philippitsch R, Winkler G (2020) The first consistent inventory of rock glaciers and their hydrological catchments of the Austrian Alps. Austrian Journal of Earth Sciences Vienna 113(1):1–23 https://doi.org/10.17738/ajes.2020.0001

Wakonigg H (1996) Unterkühlte Schutthalden. Arb Inst Geogr Univ Graz 33:209–223

Wakonigg H (2001) Ergebnisse von Temperatur-Dauerregistrierungen am Toteisboden im Schladminger Untertal. Mitt Naturwiss Ver Steiermark 131:41–56

Warburton J (1990) Secondary sorting of sorted patterned ground. Permafr Periglac Process 1:313–318

Wahrhaftig C, Cox A (1959) Rock glaciers in the Alaska Range. Geol Soc Am Bull 70:383–436

Washburn AL (1979) Geocryology: a survey of periglacial processes and environments. Edward Arnold, London, p 406

Wolff H (1986) Goethes Kenntnisse der Alpen im Lichte der modernen Geologie. Sudhoffs Archiv 70(2):143–152. http://www.jstor.org/stable/20777079

Žebre M, Stepišnik U (2016) Glaciokarst geomorphology of the northern Dinaric Alps: Snežnik (Slovenia) and Gorski Kotar (Croatia). J Maps 12:873–881. https://doi.org/10.1080/17445647.2015.1095133

Zückert G (1996) Versuch einer landschaftsökologischen Gliederung der Hochflächen der südlichen Hochschwabgruppe. Mitt Naturwiss Ver Steiermark 125:55–72

The Central European Variscan Ranges

Piotr Migoń and Jarosław Waroszewski

1 Geographical Framework

The area named as the Central European Variscan ranges refers to the latitudinal belt of medium–high mountain terrains and intervening uplands that stretches between the River Rhine in the west and the Carpathians in the east (Fig. 1). Geologically, they are predominantly built of Proterozoic and Early Palaeozoic rocks of different origin and belonging to different terranes, which were later altered to form large metamorphic complexes, amalgamated, and intruded by magmatic bodies of various size, mainly granites. As the final structural shape of the basement was acquired during the Variscan orogeny in the Devonian and the Carboniferous, the name "Variscan ranges" applies. However, the Variscan mountainous topography was subsequently eroded and the basement was partly, or completely, buried under younger sediments of Permian and Mesozoic ages. The contemporary gross topography originated in the Cenozoic through an interplay of differential uplift that occurred as a crustal response to the orogenic processes in the Alps and the Carpathians, and rock-controlled erosion (Ziegler and Dèzes, 2007). In this way, Variscan basement complexes were brought to the present-day altitudes and subject to the activity of periglacial processes, particularly intense during cold stages of the Pleistocene, but still ongoing in the most elevated parts of the Variscan belt.

The topography of the region covered in this chapter is complex and includes both well-defined mountain ranges, some exceeding 1000 m a.s.l., and extensive tracts of

P. Migoń (✉)
Institute of Geography and Regional Development, University of Wrocław, pl. Uniwersytecki 1, 50–137, Wrocław, Poland
e-mail: piotr.migon@uwr.edu.pl

J. Waroszewski
Institute of Soil Science, Plant Nutrition and Environmental Protection, Wroclaw University of Environmental and Life Sciences, Grunwaldzka, 53, 50–357, Wroclaw, Poland
e-mail: jaroslaw.waroszewski@upwr.edu.pl

M. Oliva et al. (eds.), *Periglacial Landscapes of Europe*,
https://doi.org/10.1007/978-3-031-14895-8_10

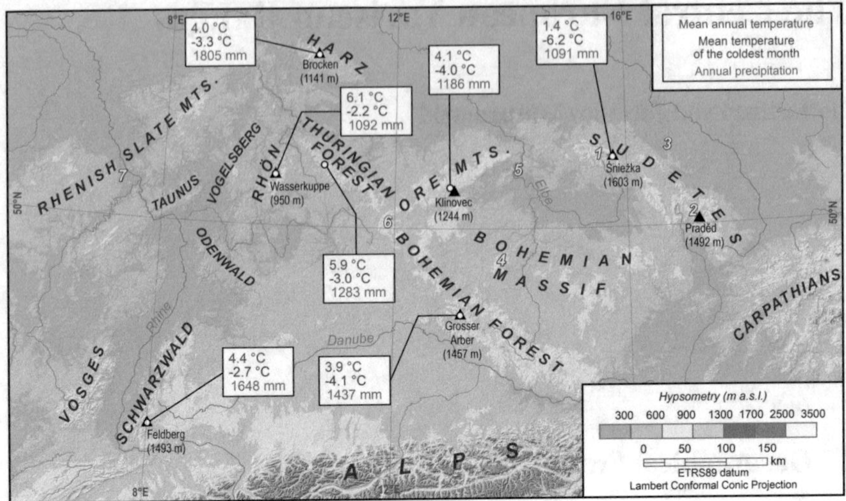

Placenames referred in the text

1 Karkonosze Mts., Mt. Wielki Szyszak, Mt. Łabski Szczyt, Mt. Luční hora, Modré sedlo; 2 Hrubý Jeseník Mts., Ztracené kameny; 3 Mt. Ślęża; 4 Brdy Highland; 5 České středohoří Mts.; 6 Fichtelgebirge Mts.; 7 Laacher See

Fig. 1 Location map

upland terrains in between. The latter may be still developed upon Permo-Mesozoic sedimentary cover, not yet eroded. The highest topographic unit within the study area is the mountain range of the Sudetes along the Czech/Polish border, itself divided into a number of lower-order geomorphic units rising to variable height. Its highest peak, Mt. Śnieżka/Sněžka in the Karkonosze Mts., is 1603 m, followed by Mt. Praděd in the Hrubý Jeseník Mts. (1492 m). The second-highest range is the Šumava/Bayerischer Wald range along the Czech/German/Austrian border (also known as the Bohemian Forest), with Mt. Grosser Arber (1457 m). In terms of altitude, it is followed by the Krušne hory/Erzgebirge (Ore Mountains) along the northern border of Czechia with Germany (Mt. Klinovec, 1244 m). Granites, gneisses and schists are dominant rock types in each, with the subordinate role of various other metamorphic rocks, including quartzites. All three ranges enclose the Bohemian Massif, with its various uplands rising to 865 m at maximum. The NW extension of the Bohemian Massif is the Thuringian Forest in Germany, where elevations do not rise above 1000 m. Further to the west are elevated massifs of Rhön (928 m) and Vogelsberg (773 m), built mainly of Cenozoic volcanic rocks, which then give away to extensive, locally deeply dissected uplands of the Rhenish Slate Mountains, crossed by the River Rhine. A northernmost outlier of the Variscan belt is the mountain group of Harz (Mt. Brocken, 1142 m), built mainly of Carboniferous granites and surrounded by younger sedimentary formations, whereas the southwesternmost part is made of Schwarzwald (Feldberg, 1493 m) and much lower Odenwald (593 m) ranges, both building the eastern shoulder of the Upper Rhine Graben.

The current climatic conditions reflect both altitude effects and decreasing maritime influence eastwards (Table 1). Nevertheless, at all stations the mean annual

Table 1 Current climate characteristics of the Central European Variscan ranges, based on data from summit and ridge top positions in individual mountain ranges (all data for 1991–2000 period)

Location	Mountain range	Altitude (m a.s.l.)	Mean annual air temperature (°C)	Mean air temperature of the coldest month (°C)	Mean annual precipitation (mm)
Śnieżka	Sudetes	1603	1.4	−6.2	1091
Feldberg	Schwarzwald	1489	4.4	−2.7	1648
Grosser Arber	Bohemian Forest	1436	3.9	−4.1	1437
Fichtelberg	Ore Mountains	1213	4.1	−4.0	1186
Brocken	Harz	1142	4.0	−3.3	1805
Schmücke	Thuringian Forest	937	5.9	−3.0	1283
Wasserkuppe	Rhön	921	6.1	−2.2	1092

Source https://ds.data.jma.go.jp/gmd/tcc/tcc/index.html

temperature is above 0 °C and except Mt. Śnieżka in the Sudetes, the mean temperature of the coldest month (typically January) does not fall below −5 °C. Higher precipitation values are recorded in exposed settings in the west and north (Brocken, Feldberg) than in the more eastern parts of the region. Snow cover can be thick and persistent during specific years, but it melts entirely during summer. Comparison of long-term records, available for Mt. Śnieżka, shows significant warming trend in the past 140 years. In the late nineteenth century mean annual air temperatures were typically just below 0 °C (Sobik et al., 2019), but have increased since to the currently observed 1.4 °C.

Regional differences notwithstanding, the topography of Central European Variscan ranges evolved under the influence of common overarching forcings, hence evident similarities between them in terms of landform inventories and the timing of major phases of landform evolution. It is very likely that at the time of highest sea-level stand in the latest Cretaceous nearly the entire Variscan belt was submerged, except a few islands supplying detrital material to the contemporaneous nearshore zones. However, tectonic stresses at the turn of the Cenozoic resulted in the origin of thrust faults, rock and surface uplift, and erosion of the Cretaceous cover to re-expose the basement. This geologically brief interval of considerable crustal instability was followed by a much longer period of limited endogenic activity, facilitating widespread planation and the origin of extensive tracts of low relief. The mechanisms, severity and time bracketing of planation remain controversial issues in regional geomorphic research, and they very likely vary between sub-regions, but it is nevertheless evident that gently undulating terrains truncating basement rocks occur widely in the Central European Variscan ranges, irrespective of altitude. Thus, high-elevation planar surfaces typify the core parts of the Šumava range, the Ore Mountains, the Sudetes, and the Harz. They are also common at lower altitudes, e.g., in the Bohemian

Highlands and the Rhenish Slate Mountains. The presence of low-relief surfaces was important for periglacial processes in the Pleistocene, defining available relief energy and limiting efficacy of various transport mechanisms. However, rock resistance to weathering and erosion was apparently an important factor in landform evolution and some bedrock types proved more resistant, giving rise to localized elevations (ridges, inselbergs) rising above the regional plains. Being higher and steeper, they developed a different suite of periglacial landforms during the Pleistocene. A good example is provided by the gabbro inselberg of Mt. Ślęża in the Sudetic Foreland (Żurawek and Migoń, 1999; see also Sect. 3.2).

Enhanced crustal instability in the Neogene (initiated earlier in some sub-regions) resulted in the origin or reactivation of numerous fault zones, fragmentation and non-uniform uplift of low-relief surfaces. In specific areas, cumulative altitude differences exceeded 1000 m (e.g., most elevated parts of the Sudetes, Schwarzwald), whereas the height of fault-generated escarpments reached 500–600 m. Increasing relief energy set the stage for fluvial downcutting and deepening of the valley system, resulting in the origin of steep slopes, locally as much as 30–40°. The close juxtaposition of planar surfaces at water divides and deeply incised valleys is among the most characteristic landmarks of the Central European Variscan ranges. Tectonic processes were accompanied by volcanism and large volcanic centres originated in various parts of the Variscan belt (e.g., Vogelsberg, Rhön, southern and eastern foreland of the Ore Mountains), whereas isolated volcanoes were even more widespread. Subsequent erosion led to the degradation of primary volcanic cones, leaving only exposed lava conduits and plugs, as well as to relief inversion upon former valley-filling lava flows, with elongated basaltic plateaus now overlooking undulating basement surfaces. The presence of well-jointed volcanic rocks in exposed topographic positions proved optimal for the efficacy of various cold-climate processes in the Pleistocene. Non-uniform uplift continued into the Quaternary and relief differentiation due to tectonic forcing is ongoing.

The Central European Variscan ranges were generally located beyond the maximum extent of the Scandinavian Ice Sheet. Only the marginal, northern parts of the Sudetes and the Ore Mountains were reached by the ice sheet, whereas the Harz was partly surrounded by the ice (Ehlers et al., 2011; Eissmann, 2002). However, local glaciation developed in the most elevated parts of the Variscan belt, including the Karkonosze Mts., Bohemian Forest, and Schwarzwald, where the geomorphic evidence is the most distinctive and includes cirques, some overdeepened and partially occupied by lakes, glacial troughs, lateral and frontal moraines (e.g., Engel et al., 2014; Hofmann et al., 2020; Mentlik et al., 2013). In southern Schwarzwald, at higher elevation and under more maritime climate, glaciation was extensive and the presence of an ice cap covering ~ 1,000 km^2 is inferred, whereas in the Bohemian Massif glaciers were confined to cirques and headwater reaches of valleys, so that large surrounding areas remained in periglacial domain. In several other ranges, the extent, effects, or even the very existence of Pleistocene glaciers are still debated, as the interpretation of landforms and deposits is uncertain. The timing of glaciation is also debated and available ^{10}Be cosmogenic exposure ages indicate the Last Glacial Maximum and subsequent phases (Engel et al., 2014; Mentlik et al.,

2013), although earlier glaciations (pre-LGM, possibly pre-Eemian) are certainly possible.

An important issue in any discussion of periglacial landforms and processes in the Central European Variscan ranges is the distinction between relict and active periglacial environment. According to the current understanding, most of periglacial landforms and deposits are inherited from the Pleistocene and hence, relict. This applies particularly to areas located below the timberline, which would be completely under forest if the natural environment was not subject to human impact. In fact, it is probably only the highest parts of the Sudetes, where the natural timberline occurs at 1250–1300 m. Otherwise, forest-free areas in the most elevated parts of Schwarzwald and the Ore Mountains are likely the effects of medieval deforestation (Treml et al., 2006). Nevertheless, in various parts of the forest belt bare blocky accumulations occur on steep slopes, interpreted as the outcome of efficient frost weathering of well-jointed, but otherwise solid rocks, that occurred in the cold phases of the Pleistocene. In specific localities such as glacial cirques, or high rock cliffs the delivery of angular material to the scree slopes is ongoing, even though the specific processes may not be periglacial (i.e. controlled by frost weathering) sensu stricto (Migoń et al., 2010). Regarding the areas above the timberline, current climatic conditions include long winter periods with temperatures below 0 °C and transitional periods with air and ground temperatures fluctuating around 0 °C, allowing for frost impact on the ground and shallow subsurface. Consequently, it is believed that certain specific cold-climate processes occur nowadays, further moulding small-scale landforms, and the contemporary environment remains partly periglacial, even though these processes no longer dominate (see Sect. 4) (Křížek, 2007; Křížek et al., 2010).

2 The History of Periglacial Studies in Central European Variscan Ranges

The Central European Variscan ranges play a special role in the history of periglacial research in general and several major concepts were developed from site-specific studies in this region. Unfortunately, nearly all early research contributions were written in languages other than English (German, Polish, Czech) and often in rather obscure publication series, no longer published. Hence, they are hardly available to the contemporary scientists in the original form, although in specific cases review papers of later date can be used. This is the case of the pioneering work of a Polish scientist, Walery Łoziński (1909, 1912), who studied blockfields in the northern Carpathians, the Holy Cross Mountains (Świętokrzyskie Mts.), and the Variscan ranges and arrived at the conclusion that they must have formed in the proximity of the marginal zone of a Scandinavian ice sheet due to harsh climatic conditions typical for the polar regions. Consequently, he coined the very term ‘periglacial’ to describe a specific geomorphic environment, moulded by cold-climate processes, but not glaciated. Investigating blockfields, he visited the Central and West Sudetes, including the Karkonosze Mts., the Harz, the Bohemian Forest, and Odenwald, acquiring a good regional overview and solid observational basis for his conclusions. The contribution of Łoziński was re-evaluated and summarized by, among others, Jahn (1954) in the inaugural issue of "Biuletyn Peryglacjalny" (although in Polish only) and recalled many years later by French (2000).

In the subsequent years, an interest in the history of blockfields continued, with important contributions by Schott (1931) and Flohr (1934) from the Karkonosze and by Hövermann (1949) from the Harz. Some of these studies suggested, however, that the origin of blockfields may be more complex and partial inheritance from pre-Quaternary weathering mantles cannot be excluded. In this scenario, large granite boulders may have emerged due to selective removal of fine sandy-gravelly regolith, facilitated by sparse vegetation in cold-climate environments. The most comprehensive approach to periglacial hillslope processes and deposits was that of Büdel (1937), who assembled observations from various parts of the Variscan belt and offered general interpretation of sequences of gravity-driven deposits, distinguishing between runoff- and solifluction-produced layers. From the first half of the twentieth century come also the first reports about patterned ground in the most elevated parts of the Variscan belt (Gellert and Schüller, 1929).

Studies of periglacial landforms resumed with vigour in the 1950s and were clearly in the mainstream of geomorphic and Quaternary research until the 1970s. Pleistocene inheritance was in focus and a characteristic feature of the time was combination of field work in high-latitude regions, where periglacial processes could have been observed *in statu nascendi*, and observations from Central European mountains and uplands. Examples can be provided from Czechia (J. Sekyra working in Antarctica and Pamir, T. Czudek and J. Demek visiting Siberia), Poland (A. Jahn working in Svalbard, Siberia and Alaska), Germany (J. Büdel, working in Svalbard, H. Richter, visiting Mongolia), to recall just a few names. Among themes, which generated particular interest, was the origin of rock cliffs, associated talus deposits and adjacent mid-slopes benches, interpreted in terms of Pleistocene cryoplanation facilitated by the presence of permafrost. This research avenue was explored mainly in the former Czechoslovakia (Czudek, 1964; Czudek and Demek, 1961, 1971; Demek, 1969), along with the related concept of cryo-pedimentation (Czudek, 1988). However, in retrospect one can also notice reduced originality of individual contributions, in which qualitative field observations were simply inserted into general conceptual models. This undoubtedly contributed to building landform inventories, but hardly resulted in a new impetus for periglacial studies. The harsh debate about the origin of tors, typical for British geomorphology of the late 1950s and 1960s, did not find its equivalent in Central Europe, where polygenesis of rock outcrops (Germ. *Weiterbildung*) and likely co-existence of rock landforms of different origins and ages was emphasized (Czudek, 1964; Demek, 1964; Jahn, 1962; Meinecke, 1957; Wilhelmy, 1958). Regarding periglacial depositional environments, inherited slope deposits remained within the sphere of interest (e.g., Czudek et al., 1963; Jahn, 1968, 1977) and became one of the central themes in Germany, where a distinctive school of studying cover beds (as they are called) emerged (Frühauf, 1991; Kleber, 1992, 1997; Semmel, 1964; Völkel et al., 2002; see recent summary in Kleber and Terhorst,

2013). This period also witnessed the appearance of numerous regional studies, in which periglacial landform inventories were described, usually in relation to large-scale landforms arising from long-term geomorphic evolution, and their origin was hypothesized. This approach used the concept of relief generations as proposed by Büdel (1977, see 1982 for English translation), within which Pleistocene periglacial processes acted upon and were re-shaping older landscape facets. One should also notice attempts to show periglacial landform inventories by means of regional maps. A notable example is the map published by Sekyra (1961) for the entire Bohemian Massif, which includes nearly 30 symbols of various periglacial landforms, deposits and structures. Another significant contribution of the time was one of the earliest experiments on frost weathering worldwide, by Martini (1967). Unfortunately, there was no follow-up and this innovative research direction vanished.

The last 20 years or so have seen another wave of increasing interest in periglacial inheritance in Central European Variscan ranges. These modern studies benefit from recent technological advances in computing, progress in sediment and landform dating, rapid expansion of GIS, as well as from the availability of new data sources such as high-resolution digital terrain models derived from airborne laser scanning (LiDAR), or very high quality ortophotomaps. Examples include determinations of ages of blocky accumulations and soil denudation rates using cosmogenic isotopes (e.g., Engel et al., 2020; Waroszewski et al., 2018b), dating of late Pleistocene hillslope deposits using tephrochronology (e.g., Schmincke et al., 1999) and improved OSL technique (e.g., Hülle et al., 2009; Waroszewski et al., 2020), modelling of palaeotemperatures (Uxa et al., 2021) and detailed mapping of periglacial landforms (e.g., Křížek et al., 2019; Migoń et al., 2020).

3 Relict Periglacial Landforms and Deposits

For the clarity of presentation, the most typical periglacial landforms and deposits were divided into six groups, but one has to bear in mind that clear-cut boundaries between these groups hardly occur. For example, blockfields may occur in close association with tors and rock cliffs, passing into more heterolithic, solifluction deposits on slopes. Likewise, the boundary between blockslopes and rock glaciers is diffuse. Patterned ground may develop upon coarse blocky accumulations, whereas solifluction mantles and other hillslope cover beds interfinger with loess deposits.

3.1 Blockfields and Related Features

Blockfields are perhaps the most conspicuous, largely relict periglacial landforms in the Central European Variscan ranges (Fig. 2). In fact, it was blockfields, which became the key piece of evidence for Łoziński (1909) to argue for cold-climate inheritance in Central Europe and to introduce the term 'periglacial'. Coarse accumulations in the region occur in considerable diversity, representing a range of dimensions, morphological settings, lithological underpinnings, structural and textural features, and vegetation coverages. Moreover, they apparently have more than one origin, some being evidently in situ, whereas downslope displacements of different kind were clearly involved in the genesis of others. Within the generic term '*blockfield*' one can distinguish, in addition to blocky mantles on low-relief surfaces, blockslopes, block streams and talus/scree slopes, understood as follows. *Blockslopes* are accumulations of angular debris on moderately steep (>15°) slopes, which are either irregular or more or less isometric in plan view. If these accumulations are markedly elongated, they are traditionally named as *block streams* or *block tongues*. Talus and scree slopes form below bedrock outcrops, due to incremental delivery of angular debris liberated by weathering from free rock faces. Their characteristic feature is surface inclination close to the angle of repose, between 30° and 40°.

Coarse blocky accumulations are associated with specific bedrock types only and the key requirements seem to include high mechanical strength and moderate to low density of jointing, facilitating breakdown into clasts of decametric or metric dimensions. Thus, they are fairly common on fine-grained granites, gabbro, volcanic rocks such as basalts, phonolites and rhyolites, quartzites, and some variants of gneiss.

Fig. 2 Blockfields in the Central European highlands. **a**—summit blockfield in fine-grained granites, Karkonosze Mts. (Czechia/Poland); **b**—blockslope on basalt, Mt. Ostrzyca (Poland); **c**—blockslope on quartzite, Brdy Highland (Czechia); **d**—blockslopes in hornfels, Mt. Śnieżka—the highest peak in the Central European Variscan ranges (Poland)

Those on granites are usually composed of blocks more than 1 m long (Fig. 2a), whereas the typical size of constituents of basaltic blockfields does not exceed 0.5 m long, in relation to much lower spacing of discontinuities in the latter (Fig. 2b). Likewise, the shape of clasts is related to textural properties of underlying bedrock, with elongated shapes more typical for metamorphic rocks.

In situ blockfields occur mainly in the most elevated parts of the Variscan belt, covering broad swells of the planar summit surfaces underlain by granite and gneiss and giving way to slope blocky accumulations as the terrain inclination increases. Numerous examples exist in the Karkonosze, Šumava and Harz. In the western part of the former area, between Mt. Wielki Szyszak (1509 m) and Mt. Łabski Szczyt, extensive granite blockfields occur, partly associated with the bedrock crag on Mt. Łabski Szczyt, which supplies the blockfield via mechanical breakdown. Northwards the blockfields become blockslopes with distinctive small-scale relief features such as treads-and-risers, shallow closed hollows and lobate arrangements of boulders, interpreted as indicators of bedrock control, ground ice melting and gravity-driven movement, respectively (Traczyk, 2007). Visually impressive are blockfields on quartzites (Fig. 2c), apparently most difficult to be colonized by vegetation because of negligible amount of fines produced by quartzite weathering. Examples can be provided from the Brdy range to the west of Prague in Czechia (Žák, 2016) and Ztracené kámeny in the Hrubý Jeseník Mts. (Stan et al., 2017).

Blocky accumulations on volcanic hills are different in that their form is adjusted to the local terrain properties, that is steepness of slopes, not uncommonly in excess of 30°. Block-covered hillslopes are typical for the volcanic edifices of the České Středohoří and nearby mountain ranges (Raška and Cajz, 2016; Růžička et al., 2012), but occur throughout the entire Variscan belt. Genetically, they comprise both clasts liberated by in situ weathering and those supplied from the rock crags above, and tend to be unstable even in the contemporary conditions. Extensive blockslopes on Mt. Śnieżka, the highest peak in the Variscan belt (Fig. 2d), composed of hard hornfels, are of similar complex origin.

Beside bare blockfields, vegetated varieties are common and the former often grade into the latter, indicating slow progress of vegetation succession. At lower elevations, vegetated blockfields are ubiquitous. A good example of extensive blocky accumulations at low altitude (<700 m) is provided by Mt. Ślęża in the foreland of the Sudetes, where gabbro boulders as much as 3 m long form densely packed clusters on slopes of variable inclination (Żurawek et al., 2005).

The timing of blockfield formation is poorly constrained. For long it was implicitly assumed that they are inherited from the last glacial cycle, although Traczyk and Migoń (2003) suggested that they may be cumulative products of weathering and downslope displacements during many successive cold periods. Engel et al. (2020) have recently dated a blockfield in the Hrubý Jeseník using cosmogenic ^{10}Be and obtained ages ranging from between 84.3 ± 3.8 and 26.8 ± 2.6 ka, whereas the tor upslope yielded an older date of 150 ± 4.8 ka. These dates are interpreted as the onset of bedrock disintegration (for the tor) and evidence of sustained supply of boulders to the block field (for the boulders), with individual ages considered as the minimum ones, allowing for further changes in boulder position.

3.2 Problems of Genetic Interpretation of Coarse Accumulations—Weathering Residues and Rock Glaciers

The origin of coarse blocky accumulations is not always easy to decipher, even though their general association with the periglacial environment seems widely accepted (at least implicitly). In one alternative interpretation, suggested particularly for granite boulder fields (Flohr, 1934; Wilhelmy, 1958), it was argued that boulders are residues of thick pre-Quaternary weathering mantles, left in situ after fine fraction (clay, silt, sand) was washed away. Supporting evidence included roundness of boulders, considered incompatible with mechanical weathering and the lack of bedrock outcrops nearby that could supply the boulders. On the other hand, timewise the phenomenon of selective erosion was placed in the Pleistocene, when cold climatic conditions limited vegetation growth, allowed permafrost to develop, and hence, facilitated surface water erosion. Thus, even if boulders themselves were not produced by periglacial processes, their exposure at the surface was linked with the periglacial environment.

Among blockfields particularly intriguing are large, tongue-like accumulations, with clear lateral boundaries separating these from the adjacent slope. Perhaps most evident are they on the gabbro slopes of Mt. Ślęża, where five such landforms exist (Fig. 3), ranging in length from 0.5 to 2 km and up to 0.9 km wide, covering the altitude span up to 250 m and being locally as much as 15 m high in respect to the surrounding terrain. Previously interpreted as moraines of the Scandinavian ice sheet or debris flow accumulations, they were comprehensively analysed by Żurawek (1999a, 1999b; Żurawek and Borowicz, 2003), who argued that these enigmatic landforms are relict rock glaciers. Their hummocky internal relief and the presence of numerous closed depressions, some occupied by ponds and marshy ground, was suggested to be compatible with degradation of ground ice. Radiocarbon dating of basal peat from one of these ponds yielded date of 12.0 ± 0.6 cal. ka BP (recalibrated using the IntCal20 dataset), interpreted as the evidence of rock glacier activity until the end of the Pleistocene (Żurawek, 2001).

As the slopes of Mt. Ślęża were covered by an ice sheet during MIS 12 (478–424 ka) and 6 (191–130 ka), except the very summit (Szczepankiewicz, 1958; Żurawek and Migoń, 1999), the origin of these accumulations was linked with the period postdating the decay of the MIS 6 ice sheet, probably including both the end of MIS 6 period and the Last Glacial period. Rock glaciers were also suspected to occur in other parts of the Central European Variscan range, for instance in the Karkonosze (Chmal and Traczyk, 1993), but some of these earlier interpretations were later disproved and the respective landforms reassessed as 'ordinary' blockfields (e.g., Engel et al., 2020).

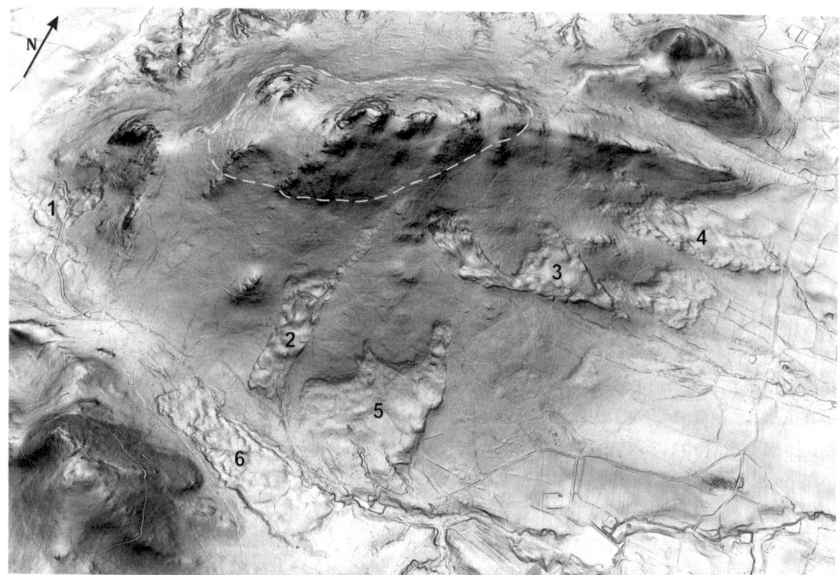

Fig. 3 Enigmatic depositional landforms (numbered 1–6) on slopes of Mt. Ślęża, Sudetic Foreland (Poland), variously interpreted, including rock glacier hypothesis, visible on high-resolution LiDAR DTM. Notice also numerous crags and rock-cut benches in the most elevated part of the hill, above c. 550 m (white broken line), which is the altitude of the Middle Pleistocene trimline related to the Scandinavian Ice Sheet advance during MIS 6. The area covered by the image is approximately 6 km × 6 km. Terrain model courtesy of Kacper Jancewicz

3.3 Tors, Rock Cliffs and Cryoplanation Terraces

Bedrock outcrops on otherwise regolith-covered summits, ridges and slopes are ubiquitous in the Central European Variscan ranges (Fig. 4), generating curiosity and scientific interest as early as 200 years ago, when Johann Wolfgang Goethe, a famous Romantic writer and naturalist, attempted to explain the origin of granite crags (German *Klippen*) in Fichtelgebirge in Germany. These bedrock protrusions occur in a variety of shapes and dimensions, with three morphological types being particularly common. Perhaps the most impressive are tors, defined as 'solid rock outcrops as big as a house rising abruptly from the smooth and gentle slopes of a rounded summit of broadly convex ridge' (Linton, 1955, p. 470). Tors have regolith-covered surfaces on all sides and are typical for summits and gently sloping surfaces. They may attain more than 20 m high and 100 m long, being composed of multiple angular or slightly rounded compartments, visually often resembling ruined buildings (castellated tors; Germ. *Felsburg*, Czech *skalní hradba*). Tors are common in granite areas such as the Karkonosze, Šumava (Fig. 4a), Harz and Fichtelgebirge, as well as in some gneiss uplands of the Bohemian-Moravian Highland. The second type includes rock cliffs, which extend perpendicularly or obliquely to the moderately or gently inclined slope, not uncommonly for tens of metres. They may be only 1–2 m high, but cliffs more than 10 m high also occur (Fig. 4b). Some have overhanging basal sections (Fig. 4c), with slots and caverns penetrating into the rock massif. Rock cliffs are typically associated with talus deposits at their base, clearly derived from the cliffs themselves, and the proportion of bedrock to talus section varies, including nearly complete burial of a rock cliff under products of its own disintegration. There seems to be no specific lithological predisposition for the formation of rock cliffs, as they are known from various rock types, except the very weak ones. Nonetheless, some are controlled by lithological change (e.g., cliffs on quartzite amidst schists), or structural dip in sedimentary successions. Finally, outcrops of various shapes occur within steep slopes. Beside rock cliffs are isolated pulpits, towers, craggy ribs and fins.

The geomorphic history of these bedrock outcrops was a popular theme of scientific debate in the Central European highlands, especially in the 1960s and 1970s and much of the debate centred on their periglacial versus non-periglacial origin. It was apparently believed that the shape provides the key to the origin, whereas the presence of angular talus next to the outcrop comprised an important supportive evidence. Thus, tors were considered as two-phase landforms, excavated from pre-Quaternary weathering mantles through removal of loose weathering products (Demek, 1964;

Fig. 4 Diversity of rock landforms in the Central European highlands. **a**—castellated tor in granite, Bohemian Forest (Germany); **b**—rock cliff in quartzite, with associated forested blockslope, Brdy Highland (Czechia); **c**—low rock cliff with basal overhang in greenschists, Kaczawskie Mts. (Poland); **d**—rock cliff in gneiss, c. 4 m high, with a cryoplanation terrace in the foreground, Hrubý Jeseník Mts. (Czechia)

Jahn, 1962). Cold periods of the Pleistocene were considered as particularly suitable periods for slope downwasting due to lack of protective vegetation cover. In this interpretation, tors emerged due to periglacial processes, but are not periglacial landforms per se. By contrast, rock cliffs were interpreted as one-phase bedrock landforms, developed by concurrent action of frost weathering and removal of regolith by solifluction (Czudek, 1964; Martini, 1969). In fact, they were usually termed as 'frost-riven cliffs' in regional studies, which clearly implied their periglacial origin, although definite evidence may have been lacking. The concept of periglacial origin was extended to bedrock landforms within steep slopes and these were described using the name of frost-riven cliffs too. This was unfortunate as the rock factor remained often ignored and there was no good explanation offered what were the controls behind the location of particular rock landforms. One recent study focusing on greenschist crags in the West Sudetes showed that their spatial distribution is not random, but reflects textural changes from massive to flaggy variants and the dip of foliation planes (Michniewicz et al., 2020).

In the context of periglacial geomorphology, the coexistence of bedrock cliffs and very low-angle slope sections is of particular interest (Fig. 4d), especially if they form repetitive patterns and give the slope stepped appearance. Such associations were described as typical periglacial hillslopes, with the benches interpreted as cryoplanation terraces separated by frost-riven cliffs (Czudek, 1964; Demek, 1964; Czudek and Demek, 1971). It was assumed that cliffs are subject to vigorous frost weathering, disintegration and retreat, leaving angular regolith at their foot. In this way a bench expands upslope, but is undercut itself by the retreating cliff in the lower slope. In the penultimate stage a residual outcrop (tump) is located on the summit, surrounded by enlarging terraces or one uniform planar surface, occasionally termed as cryo-plain (Fig. 5a). This direction of research was particularly championed in the Bohemian Massif and regional literature abounds in studies containing references to these alleged periglacial rock landforms, although illustrations were typically photos and schematic cross-sections rather than detailed cartographic presentations, capable to show the actual extent of stepped hillslopes. One classic locality, clearly visible due to its location above the timberline, is Mt. Luční hora in the Karkonosze (Fig. 5b), where a sequence of at least four treads and separating risers occurs within its slopes (Křížek, 2007). Within cryoplanation terraces one can observe evidence of material sorting in downslope direction, from large-size angular talus at the base of the cliffs to loamy materials with occasional boulders in the distal part, consistent with solifluction transport of regolith. However, unequivocal recognition of cryoplanation terraces within slopes steeper than c. 10° is problematic due to their small dimensions. Moreover, it is by no means certain that they are climate-related features, as local lithological and structural properties may play a decisive role.

Fig. 5 High-altitude planar topography of Central European highlands is usually considered to reflect combined effects of long-term planation prior to the Late Cenozoic uplift and further terrain levelling due to cryoplanation in the Pleistocene. **a**—cryo-plain with residual rock landforms, Hrubý Jeseník Mts. (Czechia); **b**—a sequence of debris-covered steps and benches, interpreted as frost-riven cliffs and cryoplanation terraces, Mt. Luční hora, Karkonosze Mts. (Czechia)

3.4 Patterned Ground

Patterned ground structures may have been ubiquitous in the Central European uplands in the Pleistocene, as they are in the contemporary high-latitude periglacial environment, but had little chance to survive climate amelioration at the beginning of the Holocene, the encroachment of the forest, with all associated bioturbations imposed on soils. In particular, tree throw was demonstrated to be an important disturbing agent, capable of reworking soils and their structures many times during the Holocene (Pawlik, 2013). Consequently, only the most elevated areas, located above the timberline, facilitated the persistence of patterned ground, although even there one needs to realize that forest communities may have reached higher altitudes during the Holocene Thermal Maximum, after ca. 7.4 ka cal BP (Treml et al., 2006).

The highest parts of the Sudetes offer the best examples of patterned ground, which occurs in several morphological variants, with the presence of particle sorting as the main classification factor (Křížek et al., 2019; Treml et al., 2010). The following categories were distinguished: (1) sorted polygons, (2) sorted nets, (3) sorted circles,

(4) sorted stripes, (5) earth hummocks, (6) peat hummocks, and (7) non-sorted stripes. Collectively, they occupy 5.23 km^2, which is c. 11.5% of the area presently located above the timberline. Among them, sorted nets are most widespread (57% of the total area), followed by sorted stripes (36% of the total). A clear dependence on slope inclination occurs, with the stripes occupying steeper slopes, whereas some control exerted by altitude was found too. Thus, sorted polygons occupy the highest areas, giving way to sorted nets as the altitude decreases. Dimensions of sorted patterns vary, with the largest polygons attaining 10 m across and the longest stripes being more than 30 m long. Most patterned ground is partly (sorted variants), or completely (non-sorted variants) vegetated, with alpine grasses and dwarf pine growing on fine regolith, which points to its largely relict nature. Patterned ground in the Sudetes has developed upon various lithologies, including granites, gneiss, schist, and hornfels. The largest polygons are known from granite bedrock, where their diameters may exceed 10 m.

3.5 *Loess and Related Deposits*

Loess, usually defined as a windblown mineral dust accumulation strictly related to the production of silt-sized particles by glacial grinding and frost shattering in a periglacial environment dominated by tundra or steppe ecosystems and product of physical and chemical weathering in desert domains, is one of the most important cold-climate terrestrial sediments. In Central Europe, the distribution of loess deposits was attributed to both glaciofluvial outwash plains existing at the margin of the Fennoscandian Ice Sheet and local mountainous sources (Central European Highlands), providing substantial amounts of silt due to mechanical weathering of outcrops of crystalline rocks (Baykal et al., 2021; Waroszewski et al., 2021). Fluvial transport and further deflation of dust from braidplains and other open surfaces was the main pathway of regional supply of silt-sized material (Badura et al., 2013). Regarding the occurrence of loess deposits in the Central European Variscan ranges they are preserved mostly in their northern forelands and, as either relatively thick domains (4–5 m, occasionally up to 10 m or so; Fig. 6a) in footslope settings and on gently rolling uplands (Moska et al., 2011) or much thinner mantles (up to 2 m) found on higher slopes, locally up to 450 m a.s.l. For instance, at Mt. Ślęża inselberg (Waroszewski et al., 2018a; Żurawek et al., 2005) and in the Rhenish Massif (Felix-Henningsen, 1991; Sauer and Felix-Henningsen, 2004) thin pedo– and bioturbated loess/aeolian silt drapes are very common, while in the Karkonosze Mts. and other mid-latitude mountain ranges the contribution of loess to hillslope sediments and the presence of thin aeolian silt bodies at higher elevations is still under debate. The main phase of loess contribution to slope deposits has been attributed to MIS 2–4. However, there is also evidence of much older (MIS 6/8) mantles of proper loess, or loess derivates on slopes and planar surfaces within the mountainous ranges (Waroszewski et al., 2020).

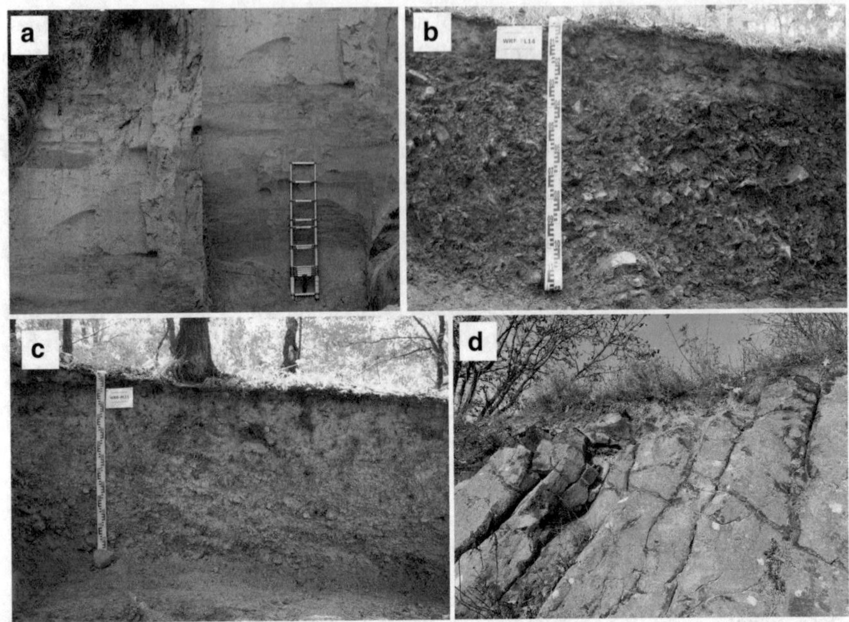

Fig. 6 Thick loess sediments and thin aeolian silt mantles in south-west Poland. **a**—9 m thick loess-paleosol sequence in Biały Kościół (Sudetic Foreland), with tundra gley soil above the last interglacial pedocomplex; **b**—thin loess over coarse serpentinite slope deposits, Oleszeńskie Hills; **c**—Late Glacial aeolian silt drape interfingering with granite regolith on the western slopes of Mt. Ślęża; **d**—thin loess covering basalt outcrop at Mt. Czartowska Skała, Kaczawa Upland

As loess deposits are susceptible to hillslope erosional processes, passing several steps of remodelling, is it often very hard to differentiate between typical loess deposits, loess derivates (being decalcified and reworked) and so-called 'loess like-sediments', which are composed of dominant silt, but of non-aeolian or complex origin. Thin loess deposits cover lithologically diverse bedrock (Fig. 6b–d) and older Pleistocene deposits, providing new properties (physical, mineralogical, and geochemical) to the substrates, with which they have been integrated. Moreover, they are widespread and therefore have significant imprint on the landscape (Kleber and Terhorst, 2013). Apart from the examples described above, loess gives rise to the formation of more complex sedimentary packages, after being incorporated into block fields, weathering residues or solifluction mantles. This complexity is accounted for in several geomorphic concepts pertinent to the palaeoenvironmental interpretation of hillslope deposits, which will be presented in detail in the following section.

3.6 Hillslope Deposits—Cover Beds and Solifluction Landforms

The most ubiquitous legacy of periglacial environments in the Central European Variscan ranges are hillslope deposits that testify to efficient downslope transport of regolith, both inherited from pre-Quaternary times and produced by various climate-related weathering mechanisms during the cold and warmer intervals of the Pleistocene. Stacked sequences of hillslope materials were noted as in the early twentieth century and recognized as potentially valuable archives of information about changing Pleistocene and Holocene environments. Hillslope deposits are crucial for understanding the 'Earth's critical zone' being an interface of soils, biota and water.

Büdel (1937, 1959), Schönhals (1957), Semmel (1964) and Jahn (1968) were among the first who described and subdivided periglacial slope deposits in mid-latitude European mountain ranges into certain types, providing also their detailed characteristics. Multi-layered periglacial slope deposits appear in various configurations depending on source material (lithology), intensity of slope and frost sorting processes, and slope properties (angle, aspect). At the most general level, one can distinguish debris-loamy deposits, with abundant angular clasts, not uncommonly as thick as 0.5 m, and finer deposits with more or less evident stratification. Genetically, the former are ascribed to predominant solifluction, whereas the latter are interpreted as indicative of milder climate, with slope wash as the dominant hillslope process. Solifluction deposits contain various materials, including coarse debris and finer fractions (mostly silt-size, but also sand) and may show typical arrangement of larger clasts more or less parallel to the slope (Fig. 7a–b), indicative of flow-like transport mode. In the Karkonosze Mts., coarse stratified materials composed of debris mixed with silty or sandy fine particles are often separated by stone lines (Fig. 7c) from older, underlying solifluction bodies (Traczyk, 1996; Waroszewski et al., 2018b), but they may be also found as layers intercalated between loamy solifluction sheets. Especially in the lower slope positions these topmost fine material (Fig. 7d) are products of slope wash and are rich in organic matter (Bogda et al., 1998). Karte (1983) provided good examples of fine-grained deposits (grèzes litées) from the Rhenish Slate Mountains, appearing predominantly between 500 and 700 m. They are well-bedded and cryoturbated, showing periglacial features like pseudomorphs of frost cracks, or ice wedges. These heterogeneous coarse and fine sediments formed by the combination of frost weathering and episodic, or periodic slope wash/debris-flow processes are frequently lacking datable materials, thus their ages may be only speculative. Recently, an interest in periglacial slope deposits was oriented towards erosion rates in such stratified materials (based on meteoric ^{10}Be). In the Karkonosze Mts. soils developed from complex hillslope deposits (mixed substrate of mica and amphibole schist) are subject to erosion varying between 80 and 410 t/km^{2}/yr, depending on the calculation model (Waroszewski et al., 2018b), which corresponds well with the values obtained for alpine and other mountainous environments.

The most comprehensive, conceptual approach to Pleistocene hillslope deposits in the Central European Variscan ranges is the cover beds concept, established in

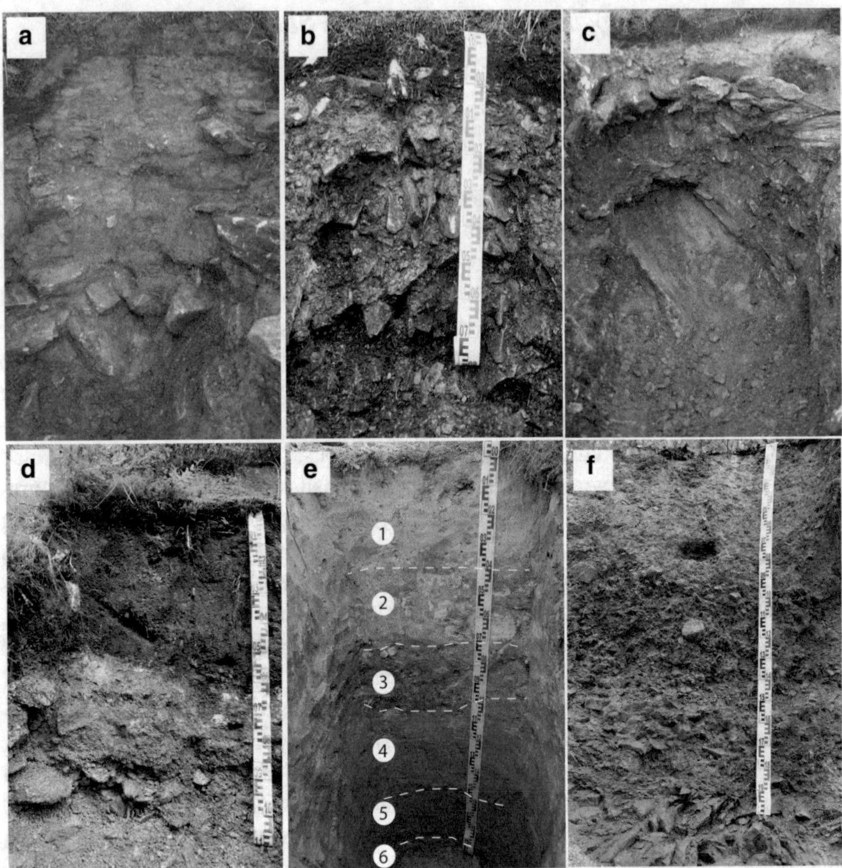

Fig. 7 Complex periglacial slope deposits. **a**—solifluction layer covered with fine-grained colluvium sediment (northern slopes of Kowarski Grzbiet, Karkonosze Mts., Poland); **b**—coarse-grained material in the proximal part of a solifluction lobe (southern slopes of Kowarski Grzbiet, Karkonosze Mts., Czechia); **c**—stone lines separating slope deposits with mica schist and amphibole schist lithologies (northern slopes of Kowarski Grzbiet, Karkonosze Mts., Poland); **d**—65 cm thick colluvial material sharply separated from the underlying coarse slope sediments (northern slopes of Mumlawski Wierch, Karkonosze Mts., Poland); **e**—complex arrangement of slope deposits including upper (1) and middle (2) periglacial cover beds (first 70 cm), glaciofluvial sediments (3), aeolian silt deposit (4), sandy material (5) and granite regolith (6) (western slopes of Mt. Ślęża, Poland); **f**—70 cm thick reworked upper periglacial cover bed in the Bardzkie Mountains (Poland)

Germany in the 1960s and 1970s (Schilling and Wiefel, 1962; Semmel, 1964) and subsequently improved by applying modern dating techniques and widely tested in and outside Europe. Periglacial cover beds are defined as deposits formed by nonlinear slope processes leading to gravity-driven displacement of upslope local material, but with a significant dose of aeolian silt (loess) admixture. They typically include several independent layers (Fig. 7e), separated by lithological boundaries (Kleber and Terhorst, 2013). In principle, their structure is subdivided into three main layers: (1) basal periglacial cover bed (*Basislage*) consisting of local, directly underlying material, with rock clasts often aligned parallel to the slope; (2) middle periglacial cover bed (*Mittellage*) derived from debris that forms the basal layer, with the addition of loess; (3) upper periglacial cover bed (*Hauptlage*) hosting mixture of bedrock material and allochthonous component, but also a stratigraphic marker, which is a tephra from the Laacher See eruption (12.9 ka, Schmincke et al., 1999). Although the upper periglacial cover bed is often overprinted by pedogenesis and exposed to near surface processes resulting in its truncation (Fig. 7f), it maintains unusually uniform thickness of 50 ± 15 cm. However, one has to be careful with the application of the three-layer concept to each locality, as hillslope sedimentary packages may be either more complex (Krzyszkowski, 1998; Migoń and Kacprzak, 2014), or less complex than the model assumes, probably indicating an important role of local lithological and topographic conditions. Moreover, it seems that the role of Holocene bioturbation in reshaping the structure of cover beds has not been given sufficient attention (Pawlik, 2013).

Recently, optical luminescence dating brought us closer into the chronology of periglacial cover beds. Ages of basal layers in the southeast Taunus Mts. cover a timespan between >43 ka to >195 ka (Hülle et al., 2009). Middle cover beds at Mt. Ślęża encompass the time range slightly after maximum dust accumulation (19–23 ka), with ages from 17.1 ± 1.1 to 19.5 ± 1.3 ka (Waroszewski et al., 2020). The OSL ages for the upper periglacial cover beds in the Taunus Mts and Ślęża Mt. span 8.7 ± 1.0 to 12.5 ± 1.4 ka (Hülle et al., 2009) and 12.0 ± 1.0 to 14.2 ± 1.1 ka (Waroszewski et al., 2020), respectively. This proves the formation of the upper cover beds during the Younger Dryas. Völkel and Mahr (2001) demonstrated more broad ranges of luminescence ages for the middle (6.0 ± 1.3 to 25.5 ± 2.6 ka) and upper cover beds (3.0 ± 0.7 to 18.3 ± 3.9 ka) in the Bayerischer Wald. This wide dispersal of dates obtained for hillslope sediments should be, however, treated with caution due to possible incomplete bleaching of mineral grains in periglacial setting and post-depositional incorporation of younger material by bio—and pedoturbation processes during slope reworking in the Holocene (Hülle et al., 2009).

Although hillslope deposits represent a ubiquitous sedimentary record of Pleistocene solifluction and occur irrespective of lithology and slope aspect, the complementary geomorphic record is relatively scarce and inherited solifluction landforms have been documented in a limited number of localities. Rather, they underlie the slopes as sheets of variable thickness, but little inherent relief. In particular, tread-and-riser topography, typical for contemporary periglacial environments, is very uncommon, probably due to Holocene pedogenesis and bioturbation. Solifluction lobes and steps can be recognized in the most elevated parts of the Karkonosze Mts., above the timberline, where they reach up to 1.5 m high (Traczyk, 1995; Traczyk and Migoń, 2003). Traczyk and Żurawek (1999) described an assemblage of solifluction lobes that moved into erosional incisions at the footslope of Mt. Ślęża. The availability of high-resolution digital terrain models has recently opened new perspectives for the identification of solifluction landforms under forest, despite their subdued appearance. An example is the study of inherited cold-climate landforms on a basalt elevation in the West Sudetes, where LiDAR DTM revealed several superimposed solifluction sheets, terminating with distinct toes (Migoń et al., 2020). Solifluction deposits are quite often integrated into sequences of stratified slope deposits, as a rule acting as the lowermost layer or basal/middle cover beds. No absolute ages exist for solifluction sheets and lobes, and it is hard to precisely reconstruct the onset of their formation and past activity.

4 Present-Day Periglacial Environments

A general consensus of opinions occurs that the great majority of periglacial landforms, deposits and structures reported from the Central European Variscan ranges is relict, even though their survival in intact form until now may be debatable. The issue of survival versus ongoing change is particularly relevant to hillslope deposits, which were certainly subject to various bioturbation factors, among which tree throw plays a crucial role (Pawlik, 2013; Pawlik et al., 2016). It is also evident that the area lacks permafrost, even in the most elevated parts of the highest mountain massifs such as the Karkonosze Mts., or the Bohemian Forest. However, occasional ground freezing and thawing is the norm in the cold part of the year, especially at higher elevations (Kasprzak et al., 2021) and in exposed places, lacking vegetation cover (Křížek, 2007; Křížek et al., 2010; Traczyk, 1992). For example, Křížek et al. (2010) reported up to thirty freeze–thaw cycles during one winter at the Modré sedlo site in the Karkonosze Mts. Consequently, modest frost heaving and frost sorting may be expected to occur nowadays and the existence of active periglacial landforms and processes is inferred. Reasoning is based on thermal measurements carried out at specific localities and observations of indirect indicators such as plants and vegetation patterns.

Active periglacial landforms include various forms of patterned ground, such as sorted circles, sorted stripes and earth hummocks. *Sorted circles* are small (0.7–1.4 m in diameter) and infrequent, limited to a few specific localities subject to strong deflation in winter. Lichen cover on individual clasts is poorly developed, which is consistent with their contemporary displacements within the circles, and contrasts with much larger (up to 4 m in diameter), but nearly completely overgrown *sorted polygons* (Křížek et al., 2010, 2019). *Earth hummocks* occur both in the Karkonosze Mts. and the Hrubý Jeseník Mts., reaching the diameter of 1.7 m and they are up to 40 cm high (Treml et al., 2010). Their compositional characteristics, including a high proportion of fine fraction (silt + fine sand) and up to 30% of organic matter in the top 60 cm of the soil profile, make them very susceptible to frost action. Cross-section through the hummocks revealed disturbed soil horizons and occasional presence of ice lenses, whereas radiocarbon dating of organic matter yielded the uncalibrated Late Holocene date of 2.1 ka. Altogether, these various pieces of evidence are consistent with the present-day activity. Likewise, repeated measurements indicate slow movement of ploughing boulders on high-altitude grass-covered slopes in both most elevated parts of the Sudetes, which is also a phenomenon associated with the periglacial environment. Thus, in the Sudetes the altitude belt above ~1400 m may be considered as marginal mountain periglacial zone, located close to its lower limit defined by the presence of the climate-controlled timberline and marginal activity of frost-related mass transport processes (Křížek, 2016; Křížek et al., 2010).

Other mountain ranges within the Central European Variscan belt are at lower elevations, do not reach the climate-controlled timberline, and do not show evidence of contemporary frost-related processes, which would have clear geomorphic impact. They do contain impressive suites of cold-climate landforms, with blockfields being the most spectacular, but they are all inherited from the Pleistocene cold stages. However, a little-known phenomenon worth mentioning in this context is the presence of thermal anomalies within scree covers in various localities in the Bohemian Massif, especially on hills built of volcanic rocks of Neogene age, such as basalt, trachyte and phonolite (Kirchner et al., 2007). Porous nature of blocky accumulation facilitates deep air circulation and persistence of very low temperatures, even below the freezing point, well into the summer. This also allows infiltrating rain- or meltwater to freeze. Ground ice may be persistent ('frozen scree') or just temporary, but temperatures at the base of the scree cover remain around 0 °C, as attested by geophysical surveying and microclimatic measurements (Gude et al., 2003; Růžička et al., 2012). These specific microclimatic conditions have important ecological implications, supporting relict plant and animal species. Gude et al. (2003) consider these localities as examples of extrazonal permafrost and define the necessary conditions in the following way: "an essential precondition for the occurrence of the cool or even ice-containing scree is a steep slope of about 25° or more, which is covered with a relatively thick layer (several metres) of primarily coarse bedrock fragments, i.e. a diameter of at least 10 cm, and an open void system between them" (p. 336).

5 The Significance of Periglacial Dynamics in the Central European Variscan Ranges Within the European Context

The Central European Variscan ranges have a peculiar position within periglacial geomorphology. Inspiring the thinking of Walery Łoziński at the beginning of the twentieth century, they can be rightly considered as one of the cradles of periglacial research, although focused on the Pleistocene rather than contemporary environments. Later on, several general concepts were proposed using evidence from the region, pertinent to the interpretation of both geomorphic (rock cliffs, cryoplanation terraces, cryopediments) and sedimentary record (hillslope and loess deposits). Nevertheless, our understanding of periglacial legacy is still incomplete, some aspects definitely require re-assessment, and chronology of periglacial remodelling of slopes and plateaus has very few solid anchoring points.

The region hosts a rich inventory of cold-climate landforms and deposits. Among the former, the most impressive are summit and slope blockfields, rock cliffs, and stepped hillslopes interpreted as cryoplanation terraces. The latter include widespread hillslope materials produced by a variety of gravity-driven processes (solifluction, soil creep, slope wash), often arranged in complex, stacked sequences. The influence of lithology is evident, especially seen in the distribution and characteristics of bedrock outcrops, blockfields and plateau regolith. However, the vast majority of these landforms and deposits is inherited from the Pleistocene and only the most elevated parts of the Sudetes (>1400 m) can be considered as an active periglacial domain, although even this is close to its lower altitudinal boundary. Moreover, inherited and contemporary landforms occur side by side, making the recognition of active versus non-active landforms a challenge. Patterned ground due to frost heaving and sorting is the most compelling evidence of ongoing periglaciation.

In the contemporary landscape, the timberline is not only an important ecological boundary, but it also separates two belts, each having its own specific features as far as periglacial geomorphology is concerned. The part above the timberline is classified as a marginal periglacial environment, with some frost-related processes at least seasonally active. In this belt, cold-climate landforms can be fairly easily recognized and some are rather impressive (summit and slope blockfields, stone circles and nets, benched hillslopes). However, this active periglacial domain essentially occurs only in the most elevated massifs of the Sudetes range. The slopes below the timberline are either forested, or converted into arable fields and pastures, which has several implications for the periglacial legacy and its ability to survive during the Holocene. First, recognition of cold-climate landforms and their spatial patterns is more difficult. Second, the degree of Holocene re-shaping and overprinting due to slope processes, pedogenesis and bioturbation is a matter of ongoing debate, but it is likely not to be trivial, at least in some settings. For example, patterned ground structures were unlikely to survive. Third, centuries of human impact may have obscured much of periglacial inheritance due to soil cultivation and associated stone removal from arable lands, forest clearance and planting.

Periglacial research in the Central European Variscan ranges long relied on detailed field mapping and investigation of purposefully made outcrops in near-surface deposits. However, recent advances in dating techniques and growing availability of remote sensing data have opened new perspectives for periglacial research in the region. Landform recognition and quantitative analyses are now much easier than before, even in forested terrains, due to an advent of high-resolution digital terrain models (LiDAR DTM). GIS tools allow the researchers to attempt objective spatial analysis, as demonstrated by several studies of patterned ground in the Karkonosze Mts. Various dating techniques help to constrain the timing of sedimentary (hillslope and aeolian) and deformation events, and the database of numerical ages for the latest Pleistocene is constantly growing.

Periglacial environments, past and present, of the Central European Variscan ranges show some similarities to those in other rejuvenated Variscan ranges of Europe, especially in the Iberian Peninsula (e.g., Serra da Estrela in Portugal—see Vieira and Nieuwendam, 2020) and the French Massif Central. They too host impressive periglacial landforms and structures on the forest-free plateaus, whereas slopes are underlain by complex depositional packages, ranging from true blockfields to fine-grained stratified deposits. Likewise, some high-elevation plateaus in Great Britain show similarities to plateaus and levelled ridges of Central Europe and they are too regarded as marginal periglacial realms (e.g. Ballantyne and Harris, 1994). However, in contrast to many other European mountain ranges, glacial remodelling during the Pleistocene was rather limited and distinctive glacial erosional landforms are restricted to a few massifs (High Sudetes, Schwarzwald, Bohemian Forest). Consequently, high-energy cold-climate hillslope processes (rock falls, scree formation) were highly localized and the respective landforms are rare. Thus, the existing inventory of periglacial landforms is most appropriately described as a cumulative result of long-term development, spanning many cold periods, but in a low-energy geomorphic setting.

Acknowledgements The authors gratefully acknowledge discussions on periglacial issues with Cezary Kabała, Andrzej Kacprzak, Marek Kasprzak, Tobias Sprafke and Andrzej Traczyk, as well as joint field work in various places of the Central European Variscan ranges. Marek Błaś kindly provided a summary of climate data and Kacper Jancewicz is thanked for preparation of Fig. 1 and help with Fig. 3. Last but not least, we are indebted to volume editors for a thorough and insightful review of the draft version of this chapter.

References

Badura J, Jary Z, Smalley I (2013) Sources of loess material for deposits in Poland and part of Central Europe: the lost Big River. Quat Int 296:15–22

Ballantyne CK, Harris C (1994) Periglaciation of Great Britain. Cambridge University Press, Cambridge

Baykal Y, Stevens T, Engström-Johansson A, Skurzyński J, Zhang H, He J, Lu H, Adamiec G, Költringer C, Jary Z (2021) Detrital zircon U-Pb age analysis of last glacial loess sources and proglacial sediment dynamics in the Northern European Plain. Quatern Sci Rev 274:107265

Bogda A, Chodak T, Szerszeń L (1998) Skład i właściwości gleb wytworzonych z granitu Karkonoszy. In: Geoekologiczne Probl Karkonoszy, Acarus, Poznań, pp 179–184

Büdel J (1937) Eiszeitliche und rezente Verwitterung und Abtragung im ehemals nicht vereisten Teil Mitteleuropas. Petermanns Geogr Mitt, Ergh. 229:5–71

Büdel J (1959) Periodische und episodische Solifluktion im Rahmen der klimatischen Solifluktionstypen. Erdkunde 13:297–324

Büdel J (1977) Klima-Geomorphologie. Gebr Borntraeger, Stuttgart - Berlin

Büdel J (1982) Climatic Geomorphology. Princeton University Press, Princeton

Chmal H, Traczyk A (1993) Plejstoceńskie lodowce gruzowe w Karkonszach. Czas Geogr 64:253–263

Czudek T (1964) Periglacial slope development in the area of the Bohemian Massif in Northern Moravia. Biul Peryglac 14:169–193

Czudek T (1988) Kryopedimente – wichtige Reliefformen der rezenten und pleistozänen Permafrostgebiete. Petermanns Geogr Mitt 132:161–173

Czudek T, Demek J (1961) Význam pleistocenní kryoplanace na vývoj povrchových tvarů České vysočiny. Anthropos 14:57–69

Czudek T, Demek J (1971) Pleistocene cryoplanation in the Česka Vysočina highlands. Institute of British Geographers, Transactions 52:95–112

Czudek T, Demek J, Panoš V, Seichterová H (1963) The Pleistocene rhytmically bedded slope sediments in the Hornomoravský úval (Upper Moravian Graben). Sborník Geologických Věd, Antropozoikum 1:75–100

Demek J (1964) Castle koppies and tors in the Bohemian Highland (Czechoslovakia). Biul Peryglac 14:195–216

Demek J (1969) Cryoplanation terraces, their geographical distribution, genesis and development. Rozprávy Československé Akad Věd, ř. MPV 79(4):1–80

Ehlers J, Grube A, Stephan HJ, Wansa S (2011) Pleistocene glaciations of North Germany—new results. In: Ehlers J, Gibbard PL, Hughes PD (eds) Quaternary Glaciations—Extent and Chronology—a Closer Look. Dev. in Quat Sci 15, Elsevier, Amsterdam, pp 149–162

Eissmann L (2002) Quaternary geology of eastern Germany (Saxony, Saxon-Anhalt, South Brandenburg, Thüringia), type area of the Elsterian and Saalian Stages in Europe. Quatern Sci Rev 21:1275–1346

Engel Z, Braucher R, Traczyk A, Leanni L (2014) Be-10 exposure age chronology of the last glaciation in the Krkonose Mountains, Central Europe. Geomorphology 206:107–121

Engel, Z., Braucher, R., Aumaitre, G., Bourles, D., Keddaouche, K., Aster Team (2020) Origin and Be-10 surface exposure dating of a coarse debris accumulation in the Hruby Jesenik Mountains. Central Europe. Geomorphology 365:107292

Felix-Henningsen P (1991) Bodentwicklung in periglazialen Deckschichten des Osthunsrücks, Rheinisches Schiefergebirge. Mitt Deutsch Bodenkundl Gesell 66:779–782

Flohr E (1934) Alter, Entstehung und Bewegungserscheinungen der Blockmeere des Riesengebirges. Veröffentl Schles Gesell Erdk 21:395–418

French HM (2000) Does Lozinski's periglacial realm exist today? A discussion relevant to modern usage of the term 'Periglacial.' Permafrost Periglac Process 11:35–42

Frühauf M (1991) Neue Befunde zur Lithologie, Gliederung und Genese der periglazialen Lockermaterialdecken im Harz: Erfassung und Bewertung postalleroedzeitlicher decksedimentbildender Prozesse. Petermanns Geogr Mitt 135:49–60

Gellert J-F, Schüller A (1929) Eiszeitböden im Riesengebirge. Z Deutsch Geol Gesell 81:444–449

Gude M, Dietrich S, Mäusbacher R, Hauck C, Molenda R, Ruzicka V, Zacharda M, (2003) Probable occurrence of sporadic permafrost in non-alpine scree slopes in central Europe. In: Proceedings of the 8th Int Conf on Permafrost, Zurich. Swets & Zeitlinger, Lisse, pp 331–336

Hofmann FM, Rauscher F, McCreary W, Bischoff J-P, Preusser F (2020) Revisiting Late Pleistocene glacier dynamics north-west of the Feldberg, southern Black Forest, Germany. E&G Quaternary Science Journal 69:61–87

Hövermann J (1949) Die Periglazial-Erscheinungen im Harz. Göttinger Geogr Abh 14:3–39

Hülle D, Hilgers A, Kühn P, Radtke U (2009) The potential of optically stimulated luminescence for dating periglacial slope deposits — a case study from the Taunus area, Germany. Geomorphology 109:66–78

Jahn A (1954) Zasługi Walerego Łozińskiego w dziedzinie badań peryglacjalnych. Biul Peryglac 1:7–18

Jahn A (1962) Geneza skałek granitowych. Czasopismo Geograficzne 33:19–44

Jahn A (1968) Peryglacjalne pokrywy stokowe Karkonoszy i Gór Izerskich. Opera Corcontica 5:9–25

Jahn A (1977) The permafrost active layer in the Sudety Mountains during the last glaciation. Quaestiones Geographicae 4:29–42

Karte J (1983) Grèzes litées as a special type of periglacial slope sediments in the German Highlands. Polarforschnung 53(2):67–74

Kasprzak M, Traczyk A, Migała K (2021) Seasonally frozen ground. In: Ref Module in Earth Syst and Environ Sci Elsevier Ref Collect., available online at: www.sciencedirect.com/science/article/pii/B9780128182345001413?via%3Dihub

Kirchner K, Máčka Z, Cílek V (2007) Scree and blocky formations in northern and central Bohemia: geologic and geomorphologic development. Acta Geographica Silesiana 2:19–26

Kleber A (1992) Periglacial slope deposits and their pedogenic implications in Germany. Palaeogeogr Palaeoclimatol Palaeoecol 99:361–372

Kleber A (1997) Cover-beds as soil parent materials in mid-latitude regions. Catena 30:197–213

Kleber A, Terhorst A (eds) (2013) Mid-Latitude Slope Deposits (Cover Beds). Dev in Sedimentol 66, Elsevier, Amsterdam
Křížek M (2007) Periglacial landforms above alpine timberline in the High Sudetes. In: Goudie AS, Kalvoda J (eds) Geomorphological Variations P3K, Praha, pp 313–337
Křížek M (2016) Periglacial landforms of the Hrubý Jeseník Mountains. In: Pánek T, Hradecký J (eds) Landscapes and Landforms of the Czech Republic. Springer, Switzerland, pp 277–289
Křížek M, Treml V, Engel Z (2010) Czy najwyższe partie Sudetów powyżej górnej granicy lasu są domeną peryglacjalną? Czas Geogr 81:75–102
Křížek M, Krause D, Uxa T, Engel Z, Treml V, Traczyk A (2019) Patterned ground above the alpine timberline in the High Sudetes, Central Europe. J Maps 15:563–569
Krzyszkowski D (1998) Late Quaternary evolution of the Czyżynka river valley, Wałbrzych Upland, Middle Sudetes Mts, southwestern Poland. Geol Sudet 31:259–288
Linton DL (1955) The problem of tors. Geogr J 121:470–487
Łoziński W, (1909) Uber die mechanische Verwitterung der Sandsteine im gemassigten Klima. Bull Int de l'Acad des Sci de Cracov 1: 1–25. On the mechanical weathering of sandstones in temperate climates. In: Evans DJ (ed) Cold Climate Landforms. Wiley, Chichester, 1994, pp 119–134 (English translation)
Łoziński W, (1912) Die periglaciale Fazies der mechanischen Verwitterung. In: Comptes Rendus XII Int Geol Congr, Stockholm, pp 1039–1053
Martini A (1967) Preliminary experimental studies on frost weathering of certain rock types from the West Sudetes. Biul Peryglac 16:147–194
Martini A (1969) Sudetic tors formed under periglacial conditions. Biul Peryglac 19:351–369
Mentlik, P., Engel, Z., Braucher, R., Leanni, L., Aster Team (2013) Chronology of the Late Weichselian glaciation in the Bohemian Forest in Central Europe. Quatern Sci Rev 65:120–128
Meinecke F (1957) Granitverwitterung, Entstehung und Alter der Granitklippen. Z Deutsch Geol Gesell 109:483–498
Michniewicz A, Jancewicz K, Migoń P (2020) Large-scale geomorphological mapping of tors – Proposal of a key and landform interpretation. Geomorphology 357:107106
Migoń P, Kacprzak A (2014) Lateral variability of hillslope regolith and soils and implications for the interpretation of Pleistocene environments. Geomorphology 221:69–82
Migoń P, Knapik R, Jała Z, Remisz J, (2010) Contemporary evolution of talus slopes in the Wielki Śnieżny Kocioł Glacial Cirque. Opera Corcontica 47/2010, Suppl. 1, pp 63–74
Migoń P, Jancewicz K, Kasprzak M (2020) Inherited periglacial geomorphology of a basalt hill in the Sudetes, Central Europe: Insights from LiDAR-aided landform mapping. Permafrost Periglac Process 31:587–597
Moska P, Adamiec G, Jary Z (2011) OSL dating and lithological characteristics of loess deposits from Biały Kościół. Geochronometria 38:162–171
Pawlik Ł (2013) The role of trees in the geomorphic system of forested hillslopes – a review. Earth Sci Rev 126:250–265
Pawlik Ł, Phillips JD, Šamonil P (2016) Roots, rock, and regolith: Biomechanical and biochemical weathering by trees and its impact on hillslopes – A critical literature review. Earth Sci Rev 159:142–159
Raška P, Cajz V (2016) Neovolcanic terrain of the České Středohoří Mountains. In: Pánek T, Hradecký J (eds) Landscapes and Landforms of the Czech Republic. Springer, Switzerland, pp 139–152
Růžička V, Zacharda M, Němcová L, Šmilauer P, Nekola JC (2012) Periglacial microclimate in low-altitude scree slopes supports relict biodiversity. J Nat Hist 46:2145–2157
Sauer D, Felix-Henningsen P (2004) Application of ground-penetrating radar to determine the thickness of Pleistocene periglacial slope deposits. J Plant Nutr Soil Sci 167:752–760
Schilling B, Wiefel H (1962) Jungpleistozäne Periglazialbildungen und ihre regionale Differenzierung in einigen Teilen Thüringens und des Harzes. Geologie 11:428–460
Schmincke H-U, Park C, Harms E (1999) Evolution and environmental impacts of the eruption of Laacher See Volcano (Germany) 12,900 a BP. Quatern Int 61:61–72

Schönhals E (1957) Ein äolisches Sediment der Jüngeren Dryas-Zeit auf dem Laacher-See Tuff. Fortschr. Geol. Rheinland U. Westfalen 4:37–340

Schott C (1931) Die Blockmeere in den deutschen Mittlegebirgen. Forschungen zur deutschen Landes- u. Volkskunde 29:1–78

Sekyra J (1961) La carte periglaciaire du Massif Bohemien. Biul Peryglac 10:43–52

SemmeI A (1964) Junge Schuttdecken in Hessischen Mittelgebirgen. Notizblatt Hess. Landesamt F. Bodenforschung 92:275–285

Semmel A, Terhorst B (2010) The concept of the Pleistocene periglacial cover beds in central Europe: A review. Quatern Int 222:120–128

Sobik M, Błaś M, Migała K, Godek M, Nasiółkowski T, (2019) Klimat. In: Knapik R, Migoń P, Raj A (eds) Przyroda Karkonoskiego Parku Narodowego. Karkonoski Park Narodowy, Jelenia Góra, pp 147–186

Stan D, Stan-Kleczek I, Kania M (2017) Geophysical approach to the study of a periglacial blockfield in a mountain area (Ztracene kameny, Eastern Sudetes, Czech Republic). Geomorphology 293:380–390

Szczepankiewicz S (1958) Peryglacjalny rozwój stoków Masywu Ślęży. Biul Peryglac 6:81–92

Traczyk A (1992) Formy współczesnego sortowania mrozowego w Karkonoszach i klimatyczne uwarunkowania ich rozwoju. Czas Geogr 63:351–359

Traczyk A (1995) Morfologia peryglacjalna Śnieżki i Czarnego Grzbietu w Karkonoszach. Czas Geogr 66:157–173

Traczyk A (1996) Geneza i znaczenie stratygraficzne rytmicznie wartwowanych osadów stokowych w Sudetach. Acta Univ Wratisl 1808. Prace Inst Geogr A8:93–104

Traczyk A (2007) Pokrywy głazowo–blokowe zachodnich Karkonoszy w rejonie Szrenica – Śnieżne Kotły. Opera Corcontica 44(1):107–116

Traczyk A, Migoń P (2003) Cold-climate landform patterns in the Sudetes. Effects of lithology, relief and glacial history. Acta Univ Carol, Geogr 35, Suppl. 2000, pp 185–210

Traczyk A, Żurawek R (1999) Pleistozäne Schuttdecken und Schuttzungen im nordwestlichenTeil des Ślęża-Massivs (Polen) und ihre Entstehung unter den Bedingungen eines Dauerfrostbodens. Petermanns Geogr Mitt 143:131–141

Treml V, Jankovská V, Petr L (2006) Holocene timberline fluctuations in the mid-mountains of Central Europe. Fennia 184:107–119

Treml V, Křížek M, Engel Z (2010) Classification of patterned ground based on morphometry and site characteristics: A case study from the High Sudetes, Central Europe. Permafrost Periglac Process 21:67–77

Uxa T, Křížek M, Hrbáček F (2021) PERICLIMv1.0: a model deriving palaeo-air temperatures from thaw depth in past permafrost regions. Geoscientific Model Development 14:1865–1884

Vieira G, Nieuwendam A (2020) Glacial and periglacial landscapes of the Serra da Estrela. In: Vieira G, Zêzere JL, Mora C (eds) Landscapes and Landforms of Portugal. Springer, Switzerland, pp 185–198

Völkel J, Mahr A (2001) Die IRSL-Datierung von periglazialen Hangsedimenten —Ergebnisse aus dem Bayerischen Wald. Z Geomorphol 45:285–305

Völkel J, Leopold M, Mahr A, Raab T (2002) Zur Bedeutung kaltzeitlicher Hangsedimente in zentraleuropäischen Mittelgebirgslandschaften and zu Fragen ihrer Terminologie. Petermanns Geogr Mitt 146:50–59

Waroszewski J, Sprafke T, Kabala C, Musztyfaga E, Labaz B, Wozniczka P (2018a) Aeolian silt contribution to soils on the mountain slopes (Mt. Ślęża, SW Poland). Quatern Res 89:702–717

Waroszewski J, Egli M, Brandová D, Christl M, Kabala C, Malkiewicz M, Kierczak J, Glina B, Jezierski P (2018b) Identifying slope processes over time and their imprint in soils of medium-high mountains of Central Europe (the Karkonosze Mountains, Poland). Earth Surf Proc Land 43:1195–1212

Waroszewski J, Sprafke T, Kabala C, Musztyfaga E, Kot A, Tsukamoto S, Frechen M (2020) Chronostratigraphy of silt-dominated Pleistocene periglacial slope deposits on Mt. Ślęża (SW, Poland): Palaeoenvironmental and pedogenic significance. Catena 190, 104549

Waroszewski J, Pietranik A, Sprafke T, Kabała C, Frechen M, Jary Z, Kot A, Tsukamoto S, Meyer-Heintze S, Krawczyk M, Łabaz B, Schultz B, Erban Kochergina YV (2021) Provenance and paleoenvironmental context of the Late Pleistocene thin aeolian silt mantles in south-west Poland e a widespread parent material for soils. Catena 204:105377

Wilhelmy H (1958) Klimamorphologie der Massengesteine. Westermann, Braunschweig

Žák K (2016) Brdy Highland: a landscape shaped in the periglacial zone of Quaternary glacials. In: Pánek T, Hradecký J (eds) Landscapes and Landforms of the Czech Republic. Springer, Switzerland, pp 73–86

Ziegler PA, Dèzes P (2007) Cenozoic uplift of Variscan Massifs in the Alpine foreland: Timing and controlling mechanisms. Global Planet Change 58:237–269

Żurawek R (1999a) Relict rock glaciers in the Central European Mid-Mountains. State-of-the-Art. Biul Peryglac 38:163–192

Żurawek R (1999b) Reliktowe lodowce skalne — nowa interpretacja form akumulacji na wschodnich i południowych stokach Ślęży. Przegl Geogr 71:77–94

Żurawek R (2001) Problem wieku reliktowych lodowców skalnych w Masywie Ślęży w świetle datowań ^{14}C i OSL oraz obserwacji geomorfologicznych. Przegl Geol 49:880–884

Żurawek R, Borowicz D (2003) Topography of a composite relict rock glacier, Slęża Massif, SW Poland. Geogr Ann 85A:31–41

Żurawek R, Migoń P (1999) Peryglacjalna morfogeneza Ślęży w kontekście długotrwałej ewolucji rzeźby. Acta Geogr Lodziensia 76:133–155

Żurawek R, Żyszkowski E, Górecki A (2005) Topographic control of periglacial slope covers, Sleza Massif, southwest Poland: A statistical approach. Permafrost Periglac Process 16:241–248

The Carpathians

Zofia Rączkowska

1 Geographical Framework

The Carpathian arc runs for a distance of 1500 km between latitudes 50°N and 44°N and longitudes 17° E and 27° E across the territory of seven countries (UNEP 2007). The area is divided into the Western Carpathians, Eastern Carpathians, Southern Carpathians and Western Romanian Carpathians (Fig. 1; Kondracki 1989). The highest elevations occur in the Tatras (Gerlachovský štít 2655 m a.s.l.) and the Făgăraş (Moldoveanu 2544 m a.s.l.).

The Carpathians are a segment of the Alpine system, with a heterogeneous lithology and folded and faulted structures, separated by Neogene and Quaternary sediment-filled intra-montane depressions. Flysch sedimentary rocks (turbidites) form the external zone present in the Western and Eastern Carpathians. The more complex, internal zone consists of crystalline, calcareous, conglomerate and volcanic rocks, which build ranges or isolated blocks along the whole Carpathian arc (Kondracki 1989; Haas 2012).

The Carpathian relief is strongly linked to the geological structure; using Starkel (1972) relief typology, it could be subdivided into two main altitudinal belts: middle mountains and high mountains. The middle mountains, lying at elevations between 300 and 1500 m, are of very varied morphology of fluvio-denudational origin and predominate in the Carpathians. High mountains (1500–2655 m) include massifs with postglacial relief originated form Pleistocene glaciations. They are well-developed in the Tatras and the Southern Carpathians. The estimated ELA (equilibrium line altitude) altitude for the Last Glacial Maximum (LGM) increases from 1580 m in the Tatras (Zasadni et al. 2018) and 1500–1700 m in the Eastern Carpathians (Kłapyta

Z. Rączkowska (✉)
Department of Geoenvironmental Research, Institute of Geography and Spatial Organisation PAS, Cracow, Poland
e-mail: raczk@zg.pan.krakow.pl

M. Oliva et al. (eds.), *Periglacial Landscapes of Europe*,
https://doi.org/10.1007/978-3-031-14895-8_11

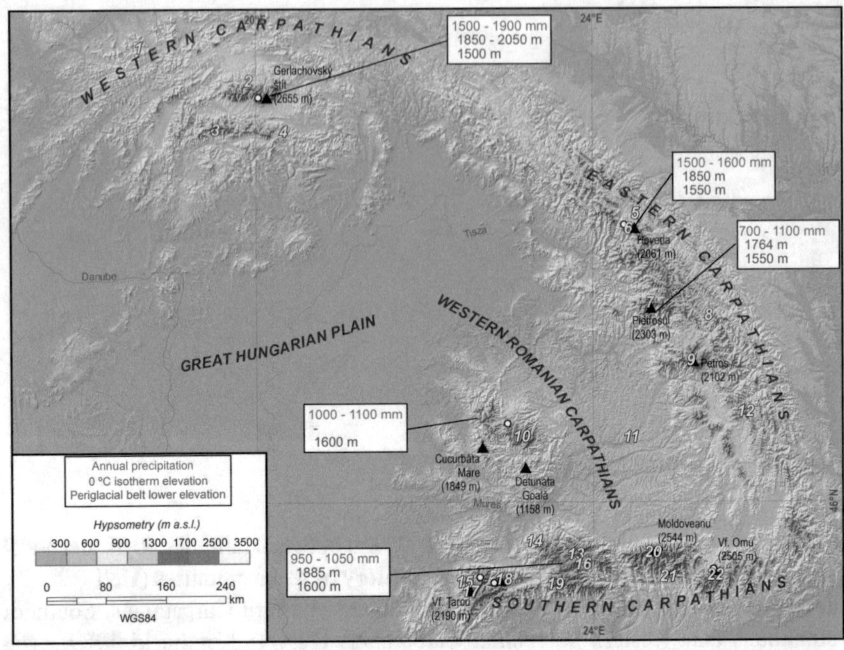

Placenames reffered in the text
1 Moravian-Silesian Beskids, *2* Tatras: Miedziany, *3* Low Tatras, *4* Slovensky raj, *5* Gorgany, *6* Chornohora, *7* Rodna, *8* Rarău *9* Călimani *10* Apuseni: Scărişoara, *11* Transylvanian Plateau, *12* Harghita, *13* Cindrel, *14* Sureanu, *15* Muntele Mic, *16* Lotru, *17* Tarcu, *18* Retezat, *19* Parâng, *20* Făgăraş, *21* Iezer, *22* Bucegi

Fig. 1 Location of the Carpathians. The map includes names of the massifs and other place names mentioned in the text. The data on mean annual precipitation, elevation of the 0 °C isotherm and the minimum altitude of the present-day periglacial belt are after Niedźwiedź (2012), Micu et al. (2015), Kłapyta et al. (2021)

et al. 2021) to 1700–1850 m in the Southern Carpathians (Ruszkiczay-Rüdiger et al. 2016).

The Carpathians are located in the transitional climate zone between the oceanic air masses in the west and the more continental ones in the east (Niedźwiedź 2012). These general characteristics vary in terms of radiation and atmospheric circulation. The mean annual air temperature (MAAT) on the summits is –3.7 °C in the Tatras, –2.5 °C in the Bucegi, –0.5 °C in the Tarcu to 0.0 °C in the Chornohora and +1.0 °C in the Apuseni (Niedźwiedź 2012; Micu et al. 2015). The mean annual precipitation amounts 600–1400 mm in the middle mountains and 1500–2000 mm in the high mountain zone, of which 30 to 85% comes as snowfall. Generally, there is an eastward and southward decrease of precipitation totals (UNEP 2007; Micu et al. 2015). The seasonal snow cover in the high mountain ranges lasts between 150 and 290 days each year, and its thickness varies from ~1.2 m in the Southern Carpathians to >2.0 m in the Tatras (Hess 1965; Micu et al. 2015). Currently, no glaciers occur in the Carpathians, only the Miedziany glacieret (~2200 m^2; Gądek and Grabiec 2008) and some small perennial snow patches occur in the Tatras (Rączkowska 1993) and in the Rodna (UNEP 2007).

Cold-climate (periglacial) geomorphological processes are currently limited to the high mountain zone of the highest Carpathian ranges. Kotarba and Starkel (1972), assigned the cryonival morphoclimatic system to geoecological belts above the upper timberline. In those locations with the most favourable topoclimatic conditions, sporadic permafrost exists (Dobiński 2005; Popescu et al. 2017c).

2 History of Periglacial Research in the Carpathians

The term "periglacial" was introduced by W. Łoziński (1912) when studying block covers on the slopes of the Gorgany in the Eastern Carpathians to refer climatic conditions in areas adjacent to Pleistocene ice sheets and mountain glaciers. Currently, this term is used in the climatic sense, for areas where the climate is cold enough to develop landforms through the activity of frost and snow related processes (French 2007). At the same time, the first observations on periglacial phenomena, though not defined as such, were also made by others (Rehman 1895; de Martonne 1907).

The significant development of periglacial research in the Carpathians, examining both the Pleistocene and contemporary phenomena and landforms, took place at the turn of the 1940s and 1950s and corresponded to the international trends of the time.

An abundance of periglacial studies followed in the Carpathians, sometimes included in regional or country syntheses, which provided a good description and explanation of the occurrence of periglacial landforms and some estimates of their ages (e.g.; Jahn 1958; Micalevich-Velcea 1961; Niculescu and Nedelcu 1961; Nedelcu 1964; Niculescu 1965; Lukniš 1973; Nemčok and Mahr 1974; Schreiber 1974; Sircu 1978; Ichim 1983; Midriak 1983; Florea 1998; Urdea 2000; Křížek 2001; Rusu 2002). The attention of researchers also focussed on the development and transformation of periglacial slope covers during the Pleistocene (e.g.; Klimaszewski 1948; Jahn 1958; Starkel 1960, 2015; Pécsi 1967; Pękala 1969; Baumgart-Kotarba 1971; Henkiel 1972; Czudek and Demek 1976; Ichim 1983; Ziętara 1989, 2004).

The analysis of contemporary periglacial processes is characterised by inconsistency. Two various periods could be highlighted. The studies were mostly descriptive at the beginning, without discussing quantitative approaches because of a lack of data (e.g.; Jahn 1958; Nedelcu 1964; Sircu 1976). Quantitative studies of the processes started in the Tatras at the end of 1950s (e.g.; Gerlach 1959, 1972; Kotarba 1976; Midriak 1983; Kotarba et al. 1987), when instrumental measurements of ground temperature and freezing depth were also initiated (Hess 1963; Gerlach 1972).

Since 1990s, the research intensified both in the Tatras (e.g.; Rączkowska 1993, 1995, 2007; Baranowski et al. 2004; Gądek et al. 2009; Lubera 2016; Rączkowska et al. 2016) and the Romanian Carpathians—mostly the Southern Carpathians (e.g.; Urdea 1992, 1993; Nagy et al. 2004; Voiculescu 2004; Urdea et al. 2008; Ardelean et al. 2015; Puţan 2015; Onaca 2017; Onaca et al. 2017b; Vasile and Vespremeanu-Stroe 2017). The broader set of processes and the spatial scope of the research were considered. The range of methods applied was extended, including monitoring of the ground thermal regime, continuous monitoring of dynamic processes and application of various geophysical methods. In effect, monitoring of the weathering rates, rockfall, talus slope formation, solifluction, nivation, patterned ground, and needle ice activity as well as the thermal regime of the ground were established there, although, such studies are still lacking in other periglacial areas of the Carpathians.

Almost simultaneously with the quantitative studies of periglacial processes, the study of permafrost and its geomorphological effect started in the 1990s—and continued in the following decades—in the Southern Carpathians (e.g. Urdea 1992, 1993; Kern et al. 2004; Urdea et al. 2008; Vespremaeanu-Stroe et al. 2012; Onaca et al. 2013b, 2015; Popescu et al. 2015, 2017a), and in the Tatras (e.g. Dobiński 1998; Kędzia 2004; Gądek and Grabiec 2008; Gądek and Kędzia 2008). Their aim was to demonstrate the presence of sporadic permafrost and its features there (Mościcki and Kędzia 2001; Dobiński 2004; Popescu et al. 2017a) as well as to make significant progress in the recognition of rock glaciers and their dynamic behaviour (e.g. Onaca et al. 2013a, 2017a; Necşoiu et al. 2016; Uxa and Mida 2017a, b).

In the last few decades, the attention of researchers also turned to the dating of periglacial landforms using dendrogeomorphological methods (e.g. Voiculescu and Onaca 2014; Lempa et al. 2016; Šilhán et al. 2011; Pop et al. 2018), lichenometry (Kędzia 2014), Schmidt hammer testing (e.g. Kłapyta 2011; Onaca 2017; Şerban et al. 2019) and cosmic-ray exposure dating (e.g. Pánek et al. 2016; Zasadni et al. 2020). Besides using geophysical techniques, the inner structure of some periglacial deposits and landforms was analysed (e.g. Urdea et al. 2008; Onaca et al. 2013a, b, 2016; Gądek et al. 2016).

3 Periglacial Landforms and Processes

The Carpathian landscape covers a wide range of periglacial landforms and phenomena, which evolved during both past and present climate conditions. Active landforms under the current climate are mostly concentrated in the highest zones of the high-mountain massifs. The widespread occurrence of relic or inactive periglacial phenomena and landforms suggests that they must have formed during a past and more severe cold climate regime. In the Polish Carpathians, mesoforms from the last cold stage occupy 90% of the area (Starkel 1986). On the other hand, in small alpine catchments of the Southern Carpathians, 80% of the deposits are of periglacial origin (Ardelean et al. 2017).

Pleistocene periglacial environment encompassed the whole of the Carpathians except the areas occupied by mountain glaciers (Ichim 1983; Demek 1984; Mojski 2005). In the non-glaciated middle mountains, it covered their entire vertical extent, and in the high-mountains two periglacial zones occurred, one above and one below the mountain glaciers. The extent of the periglacial domain during Pleistocene glacial cycles was dependent on the intensity of cold conditions. Permafrost underlaid the area in the Pleistocene (Gierasimov and Vielichko 1982; Ichim 1983) conditioning periglacial processes and landform development. It melted during interglacial periods, and the Interpleniglacial (58–28 ka BP; MIS 3; Starkel 2015). The last

Weichselian permafrost melted about 10–9 ka BP (Demek 1984; Starkel 2015). The Carpathians were in the loess-tundra belt during the cold stages of the Pleistocene (Büdel 1951; Starkel 2015). In the Polish Carpathians the upper timberline has fluctuated between 500 and 1000 m in glacial and interglacial periods, respectively. A polar desert was located above 1,000–1,200 m (Mojski 2005).

The current periglacial environment, in the Carpathians is confined to areas above the natural upper timberline (Rączkowska 2007; Onaca et al. 2017b). The 0 °C isotherm is located at an altitude 1885 m in the Southern Carpathians, 1764–1850 m in the Eastern Carpathians (Niedźwiedź 2012; Micu et al. 2015) and at 1850 m in the Western Carpathians (Niedźwiedź 2012). The –2 °C isotherm appears at 2100–2200 m (Micu et al. 2015) and at 2150–2300 m (Hess 1965) in the Southern Carpathians and Tatras, respectively. The +2 °C isotherm, corresponding roughly with the upper timberline, is located at 1500 m in the Western and the Eastern Carpathians (Hess 1965; Niedźwiedź 2012; Micu et al. 2015) and at 1600 m in the Southern and the Western Romanian Carpathians (Micu et al. 2015). The +3 °C isotherm is located at 1300 m in the Southern and Western Romanian Carpathians and at 1450 m in the Eastern Carpathians (Micu et al. 2015), which is much lower than the upper timberline.

The activity of periglacial processes is currently driven by seasonal and diurnal frost conditions (Rączkowska 2007; Onaca et al. 2017b). The length of the freezing season is from 150 to >200 days at 2000 m (Hess 1965; Micu et al. 2015). The freeze–thaw cycles (FTC) mainly occur in the spring and autumn seasons (Rączkowska 2007; Onaca et al. 2017b). In the Southern Carpathians the total number of FTC of air temperature ranges between 80 and 100 days per year (Vasile and Vespremeanu-Stroe 2017), and it varies from 78 to 93 in the Tatras (Rączkowska 2007).

Generally, the ground remains frozen during the cold half of the year in areas of the periglacial domain, although monitoring showed that the thermal regime of the ground varies greatly. Its dynamics is mostly influenced by slope exposure, local topography, local air circulation over the surface (Gądek and Kędzia, 2008), ground structure and humidity (Mościcki 2008) and snowpack (thickness, development, extent), which is unstable during the winter seasons of recent decades (Gądek 2014).

In the Southern Carpathians, at altitudes >2,000 m the ground is frozen to the depth of 50 cm during 130–200 days per season (Onaca et al. 2017b). In the Bucegi, at altitudes >2000 m, the annual number of FTC of soil temperature is 22–23, similar to the N-oriented rockwalls, where it counts to 24. On S-oriented rockwalls, the number is 110 (Vasile et al. 2014), a similar number to the Tatras (91.8 cycles yr^{-1}) at an altitude of 1645 m (Lubera 2016). At a similar altitude (1770 m) in Muntele Mic, only a few daily FTC per year were recorded in the ground temperature at a depth down to 20 cm within earth hummocks, while none were recorded within the solifluction lobes. On rock glaciers (permafrost terrain), daily FTC occurred the whole year, but mainly in spring and autumn (Onaca et al. 2013b).

In the Tatras, the ground temperature remains negative at 1950 m during 8–9 months at 50 cm and 100 cm depth; at 1720 m during 6–7 months at 50 cm and 25 cm depth; and at 1520 m during 3 months (Baranowski et al. 2004). Seasonal, daily and hourly FTC of the ground temperature occur at depths down to 50 cm; the depth differs from a few cm to a few tens of cm in individual years. The FTC are the most frequent in autumn (Rączkowska 2007). On debris slopes with sporadic discontinuous permafrost, the mean annual ground surface temperature varies between –1.6 °C and +2.3 °C (Mościcki and Kędzia 2001; Gądek and Kędzia 2008), and gradually increases in recent decades (Gądek and Leszkiewicz 2012). The bottom temperature of the snow cover (BTS) is negative from –9.6 °C to –0.2 °C (Gądek and Kędzia 2008). The BTS in the Retezat is higher, from –4.1 °C to –2.3 °C on average (Ardelan et al. 2015).

Dobiński (2005) found climatic conditions sufficient for permafrost at altitudes >1930 m in the Tatras and >2000 m in the Southern Carpathians. Patches of sporadic permafrost exist in the Tatras on talus slopes at ~1950–2000 m (Gądek and Szypuła 2015), as proved by permafrost mapping using BTS, thermal regime monitoring, geophysical data and infrared imaging (e.g.; Dobiński 1998; Kędzia 2004; Gądek and Grabiec 2008; Gądek and Kędzia 2008). Contemporary permafrost includes both interstitial ice and massive ice (e.g.; Kędzia 2004; Gądek and Grabiec 2008). The thickness of the active layer ranges from 0.5 to 6 m (Dobiński 1998), and the thickness of identified permafrost patches from 0.5 to 25 m (Dobiński, 1998; Mościcki and Kędzia 2001; Gądek and Grabiec, 2008). The presence of relict (Pleistocene) permafrost in the Tatras is also not excluded (Dobiński et al. 2008).

In the Southern Carpathians (the Retezat, the Parâng, the Făgăraş, the Iezer and the Bucegi massifs), sporadic permafrost exists in rock glaciers (>1950 m), talus slopes and shaded rock walls (>2400 m; Popescu et al. 2017a) on N-facing slopes, as was proved by thermal monitoring of BTS on rock glaciers, summer temperature of the springs, and geophysical data (Urdea 1993; Kern et al. 2004; Vespremeanu-Stroe et al. 2012; Onaca et al. 2013b, 2015). Geophysical data indicate a thin (<10 m) undersaturated contemporary permafrost layer located under a thick (8–10 m) active layer at altitudes 1950–2100 m; and also, a thicker (>10–20 m) layer, sometimes supersaturated in ice, at altitudes exceeding 2100 m (Popescu et al. 2015, 2017a). A lower-lying relict permafrost has been detected in a talus derived rock glacier on Detunata Goală Mt. in the Apuseni, in the forest belt (Popescu et al. 2017c). A specific case is the underground permafrost in ice caves with perennial ice bodies accompanied by cryospeleothems (Siarzewski 1994; Bella 2008, Persoiu and Onac 2019).

The periglacial environments described above determine the development and distribution of periglacial landforms and the dynamics of periglacial processes in the Carpathians.

3.1 Rock Weathering and Derived Landforms

The efficiency of frost weathering depends on the thermal and humidity regime of rocks (e.g. Matsuoka 2011), whose variability followed temporal environmental variability in the Carpathians. In the past, in more severe environmental conditions, rock weathering was a common process in the Carpathians as can be inferred from a widespread occurrence and diversity of relict landforms related to this process. At present, frost weathering is the principal type of weathering in the Carpathian high-mountains above the upper timberline. It is effective, particularly in autumn and spring when the freeze–thaw cycles are most frequent. However, low rates of rockwall retreat of 0.002–3.0 mm yr^{-1} (Table 1) indicate that its intensity is rather low despite the lithology. Similar rates were found for crystalline and carbonate rocks (Rączkowska 2007; Onaca et al. 2017b). The intensity of weathering and rockfall was higher during the Little Ice Age (LIA). Lichenometric dating of talus slopes in the Tatras indicated more intensive physical weathering and rockfall during last 300 years in the period 1810–1910 (Kotarba and Pech 2002).

Frost weathering contributes to the development (lowering) of ridges and slopes, formation of blockfields, cryoplanation terraces and in the interaction with other processes is involved in the development of talus slopes, rock glaciers and protalus ramparts (Rączkowska 2007; Kravčuk 2008; Onaca et al. 2017b; Kłapyta et al. 2020). In the Southern Carpathians, the rates of periglacial denudation range between 0.26 and 0.90 mm yr^{-1} in Făgăraş and Retezat, respectively (Table 1).

Blockfields are relict Pleistocene periglacial landforms, occurring high on the ridges and slopes (Fig. 2a), both in massifs not glaciated in the Pleistocene and above the Pleistocene trimline in glaciated mountain massifs (e.g. Klimaszewski 1948; Jahn 1958; Niculescu and Nedelcu 1962; Pękala 1969; Lukniš 1973; Florea 1998; Urdea 2000; Kłapyta 2006, 2020a, b; Kravčuk 2008; Kłapyta et al. 2020). They result from in situ weathering, favoured by the polar desert that existed in the Carpathians during the Pleistocene cold stages (Mojski 2005). They are more common in the Eastern and Southern Carpathians, because of their more continental climate. Blockfields mostly comprise large boulders covered with lichens, often located covering cryoplanation terraces, and on lower located terraces buried by a soil layer few-tens cm thick (Křížek 2001). Although some authors have assumed that the present activity of blockfields is weak (Jahn 1958; Urdea 1995), they tend to become overgrown rather slowly.

On steeply-graded slopes, elongated *blockstreams* have developed as a result of slow frost creep or solifluction (Fig. 2b; Henkiel 1972; Křížek 2001; Kłapyta 2006, 2020a, b; Rączkowska 2007; Puţan 2015; Şerban et al. 2019). Now they are relict landforms, occurring mostly in the high mountains, but also in some middle-mountain ranges. In the Southern Carpathians, they very often occur on south-facing slopes, where they were developed at the end of Pleistocene under permafrost conditions (Puţan 2015; Şerban et al. 2019). Their measured annual rates of movement are lower than the uncertainty of the measuring instrument (Onaca 2017; Onaca et al. 2017b). Lichenometic dating suggests that the formation of blockstreams was re-initiated above 2100 m a.s.l during the cold Holocene episodes, e.g. in the LIA (Puţan 2015; Şerban et al. 2019).

Step-like *cryoplanation terraces* with gently sloping treads and debris-covered scarps related to outcrops of the sedimentary beds were identified in the upper slope sectors or on summit surfaces, mostly within the flysch Carpathians (Fig. 2d; e.g. Demek 1969; Baumgart-Kotarba 1971; Ziętara 1989, 2004; Florea 1998; Křížek 2000; Kłapyta 2006, 2020a, b). They developed due to intense frost shattering, nivation, cryoturbation and gelifluction in cold Pleistocene phases and now represent

Table 1 Dynamic of present-day periglacial processes in the Carpathians

Geomorphological process	Rates of the process	Study site	Author
Chemical denudation ($m^3\ km^{-2}\ yr^{-1}$)	36–38	Limestones, dolomites, Tatras	Kotarba (1972)
Denudation ($mm\ yr^{-1}$)	0.26	Crystalline rocks, Făgăraş	Ardelean et al. (2017)
	0.90	Crystalline rocks, Retezat	Vespremeanu-Stroe et al. (2012)
Retreat of rockwalls ($mm\ yr^{-1}$)	0.7	Metamorphic rocks and granites, Tatras	Kotarba et al. (1987)
	0.003	Granites, Tatras	Lubera (2016)
	0.002	Dolomites, Tatras	Lubera (2016
	0.1–3.0	Limestones, dolomites, Tatras	Kotarba (1972)
	0.003–0.019	Limestones, dolomites, Tatras	Midriak (1983)
Permafrost creep (cm yr^{-1})	Few	Rock glaciers, Retezat	Necsoiu et al. (2016)
Frost creep (cm yr^{-1})	<1	Block streams, Southern Carpathians	Onaca (2017), Onaca et al. (2017b)
	0.6–1.7	Tatras	Kotarba (1976)
	few–18	Retezat	Urdea (2000), Onaca et al. (2017b)
Frost creep and solifluction (cm yr^{-1})	3–5	Rarău	Rusu (2002)
	0.1–2.0	Metamorphic rocks, Tatras	Baranowski et al. (2004)
Solifluction (cm yr^{-1})	1.0–2.0 (14.1)	turf-banked lobes, Făgăraş	Onaca et al. (2017b)
	9.7	turf-banked lobes, Izer	Onaca et al. (2017b)
	17.4	turf-banked lobes, Cindrel	Onaca et al. (2017b)
	9	turf-banked lobes, Parâng	Onaca et al. (2017b)
	3.3	turf-banked lobes, Şureanu	Onaca et al. (2017b)
	1–4	stone-banked lobes, Parâng	Urdea et al. (2004)
Ploughing boulders (cm yr^{-1})	0.1–3.3	Limestones, dolomites, Tatras	Kotarba (1976)
	2.7	Granites, Tatras	Kotarba et al. (1987)

(continued)

Table 1 (continued)

Geomorphological process	Rates of the process	Study site	Author
	11.1	Făgăraș	Onaca et al. (2017b)
	6.6	Parâng	Onaca et al. (2017b)
	8.1	Lotru	Onaca et al. (2017b)
	10.4	Muntele Mic	Onaca et al. (2017b)
Needle ice (cm yr^{-1})	3.8	Granites, Tatras	Gerlach (1959)
Frost heave (cm yr^{-1})	0.28–0.69	Muntele Mic	Onaca (2017)
Frost heave (cm)	2.0	Retezat	Urdea (2000)
	3.1–4.0	Tatras	Midriak (1983)
Max accretion of talus slopes (cm yr^{-1})	2.0–10.0	Carbonate rocks, Tatras	Kotarba (1976)
Accumulation on talus slopes around snow patches (cm yr^{-1})	0,03–0,06	Granites, Tatras	Rączkowska (1995)
Nival erosion (cm yr^{-1})	0.05–5.0	Tatras	Rączkowska (1993)

Blockfield (Eastern Carpathians)
Author: Piotr Kłapyta

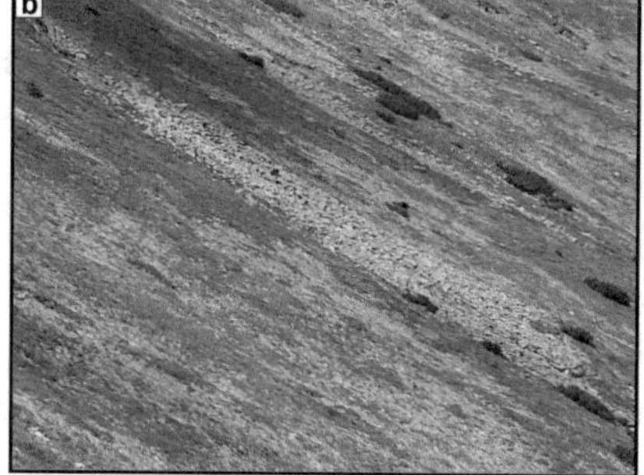

Blockstream (Western Carpathians)
Author: Zofia Rączkowska

Tor (Eastern Carpathians)
Author: Piotr Kłapyta

Cryoplanation terrace (Western Carpathians)
Author: Zofia Rączkowska

Fig. 2 Rock weathering derived landforms

relict landforms (e.g.; Demek 1969; Henkiel 1972; Křížek 2000). Their scarps (*frost riven cliffs*) are a few to 10 m high, and their ledges from a few tens to more than 100 m in width. The sandstone scarps of terraces were remodelled by weathering in the Holocene, as indicated by microforms on their surface (pits, honeycombs etc.; Křížek 2001).

Tors are landforms originating from differential weathering of bedrock over thousands of years (Fig. 2c). They are widespread outside the areas glaciated during the last glaciations, both in the middle and the high mountains of the Carpathians. Their sizes vary from few to several tens of metres, depending on the lithology. These mostly relict landforms are preserved both on the ridges and the upper slope sections (e.g.; Micalevich-Velcea 1961; Niculesu and Nedelcu 1961; Alexandrowicz 1970; Starkel 1986; Florea 1998; Kravčuk 2008; Lubera 2011). However, data from thermal regime monitoring of tor rockwalls at the highest altitudes indicate a very weak current weathering activity (Lubera 2011; Vasile and Vespremanu-Stroe 2017).

High-mountain calcareous massifs or ranges, where chemical weathering and karst processes affect the landscape are also present in the Carpathians (e.g. Goran 2012; Szczygieł et al. 2015). These areas present a wide diversity of karst landforms, such as *niveo-karst* landforms like karren, chimneys, dissolution arches and sinkholes. Through monitoring, Kotarba (1972) showed that rillkarren and rinnenkarren may now develop in the periglacial zone in the Tatras (1600–1900 m) by surface chemical erosion by meltwater. Its intensity is 0.38 mm year^{-1} (Table 1).

3.2 Talus Slopes and Related Landforms

Talus cones and *sheets* formed on talus slopes are the most widespread periglacial features in the Carpathian high-mountains, located below rocky slopes and rockwalls of glacial cirques and U-shaped valleys (Fig. 3a; e.g. Kotarba et al. 1987; Urdea 2000; Rączkowska 2007). Frost weathering is the major process responsible for the supply of debris, which may be transported by rockfall, debris flows and snow avalanches and subsequently accumulates on talus slopes. Ground penetrating radar (GPR) data has shown that the maximum thickness of the talus slopes in the Tatras ranges from 20 to 35 m (Gądek et al. 2016), and is >15 m thick in the Făgăraş (Onaca et al. 2016). The development of the talus slopes began in paraglacial conditions, once glacial ice had retreated in the Late Glacial, around 12.0–12.5 ka BP (Makos 2015; Popescu et al. 2017b). At present a low intensity of rockwall retreat and accretion on talus slopes indirectly indicates weak activity of weathering and rockfall. In the Tatras, rates of rockwall retreat are from 0.02 to 3.0 mm yr^{-1}, and of talus slope accretion from 0.03 to 8.0 cm yr^{-1} (Table 1). Lichenometric dating of talus slopes suggests a higher rockfall intensity during the LIA (Kotarba and Pech 2002). Dendrogeomorphological studies in the Moravian-Silesians Beskids also confirmed probable recent rockfall activity in the middle-mountain ranges (Šilhán et al. 2011).

Talus cones are mainly reworked by *debris flows* triggered by intense rainfall, which started with the warming of the climate at the beginning of the Holocene

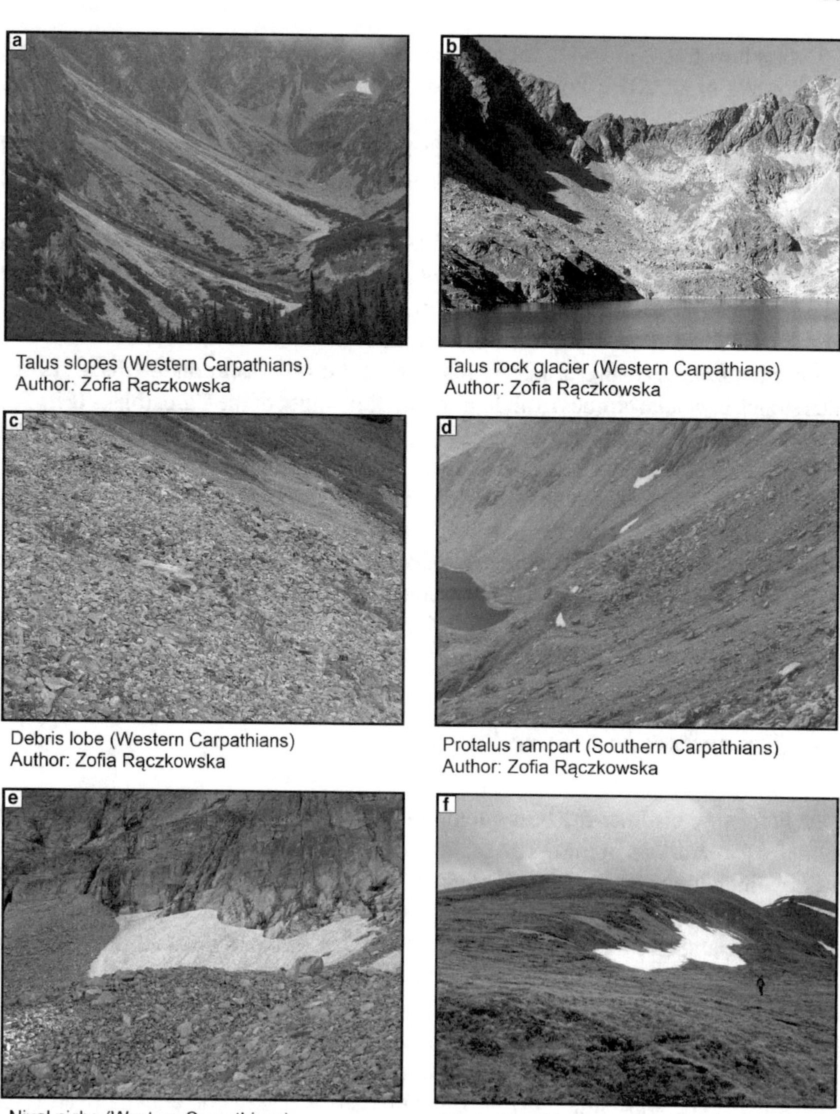
a
Talus slopes (Western Carpathians)
Author: Zofia Rączkowska

b
Talus rock glacier (Western Carpathians)
Author: Zofia Rączkowska

c
Debris lobe (Western Carpathians)
Author: Zofia Rączkowska

d
Protalus rampart (Southern Carpathians)
Author: Zofia Rączkowska

e
Nival niche (Western Carpathians)
Author: Zofia Rączkowska

f
Nival niche (Western Carpathians)
Author: Zofia Rączkowska

Fig. 3 Talus slopes and associated landforms and the nival landforms

(Onaca et al. 2016; Senderak et al. 2019). In effect, *gullies* and *levées* occur on talus cone surface. Terrestrial laser scanning (TLS) monitoring proved debris flow tracks are the only active sectors of the talus cones at present (Rączkowska and Cebulski 2020). Debris flow activity was significantly higher during the LIA, as was proved by lichenometric dating and lake deposits in the Tatras (Kotarba 1995, 1996).

Debris flow tracks are often followed by *snow avalanches* (e.g.; Voiculescu 2004; Rączkowska et al. 2016; Pop et al. 2018), which contribute to reshaping of talus slopes, and even to the formation of specific *avalanche talus cones* (Lempa et al. 2015). However, a diminution of snow avalanche activity during the latter half of the twentieth century has been demonstrated (Lempa et al. 2016; Gądek et al. 2017; Meseșan et al. 2019).

Frost creep and *gelifluction* on shaded talus slopes above 2000 m has resulted in the development of small *sorted debris lobes* a few metres in size, built with coarse debris, and observed on talus slopes in the Southern Carpathians and the Tatras (Fig. 3c; Nedelcu 1964; Rączkowska 2004, 2007).

Protalus ramparts several metres high, often arcuate, occur on the distal part of talus slopes in glacial cirques in all high-mountain ranges of the Carpathians (Fig. 3d; Niculescu and Nedelcu 1961; Sircu 1978; Florea 1998; Urdea 2000; Rączkowska 2007). These are mostly relict landforms, often covered by shrubs or meadows. However, even today they are weakly built up from the inside, with debris delivered by rockfall and sliding on snow patches. Their outer slopes are transformed by frost action and solifluction as well as other erosional processes. Active, smaller, not higher than 1.5 m, protalus ramparts were identified in the Tatras in front of perennial snow patches above 2000 m (Rączkowska 2007).

3.3 Permafrost Indicative Landforms

Rock glaciers constitute the best geomorphological indicator of the occurrence of permafrost conditions in European mountains (Barsch 1996). They are concentrated in Carpathian high-mountain massifs glaciated during the Pleistocene (Nemčok and Mahr 1974; Ichim 1978; Kotarba 1991/1992; Urdea 1992). There is an abundance of rock glaciers, and more than 300 were identified in the Tatras and the Southern Carpathians (Onaca et al. 2017a; Uxa and Mida 2017b). They were also found in the Rodna and the Călimani (Nagy et al. 2004). These are talus-derived rock glaciers and debris rock glaciers with a size of 0.02–0.03 and 0.06–0.08 km^{-2} respectively (Fig. 3b; Onaca et al. 2017a; Uxa and Mida 2017b). In the Tatras, debris rock glaciers were found to be substantially (four times) larger in size, while talus-derived rock glaciers predominate (Uxa and Mida 2017b). Regarding their shape, lobate rock glaciers, protalus lobes and tongue-shaped glaciers occur. They are located in the upper parts of glacial valleys, at the base of talus slopes. The elevations of the relict rock glacier fronts range from 1400 m in the Tatras to 1700 m in the Southern Carpathians. In the case of the intact rock glacier, fronts elevations are between 2000 m and 2090 m in the Tatras and Southern Carpathians, respectively (Urdea 1998; Onaca et al. 2017a; Uxa and Mida 2017b). Rock glaciers in the Carpathians are predominantly relict landforms (85% and 74% of forms in the Tatras and the Southern Carpathians, respectively), the YD being most frequently indicated as the period of their development (Kotarba 1991–1992; Urdea 1998; Kłapyta 2013; Uxa and Mida 2017b; Zasadni et al. 2020). Schmidt hammer relative age dating has proved that rock glaciers developed in the period 12.9–11.5 ka BP, during the final phase of cirque glacier recession (Kłapyta 2013; Onaca et al. 2017a; Popescu et al. 2017b). Urdea (1998) extended their formation in the Southern Carpathians back to the Older Dryas. On the other hand, Vespreameanu-Stroe et al. (2012), indicated that rock glaciers could developed there even in the Early Holocene, and Urdea (1998) suggested that their reactivation took place at an altitude >2050 m also during the LIA. In the Tatras, stabilisation of the rock glaciers took place in the Early Holocene (11.8–10.04 ka BP) was proved (Zasadni et al. 2020).

Recent geophysical, BTS and spring temperature monitoring data indicate that permafrost is likely to occur within a few rock glaciers in the Southern Carpathians (Urdea 1993; Onaca et al. 2017b). Displacements of a few cm yr^{-1} were recorded in three rock glaciers in Retezat using optical and radar remote sensing techniques (Necşoiu et al. 2016). Recent monitoring of ground surface temperature and BTS on a rock glacier in the Tatras also suggested the probable existence of permafrost within it (Uxa and Mida 2017a), but lichenometric data indicated no activity of rock glaciers at the present time and nor in the LIA (Kędzia 2014).

Thermokarst can be also an indicator of permafrost. This single phenomenon occurred at an altitude of ~ 2000 m and was related to thawing of modern permafrost within a moraine ridge in the Tatras following the extraordinarily hot summer of 2003 (Gądek et al. 2009).

Ice caves exist in the Apuseni (Persoiu and Onac 2019), in the Tatras (Siarzewski 1994; Szczygieł et al. 2015) and in the Lower Tatras and Slovenský ráj (Bella 2008) below the lower limit of the present-day periglacial zone and the elevation of the 0 °C isotherm. They are inherited from past cold periods. In the Scărişoara ice cave (Apuseni Mts.), ice older than 10 ka BP was found (Perșoiu and Onac 2019). Currently an ice mass loss is being observed, related to climatic warming (Rachlewicz and Szczuciński 2004; Perșoiu et al. 2021).

3.4 Frost Derived Landforms

Seasonal and diurnal freezing and thawing of the ground and frost heave act as a sorting mechanism leading to the development of patterned grounds (Ballantyne 2018), which are not very common in the Carpathians. However, they occur in both relict and active forms, as well as sorted and non-sorted. *Patterned grounds* are found on wide ridges and passes and in the bottom of glacial cirques or heads of U-shaped valleys in the Southern Carpathians and in the Tatras, where slope covers are relatively rich in fine material (e.g.; Ksandr 1954; Jahn 1958; Sekyra 1960; Niculescu and Nedelcu 1961; Lukniš 1973; Sircu 1978; Rączkowska 2007; Onaca 2017).

Relict *sorted polygons* (Fig. 4c) and circles are relatively large features with diameters of 2–6 m formed in past cold periods (e.g.; Jahn 1928; Lukniš 1973; Urdea et al. 2004; Onaca 2017). The relict *unsorted polygons* occur much more rarely (Lukniš 1973; Rączkowska 2007; Onaca 2017). Lichenometric data suggests their probable activity during the LIA. Currently, some of these landforms located on ridges or passes are being transformed by deflation (Rączkowska 2007). Additionally, in surface sediments on the bottoms of some Carpathian basins (~600–800 m), networks of fossil frost polygons were identified, originating probably from the Weichselian period when permafrost existed there (Ichim 1983).

Small patterned grounds are being developed nowadays by frost heave and sorting in glacial cirque bottoms in the Tatras at elevations >1940–2000 m (Rączkowska

a Sorted circle (Western Carpathians) Author: Zofia Rączkowska

b Sorted polygons (Western Carpathians) Author: Zofia Rączkowska

c Relict polygons (Western Carpathians) Author: Zofia Rączkowska

d Miniature circle (Western Carpathians) Author: Zofia Rączkowska

e Sorted strips (Western Carpathians) Author: Zofia Rączkowska

f Earth hummocks (Southern Carpathians) Author: Zofia Rączkowska

Fig. 4 Frost related landforms

2007), and in the Southern Carpathians at elevations >2300 m (Niculescu and Nedelcu 1961; Rączkowska 2007; Onaca et al. 2017b). Their activity is related to seasonal and diurnal FTC (Rączkowska 2007). Sorted circles (Fig. 4a) and polygons (Fig. 4b) develop in wet places such as the immediate surroundings of small lakes, depressions within the rocky floors of glacier cirques, or the foregrounds of perennial snow patches (e.g.; Ksandr 1954; Jahn 1958; Lukniš 1973; Rączkowska 2004, 2007). Their centre of finer-grained debris is surrounded by coarse debris up to a few tens of cm in diameter. Circles are smaller, 0.7–2.5 m in diameter, and the size of polygons reaches 4 m (Rączkowska 2007). The most widespread are miniature patterned grounds (Fig. 4d), a few tens cm in size. In such landforms, a circle of a small-debris, or turf usually surrounds their centre of fine-grained material. They mostly develop by needle ice activity, mainly in spring and autumn (Urdea 2000; Rączkowska 2004, 2007; Onaca et al. 2017b). Frost heave by *needle ice* reaches up to 13.5 cm in Retezat (Urdea 2000) and 0.3–0.4 cm in the Tatras (Midriak 1983), and displacement by needle ice totals 38 cm year^{-1} (Gerlach 1959).

On a sloping surface in the Tatras (15–35°), *sorted stripes* (Fig. 4e) develop, consisting of alternating stripes of debris on average 0.4 m wide and turf steps about 0.3 m high (Midriak 1983; Rączkowska 2007), similar to other high-mountain massifs in the Carpathians (Onaca et al. 2017b). They occur on ridges or near ridge slopes, but not very often. At present sorted stripes are an active landform affected by frost heave and frost creep supported by deflation (Midriak 1983; Rączkowska 2004, 2007).

On flat or gently sloping poorly drained areas above the upper timberline, particularly on wide passes and ridges in the high-mountains of the Carpathians, fields with tens to hundreds of *earth hummocks* (Fig. 4f) are common. These vegetation mounds of semi-circular or slightly elongated shape with a height of 0.1–0.6 m and a diameter of 0.3–1.2 m, are mostly composed of an organic-rich layer over a mineral core. They have been observed also in the Southern Carpathians (Urdea et al. 2004; Onaca 2017; Onaca et al. 2017b), in the Harghita (Schreiber 1994) and in the Tatras (Rączkowska 2007). Nowadays, these have become active landforms because of frost heave and needle ice activity (Gerlach 1972; Urdea et al. 2004; Rączkowska 2007; Onaca 2017; Onaca et al. 2017b). The frost heave rates in earth hummocks in the Tatras were found to be few mm per year (Table 1).

3.5 Periglacial Mass Movements and Derived Landforms

Solifluction could be considered the most common among the periglacial processes on the Carpathian slopes, both in past cold periods and at present, and it forms the dominant mechanism of sediment transport in the Carpathian periglacial environment.

Regolith and slope cover deposits in middle-mountain ranges of the Carpathians, mainly built of silt and clay deposits, have undergone displacement by solifluction to the foot of the slopes during cold stages of the Pleistocene (Klimaszewski 1971;

Henkiel 1972; Starkel 1986). In effect several tens meters of solifluction covers were formed at the foot of the slopes, often consisting of several layers of different ages, which are interbedded with fluvial deposits, frequently with organic material, when entering the river terrace. The post-Brörup or Odderade early Pleniglacial beds (75–58 ka BP; MIS 4), various Interpleniglacial beds (58–28 ka BP; MIS 3), late Pleniglacial (28–14 ka; MIS 2) and Late Glacial members were identified (Klimaszewski 1971; Starkel 2015). Solifluction was also active in past cold periods on high-mountain slopes, besides areas occupied by glaciers, as indicated by the presence of solifluction lobes (Jahn 1958), or the presence of lobe-like landforms within the blockfields (Henkiel 1972; Kłapyta 2006, 2020a, b).

Solifluction and other cryogenic processes, which took place in cold stages of the Pleistocene, caused the development of *cryopediments*. They are formed by the gently inclined surfaces at the foot of valley sides, or of the marginal slopes of mountains that were identified at many localities in the middle mountains of the Western Carpathians (Klimaszewski 1971; Czudek and Demek 1976; Starkel 2015). Although the last studies in the Moravian-Silesian Beskids showed cryopediment being the levelled surface of old landslides from the MIS 3 period (51–56 to < 36 ka cal. BP; Pánek et al. 2014).

At present, solifluction is the most widespread process on debris mantled slopes above the upper timberline in the high mountains of the Carpathians (Nagy et al. 2004; Rączkowska 2007; Onaca 2017; Onaca et al. 2017b). It was also identified on the uppermost slopes in some middle mountains (Pękala 1969; Henkiel 1972; Kravčuk 2008; Kłapyta et al. 2020). This has produced a wide range of solifluction landforms that are found there such as: lobes, terraces, garlands, steps (*terracettes*), and ploughing boulders identified in the Tatras above 1500 m and in the Southern Carpathians above 1600 m (Urdea 2000; Rączkowska 2007; Onaca et al. 2017b).

Monitoring of present-day solifluction movement has shown annual rates from <1 cm to >15 cm (Table 1). Solifluction is now driven by daily and seasonal FTC as well as the texture and moisture of the slope covers, but also depends on slope aspect (Urdea 2000; Rączkowska 2007; Onaca 2017; Onaca et al. 2017b).

Solifluction lobes are longer (a few to ten metres) and wider (a few metres) and their fronts are around 1 m high (Fig. 5a; Rączkowska 2007). These are mostly *turf-banked lobes*, situated near ridges, although often with a bare debris surface modified by different frost processes. *Stone-banked lobes* are inactive at present. Turf-banked lobes are mostly found to be active. Monitoring showed displacement of their surface layer from several to tens of millimetres per year in the Southern Carpathians (Urdea 2000; Onaca 2017). Stone lobes of free solifluction occur rather rarely and are small—a few metres long (Rączkowska 2007; Onaca et al. 2017b).

The specific solifluction landforms are transversal *steps*, a few tens metres long and up to 1 m high, with flat surfaces few metres wide known as terraces or garlands (Fig. 5b). Mostly these are separated by turf-banked steps. Some are even completely vegetated with alpine meadows. They develop on relatively gentle summit surfaces, with slope covers containing a large proportion of fine-grained particles. At present they seem to show only weak activity (Niculescu and Nedelcu 1961; Florea 1998; Rączkowska 2007).

Solifluction lobe (Western Carpathians)
Author: Zofia Rączkowska

Solifluction terraces (Southern Carpathians)
Author: Zofia Rączkowska

Solifluction terracete (Western Carpathians)
Author: Zofia Rączkowska

Ploughing boluder (Western Carpathians)
Author: Zofia Rączkowska

Fig. 5 Periglacial mass movements

The most common and widespread active features are solifluction steps (*terracettes*) of <1 m in size, flat surfaces often with bare debris, and turf-banked fronts (Fig. 5c; Rączkowska 2004, 2007).

Solifluction is also the agent of movement of *ploughing boulders* (Fig. 5d), which are currently widespread on high-mountain slopes (e.g.; Kotarba 1976; Sircu 1978; Urdea 2000; Rusu 2002; Rączkowska 2007; Onaca et al. 2017b). Their mean rate of movement is 1.3–2.7 cm yr^{-1} in the Tatras and 6.6–11.1 cm yr^{-1} in the Southern Carpathians (Table 1), and is twice as fast in areas built of carbonate rocks than in those of crystalline rocks. As a result, characteristic mounds develop in front of the boulders with hollows behind them.

During the Late Glacial-Holocene transition (~11.7 ka), permafrost degradation combined with the establishment of groundwater circulation caused numerous large *landslides* in the Carpathian middle mountains (Margielewski 2006; Pánek et al. 2014). In the high mountains, the occurrence of *slope failures* and large *rock avalanches* in the paraglacial period was proven by cosmic-ray exposure dating (Pánek et al. 2016). At present, landslides are not specific of the periglacial environment in the Carpathians, but are among the most significant factors transforming slopes of the middle mountains.

3.6 Nival and Aeolian Landforms

Processes of nival erosion related to long lasting or perennial snow patches lead to the fragmentation of the slopes and the development of the *nival hollows* in high-mountain massifs of the Carpathians (Fig. 3f; Rączkowska 1993, 1995, 2007; Kłapyta et al. 2020). These landforms even attain a few hundred metres in length and a few tens metres in width. *Boulder pavements* have developed at the bottom of most of them due to the effect of snow pressure and creep. Field experiments in the Tatras showed that the edge of the nival hollow developed by debris mantled slopes retreat at a rate of 1–5 cm yr^{-1}, while meltwater creates 20 cm deep gullies within the hollow bottom, or on the slope below. On talus slopes, nival hollows develop as an effect of accretion on a snow free slope, which doesn't happen on a slope section under a snow patch. Rates of accumulation, however, are less than 0.1 mm yr^{-1} (Fig. 3e; Table 1).

Wind action in the periglacial environment of the Pleistocene cold phases was recorded by *loess deposits* and *ventifacts* (Gerlach et al. 1991). The silty-sandy loess, of a few to up to ten metres thick, covers low elevated ridges and slopes (at altitudes 300–400 m) along the northern and southern rim of the Western and the Eastern Carpathians and in some intramountain basins (Pécsi 1967; Łańczont 1995; Haase et al. 2007).

At present, wind action in high mountains contributes to the development of frost or solifluction landforms. On windward slopes with a not very dense vegetation cover, *deflation* together with frost processes causes the development of small *scars* with bare soil in the bottom, and even larger *deflation depressions* (Lukniš 1973; Midriak 1983; Rączkowska 2007).

4 The Significance of Periglacial Dynamics in the Carpathians within the European Context

The more than a hundred years of research into the periglacial relief of the Carpathians has allowed the distribution of the periglacial features and landforms to be relatively well known. The studies have shown that such landforms are present both in the landscapes of the middle and high mountain landscapes. However, the nature of the periglacial relief in these areas is very varied. In the high mountains, there are currently active, inactive and relict forms, while in the middle mountains, only periglacial relict forms occur. The main extant landforms are those originated from frost weathering and mass movements (Fig. 6). Other forms shaped by the periglacial transformation of the middle mountains in the Pleistocene include aeolian and solifluction slope covers. The above differences in relief stem from the change in the extent of the periglacial environment during cold Pleistocene stages and at present.

In the Pleistocene, a compact periglacial zone covered the entire Carpathian region, although the conditions varied in time and space, especially regarding the

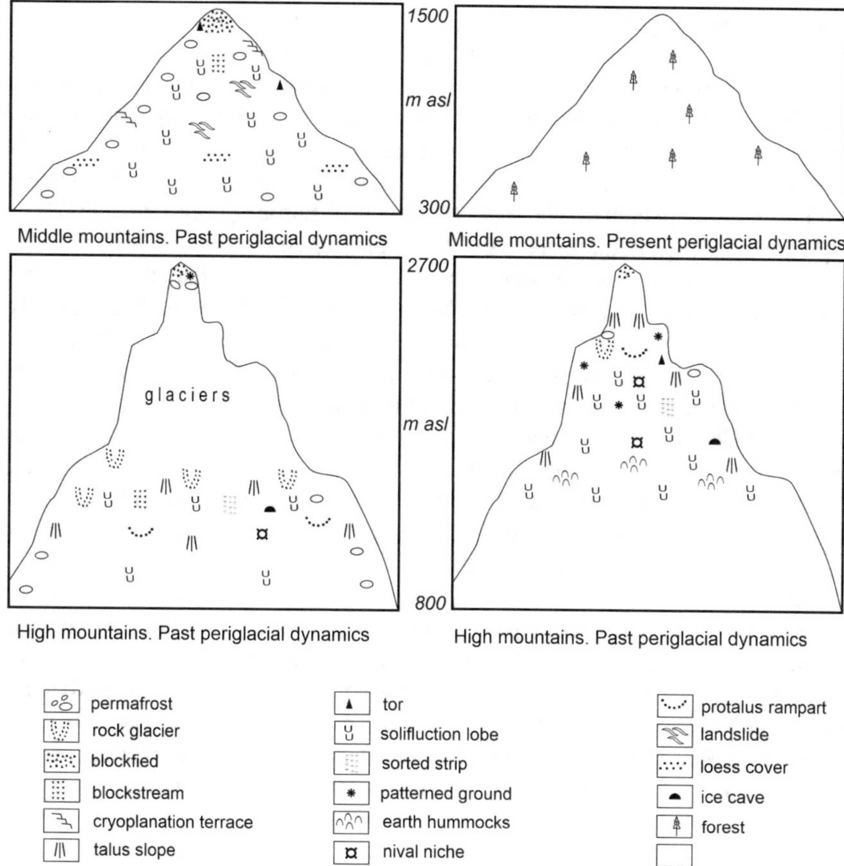

Fig. 6 Scheme of the past and present periglacial phenomena in the Carpathians

presence of permafrost, and were pronouncedly different during glacial and interglacial periods. Currently, the periglacial (cold climate) zone is discontinuous and has the character of patches located in the highest massifs with alpine relief, mainly in the Western and Southern Carpathians. In addition, its vertical reach has shrunk to areas above 1500–1800 m, which means that, currently, this zone is nearly completely absent in the middle mountains.

In the areas currently covered by the periglacial zone, periglacial phenomena and processes are active, which is confirmed and parameterised, *inter alia*, through monitoring studies. However, the intensity of the periglacial processes is low, and the duration of their activity throughout the year is limited, mainly to spring and autumn (Rączkowska 2007; Onaca et al. 2017b). As a result, only some types of landforms of periglacial relief develop, especially the frost-derived ones. Moreover, the periglacial features are often not fully developed and their sizes are small. Permafrost-indicative landforms, including numerous rock glaciers, do exist, but they are inactive or relict, despite the residual presence of permafrost (Fig. 3). Active periglacial landforms are found (present) above the upper timberline (Rączkowska 2007; Onaca et al. 2017b), but they are prevalent above the altitude of 1850–1900 m, which roughly corresponds to the 0 °C isotherm elevation. Their distribution is controlled by the local topographic and soil conditions, which is another factor adding to the regional variations of the contemporary periglacial relief in the Carpathians. In essence, such relief is much less developed here than in other high mountains of Europe (Rączkowska 2007).

The increasing amount of data on the chronology of the periglacial landforms in recent decades, mainly in the high mountains, indicates that the end of the Last Glacial Cycle was the period when the relief developed most intensively. Ever since, the periglacial dynamics in the Carpathians have been declining, with breaks in the cold periods of the Holocene, especially in the LIA, when certain phenomena/processes intensified, to become significantly limited today. This is conditioned by a changing climate, which means that we can expect a further decline in periglacial processes, although the observable instability of the snow cover and of the thermal regime (Gądek 2014; Micu et al. 2015) may, paradoxically, foster the activity of periglacial processes, which is associated with both diurnal and seasonal temperature fluctuations in the Carpathians.

Although monitoring studies and a number of new research methods have provided valuable insight into periglacial processes and relief, some processes, e.g. physical or chemical weathering, are still only known to a limited extent. In addition, the dynamics and the chronology of periglaciation have been unevenly explored on the long Carpathians arc. The current very low activity of periglacial processes and relief makes periglacial landforms sensitive to climate change and means that they are an attractive subject for research in terms of the various climate scenarios.

References

Alexandrowicz Z (1970) Skałki piaskowcowe w okolicach Ciężkowic nad Biała. Ochrona Przyrody 35:281–235

Ardelean AC, Alexandru L, Onaca A, Urdea P, Şerban RD, Sîrbu F (2015) A first estimate of permafrost distribution from BTS measurements in the Romanian Carpathians (Retezat Mountains). Géomorphol Relief Process Environ 21(4):297–312. https://doi.org/10.4000/geomorphologie.11131

Ardelean AC, Onaca A, Urdea P (2017) Quantifying postglacial sediment storage and denudation rates in a small alpine catchment of the Făgăraş Mountains (Romania). Sci Total Environ 599–600:1756–1767. https://doi.org/10.1016/j.scitotenv.2017.05.131

Ballantyne CK (2018) Periglacial Geomorphology. Wiley Blackwell, 472 p

Baranowski J, Kędzia S, Rączkowska Z (2004) Soil freezing and its relation to slow soil movements on Alpine slopes (of the Tatra Mountains, Poland). Ann Univ De Vest Din Timisoara Geogr 14:169–179

Barsch D (1996) Rock Glaciers: indicators for the present and former geoecology in high mountain environments. Springer, Berlin and New York, p 331

Baumgart-Kotarba M (1971) Cryonival features of flysch ridge crests in the Carpathians. Stud Geomorphol Carpatho Balc 5:199–211

Bella P (2008) Geographical distribution of ice-filled caves in Slovakia. In: Kadebskaya O, Mauvludov B, Pyatunin M (eds) 3th International workshop on ice caves, Proceedings. Kungur, p 33–37

Büdel J (1951) Die Klimazonen des Eiszeitalters. Eiszeit Gegenw 1:16–26

Czudek T, Demek J (1976) The slopes of Central Moravian Carpathians: periglacial or temperate? Stud Geomorphol Carpatho Balc 10:3–14

de Martonne E (1907) Recherches sur l'évolution morphologique des Alpes de Transylvanie (Karpates Méridionales). Revue De Géographie Annuelle I(1906–1907):286

Demek J (1969) Cryoplanation terraces, their geomorphological distribution, genesis and development. Rozpravy ČSAV, Academia, Praha, 80 p

Demek J (1984) Fossil periglacial phenomena in Czechoslovakia and their paleoclimatic evaluation. Scripta Facultatis Scientiarum Naturalium, Universitatis Purkynianae Brunensis 14:343–348

Dobiński W (1998) Permafrost occurrences in the alpine zone of the Tatra Mountains, Poland. In: Proceedings of the 7th International Conference Permarost. Yellowknife, Quebec City, Canada, p 231–237

Dobiński W (2004) Wieloletnia zmarzlina w Tatrach: geneza, cechy, ewolucja. Prz Geogr 76(3):327–343

Dobiński W (2005) Permafrost of the Carpathian and Balkan Mountains, Eastern and Southern Europe. Permafr Periglac Process 16:395–398

Dobiński W, Żogała B, Wziętek K, Litwin L (2008) Results of geophysical surveys on Kasprowy Wierch, the Tatra Mountains, Poland. In: Hauck C, Kneisel C (eds) Applied geophysics in periglacial environments. Cambridge University Press, Cambridge, pp 126–136

Florea M (1998) Munti Fagarasului. Studiu geomorfologic. Editura Foton, Brasov, 114 p (in Romanian)

French HM (2007) The periglacial environment. Wiley, 544 p

Gądek B (2014) Climatic sensitivity of the non-glaciated mountains cryosphere (Tatra Mts., Poland and Slovakia). Glob Planet Change 121:1–8

Gądek B, Grabiec M (2008) Glacial ice and permafrost distribution in the Medena kotlina (Slovak Tatras): mapped with application of GPR and GST measurements. Stud Geomorphol Carpatho-Balc 42:5–2

Gądek B, Kędzia S (2008) Winter surface thermal regimes in the zone of sporadic discontinuous permafrost, Tatra Mountains (Poland and Slovakia). Permafr Periglac Process 19:315–321

Gądek B, Leszkiewicz J (2012) Impact of climate warming on the ground surface temperature in the sporadic permafrost zone of the Tatra Mountains, Poland and Slovakia. Cold Reg Sci Technol 79–80:75–83. https://doi.org/10.1016/j.coldregions.2012.03.006

Gądek B, Rączkowska Z, Żogała B (2009) Debris slope morphodynamics as an permafrost indicator in its sporadic occurrence zone, High Tatra Mts., Slovakia. Z Geomorph NF53 (Suppl2):79–100

Gądek B, Grabiec M, Kędzia S, Rączkowska Z (2016) Reflection of climate changes in the structure and morphodynamics of talus slopes (the Tatra Mountains, Poland). Geomorphology 263:39–49

Gądek B, Kaczka RJ, Rączkowska Z, Rojan E, Casteller A, Bebi P (2017) Snow avalanche activity in Żleb Żandarmerii in a time of climate change (Tatra Mts., Poland). CATENA 158:201–212. https://doi.org/10.1016/j.catena.2017.07.005

Gądek B, Szypuła B (2015) Contemporary cryosphere. Map in scale 1:250,000. In: Dąbrowska K, Guizk M (eds) Atlas of the Tatra Mountains-Abiotic Nature. Tatrzański Park Nar., Zakopane

Gerlach T (1959) Lód włóknisty i jego rola w przemieszczaniu pokrywy zwietrzelinowej w Tatrach (Needle ice and its role in the displacement of the cover of waste material in the Tatra Mts.). Prz Geogr. 31:590–605 (in Polish)

Gerlach T (1972) Contribution à la connaissance du dévelopment actuel des buttes gazonnées (thufurs) dans les Tatras polonaises. In: Processus périglaciaires, étudiés sur le terrain. Les Congres et Colloques de l'Université de Liege 67:57–74

Gerlach T, Krysowska-Iwaszkiewicz M, Szczepanek K, Alexandrowicz SW (1991) Karpacka odmiana lessów w Humniskach koło Brzozowa na Pogórzu Dynowskim w polskich Karpatach fliszowych. Kwartalnik AGH, Geologia 17:193–219 (in Polish)

Gerasimov IP, Velichko AA (eds) (1982) Paleogeografiya Evropy za poslednye sto tysiac let: Atlas-monografiya (Paleogeography of Europe during the last one hundred thousand years: Atlas-monograph). Nauka, Moscow (in Russian)

Goran C (2012) Karst In: Lóczy D, Stankoviansky M, Kotarba A (eds) Recent landform evolution. The Carpatho-Balkan-Dinaric Region. Springer, Dordrecht/Heidelberg/London/New York, p 362–265. https://doi.org/10.1007/978-94-007-2448-8_10

Haas J (2012) Geological and tectonic setting. In: Lóczy D, Stankoviansky M, Kotarba A (eds) Recent landform evolution. The Carpatho-Balkan-Dinaric Region. Springer, Dordrecht/Heidelberg/London/New York, p 3–18. https://doi.org/10.1007/978-94-007-2448-8_1

Haase D, Fink J, Haase G, Ruske R, Pecsi M, Richter H, Altermann M, Jager KD (2007) Loess in Europe—its spatial distribution based on a European Loess Map, scale 1:2,500,000. Quat Sci Rev 26(9–10):1301–1312. https://doi.org/10.1016/j.quascirev.2007.02.003

Henkiel A (1972) Soliflukcja w polskich Karpatach. Czas Geogr 43:295–305 (in Polish)

Hess M (1963) Problems of the perinival climate in the Tatra Mountains. Bull Pol Acad Sc Ser Geol Geogr 1(4):247–251

Hess M (1965) Piętra klimatyczne w polskich Karpatach Zachodnich. Zesz Nauk Uniw Jagiell Pr Geogr 11:1–267 (in Polish)

Ichim I (1978) Preliminary observations on the rock glaciers phenomenon in the Romanian Carpathians. Revue Roumaine Géol Géophys Géogr, Serie Géographie 23:295–299

Ichim I (1983) Relieful periglaciar. In: Geografia Romåniei, Vol. I. Geografia fizicâ. Bucuresti, Editura Academiei, p 141–144, (in Romanian)

Jahn A (1958) Mikrorelief peryglacjalny Tatr i Babiej Góry. Biul Peryglac 6:57–80 (in Polish)

Kern Z, Balogh D, Nagy B (2004) Investigations for the actual elevation of the mountain permafrost zone on postglacial landforms in the head of Lăpuşnicu Mare Valley, and the history of deglaciation of Ana Lake—Judele Peak region, Retezat Mountains, Romania. Ann Univ De Vest Din Timisoara Geogr 14:119–132

Kędzia S (2004) Klimatyczne i topograficzne uwarunkowania występowania wieloletniej zmarzliny w Tatrach Wysokich (na przykładzie Koziej Dolinki). PhD thesis, Inst Geogr Przestrz Zagosp PAN, Warszawa, 124 p (in Polish)

Kędzia S (2014) Are there any active rock glaciers in the Tatra Mountains? Stud Geomorphol Carpatho Balc 48:5–16. https://doi.org/10.1515/sgcb-2015-0001

Klimaszewski M (1948) Polskie Karpaty Zachodnie w okresie dyluwialnym. Prace Wrocławskiego Tow Nauk B7:1–233 (in Polish)

Klimaszewski M (1971) The effect of solifluction processes on the development of mountain slopes in the Beskidy (Flysch Carpathians). Folia Quat 28:3–18

Kłapyta P (2006) Rzeźba południowych stoków Czarnohory (Karpaty Ukraińskie) pomiędzy Howerlą a Turkułem. In: Troll M (ed) Czarnohora. Przyroda i człowiek. Instytut Geografii i Gospodarki Przestrzennej UJ, Kraków, p 27–46 (in Polish)

Kłapyta P (2011) Relative surface dating of rock glacier systems in the Žiarska Valley, the Western Tatra Mountains, Slovakia. Stud Geomorphol Carpatho Balc 45:89–106

Kłapyta P (2013) Application of Schmidt hammer relative age dating to Late Pleistocene moraines and rock glaciers in the Western Tatra Mountains, Slovakia. CATENA 111:104–121. https://doi.org/10.1016/j.catena.2013.07.004

Kłapyta P (2020a) Geomorphology of the high-elevated flysch range—Mt. Babia Góra Massif (Western Carpathians). J Maps 16(2):689–701. https://doi.org/10.1080/17445647.2020.1800530

Kłapyta P, Dubis L, Krzemień K, Gorczyca E, Krąż P (2020b) Rzeźba i współczesne procesy morfogenetyczne wysokogórskiego masywu Świdowca (Karpaty Wschodnie, Ukraina). Roczniki Bieszczadzkie 28:159–188 (in Polish)

Kłapyta P, Zasadni J, Dubis L, Świąder A (2021) Glaciation in the highest parts of the Ukrainian Carpathians (Chornohora and Svydovets massifs) during the local last glacial maximum. CATENA 203:105346. https://doi.org/10.1016/j.catena.2021.105346

Kondracki J (1989) Karpaty. Wydawnictwo Szkolne i Pedagogiczne, Warszawa, 261 p (in Polish)

Kotarba A (1972) Powierzchniowa denudacja chemiczna w wapienno-dolomitowych Tatrach Zachodnich. Pr Geogr IG PAN 96:1–119 (in Polish)
Kotarba A (1976) Współczesne modelowanie węglanowych stoków wysokogórskich na przykładzie Czerwonych Wierchów w Tatrach Zachodnich. Pr Geogr IGiPZ PAN 120:1–128 (in Polish)
Kotarba A (1991–1992) Reliktowe lodowce gruzowe jako element deglacjacji Tatr Wysokich. Stud Geomorphol Carpatho Balc 25–26:133–150 (in Polish)
Kotarba A (1995) Rapid mass wasting over the last 500 years in the High Tatra Mountains. Quest Geogr Spec Issue 4:177–183
Kotarba A (1996) Sedimentation rates in the High Tatra lakes during the Holocene - geomorphic interpretation. Stud Geomorphol Carpatho Balc 30:51–61
Kotarba A Kaszowski L, Krzemień K (1987) High-mountain Denudational Systemin the Polish Tatra Mountains. Geogr Stud IGiPZ PAN, Spec issue 3, 1–106
Kotarba A, Pech P (2002) The recent evolution of talus slopes in the High Tatra Mountains (with the Pańszczyca valley as example). Stud Geomorphol Carpatho Balc 36:69–76
Kotarba A, Starkel L (1972) Holocene morfogenetic altitudinal zones in the Carpathians. Stud Geomorphol Carpatho-Balc 6:21–35
Kravčuk A (2008) Geomorfologia Polonińs′ko-Čarnogirs′kih Karpat. Wyd. Centr LNU im. I. Franka, Lviv, 187 p (in Ukrainian)
Křížek M (2000) Frost-riven cliffs and cryoplanation terraces in the Hostýnské vrchy Hills (east Moravia, Czech Republic). Acta Univ Carol Geogr Suppl 35:239–245
Křížek M (2001) The Quaternary sculpturing of sandstones in the Rusavská hornatina Mts. Acta Univ Carol Geogr 36:99–109
Ksandr J (1954) Mrazové půdní formy v Tatrách. Ochrana Přírody 9(4):97–108
Lempa M, Kaczka RJ, Rączkowska Z (2015) Morphological and morphometrical analyses reveal the avalanche influence over the talus cones in the Rybi Potok Valley, Tatra Mountains. Stud Geomorphol Carpatho Balc 49:15–33
Lempa M, Kaczka RJ, Rączkowska Z, Janecka K (2016) Combining tree-ring dating and geomorphological analyses in the reconstruction of spatial patterns of the runout zone of snow avalanches, Rybi Potok Valley, Tatra Mountains (Poland). Geogr Pol 89(1):31–45
Lubera E (2011) Tors of the Chochołowska Valley (Western Tatra Mountains). Geogr Pol 84(1):75–93
Lubera E (2016) Wietrzenie mrozowe i odpadanie ze ścian skalnych w obszarze wysokogórskim, na przykładzie Tatr Zachodnich, PhD Thesis, Inst Geogr i Gosp Przestrz Uniw Jagiell, Kraków, 157 p
Lukniš M (1973) Reliéf Vysokých Tatier a ich predpolia. Veda, Bratislava, 375 p (in Slovak)
Łanczont M (1995) Regional stratigraphy and lithology of the Carpathians loess deposits in Przemyśl environs. Bull Pol Acad Sc Earth Sci 43(1):43–56
Łoziński W (1912) Die periglaziale Fazies der mechanischen Verwitterung. In: Comptes Rendus. XI Congres Internationale Geologie, Stockholm. p 1039–1053
Makos M (2015) Deglaciation of the High Tatra Mountains. Cuadernos De Investigación Geográfica 41(2):317–335
Margielewski W (2006) Records of the Late Glacial-Holocene palaeoenvironmental changes in landslide forms and deposits in the Beskid Makowski and Beskid Wyspowy Mts. Area (Polish Outer Carpathians). Folia Quat 76:1–149
Matsuoka N (2011) Climate and material controls on periglacial soil processes: toward improving periglacial climate indicators. Quatern Res 75:356–365
Meseșan F, Man TC, Pop OT, Gavrilă IG (2019) Reconstructing snow-avalanche extent using remote sensing and dendrogeomorphology in Parâng Mountains. Cold Reg Sci Technol 157:97–109
Micalevich-Velcea V (1961) Masivul Bucegi. Studiu geomorfologic. Edit, Academiei, Bucureşti (in Romanian), 151 p
Micu DM, Dumitrescu A, Cheval S, Birsan MV (2015) Climate of the Romanian Carpathians. Variability and Trends. Springer International Publishing Switzerland, Springer Cham Heidelberg New York Dordrecht London, 212 p. https://doi.org/10.1007/978-3-319-02886-6

Midriak R (1983) Morfogenéza povrchu vysokých pohorí. Veda, Bratislava, 516 p (in Slovak)
Mojski JE (2005) Ziemie polskie w czwartorzędzie. Zarys morfogenezy. Państwowy Instytut Geologiczny, Warszawa 404 p
Mościcki JW, Kędzia S (2001) Investigation of mountain permafrost in the Kozia Dolinka valley. Tatra Mountains, Poland, Norsk Geogr Tidsskr 55:235–240
Mościcki J (2008) Temperature regime on northern slopes of Hala Gąsienicowa in the Polish Tatra Mountains and its relationship to permafrost. Stud Geomorphol Carpatho Balc 42:23–40
Nagy B, Kern Z, Bugya É, Kohán B (2004) Investigation of postglacial surface-evolution in the alpine region of the Călimani Mountains—with an outlook to the cirque region of the Rodnei Mountains. Ann Univ De Vest Din Timisoara Geogr 14:101–118
Necşoiu M, Onaca A, Wigginton S, Urdea P (2016) Rock glacier dynamics in Southern Carpathian Mountains from high-resolution optical and multi-temporal SAR satellite imagery. Remote Sens Environ 177:21–36
Nedelcu E (1964) Sur la cryo-nivation actuelle dans les Carpates Méridionales entre les rivieres Ialomiţa et Olt. Rev Roum Géol, Géophys Géogr Géographie 8:121–128
Nemčok A, Mahr T (1974) Kamenné ladovce v Tatrách. Geografický Časopis 36(4):359–374 (in Slovak)
Niculescu G (1965) Munţii Godeanu. Studiu geomorfologic. Edit, Academiei, Bucureşti (in Romanian)
Niculescu G, Nedelcu E (1961) Contribuţii la studiul microreliefului crionival din zona înaltă a munţilor Retezat–Godeanu–Ţarcu şi Făgăraş-Iezer. Probleme De Geografie 8:87–121
Niedźwiedź T (2012) Climate. In: Lóczy D, Stankoviansky M, Kotarba A (eds) Recent landform evolution. The Carpatho-Balkan-Dinaric Region, Springer geography. Springer, Dordrecht/Heidelberg/London/New York, 19–29, https://doi.org/10.1007/978-94-007-2448-8_2
Onaca A (2017) Procese și forme periglaciare din Carpații Meridionali. Abordare geomorfologică și geofizică. Editura Universităţii de Vest, Timișoara, 264 p (in Romanian)
Onaca A, Urdea P, Ardelean A (2013a) Internal structure and permafrost characteristics of the rock glaciers of Southern Carpathians (Romania) assessed by geoelectrical soundings and thermal monitoring. Geogr Ann Ser A Phys Geogr 95:249–266
Onaca A, Urdea P, Ardelean A, Șerban R (2013b) Assessment of internal structure of periglacial landforms from Southern Carpathians (Romania) using DC resistivity tomography. Carpathian J Earth Environ Sci 8(2):113–122
Onaca A, Ardelean AC, Urdea P, Ardelean F, Sîrbu F (2015) Detection of mountain permafrost by combining conventional geophysical methods and thermal monitoring in the Retezat Mountains, Romania. Cold Reg Sci Technol 119:111–123
Onaca AL, Ardelean AC, Urdea P, Ardelean F, Sărăs A (2016) Genetic typologies of talus deposits derived from GPR measurements in the alpine environment of the Făgăras. Mountains. Carpathian J. Earth Environ. Sci. 11(2):609–616
Onaca A, Ardelean F, Urdea P, Magori B (2017a) Southern Carpathian rock glaciers: Inventory, distribution and environmental controlling factors. Geomorphology 293:391–404. https://doi.org/10.1016/j.geomorph.2016.03.032
Onaca A, Urdea P, Ardelean AC, Șerban R, Ardelean F (2017b) Present-Day Periglacial Processes in the Alpine Zone. In: Rădoane M, Vespremeanu-Stroe A (eds) Landform Dynamics and Evolution in Romania. Springer Geography, p 147–176. https://doi.org/10.1007/978-3-319-32589-7_7
Pánek T, Engel Z, Mentlík P, Braucher R, Břežný M, Škarpich V, Zondervan A (2016) Cosmogenic age constraints on post-LGM catastrophic rock slope failures in the Tatra Mountains (Western Carpathians). CATENA 138:52–67. https://doi.org/10.1016/j.catena.2015.11.005
Pánek T, Hartvich F, Jankovská V, Klimeš J, Tábořík P, Bubík M, Smolková V, Hradecký J (2014) Large Late Pleistocene landslides from the marginal slope of the Flysch Carpathians. Landslides 11:981–992
Pécsi M (1967) Relationship between slope geomorphology and Quaternary slope sedimentation. Acta Geol 11(1–3):307–321

Perşoiu A, Onac BP (2019) Ice caves in Romania. In: Ponta GML, Onac BP (eds) Cave and karst systems of Romania. Springer, Switzerland, p 455–466. https://doi.org/10.1007/978-3-319-90747-5-52

Perşoiu A, Buzjak N, Onaca A, Pennos C, Sotiriadis Y, Ionita M, Zachariadis S, Styllas M, Kosutnik J, Hegyi A, Butorac V (2021) Record summer rains in 2019 led to massive loss of surface and cave ice in SE Europe. Cryosphere 15:2383–2399

Pękala K (1969) Rumowiska skalne i współczesne procesy geomorfologiczne w Bieszczadach Zachodnich. Annales UMCS Sectio B 24:47–98 (in Polish)

Pop OT, Munteanu A, Meseşan F, Gavrilă I-G, Timofte C, Holobâcă I-H (2018) Tree-ring-based reconstruction of high-magnitude snow avalanches in Piatra Craiului Mountains (Southern Carpathians, Romania). Geogr Ann Ser A Phys Geogr 100:99–115. https://doi.org/10.1080/04353676.2017.1405715

Popescu R, Onaca A, Urdea P, Vespremeanu-Stroe A (2017a) Spatial Distribution and Main Characteristics of Alpine Permafrost from Southern Carpathians, Romania. In: Rădoane M, Vespremeanu-Stroe A (eds) Landform Dynamics and Evolution in Romania. Springer Geography, p 117–146. https://doi.org/10.1007/978-3-319-32589-7

Popescu R, Urdea P, Vespremeanu-Stroe A (2017b) Deglaciation history of high massifs from the Romanian Carpathains: Towards an Integrated View. In: Rădoane M, Vespremeanu-Stroe A (eds) Landform Dynamics and Evolution in Romania. Springer Geography, p 87–116. https://doi.org/10.1007/978-3-319-32589-7_7

Popescu R, Vespremeanu-Stroe A, Onaca A, Cruceru N (2015) Permafrost research in the granitic massifs of Southern Carpathians (Parâng Mountains). Z Geomorph 59(1):1–20

Popescu R, Vespremeanu-Stroe A, Onaca A, Vasile M, Cruceru N, Pop O (2017c) Low-altitude permafrost research in an overcooled talus slope–rock glacier system in the Romanian Carpathians (Detunata Goală, Apuseni Mountains). Geomorphology 295:840–854. https://doi.org/10.1016/j.geomorph.2017.07.029

Puţan R (2015) Analiza unor procese și forme periglaciare din bazinul superior al văii Capra din Munții Făgăraș. Unpublised PhD Thesis, Universitatea de Vest din Timişoara, 36 p (in Romanian)

Rachlewicz G, Szczuciński W (2004) Seasonal and decadal ice mass balance change in the ice cave Jaskinia Lodowa w Ciemniaku, the Tatra Mountains, Poland. Theor Appl Karstol 17:11–18

Rączkowska Z (1993) Ilościowe wskaźniki niwacji w Tatrach Wysokich. Dokum Geogr Inst Geogr Przestrz Zagospod PAN 4–5:63–81 (in Polish)

Rączkowska Z (1995) Nivation in the High Tatras, Poland. Geogr Ann Ser A Phys Geogr 77:251–258

Rączkowska Z (2004) Considerations on periglacial landforms and slope morphodynamic in periglacial zone of Tatra Mountains. Ann Univ De Vest Din Timisoara Geogr 19:35–50

Rączkowska Z (2007) Współczesna rzeźba peryglacjalna wysokich gór Europy. Pr Geogr Inst Geogr Przestrz Zagospod PAN 212:1–252 (in Polish)

Rączkowska Z, Cebulski J (2020) Morfodynamika stoków gruzowych Tatr w świetle wieloletniego monitoringu metodą skaningu laserowego. In: VI Konferencja Przyroda Tatrzańskiego Parku Narodowego a Człowiek. Zmiany w Tatrach—zagrożenia istniejące i potencjalne. Tatrzański Park Nar., Zakopane, p 9 (in Polish)

Rączkowska Z, Długosz M, Rojan E (2016) Geomorphological conditions of snow avalanches in the Tatra Mountains. Z Geomorph 60(4):285–297. https://doi.org/10.1127/zfg/2016/0289

Rehman A (1895) Ziemie dawnej Polski i sąsiednich krajów słowiańskich, opisane pod względem fizyczno-geograficznym. I. Karpaty, p 1–657 (in Polish)

Rusu C (2002) Masivul Rarău. Studiu de geografie fizică. Edit, Academiei Române, Bucureşti (in Romanian)

Ruszkiczay-Rüdiger Z, Kern Z, Urdea P, Braucher R, Balazs M, Schimmelpfennig I, Aster Team (2016) Revised deglaciation history of the Pietrele-Stanisoara glacial complex, Retezat Mts, Southern Carpathians, Romania. Quat Int 415:2016–2029

Schreiber WE (1974) Das Periglazialrelief des Harghita-Gebirges. Revue Roumaine Géol Géophys Géogr, Serie Géographie 18:179–187

Schreiber WE (1994) Munţii Harghita. Studiu geomorfologic. Edit, Academiei, Bucureşti (in Romanian)
Sekyra J (1960) Působení Mrazu Na Půdu. Geotechnica 27:1–64
Senderak K, Kondracka M, Gądek B (2019) Postglacial talus slope development imaged by the ERT method: comparison of slopes from SW Spitsbergen, Norway and Tatra Mountains, Poland. Open Geosciences 11:1084–1097. https://doi.org/10.1515/geo-2019-0084
Şerban RD, Onaca A, Şerban M, Urdea P (2019) Block stream characteristics in Southern Carpathians (Romania). CATENA 178:20–31. https://doi.org/10.1016/j.catena.2019.03.003
Siarzewski W (1994) Jaskinia Lodowa w Ciemniaku. In: Grodzicki J (ed), Jaskinie Tatrzańskiego Parku Narodowego, 5. Pol. Tow. Przyjaciół Nauk o Ziemi—Tatrzański Park Nar., Warszawa, p 142–153 (in Polish)
Sîrcu I (1976) Frost heaving stones in the Godeanu Mountains. Anal. şt. Univ."Al. I. Cuza" Iaşi, secţ. 2 Geologie-Geografie 21:126
Sîrcu I (1978) Munţii Rodnei. Studiu morfogeografic. Edit, Academiei, Bucureşti (in Romanian)
Šilhán K, Brázdil R, Pánek T, Dobrovolný P, Kašičková L, Tolasz R, Turský O, Václavek (2011) Evaluation of meteorological controls of reconstructed rockfall activity in the Czech Flysch Carpathians. Earth Surf Process Landf 36:1898–1909
Starkel L (1960) Periglacial covers in the Beskid Wyspowy (Carpathians). Biul Peryglac 8:155–170
Starkel L (1972) Charakterystka rzeźby polskich Karpat i jej znaczenie dla gospodarki ludzkiej. Probl Zagospod Ziem Górskich 10:75–150 (in Polish)
Starkel L (1986) The role of inherited forms in present-day relief of the Polish Carpathains. In: Gardiner V (ed) International Geomorphology II. Wiley, London, pp 1033–1045
Starkel L (2015) Climatic fluctuations reflected in slope and fluvial systems of Polish Carpathians and their foreland during upper Quaternary. Stud Geomorphol Carpatho Balc 49:55–69
Szczygieł J, Gradziński M, Pavlarcik S, Holubek P, Kicińska D, Barczyk G, Dąbrowska K, Michalec V (2015) Caves and karst phenomena—part 1. Map in scale 1:100,000. In: Dąbrowska K, Guizk M (eds) Atlas of the Tatra Mountains-Abiotic Nature. Tatrzański Park Nar., Zakopane
UNEP (2007) Carpathian environmental outlook. United Nations Environment Programme, Geneva, p 236
Urdea P (1992) Rock glaciers and periglacial phenomena in the Southern Carpathians. Permafr Periglac Process 3:267–273
Urdea P (1993) Permafrost and periglacial forms in the Romanian Carpathians. In: Proceedings of sixth international conference on permafrost, Beijing, University of Technology Press 1, p 631–637
Urdea P (1995) Quelques considerations concernant des formations de pente dans les Carpates Meridionales. Permafr Periglac Process 6:195–206
Urdea P (1998) Rock glaciers and permafrost reconstruction in the Southern Carpathians Mountains, Romania. In: Permafrost—seventh international conference proceedings, Yellowknife, Canada. Collection Nordicana, 57, Univ. Laval, p 1063–1069
Urdea P (2000) Munţii Retezat. Studiu geomorfologic. Edit, Academiei Române, Bucureşti (in Romanian)
Urdea P, Vuia F, Ardelean M, Voiculescu M, Török-Oance M (2004) Investigations of some present-day geomorphological processes in the alpine area of Southern Carpathians (Transylvanian Alps). Geomorphologia Slovaca 4(1):5–11
Urdea P, Ardelean M, Onaca A, Ardelean F (2008) An outlook on periglacial of the Romanian Carpathians. Ann Univ De Vest Din Timisoara Geogr 18:5–30
Uxa T, Mida P (2017a) Ground surface thermal regime of rock glaciers in the High Tatra Mts., Slovakia. Geophysical Research Abstracts, 19, EGU2017a–1740
Uxa T, Mida P (2017b) Rock glaciers in the Western and High Tatra Mountains, Western Carpathians. J Maps 13(2):844–857. https://doi.org/10.1080/17445647.2017.1378136
Vasile M, Vespremeanu-Stroe A, Popescu R (2014) Air versus ground temperature data in the evaluation of frost weathering and ground freezing. Examples from the Romanian Carpathians. Revista De Geomorfologie 16:61–70

Vasile M, Vespremeanu-Stroe A (2017) Thermal weathering and distribution of mountain rockwalls. In: Rădoane M, Vespremeanu-Stroe A. (eds), Landform Dynamics and Evolution in Romania. Springer Geography, p 117–146. https://doi.org/10.1007/978-3-319-32589-8

Vespremeanu-Stroe A, Urdea P, Popescu R, Vasile M (2012) Rock glacier activity in the Retezat Mountains, Southern Carpathians, Romania. Permaf Permafr Periglac Process 23:127–137. https://doi.org/10.1002/ppp.1736

Voiculescu M (2004) Types of avalanches and their morphogenetic impact in Făgăraş Massif—Southern Carpathians (Romania). Geomorphologia Slovaca 4(1):72–81

Voiculescu M, Onaca A (2014) Spatio-temporal reconstruction of snow avalanche activity using dendrogeomorphological method in Bucegi Mountains-Romanian Carpathians. Cold Reg Sci Technol 104–105:63–75

Zasadni J, Kłapyta P, Świąder A (2018) Predominant western moisture transport to the Tatra Mountains during the last glacial maximum, inferred from glacier palaeo- ELAs. XXI International Congress of the CBGA, Salzburg, Austria, September 10–13, 2018. Abstract 237

Zasadni J, Kłapyta P, Broś E, Ivy-Ochs S, Świąder A, Christl M, Balážovičová L (2020) Latest Pleistocene glacier advances and post-Younger Dryas rock glacier stabilization in the Mt. Kriváň group, High Tatra Mountains, Slovakia. Geomorphology 358, https://doi.org/10.1016/j.geomorph.2020.107093

Zietara T (1989) Rozwój teras krioplanacyjnych w obrębie wierzchowiny Babiej Góry w Beskidzie Wysokim. Folia Geogr Ser Geogr Phys 21:79–82

Ziętara T (2004) Modifying of cryoplanation terraces in the flysch Carpathians. Geografický Časopis 56:85–97

The North European Plain

Barbara Woronko and Maciej Dąbski

1 Introduction: Geographical and Chronological Framework

North European Plain (NEP) spreads NE from Variscan massifs of Adrennes, Harz, Erzgebirge and Sudetes. The south-eastern boundary of NEP is made by the Meta-Carpathian Arch (Małopolska Upland and Lubelska Upland), uplifted in Alpine orogeny, while the eastern boundary is less obvious as it merges with East European Plain, developed on the Precambrian Platform (Mydel and Groch 2000). Northern reaches of NEP include the coast of North Sea and Baltic Sea, both sees being separated by the Jutland Peninsula (Fig. 1A).

NEP occupies a tectonic depression filled with Mesozoic and Cainozoic strata and contemporary relief was shaped by several Pleistocene glaciations and postglacial processes, most importantly: glacial and glaciofluvial accumulation and erosion, thermokarstic denudation and fluvial and aeolian processes (Stankowski 1996; Koster 2009; Mojski 2005). Elevations are generally below 200 m a.s.l., the highest point of NEP being Wieżyca summit located at 329 m within end moraines of the Pomeranian Stage of the Weichselian Glaciation (Beniuszys 1968). Characteristic landforms include: morainic uplands, sandar plains, subglacial valleys, east-to-west running sets of end moraine ridges (in Jutland the direction is changing to north-south) and urstromtals (ice-marginal valleys), which transported meltwaters along the ice sheet front in the general inclination of the area—towards WNW. Large areas of coversands and inland dune fields are developed within the so called the 'European sand

B. Woronko (✉)
Faculty of Geology, University of Warsaw, Żwirki i Wigury 93, 02–089 Warsaw, Poland
e-mail: bworonko@uw.edu.pl

M. Dąbski
Faculty of Geology and Regional Studies, University of Warsaw,
Krakowskie Przedmieście 30, 00-927 Warsaw, Poland
e-mail: mfdabski@uw.edu.pl

M. Oliva et al. (eds.), *Periglacial Landscapes of Europe*,
https://doi.org/10.1007/978-3-031-14895-8_12

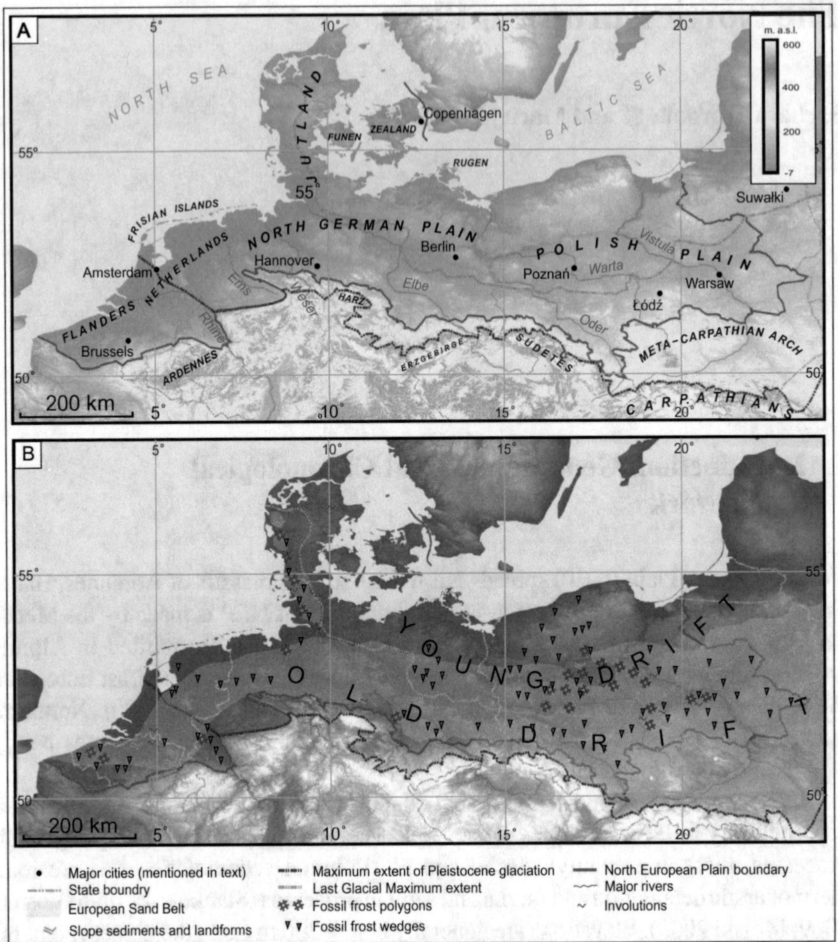

Fig. 1 **A.** Location of North European Plain with Scandinavian Ice Sheet maximum extents during cold stages of the Pleistocene (DEM based on SRTM satellite images provided by EarthExplorer, mosaic elaborated in ESRI ArcGIS), **B.** Location of relict permafrost related features and the European 'sand belt' (ESB) with Scandinavian Ice Sheet maximum extents during cold stages of the Pleistocene (Zeeberg 1998)

belt' (Koster 2009; Fig. 1B). Apart from urbanized areas, most of NEP is used as farmland, but there are also significant areas of wetlands, postglacial lakes and forest (especially in NE part of NEP).

Western parts of NEP coastline are characterized by Dutch polders, Frisian Islands, Wadden Sea and large estuaries of Ems, Weser and Elbe rivers. Danish, German and Polish coastline is predominantly flat but some parts (e.g., Rügen Island) have relatively high cliffs. There are numerous sand spits, barrier beaches and lagoons (e.g., Szczecin Lagoon). Largest archipelago belongs to Denmark, e.g. islands of Zealand,

Vendsyssel-Thy and Funen, which are located in Danish straits of Skagerrak and Kattegat. The NEP can be subdivided into the following regions: Flanders (northern Belgium and norther-eastern tip of France), the Netherlands, Northern Germany (North German Plain), Denmark (Jutland and Danish islands), and most of central-western Poland (Polish Plain). Major river-drainage basins include, from the west to the east: the Rhine, Ems, Weser, Elbe, Oder and Vistula rivers (Fig. 1A).

Contemporary long-term climatic conditions place the NEP in the temperate zone with mean annual air temperature between 7 °C and 10 °C, and mean annual precipitation between 450 and 1000 mm (Steinhauser 1970), which mean that the NEP lies outside of the current permafrost zone. However, Szewczyk and Nawrocki (2011) recently discovered deep-seated relict permafrost in NE Poland near Suwałki (slightly outside of NEP, but within the European lowlands, see Fig. 1B), which is linked to the permafrost aggradation after deglaciation of Weichselian ice sheet and deep anorthosite rocks responsible for a very low geothermal heat. The contemporary global climate change in Europe results in warming, decrease in the number of frost days and slight increase in wet extremes (Schuurmans 2005).

During the Pleistocene, the NEP underwent glaciation several times due to the build-up and retreat of the Scandinavian Ice Sheet (SIS), which exerted a profound influence on the superficial geological structure and relief of the area. Traces of the older mid-Pleistocene glaciations (Cromerian Glacial b and Elsterian, MIS 16 and MIS 12 respectively) are usually limited to geological profiles, but the following glaciations of Saalian and Weichselian (MIS 8, 6, and especially MIS 2) left characteristic glacial and glaciofluvial land systems (Stankowski 1996; Koster 2009; Mojski 2005).

The NEP can be divided between **the Old Drift** (or old-glacial landscape) and **Young Drift** (or young-glacial landscape) belts (Fig. 1B). The former was is inherited from pre-Weischelian glaciations and therefore long periods of denudation resulted in domination of relatively dry flatlands. On the other hand, the Young Drift landscape was formed due to the retreat of the last SIS, between 22.5 ka cal. BP and 14.4 ka cal. BP (Stankowski 1996; Koster 2009; Mojski 2005; Puzachenko et al. 2017; Marks et al. 2019; Tylmann and Uścinowicz 2022). Its relatively young age allows for more undulating landscape with numerous moraine hills and ridges, drumlins, eskers, kames and deep subglacial channels or inter-morainic depression, frequently filled with lakes.

Each time the SIS developed, a periglacial zone with permafrost, desert or tundra-like vegetation, usually with significant aeolian activity and characteristic river channel patterns formed in front of the ice sheet (Fig. 2). As the ice sheet advanced and covered frozen glaciofluvial sediments or former glacial till, the geothermal heat and glacial pressure degraded overridden permafrost allowing only proglacial permafrost to maintain (French 1996; Waller et al. 2012; Narloch et al. 2013; Woronko et al. 2021; Hanáček et al. 2021; Mleczak et al. 2021). During deglaciation, the receding ice sheet margin was followed by re-establishment of permafrost in the recently deglaciated area, development of frost polygons, pingos and active layer regelation cycles features such as cryoturbation and solifluction landforms as well as increased fluvial accumulation. In front of the ice sheet, cold desert (or semi-desert) conditions facilitated intense aeolian processes leading to the formation of extensive coversands and dune fields. As the climatic conditions ameliorated, permafrost underwent degradation, thermokarst landforms developed, aeolian processes weakened and rivers displayed meandering patterns. The final disappearance of superficial permafrost in eastern part of NEP probably occurred at the end of Preboreal—ca. 10,544 ± 78 – 11,223 ± 23 cal. BP (Błaszkiewicz 2011; Błaszkiewicz et al. 2015; Słowiński et al. 2015). In interglacial times, the thermokarst relief underwent further degradation with colluvium deposits accumulating at foothills, active fluvial processes in the valley floors and dunes and coversand becoming fully stabilised by forest.

2 The Development of Periglacial Research Focusing on North European Plain

Periglacial research performed in NEP concentrated on its characteristic Pleistocene landforms and sediments such fossil frost polygons (ice-wegde pseudomorphs, sand wedges and composite wedges), involutions associated with active layer dynamics, slope sediment associated with solifluction and surface wash, asymmetrical valleys, extensive coversands or inland dunes of aeolian origin (French 2003). Fluvial sediments and landforms developed under permafrost conditions were also studied (see Sect. 5). Due to a low latitude and altitude, active permafrost is absent in NEP and contemporary freeze-thaw processes are insignificant, limited to a superficial soil layer. Analyses of Pleistocene periglacial sediment structures and landforms have played an important role in Quaternary palaeogeographic reconstructions.

2.1 Beginning of Modern Periglacial Research

The modern approach to periglacial studies on Pleistocene sediments found in NEP, especially aeolian sediments found in coversand and dunes, was developed in France by André Cailleux (1942). His study showed that quartz sand grains deposited during the cold stages of Pleistocene have characteristic textural features: the sand quartz grains are well-rounded and matt ("frosted") and coarse gravel (pebbles and cobbles) possess characteristic three ridges due to wind abrasion (ventifacts, or "dreikanters"). André Cailleux leadership of the European periglacial community just after the World War II was emphasized by the fact that he became the Secretary of the first Periglacial Commission of the International Geographical Union in 1949.

The first years of post-war periglacial research was also significantly influenced by the development of the German school of climatic geomorphology represented by Carl Troll and Julius Büdel, who had reconstructed climatic conditions during Pleistocene cold stages recorded in contemporary soils (Büdel 1944; Troll 1944). The studies concluded that the wide belt adjacent to the continental margin of the

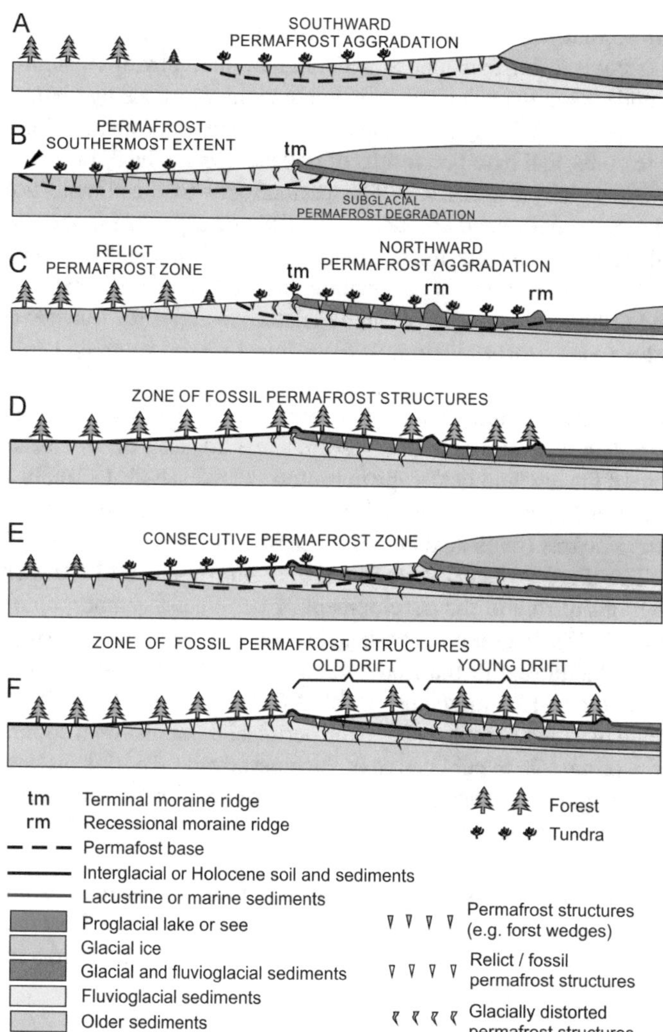

Fig. 2 Simplified model of North European Plain periglaciation during the Pleistocene. **A.** Periglacial zone developed adjacent to the advancing front of the Scandinavian Ice Sheet, **B.** Glaciation and the periglacial zone reach maximum extent in Pleistocene; northern part of permafrost degrades under the ice sheet (in most of the NEP this situation occurred in the Elsterian glacial), **C.** Recession of the ice sheet allows for re-establishment of permafrost in the northern part of the NEP (in deglaciated area), **D**. Fossilisation or degradation of permafrost structures in interglacial period **E**. Consecutive glaciation and adjacent periglacial zone do not reach as south as in the previous cold period (this does not apply to early- to mid-Pleistocene glaciations); note that the former permafrost structures could have been reactivated, **F**. Simplified situation in Holocene with Young Drift landscape (inherited from the last glaciation) and Old Drift landscape created in area covered by older advances of the Scandinavian Ice Sheet

ice sheet in nowadays NEP was characterised by intense aeolian processes operating over a cold desert and tundra underlain by permafrost. A significant step forward to understand the palaeoclimatic fluctuations was done by German geographer Hans Poser (1948). By synthesising the knowledge on the distribution of various periglacial features, and based on results of pollen analyses, he draw a map of Pleistocene climatic regions (Poser 1948). The 'permafrost—tundra climate' zone which, according to Poser, developed adjacent to the SIS corresponded to the concept of a "periglacial zone"—for the first time formulated by a Polish scientist Walery Łoziński (1912). Further developments in palaeoclimatic reconstructions came quickly from Julius Büdel (1950), who identified three Pleistocene zones for Europe: (i) a frost-rubble tundra zone ('frostschüttzone'), (ii) a forest-tundra zone, and (iii) a steppe zone.

In 1956 Jan Dylik from Poland took over the Presidency of the Periglacial Commission. His post-war scientific research was focused on understanding the development of the lowland landscape in central Poland (Dylik 1951a, b, 1953a, b). Dylik founded the first world famous scientific journal dedicated to periglacial studies *Biuletyn Peryglacjany* (published in Polish, English and French in years 1954–2000 by Łódzkie Towarzystwo Naukowe) and was the editor-in-chief till 1972. The journal played a significant role in the development of periglacial geomorphology worldwide (French 2003). It is not surprising that the findings about fossil periglacial sediments and landforms in the Łódź area (central Poland) were especially well represented in *Biuletyn Peryglacjalny*. In 1990, *Permafrost and Periglacial Processes* journal, edited by Hugh French, entered the periglacial international community and soon took the prime role in publication of the most advanced periglacial studies. The scope of the periglacial research slowly turned from palaeoenvironmental issues to contemporary high-latitude and high-altitude research, accompanied by experimental studies.

More recent and novel studies in NEP usually focus on detailed dating of relict periglacial sediments or reinterpretation of certain landforms, e.g. frost mounds (see following subsections). However, there are also novel untypical works in terms of findings or methods used (see Sect. 2.4).

2.2 *Literature Overview on Specific Sedimentological Structures*

There is a wealth of literature focusing on sediments and landforms inherited from Pleistocene periglacial times and developed in NEP, including reports on: frost fissures and polygons, involutions, periglacial slope covers (solifluction and slope wash sediments), asymmetrical valleys, aeolian coversand, ventifacts and inland dunes, pingo scars (Fig. 1B). Examples of the mentioned periglacial structures, or landforms are given below (restricted to studies on periglacial structures developed in the NEP).

Frost fissures and polygons found in Belgium and the Netherlands were described by Van der Tak-Schneider (1968), Heyse (2000), Heyse and Ghysels (2003), Buylaert et al. (2009), Beerten et al. (2017, 2021), in Denmark by Nörvang (1939, 1942), Svensson (1963, 1984), Christensen (1974), Kolstrup (1986, 2004), in Germany by Wortmann (1956), Kolstrup (1980), Karte (1981, 1987), Grube (2015), in Poland by Dylik (1951a), Dylikowa (1956), Gołąb (1956), Laskowska (1960), Goździk (1967, 1986, 1994), Kozarski (1986), Dąbski et al. (2007), and Ewertowski et al. (2017).

Involutions in active layer were described by: Klatka (1954), Dudkiewicz (1956), Mojski (1958), Kozarski and Rotnicki (1964), Christensen (1968), Paepe and Pissart (1969), Gullentops et al. (1981), De Moor (1983), Vandenberghe (1983), Kozarski (1986), Heyse (2000), Kolstrup (2004), and Dąbski et al. (2017). An interesting example of upfreezing of stones in relict active layer developed on coversand was reported by Goździk and French (2004). Genetic interpretation of certain periglacial structures, especially involutions and, to a lesser degree, also fissure-like features (followed by palaeoclimatic reconstructions) were sometimes problematic, because in certain conditions frost action is not required as a triggering factor (van Loon et al. 2020). This was raised mainly by Schenk (1955), Halicki (1960), Słowański (1963), Butrym et al. (1964), and Kasse (1999a).

Periglacial slope covers were reported by Dylik (1953a, 1955, 1961), Klatkowa (1954), Olchowik-Kolasińska (1955), Kozarski (1958), Rotnicki (1964), Urbaniak (1969), Woldstedt and Duphorn (1974), De Moor (1981), Turkowska and Wieczorkowska (1986), Kleber (1992), and Turkowska (1994) and Klosterman (1995). This issue was frequently linked with the development of asymmetrical valleys (Edelman and Maarleveld 1949; Weise 1983; Gullentops et al. 1993; Christiansen 1996). In Germany, contemporary soil cover development has been associated with periglacial deposits (Kopp 1965, 1970; Kleber 1992).

In the study area, the most famous site for pingo scars or rather lithalsa is Hautes Fagnes plateau in Adrenne (Pissart 1956, 1983, 2000), which lies beyond the NEP, but close to its westernmost margin. Hypothetic post-pingo depressions in the territory of Belgium and the Netherlands were described by Ploeger and Groenman-van Waateringe (1964), De Gans (1981), Kluiving et al. (2010), in Germany by Dahms (1972), Liedtke (1993), in Denmark by Svensson (1984) and in Poland by Dylik (1964) and Błaszkiewicz and Danel (2019). Similarly, to the Hautes Fagnes site (located beyond the W boundary of NEP), the Petronajć Lake, interpreted as post-pingo lake by Rutkowski et al. (1998) lay outside NEP, in NE Poland. At places, the "pingo hypothesis" of certain rimmed depressions did not stand the test of time (see Sect. 3.1.4).

Aeolian sand covers, or ventifacts in NEP were studied by Dylik (1951b, 1969), Cailleux (1960), Maarleveld (1960a), Schönhage (1969), Urbaniak (1969), Christiansen and Svensson (1998), Antczak-Górska (1999), Bateman and Huissteden (1999), Vandenberghe et al. (1999), and Berteen et al. (2017). A modern summary of periglacial aeolian processes and sediments from the eastern part of NEP was published by Goździk (2007). A comprehensive article on luminescent dating of periglacial aeolian sediments, mostly coversand and inland dunes (including samples from NEP) was performed by Bateman (2008). Most recent sediment-texture-based approach to periglacial aeolian studies in central Poland can be found in publications of Zieliński et al. (2015, 2016a, b), Zieliński (2017), Kalińska-Nartiša et al. (2017), and Moska et al. (2020, 2021). During the periglaciation of NEP numerous cold inland dunes developed and those from the Late Weichselian are very well visible in contemporary landscape, especially in Poland (e.g., Poser 1950; Kamińska et al. 1986; Nowaczyk 1986, Andrzejewski and Weckwerth 2010). The dunes frequently include fossil soil horizons from the Bølling and Allerød warm periods (e.g., Manikowska 1986; Zieliński et al. 2015, 2016a; Moska et al. 2020, 2021).

Among aeolian sediments associated with Pleistocene periglacial zone one should mention also loess. The loess covers were primarily deposited in upland parts of Europe, south of NEP (Radtke and Janotta 1998; Mojski 2005; Meszner et al. 2011; Badura et al. 2013; Jary and Ciszek 2013; Rousseau et al. 2017) in the so called the 'European loess belt', therefore we will not discuss the issue of loess in depth. However, within NEP there are also limitted areas with loess or sandy-loess deposits, e.g. near Hannover (Liedtke 2003) and in western Poland (Issmer 2001). They frequently contain palaeosoil horizons from interglacial times and very well developed fossil frost wedges (Krauss et al. 2016).

Periglaciation of NEP had a profound influence on the fluvial system and fluvial sediments in NEP (e.g., Turkowska 1995; Stankowski 1996; Starkel 1996; Mol et al. 2000; van Huissteden et al. 2001a; Mojski 2005; Weckwerth and Pisarska-Jamroży 2015). Studies on fluvial sediments developed under periglacial conditions can be found in Klatkowa (1965), Van Huissteden et al. (1986, 2001a), Kasse et al. (1995a), Mol et al. (2000), Bogaart et al. (2003), Marks et al. (2016), and Dzieduszyńska et al. (2020). We dedicate separate sections for better description of aeolian and fluvial sediments and landforms due their profound significance in the NEP.

2.3 Synthetic Studies

There are many synthetic studies focusing on periglacial landforms and sediments with relevant palaeogeographic or palaeoclimatic discussion with numerous information about the NEP. Palaeoclimatic interpretation of periglacial structures, including aeolian sediment for the whole NEP can be found in Böse (1991) and Vandenberghe et al. (1998). Synthetic studies focusing on Belgium and the Netherlands were published by Maréchal (1956), Maarleveld (1956, 1960b, 1976), van Leckwijck and Macar (1960), Paepe and Pissart (1969), Paeper and Vanhoorne (1970), Gullentops et al. (1981), Langohr and Van Vliet (1981), De Moor (1983), Vandenberghe (1983), Pissart (1987), Kasse (1993), and Van Vliet-Lanoë (1994). Similar works, but with the focus on Germany were published by e.g. Richter (1973), Eissmann (2002), and Lippstreu et al. (2003). Germans emphasised the periglacial conditions in the formation of contemporary soils in postglacial areas (Kopp 1965, 1970). Synthetic approach to reconstruction of periglacial conditions for parts of Denmark were published by

Svensson (1984), Houmark-Nielsen (1994, 2007), Berglund et al. (1994), and Christiansen (1993, 1997). Houmark-Nielsen (2004) published a review of stratigraphy and glaciation history, including periglacial conditions for Denmark. Synthetic works on periglacial structures, sediments, permafrost extent and and palaeoclimatic interpretations with many information on Polish part of NEP can be found in Dylik (1953b, 1956, 1960, 1967), Jahn (1956, 1960, 1970), Goździk (1973, 1995a), Brodzikowski and van Loon (1987), Starkel (1988), Klatkowa (1994, 1996), Kozarski (1995), Kasse et al. (1998), Marks et al. (2019), and Dzieduszyńska et al. (2020).

2.4 Recent Untypical Studies

Recently some innovative studies to reconstruct past periglacial conditions in NEP were published. For example, an interesting study of permafrost conditions in lowland Belgium was performed by Blaser et al. (2010), who utilised noble gas measurements in order to study the climatic evolution of the region during the Last Glacial Maximum (LGM). An unexpected discovery was published by Szewczyk and Nawrocki (2011) who proved existence of a deep-seated perennially cryotic rocks (**relict permafrost**) in Suwałki lake district (NE Poland), slightly beyond NEP, but inside the Young Drift belt. In Flanders, a novel approach to detect near-surface frost polygons was successfully tested by Meerschman et al. (2011), who used an Electromagnetic Induction Sensor mounted on a sled pulled by an all-terrain vehicle. This method occurred to work well and offered detailed information about the morphology of frost polygons (Ewertowski et al. 2017). However, the structures should be rather no deeper that 0.6 m below the ground surface and there should be a significant granulometric contrast between the host material and the wedge infill.

3 Distribution and Chronology of Periglacial Landforms

3.1 Permafrost: Associated Sediments and Landforms

3.1.1 Frost Wedges and Polygons

Ice, or composite wedge pseudomorphs, relict sand wedges or soil wedges, which are thin and shallow wedges filled with soil, sometimes limited to active layer or seasonally frozen ground (Pissart 1987), developed in pre-Weichselian periglacial zones have been reported from a variety of sites in NEP (e.g. Figs. 1 and 3; Pissart 1987; Mojski 2005), frequently in gravel or sand pits and deep open-cast mines (e.g. Kasse et al. 1998). However, the largest collection of such structures are dated to the last glacial cycle (Nörvang 1942; Höfle 1983; Pissart 1987). Some fissures could have developed and decayed during the Saalian (MIS 6) and reactivated during the Early and Late Pleniglacial (MIS 4–2; Kolstrup 2004).

Frost fissures found in the NEP developed in fluvial or glaciofluvial sediments, aeolian coversands, or glacial till can be found in a variety of geomorphological settings, but most commonly in relatively flat areas such as upper fluvial terraces, sandur plains or uplands overlain by basal glacial till. They are distributed within Young Drift, Old Drift or even outside of the maximum Pleistocene ice sheet extent, as is the case in the western part of NEP (Figs. 1B, and 3; Paepe and Pissart 1969; Christensen 1974; Gullentops et al. 1981; Kolstrup 1986; De Moor 1983; Goździk 1994; Kozarski 1995; Mojski 2005; Grube 2015; Ewertowski et al. 2017). The majority of their ages indicate extensive thermal contraction cracking and frequent filling with aeolian material during the Late Pleniglacial–Late Glacial (MIS 2 to early MIS 1) (Buylaert et al. 2009), which show evidence of the occurrence of continuous permafrost conditions (French 1996; Vandenberghe et al. 1998), and during the Middle Pleniglacial (MIS 3) to Late Glacial times in the NEP (Dzierżek and Stańczuk 2006).

Frost polygons occurring immediately below contemporary soil cover developed on various sediments and in different geomorphological settings can be visible on aerial images as crop marks. This is due to differential growth of crops (preferably: barley, oat, alfalfa) on host material and wedge infill. Environmental conditions, which allow for photographic documentation of crop marks were described by Svensson (1982) and since then periglacial studies in NEP based on photointerpretation of aerial images have been frequently performed. A selection of such superficial polygonal network sites (ice-wedge pseudomorphs, sand wedges or composite wedges) in northwest Flanders can be found in works of Heyse and Ghysels (2003) and Meerschman et al. (2011), in Denmark—in works of Christensen (1974), Svennson (1984), in Germany by Liedtke (2003), and in Poland by Gawlik (1970), Bogdański and Kijowski (1990), Kozarski (1995), Dąbski et al. (2007), and Ewertowski et al. (2017).

Sand wedges in Poland were thoroughly investigated by Goździk (1986), who found that there is a gradual downward decrease in grain size of the wedge infill, most of the wedges developed in glacial till and there is usually a gravel horizontal layer (periglacial pavement) at the uppermost part of the wedge. He also found that wedges are most frequent in the Old Drift area, and attributed it to the prolonged duration of periglacial conditions and the more severe climatic conditions during the LGM (28.6–22.5 ka cal. BP; time scale after Puzachenko et al. 2017) (Fig. 3A, E, F). The northernmost zone (north of the Pomeranian phase limit) have lower density of frost wedges, including **ice-wedge pseudomorphs**. Shallow frost wedges, developed immediately below contemporary soil in glacial till of the Old Drift were frequently associated with the Weichselian periglaciation, however Saalian age is not excluded (Laskowska 1960; Goździk 1986; Dąbski et al. 2017). Aerial photographs (and interpretation of crop marks) have revealed that frost polygons are also very common within the Young Drift (but south of Pomeranian phase) (Kozarski 1995; Dąbski et al. 2007; Ewertowski et al. 2017) and suggest a northward expansion of permafrost following the retreat of SIS.

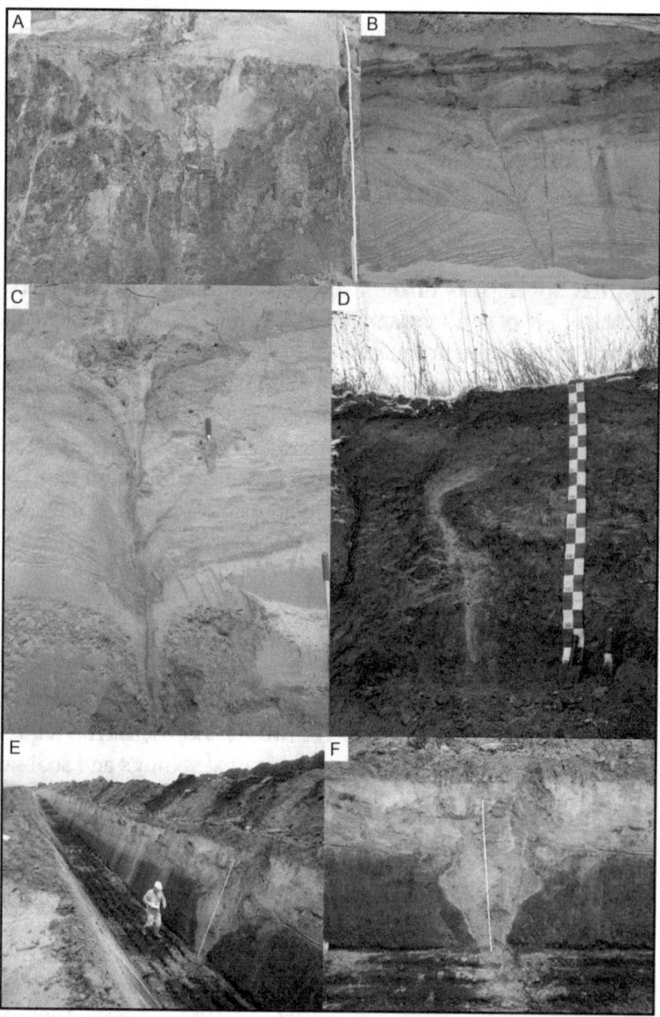

Fig. 3 Main periglacial structures in North European Plain: **A**. Sand wedge on the roof of the Saalian glacial till overlaying by the Late Weichselian cover sand, Poland, **B**. ice wedge in Weichselian glaciofluvial sands generated under continuous permafrost conditions, most probably after Frankfurt/Poznan phase (Pomeranian phase), NEL gas pipeline trench (Nordeuropäische Erdgasleitung), near Dobbertin Easter Germany, photo by Andreas Börner, Geological Survey Mecklenburg-Vorpommern, **C**. Ice wedge cutting Allerød calcareous limnic gyttja (grey) into Weichselian glaciofluvial gravelly sand generated under discontinuous permafrost conditions most probably during Younger Dryas. NEL gas pipeline trench Eastern Germany, photo by Andreas Börner, Geological Survey Mecklenburg-Vorpommern, D. Sand wedge modified by solifluction, developed in Saalian glacial till, **E**. and **F**. Saalian till morainic uplands: big ice wedge (Weichselian) cutting Miocene organic silt (rafts) filled with Weichselian coversand generated under continuous permafrost conditions, most probably during MIS 2, NEL gas pipeline trench, Eastern Germany, photo by Karin Schätzchen & Hans-Werner Lübcke, Geological Survey Mecklenburg-Vorpommern

An interesting and rare example of an interplay between thermal **contraction cracking** of glacial till (Fig. 3A, D), ice wedge development and weathering of underlying glacial pebbles and cobbles was described by Dylikowa (1956) from central Poland (Old Drift area). However, not all wedges are interpreted as indicative of thermal contraction in permafrost. Paepe and Pissart (1969) described unusual sets of small-scale wedges (up to 0.05 m wide, max. 1 m deep and max. 1 m spacing between), which are filled with sand and developed in loamy and sandy palaeosoil horizons in Belgium. They are interpreted as cryoturbation desiccation structures.

Frequently, the upper parts of thermal-contraction wedges are tilted, which is interpreted as the effect of solifluction (Goździk 1967), especially during permafrost degradation stage or glacial deformation (Narloch et al. 2013). Frost polygons and wedges described by Heyse and Ghysels (2003) in Flanders are interpreted as composite, or ice-wedge pseudomorphs due to wide and irregular shape suggesting significant thaw consolidation upon permafrost degradation. The age of the Flandrian frost polygons is constrained to the period between the Elsterian and Weichselian (MIS 12 and MIS 2 respectively), but the structures most probably developed during the Late Pleistocene (Heyse and Ghysels 2003).

3.1.2 Cryoturbations (Involutions) and Upfreezing of Stones

Deformed sedimentary layers, interpreted as an effect of the Pleistocene cold stages frost processes within the active layer (**cryoturbations**), have frequently been recorded in the NEP in a variety of geomorphological settings and sediments, most commonly in gravels, sands, loams and clays, sometimes as a single cryoturbated horizon (Fig. 4; Klatka 1954; Dudkiewicz 1956; Mojski 1958; Christensen 1968; Petera-Zganiacz 2016), sometimes as several horizons (Kozarski and Rotnicki 1964; De Moor 1983; Vandenberghe 1983, 1992). They were usually associated with load-casting and/or frost heaving, sometimes also associated with solifluction, and interpreted by Polish researchers as "struktury tundrowe" (tundra structures) in 1950's. However, such interpretation was criticised, because similar structures can be developed without any frost-induced processes (Schenk 1955; Halicki 1960; Słowański 1963; Butrym et al. 1964; Kasse 1999a; Van Loon et al. 2020). Nevertheless, liquefaction, increased plasticity of sediments, or reversed density gradients, required for involutions, frequently occur due to freeze–thaw processes within the active layer (Van Vliet-Lanoë 1991; French 1996) and therefore, such palaeo-climatic interpretation of involutions is still used in studies of Pleistocene structures.

Small-scale involutions (up to several decimeters) are widespread, with large-scale involutions (with an amplitude exceeding 1 m) observed in certain regions, e.g. in Flanders. They are associated with thermokarst processes and lowering of permafrost table at the end of Pleistocene (Vandenberghe 1983). On the other hand, some large-scale involutions are interpreted as structures from cold-climate developments during the Saalian Glaciation, (MIS 6) e.g. in Schöningen, Germany (Urban et al. 1991) or in Tjæreborg, Jutland (Kolstrup 2004).

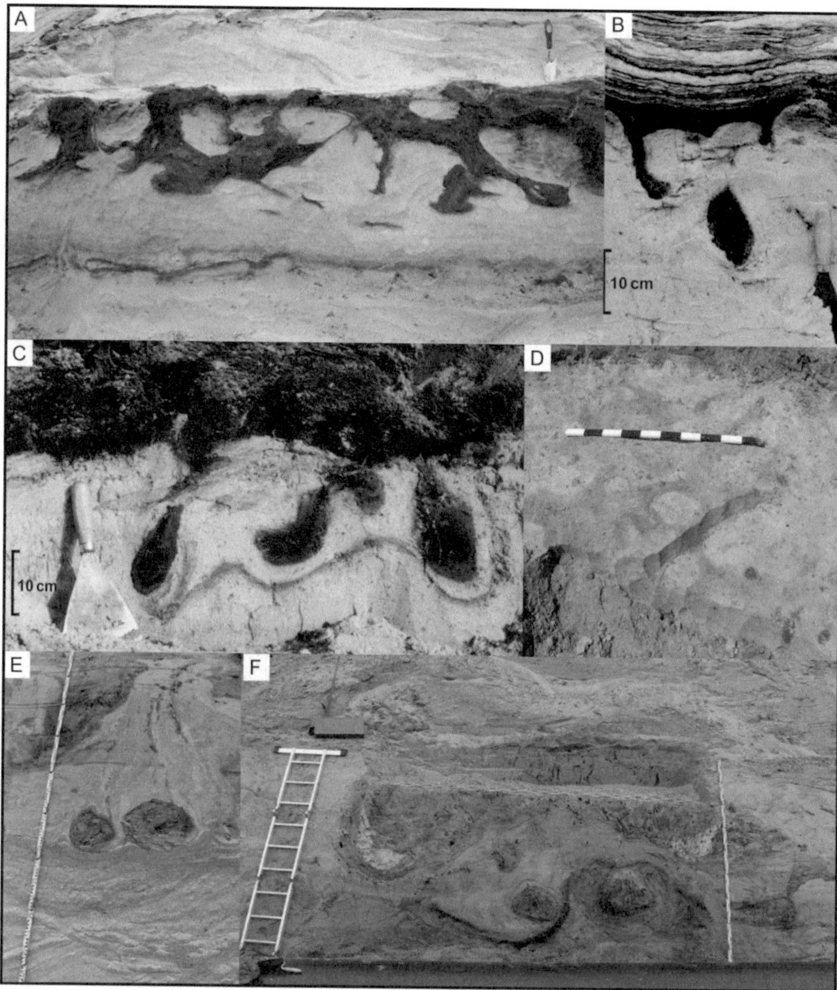

Fig. 4 **A**. Diapiric involutions, Pleniweichselian, Adamów open coal mine, Poland (photo by J. Petera-Zganiacz), **B**. Drop-like involutions, Younger Dryas, Koźmin site, Poland (photo by J. Petera-Zganiacz 2016), **C**. Drop-like involutions, Younger Dryas, Koźmin site, Poland (photo by J. Petera-Zganiacz), **D**. Regular involution in the Łowicko-Błońska Plane, Poland, **E** and **F**. Ball and Pillow soft-sediment deformation structures (SSDS) of Allerød calcareous limnic gyttja developed into Weichselian glaciofluvial gravelly sand under discontinuous permafrost conditions during Younger Dryas, OPAL gas pipeline trench (Ostsee Pipeline Anbindungs-Leitung) (photo by Andreas Börner, Geological Survey Mecklenburg-Vorpommern)

Frost-heave operating in the active layer resulted also in **upfreezing of stones** (French 1996). This process resulted in vertical orientation and sorting of clasts and finally, together with aeolian deflation, sometimes periglacial pavement was created with significant share of tilted stones (Dąbski et al. 2017). Frequently, a stone horizon is found at the uppermost level of fossil frost wedges (Goździk 1986; Fig. 3D). A peculiar case study was described by Goździk and French (2004) from central Poland of apparently upfreezing of stones within non-frost susceptible cover sands, which poses a significant challenge in terms of process reconstruction. Some stones of the periglacial layer exhibit clear signs of aeolian abrasion.

3.1.3 Assymetric Valleys and Slope Deposits

Fine-grained laminated sediments covering slopes of small valleys near Łódź were extensively studied by by Dylik (1953a, 1955, 1961), Klatkowa (1954), Olchowik-Kolasińska (1955), Turkowska and Wieczorkowska (1986), and interpreted as slope wash or solifluction deposits from Weischelian age. Backwashing of slopes under periglacial conditions during Saalian glaciation, sometimes associated with nivation, was responsible for the formation of asymmetrical valleys in Belgium (Gullentops et al. 1993).

Contemporary soil cover and soil parent material are described in German literature as periglacial cover-beds (Kleber 1992; Semmel and Terhorst 2010). They are composed of three layers: (1) the upper one, which has relatively constant thickness and occurs on almost all slopes, (2) the intermediate layer that is restricted to flat relief, to hollows, and to leeward facing slopes, and (3) the basal layer that is rather widespread again. These layers were mainly formed by periglacial slope processes (solifluction and surface wash) and have varying portions of loess and local rock fragments transported downslope.

3.1.4 Frost Mounds

Landforms interpreted as **post-pingo** structures in the NEP are usually peat-filled shallow depressions, surrounded by a relatively circular rampart built with solifluction or surface wash sediments and developed during Weichselian glaciation. They were predominantly, but not exclusively, found in the Old Drift zone of Belgium and the Netherlands (Ploeger and Groenman-van Waateringe 1964; De Gans 1981; Kluiving et al. 2010), in Germany (Dahms 1972; Liedtke 1993) in Denmark (Svensson 1984), but also in Poland (Dylik 1964; Błaszkiewicz and Danel 2019).

Today, following the famous case of periglacial structures in Hautes Fagnes (Belgium) and the change of their interpretation from pingo (Pissart 1956) to another frost mount—e.g. lithalsa (Pissart 1983, 2000), the "pingo hypothesis" of certain structures did not stand the test of time. This is also the case of small depressions near Łódź previously interpreted by Dylik (1964) as post-pingo, but Majecka et al.

(2021) provided arguments in favour of a different (yet undefined) periglacial mound evolution.

New precise lidar-based digital elevation models allow new geomorphological discoveries, a good example being the study of Błaszkiewicz and Danel (2019) undertaken in northern Poland (Żarnowicka Highland) in the Young Drift zone. Over 80 very well-developed ring rampart ridges (3–7 m high) have been distinguished with central circular depressions (60–80 m wide), filled with lake gyttja and peat with a total thickness of 6–7 m. Preliminary geomorphological observations indicate that the genesis of these landforms was associated with the development of pingo mounts under the conditions of discontinuous permafrost and their subsequent degradation.

3.2 Aeolian Sediments and Landforms in Periglacial Environments

One of the most characteristic aspect of NEP periglaciation is a widespread development of aeolian sediments and landforms (Tricart 1960; Koster 1988; Kozarski and Nowaczyk 1991; Mycielska-Dowgiałło 1993; Zeeberg 1998; Schirmer 1999a; Kasse 2002; Seppälä 2004; French 2017). They constitute the 'European sand belt' (ESB) running from Belgium and the Netherlands in the west, across Germany, southern Denmark, Poland into Russia in the east (Fig. 1B; Koster 1988, 2009; Zeeberg 1998; Kasse 2002).

Cold-climate inland aeolian sediments and landforms of ESB were subject to numerous detailed studies, including grain-size characteristic, micro-morphology of quartz grain, palaeowind direction and dating of their deposition. In the Netherlands they were studied by van der Hammen (1971), van Geel et al. (1989), Bateman and Van Huissteden (1999), Van Huissteden et al. (2000), Kasse (1999b, 2002) and Vandenberghe (1991, 2013), in Denmark by Kolstrup (1982), Christiansen and Svensson (1998), Vandenberghe et al. (1999) in Germany by Alisch (1995), Bussemer et al. (1998), Schirmer (1999a), Kaiser et al. (2006), Hilgers (2007), Hirsch et al. (2017), and in Poland by Urbaniak-Biernacka (1976), Manikowska (1985, 1993), Nowaczyk (1986), Kozarski and Nowaczyk (1991), Kalińska (2012), Hirsch et al. (2015), Woronko et al. (2015), Zieliński et al. (2015, 2016a), Zieliński (2017), Rychel et al. (2018), Łapcik et al. (2021), and Moska et al. (2021).

The European 'sand belt' (ESB) (Fig. 1B) is composed of cold-climate aeolian coversand, inland dunes, wind-abraded quartz grains, ventifacts and deflation depressions. Intensive aeolian processes are also recorded by sand wedges filled with aeolian sediments (Goździk 2007). Cold inland dunes are found within large dune fields, e.g. in northern Belgium and the Netherlands, northern Germany, or Polish sections of urstromtals (Fig. 5; Nowaczyk 1986; Kozarski and Nowaczyk 1991; Andrzejewski and Weckwerth 2010), but also there are tens and hundreds of smaller scattered dunes (Fig. 6). However, the most dominant features of ESB are aeolian coversands (Koster 1988, 2009; Kasse 2002).

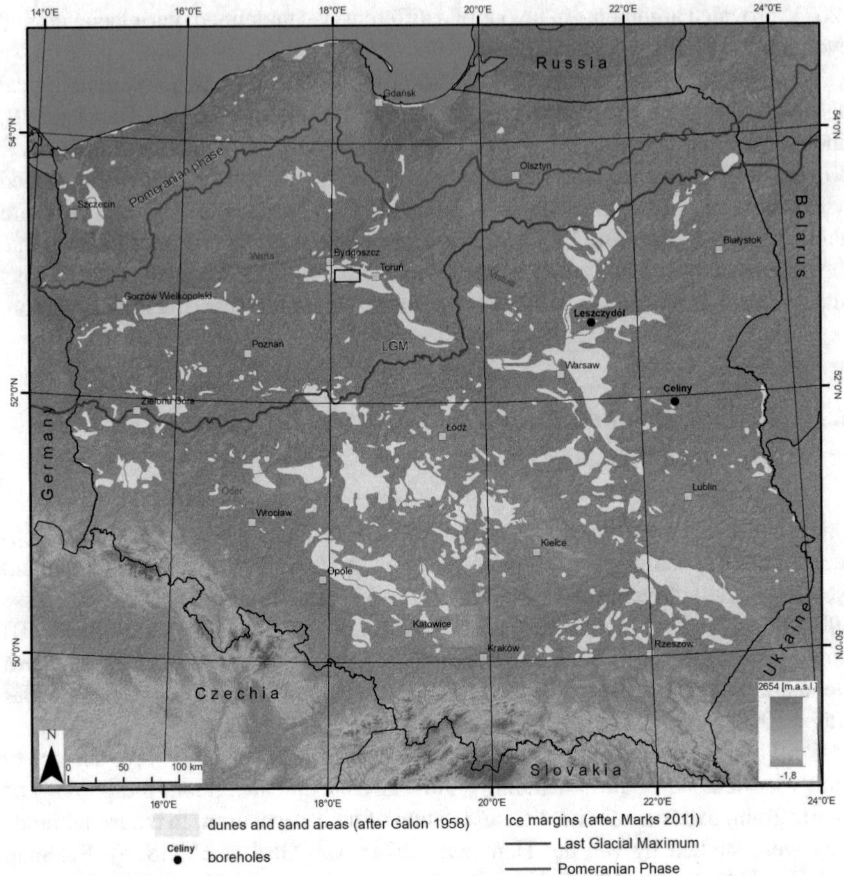

Fig. 5 The Late Weichselian inland dunes developed in Poland, in the eastern part of the ESB according to Galon (1958)

The norther boundary of ESB in the Netherland and western Germany is determined by marine deposits of the North Sea Holocene transgression. In Jutland and Schleswig-Holstein (N Germany), the ESB northern limit is the SW limit of the SIS during the LGM. In eastern Germany and in Poland, ESB is much wider than in its western part. Dune fields and coversands are well developed from the limit of the Pomeranian phase (MIS 2) in the north to the maximum extent of SIS, the Meta-Carpathian Arch or, in places, to the Carpathians in the south (Fig. 1B; Koster 1988, 2009). South of ESB, well developed loess covers ('European loess belt') are found (Haase et al. 2007).

During the Late Pleniglacial (MIS 2), aeolian processes were facilitated by periglacial climate with low precipitation and strong winds, scarcity of vegetation and relatively flat relief with no significant topographic barriers (Zeeberg 1998; Kasse 1997, 2002). Aeolian landforms were fed by enormously vast fine-grained

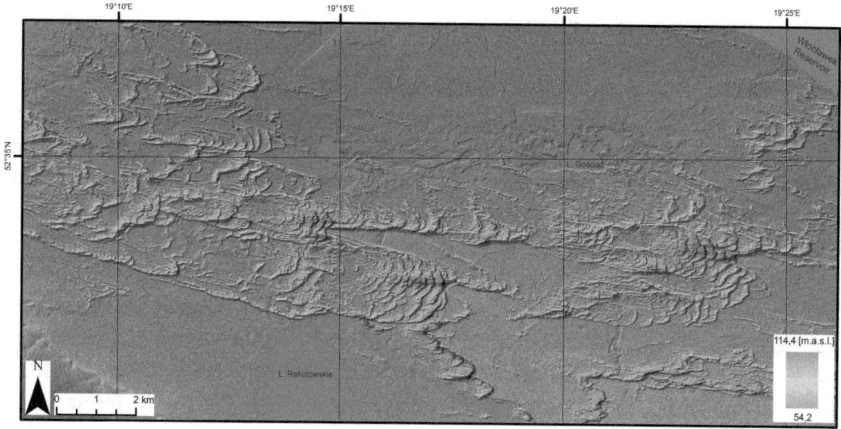

Fig. 6 Complex aeolian dunes on the left bank of the Vistula Valley near Płock (site 1). For the location of the site 1, see rectangle in Fig. 5

fluvial and glaciofluvial deposits, but also by widely available frost-shattered, or glacially crashed material (Kasse 1997, 2002; van Huissteden et al. 2001b; Seppälä 2004; Bertrant et al. 2018; Kasse and Aalbersberg 2019). According to Böse (1991), aeolian activity occurred especially under dry permafrost conditions and in upland areas with better drainage. In some places, frozen ground limited aeolian deflation, but also preserved aeolian sediments (Maarleved 1976; Kozarski 1990; Manikowska 1991a, b; Kasse 1997, 2002). Sedimentological analysis, including fossil faunal and botanical records, show that during the LGM (28.6–22.4 ka cal. BP) and Late - Pleniglacial (22.4 to ~14.7 ka cal. BP; time scale after Puzachenko et al. 2017) large areas of the NEP were represented by sparsely vegetated cold desert and aeolian processes acted intensively (Kozarski 1980; Krzyszkowski 1990; Mol et al. 2000).

Zeeberg (1998) determined three zones of European aeolian environments regarding timing and duration of aeolian processes, which was translated to different development of aeolian sediments:

i. a zone of cold periglacial environment south of SIS LGM extent (Old Drift, zone) with continuous and discontinuous permafrost (28.6–22.4 ka cal. BP), where aeolian processes operated very long time and had the highest 'aeolian effectiveness';
ii. an intermediate zone between LGM and the ice retreat to 14.4 ka cal. BP), constituted by numerous spillways and glacial landforms within the Young Drift zone;
iii. a deglaciation zone (14.4–10.2 ka cal. BP ka), where significant aeolian processes operated for the shortest period on freshly deglaciated areas and were limited to sections of urstromtals or small sandar (Young Drift zone, Fig. 6).

3.2.1 Chronology of Aeolian Activity

Intense aeolian processes operated in front of the SIS during Pleistocene cold periods (Zeeberg 1998; van Huissteden et al. 2001b; Woronko 2012; Woronko and Bujak 2018; Dzieduszyńska et al. 2020), which is well recorded in loess sequences south of ESB (Haase et al. 2007; Solarska et al. 2013). However, aeolian activity is mainly detected in sediments of the last glacial cycle (Nowaczyk 1986; Zeeberg 1998; Kasse 1999b, 2002; Hilgers 2007; Zieliński et al. 2016a, b; Moska et al. 2021) and records of pre-Weischelian aeolian activity are very sparse and allow only for limited reconstruction of palaeo-environments (Goździk 1980, 2007; Goździk and Maruszczak 1998; Maruszczak and Goździk 2001; Mycielska-Dowgiałło 1993; Mycielska-Dowgiałło and Woronko 2004; Goździk and Zieliński 2007; Zieliński 2007; Woronko 2012; Woronko and Bujak 2018). Some studies show the existence of five different layers of sediments bearing imprint of aeolian processes in central Poland (e.g., Goździk 1980, 2001), but later studies on full Pleistocene chronostratigraphy argued for eight aeolian horizons correlated with advances of SISs (e.g., Woronko 2012; Woronko and Bujak 2018), which favoured the development of periglacial phenomena (Lindner and Bogucki 2002). Sediments subject to aeolian transport contain high shares of well-rounded and matt grains (RM; Fig. 7A, B), or semi-rounded ones, which represent abraded protruding fragments (EM/RM; Fig. 7C, D) (Mycielska-Dowgiałło and Woronko 1998; Woronko 2012; Woronko and Bujak 2018).

Sandy sediments from the Mid-Pleistocene (Elsterian, MIS 12) have a relatively high share of RM grains, gradually increasing upwards, which allow to infer that intensity of aeolian processes was increasing together with sedimentation (Fig. 8A). On the other hand, sediments from Saalian Glacial Complex (MIS 8–6) from a variety of locations in Poland are dominated by EM/RM grains and the share of RM is low (Fig. 8B). This is interpreted as a signature of a massive activation of aeolian processes, but for a relatively short time. However, the duration of aeolisation was longer in southern parts of Poland (Woronko 2012), which is in accordance with the pattern observed for the Weischelian SIS (Goździk 1980, 2001). Throughout the Pleistocene, aeolian sediments were frequently subject to fluvial erosion, which is manifested in high share (sometimes up to 90–100% of sand grains) of wind-abraded grains in fluvial sediments (Zieliński 2007; Woronko 2012).

Several phases of aeolian activity can be distinguished in ESB, which were interrupted by ground surface stabilisation and development of soil covers (Fig. 9; Kasse and Aalbersberg 2019). The difference between aeolisation of sediments within the Young Drift and Old Drift zone is significant (Goździk 1991; ; Zeeberg 1998; Hilgers 2007; Zieliński et al. 2015, 2016a; Kasse and Aalbersberg 2019; Andrzejewski and Weckwerth 2010; Moska et al. 2021). This gradient is also clearly visible in the west-east profile through ESB (Galon 1958; Mycielska-Dowgiałło 1993; Isarin et al. 1997). Aeolian processes reworked sediments within the Young Drift zone only after the retreat of the Weischelian SIS (MIS 2). On the other hand, the Old Drift zone sediments experienced aeolisation already from the Saalian SIS (MIS 6) retreat (in each cold period of the last 140 ka). In the eastern part of the ESB, wind deflation could have removed up to 0.5 m-thick layer of sediments on certain morainec

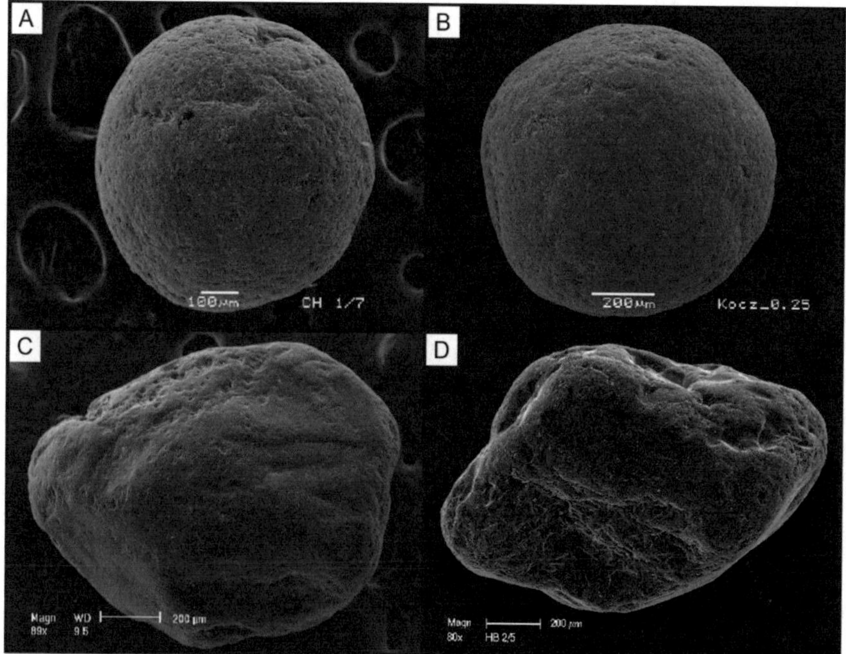

Fig. 7 SEM microphotograph of a wind-abraded sand-sized quartz grain surfaces: **A**, and **B**. Well-rounded grain with matt surface (type RM according to Mycielska-Dowgiałło and Woronko 1998), **C**, and **D**. Semi-rounded grain with matt surface only on the most concave part of grain (type EM/RM according to Mycielska-Dowgiałło and Woronko 1998)

terrain (Woronko et al. 2022). Aeolian processes are recorded in eastern part of ESB in sediments of Middle Pleniglacial (MIS 4) (Woronko et al. 2021; Dzieduszyńska et al. 2020) and in the later Late Pleniglacial (MIS 3) as an infill of sand wedges (Dzierżek and Stańczuk 2006). Additionally, Wojtanowicz (1999) found based on TL dating (Thermoluminescence method) of sediments from eastern Poland inland dunes 32 ± 5.4 and 22 ± 3.7 ka old. However, aeolian deposits from the period >17–18 ka (OSL—optically stimulated luminescence) are rather rare (Kozarski 1990; Manikowska 1991a, b; Kasse 1997; Hilgers 20072007; Fig. 9).

Schirmer (1999a) summarised results of investigations of aeolian activity from the Netherlands in the west, to Poland in the east. He distinguished four periods of aeolian activity since the Late Pleniglacial to the beginning the Holocene: period 1—the period around LGM (28–18 ka cal. BP), period 2—17.2–14.2 ka cal. BP which ends in phase of the Bølling Interstadial (BI), period 3—13.5–13.25 ka cal. BP (Dryas 2)—the main dune phase ending in the phase the Allerød Interstadial and period 4—12.9–11.7 ka cal. BP (Dryas 3) which ends in the Holocene soil formation (time scale after Rasmussen et al. 2014). During the Period 1 and 2, the superficial processes developed under continuous permafrost conditions, whereas in the period 3 permafrost seems to have already disappeared. Kasse (1999b, 2002) distinguished

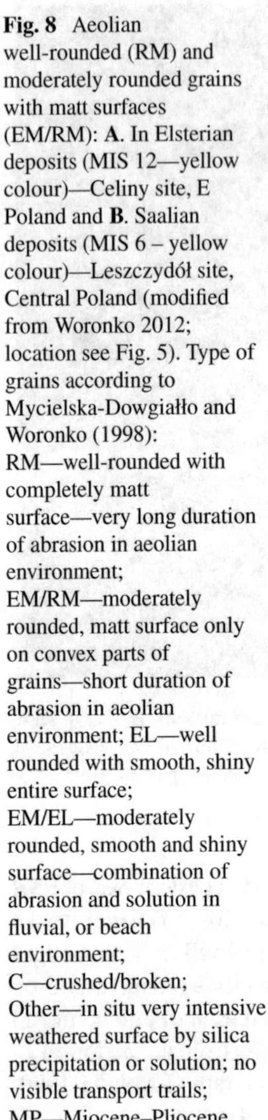

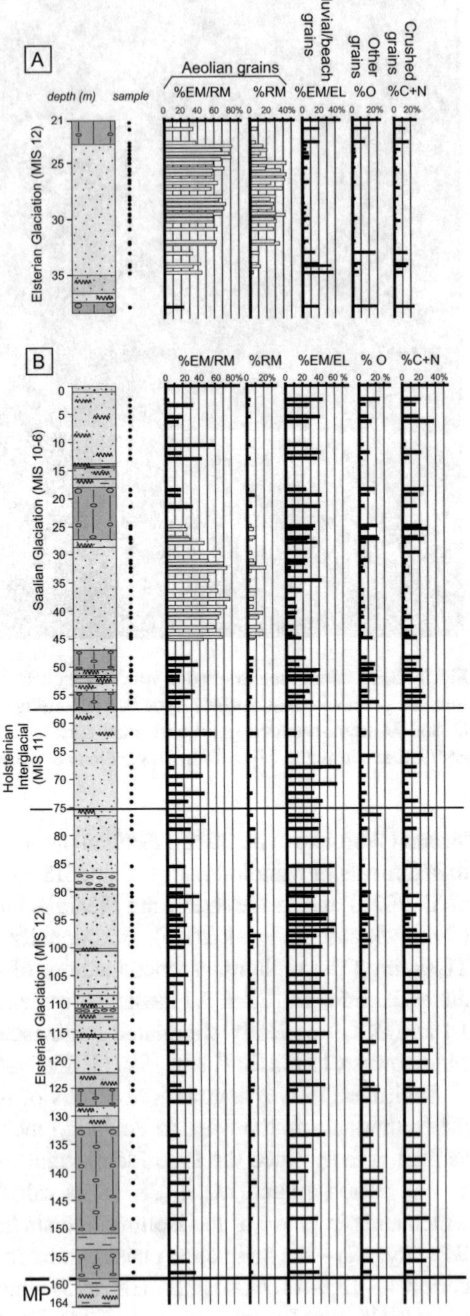

Fig. 8 Aeolian well-rounded (RM) and moderately rounded grains with matt surfaces (EM/RM): **A**. In Elsterian deposits (MIS 12—yellow colour)—Celiny site, E Poland and **B**. Saalian deposits (MIS 6 – yellow colour)—Leszczydół site, Central Poland (modified from Woronko 2012; location see Fig. 5). Type of grains according to Mycielska-Dowgiałło and Woronko (1998): RM—well-rounded with completely matt surface—very long duration of abrasion in aeolian environment; EM/RM—moderately rounded, matt surface only on convex parts of grains—short duration of abrasion in aeolian environment; EL—well rounded with smooth, shiny entire surface; EM/EL—moderately rounded, smooth and shiny surface—combination of abrasion and solution in fluvial, or beach environment; C—crushed/broken; Other—in situ very intensive weathered surface by silica precipitation or solution; no visible transport trails; MP—Miocene–Pliocene

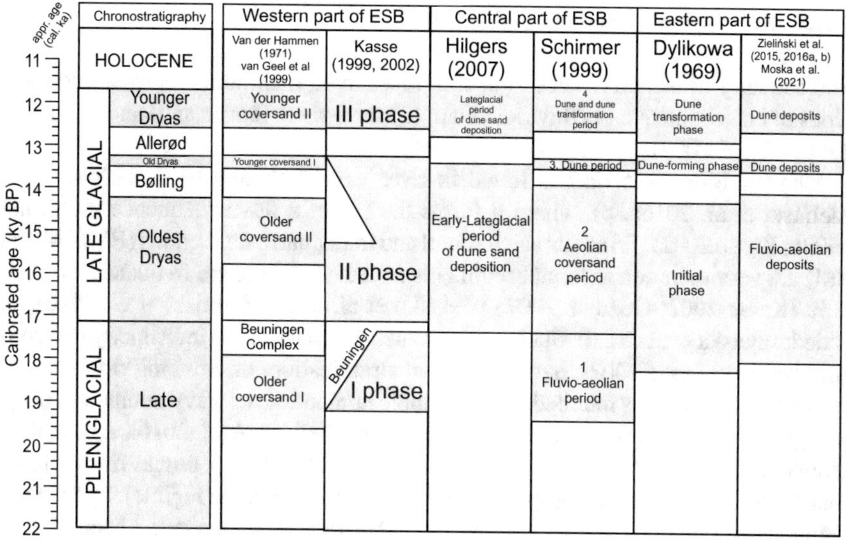

Fig. 9 Synthesis of the Late Glacial aeolian activity phases in European sand belt (yellow color –aeolian activity, gray color—cold periods)

three phases of activity of aeolian processes from the LGM to the beginning of the Holocene: (1) phase I between ca. 28 and 18 ka cal BP related to permafrost conditions of the LGM (2) phase II between 18 and 14 ka cal. BP during the Late Pleniglacial and early Late Weichselian and (3) phase III between ca. 13 and 10 ka cal BP during the later part of Younger Dryas (YD) and the early Holocene. On the other hand, Vandenberghe et al. (2013) aeolian sand overlaying the Beuningen Gravel Bed took place between 15.8 ± 1.4 ka and 12.3 ± 1.0 ka OSL (Fig. 9). According to results of OSL data, Moska et al. (2021) in the Old Drift areas of western-central Poland, claimed that aeolian deposition started in the Oldest Dryas (OsD) and continued throughout the Older Dryas (OD, main the phase of the dune build-up period) with limited deposition in the YD. Whereas in the Young Drift areas, aeolian deposition started in the YD and continued to the early Boreal. According to Hilgers (2007), in north-eastern Germany and adjacent parts of Poland (middle part of ESB), it is possible to determine two phases of dune formation: the first one around 16.7 ka OSL, 2) and the second one, the reactivation period from 12.9 to 11.1 ka OSL. On the other hand, in the easternmost part of ESB, the first phase is usually dated to the OsD (the initial period of dune formation), the second to the Older Dryas (OD; the main phase of dune formation) and the third one to the YD (Dylikowa 1969; Kozarski et al. 1969; Nowaczyk 1986; Kozarski 1990; Manikowska 1991a, b, 1994; Zieliński et al. 2015, 2016a, b; Fig. 9).

3.2.2 Succession of Aeolian Sediments

The aeolian sedimentary succession comprises three main units representing three types of environments: (1) fluvio-aeolian, (2) sand-sheet (coversand) and (3) inland dunes (Kasse 2002).

The fluvio-aeolian unit is found in river valleys (Kasse 1997, 1999b, 2002; Zieliński et al. 2016a, b), where it forms the top of a fluvial sequence (Schirmer 1999b; Kasse 2002). Wind-abraded, round and matt quartz sand grains (EM/RM and RM) are very common in fluvial sediments of the NEP, its share frequently exceeds 50% (Kasse 2002; Goździk 2007; Woronko et al. 2015; Zieliński et al. 2016a, b; Dzieduszyńska et al. 2020). Such sands are usually massive, or indistinctly horizontally bedded (Kasse 2002), have translatent stratification, or climbing ripple cross-lamination, sometimes interbedded with ripple lamination, or wavy lamination (silty sands). Cut-and-fill structures are quite common (Zieliński et al. 2016a, b). Aeolian sediments accumulated in river valleys were incorporated to fluvial transport by shallow flowing water during spring snow melt (nival discharge regime). Syngenetic pseudomorphs of ice wedges, sand wedges, and load casts are frequent. Accumulation of fluvio-aeolian deposits occurred under continuous permafrost cold-arid conditions (Van Huissteden et al. 1986; French and Goździk 1988; Mol et al. 1993; Manikowska 1995; Goździk 1998; Schirmer 1999a; Kasse 2002; Zieliński et al. 2015, 2016a, b). However, in the eastern part of the NEP, the fluvio-aeolian sedimentation could last till the OsD (Zieliński et al. 2015, 2016a, b). In the Netherlands such sediments are typical for Older Coversand and Beverlorg Member as they are usually found below Beuningen Gravel Bed (Van Huissteden 1990; Kasse 2002). On morainic uplands this period is recorded in the presence of ventifacts (Fig. 10A–C).

Aeolian coversand deposits constitute flat and widespread sand veneer with small dunes, covering pre-existing landscape of the NEP in interfluve areas as well as in river valleys (Kolstrup 1982; Goździk 1991; Manikowska 1991a, b 1994; Schirmer 1999a; van Huissteden et al. 2001b; Kasse 2002; Kalinska 2012; Woronko et al. 2015; Zieliński et al. 2016a, b). Internal structure of coversands reveals alternating bedding of sand and silt/fine and sand with wavy lamination (Fig. 10D), whereas bedding attributable to dune slip face progradation is very rare. Cut-and-fill structures, different scale plastic deformations and ice wedge casts are relatively common (Kolstrup 1982; Goździk 1998; van Huissteden et al. 2001b; Kasse 2002; Zieliński et al. 2016a, b). Coversands were preceded in places by intense wind deflation, which lead to the development of stone pavements. In the eastern part of the ESB, the coversands are characterized by the highest share of wind-abraded grains (RM and EM/RM), which indicates numerous redeposition and simultaneous strong abrasion with limited supply of new material in increasingly arid conditions (Goździk 2007; Woronko et al. 2015; Zieliński et al. 2015, 2016a). According to Kasse (1997), coversands resulted from high ratio between wind energy and sand availability. In the eastern part of ESB, lithostratigraphic tripartition is observed within the coversands. The lower unit consists of medium sand, the middle unit contains medium sand alternating with layers of silt and fine sand and the upper one is built of medium

Fig. 10 **A**. Ventifact from Obórki, Young Drift landscape (photo by Wojciech Wysota), **B**. Ventifact on the roof of the Saalian glacial till, Łowicko-Błońska Plain, Poland, **C**. Wind abraded big stone from Old Drift landscape (SE Poland), **D**. Alternating bedding of sand and silt in coversand, **E**. Slip face large-scale cross-bedded sand of the parabolic dune from the Late Glacial in the eastern part of ESB, **F**. Bedded aeolian sand of a dune from the Late Glacial in Garwolin site, eatern part of ESB (central Poland)

and fine sand (Goździk 1980, 2007). In the western part of the NEP (the Netherlands), such sediments are known as Older Coversand II of the older phase and the Younger Coversand I form the younger phase (Van Huissteden 1990; Kasse 2002). They originated during the Late Weichselian at 14.7 ka cal BP and can be extended to the Bølling Interstadial and OD periods (Kasse 2002; Zieliński et al. 2015, 2016a, b).

Reconstruction of dominant wind direction during formation of coversands is very difficult due to insignificant surface and because they form a more or less continuous cover over the pre-existing morphology (Renssen et al. 2007). However, the succession of sediments in the coversand ridges in the eastern Netherlands allows to infer that the Late Pleniglacial Older Coversand II unit was deposited by dominating westerly winds (Kasse 1999b; Renssen et al. 2007).

Inland dunes are landforms usually developed at the upper end of the aeolian succession in NEP (Zieliński et al. 2016a, b). According to Koster (2009), the proportion of sediments in coversands and in dunes gradually change in the west-east direction: coversands dominate in the western part of NEP, while large dunes fields are typical for its eastern part. Parabolic dunes dominate (Koster 1988; Fig. 6), they are frequently up to 15 m high, sometimes reaching 30 m and are over 5000 m long. Their internal structure is dominated by slip face large scale cross-bedding of sand (Fig. 10E, F). The dunes were formed from the OsD to the beginning of the Holocene (Manikowska 1991a, b; Zieliński et al. 2015, 2016a, b; Zieliński 2017; Moska et al. 2021). Dunes formation during cold phases of Late Glacial was interrupted by warm phases of Bølling and Allerød allowing for soil formation and dunes stabilization by vegetation (Derese et al. 2009, 2012; Vandenberghe et al. 2013; Kasse and Aalbersberg 2019). The major phase of dune formation occurred in the OD (Dylikowa 1968; Manikowska 1991a, b; Schirmer 1999a; Hilgers 2007; Zieliński et al. 2016a, b; Moska et al. 2021). In the western part of ESB, wind directions changed from WNW in the OD towards WSW during the YD (Maarleveld 1960a). Local permafrost conditions, or seasonally frozen ground during the YD have been reported in the Netherlands (Kasse 1995, 2002; Isarin 1997; Kasse and Aalbersberg 2019). The central part of ESB was characterised by north-westerly or westerly winds (Nowaczyk 1986; Hilgers 2007), but in the eastern part the direction of wind was more diverse, from SW to NW (Dylikowa 1968, 1969; Nowaczyk 1986; Zieliński 2001, 2017). The wind must have changed between different dune forming phases: W in the OSD, NW–W in the OD and SW–W in the YD (Wojtanowicz 1965; 1969; Dylikowa 1968, 1969; Szczypek 1977; Böse 1991; Zieliński 2001, 2017).

3.3 Fluvial Activity, Sediments and Landforms in Periglacial Environments

Functioning of river systems under periglacial environmental conditions has been a subject of numerous studies performed in different parts of the NEP, especially in relatively small river catchments. In a period of MIS 5a–d, when sedimentological records show episodic deep seasonal frost action and discontinuous permafrost, through continuous permafrost in MIS 4 and second part of MIS 3 and MIS 2, up to discontinuous permafrost in late MIS 2 and the beginning of MIS 1 (Kasse et al. 2003), different phases of river incision and aggradation, changes of river channel patterns, changes of source sediments for alluvia and rates of aeolian supply to rivers occurred. Mol et al. (2000) emphasize that in this period the river catchments in western and central parts of Europe exhibited several similarities.

3.3.1 The Early Pleniglacial (MIS 4)

Strong cooling at the Early Pleniglacial (MIS 4), comparable with MIS 2, leading to continuous permafrost aggradation, is recorded in frequent involutions and large ice-wedge casts and frozen sand in alluvial sediments, which resulted in changes in the water and sediment budget, and therefore in a significant change of river channels (Klatkowa 1965; Huissteden et al. 1986, 2001, Mol et al. 2000; Jary 2007; Marks et al. 2016). However, preserved fluvial sediments deposited in this period are limited. In western parts of the NEP (e.g., Vandenberghe 1993; Mol et al. 2000; Kasse et al. 1998, 2003; Gerfield et al. 2001; van Huissteden et al. 2001a) and in its eastern parts (Turkowska 1988, 2007; Goździk and Zieliński 1996; Marks 2004; Krzyszkowski and Kuszell 2007, Wachecka-Kotkowska et al. 2014; Woronko et al. 2018, 2021; Dzieduszyńska et al. 2020) deep fluvial incision in part of river valleys occurred. In most river valleys erosion removed the entire Early Pleniglacial sequence and a part of the Middle Pleniglacial sediments (Fig. 11). Erosion also acted in dry valleys in eastern parts of the NEP. Shallow pools, which functioned since the end of Saalian Glacial Complex (MIS 6), were incorporated into river systems (Woronko et al. 2021). In the eastern parts of the NEP, deep incision responsible for erosion of MIS 4 sediments most probably occurred around 40 ka BP OSL (Goździk and Zieliński 1996; Wachecka-Kotkowska et al. 2014; Moskalewicz et al. 2020). In the Nochten mine, eastern Germany, sedimentary structures indicate erosion which could have happened during MIS 4 or at the transition of MIS 4 to MIS 3. Whereas, in Vecht and Dinkel valleys in the Netherlands, a phase of incision took place after 36 ka BP which correlates with Hasselo Stadial (MIS 3; 43.3–48.3 ka cal. BP) (Huisink 2000). However, in MIS 4, some valleys experienced deposition and recorded faster-flowing braided systems (van Huissteden et al. 1986; Moskalewicz et al. 2020). After the phase of incision, high sediment supply from surrounding upland, including significant aeolian input resulted in intensive aggradation and coarse-grained braided patterns (van Huissteden et al. 1986; Zieliński et al. 2015).

3.3.2 The Middle Pleniglacial (MIS 3)

Fluvial erosion is documented in sediments dated at MIS 4–3 transition (Mol et al. 2000). Aggradation started after 40 ka ^{14}C BP (Huissteden et al. 1986, 2001; Turkowska 1988; Goździk 1995b; Goździk and Zieliński 1996; Van Huissteden and Kasse 2001; Mol et al. 2000; Starkel et al. 2007; Dzieduszyńska et al. 2020; Woronko et al. 2022). The MIS 3 can be divided into an older part (>45 ka cal BP) with relatively stable landscape and younger part with gradual change towards less stable environment (van Huissteden and Pollard 2003). In the Middle Pleniglacial (MIS 3), rivers adapted to changing climatic conditions and to discontinuous permafrost

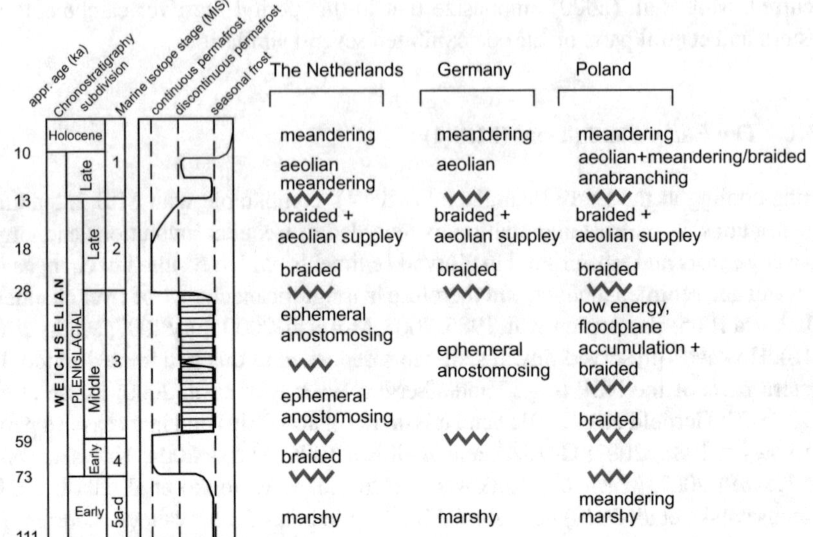

Fig. 11 Synthesis of fluvial activity from West to East of the NEP during the Early Pleniglacial—Late Weichselian based on composition diagrams of Mol et al. (2000) and Dzieduszyńska et al. (2020)

(Fig. 11; Huisink 2000; Mol et al. 2000). Fluvial rhythmically bedded fine-grained sand and silt, characteristic for younger part of MIS 3, are inherited from low-energy fluvial environment and overbank sedimentation (van Huissteden et al. 1986, 2001; Turkowska 1988; van Huissteden and Pollard 2003; Dzieduszyńska et al. 2020). In the Łódź Plateau (central Poland), the bottom layer of theses deposits was ^{14}C and TL dated to about 43 ka cal. BP and its top to ca 24 ka cal. BP (Turkowska 1994, 1995; Starkel et al. 2007). Similar fine sediments (depleted of organic matter) were found in small depressions within dry valleys in the second part of MIS 3 (Woronko et al. 2022). River channels form this period were meandering or anastomosing (Fig. 11; Van Huissteden et al. 1986, 2001; Mol et al. 2000). Syngenetic cryoturbation and frost fissures are found in these sediments in the Netherlands testifying for periglacial conditions during their origin (Van Huissteden et al. 2001). In eastern Germany involutions are common, but ice-wedge casts are absent, which probably resulted from cool climatic conditions with deep seasonal freezing. In the Wartha valley (Poland), in such positions peat is found, dated at 45.3–37.6 ka cal. BP (Petera 2002; Forysiak 2005; Dzieduszyńska et al. 2020), which is interpreted as a record of shallow pools and peat bogs (Van Huissteden et al. 1986, 2001; Turkowska 1988; Krzyszkowski 1990).

In the western part of the NEP, lateral immobility of river channels is recorded, resulting most probably from anastomosing river style with channel avulsion responsible for deposition of flood basin sediments (van Huissteden et al. 1986, 2001).

Huisink (2000) argued that an increase in the spring snowmelt discharge was responsible for an increase in the frequency of avulsion. Aeolian processes activated in this period supplied fluvial system with additional sediments (van Huissteden et al. 1986, 2001; Zieliński et al. 2015). However, deep cooling around 43 ka cal. BP resulted in a change of river channels from anabranching to sheet-flood and braided river systems in certain valleys. This could stem from the development of continuous permafrost, which triggered overland flow, unstable runoff and higher sediment supply to river beds (Woronko et al. 2021).

3.3.3 The Late Pleniglacial (MIS 2)

The MIS 3–2 transition in marked with an incision in most of the river channels (Mol et al. 2000). In the Late Pleniglacial (MIS 2) in the Old Drift belt, fluvial system functioned under the coldest climate conditions with continuous permafrost, in increased continentality and aridity (Huisink 2000; Mol et al. 2000; Van Huissteden et al. 2000; Dzieduszyńska et al. 2020). The lack of organic matter indicates polar desert conditions (Goździk 2007). Large ice-wedge casts and pingo mounts were frequent and aeolian processes were very active around the entire NEP (Goździk 1986; Kozarski 1993; Mol et al. 2000; Petera-Zganiacz 2011; Zieliński et al. 2016a, b; Dzieduszyńska et al. 2020; Moska et al. 2020, 2021). In the fluvial succession, this period is marked with a significant sediment aggradation (Fig. 11). Alluvium is usually sandy-gravelly indicating high-energetic environment of the braided river system with a significant admixture of aeolian-abraded quartz grains in the fluvio-aeolian series (Goździk 1995, 2007; Mol et al. 2000; Zieliński and Goździk 2001; Woronko 2012; Woronko et al. 2015; Zieliński et al. 2016; Dzieduszyńska et al. 2020). Floodplains were especially remodelled by aeolian processes (Van Huissteden and Kasse 2001; Kasse et al. 2003). Fluvial reworking of aeolian sediments took place during spring snow melts (nival regime). Syngenetic ice-wedge pseudomorphs are frequent in alluvial sediments form this period. An increase in aridity was accompanied by a decrease in fluvial processes, or even complete cessation of river activity (Mol et al. 1993; Kasse 1997; Kasse et al. 2003). This is recorded by the common transition from fluvial series into aeolian coversands (Fig. 11; Kasse 1997; Van Huissteden and Kasse 2001; Woronko et al. 2015; Zieliński et al. 2016).

3.3.4 Late Glacial (MIS 2–1 Transititon)

First signals of climatic amelioration are dated at 18.4 cal. ka BP—a shift from continuous to discontinuous permafrost, which resulted in a change of river channel pattern (Dzieduszyńska et al. 2020).

Several abrupt climatic oscillations occurred in the Late Weichselian (14.7–11.7 ka cal. BP) (Rasmussen et al. 2014). In cold phases (OsD, OD and YD) aeolian processes reactivated (see previous section). The braided river system was abandoned just before the end of the Bølling warm period (Van Huissteden and Kasse

2001). In the eastern part of the NEP during the Bølling warm period, rivers were already meandering (Turkowska 1994; Zieliński et al. 2015) just like in Allerød warm period in other parts of the NEP (Van Huissteden and Kasse 2001; Zieliński et al. 2015). Many rivers maintained their meandering character, despite the climatic deterioration of YD, but some other changed into braided pattern (Fig. 11; Starke 1995; Mol et al. 2000; Dzieduszyńska et al. 2020). The discontinuous permafrost conditions returned during the YD (Isarin 1997; Huijzer and Vandenberghe 1998; Huisink 2000; Mol et al. 2000), or sporadic permafrost in the eastern part of the NEP with thermal contracting structures and small scale involution (Zieliński et al. 2016a, b; Dzieduszyńska and Petera-Zganiacz 2018; Dzieduszyńska et al. 2020). The presence of permafrost is recorded in sediments of the Meuse River valley in the Netherlands (Bohncke et al. 1993) and in northeast Germany (Börner et al. 2011). This cooling directly resulted in incision and change in the river style (Mol et al. 2000; Van Huissteden and Kasse 2001; Zieliński et al. 2015; Dzieduszyńska et al. 2020). Increased melt-water runoff during spring and more frequently peak discharges resulted in multichannel or braided patterns (Kasse et al. 1995; Van Huissteden and Kasse 2001; Starkel et al. 2007; Petera-Zganiacz et al. 2015; Dzieduszyńska et al. 2020).

4 The Significance of Periglacial Dynamics in the North European Plain Within the European Context

The NEP lies south of the contemporary latitudinal periglacial zone and below the alpine permafrost zone. However, during the Pleistocene periglacial processes extended over large areas, deep permafrost developed, and the periglacial domain expanded and retreated several times during glacial-interglacial cycles. As a result, periglacial landforms and deposits were significantly reshaped and remained relativelly well preserved from the last glacial-interglacial cycle.

Some of the first paleoclimatic reconstructions of the Northern Hemisphere periglacial conditions and processes were based on analogies between the contemporary polar and subpolar periglacial zones and those, which functioned during the Pleistocene cold stages in the NEP. These comparisons provided a better general understanding of periglacial processes, particularly for Quaternary cold periods. The interaction of the past glacial, fluvial, aeolian, and periglacial processes makes the NEP a classical area of periglacial research throughout the last century, where numerous geomorphological concepts evolved.

The current landscape of the NEP can be divided into two belts depending on the different extent of Pleistocene ice sheets: the Old Drift and Young Drift. Deposits of the former one were accumulated by pre-Weischelian ice sheets, therefore the landforms of the Old Drift have been significantly modified. On the other hand, the Young Drift landscape was exposed during the retreat of the last Scandinavian Ice Sheet between 22.5 and 14.4 ka cal. BP. Therefore, periglacial processes were fully operative only after deglaciation for a relatively short time, and their impact in the landscape was moderate. Large areas of permafrost and glacial dead ice disappeared during the Preboreal on most of the NEP.

Evidences of intense Pleistocene periglacial processes were recorded in the NEP in forms of ice- and composite wedge pseudomorphs, sand wedge casts, soil wedges, cryoturbations (involutions), ventifacts, solifluction deposits and dry (asymmetric) valleys. Intense aeolian processes operating in the periglacial desert or semi-desert produced abundant coversands and inland dunes, which create the 'European sand belt', one of the most spectacular features of NEP's landscape. Rivers flowing in the permafrost affected areas developed characteristic channel patterns and alluvial sediments acquired characteristic texture as lot of aeolian grains were incorporated into the fluvial systems.

Periglacial landforms are now mostly stabilized by grassland or forest, and some of them are protected by law, such as the dune fields in the Kampinos National Park in Poland. However, in many places in the "European sand belt" (from the Netherlands to Poland) the sediments are remobilised due to deforestation, overgrazing, fires or other human actions. New measures to nature management and geoconservation have led to the reactivation of some dune areas, which thus increasing their geomorphological, ecological, and educational importance. On the other hand, periglacial sediments and structures like laminated slope sediments, involutions or frost wedges continue to be target features that need to be further investigated in the future.

Acknowledgements The authors would like to thank Andreas Börner from Geological Survey Mecklenburg-Vorpommern (Germany), Joanna Petera-Zganiacz from University of Łódź (Poland) and Wojciech Wysota from Nicolaus Copernicus University in Toruń (Poland) for providing images to this publication.

References

Alisch M (1995) Das äolische Relief der mittleren Oberen Allerniederung (Ost-Niedersachsen); spät- und postglaziale Morphogenese, Ausdehnung und Festlegung historischer Wehsande, Sandabgrabungen und Schutzaspekte. – Kölner Geographische Arbeiten 62

Andrzejewski L, Weckwerth P (2010) Dunes of the Toruń Basin against palaeogeographical conditions of the Late Glacial and Holocene. Ecol Questions 12, Special Issue:9–1

Antczak-Górska B (1999) Wind-faced stones within the maximum limit of the last ice-sheet in southern Wielkopolska and the periglacial phenomena concurrent with them. Biul Peryglac 38:63–72

Badura J, Jary Z, Smalley I (2013) Sources of loess material for deposits in Poland and parts of Central Europe: the lost Big River. Quat Int 296:15–22

Bateman MD (2008) Luminescence dating of periglacial sediments and structures. Boreas 37:574–588

Bateman MD, van Huissteden J (1999) The timing of last-glacial periglacial and aeolian events, Twente, eastern Netherlands. J Quat Sci 14(3):277–283

Beerten K, Heyvaert VMA, Vandenberghe DAG, Van Nieuland J, Bogemans F (2017) Revising the Gent Formation: a new lithostratigraphy for Quaternary wind-dominated sand deposits in Belgium. Geologica Baltica 20(1–2):95–102

Beerten K, Meylemans E, Kasse C, Mestdagh T, Van Rooij D, Bastiaens J (2021) Networks of unusually large fossil periglacial polygons, Campine area, northern Belgium. Geomorphol 377:107582

Berglund BE, Björck S, Lemdahl G, Bergsten H, Nordberg K, Kolstrup E (1994) Late Weichselian environmental change in southern Sweden and Denmark. J Quat Scien 9(2):127–132

Beniuszy S (1968) Budowa geologiczna wzgórza Wieżyca w świetle najnowszych badań. Geol Q 12(2):403–409

Bertran P, Andrieux E, Bateman M, Font M, Manchuel K, Sicilia D (2018) Features caused by ground ice growth and decay in Late Pleistocenefluvial deposits, Paris Basin, France. Geomorphology 310:84–101.

Blaser PC, Kipfer R, Loosli HH, Walraevens K, van Camp M Aeschbach-Hertig W (2010) A 40 ka record of temperature and permafrost conditions in northwestern Europe from noble gases in the Ledo-Paniselian Aquifer (Belgium). J Quat Sci 25:1038–1044.

Błaszkiewicz M (2011) Timing of the final disappearance of permafrost in the central European Lowland, as reconstructed from the evolution of lakes in N Poland. Geol Q 55(4):361–374

Błaszkiewicz M, Danel W (2019) Formy pierścieniowe w rejonie Wejherowa jako prawdopodobne pozostałości po-pingo i ich znaczenie dla paleogeografii późnego glacjału w północnej Polsce (ang. Ring forms in the area of Wejherowo as likely remnants of pingos, and their significance for Late-Glacial paleogeography in Northern Poland). Przegląd Geograficzny 91(3):405–419 (in Polish with English summary).

Błaszkiewicz M, Piotrowski J, Brauer A, Gierszewski P, Kordowski J, Kramkowski M, Lamparski P, Lorenz S, Noryśkiewicz A, Ott F, Słowiński M, Tyszkowski S (2015) Climatic and morphological controls on diachronous postglacial lake and river valley evolution in the area of Last Glaciation, northern Poland. Quat Sci Rev 109:13–27

Bogaart PW, Van Balen RT, Kasse C, Vandenberghe J (2003). Process-based modelling of fluvial system response to rapid climate change II. Application to the River Maas (The Netherlands) during the Last Glacial–Interglacial Transition. Qua. Sci Rev 22:2097–2110

Bogdański P, Kijowski A (1990) Photointerpretation of geometry of Vistulian ice-wedge polygons: the Grabianowo and Sulejewo sites, south of Poznań. Questiones Geographicae 11(12):39–52

Bohncke S, Vandenberghe J, Huijzer AS (1993) Periglacial environments during the Weichselian Late Glacial in the Mass valley, the Netherlands. Geol Mijnbouw 72:193–210

Börner A, Janke W, Lampe R, Lorentz S, Obst K, Schütze K (2011) Geowissenschaftliche Untersuchungen an der OPAL-Trasse in Mecklenburg-Vorpommern – Geländearbeiten und erste Ergebnisse. Brandenburgische Geowissenschaftliche Beiträge 18:9–28 (in German with English summary)

Böse M (1991) A palaeoclimatic interpretation of frost-wedge casts and aeolian sand deposits in the lowlands between Rhine and Vistula in the Upper Pleniglacial and Late Glacial. Z Geomorphol, Suppl 90:15–28

Brodzikowski K, van Loon AJ (1987) A systematic classification of glacial and periglacial environments, facies and deposits. Earth Sci Rev 24(5):297–381

Bussemer S, Gärtner P, Schlaak N (1998) Stratigraphie, Stoffbestand und Reliefwirksamkeit der Flugsande im brandenburgischen Jungmoränenland. Petermanns Geographische Mitteilungen 142:115–125

Büdel J (1944) Die morphologischen Wirkungen des Eiszeitklimas im gletscherfreien Gebiet (Beiträge zur Geomorphologie der Klimazonen und Vorzeitklimate I). In: Geol. Rundschau 34:482–519

Büdel J (1950) Das System der klimatischen Morphologie. Beiträge zur Geomorphologie der Klimazonen und Vorzeitklimate (V). Deutscher Geographentag München: 65–100

Butrym J, Cegła J, Dżułyński S, Nakonieczny S (1964) New interpretation of 'periglacial' structures. Folia Quaternaria, Polska Akademia Nauk, Kraków 17:34

Buylaert JP, Ghysels G, Murray AS, Thomsen KJ, Vandenberghe D, De Corte F, Heyse I, Van den haute P (2009) Optical dating of relict sand wedges and composite-wedge pseudomorphs in Flanders, Belgium. Boreas 38:160–175

Cailleux A (1942) Les actions éoliennes périglaciaires en Europe. Paris—Société Géologique de France. Mémoires 21 (46)
Cailleux A (1960) Action du vent quaternaires préwurmiennes en Europe. Biul Peryglac 9:73–82
Chmielewska M, Chmielewski W (1960) Stratigraphie et chronologie de la dune de Witów, distr. De Łęczyca. Biul Peryglac 8:133–142
Christensen L (1968) An occurrence of periglacial struktures at Langå. Jylland. Medd. Dansk Geol. Foren. 18:46–56
Christensen L (1974) Crop-marks revealing large-scale patterned ground structures in cultivated areas, southwestern Jutland, Denmark. Boreas 3:153–180
Christiansen HH (1993) Late Weichselian periglacial landforms in the Bjergsted area, north-western Zealand. Denmark. Geografisk Tidsskrift 93:39–48
Christiansen HH (1996) Effects of nivation on periglacial landscape evolution in western Jutland, Denmark. Permafrost Periglac Process 17:111–138
Christiansen HH (1997) Periglacial sediments in an Eemian-Weichselian succession at Emmervel Klev, southwestern Jutland, Denmark. Palaeogeogr Palaeoclimatol Palaeoecol 138:245–258
Christiansen HH, Svensson H (1998) Windpolished boulders as indicators of a Late Weichselian wind regime in Denmark in relation to neighbouring areas. Permafrost Periglac Process 9:1–21
Dąbski M (2011) Grunty Strukturalne w Polsce. Przegląd Geograficzny 83:307–321
Dąbski M, Łapaj K, Pudłowski M (2007) Identyfikacja poligonów mrozowych na podstawie analizy zdjęć lotniczych z rejonu Kruszwicy. In: Smolska E, Giriat D (eds) Rekonstrukcja dynamiki procesów geomorfologicznych – formy rzeźby i osady. WGiSR UW, Komitet Badań Czwartorzędu PAN, pp 117–124
Dąbski M, Zawadzka-Pawlewska U, Greń K (2017) Struktury peryglacjalne na stanowisku Łubienica-Superunki (Wysoczyzna Ciechanowska) – wstępne wyniki badań. Land Anal 33:17–24
Dahms E (1972) Limnogeologische Untersuchungen im Dümmer-Becken. Freie Universität, Berlin, Geologische Untersuchungen an Niedersächsischen Binnenseen
De Gans W (1981) The Drentsche Aa valley system. a study in quaternary geology, Vrije Universiteit, Academisch Proefschrift, Amsterdam.
De Moor G (1981) Periglacial deposits and sedimentary structures in the Upper Pleistocene infilling of the Flemish Valley (NW Belgium). Biul Peryglac 28:277–290
De Moor G (1983) Cryogenic structures in the Weichselian deposits of northern Belgium and their significance. Polarforschung 53(2):79–86
De Ploey J (1977) Some experimental data on slopewash and wind action with reference to quaternary morphogenesis in Belgium. Earth Surf Process 2:101–105
Derese C, Vandenberghe D, Paulissen E, Van den haute, P (2009) Revisiting a type locality for Late Glacial aeolian sand deposition in NW Europe: optical dating of the dune complex at Opgrimbie (NE Belgium). Geomorphol 109:27–35
Derese C, Vandenberghe DAG, Van Gils M, Mees F, Paulissen E, Van den haute P (2012) Final Palaeolithic settlements of the Campine region (NE Belgium) in their environmental context: optical age constraints. Quat Int 251:7–21
Dudkiewicz L (1956) Struktury Tundrowe w Patokach. Biul Peryglac 3:39–46
Dylik J (1951a) Some periglacial structures in Pleistocene deposits of Middle Poland. Bull Soc Sci Lettres De Łódź 3(2):1–6
Dylik J (1951b) The loess-like formations and the windworn stones in Middle Poland. Bull Soc Sci Lettres De Łódź 3(3):1–17
Dylik J (1953a) Premières notions sur les formations de couverture dans la Pologne centrale. Bull Soc Sci Lettres De Łódź 4(1):1–20
Dylik J (1953b) O peryglacjalnym charakterze rzeźby środkowej Polski, vol 4. Acta Geographica Universitatis, 109 p (in Polish)
Dylik J (1955) Peryglacjalne osady stokowe rytmicznie Warstowane (Rhythmically stratified periglacial slope deposits). Biul Peryglac 2:15–32
Dylik J (1956) Coup d'oeil sur la Pologne périglaciaire. Biul Peryglac 4:195–238

Dylik J (1960) Sur le système triparti de la stratigraphie du Pléistocène dans les pays d'accumulation glaciaire. Biul Peryglac 9:25–40

Dylik J (1961) Analyse sedimentologique des formations de versant remplissant les depressions fermées aux environs de Łódź. Biul Peryglac 10:57–74

Dylik J (1964) Eléments essentiels de la notion de "périglaciaire" – réponse a l'enquete. Biul Peryglac 14:111–132

Dylik J (1967) The main elements of Upper Pleistocene paleogeography in Central Poland. Biul Peryglac 16:85–116

Dylik J (1969) L'action du vent pendant le dernier âge froid sur le territoire de la Pologne Centrale. Biul Peryglac 20:29–44

Dylikowa A (1956) Kliny Zmarzlinowe w Stawęcinie. Biul Peryglac 3:47–60

Dylikowa A (1968) Fazy rozwoju wydm w środkowej Polsce w schyłkowym plejstocenie. Folia Quat 29:119–126

Dylikowa A (1969) Le problème des dunes intérieures en Pologne à la lumière des études de structure. Biul Peryglac 20:45–80

Dylikowa A, Klatkowa H (1956) Example du modele périglaciaire du Plateau de Łódź. Biul Peryglac 4:239–254

Dzieduszyńska D, Petera-Zganiacz J (2018) Small-scale geologic evidence for Vistulian decline cooling periods: case studies from the Łódź Region (Central Poland). Bull Geol Soc Finland 90(2):209–222

Dzieduszyńska D, Petera-Zganiacz J, Roman M (2020) Vistulian periglacial and glacial environments in central Poland: an over view. Geol Q 64(1):54–73

Dzierżek J, Stańczuk D (2006) Record and palaeogeographical implications of Pleistocene periglacial processes in the Drohiczyn Plateau, Podlasie Lowland (Eastern Poland). Geol Q 50:219–228

Edelman CH, Maarleveld GC (1949) De asymmetrische dalen van de Veluwe. Tjdschrift Koninklijk Nederlans Aardrijkskundig Genoot 66:143–146

Eissmann L (2002) Quaternary geology of eastern Germany (Saxony, Saxon-Anhalt, South Brandenburg, Thuringia), type area of the Elsterian and Saalian stages in Europe. Quat Sci Rev 21:1275–1346

Ewertowski MW, Kijowski A, Szuman I, Tomczyk AM, Kasprzak L (2017) Low-altitude remote sensing and GIS-based analysis of cropmarks: classification of past thermal-contraction-crack polygons in central western Poland. Geomorphol 293, B: 418–432

Forysiak J (2005) Rozwój doliny Warty między Burzeninem a Dobrowem po zlodowaceniu warty. Acta Geographica Lodziensia 90:116 p (in Polish with English summary)

French H (1996) The periglacial environment. Longman, Harlow

French H (2003) The development of periglacial geomorphology: 1- up to 1965. Permafrost Periglac Process 14:29–60

French HM (2017) The Periglacial Environment, 4th Edition. Wiley Blackwell

French HM, Goździk JS (1988) Pleistocene epigenetic and syngenetic frost fissures, Belchatów, Poland. Can J Earth Sci 25:2017–2027

Galon R (1958) Z problematyki wydm śródlądowych w Polsce. In: Galon R (ed) Wydmy śródlądowe Polski. Państwowe Wydawnictwo Naukowe, Warszawa, pp 13–30

Gawlik H (1970) Wieloboki peryglacjalnych szczelin mrozowych na obszarze Polski Środkowej w świetle interpretacji zdjęć lotniczych. Fotointerpret Geogr 8:71–79

Geukens F (1947) De asymmetrie der droge dalen van Haspengouw. Natuurw. Tijdschr 29:13–18

Gołąb JS (1956) Kliny zmarzlinowe jako drogi przewodzące wód gruntowych. Biul Peryglac 3:61–64

Goździk JS (1967) Fauchage des fentes en coin du aux mouvements de masses sur des pentes douces. Biul Peryglac 16:133–146

Goździk JS (1973) Origin and stratigraphic position of periglacial structures in Middle Poland. Acta Geographica Lodzensia 31:1–119

Goździk JS (1980) Zastosowanie morfometrii i graniformametrii do badań osadów w kopalni węgla brunatnego Bełchatów. Studia Regionalne, IV (IX). PWN, Warszawa-Łódź pp 101–114
Goździk JS (1986) Structures de fentes à remplissage primaire sableaux du Vistulien en Pologne et leur importance paléogéographique. Biul Peryglac 31:71–106
Goździk JS (1991) Sedimentological record of aeolian processes from the upper Plenivistulian at the turn of Pleni- and Late Vistulian in Central Poland. Z Geomorphol, Suppl 9:51–60
Goździk JS (1994) Études des fentes de gel en Pologne Centrale. Biul Peryglac 33:49–78
Goździk JS (1995a) A permafrost evolution and its impact on some depositional conditions between 20 and 10 ka in Poland. Biul Peryglac 34:53–72
Goździk JS (1995b) Vistulian sediments in the Bełchatów open cast mine, central Poland. Quat Stud Pol 13:13–26
Goździk J (2001) Stratygrafia i paleogeografia osadów czwartorzędowych z środkowo–zachodniej części kopalni Bełchatów z wykorzystaniem morfoskopii ziarn kwarcowych. In Mycielska-Dowgiałło E. (ed), Eolizacja osadów jako wskaźnik stratygraficzny czwartorzędu. Warszawa, pp 81–93
Goździk JS (2007) The Vistulian aeolian succession in central Poland. Sed Geol 193:211–220
Goździk J, Zieliński T (1996) Sedymentologia vistulianskich osadów malych dolin srodkowej Polski– przyklad z kopalni Bełchatów. Biul Panstw Inst Geol. 373:67–77
Goździk JS, French H (2004) Apparent upfreezing of stones in Late-Pleistocene coversand, Bełchatów vicinity, Central Poland. Permafrost Periglac Process 15:359–366
Goździk JS, Maruszczak H (1998) Evidence of strong aeolian abrasion in fluvial deposits immediately before the Odranian ice sheet advance in an area of the Middle Vistula river. Biul Peryglac 37:101–114
Grube A (2015) Periglaziäre, polygonal-verzweigte rinnenförmige Bildungen und glazitektonische Strukturen in Saale-Till am Elbe-Urstomtalrand bei Wedel (Schleswig-Holstein). E&G Quat Sci J 64(1):3–13
Gullentops F, Jansen J, Paulissen E (1993) Saalian nivation activity in the Bosbeek valley, NE Belgium. Geol Mijnbouw 72:125–130
Gullentops F, Paulissen E, Vandenberghe J (1981) Fossil periglacial phenomena in NE-Belgium. Biul Peryglac 28:345–365
Guy De Moor (1983) Cryogenic structures in the Weichselian deposits of Northern Belgium and their significance. Polarforschung 53(2):79–86
Haase D, Fink J, Haase G, Ruske R, Pécsi M, Richter H, Altermann M, Jäger KD (2007) Loess in Europe—its spatial distribution based on a European Loess Map, scale 1:2,500,000. Quat Sci Rev 26:301–1312
Halicki B (1960) O różnej genezie strukturalnych deformacji osadów w środowisku hydroplastycznym. Biul Peryglac 7:35–58
Hanáček M, Nývlt D, Jennings SJA (2021) Thermal basal regimeof the Elsterian ice-sheet marginal zone in a hilly mountain foreland, Rychleby Mts. Eastern Sudetes. Boreas 50:582–605
Harry DG, Goździk JS (1988) Ice wedges: growth, thaw transformation, and palaeoenvironmental significance. J Quat Sci 3:39–55
Heyse I (2000) Fossil periglacial remnants in the Beernem-Mouton excavation in Flanders (Belgium). Biul Peryglac 39:53–68
Heyse I, Ghysels G (2003) Fossil polygonal periglacial structures in Flanders (Belgium). In: Phillips M, Springman SM, Arenson LU (eds) Permafrost. Swets & Zeitlinger, Lisse, pp 395–400
Hilgers A (2007) The chronology of Late Glacial and Holocene dune development in the northern Central European lowland reconstructed by optically stimulated luminescence (OSL) dating. Diss. Univ. Köln
Hirsch F, Schneider A, Nicolay A, Błaszkiewicz M, Kordowski J, Noryskiewicz AM, Tyszkowski S, Raab A, Raab T (2015) Late Quaternary landscape development at the margin of the Pomeranian phase (MIS 2) near Lake Wygonin (Northern Poland). Catena 124:28-44. https://doi.org/10.1016/j.catena.2014.08.018

Hirsch F, Spröte R, Fischer T, Forman SL, Raab T, Bens O, Schneider A, Hüttl RF (2017) Late Quaternary aeolian dynamics, pedostratigraphy and soil formation in the North European Lowlands—new findings from the Baruther ice-marginal valley. DIE ERDE 148(1):58–73

Houmark-Nielsen M (1994) Late Pleistocene stratigraphy, glaciation chronology and Middle Weichselian environmental history from Klintholm, Møn, Denmark. Bull Geol Soc Den 41:181–202

Houmark-Nielsen M (2004) The Pleistocene of Denmark: a review of stratigraphy and glaciation history In: Ehlers J, Gibbard PL (ed) Developments in Quaternary Sciences, Elsevier, 2, Part 1, pp 35–46

Houmark-Nielsen M (2007) Extent and age of Middle and Late Pleistocene glaciations and periglacial episodes in Southern Jylland, Denmark. Bull Geol Soc Den 55:9–35

Höfle H-C (1983) Periglacial phenomena. In: Ehlers J (ed) Glacial deposits in North-west Europe. Balkema, Rotterdam, pp 297–298

Huijzer AS, Vandenberghe J (1998) Climatic reconstruction of the Weichselian Pleniglacial in north-western and central Europe. J Quat Sci 13:391–417

Huisink M (2000) Changing river styles in response to Weichselian climate changes in the Vecht valley, eastern Netherlands. Sed Geol 133:115–134

Isarin RFB (1997) Permafrost Distribution and Temperatures in Europe During the Younger Dryas. Permafr Periglac Process. 8:313–333

Isarin RFB, Renssen H, Koster EA (1997) Surface wind climate during the Younger Dryas in Europe as inferred from aeolian records and model simulations. Palaeogeogr, Palaeoclimatol, Palaeoecol 134:127–148

Issmer K (2001) Vistulian loess deposits in western Poland and their palaeoenvironmental implications. Quat Int 76(77):129–139

Jahn A (1956) Some periglacial problems in Poland. Biul Peryglac 4:169–184

Jahn A (1960) The oldest periglacial period in Poland. Biul Peryglac 9:159–162

Jahn A (1970) Zagadnienia strefy peryglacjalnej. PWN, Warszawa, p 200

Jary Z, Ciszek D (2013) Late Pleistocene loess–palaeosol sequences in Poland and western Ukraine. Quat Int 296:37–50

Kaiser K, Barthelmes A, Czakó Pap S, Hilgers A, Janke W, Kühn P, Theuerkauf M (2006) A Lateglacial palaeosol cover in the Altdarss area, southern Baltic Sea coast (Northeast Germany): investigations on pedology, geochronology, and botany. Netherlands J Geosci 85 (3):199–222. https://doi.org/10.1017/S0016774600021478

Kalinska E (2012) Geological setting and sedimentary characteristics of the coversands distributed in the western part of the Blonie glaciolacustrine basin (Central Poland)—preliminary results. Bull Geol Soc Finland 84: 33–44

Kalińska-Nartiša E, Woronko B, Ning W (2017) Microtextural inheritance on Quartz Sand Grains from Pleistocene periglacial environments of the Mazovian Lowland, Central Poland. Permafrost Periglac Process 28:741–756

Kamińska R, Konecka-Betley K, Mycielska-Dowgiałło E (1986) The Liszyno dune in the Vistula valley (east of Płock). Biul Peryglac 31:141–162

Karte J (1981) Zur Rekonstruktion des weichselhochglazialen Dauerfrostbodens im westlichen Mitteleuropa. Bochumer Geographische Arbeiten 40:59–71

Karte J (1987) Pleistocene periglacial conditions and geomorphology of in north and central Europe. In: Boadrman J (ed) Periglacial processes and landforms in Britain and Ireland. Cambridge University Press, Cambridge, pp 67–75

Kasse C (1993) Periglacial environments and climatic development during the early Pleistocene Tiglian stage (Beerse Glacial) in northern Belgium. Geol Mijnbouw 72:107–123

Kasse C (1995) Younger Dryas climatic changes and aeolian depositional environments. In Troelstra SR, Van Hinte, J. E. and Ganssen, G. M. (eds), The Younger Dryas. Koninklijke Nederlandse Akademie van Wetenschappen, Afdeling Natuurkunde, Eerste Reeks 44:27–31

Kasse C, Bohncke SJP, Vandenberghe J (1995a) Fluvial periglacial environments, climate and vegetation during the middle Weichselian in the northern Netherlands, with special reference to the Hengelo Interstadial. Mededelingen—Rijks Geologische Dienst 52:387–413.

Kasse K, Vandenberghe J, Bohncke S (1995b) Climatic change and fuvial dynamics of the Maas during the late Weichselian and early Holocene. In: Frenzel B (ed) European River Activity and Climatic Change during the Late Glacial and Early Holocene. PalaÈ okli-maforschung/Palaeoclimate Research vol. 14. Stuttgart, Gustav Fisher, pp 123–150

Kasse C (1997) Cold-climate sand-sheet formation in North-Western Europe (c. 14–12.4 ka); a response to permafrost degradation and increased aridity. Permafrost Periglac Process 8:295–311

Kasse C (1999a) Can involutions be used as palaeotemperature indicators? Biul Peryglac 38:95–110

Kasse C (1999b) Late Pleniglacial and Late Glacial aeolian phases in The Netherlands. GeoArchaeoRhein 3:61–82

Kasse C (2002) Sandy aeolian deposits and environments and their relation to climate during the Last Glacial Maximum and Lateglacial in northwest and Central Europe. Prog Phys Geogr 26:507–532

Kasse C, Aalbersberg G (2019) A complete Late Weichselian and Holocene record of aeolian coversands, drift sands and soils forced by climate change and human impact, Ossendrecht, the Netherlands. Neth J Geosci 98:e4

Kasse C, Huijzer AS, Krzyszkowski D, Bohncke SJP, Coope GR (1998) Weichselian Late Pleniglacial and Late-glacial depositional environments, Coleoptera and periglacial climatic records from central Poland (Bełchatów). J Quat Sci 13:455–469

Klatka T (1954) Peryglacjalne struktury tundrowe w Tychowie. Biul Peryglac 1:63–68

Klatkowa H (1954) Niecki korazyjne w okolicach Łodzi. Biul Peryglac 1:69–75

Klatkowa H (1965) Niecki i doliny denudacyjne w okolicach Łodzi. Acta Geographica Lodzensia 19

Klatkowa H (1994) Évaluation du role de l'agent périglaciaire en Pologne Centrale. Biul Peryglac 33:79–106

Klatkowa H (1996) Symptoms of permafrost presence in Middle Poland during the last 150 000 years. Biul Peryglac 34:45–86

Kleber A (1992) Periglacial slope deposits and their pedogenic implications in Germany. Palaeogeogr Palaeoclimatol Palaeoecol 99:361–371

Klosterman J (1995) Nordrhein-Westfalen. In: Benda L (ed) Das Quartär Deutschlands. Borntrager, Stuttgart, pp 59–94

Kluiving SJ, Verbers ALLM, Thijs WJF (2010) Lithological analysis of 45 presumed pingo remnants in the northern Netherlands (Friesland): Substrate control and fill sequences. Neth J Geosci Geol Mijnbouw 89:61–75

Kolstrup E (1980) Climate and stratigraphy in northwestern Europe between 30,000 BP and 13,000 BP, with special reference to the Netherlands. Meded Rijks Geol Dienst 32(15):181–238

Kolstrup E (1982) Cover sand and cover sand stratigraphy in southern Denmark. Geogr Tidsskr 82:88–91

Kolstrup E (1986) Reappraisal of the upper Weichselian periglacial environment from Danish frost wedge casts. Palaeogeogr Palaeoclimatol Palaeoecol 56:237–249

Kolstrup E (1994) Late Weichselian environmental change in southern Sweden and Denmark. J Quat Sci 9:127–132

Kolstrup E (2004) Stratigraphic and environmental implications of a large Ice-wedge cast at Tjæreborg, Denmark. Permafrost Perigla Process 15:31–40

Kolstrup E (2005) Periglacial geomorphology. In: Koster EA (ed) The physical geography of Western Europe Oxford University Press, New York

Kopp D (1965) Die periglaziare Deckzone (Geschiebedecksand) im nordost -deutschen Tiefland und ihre bodenkundliche Bedeutung. Berichte Der Geologischen Gesellschaft in Der DDR 10:39–771

Kopp D (1970) Kryogene Perstruktion und ihre Beziehung zur Bodenbildung im Moränengebiet. In: Richter H, Haase G, Lieberoth I, Ruske R (eds) Periglazial – Löß – Paläolithikum im Jungpleistozän der DDR. Erg.-Heft 274 Petermanns. Geographische Mitteilungen, pp 269–279

Koster EA (1988) Ancient and modern cold-climate eolian sand deposition: a review. J Quat Sci 3:69–83

Koster EA (2009) The "European aeolian sand belt": Geoconservation of drift sand landscapes. Geoheritage 1:93–110

Kozarski S (1958) Warstwowe osady stokowe w okolicy Chodzieży. Biul Peryglac 6:137–144

Kozarski S (1980) An outline of Vistulian stratigraphy and chronology of the Great Poland Lowland. Quat St in Poland 2:21–35

Kozarski S (1986) Early Vistulian permafrost occurrence in north-west Poland. Biul Peryglac 31:163–170

Kozarski S (1990) Pleni and Late Vistulian aeolian phenomena in Poland: new occurrences, palaeoenvironmental and stratigraphic interpretations. Acta Geographica Debrecina 1987–1988, Tomus XXVI–XXVII:31–45

Kozarski S (1993) Morfostratygraficzna i litostratygraficzna pozycja subfazy chodzieskiej w północno-wschodniej Wielkopolsce. II Seminarium—Geneza, litologia i stratygrafia utworów czwartorzędowych (Poznań, 14-15 października 1993), Streszczenia referatów i opisy posterów: 35–36

Kozarski S (1995) The periglacial impact on the deglaciation areas of northern Poland after 20 kyr BP. Biul Peryglac 34:73–102

Kozarski S, Nowaczyk B (1991) Lithofacies variation and chronostratigraphy of Late Vistulian and Holocene aeolian phenomena in northwestern Poland. Z Suppl 90:107–122

Kozarski S, Nowaczyk B, Rotnicki K, Tobolski K (1969) The eolian phenomena in west-central Poland with special reference to the chronology of phases of eolian activity. Geogr Pol 17:231–248

Kozarski S, Rotnicki K (1964) Inwolucje w sandrze stadium poznańskiego na południe od Gniezna. Biul Peryglac 13:15–52

Krauss L, Zens J, Zeeden C, Schulte P, Eckmeier E, Lehmkuhl F (2016) A multi-proxy analysis of two Loess-Paleosol sequences in the Northern Harz Foreland, Germany. Palaeogeogr Palaeoclimatol Palaeoecol 461:401–417

Krzyszkowski D (1990) Middle and Late Weichselian stratigraphy and palaeoenvironments in central Poland. Boreas 19:333–350

Krzyszkowski D, Kuszell T (2007) Middle and Upper Weichselian Pleniglacial fluvial erosion and sedimentation phases in southwestern Poland, and their relationship to Scandinavian ice-sheet built-up and retreat. Ann Soc Geol Pol. 77:17–38

Langohr R, Van Vliet N (1981) Properties and distribution of Vistulian permafrost traces in today surface soil of Belgium, with special reference to the data provided by Soil Survey. Biul Peryglac 28:137–148

Łapcik P, Ninard K, Uchman A (2021) Extra-large grains in Late Glacial—early Holocene aeolian inland dune deposits of cold climate, European Sand Belt, Poland: an evidence of hurricane-speed frontal winds. Sed Geol 415:105847

Laskowska W (1960) Kopalne struktury poligonalne na glinach zwałowych. Biul Peryglac 7:73–88

Liedtke H (1993) Phasen periglaziär-geomorphologischer Prägung während der Weichseleiszeit im norddeutschen Tiefland. Z Geomorphol 93:69–94

Liedtke H (2003) Deutschland zur letzen Eiszeit. In: Liedtke H, Mäusbacher R, Schmidt KH (eds) Nationalatlas Bundesrepublik Deutschland. Relief, Boden und Wasser. Spectrum Akademischer Verlag Heidelberg, Berlin.

Lindner L, Bogucki A (2002) Pozycja wiekowa środkowo- i późnoplejstoceńskich zjawisk peryglacjalnych we środkowo-wschodniej Europie. Zagadnienia Peryglacjału Polski i obszarów sąsiednich. Prace Instytutu Geografii AŚ w Kielcach 8:81–106 (in Polish with English summary)

Lippstreu L, Müller U, Stephan HJ, Hammer J, Meyer KD, Bremer F (2003) Fazies, Stratigraphie, Lagerungsverhältnisse, Verbreitung und Genese der Quartärsedimente in Norddeutschland. In: Handbuch zur Erkundung des Untergrundes von Deponien und Altlasten. Springer, Berlin, Heidelberg

Łoziński W (1912) Die periglaziale Fazies der mechanischen Verwitterung. [In:] XI Congr. Geol. Intern. Stockholm 1910: 1039–1053

Maarleveld GC (1956) Sur les sédiments périglaciaires en Hollande: formes et phénomènes. Biul Peryglac 4:73–82
Maarleveld GC (1960a) Wind directions and cover sands in the Netherlands. Biul Peryglac 8:49–58
Maarleveld GC (1960b) Glacial deposits in the Netherlands transformed under periglacial conditions. a review. Biul Peryglac 8:13–20
Maarleveld GC (1976) Periglacial phenomena and the mean annual temperature during the last glacial time in the Netherlands. Biul Peryglac 26:57–78
Majecka A, Forysiak J, Moska P, Okupny D, Marks L, Tołoczko-Pasek A (2021) Wiek bezwzględny i charakter procesów peryglacjalnych zapisanych w osadach z Józefowa, Wysoczyzna Łódzka (in Polish). In: Dobrowolski R, Orłowska A, Hołub B, Janicki G (eds) Glacjał i Peryglacjał Europy Środkowej. Wyd. Uniw. Marii Curie-Skłodowskiej, Lublin, pp 229–232
Manikowska B (1985) O glebach kopalnych, stratygrafii i litologii wydm środkowej Polski (summary: On the fossil soils stratigraphy and lithology of the dunes in central Poland). Acta Geographica Lodziensia 52
Manikowska B (1986) Sol fossile de la phase de transition pléistocéne-holocène dans les dunes continentales de la Pologne Centrale. Biul Peryglac 31:199–212
Manikowska B (1991a) Vistulian and Holocene aeolian activity, pedostratigraphy and relief evolution in Central Poland. Z Geomorphol 90:131–141
Manikowska B (1991b) Dune processes, age of dune terrace and Vistulian decline in the Vistlulian decline in the Vistlula Valley near Wyszogród, Central Poland. Bull Pol Acad Sci Earth Sci 39(2):137–148
Manikowska B (1993) Mineralogy and abrasion of sand grains due to Vistulian (Late Pleistocene) aeolian processes in Central Poland. Geol Mijnbouw 72:167–177
Manikowska B (1994) État des études des processus éoliens dans la région de Łódź (Pologne Centrale). Biul Peryglac 33:107–132
Manikowska B (1995) Aeolian activity differentiation in the area of Poland during the period 20-8 ka BP. Biul Peryglac 34:125–166
Maréchal R (1956) L'étude de phénomènes périglaciaires en Belgique. Biul Peryglac 4:83–98
Marks L, Gałązka D, Woronko B (2016) Climate, environment and stratigraphy of the last Pleistocene glacial stage in Poland. Quat Int 420:259–271
Marks L, Makos M, Szymanek M, Woronko B, Dzierżek B, Majecka A (2019) Late Pleistocene climate of Poland in the mid-European context. Quat Int 504:24–39
Maruszczak H, Goździk J (2001) Znaczenie paleogeograficzne osadów fluwioperyglacjalnych poprzedzających nasunięcie lądolodu odrzańskiego w dolinie Wisły środkowej. In: Mycielska-Dowgiałło E (ed), Eolizacja osadów jako wskaźnik stratygraficzny czwartorzędu. Warszawa, pp 65–81
Meerschman E, Van Meirvenne M, De Smedt P, Saey T, Islam MM, Meeuws F, Van De Vijver E, Ghysels G (2011) Imaging a polygonal network of Ice-Wedge casts with an electromagnetic induction sensor. Soil Sci Soc Am J 75:2095–2100
Meszner S, Fuchs M, Faust D (2011) Loess-Palaeosol-sequences from the loess area of Saxony (Germany). E&G Quat Sci J 60:47–65
Meszner S, Kreutzer S, Fuchs M, Faust D (2013) Late Pleistocene landscape dynamics in Saxony, Germany: paleoenvironmental reconstruction using loess-paleosol sequences. Quat Int 296:94–107
Mleczak M, Woronko B, Pisarska-Jamroży M, Bujak Ł (2021) Permafrost as the main factor controlling the fluvial sedimentation style on glaciomarginal fans. Sed Geol 422:105971
Mojski JE (1958) Struktury krioturbacyjne na tarasach Wisły w okolicy Włocławka. Biul Peryglac 6:145–152
Mojski JE (2005). Ziemie Polskie w czwartorzędzie. Zarys morfogenezy. Państwowy Instytut Geologiczny, Warszawa
Mol J, Vandenberghe J, Kasse K, Stel H (1993) Periglacial microjointing and faulting in Weichselian fluvio-aeolian deposits. J Quat Sci. 8:15–30

Mol J, Vandenberghe J, Kasse C (2000) River response to variations of periglacial climate in mid-latitude Europe. Geomorphol 33:131–148

Moska P, Jary Z, Sokołowski RJ, Zieliński P, Poręba G, Raczyk J, Krawczyk M, Skurzyński J, Zieliński P, Michczyński A, Tudyka K, Adamiec G, Piotrowska N, Pawełczyk F, Łopuch M, Szymak A, Ryzner K (2020) Chronostratigraphy of Late Glacial aeolian activity in SW Poland—a case study from the Niemodlin Plateau. Geochronometria 47:124–137

Moska P, Sokołowski RJ, Jary Z, Zieliński P, Raczyk J, Szymak A, Krawczyk M, Skurzyński J, Poręba G, Łopuch M, Tudyka K (2021) Stratigraphy of the Late Glacial and Holocene aeolian series in different sedimentary zones related to the Last Glacial maximum in Poland. Quat Int. https://doi.org/10.1016/j.quaint.2021.04.004

Moskalewicz D, Tylmann K, Woźniak PP, Kopyść N, Moska P (2020) Fluvial response to environmental conditions during MIS 4–3: a sedimentary record at the Brześnica site, central-western Poland. Geol Q 64(4). https://doi.org/10.7306/gq.1560

Mycielska-Dowgiałło E (1993) Estimates of Late Glacial and Holocene aeolian activity in Belgium, Poland and Sweden. Boreas 22:165–170

Mycielska-Dowgiałło E, Woronko B (1998) Analiza obtoczenia i zmatowienia powierzchni ziarn kwarcowych frakcji piaszczystej i jej wartość interpretacyjna. Przeg. Geol 46:1275–1281

Mycielska-Dowgiałło E, Woronko B (2004) The degree of aeolization of Quaternary deposits in Poland as a tool for stratigraphic interpretation. Sed Geol 168:149–163

Mydel R, Groch J (2000) Przeglądowy Atlas Świata. Europa, część 1. Fogra Oficyna Wydawnicza, Kraków

Narloch W, Wysota W, Piotrowski JA (2013) Sedimentological record of subglacial conditions and ice sheet dynamics of the Vistula Ice Stream (north-central Poland) during the Last Glaciation. Sed Geol 293:30–44

Nörvang A (1939) Stenringe og frostspalter I Danmark. Natur-hist. Tidende 3.

Nörvang A (1942) Frostspalter I Jylland. Medd. Dansk Geol. Fren 10:2

Nowaczyk B (1986) Wiek Wydm w Polsce. Uniwersytet im. Adama Mickiewicza w Poznaniu, Seria Geografia 28. Poznan:230–45. English abstract: The age of dunes, their textural and structural properties against atmospheric circulation pattern of Poland during the Late Vistulian and Holocene

Olchowik-Kolasińska J (1955) Struktury kongeliflukcyjne w okolicach Łodzi. Biul Peryglac 2:109–116

Paepe R, Pissart A (1969) Periglacial structures in the Late-Pleistocene stratigraphy of Belgium. Biul Peryglac 20:321–336

Paeper R, Vanhoorne R (1970) Stratigraphical position of periglacial phenomena in the Campine Clay of Belgium, based on palaeobotanical analysis and palaeomagnetic dating. Bull Soc Ver Geol Paleont Hydrol 79(3–4):201–211

Petera J (2002) Vistuliańskie osady dolinne w basenie uniejowskim i ich wymowa paleogeograficzna. Acta Geographica Lodziensia 83:174 p (in Polish with English summary)

Petera-Zganiacz J (2011) Changes in the development of frost-wedges in the middle Warta valley deposits (Central Poland). Geologija 53 (1) (73):15–20

Petera-Zganiacz J (2016) Czynniki determinujące zróżnicowanie inwolucji późnovistuliańskich w północno zachodniej części regionu łódzkiego Folia Geographica. Physica 15:45–54

Petera-Zganiacz J, Dzieduszyńska DA, Twardy J, Pawłowski D, Płóciennik M, Lutyńska M, Kittel P (2015) Younger Dryas flood events: a case study from the middle Warta River valley (Central Poland). Quat Int 386:55–69

Pissart A (1956) L'origine périglaciaire des vivieres des Hautes Fagnes. Ann Phys Quat 33:339–357

Pissart A (1983) Remnants of periglacial mounts in the Hautes Fagnes (Belgium): structure and age of ramparts. In: Terwindt JHJ, Van Steijn H (eds) Development in physical geography—a tribute to J.I.S. Zonneveld. Geol Mijnbouw 62:551–555

Pissart A (1987) Weichselian periglacial structures and their environmental significance: Belgium, the Netherlands and northern France. In: Boardman J (ed) Periglacial processes and landforms in Britain and Ireland. Cambridge University Press, Cambridge, pp 77–85

Pissart A (2000) Remnants of Lithalsas of the Hautes Fagnes, Belgium: a summary of present-day knowledge. Permafrost Periglac Process 11:327–355

Ploeger PL, Groenman-van Waateringe W (1964) Late glacial pingo and valley development in the Boorne region near Wijnjeterp, province of Friesland, Netherlands. Biul Peryglac 13:199–234

Poser H (1948) Boden- und Klimaverhältnisse in Mitteleuropa während der Würm-Eiszeit. Erdkunde 2:53-68

Poser H (1950) Zur Rekonstruktion der Spätglazialen Luftdruckverhältnisse in Mittel- und Westeuropa auf Grund der vorzeitlichen Binnendünen. Erdkunde 4:81–88

Poser H (ed) (1954) Studien über die Periglazial-Erscheinungen in Mitteleuropa. Gottinger Geographische Abhandlungen, pp 1–4

Puzachenko AY, Markova AK, Kosintsev PA, van Kolfschoten T, van der Plicht J, Kuznetsova TV, Tikhonov AN, Ponomarev DV, Kuitems M, Bachura OP (2017) The Eurasian mammoth distribution during the second half of the Late Pleistocene and the Holocene: regional aspects. Quat Int 445:71–88

Radtke U, Janotta A (1998) Ein Beitrag zur Beurteilung der Aussagekraft von Lumineszenzaltern fűr die Datierung von spätpleistozänen und holozänen Dűnen anhand des Laacher See-Tuffes (12.900 J.v.h.): Der Testfall Dűne Mainz-Gonsenheim. Kőlner Geographische Arbeiten 70:1–18

Rasmussen SO, Bigler M, Blockley SP, Blunier T, Buchardt SL, Clausen HB, Cvijanovic I, Dahl-Jensen D, Johnsen SJ, Fischer H, Gkinis V, Guillevic M, Hoek WZ, Lowe JJ, Pedro JB, Popp T, Seierstad IK, Steffensen JP, Svensson AM, Vallelonga P, Vinther BM, Walker MJC, Wheatley JJ, Winstrup M (2014) A stratigraphic framework for abrupt climatic changes during the Last Glacial period based on three synchronized Greenland ice-core records: refining and extending the INTIMATE event stratigraphy. Quat Sci Rev 106:14–28

Renssen H, Kasse C, Vandenberghe J, Lorenz SJ (2007) Weichselian Late Pleniglacial surface winds over northwest and central Europe: a model—data comparison. J Quat Sci 22:281–293

Richter HG (1973) Main lines of the regional structure of periglacial facies on the territory of G.D.R. Biul Peryglac 23:125–134

Rotnicki K (1964) Periglacial slope planations in the south-eastern part of the Ostrzeszów hills. Biul Peryglac 13:235–260

Rousseau DD, Boers N, Sima A, Svensson A, Bigler M, Lagroix F, Taylor S, Antoine P (2017) (MIS3 & 2) millennial oscillations in Greenland dust and Eurasian aeolian records—a paleosol perspective. Quat Sci Rev 169(1):99–113

Rutkowski J, Król K, Lemberger M (1998) The pingo remnant in the Suwalki Lake region (NE Poland). Quat Stud Pol 15:55–60

Rychel J, Woronko B, Błaszkiewicz M, Karasiewicz T (2018) Aeolian processes records within last glacial limit areas based on the Płock Basin case (Central Poland). Bull Geol Soc Finland 90:223–237

Schenk E (1955) Die periglazialen strukturbodenbildungenen als folgen der hydratationsvorgange in bödem. Eiszeit Gegenw 6:170–184

Schirmer W (1999a) Dune phases and fossil soils in the European sand belt. In: Schirmer W (ed) Dunes and fossil soils. - GeoArchaeoRhein 3:11–42

Schirmer W (ed) (1999b) Dunes and fossil soils. GeoArchaeoRhein 3:190.

Schönhage W (1969) Note on the ventifacts in the Netherlands. Biul Peryglac 20:355–360

Schuurmans C (2005) Climate: mean state, variability, and change. In: Koster EA (ed) The physical geography of western Europe. Oxford University Press, New York

Semmel A, Terhorst B (2010) The concept of the Pleistocene periglacial cover beds in central Europe: a review. Quat Int 222:120–128

Seppälä M (2004) Wind as a geomorphic agent in cold climates. Cambridge University Press, 358pp

Svensson H (1982) A low-lying polygon locality on the Laholm plain, Swedish west coast. Geologiska Föreningen i Stockholm Förhandlingar 104(1):69–73

Słowański W (1963) Pseudozmarzlinowe formy szczelinowe w kredzie jeziornej okolic Laski, koło Brus. Biul Peryglac 12:165–172

Słowiński M, Błaszkiewicz M, Brauer A, Noryśkiewicz B, Ott F, Tyszkowski S (2015) The role of melting dead ice on landscape transformation in the early Holocene in Tuchola Pinewoods. North Poland, Quat Int 388:64–75

Solarska A, Hose TA, Vasiljević DA, Mroczek P, Jary Z, Marković SB, Widawski K (2013) Geodiversity of the loess regions in Poland: inventory, geoconservation issues, and geotourism potential. Quat Int 296:68–81

Stankowski W (1996) Wstęp do geologii kenozoiku ze szczególnym odniesieniem do terytorium Polski. Wydawnictwo Naukowe UAM, Poznań

Starkel L (1988) Paleogeography of the periglacial zone in Poland during the maximum advance of the Vistulian ice sheet. In: Kozarski S (ed) Environmental changes in Poland and Sweden after the maximum of the last glaciation. Proceedings of the First Polish—Swedish Seminar Poznan, Poland, Oct 9–15, 1986. Geographia Polonica 55:151–164

Starkel L (1996) Cykle glacjalno-interglacjalne w ewolucji systemu rzecznego Wisły. In: Kostrzewski A. (Ed.) Geneza, litologia i stratygrafia utworów czwartorzędowych 2, Wyd. UAM, Poznań, pp 297–305

Starkel L, Gębica P, Superson J (2007) Last Glacial-Interglacial cycle in the evolution of river valleys in southern and central Poland. Quat Sci Rev 26:2924–2936

Steinhauser F (ed) (1970) Climatic atlas of Europe. WNO, Unesco, Cartographia. Geneva, Budapest

Svensson H (1963) Some observations in West-Jutland of a polygonal pattern in the ground. Geogr Tidsskr 62:122–124

Svensson H (1984) The periglacial form group of southwestern Denmark. Geogr Tidsskr 84:25–34

Szczypek T (1977) Utwory i procesy eoliczne w północnej części Wyżyny Śląskiej. Prace Naukowe UŚ 190:116

Szewczyk J, Nawrocki J (2011) Deep-seated relict permafrost in northeastern Poland. Boreas 40:385–388

Tricart J (1960) Klimat i geomorfologia. In: Zagadnienia geomorfologiczne, PWN, Warszawa, pp 25–50

Troll C (1944) Strukturböden, Solifluktion und Frostklimate der Erde, Geol. Rundschau 34:545-694

Turkowska K (1994) La morphogenèse périglaciaire dans les vallées fluviales du Plateau de Łódź et sa differentiation dans le temps et dans l'espace. Biul Peryglac 33:153–164

Turkowska K (1995) Recognition of valleys evolution during the Pleistocene-Holocene transition in non-glaciated regions of Polish lowland. Biul Peryglac 34:209–227

Turkowska K (1988) Évolution des vallées fluviatiles sur le Plateau Plateau du Łódź au cours du Quaternaire tardif. Acta Geogr Lodz.57:1–157 (in Polish with French summary)

Turkowska K (1999) Some relfections on the Łódź contribution to periglacial research and evolution criteria of periglacial morphogeny in middle Poland. Biul Peryglac 38:43–61

Turkowska K (2007) Morphology and infillings of upper Mroga and Mro˙ zyca Valleys as evidence of post-glacial stages of the watershed area evolution (in polish with English summary). Acta Geogr. Lodz. 93:87–105

Turkowska K, Wieczorkowska J (1986) L'influence du relief sur le caractères des depots de versant périglaciaire dans la région de Łódź. Biul Peryglac 31:293–309

Tylmann K, Uścinowicz S (2022) Timing of the last deglaciation phases in the southern Baltic area inferred from Bayesian age modeling. Quat Sci Rev. 287:1–11

Urban B, Lenhard R, Mania D, Albrecht B (1991) Mittelpleistozän im Tagebau Schöningen. Ldkr. Helmstedt. Zeitschrift Deutscher Geologischer Geselschaft 142:351–372

Urbaniak U (1969) Les sables de couverture; les cryoturbations et les fractures dans les dunes du Bassin de Płock. Biul Peryglac 19:399–422

Urbaniak-Biernacka U (1976) Badania wydm środkowej Polski zwykorzystaniem metod statystycznych. Prace Naukowe, Geodezja, 17, Wyd. Polit. Warszaw

van der Tak-Schneider U (1968) Craks and fissures of post-Alleröd age in the Netherland. Biul Peryglac 17:221–226

Van Huissteden J (1990) Tundra rivers of the Last Glacial: sedimentation and geomorphological processes during the middle pleniglacial (eastern Netherlands). Meded Rijks Geol Dienst 44:1–138

van Huissteden J, Gibbard PL, Briant RM (2001a) Periglacial fluvial systems in northwest Europe during marine isotope stages 4 and 3. Quat Int 79:75–88

Van Huissteden J, Schwan JCG, Bateman MD (2001b) Environmental conditions and paleowind directions at the end of the Weichselian Late Pleniglacial recorded in aeolian sediments and geomorphology (Twente, Eastern Netherlands). Geol Mijnbouw 80:1–18

van Huissteden J, Pollard D (2003) Oxygen isotope stage 3 fluvial and eolian successions in Europe compared with climate model results. Quat Res 59:223–233

Van Huissteden J, Vandenberghe J, Van der Hammen T, Laan W (2000) Fluvial and aeolian interaction under permafrost condition: Weichselian Late Pleniglacial, Twente, eastern Netherlandes. CATENA 40:307–321

Van Huissteden K, Vandenderghe J, van Gel B (1986) Late Pleistocene stratigraphy and fluvial history of the Dinkel Basin (Twente, Eastern Netherlands). Eiszeit Gegenw 36:43–59

van Leckwijck W, Macar P (1960) Les structures périglaciaires antérieures au Wurm en Belgique. Biul Peryglac 9:47–60

Van Vliet-Lanoë B (1991) Differential frost heave, load casting and convection: converging mechanisms; a discussion of the origin of cryoturbations. Permafrost Periglac Process 2:123–139

Van Vliet-Lanoë B (1994) Histoire et dynamique du pérgelisol européen au Weichselien. Biulety Peryglacjalny 33:165–176

Vandenberghe J (1983) Some periglacial phenomena and their stratigraphic position in Weichselian deposits in the Netherlands. Polarforschung 53(2):97–107

Vandenberghe J (1993) Changing fluvial processes under changing periglacial conditions. Zeitschrift für Geomorphologie Neue Folge Supplement-Band 88:17–28

Vandenberghe J (1992) Cryoturbations: a sediment structural analysis. Permafrost Periglac Process 3:343–352

Vandenberghe J (2013) Cryoturbation structures. In: Elias S.A. (ed.) The Encyclopedia of Quaternary Science. Elsevier, Amsterdam, pp 430–435

Vandenberghe DAG, Derese C, Kasse C, Van den haute P (2013) Late Weichselian (fluvio-)aeolian sediments and Holocene drift-sands of the classic type locality in Twente (E Netherlands): a high-resolution dating study using optically stimulated luminescence. Quat Sci Rev 68:96–113

Vandenberghe J, Isarin RFB, Renssen H (1999) Comments on windpolished boulders as indicators of a Late Weichselian wind regime in Denmark in relation to neighbouring areas by Christiansen and Svensson 9(1), 1–21. Permafrost Periglac Process 10:199–201

Vandenberghe J, Kasse K, Coope R (eds) (1998) Climatic reconstruction of the Last interglacial-Glacial cycle in Western & Central Europe (Special Issue of the J Quat Sci 13(5), 361–497). John Wiley & Sons, Chichester

Van der Hammen T (1971) The Upper Quaternary stratigraphy of the Dinkel valley. In Van der Hammen T and Wijmstra TA (eds), The Upper Quaternary of the Dinkel Valley. Mededelingen Rijks Geologische Dienst Nieuwe Serie 22, pp 59–72

Van Geel B, Coope GR, Van der Hammen T (1989) Palaeoecology and stratigraphy of the Lateglacial type section at Usselo (The Netherlands). Rev Palaeobot Palynol 60:25–129

Van Loon AJ (Tom), Pisarska-Jamroży M, Woronko B (2020) Sedimentological distinction in glacigenic sediments between load casts induced by periglacial processes from those induced by seismic shocks. Geol Q 64(3):626–640

Wachecka-Kotkowska L, Krzyszkowski D, Król E, Klaczak K (2014) Middle Weichselian Pleniglacial fluvial erosion and sedimentation in the Krasówka river valley, Szczerców field, Bełchatów open cast mine, central Poland. Ann Soc Geol Pol. 84(4):323–340

Waller RI, Murton JB, Kristensen L (2012) Glacier–permafrost interactions: processes, products and glaciological implications. Sed Geol 255:1–28

Weckwerth P, Pisarska-Jamroży M (2015) Periglacial and fluvial factors controlling the sedimentation of Pleistocene breccia in NW Poland. Geogr Ann 97(A):415–430

Weise OR (1983) Das Periglazial. Geomorphologie und Klima in gletscherfreien, kalten Regionen. Gebr. Borntraeger, Berlin

Wojtanowicz J (1965) Wydmy międzyrzecza Sanu i Łęgu. Annales UMCS, B 20:89–124

Wojtanowicz J (1969) Typy genetyczne wydm Niziny Sandomierskiej. Annales UMCS, B 24:1–45

Wojtanowicz J (1999) Problem of occurrence and age (TL) of inland Plenivistulian dunes in Poland (on the example of Sandomierz Basin). In Schirmer W (ed) Dunes and fossil soils, GeoArchaeoRhein 3; 43–53.

Woldstedt P, Duphorn K (1974) Norddeutschland und angrenzende Gebiete im Eiszeitalter. Koehler, Stuttgart, p 500

Woronko B (2012) Zapis procesów eolicznych w piaszczystych osadach plejstocenu na wybranych obszarach Polski środkowej i północno-wschodniej. Wydział Geografii i Studiów Regionalnych, Warszawa, pp 130

Woronko B, Bujak Ł (2018) Quaternary aeolian activity of Eastern Europe (a Poland case study). Quat Int 478:75–96

Woronko B, Karasiewicz TM, Rychel J, Kupryjanowicz M, Fiłoc M, Moska P, Adamczyk A, Demitroff MN (2022). A palaeoenvironmental record of MIS 3 climate change in NE Poland—Sedimentary and geochemical evidence. Quat Int. 617:80–100

Woronko B, Rychel J, Honczaruk M (2021) Quartz-grain microweathering amid Pleistocene-aged deep-seated relict permafrost in Central Europe, NE Poland. Quat Int. 605–606: 65–80

Woronko B, Zieliński P, Sokołowski RJ (2015) Climate evolution during the Pleniglacial and Late Glacial as recorded in quartz grain morphoscopy of fluvial to aeolian successions of the European Sand Belt. Geologos 21(2):89–103

Wortmann H (1956) Ein erstes sicheres Vorkommen von periglazialem Steinnetzboden im Norddeutschen Flachland. – Eiszeitalter & Gegenwart Quat Sci 7(1):119–126

Zeeberg J (1998) The European sand belt in eastern Europe—and comparison of Late Glacial dune orientation with GCM simulation results. Boreas 27:127–139

Zieliński T (2007) The Pleistocene climate-controlled fluvial sedimentary record in the Bełchatów mine (central Poland). Sediment Geol 193:203–209.

Zieliński P (2017) Regional and local conditions of the Late Vistulian aeolian deposition in the central part of the European sand belt. Wyd. Uniwersytetu Marii Curie-Skłodowskie, Lublin, pp 235

Zieliński P (2001) Procesy eoliczne w Kotlinie Chodelskiej (Wyżyna Lubelska) – ich natężenie i fazy rozwoju. Prz Geogr. 73:57–73

Zieliński T, Goździk J (2001) Palaeoenvironmental interpretation of a Pleistocene alluvial succession in central Poland: sedimentary facies analysis as a tool for palaeoclimatic inferences. Boreas 30:240–253

Zieliński P, Sokołowski RJ, Fedorowicz S, Woronko B, Hołub B, Jankowski M, Kuc M, Tracz M (2016a) Depositional conditions on an alluvial fan at the turn of the Weichselian to the Holocene—a case study in the Żmigród Basin, southwest Poland. Geologos 22(2):105–120

Zieliński P, Sokołowski RJ, Woronko B, Jankowski M, Fedorowicz S, Standzikowski K (2016b) Sandy deposition in a small dry valley in the periglacial zone of the Last Glacial maximum: a case study from the Józefów site, SE Poland. Quat Int 399:58–71

Zieliński P, Sokołowski RJ, Woronko B, Jankowski M, Fedorowicz S, Zaleski I, Molodkov A, Weckwerth P (2015) The depositional conditions of the fluvio-aeolian succession during the last climate minimum based on the examples from Poland and NW Ukraine. Quat Int 386:30–41

Periglacial Landforms in Northern Europe

Great Britain and Ireland

Colin K. Ballantyne and Julian B. Murton

1 Introduction

1.1 Topography and Geology

Great Britain and Ireland (together known as the British Isles) are located on the NE margin of the Atlantic Ocean (50–61°N, 10°W–02°E; Fig. 1) and contain an outstanding variety of periglacial landscapes, some of which evolved under successive glacial stages during the Quaternary, whilst others developed no earlier than the Lateglacial or Holocene. In considering the nature and distribution of periglacial landscapes in Britain and Ireland, it is useful to distinguish between lowland periglacial environments and upland periglacial environments of high relief. This is because lowland areas not only contain periglacial landsystems (landform assemblages) that differ from those of upland areas, but also because all lowland periglacial landforms are relict features of pre-Holocene age, whereas most upland periglacial landforms are no older than the Lateglacial period, and some are active at present.

The main areas of lowland periglaciation are located on Permian and younger sedimentary rocks in central, eastern and southern England and on the Carboniferous sedimentary rocks of the Central Plain of Ireland and Midland Valley of Scotland. Most of these areas lie below 400 m, but are interrupted by low hills and escarpments, such as the Cotswold and Chiltern Hills in England, the Slieve Bloom in Ireland and the Ochil Hills in central Scotland. The lowlands largely developed on low-relief

C. K. Ballantyne (✉)
School of Geography and Sustainable Development,
University of St Andrews, St Andrews KY16 9AL, Scotland, UK
e-mail: ckb@st-andrews.ac.uk

J. B. Murton
Permafrost Laboratory, Department of Geography, University
of Sussex, Brighton BN1 9QJ, UK
e-mail: j.b.murton@sussex.ac.uk

M. Oliva et al. (eds.), *Periglacial Landscapes of Europe*,
https://doi.org/10.1007/978-3-031-14895-8_13

Fig. 1 The British Isles (Great Britain and Ireland), showing key locations mentioned in the text

land surfaces formed by Palaeogene and Neogene erosion, surfaces now represented by horizontal, gently undulating (<5°) plains or low, dissected plateaux.

The uplands and mountains of Britain and Ireland are, by alpine standards, of modest elevation: the highest mountains in Scotland, Wales, Ireland and England are, respectively, Ben Nevis (1344 m), Snowdon (1085 m), Carrauntoohil (1039 m) and Scafell Pike (978 m); only in the Scottish Highlands are there extensive areas of ground over 900 m. The upland and mountain areas are underlain by a remarkable variety of igneous, metamorphic and resistant sedimentary rocks, that range in age from Archaean to Palaeogene, though Neoproterozoic and Palaeozoic rocks predominate. Most mountains in the British Isles were formed through differential uplift of a land surface of relatively low relief during the Cenozoic, and their general form has been strongly influenced by glacial erosion during the Pleistocene: those in the west have been strongly dissected to form alpine-type landscapes of glacial troughs, deep rock basins, cirques and arêtes; those in the east, particularly in Scotland, are dominated by extensive residual Cenozoic palaeosurfaces (high plateaux) incised by glacial troughs.

1.2 *Quaternary Landscape Evolution*

During successive glacial stages, the Quaternary Period in the British Isles was characterised by glaciation and periglaciation under a climate of arctic severity. Each return to temperate conditions during successive interglacial stages was dominated by paraglacial landscape modification and the operation of weathering, fluvial, slope, aeolian and coastal processes. These changes were accompanied by alternation of vegetation cover, from temperate deciduous forest during interglacials to boreal forest, tundra and cold desert during glacial stages. During Middle and Late Pleistocene glacial stages, relative sea level fell by up to 130 m around the British Isles, exposing much of the adjacent continental shelf.

There is evidence that glacial stages in the British Isles had two climatic components: (i) periods of cold maritime conditions, when the NE Atlantic Ocean lacked sea ice; and (ii) periods of cold continental conditions, when the Atlantic oceanic polar front extended far to the south, allowing extensive sea-ice cover to form across the NE Atlantic Ocean and causing arid conditions with extremely cold winters. Rose (2010) has argued that the cold maritime regime favoured rapid expansion of glaciers, highly active periglacial slope processes and enhanced fluvial activity, whereas the onset of cold continental conditions resulted in glacier retreat, renewed permafrost aggradation and prevalence of wind erosion and associated deposition of coversand and loess.

The last glacial stage in the British Isles is termed the Devensian in Great Britain (Fig. 2) and Midlandian in Ireland, and is equivalent to the Weichselian in NW Europe; the preceding interglacial is termed the Ipswichian in Great Britain (Knocknacran in Ireland) and is equivalent to the Eemian in NW Europe. In broad terms, the Devensian is conventionally subdivided into the Early Devensian (Marine Isotope Stages (MIS) 5d–4; ~116–58 ka), Middle Devensian (MIS 3; ~58–31 ka) and Late Devensian (MIS 2; ~31–11.7 ka). The Late Devensian is subdivided into the Dimlington Stade (~31–14.7 ka) and Loch Lomond (≈ Younger Dryas) Stade (~12.9–11.7 ka), separated by the cool temperate Lateglacial (≈ Bølling/Allerød) Interstade (~14.7–12.9 ka). Pre-Ipswichian stages are poorly differentiated in the British Isles, apart from the Anglian (≈ Elsterian in NW Europe) Glacial Stage (MIS 12), the preceding Cromerian Interglacial (MIS 13) and the succeeding Hoxnian Interglacial (MIS 11). The long period of time (~260 ka) between the end of the Hoxnian and

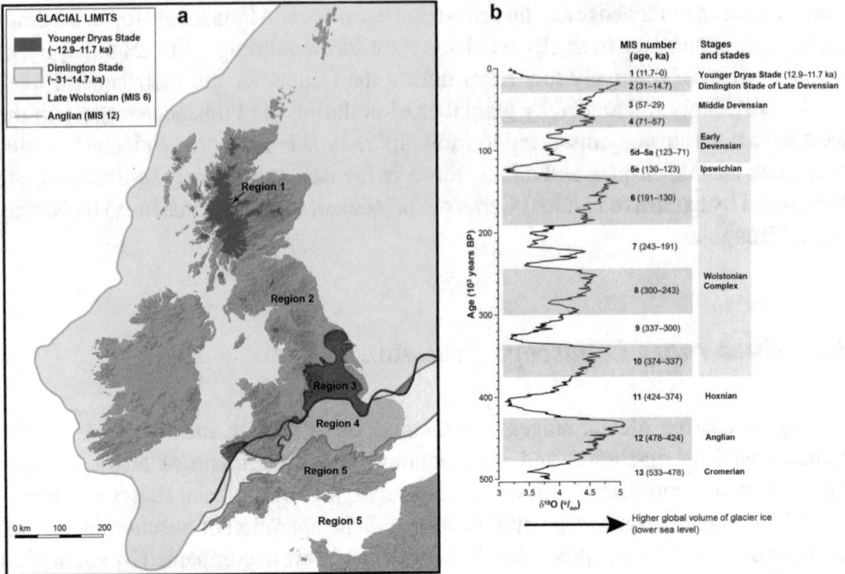

Fig. 2 **a** Periglacial regions of Great Britain and Ireland. For explanations, see text. **b** Timescale for periglacial conditions in the British Isles during the past 500 ka. Blue shading indicates cold (even-numbered) marine isotope stages (MIS), during which periglacial conditions with extensive permafrost prevailed; periglacial conditions with at least localised permafrost also occurred during MIS 3. Adapted from Murton JB, Ballantyne CK (2017) Geol Soc Lond Eng Geol Spec Publ 28:501–597 © The authors

the beginning of the Ipswichian Interglacials, which encompasses three cold stages (MIS 10, 8 and 6) and two intervening warm stages (MIS 9 and 7), is termed the Wolstonian Stage or Wolstonian Complex (≈ Saalian of NW Europe; Fig. 2).

The British Isles can be subdivided into periglacial regions (Fig. 2a) based on the extent of glacier ice at different times (Murton and Ballantyne 2017). Region 1 comprises uplands that were glaciated during the Loch Lomond Stade, and in this region periglacial processes have affected the landscape only after ~12.5–11.5 ka as these glaciers retreated. Region 2 is the most extensive and represents all other areas that were covered by the last British-Irish Ice Sheet, which reached its maximum extent diachronously over the period ~30–21 ka. Within this region, permafrost formed and periglacial conditions prevailed both in the wake of the retreating ice sheet prior to ~14.7 ka and again during the Loch Lomond Stade. Region 3 was probably last glaciated at ~160 ka (MIS 6), and region 4 during the Anglian Glacial Stage (MIS 12; ~478–424 ka): both these regions were affected by severe periglacial conditions during all subsequent cold stages, with associated development of permafrost and, during intervening warmer episodes, thermokarst. Finally, region 5 represents the 'never-glaciated' part of England, where periglacial conditions of varying severity affected the landscape during all Pleistocene cold stages. Within this zone there is evidence for former glaciation only on the higher parts of Dartmoor in SW England, which probably supported a small ice cap during the Dimlington Stade and possibly earlier.

This regional model, however, identifies only the minimum timescale (i.e. since the most recent glacier ice cover) during which periglacial processes have affected regions 1–4, as within these regions there is stratigraphic evidence for earlier episodes of permafrost development or other forms of periglacial activity. East Anglia (region 4), for example, contains evidence for at least three episodes of permafrost development prior to the Anglian Stage (Lee et al. 2003), and in NE Scotland (region 2), Late Devensian till overlies a succession of earlier glacigenic and periglacial facies that date back to MIS 8 (300–243 ka; Hall et al. 2019). The present landscape north of the Anglian glacial limit is therefore a palimpsest of past glacial, periglacial and interglacial landforms and sediments. Because successive Middle and Late Pleistocene ice sheets have largely erased the imprint of earlier events, however, the effects of periglacial activity are usually only represented at the surface following the most recent deglaciation, though locally the record of much earlier periglacial conditions can be reconstructed from stratigraphic evidence (e.g. Rose et al. 1985; Candy et al. 2011; Murton et al. 2015).

2 The Development of Periglacial Research in Great Britain and Ireland

The development of periglacial research in Great Britain prior to 1990 has been summarised by Ballantyne and Harris (1994), and early research in Ireland has been outlined by Lewis (1985). The first accounts of periglacial phenomena in the British Isles appeared over a century ago, and by 1940 a wide range of periglacial features had been recognised and investigated. Three early accounts were particularly prescient: Hollingworth (1934) discussed the origin of active patterned ground and solifluction landforms in the English Lake District; Paterson (1940) interpreted the frost polygons, ice-wedge pseudomorphs and involutions of eastern England through comparison with their arctic analogues; and Dines et al. (1940) described how the head (gelifluctate) deposits of southern England could be interpreted as climatostratigraphic units in Quaternary successions.

The following quarter century (1940–1965) provided much of the basis for the present understanding of periglacial phenomena in Britain and Ireland, with the publication of papers on features indicative of past permafrost (ice-wedge pseudomorphs, involutions, frost polygons, thermokarst depressions and fragipans), cryoplanation and nivation features, head deposits, cambered bedrocks, stratified slope deposits, coversands, loess, tors, talus and active periglacial features on high ground. A particularly important contribution was made by Williams (1965), who demonstrated that continuous permafrost formerly underlay almost all of southern England during the last (Late Devensian) glaciation, and who subsequently employed this evidence to reconstruct past temperature conditions (Williams 1969).

Periglacial research during the period 1965–1990 extended the range of periglacial phenomena identified in Britain and Ireland to include relict active-layer detachment slides, relict sand wedges, periglacial valley-fill deposits and, on high ground, such features as ploughing boulders, earth hummocks (thúfur) and niveo-aeolian deposits. Key developments during this period included application of radiocarbon and luminescence dating to establish the age of periglacial deposits, together with detailed sedimentological and stratigraphic analysis of such sediments, and increased involvement of engineering geologists in analysis of the geotechnical significance of periglacial deposits (e.g. Higginbottom and Fookes 1971; Skempton and Weeks 1976). Our understanding of the role of periglaciation in fashioning the landscapes of the British Isles towards the end of the twentieth century was encapsulated in two books: *Periglacial processes and landforms in Great Britain and Ireland* (Boardman 1987) and *The periglaciation of Great Britain* (Ballantyne and Harris 1994).

Subsequent research on periglacial features in Britain and Ireland is described in recent reviews (Murton and Ballantyne 2017; Wilson 2017; Ballantyne 2018). Several trends are evident. There has been increased emphasis on placing relict periglacial deposits and structures within their chronostratigraphic and palaeoclimatic context. In particular, the burgeoning literature on periglacial processes, landforms and sediments in present permafrost environments has informed much more forensic interpretation or reinterpretation of Middle and Late Pleistocene periglacial stratigraphy in lowland England (zones 3–5), often complemented by the employment of biotic evidence (such as subfossil insect remains, pollen assemblages and/or plant microfossils) to yield fuller interpretations of changing environmental conditions (e.g. Murton et al. 2003, 2015). Also in lowland Britain there has been increased focus on periglacial fluvial deposits and river terraces and their implications for palaeoenvironmental changes (e.g. Briant et al. 2004; Brown et al. 2010; Lewin and Gibbard 2010; Murton and Belshaw 2011). Similarly, an increased emphasis on periglacial aeolian deposits has permitted these to be placed within the context of wider European loess and coversand sequences (e.g. Antoine et al. 2003; Baker et al. 2013). Laboratory experiments have been employed to throw light on the origins of several periglacial features, such as brecciated bedrocks in southern England (Murton and Lautridou 2003; Murton et al. 2006) and solifluction features (e.g. Harris et al. 2008). In mountain areas, field monitoring of solifluction lobes, ploughing boulders and frost-sorted patterned ground has thrown light on the nature and rates of present activity (Ballantyne 2001a, b, 2013a). The advent of cosmic-ray exposure (CRE) dating has enabled the ages of such features as tors, blockfields and postglacial rock-slope failures to be determined (e.g. Phillips et al. 2006; Fabel et al. 2012; Gunnell et al. 2013; Ballantyne et al. 2014). A recurrent theme of research throughout the past 30 years is growing involvement of engineering geologists in investigation of periglacial sediments and structures, and their implications for foundations and slope stability, recently reviewed by Culshaw et al. (2017), de Freitas et al. (2017), and Giles et al. (2017).

Murton and Ballantyne (2017) have recently introduced the concept of *periglacial landsystems* to encapsulate the nature of periglacial landforms, sediments and subsurface structures associated with particular terrain types. Both lowland and upland terrains are characterised by four periglacial landsystems: (i) plateau landsystems; (ii) sediment-mantled hillslope landsystems; (iii) rock-slope landsystems; and (iv) slope-foot landsystems. Two additional types, valley landsystems and buried landsystems, also occur in lowland areas where thick sequences of periglacial deposits are common. Some examples of such landsystems are illustrated in Sects. 3–5 below. Periglacial sediments, structures and landforms found within these landsystems are systematically described and illustrated by Giles et al. (2017) in a review that also outlines the characteristics and formal engineering properties of periglacial sediments.

3 Relict Periglacial Features in Lowland Areas

Relict periglacial features are common in many lowland areas of the British Isles, highlighting periglaciation as a major component of Quaternary landscape evolution. Periglacial processes likely began to dominate landscape change and produce coarse-grained sediments in susceptible regions from the onset of the Quaternary at ~2.6 Ma (MIS 104; Rose 2010). The effects of periglacial processes on lowland landscapes intensified after the Mid-Pleistocene Transition (1.25–0.7 Ma) with the emerging dominance of eccentricity-driven orbital cycles that lasted ~100 ka and led to the development of extensive ice sheets across much of the British Isles and periods of prolonged permafrost development in unglacierized terrain. As a result of periglacial weathering and incision by rivers during cold stages of the Pleistocene, the flattish land surfaces inherited from the Palaeogene and Neogene in lowland regions experienced significant relief modification, leading to the development of numerous vales and scarps (Fig. 1). The impact of periglaciation in shaping the topography is particularly evident in unglaciated lowlands underlain by frost-susceptible bedrock, such as chalk in southern England and schist and slate in SW England (Murton 2021). Such areas are characterised by convexo–concave hillslopes and by low plateaux indented with valleys. The topography of such terrain is attributed to 'recurrent development of a palaeo-transition zone of ice-rich brecciated bedrock at the top of past permafrost – supplemented by deep seasonal freezing during milder periglacial conditions – coupled with periglacial mass wasting and stream incision operating on a tectonic background of Pleistocene uplift' (Murton 2021, p. 207).

During cold stages, periglacial ground thermal regimes varied between cold (continuous) permafrost, warm (discontinuous) permafrost and seasonally frozen ground (Murton and Ballantyne 2017). Such variation of periglacial regimes in part reflects Britain's exposure to airmasses moving eastwards across the North Atlantic Ocean, which experienced major and rapid changes in sea-ice conditions during the Pleistocene (Rose 2010). Vegetation communities also responded to stadial-interstadial climatic changes, and comprised complex and dynamic mosaics determined by local or regional factors such as geology, soil, relief, drainage and climate. During cold periods the lower latitudes of the British Isles supported (at different times and locations) grassland, meadows, rocky areas, dunes, heaths, boreal woodland, scrub, fens, marshes, and saline and aquatic vegetation cover (West 2000). The legacy of lowland periglaciation is a rich array of weathering features, indicators of past permafrost, mass-movement features, aeolian features and evidence of periglacial fluvial activity (Giles et al. 2017).

3.1 Periglacial Weathering: Brecciated Bedrocks and Periglacial Karst

Periglacial weathering formed highly fractured (brecciated) weathering profiles referred to as *brecciated bedrocks* and contributed to the formation of irregular and pitted profiles termed *periglacial karst*. Brecciated bedrock commonly underlies low-relief plateaux and lowlands underlain by frost-susceptible (fine-grained and porous) bedrock such as chalk, Mesozoic and Cenozoic clays, shale, slate, schist and silty sandstone. The brecciation extends to depths of several tens of centimetres to a few tens of metres, though depths of 1–5 m are typical (Murton 1996), and can be subdivided into two types. Type 1 is texturally a breccia of angular to subangular rock fragments separated by unfilled fractures with matched sides (Fig. 3a), whereas type 2 is texturally a diamict comprising subangular to rounded lumps of rock (lithorelicts) within a fine-grained matrix of similar but softened and remoulded material (Fig. 3b; Murton and Ballantyne 2017). Type 1 brecciation of chalk has been shown experimentally to result from ice segregation during perennial and/or seasonal freezing (Murton et al. 2006). Type 2 is more difficult to interpret but may have involved a combination of frost weathering and limited vertical or lateral ground movement. A gradation exists between types 1 and 2, and also probably between type 2 brecciation (where the diamict is more or less in situ) and head deposits (Sect. 3.3) where the diamict has moved substantially downslope.

Periglacial karst features in limestone terrain include epikarst, solutional unconformities, clay-with-flints and periglacially influenced cave deposits. *Epikarst* comprises strongly weathered carbonate bedrock characterised by numerous enlarged rock joints and/or involutions, and sediment-filled pockets or pipes (Fig. 3c) within the upper few to several metres of limestone. *Solutional unconformities* tend to form sharp boundaries marking the current or former position of solution fronts, for example in calcareous deposits such as loess and chalky coversand (Sect. 3.4). *Clay-with-flints* is a diamict consisting of flint clasts within a matrix of clay to sandy clay loam (Fig. 3d) that forms a discontinuous mantle up to 2–15 m thick on plateaux and interfluves developed on the Chalk or Upper Greensand Formation in southern England (Gallois 2009; Catt 2010). Clay-with-flints is interpreted as a weathered residual material that has experienced weathering and pedogenesis (decalcification, rubification and clay illuviation) during relatively warm periods and periglacial processes (cryoturbation, frost weathering and gelifluction) during cold periods. *Periglacially influenced cave deposits* include speleothems fractured by frost

Fig. 3 Periglacial weathering features. **a** Type 1 brecciation (breccia) in chalk, Thanet, Kent. **b** Type 2 brecciation (diamict) in chalk, Thanet, Kent. **c** Epikarst features in the form of sediment-filled pipes beneath patterned ground, Breckland, Norfolk. **d** Clay-with-flints above chalk, Kent (Images: **a–c** Julian Murton; **d** Ursula Lawrence)

heave or mass-movement, and gelifluction and/or debris flow deposits transported into caves (Lundberg and McFarlane 2007, 2012). In general, karst processes such as solution and speleothem formation tend to operate only in unfrozen ground, as movement of liquid water within permafrost is very limited (Ballantyne 2018).

3.2 Indicators of Past Permafrost

Indicators of past permafrost in lowland Britain and Ireland include ice-wedge pseudomorphs, composite-wedge pseudomorphs and frost polygons, relict ground-ice features, active-layer phenomena and glacially deformed past permafrost. *Ice-wedge pseudomorphs* commonly take the form of downward-tapering sediment structures (Fig. 4a) that represent melting of a former ice wedge in permafrost and infilling of the space occupied by ice by sediment slumping from the sides and/or above

(Murton 2013). Melt of ice within wedges that originally had a composite filling of ice and sediment may produce a *composite-wedge pseudomorph*. Both types of pseudomorph are abundant in river terrace deposits and glacigenic deposits of lowland Britain, particularly of Late Devensian age, but some are much older. Wedge structures that formed during deposition of a host stratum are termed *intraformational* and are identified by their tops occurring within a stratum, whereas wedges that formed after deposition of the host stratum are termed *interformational* and extend down from bedding surfaces or erosional contacts between stratigraphic units. Examples of wedges dating from the Early, Middle and Late Devensian occur in the Fenland of eastern England (West et al. 1999; Briant et al. 2004, 2005). Such wedges are best viewed in vertical cross section in pits, quarries and excavations, whereas viewed from above near-surface wedges may form distinctive polygonal structures. *Polygons* with diameters of commonly 10–15 m may be seen on aerial photographs and high-resolution satellite images, especially at times when vegetation growing within and between the underlying wedges experiences differential water stress (Fig. 4b). Such polygons formed by thermal contraction cracking of past permafrost and resemble actively cracking polygons in many arctic regions.

Relict ground-ice features indicative of past permafrost comprise the remnants of frost mounds as well as thermokarst landforms and structures. *Remnant frost mounds* take the form of circular to elliptical depressions, often enclosed by annular ring-ridges or ramparts (Fig. 4c). Such depressions occur singly or in clusters, notably on valley floors and scarp-foot locations in central and west Wales, in Norfolk (Ballantyne and Harris 1994, pp. 74–80) and in southern Ireland (Coxon and O'Callaghan 1987). The depressions are thought to result from a radial outward movement of sediment by mass wasting from the elevated surface of the former frost mound. After the excess ice in the core of the frost mound has melted, a depression may remain, surrounded by a rampart of mass-movement deposits. Such depressions, however, are often difficult to interpret; although many were initially interpreted as pingo scars, increased understanding of modern frost mounds has led to the re-evaluation that some represent the sites of former lithalsas (Gurney 2000; Pissart 2003). Other interpretations of these depressions in the British Isles include seasonal frost mounds or frost blisters (West 2015), frost mounds transitional between hydraulic pingos and lithalsas (Ross et al. 2011), and glacigenic features (Ross et al. 2007). This range of interpretations reflects in part to the genetic complexity of frost mounds on permafrost terrain (Gurney 1998, 2001) compounded by the limited data on stratigraphic conditions at depth within both modern and relict structures.

Thermokarst landforms and sedimentary structures include large relict thermokarst depressions and thermokarst involutions. Landforms interpreted as large relict *thermokarst depressions* have been identified in the Fenland Basin and the Vale of York, both in eastern England. Those in the Fenland form near-circular features over 1 km in major diameter, with depths of several metres, flat to very gentle concave floors, and enclosed by higher ground, except for narrow drainage outlets (Ballantyne and Harris 1994, pp. 80–82). These depressions are inset into sand and gravel terrace deposits that contain a polygonal network of ice-wedge pseudomorphs, indicating the former occurrence of ground ice in past permafrost. Although periglacial

Fig. 4 Indicators of past permafrost. **a** Ice-wedge pseudomorph, north Yorkshire. **b** Relict frost polygons, Staffordshire. **c** Ramparted ground-ice depression, west Wales. **d** Thermokarst involutions, Thanet, Kent. **e** Large-scale periglacial stripes, Breckland, Norfolk. **f** Intraclasts of stratified sand in glacially deformed past permafrost, north Norfolk. **g** Chalk raft, north Norfolk (Images: **a** Dave Giles; **b, e** Google Earth; **c, d, f, g** Julian Murton;)

processes have likely influenced the development of these landforms (West et al. 2000), the amount of past excess ice within the enclosing terraces appears to have been rather limited, and the role of thermokarst development in the initiation and growth of these depressions remains uncertain (Murton and Ballantyne 2017).

Thermokarst involutions are contorted strata (Fig. 4d) that result from soft-sediment deformation due to thaw of ice-rich permafrost, as simulated experimentally under laboratory conditions (Harris et al. 2000). Prominent involuted layers of Anglian and Devensian age in lowland Britain have been interpreted as thermokarst involutions at sites where there is strong evidence (such as brecciated bedrock or platy structures in sediments) for the past occurrence of an ice-rich transition zone at the top of past permafrost (Murton et al. 1995; Murton 2022a, b). Such involutions are believed to have formed in a deepening active layer or residual thaw layer as permafrost degraded.

Relict active-layer phenomena include periglacial involutions (cryoturbation structures) and large-scale patterned ground. *Periglacial involutions* are polygenetic sedimentary structures, and some within widespread involuted layers may have formed or been modified by cryoturbation and/or differential frost heave as well as thermokarst activity, as inferred for Thanet, east Kent (Murton et al. 2003). The maximum depth of involutions may in some cases approximate the minimum depth of a former active layer at some time in the past, as indicated at sites where involutions in sand or gravel terminate downwards at or near the tops of ice-wedge pseudomorphs. In frost-susceptible materials such as chalk, shale or silt, however, it is likely that differential frost heave within the ice-rich transition zone at the top of past permafrost has also produced involutions, as suggested by cryoturbated shale in near-surface permafrost in arctic Canada (French et al. 1986; Murton 1996). Similar involutions occur in the upper metre or so of slate, slatey breccia and chalk in southern England (Scourse 1987; Murton 1996).

Large-scale patterned ground in the form of circles, polygons, irregular networks or (on slopes) stripes several metres in width is rarely preserved in most lowland areas. However, distinctive stripes and polygons spaced ~7.5 m and ~10.5 m apart, respectively, are widespread in the chalklands of East Anglia and Thanet (Fig. 4e). These patterns are now represented by differences in vegetation that reflect contrasts in the texture and composition of the underlying finer-grained chalk-rich centres and the coarser sandy marginal troughs, and are thought to have developed through soil circulation driven by buoyancy and differential frost heave in a former active layer ~2 m thick (Murton et al. 2003). Three episodes of patterned ground (and permafrost) development during the Wolstonian and Devensian stages have been inferred from stratigraphic evidence in west Norfolk (West 2015), and luminescence dating of stripes and polygons elsewhere in East Anglia has established that multiple phases of pattern development occurred during the Devensian (Bateman et al. 2014).

Glacially deformed past permafrost likely developed beneath and/or at the margin of successive ice sheets during the Pleistocene. In north Norfolk, for example, pervasively deformed sediments >10 m thick that contain intraclasts of stratified sand (Fig. 4f) have been attributed to deformation of partially frozen ground at temperatures just below the pressure melting point of ice, and therefore indicate the former occurrence of warm permafrost (Waller et al. 2011). Moreover, the occurrence of rafts of chalk bedrock below undeformed sand and gravel within glacitectonic sequences (Fig. 4g) suggests that these were probably emplaced in a frozen state (Burke et al. 2009), again implying glacier advance over frozen ground. As north Norfolk lies in region 4 in Fig. 2a, the substantial glacitectonic deformation described above probably occurred during the Anglian glaciation of MIS 12 (478–424 ka).

3.3 Mass-Movement Features

Various periglacial mass-movement structures, sediments and landforms are represented on hillslopes in lowland regions. Deformation structures in periglacially weathered bedrock include downslope-deflected (overturned) brecciated rock, relict periglacial shears and periglacial creep folds. *Downslope-deflected brecciated rock* is represented by a deformed layer up to ~1 m thick that occurs above in situ bedrock and is overlain by granular head deposits (Fig. 5a). Many fine examples are exposed in coastal sections through slates and shales in Devon, Cornwall and SW Ireland. The downslope deflection indicates gravity-driven displacement of fractured rock, but the processes responsible for deformation are uncertain, and may have involved frost creep, gelifluction, shearing or permafrost creep acting alone or in some combination (Murton and Ballantyne 2017). *Relict periglacial shears* are planar to undulating slip surfaces (Fig. 5b) that typically occur where gentle slopes are underlain by weathered (usually brecciated) clay bedrock (Spink 1991). Weathering profiles in clay bedrock reach depths of several metres or more in the London Clay, Gault Clay and Reading Formations of southeast England, and though relict shears may occur at any depth within them, they tend to be particularly common at depths of ~1.5–3.0 m and run parallel to the ground surface. Such shallow periglacial shears are thought to indicate shallow translational landsliding, analogous to active-layer detachment slides triggered by high pore-water pressures during thaw consolidation of ice-rich soil in arctic active layers (Hutchinson 1991; Harris 2013). *Large-scale folding* in mudstones (such as the Lias Group and Kimmeridge Clay of Dorset) that underlie gentle hillslopes has been interpreted as periglacial creep folds (Gallois 2010). These structures include tight and kink folds, are locally overturned, and are thought to indicate former gradual downslope deformation due to permafrost creep.

Relict periglacial slope deposits are abundant in lowland regions of the British Isles, where they are commonly termed *head*. This term encompasses a wide textural range of clastic sediments, typically poorly sorted, with two main types distinguished: granular head deposits and clay-rich head deposits (Harris 1987; Hutchinson 1991).

Granular head deposits derive from non-argillaceous bedrock such as sandstone, limestone, granite, mudstone and slate, and typically consist of angular to sub-angular clasts set within a silty or sandy frost-susceptible matrix (Fig. 5c). Texturally, such deposits are commonly matrix-supported diamicts, massive to crudely stratified,

Fig. 5 Lowland mass-movement features. **a** Downslope deflected (overturned) brecciated slatey bedrock, SW Ireland. **b** Relict periglacial shear surface (polished and striated) in clay-rich head deposits, Gloucestershire. **c** Granular head deposits, south Devon. **d** Clay-rich head deposits in trial pit, East Sussex. **e** Relict active-layer detachment slide, Kent. Arrows indicate front of lobate slide. **f** Cambered strata, Kent. **g** Gulls, Gloucestershire. **h** Valley bulge, East Sussex (Images: **a** Derek Moss; **b, d** Adrian Humpage; **c, e, f** Julian Murton; **g** Andrew Farrant; **h** JM Pulsford, BGS Geoscenic)

within which elongate clasts are typically aligned downslope and imbricated. Laboratory experiments have suggested that granular head deposits accumulated by solifluction (needle-ice creep, frost creep and/or gelifluction) and debris-flow processes. Such experiments have shown that gelifluction represents pre-failure elasto-plastic deformation operating in frictional soils due to the development of excess pore-water pressures at the thaw front during thaw consolidation of ice-rich ground (Harris et al. 2001, 2003; Harris 2013). The depth at which segregated ice is concentrated determines the depth where strain is focused (Harris et al. 2008). The nature of solifluction processes associated with granular head deposits was conditioned by soil texture: sandy granitic soil and gravelly limestone soil were displaced mainly by needle-ice creep and frost creep, whereas more fine-grained soil derived from slate or mudstone promoted ice segregation and thus the operation of gelifluction (Harris et al. 1993, 1997). On steep hillslopes, shallow debris flows probably contributed to the accumulation of head deposits in locations where increased water content and pore-water pressure developed during active-layer thaw (Font et al. 2006; Védie et al. 2010). In formerly glaciated lowlands, paraglacial mudflows and slumping of glacigenic sediments also resulted in the accumulation of thick diamicts very similar to granular head deposits (Harris 1998).

Clay-rich head deposits consist of reworked (partially remoulded) clayey sediments that contain lithorelicts of the original clay fabric and exhibit relict shear surfaces within and/or beneath them (Fig. 5d; Murton and Ballantyne 2017). Clay-rich head deposits are widespread beneath hillslopes developed on argillaceous strata such as the London Clay, and are distinguished from the underlying unweathered in situ clay by the random orientation of lithorelicts, the soft clay matrix between them and sometimes by small (<1 mm diameter) air bubbles in the matrix (Hutchinson 1991, 2010). The relict shears within clay-rich head deposits (Fig. 5b) represent planes of weakness, usually at residual shear strength, and have determined the low gradients of many hillslopes. Clay-rich head deposits typically form smooth sheets, sometimes partly eroded, mantling low-gradient slopes; less commonly they form distinct lobate landforms (Fig. 5e). Genetically, clay-rich head deposits probably developed as active-layer detachment slides (Harris 2013) similar to those that presently occur in arctic and subarctic regions (Ballantyne 2018).

Where valleys dissect stratigraphic sequences comprising a relatively strong caprock overlying weaker (argillaceous) strata, periglacial mass-movement features may take the form of cambered strata, gulls and valley bulges. *Cambered strata* exhibit 'large-scale flexing and stretching of competent caprocks over the upper parts of valley-side slopes' (Ballantyne and Harris 1994, p. 137), and characteristically dip towards the valley axis (Fig. 5f), irrespective of the regional dip of the rocks. Cambered caprocks include limestone, sandstone or ironstone overlying an incompetent mudrock (usually clay), a common stratigraphy in central and southern England. Cambering has involved gradual mass movement of bedrock, with the weight of the caprock driving the underlying clay towards the valley axis. Such movement has taken the form of flexing, stretching and slumping downward and outward of the caprock, and its subsequent gradual breakup into separate blocks (Hobbs and Jenkins 2008; Hobbs et al. 2012). Following initial valley incision by periglacial rivers, cambering is thought to have been caused by aggradation of ice-rich permafrost in the argillaceous rocks, which experienced a reduction in shear strength during subsequent permafrost thaw, permitting the overlying cambered caprock to settle and rotate downslope (Parks 1991).

Cambered strata are often associated with *gulls*, which represent vertically widened joints in competent caprock (Fig. 5g). Gulls vary in width from centimetres to a few metres, and in height from ~1 m to at least 15 m. Commonly, they are aligned vertically across the surface hillslope and more or less parallel to valley sides. Gulls have developed where a mechanically strong caprock has undergone extension, typically on steep hillslopes where jointed rocks are unsupported on their downslope side (Self 1995). Extension of caprock and related opening of joints to form gulls has often taken place along bedding planes (Hawkins and Privett 1981); shear has often been focused along a basal plane, but in some instances also within a caprock to produce gulls with intact roofs, a feature common in the thick limestone caprocks of the Cotswold Hills (Self and Farrant 2013).

Valley bulges are anticlinal folds or diapirs, rim synclines and décollement planes that underlie the centre and margins of valleys incised through a cap of competent, jointed rock, such as sandstone or limestone, that overlies stiff clays or shales (Fig. 5h). The bulges have formed by the plastic deformation of clays and mudrocks under the weight of the overlying caprock, with the greatest upward deformation usually concentrated along the valley axis (Hobbs et al. 2011). Such deformation has been attributed, at least in part, to 'softening' of the argillaceous rocks through generation of excess pore-water pressures during thaw of ice-rich past permafrost (Hutchinson 1991).

3.4 Aeolian Features: Coversands, Loess and Ventifacts

Periglacial aeolian deposits (loess and coversands) are common on low plateaux and within valleys, particularly (but not exclusively) in lowland areas that remained outside the limits of the last (Late Devensian) ice sheet. *Loess* is a silty aeolian sediment that was transported under periglacial conditions and deposited in a cold steppe-tundra environment (Murton and Ballantyne 2017). The loess deposits in England and Wales tend to contain abundant coarse silt (20–50 μm), have a median particle size of 25–35 μm and are typically unstratified (Fig. 6a; Antoine et al. 2003). Two loess provinces have been identified: the main, eastern one extends from Yorkshire to Kent and westwards to east Devon, and a second, more limited western loess occurs in Cornwall, and inside the Late Devensian ice-sheet limit in NW England and north Wales. Sediment sources for the eastern and western provinces are thought to have been the dry, exposed floors of the southern North Sea Basin and the southern Irish Sea Basin, respectively. Loess has also been identified on the karstic uplands of the Burren in western Ireland (Moles et al. 1995). The loess probably accumulated where vegetation such as grasses, sedges and forbs as well as biological soil crust communities trapped and stabilized windblown silt. Luminescence dating of loess deposits in England suggests that most accumulated during the Late Devensian to Early Holocene (~25–10 ka; Stevens et al. 2020), though older Devensian and pre-Devensian luminescence ages back to ~229–175 ka have also been obtained (Parks and Rendell 1992; Rose et al. 2000; Murton et al. 2015), suggesting recurrent accumulation of loess during successive periglacial intervals. Much of the loess has been modified by interglacial or postglacial weathering or erosion, and redeposited on lower slopes and valley floors, or within karstic depressions (Vincent et al. 2011).

The term *coversand* describes a near-surface sedimentary unit dominated by aeolian sand, usually <5 m thick, that covers a low-relief landscape (Bateman 2013). Depositionally, coversand represents periglacial aeolian sand sheets (Bateman 2008). In the British Isles the distribution of coversand deposits is more limited than that of loess, and concentrated in the Vale of York, north Lincolnshire, the Breckland of East Anglia and SW Lancashire, with some local occurrences elsewhere. Texturally, coversand is often well sorted, with a mean grain size of ~150–200 μm (Bateman 1995), and usually well-stratified, with horizontal to gently dipping strata typically a few millimetres to a few centimetres thick (Fig. 6b). Sediment sources for the coversand included sandy outwash deposits as well as other Pleistocene sands and sandstone bedrock. Multiple episodes of coversand deposition occurred during the Devensian, most recently during the Loch Lomond Stade of ~12.9–11.7 ka (Bateman et al. 2014). The episodes of coversand deposition in East Anglia are thought to coincide with those associated with the Older Coversand I and II (~26–15 ka), and Younger Coversand I (~14 ka) in mainland NW Europe. Earlier episodes of coversand deposition have been inferred to have occurred during MIS 6 (~191–130 ka) in south-central England (Murton et al. 2015) and MIS 12 (~478–424 ka) in East Anglia (Rose et al. 1985), both also consistent with coversand deposition at these times in mainland NW Europe.

Wind erosion under former periglacial conditions is recorded by the occurrence of *ventifacts*, stones that have been shaped, facetted, indented, polished or grooved by the abrasive action of windblown sand or silt particles. In a few locations, a combination of deflation and aeolian abrasion has produced gravelly lag deposits on wind-scoured surfaces. An example of one such ventifact pavement is shown in Fig. 6c, and is thought to have developed during the Loch Lomond Stade (Hoare et al. 2002).

3.5 Periglacial Fluvial Activity

Periglacial fluvial activity was widespread in lowland areas during multiple Pleistocene cold stages, as recorded by the abundance of dry valleys, slopewash and fluvio-colluvial deposits, and by extensive periglacial fluvial deposits and river terraces. *Dry valleys* are common on low plateaux underlain by limestone, sandstone and other porous and permeable rocks. Some of the best-known examples are incised onto the chalk plateaux and cuestas of southern and eastern England (Fig. 6d; Ballantyne and Harris 1994). Although there is some debate regarding the origin of dry valleys,

Fig. 6 Lowland aeolian and fluvial features. **a** Loess (brown), Thanet, Kent. **b** Coversand containing a relict sand vein, Lincolnshire. **c** Ventifact pavement, west Norfolk. **d** Dry valley East Sussex. **e** Stratified slopewash deposits, East Sussex. **f** Fluvial gravel deposits containing an ice-wedge pseudomorph, Oxfordshire. **g** Erosion surface between underlying brecciated clay bedrock (grey) and overlying fluvial terrace gravel (brown), Wiltshire (Images: **a, b, d–g** Julian Murton; **c** Peter Hoare)

it seems likely that most were incised under periglacial conditions, when frozen ground favoured surface runoff (Murton and Ballantyne 2017). The sides and floors of dry valleys are often underlain by sediments attributed to deposition by solifluction (head deposits), slopewash, or fluvio–colluvial processes. *Lowland slopewash deposits* tend to be stratified and often have a fine gravel component together with varying amounts of finer sediments (Fig. 6e) eroded from hillslopes and redeposited by sheets or threads of water flowing over the surface during periods of snowmelt, ground-ice melt or rainstorms (Preece et al. 1995). *Fluvio–colluvial deposits* may include a coarse gravel component and tend to consist of sorted gravel or sandy gravel derived from solifluction or other colluvial deposits then reworked and sorted episodically during periods of high stream discharge (West 2015; Giles et al. 2017).

Periglacial fluvial deposits and river terraces occur in many major lowland river valleys, such as those of the rivers Thames, Severn and Trent, and are particularly well developed outside the limits of the Late Devensian ice sheet. During floods, channel migration and avulsion operating within such trunk valleys spread gravel across floodplains. Subsequent long-term uplift and floodplain incision has created flights of river terraces that flank the active floodplain and provide a valuable stratigraphic archive of former periglacial conditions, though the timing of valley incision, valley widening and episodes of sediment aggradation is disputed (Murton and Belshaw 2011; Murton and Ballantyne 2017). Periglacial fluvial deposits within the river terraces comprise mainly sorted gravels and sands. Massive or horizontally and crudely bedded gravel, commonly containing broad, shallow channel scours, is the most abundant facies (Fig. 6f; Lewin and Gibbard 2010), though in some locations stacked sequences of channel and bar deposits record channel migration across former floodplains (Briant et al. 2004, 2005). Gravel and sand typically form a unit 2–5 m thick that sharply overlies brecciated clay bedrock, producing a characteristic terrace stratigraphy (Fig. 6g). Ice-wedge pseudomorphs and involutions often occur within the terrace stratigraphy, and periglacial fluvial deposits often interdigitate with colluvial deposits at floodplain margins. Palaeochannels on terrace surfaces demonstrate that fluvial deposition of the gravel and sand was accomplished by wandering or braided rivers.

3.6 Lowland Periglacial Landsystems

The characteristics of lowland periglacial landsystems in the British Isles can be encapsulated as ground models that illustrate the influences of bedrock lithology, stratigraphy and relief on the distribution and characteristics of relict periglacial and permafrost features (Murton and Ballantyne 2017). A *limestone plateau–clay vale ground model* (Fig. 7a) depicts limestone or fine-grained sandstone (~20–100 m thick) above a clay formation and characterises the periglacial landform assemblage associated with major escarpments, plateaux and clay vales of southern and central England. Specifically, it highlights the geomorphic significance of brecciated bedrock, the incision of dry valleys in limestone plateaux and the development of river terraces in the major valleys. The second landsystem model (Fig. 7b) depicts the characteristic *periglacial landscape associated with chalk terrain* in southern and eastern England. The third model (Fig. 7c) illustrates *features associated with a thin caprock* of limestone, sandstone or ironstone (≤ ~20 m thick) overlying clay (Fig. 7c) and shows the geomorphic impact of valley incision through a competent caprock and development of cambering of the caprock and valley bulging in the clay, particularly during former periods of thaw of ice-rich brecciated clay.

4 Relict (Lateglacial) Periglacial Features on British and Irish Mountains

The Lateglacial period in Britain and Ireland extends from the timing of local ice-sheet deglaciation until the beginning of the Holocene. It incorporates: (i) the final millennia of the Dimlington Stade, when permafrost extended to sea level on terrain vacated by retreat of the last ice sheet; (ii) the Lateglacial Interstade of ~14.7–12.9 ka, when summer temperatures approached those of the present; and (iii) the Loch Lomond Stade (LLS) of ~12.9–11.7 ka, when renewed permafrost aggradation occurred in some parts of the British Isles. The Lateglacial period was therefore much more protracted in areas that emerged earliest as the last ice sheet thinned and retreated. Mountains in southern Ireland and south Wales, for example, emerged from the last ice sheet as early as ~20 ka, whereas summits in NW Scotland did not emerge from the shrinking ice sheet until ~16 ka, and some mountains in the Grampian Highlands may have retained ice cover as late as ~14 ka, or even throughout the Lateglacial Interstade (Ballantyne and Small 2019). Most relict periglacial features in mountainous areas formed during the Lateglacial period, apart from blockfields and tors, which are of much greater antiquity.

4.1 *Blockfields and Tors*

Many high plateaux and mountain summits support in situ periglacial regolith covers (*blockfields*), typically up to a metre deep. Some comprise mainly openwork boulders with an infill of fine sediment at depth, but others are sandy or silt-rich diamicts (Fig. 8a). Research on the sedimentary characteristics of blockfields on Scottish mountains indicates that they formed mainly through frost weathering of jointed bedrock (Ballantyne 2010; Hopkinson and Ballantyne 2014). CRE dating of erratics resting on blockfields, however, has demonstrated that mature blockfields pre-date the Late Devensian, implying that they escaped erosion by the last ice sheet because the ice cover on high ground was persistently cold-based and non-erosive (Fabel et al. 2012; Ballantyne and Stone 2015). Wilson and Matthews (2016) have argued that a blockfield in NW Ireland remained 'active' during brief cold intervals in the

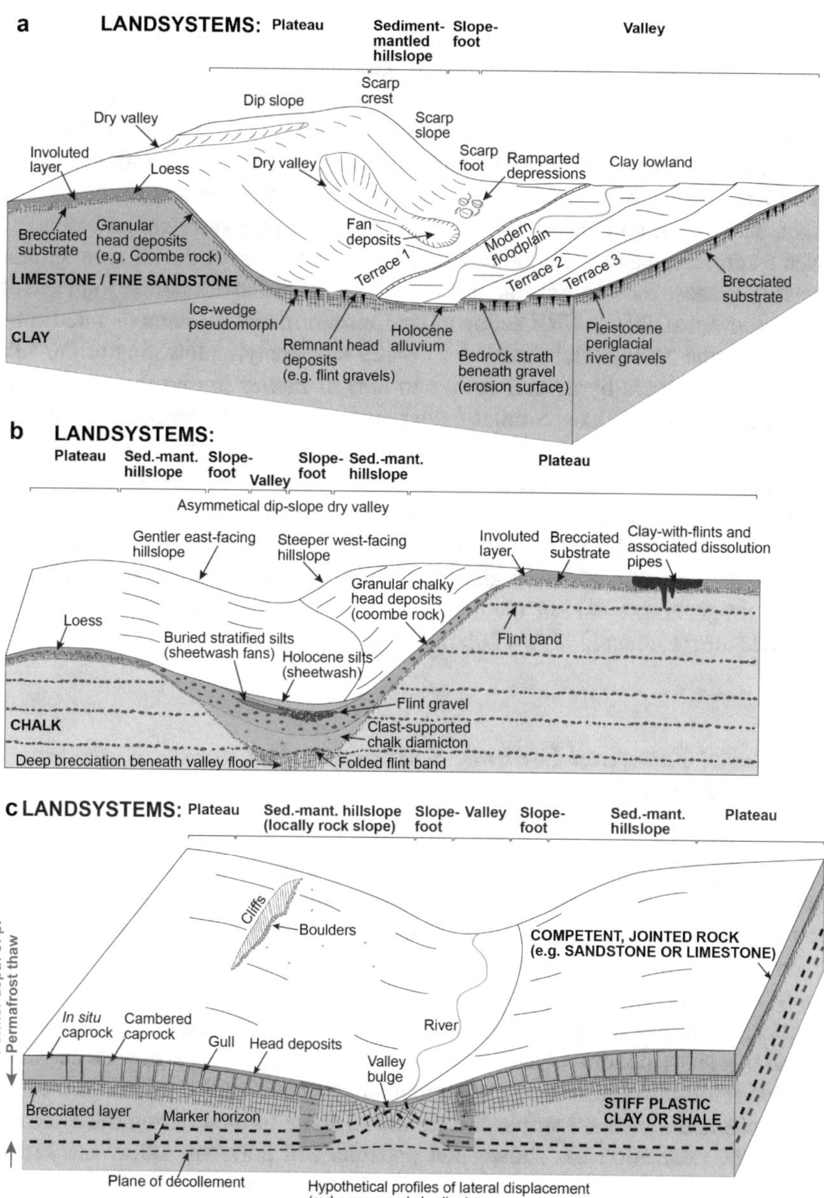

Fig. 7 Lowland periglacial landsystems of Britain and Ireland. **a** Ground model of limestone plateau–clay vale. **b** Landsystems in the chalklands, where a dry valley is incised into a plateau. Note the greater depth of brecciation beneath the valley floor compared to the plateau. **c** Ground model of caprock plateau–mudstone valley, showing cambered caprock and gulls beneath valley sides and valley bulge structure in clay beneath the valley centre. Modified from Murton JB, Ballantyne CK (2019) Geol Soc Lond Eng Geol Spec Publ 28:501–597 © The authors

Early Holocene, but their results rest on the questionable assumption that Schmidt hammer rebound values scale linearly with exposure age.

Tors are present on high ground underlain by a range of lithologies, but the most spectacular examples are those on the granite of the Cairngorm Mountains in Scotland and Dartmoor in SW England (Fig. 8b). Some of those on the Cairngorms rise over 15 m above the surrounding plateau, but few Dartmoor tors exceed 5 m in height. The tors in both locations have 'emerged' from the surrounding blockfield-covered surfaces due to preferential weathering and erosion of the adjacent ground, primarily because the tors are composed of massive rock and have lower joint density (Goodfellow et al. 2014). CRE dating of the Cairngorm tors has demonstrated emergence since the Middle Pleistocene (~774–129 ka), implying lowering of the adjacent plateau surfaces by several metres to tens of metres during the past million years (Phillips et al. 2006). Similar dating applied to the Dartmoor tors, however, suggests that most of these are much younger, and represent rapid stripping of overlying regolith cover after ~115 ka, under severe periglacial conditions during the Devensian (Gunnell et al. 2013). This difference in age suggests that the oldest Cairngorm tors may have been shielded from periglacial weathering and erosion under a persistent cover of cold-based glacial ice during successive cold periods, though some Cairngorm tors exhibit modification by glacier ice and consequent loss of superstructure (Hall and Phillips 2006).

4.2 Relict Patterned Ground, Solifluction Features and Periglacial Valley Fills

Relict Lateglacial *frost-sorted patterned ground* up to 1.5–5.0 m wide occurs on some blockfields and is indicative of formation above former permafrost (Ballantyne 2018). On plateaux these patterns take the form of circles of boulders around vegetated cells of finer sediment (Fig. 8c); on slopes, they are represented by alternating vegetation-covered and bouldery *sorted stripes*. Nonsorted patterned ground of probable Lateglacial age is common on frost-susceptible regolith covers; on level ground this takes the form of earth hummocks (*thúfur*) up to ~2.0 m wide and 0.6 m high that grade into relief stripes with similar dimensions (Fig. 8d). Although the relict hummocks and nonsorted stripes are similar to some formed during the Holocene (Sect. 5.2), excavation has shown that podzolic soil horizons follow the surface microtopography, implying prolonged inactivity (Ballantyne and Harris 1994).

Lateglacial mass-movement of periglacial regolith has resulted in the formation of large *terraces* and *lobes* of coarse bouldery debris. These features are best developed on terrain underlain by granitic rocks (e.g. Cunningham and Wilson 2004), but also occur on mountain slopes underlain by gneiss, sandstone and most igneous and metasedimentary rocks. Terraces consisting of 1–4 m high bouldery risers that mark the downslope termination of vegetated treads extend for hundreds of metres along slope crests (Fig. 8e). As the gradient increases downslope, the terraces become

Fig. 8 Relict periglacial landforms in mountain areas. **a** Blockfield at 1250 m on Ben Nevis. **b** Granite tor on the Cairngorm Mountains. **c** Relict large-scale sorted circles at 1050 m in the Grampian Highlands; vegetation covers the fine centres. **d** Sinuous nonsorted relief stripes at 900 m in the NW Highlands. **e** Stone-banked terraces at 850 m, NW Highlands. **f** Stone-banked lobes (boulder lobes) at 820 m, Grampian Highlands. **g** Relict Lateglacial talus accumulations, NW Highlands. **h** Runout from a Lateglacial rock-slope failure (previously interpreted as a rock glacier) that occurred at ~17.7 ka in NW Ireland (Images: **a–g** Colin Ballantyne; **h** Pete Coxon)

increasingly crenulate in planform, and locally over-riding of terraces has resulted in the isolation of *stone-banked banked lobes* (Fig. 8f). In the Cairngorm Mountains, such lobes occur at altitudes as low as 540 m but are absent from areas occupied by LLS glaciers, suggesting that downslope movement was terminated as permafrost degraded under the rapidly warming conditions after ~11.7 ka. Some authors have suggested that such bouldery terraces and lobes were formed by downslope creep of ice-rich permafrost, but it is perhaps more plausible that a near-surface cover of frost-heaved boulders was rafted downslope by solifluction operating in underlying frost-susceptible fine sediment (Ballantyne and Harris 1994). Vegetation-covered, nonsorted Lateglacial terraces and lobes with risers typically 1–2 m high also occur on some mountains and can be distinguished from their active counterparts (Sect. 5.2) by their degraded, low-gradient risers.

Lateglacial solifluction has also been implicated in the deposition of periglacial slope deposits that occupy lower slopes and valley floors in some upland areas, notably Wales, southern Scotland and the Cheviot Hills of northern England. Such periglacial valley fills form gently sloping terraces up to 300 m wide with steep frontal bluffs produced by river incision. Some consist of reworked till or regolith; others of in situ till overlain by reworked till and frost-weathered debris derived from adjacent slopes. Some are massive, some are crudely stratified, and others are intercalated with or overlain by slopewash gravels. Luminescence dating suggests that the most recent period of sediment accumulation occurred during the LLS (Harrison et al. 2010). Whether such thick sediment accumulations represent the products of solifluction alone is uncertain; some may have been deposited by active-layer detachment failures that occurred over former permafrost on the adjacent valley-side slopes (Harrison 2002).

4.3 *Lateglacial Modification of Steep Rock Slopes*

Following retreat of the last ice sheet, steep rock slopes have exhibited four types of *paraglacial* behaviour in response to deglacial unloading, hydromechanical fatigue and thermomechanical effects (Ballantyne 2019, 2021): (i) many mountain slopes have remained essentially stable; (ii) others have progressively degraded through incremental rockfall activity, with talus accumulating below cliffs; (iii) others still have experienced gradual large-scale rock-slope deformation; and (iv) some have failed catastrophically as rockslides, major rockfalls or topples, or rock avalanches.

Outside the limits of LLS glaciation, most *talus* accumulations are essentially relict features that accumulated rapidly during the Lateglacial as a result of paraglacial stress release and periglacial frost wedging operating on steep rockwalls (Hinchliffe and Ballantyne 1999, 2009; Curry and Morris 2004; Anderson and Harrison 2006). Lateglacial talus slopes are typically vegetation-covered and incised by gullies eroded by translational *debris slides*, torrential runoff and *debris flows*, resulting in the deposition of *debris cones* at the slope foot (Fig. 8g). At a few locations, possible Lateglacial *protalus ramparts* in the form of arcuate or linear debris ridges at the foot of talus slopes have been identified, but some have subsequently been reinterpreted as rockslide runout deposits, and the status of others is uncertain. The same is true for putative Lateglacial 'rock glaciers' (Fig. 8h) that have been reinterpreted as rock avalanche deposits; there is no conclusive evidence for rock-glacier formation in the British Isles since retreat of the last ice sheet (Ballantyne 2019). A few relict avalanche boulder tongues of probable Lateglacial age have been identified in Scotland; it is likely that these were once more widespread but have been obliterated by Holocene debris flows (Luckman 1992).

Paraglacial *rock-slope failures* (RSFs) in the form of *rock-slope deformations* and *catastrophic rock-slope failures* are common on mountains throughout Britain and Ireland (Fig. 8h). Jarman and Harrison (2019) have estimated that over 1000 post-glacial RSFs >0.01 km^2 in area (and over 200 >0.25 km^2) are present in the Scottish Highlands, Southern Uplands of Scotland, English Lake District and mountains of Wales, and many others have been recorded in Ireland (Wilson 2017), the Hebrides and other upland areas. CRE dating of catastrophic RSFs in NW Ireland and Scotland has shown that 95% of those located outside the limits of the LLS glaciers occurred during the Lateglacial period, with peak activity occurring 1.6–1.7 ka after local deglaciation. Ballantyne et al. (2014) attributed this time lag to progressive reduction in rock-mass integrity (joint propagation) following deglaciation, but noted that peak RSF activity coincided with the timing of maximum rates of glacio-isostatic uplift, suggesting that failure in some (possibly many) cases was triggered by uplift-driven seismic activity. The role of periglacial conditions (thermal changes, hydrostatic pressures and frost wedging) in conditioning or triggering Lateglacial RSFs is uncertain. Very few dated RSFs occurred during or soon after periods of rapid warming at ~14.7 ka (the start of the Lateglacial Interstade) and ~11.7 ka (the start of the Holocene), however, suggesting that warming and thaw of permafrost ice in joints was not a widespread trigger of failure. Inside the limits of LLS glaciation, especially in the Scottish Highlands, numerous rockslide scars mark the sites of Lateglacial RSFs where runout debris has been removed by LLS glaciers and incorporated in moraines (Ballantyne 2013b; Cave and Ballantyne 2016).

4.4 *Relict Periglacial Landsystems on British and Irish Mountains*

The range of relict periglacial features present on mountains is summarised in Fig. 9a, which depicts a hypothetical mountain area during the LLS, but not all the landforms shown are of the same age. As noted above, the tors and blockfields are essentially pre-Late Devensian features, though periglacial weathering continued during the Lateglacial, and most rockslides and talus accumulation probably occurred prior to the LLS. Moreover, the nature of relict upland periglacial landsystems is conditioned by lithology: tors are mainly restricted to granite mountains, for example, and large-scale sorted patterned ground occurs only on terrain underlain by frost-susceptible regolith.

5 Active (Holocene) Periglacial Features on British and Irish Mountains

5.1 Holocene and Present-Day Periglacial Environments on High Ground

Climatic conditions similar to those of the present were established in Britain and Ireland by ~11 ka. During much of the Early Holocene, the climate of the British Isles was characterised by relatively warm summers, but the past 5000 years have witnessed gradual cooling and oscillation of average temperatures by up to ± 2 °C over decadal timescales (Briffa and Atkinson 1997), with at least local shifts in wetness (Langdon and Barber 2005). The coldest climatic events during the past ~10 ka occurred during short-lived cooling at ~8.2 ka and the 'Little Ice Age' of the sixteenth–nineteenth centuries, a period of exceptional storminess, cool summers, and persistence of perennial snowcover on some mountains in the Scottish Highlands. During the Little Ice Age, summer temperatures in the Cairngorms were about 1 °C lower than those of the period 1961–1990 (Rydval et al. 2017).

The limited variation of climatic conditions during the Holocene suggests that present-day environmental conditions on mountain summits in Britain and Ireland are representative of those that prevailed over much of the past ~10 ka. The present maritime periglacial environment above 700–800 m altitude is characterised by strong, gusty winds and extreme wetness but only shallow ground freezing. Even on the highest summits, mean annual air temperatures exceed 0 °C: those recorded for the Scottish summits of Cairn Gorm (1245 m), Aonach Mòr (1130 m) and The Cairnwell (933 m) for the period 1981–2010 were + 0.9 °C, + 1.7 °C and + 2.7 °C, respectively, and thus fall marginally within the + 3.0 °C threshold recommended by French (2017) for identifying periglacial environments. Farther south, the mean annual air temperature on all mountains exceeds this threshold: that on Snowdon (1085 m), the highest mountain in Wales, is roughly + 5.0 °C. Seasonal ground freezing reaches depths of less than 0.4–0.5 m, even in high, shaded sites, and permafrost is absent. Precipitation on all but the most easterly mountains exceeds 2000 mm a^{-1}; and although annual snow-lie (>50% cover) at 900 m altitude exceeded 100–180 days during much of the last century, in recent years snowcover has been less persistent because of frequent winter thaw events. All mountain areas experience frequent gale-force (≥80 km h^{-1}) winds associated with the passage of winter cyclones; a gust of 277 km h^{-1} has been recorded at the summit of Cairn Gorm.

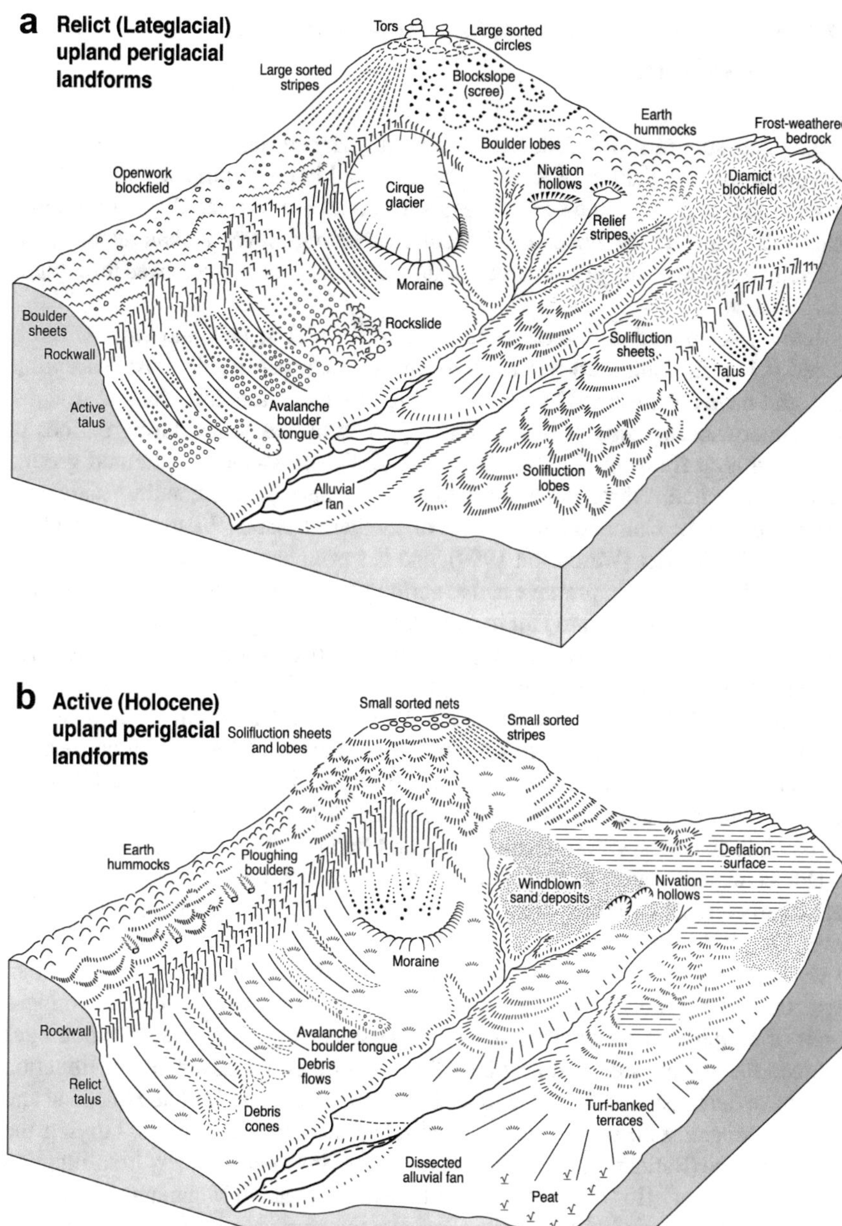

Fig. 9 Upland periglacial landsystems for British and Irish Mountains. **a** Relict (Lateglacial) periglacial features. **b** Active (Holocene) periglacial features. Although the topography depicted is similar for the two landsystems, the distribution of active (Holocene) features is unrelated to that of features depicted in the top diagram. Both diagrams take no account of the influence of underlying geology in determining the distribution of periglacial features. Reproduced from Murton JB, Ballantyne CK (2019) Geol Soc Lond Eng Geol Spec Publ 28:501–597 © The authors

5.2 Holocene Frost Weathering, Patterned Ground and Solifluction

On mountain summits, Holocene frost weathering has been largely limited to granular disaggregation and flaking of exposed bedrock and boulder surfaces, so that those on some lithologies (such as granite and sandstone) are characteristically rounded rather than angular. The widespread occurrence of high-level aeolian deposits and aeolisols (see below) suggests that such granular weathering has persisted throughout the Holocene.

Active frost-sorted patterned ground features (circles, nets and stripes) rarely exceed 0.7 m in width, are restricted to unvegetated, permeable frost-susceptible soils, and have been formed by lateral sorting of small clasts into shallow troughs. The primary sorting process is differential growth of needle ice during periods of shallow ground freezing (Ballantyne 1996). Active frost-sorted patterned ground occurs on mountains throughout the British Isles where ground conditions are suitable, including the Comeragh Mountains of southern Ireland (Wilson 1992) and the English Lake District (Warburton 1999), and has even been recorded near sea level in Shetland. The most impressive active sorted patterns are *stone stripes* at ~640 m on the felsite regolith of Tinto Hill in southern Scotland (Fig. 10a). At this site, well-developed sorted stripes have developed over disturbed ground within one to three winters, and clasts have moved downslope through needle-ice creep on a gradient of 23° at rates of 24–62 cm a^{-1} (Ballantyne 2001a). Although most nonsorted patterned ground features in Britain and Ireland are relict features of Lateglacial age, extensive fields of *earth hummocks* that have demonstrably formed in the Late Holocene occur in parts of Dartmoor, where their formation is limited to areas of silt-rich, frost-susceptible loessic soils (Killingbeck and Ballantyne 2012; Fig. 10b).

Holocene solifluction landforms occur on many mountain slopes above 600 m, particularly on rocks that have weathered to form a silt-rich, frost-susceptible regolith cover. On gentle slopes, Holocene solifluction has resulted in the formation of flights of vegetated *solifluction terraces* that terminate downslope at steep risers 0.2–1.0 m high. On steeper slopes, such terraces become crenulate in planform, forming a series of *solifluction lobes* arrayed across the slope (Fig. 10c). Radiocarbon ages obtained for organic soils and peat buried by the downslope advance of solifluction lobes have demonstrated that such lobes been active since the Middle Holocene and probably longer, with some evidence that lobes occasionally rupture, burying the soil downslope (Ballantyne 2019). Segmented tubes inserted vertically in solifluction lobes on slopes of ~15° at over 900 m altitude in the Fannich Mountains in northern Scotland and excavated after 35 years indicate that surface velocities average 7.8–10.6 mm a^{-1}, declining exponentially to zero at depths of 29–45 cm (Ballantyne 2013a).

Commonly associated with active solifluction terraces and lobes are *ploughing boulders* that have moved downslope faster than the surrounding soil, leaving a vegetated furrow upslope and sometimes pushing up a ridge of soil downslope (Fig. 10d). Ploughing boulders are amongst the most widely recorded periglacial phenomena on

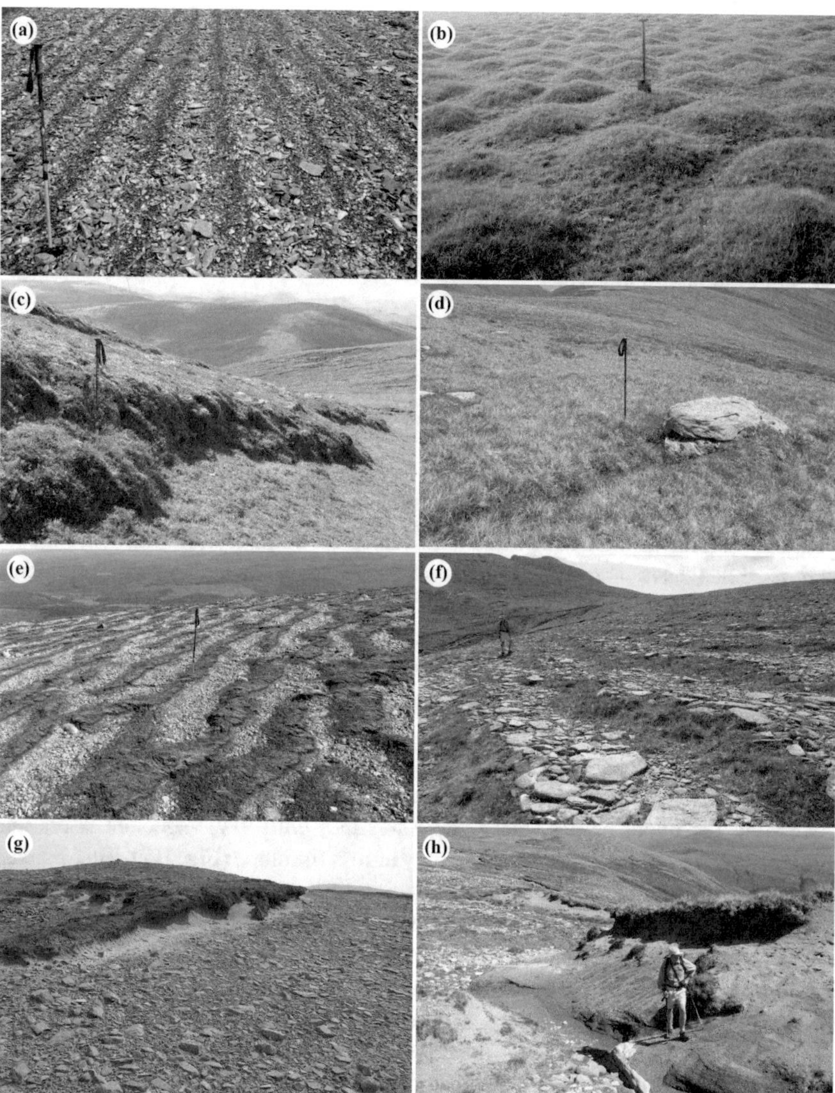

Fig. 10 Holocene and active periglacial features on high ground. **a** Active stone stripes, Tinto Hill, southern Scotland. **b** Earth hummocks (thúfur) of probable Holocene age on loessic soils, Dartmoor, SW England. **c** Active solifluction lobes, NW Highlands. **d** Ploughing boulder, NW Highlands. **e** Wind stripes, NW Highlands. **f** Turf-banked terraces, An Teallach, NW Highlands. **g** Deflation surface and residual 'island' of windblown sand, Ward Hill, Orkney Islands. **h** Niveo-aeolian sand deposits up to 4 m thick on lee slopes, An Teallach, NW Highlands (Images: Colin Ballantyne)

British and Irish mountains (e.g. Wilson 1993; Allison and Davies 1996). Most occur on vegetated slopes of 8–30°, and recent activity is indicated by a niche that extends to the base of the boulder at its upslope end; measured rates of recent ploughing boulder movement on Scottish mountains extend over two orders of magnitude (0.3–30 mm a^{-1}) and tend to increase with gradient. Boulder movement has been attributed by Ballantyne (2001b) to formation of ice lenses under boulders during periods of ground freezing; the resulting development of high pore-water pressures in the underlying soil during the ensuing thaw is thought to cause boulders to slide downslope over liquified or softened soil.

5.3 Holocene Aeolian Landforms and Deposits on British and Irish Mountains

The gale-force winds that sweep across the mountains of the British Isles have produced several distinctive aeolian landforms. On some lithologies, wind-driven sand particles have abraded boulders and low-lying rock outcrops, producing *ventifacts* with pitted, facetted and/or polished surfaces (Christiansen 2004). Over extensive areas, the interaction of wind stress, frost action (particularly needle-ice growth) and vegetation cover rooted in friable, cohesionless soils has produced *wind-patterned ground* on high cols, upper slopes and exposed plateaux. Wind-related ground patterns include *deflation scars* (pockets of bare ground on vegetated terrain), *wind stripes* (straight or wavy lines of bare ground alternating with bands of vegetation; Fig. 10e) and arcuate clumps of vegetation (*wind crescents*) that migrate slowly downwind. On upper slopes, *turf-banked terraces* with low, vegetated risers and bare treads extend horizontally or obliquely across the slope (Fig. 10f). Turf-banked terraces are probably attributable to retardation of downslope-moving debris cover by wind stripes (Ballantyne and Harris 1994).

The most widespread landforms produced by wind erosion of high plateaux are *deflation surfaces*, extensive areas of bare ground where removal of vegetation cover has exposed the soil to wind scour. At such sites, strong gusty winds have removed all fine soil particles (up to 4–6 mm), leaving unvegetated surfaces carpeted by gravel and boulders. The most extensive deflation surfaces occur on sandstone and granite plateaux, though similar features are present, if more localised, on most other lithologies. Most deflation surfaces occur above 600 m, but they are present down to 350 m on the windswept northern islands of Orkney and Shetland. Many are interspersed with remnant 'islands' of vegetated sand or soil cover that demonstrate that soil or sand cover was formerly more extensive (Fig. 10g).

On some mountains, the sand and soil particles stripped from deflation surfaces have accumulated as *vegetated sand sheets* on adjacent lee slopes. Those on the sandstone mountains of NW Scotland are of predominantly niveo-aeolian origin, and up to 4.0 m thick on plateau margins but thin downwind (Ballantyne and Whittington 1987; Ballantyne and Morrocco 2006; Fig. 10h). Many contain a lower unit of weathered sand overlain (sometimes unconformably) by an upper unit of fresh, unweathered sand. Radiocarbon dating of organic horizons has confirmed that the lower unit began to accumulate in the Early Holocene, and optically-stimulated luminescence (OSL) dating of sand immediately above and below the contact between the two units has demonstrated that the upper unit began to accumulate in AD 1550–1700, suggesting that it represents catastrophic wind erosion of upwind plateau surfaces during the Little Ice Age (Morrocco et al. 2007).

Other vegetated aeolian sand deposits have accumulated on high ground as a result of particles released by weathering from adjacent rockwalls being blown upwards and accumulating on plateau surfaces, where they are anchored by vegetation. Wilson (1989) has shown that sand deposits have accumulated in this way on Muckish Mountain (667 m) in NW Ireland during the Middle and Late Holocene. On the summit of The Storr (719 m) on Skye, plateau-top aeolian sands achieve a maximum thickness of 2.9 m and thin away from the edge of adjacent cliffs. Radiocarbon dating of organic soil buried by the sand demonstrates that accumulation was initiated by exposure of the cliff by a rockslide at ~6.1 ka, and that sand has subsequently accumulated at an average rate of 0.6 mm a^{-1} close to the cliff crest (Ballantyne 1998).

5.4 *Holocene Landslides, Debris Flows and Snow-Avalanche Landforms*

Although most paraglacial RSFs occurred during the Lateglacial period, CRE dating of catastrophic rockslides and rock avalanches shows that these continued to occur throughout the Holocene, particularly on steep mountain slopes inside the limits of Loch Lomond Stadial glaciation. Notable Holocene RSFs include the Beinn Alligin rock avalanche in NW Scotland, which occurred at ~4.3 ka, involved the failure of 9 Mt of sandstone bedrock and extended for over 1.2 km downvalley, the Storr landslide on Skye (~6.1 ka), which involved translational failure of stacked Palaeocene lava flows over Jurassic shales, and a rock avalanche near Glen Coe, where 0.6 Mt of rhyolitic ignimbrite failed at ~1.7 ka, blocking the mouth of a tributary hanging valley (Ballantyne and Stone 2013; Ballantyne et al. 2014).

Accumulation of *rockfall talus* continued inside the limits of LLS glaciation during the Early Holocene, but dating of buried palaeosols within talus accumulations suggests that rockfall activity had declined markedly by the Middle Holocene, and that many Holocene talus slopes thereafter experienced net erosion by shallow landslides, debris flows and flood torrents (Hinchliffe and Ballantyne 2009). Reworking of talus, debris-mantled slopes and valley-side till deposits by debris flows triggered by exceptional rainstorms has been the dominant process affecting most mountain slopes during the Holocene, and continues to represent a hazard at present, frequently affecting roads routed along the foot of steep mountain slopes (Winter 2020). Conversely, although over 100 snow avalanches are recorded on British mountains in most years, most have negligible geomorphological impact, though a few active *avalanche boulder tongues* are present high-level sites in the Cairngorm Mountains (Luckman 1992; Ballantyne 2019).

5.5 The Holocene Upland Periglacial Landsystem

A terrain model encapsulating the characteristic features of active (Holocene) periglaciation on British and Irish mountains is depicted in Fig. 9b but represents a composite that takes no account of regional variations introduced by lithology. Extensive aeolian landforms (deflation surfaces and aeolian sand deposits), for example, are much more common on lithologies that have weathered to produce cohesionless sandy regolith, such as sandstone and granite. Conversely, frost-sorted patterned ground and solifluction landforms are restricted to lithologies that have weathered to form silt-rich frost-susceptible regolith, such as schist. The full range of periglacial features depicted in Fig. 9b is therefore rarely, if ever, encountered on a single mountain.

6 The Significance of Periglacial Dynamics in Great Britain and Ireland within the European Context

Periglacial landscapes in Great Britain and Ireland fall into two broad categories distinguished by different landform assemblages and by age: (i) lowland landscapes in which all periglacial landforms, sediment accumulations and structures are relict features of pre-Holocene age, with some dating back to the Middle Pleistocene or earlier; and (ii) upland or montane landscapes in which most periglacial features are of Lateglacial or Holocene age. Lowland periglacial regions may also be subdivided into those occupied by successive Middle and Late Pleistocene ice sheets, and the 'never-glaciated' parts of southern and eastern England, where periglacial activity dominated landscape evolution throughout the Quaternary.

Lowland periglacial landscapes of the British Isles contain a range of periglacial features similar to those found in neighbouring continental Europe. Long-term periglacial weathering is evident in widespread brecciation of frost-susceptible bedrock, particularly the Mesozoic and Cenozoic strata that underlie central, southern and eastern England, and in periglacial karst landscapes on carbonate rocks in England and Ireland. The abundant evidence for past permafrost in lowland regions includes ice-wedge or composite-wedge pseudomorphs in both near-surface and buried stratigraphic successions (in some cases dating back to the Early or Middle Pleistocene), frost polygons, relict frost mounds (lithalsas, and possible pingos or pingo-lithalsa hybrids), thermokarst features, relict active-layer phenomena (cryoturbation structures and large-scale patterned ground) and glacially deformed past permafrost. Former periglacial mass-movement is represented by deformation structures at rockhead, periglacial shears that are believed to represent former active-layer detachment slides, in the accumulation of both granular and clayey head (mainly gelifluctate) deposits that mantle gentle slopes, locally in the form of lobate landforms, and in the development of cambered strata, gulls and valley bulges in areas where valleys are incised through a resistant caprock overlying argillaceous bedrock. Extensive loess deposits, mainly of Late Devensian (Dimlington Stadial) age, mantle the Mesozoic rocks of central, southern and eastern England, with smaller pockets elsewhere, though British loesses are generally thinner and younger than those of continental Europe. Coversand deposits of Middle to Late Pleistocene age are more restricted in distribution; reported examples are limited to just four main areas, all in England. Finally, periglacial fluvial activity has played a major role in modifying the landscape of lowland England, particularly outside the limit of the last ice sheet: it has been responsible for the evolution of dry valleys mantled by solifluction, slopewash and fluvio-colluvial deposits, and for the development of the flights of terraces that flank major trunk valleys.

In upland areas of Great Britain and Ireland, the oldest periglacial landforms are plateau blockfields and summit tors, both of which pre-date the Late Devensian and have been preserved under a cover of cold-based ice during the last (and in some cases earlier) ice-sheet glaciation(s). Lateglacial periglacial features in upland areas include large-scale sorted patterned ground and bouldery 'stone-banked' terraces and lobes on high ground, and periglacial valley-fill deposits. The Lateglacial period also witnessed widespread paraglacial rock-slope failures on British and Irish mountains, and the development of (now relict and eroded) talus slopes through rockfall from cliffs. Under the maritime periglacial climate of the Holocene, periglacial activity on high ground has been limited to granular weathering of exposed bedrock and boulder surfaces, formation of miniature sorted patterned ground by needle-ice, and shallow solifluction, the last represented by active terraces, lobes and ploughing boulders on slopes above ~600 m. Wind erosion has created deflation surfaces and wind-patterned ground, and some mountains support Holocene (niveo-)aeolian sand sheets; a phase of enhanced aeolian erosion and deposition on high ground occurred under the stormy conditions of the Little Ice Age. Paraglacial rock-slope failure has continued to operate sporadically during the Holocene, though debris flows now constitute the dominant mass-movement activity operating on most British and Irish mountains.

As the above summary demonstrates, the range of periglacial features present in the British Isles encompasses, in relict or active form, most of the landforms, sediments and soil structures of the periglacial realm. Although most of Great Britain and Ireland were covered by successive ice sheets, the 'never-glaciated' parts of England essentially evolved under periglacial conditions, whilst even those areas covered by the last ice sheet have experienced (and on high ground, continue to experience) modification by Lateglacial and Holocene periglacial processes.

References

Allison RJ, Davies KC (1996) Ploughing blocks as evidence of down-slope sediment transport in the English Lake District. Z Geomorph 106:199–219

Anderson E, Harrison S (2006) Late Quaternary paraglacial sedimentation in the Macgillycuddy Reeks, southwest Ireland. Irish Geogr 39:69–77

Antoine P, Catt J, Lautridou J-P, Sommé J (2003) The loess and coversands of northern France and southern England. J Quat Sci 18:309–318

Baker CA, Bateman M, Bateman P, Jones H (2013) The aeolian sand record in the Trent valley. Mercian Geol 18:108–118

Ballantyne CK (1996) Formation of miniature sorted patterns by shallow ground freezing: a field experiment. Permafrost Periglac Process 7:409–424

Ballantyne CK (1998) Aeolian deposits on a Scottish mountain summit: characteristics, provenance, history and significance. Earth Surf Proc Landf 23:625–641

Ballantyne CK (2001a) The sorted stone stripes of Tinto Hill. Scot Geogr J 117:313–324

Ballantyne CK (2001b) Measurement and theory of ploughing boulder movement. Permafrost Periglac Process 12:267–288

Ballantyne CK (2010) A general model of autochthonous blockfield evolution. Permafrost Periglac Process 21:289–300

Ballantyne CK (2013a) A 35-year record of solifluction in a maritime periglacial environment. Permafrost Periglac Process 24:56–66

Ballantyne CK (2013b) Lateglacial rock-slope failures in the Scottish Highlands. Scot Geogr J 129:67–84

Ballantyne CK (2018) Periglacial geomorphology. Wiley-Blackwell, Chichester

Ballantyne CK (2019) After the ice: Lateglacial and Holocene landforms and landscape evolution in Scotland. Earth Env Sci Trans R Soc Edinb 110:133–171

Ballantyne CK (2021) Rock-slope failures in the North West Highlands. In: Ballantyne CK, Gordon JE (eds) Landscapes and landforms of Scotland. Springer, Switzerland

Ballantyne CK, Harris C (1994) The periglaciation of Great Britain. Cambridge University Press, Cambridge UK

Ballantyne CK, Morrocco SM (2006) The windblown sands of An Teallach. Scot Geogr J 122:149–159

Ballantyne CK, Sandeman GF, Stone JO, Wilson P (2014) Rock-slope failure following Late Pleistocene deglaciation on tectonically stable mountainous terrain. Quat Sci Rev 86:144–157

Ballantyne CK, Small D (2019) The last Scottish ice sheet. Earth Env Sci Trans R Soc Edinb 110:133–171

Ballantyne CK, Stone JO (2013) Timing and periodicity of paraglacial rock-slope failures in the Scottish Highlands. Geomorphol 186:150–161

Ballantyne CK, Stone JO (2015) Trimlines, blockfields and the vertical extent of the last ice sheet in southern Ireland. Boreas 44:277–287

Ballantyne CK, Whittington GW (1987) Niveo-aeolian sand deposits on An Teallach, Wester Ross, Scotland. Trans R Soc Edinb Earth Sci 78:51–63

Bateman MD (1995) Thermoluminescence dating of the British coversand deposits. Quat Sci Rev 14:791–798

Bateman MD (2008) Luminescence dating of periglacial sediments and structures. Boreas 37:574–588

Bateman MD (2013) Aeolian processes in periglacial environments. In: Shroder JF, Giardino R, Harbor J (eds) Treatise on geomorphology, vol 8. Glacial and periglacial geomorphology. Academic Press, San Diego, pp 416–429

Bateman MD, Hitchens S, Murton JB et al (2014) The evolution of periglacial patterned ground of central East Anglia, UK. J Quat Sci 29:301–317

Boardman J (ed) (1987) Periglacial processes and landforms in Britain and Ireland. Cambridge University Press, Cambridge UK

Briant RM, Bateman MD, Coope GR, Gibbard PL (2005) Climatic control on Quaternary fluvial sedimentology of a Fenland Basin river, England. Sedimentol 52:1397–1423
Briant RM, Coope GR, Preece RC et al (2004) Fluvial system response to Late Devensian (Weichselian) aridity, Baston, Lincolnshire, England. J Quat Sci 19:479–495
Briffa KR, Atkinson TC (1997) Reconstructing Late-glacial and Holocene climates. In: Hulme M, Barrow E (eds) Climates of the British Isles: present, past and future. Routledge, London, pp 84–111
Brown AG, Basell LS, Toms PS et al (2010) Later Pleistocene evolution of the Exe valley: a chronostratigraphic model of terrace formation and its implications for Palaeolithic archaeology. Quat Sci Rev 116:106–121
Burke H, Phillips E, Lee JR, Wilkinson IP (2009) Imbricate thrust stack model for the formation of glaciotectonic rafts: an example from the Middle Pleistocene of north Norfolk, UK. Boreas 38:620–637
Candy I, Silva B, Lee J (2011) Climates of the early Middle Pleistocene in Britain: environments of the earliest humans in northern Europe. In: Ashton N, Lewis SG, Stringer C (eds) The ancient human occupation of Britain. Elsevier, Amsterdam, pp 11–22
Catt J (2010) Neogene deposits and early landscape development. In: Catt J (ed) Hertfordshire geology and landscape. Herts Natural Hist Soc, Welwyn Garden City, pp 79–117
Cave JAS, Ballantyne CK (2016) Catastrophic rock-slope failures in NW Scotland: quantitative analysis and implications. Scot Geogr J 132:185–209
Christiansen HH (2004) Windpolished boulders and bedrock in the Scottish Highlands: evidence and implications of Late Devensian wind activity. Boreas 33:82–94
Coxon P, O'Callaghan P (1987) The distribution and age of pingo remnants in Ireland. In: Boardman J (ed) Periglacial processes and landforms in Britain and Ireland. Cambridge University Press, Cambridge, pp 195–202
Culshaw MG, Entwisle DC, Giles DP et al (2017) Material properties and geohazards. Geol Soc Lond Eng Geol Spec Publ 28:599–740
Cunningham A, Wilson P (2004) Relict periglacial boulder sheets and lobes on Slieve Donard, Mountains of Mourne, Northern Ireland. Irish Geogr 37:187–201
Curry AM, Morris CJ (2004) Lateglacial and Holocene talus slope development and rockwall retreat on Mynedd Du, UK. Geomorphol 58:85–106
de Freitas MH, Griffiths JS, Press N et al (2017) Engineering investigation and assessment. Geol Soc Lond Eng Geol Spec Publ 28:741–830
Dines HG, Hollingworth SE, Edwards H et al (1940) The mapping of head deposits. Geol Mag 77:198–226
Fabel D, Ballantyne CK, Xu S (2012) Trimlines, blockfields, mountain-top erratics and the vertical dimensions of the last British-Irish Ice sheet in NW Scotland. Quat Sci Rev 55:91–102
Font M, Lagarde J-L, Amorese D et al (2006) Physical modelling of fault scarp degradation under freeze/thaw cycles. Earth Surf Proc Landf 31:1731–1745
French HM (2017) The periglacial environment, 4th edn. Wiley-Blackwell, Chichester
French HM, Bennett L, Hayley DW (1986) Ground ice conditions near Rea Point and on Sabine Peninsula, eastern Melville Island. Can J Earth Sci 23:1389–1400
Gallois RW (2009) The origin of the Clay-with-flints: the missing link. Geosci SW Engl 12:153–161
Gallois RW (2010) Large-scale periglacial creep folds in Jurassic mudstones on the Dorset coast, UK. Geosci SW Engl 12:223–232
Giles DP, Griffiths JS, Evans DJA, Murton JB (2017) Geomorphological framework: glacial and periglacial sediments, structures and landforms. Geol Soc Lond Eng Geol Spec Publ 28:59–368
Goodfellow BW, Skelton A, Martel SJ et al (2014) Controls of tor formation, Cairngorm Mountains, Scotland. J Geophys Res Earth Surf 119:225–246
Gunnell Y, Jarman D, Braucher R et al (2013) The granite tors of Dartmoor, Southwest England: rapid and recent emergence revealed by Late Pleistocene cosmogenic apparent exposure age. Quat Sci Rev 61:62–76

Gurney SD (1998) Aspects of the genesis and geomorphology of pingos: perennial permafrost mounds. Prog Phys Geog 22:307–324

Gurney SD (2000) Relict cryogenic mounds in the UK as evidence of climate change. In: McLaren SJ, Kniveton DR (eds) Linking climate change to land surface change. Kluwer Academic, Dordrecht, pp 209–229

Gurney SD (2001) Aspects of the genesis, geomorphology and terminology of palsas: perennial permafrost mounds. Prog Phys Geog 25:249–260

Hall AM, Merritt JW, Connell ER, Hubbard A (2019) Early and Middle Pleistocene environments, landforms and sediments in Scotland. Earth Env Sci Trans R Soc Edinb 110:5–37

Hall AM, Phillips WM (2006) Glacial modification of granite tors in the Cairngorms, Scotland. J Quat Sci 21:811–830

Harris C (1987) Solifluction and related deposits in England and Wales. In: Boardman J (ed) Periglacial processes and landforms in Britain and Ireland. Cambridge University Press, Cambridge, pp 209–224

Harris C (1998) The micromorphology of paraglacial and periglacial slope deposits: a case study from Morfa Bychan, west Wales, UK. J Quat Sci 13:73–84

Harris C (2013) Slope deposits and forms. In: Elias SA, Mock CJ (eds) Encyclopedia of Quaternary science, vol 3, 2nd edn. Elsevier, Amsterdam, pp 481–489

Harris C, Davies MCR, Coutard J-P (1997) Rates and processes of periglacial solifluction: an experimental approach. Earth Surf Proc Landf 22:849–868

Harris C, Davies MCR, Rea BR (2003) Gelifluction: viscous flow or plastic creep? Earth Surf Proc Landf 28:1289–1301

Harris C, Gallop M, Coutard J-P (1993) Physical modelling of gelifluction and frost creep: some results of a large-scale laboratory experiment. Earth Surf Proc Landf 18:383–398

Harris C, Kern-Luetschg M, Murton J et al (2008) Solifluction processes on permafrost and non-permafrost slopes: results of a large-scale laboratory simulation. Permafrost Periglac Process 19:359–378

Harris C, Murton JB, Davies MCR (2000) Soft-sediment deformation during thawing of ice-rich frozen soils: results of scaled centrifuge modelling experiments. Sedimentol 47:687–700

Harris C, Rea B, Davies MCR (2001) Scaled physical modelling of mass movement processes on thawing slopes. Permafrost Periglac Process 12:125–136

Harrison S (2002) Lithological variability of Quaternary slope deposits in the Cheviot Hills, UK. Proc Geol Assoc 113:121–138

Harrison S, Bailey RM, Anderson E et al (2010) Optical dates from British Isles 'solifluction sheets' suggest rapid landscape response to Late Pleistocene climate change. Scot Geogr J 126:101–111

Hawkins AB, Privett KD (1981) A building site on cambered ground at Radstock, Avon. Quart J Eng Geol 14:151–167

Higginbottom IE, Fookes PG (1971) Engineering aspects of periglacial features in Britain. Q J Eng Geol 3:85–117

Hinchliffe S, Ballantyne CK (1999) Talus accumulation and rockwall retreat, Trotternish, Isle of Skye, Scotland. Scot Geogr J 115:53–70

Hinchliffe S, Ballantyne CK (2009) Talus structure and evolution on sandstone mountains in NW Scotland. The Holocene 19:477–486

Hoare PG, Stevenson CR, Godby SP (2002) Sand sheets and ventifacts: the legacy of aeolian action in west Norfolk, UK. Proc Geol Ass 113:301–317

Hobbs PRN, Dobbs MR, Cuss RJ (2011) Desk based study and literature review of diapirism in plastic clays and an analysis of the critical state of Boom Clay. Brit Geol Surv Commercial Report CR/11/012

Hobbs PRN, Entwisle DC, Northmore KJ et al (2012) Engineering geology of British Rocks and soils—Lias Group. Brit Geol Surv Internal Report OR/12/032

Hobbs PRN, Jenkins GO (2008) Bath's 'foundered strata'—A re-interpretation. Brit Geol Surv Research Report OR/08/052

Hollingworth SE (1934) Some solifluction phenomena in the northern part of the Lake District. Proc Geol Assoc 2:167–188

Hopkinson C, Ballantyne CK (2014) Age and origin of blockfields on Scottish mountains. Scot Geogr J 130:116–141

Hutchinson JN (1991) Periglacial slope processes. Geol Soc Lond Eng Geol Spec Publ 7:283–331

Hutchinson JN (2010) Relict sand wedges in soliflucted London Clay at Wimbledon, London, UK. Proc Geol Ass 121:444–454

Jarman D, Harrison S (2019) Rock slope failure in the British mountains. Geomorphol 340:202–233

Killingbeck J, Ballantyne CK (2012) Earth hummocks in west Dartmoor, SW England: characteristics, age and origin. Permafrost Periglac Process 23:152–161

Langdon DG, Barber KE (2005) The climate of Scotland over the last 5000 years inferred from multiproxy peatland records: inter-site correlations and regional variability. J Quat Sci 20:549–566

Lee J, Brown EJ, Rose J et al (2003) A reply to 'Implications of a Middle Pleistocene ice-wedge cast at Trimingham, Norfolk, Eastern England' (Whiteman, 2002). Permafrost Periglac Process 14:75–77

Lewin J, Gibbard PL (2010) Quaternary river terraces in England: forms, sediments and processes. Geomorphol 120:293–311

Lewis CA (1985) Periglacial features. In: Edwards KJ, Warren WP (eds) The Quaternary history of Ireland. Academic Press, London, pp 95–113

Luckman BH (1992) Debris flows and snow avalanche landforms in the Lairig Ghru, Cairngorm Mountains, Scotland. Geogr Annr 74A:109–121

Lundberg J, McFarlane D (2007) Pleistocene depositional history in a periglacial terrane: a 500 ka record from Kents Cavern, Devon, United Kingdom. Geosphere 3:199–219

Lundberg J, McFarlane D (2012) Cryogenic fracturing of calcite flowstone in caves: theoretical considerations and field observations in Kents Cavern, Devon, UK. Int J Speleol 41:307–316

Moles N, Moles R, O'Donovan G (1995) Evidence for the presence of Quaternary loess-derived soils in the Burren karstic area, western Ireland. Irish Geogr 28:48–63

Morrocco SM, Ballantyne CK, Spencer JQ, Robinson RAJ (2007) Age and significance of aeolian sediment reworking on high plateaux in the Scottish Highlands. The Holocene 17:349–360

Murton JB (1996) Near-surface brecciation of Chalk, Isle of Thanet, southeast England: a comparison with ice-rich brecciated bedrocks in Canada and Spitsbergen. Permafrost Periglac Process 7:153–164

Murton JB (2013) Ice wedges and ice-wedge casts. In: Elias SA, Mock CJ (eds) Encyclopedia of Quaternary science, 2nd edn. Elsevier, Amsterdam, vol 3, pp 436–451

Murton JB (2021) What and where are periglacial landscapes? Permafrost Periglac Process 32:186–212

Murton JB (2022a) Ground ice. In: Shroder JF (ed-in-chief) Treatise on geomorphology, 2nd edn. 1146 Haritashya U (ed) vol 4 Cryospheric geomorphology. Elsevier, Academic Press, pp 428–457

Murton JB (2022b) Cryostratigraphy. In: Shroder JF (ed-in-chief) Treatise on geomorphology, 2nd edn. 1146 Haritashya U (ed) vol 4 Cryospheric geomorphology. Elsevier, Academic Press, pp 458–490

Murton JB, Ballantyne CK (2017) Periglacial and permafrost conceptual ground models for Great Britain. Geol Soc Lond Eng Group Spec Publ 28:501–597

Murton JB, Bateman MD, Baker CA et al (2003) The Devensian periglacial record on Thanet, Kent, UK. Permafrost Periglac Process 14:217–246

Murton JB, Belshaw RK (2011) A conceptual model of valley incision, planation and terrace formation during cold and arid permafrost conditions of Pleistocene southern England. Quat Res 75:385–394

Murton JB, Bowen DQ, Candy I et al (2015) Middle and Late Pleistocene environmental history of the Marsworth area, south-central England. Proc Geol Assoc 126:18–49

Murton JB, Lautridou J-P (2003) Recent advances in the understanding of Quaternary periglacial features of the English Channel coastlands. J Quat Sci 18:301–307

Murton JB, Peterson R, Ozouf J-C (2006) Bedrock fracture by ice segregation in cold regions. Sci 314:1127–1129

Murton JB, Whiteman CA, Allen P (1995) Involutions in the Middle Pleistocene (Anglian) Barham Soil, Eastern England: a comparison with thermokarst involutions from arctic Canada. Boreas 24:269–280

Parks CD (1991) A review of the possible mechanisms of cambering and valley bulging. Geol Soc Lond Eng Geol Spec Publ 7:373–380

Parks DA, Rendell HM (1992) Thermoluminescence dating and geochemistry of loessic deposits in south-east England. J Quat Sci 7:99–107

Paterson TT (1940) The effects of frost action and solifluction around Baffin Bay and the Cambridge district. Q J Geol Soc Lond 96:99–130

Phillips WM, Hall AM, Mottram R et al (2006) Cosmogenic ^{10}Be and ^{26}Al exposure ages of tors and erratics, Cairngorm Mountains, Scotland: timescales for the development of a classic landscape of selective linear erosion. Geomorphol 73:222–245

Pissart A (2003) The remnants of Younger Dryas lithalsas on the Hautes Fagnes Plateau in Belgium and elsewhere in the world. Geomorphol 52:5–38

Preece RC, Kemp RA, Hutchinson JN (1995) A Late-glacial colluvial sequence at Watcombe Bottom, Ventnor, Isle of Wight. J Quat Sci 10:107–121

Rose J (2010) The Quaternary of the British Isles: factors forcing environmental change. J Quat Sci 25:399–418

Rose J, Allen P, Kemp RA et al (1985) The Early Anglian Barham Soil of Eastern England. In: Boardman J (ed) Soils and Quaternary landscape evolution. Wiley, Chichester, pp 197–228

Rose J, Lee JA, Kemp RA, Harding PA (2000) Palaeoclimate, sedimentation and soil development during the last Glacial Stage (Devensian), Heathrow Airport, London, UK. Quat Sci Rev 19:827–847

Ross N, Harris C, Brabham P (2007) Internal structure and origins of Late Devensian 'ramparted depressions', Llanio Fawr, Ceredigion. Quat Newsletter 112:6–21

Ross N, Harris C, Brabham P, Sheppard TH (2011) Internal structure and geological context of ramparted depressions, Llanpumsaint, Wales. Permafrost Periglac Process 22:291–305

Rydval M, Loader NJ, Gunnarson BE et al (2017) Reconstructing 800 years of summer temperatures in Scotland from tree rings. Clim Dyn 49:2951–2974

Scourse JD (1987) Periglacial sediments and landforms in the Isles of Scilly and West Cornwall. In: Boardman J (ed) Periglacial processes and landforms in Britain and Ireland. Cambridge University Press, Cambridge, pp 225–236

Self CA (1995) The relationship between the gull cave Sally's Rift and the development of the River Avon east of Bath. Proc Univ Bristol Spelaeolog Soc 20:91–108

Self CA, Farrant AR (2013) Gulls, gull-caves and cambering in the southern Cotswold Hills, England. In: Filippi M, Bosak P (eds) 16th Int Cong Speleol. Czech Speleolog Soc, Brno 3:132–136

Skempton AW, Weeks AG (1976) The Quaternary history of the Lower Greensand escarpment and Weald Clay vale near Sevenoaks, Kent. Phil Trans R Soc Lond A283:493–526

Spink TW (1991) Periglacial discontinuities in Eocene Clay near Denham, Buckinghamshire. Geol Soc Lond Eng Geol Spec Publ 7:389–396

Stevens T, Sechi D, Bradák B et al (2020) Abrupt last glacial dust fall over southeast England associated with dynamics of the British-Irish ice sheet. Quat Sci Rev 250:106641

Védie E, Lagarde J-L, Font M (2010) Physical modelling of rainfall- and snowmelt-induced erosion of stony slopes underlain by permafrost. Earth Surf Proc Landf 36:395–407

Vincent PJ, Lord TC, Telfer MW, Wilson P (2011) Early Holocene loessic colluviation in northwest England: new evidence for the 8.2 ka event in the terrestrial record? Boreas 40:105–115

Waller RI, Phillips E, Murton JB et al (2011) Sand intraclasts as evidence of subglacial deformation of Middle Pleistocene permafrost, north Norfolk, UK. Quat Sci Rev 30:3481–3500

Warburton J (1999) Sorted patterned ground in the English Lake District. Permafrost Periglac Process 10:193–197

West RG (2000) Plant life of the Quaternary cold stages: evidence from the British Isles. Cambridge University Press, Cambridge

West RG (2015) Evolution of a Breckland landscape: Chalkland under a cold climate in the area of Beachamwell. Norfolk. Suffolk Naturalists's Society, Ipswich

West RG, Andrew R, Catt J et al (1999) Late and Middle Pleistocene deposits at Somersham, Cambridgeshire, UK: a model for reconstructing fluvial/estuarine depositional environments. Quat Sci Rev 18:1247–1314

West RG, Burton RGO, Andrew R, Pettit ME (2000) Evolution of a periglacial landscape in the Late Devensian: environments and palaeobotany of the Mepal area, Cambridgeshire, England. J Quat Sci 17:31–50

Williams RBG (1965) Permafrost in England during the last glacial period. Nat 205:1304–1305

Williams RBG (1969) Permafrost and temperature conditions in England during the last glacial period. In: Péwé TL (ed) The periglacial environment. McGill-Queens University Press, Montreal, pp 399–410

Wilson P (1989) Nature, origin and age of Holocene aeolian sand on Muckish Mountain, Co., Donegal. Ireland. Boreas 18:159–168

Wilson P (1992) Small-scale patterned ground, Comeragh Mountains, southeast Ireland. Permafrost Periglac Process 3:63–70

Wilson P (1993) Ploughing boulder characteristics and associated soil properties in the Lake District and southern Scotland. Scot Geogr J 109:18–26

Wilson P (2017) Periglacial and paraglacial processes, landforms and sediments. In: Coxon P, McCarron S, Mitchell F (eds) Advances in Irish Quaternary Studies. Atlantis, Paris, pp 217–254

Wilson P, Matthews JA (2016) Age assessment and implications of Late Quaternary periglacial and paraglacial landforms on Muckish Mountain, northwest Ireland, based on Schmidt-hammer exposure dating (SHD). Geomorphol 270:134–144

Winter MG (2020) Debris flows. Geol Soc Lond Eng Geol Spec Publ 29:163–185

Scandinavia

John A. Matthews and Atle Nesje

1 Introduction

For purposes of this chapter, Scandinavia is defined as four countries—Norway, Sweden, Finland and Denmark—which occupy a total area of ~1.2 million km^2, which is over one tenth of the area of Europe. By virtue of their northerly location, latitudinal extent and topography, the present-day periglacial landscapes of Scandinavia exhibit characteristics of both Arctic and alpine regions (Fig. 1). Most of these periglacial landscapes are located in the 'Scandes', the Scandinavian mountain chain that extends from southern Norway to northern Finland. The Scandes occupy more than one-third of the area of Norway, almost one-tenth of the area of Sweden and a very small part of northern Finland (Corner 2005a). They rise to 2469 m above sea level in Jotunheimen, southern Norway, and 2097 m a.s.l. in Kebnekaise, northern Sweden, and include extensive plateau areas that lie generally between 1000 and 1500 m a.s.l. Together with high-level basins, valleys and cols, the plateaux have been seen as macro-remnants of 'palaeic' landscapes, which retain pre-Pleistocene characteristics little modified by subsequent uplift and glaciation (Holtedahl 1960; Gjessing 1967; Lidmar-Bergström et al. 2000). According to Etzelmüller et al. (2007) such 'palaeic' landscapes may account for ~12% of the surface area of southern Norway.

J. A. Matthews (✉)
Department of Geography, College of Science, Swansea University, Singleton Park, Wales SA2 8PP, UK
e-mail: J.A.Matthews@Swansea.ac.uk

A. Nesje
Department of Earth Science and Bjerknes Centre for Climatic Research, University of Bergen, NO-5020 Bergen, Norway
e-mail: Atle.Nesje@uib.no

M. Oliva et al. (eds.), *Periglacial Landscapes of Europe*,
https://doi.org/10.1007/978-3-031-14895-8_14

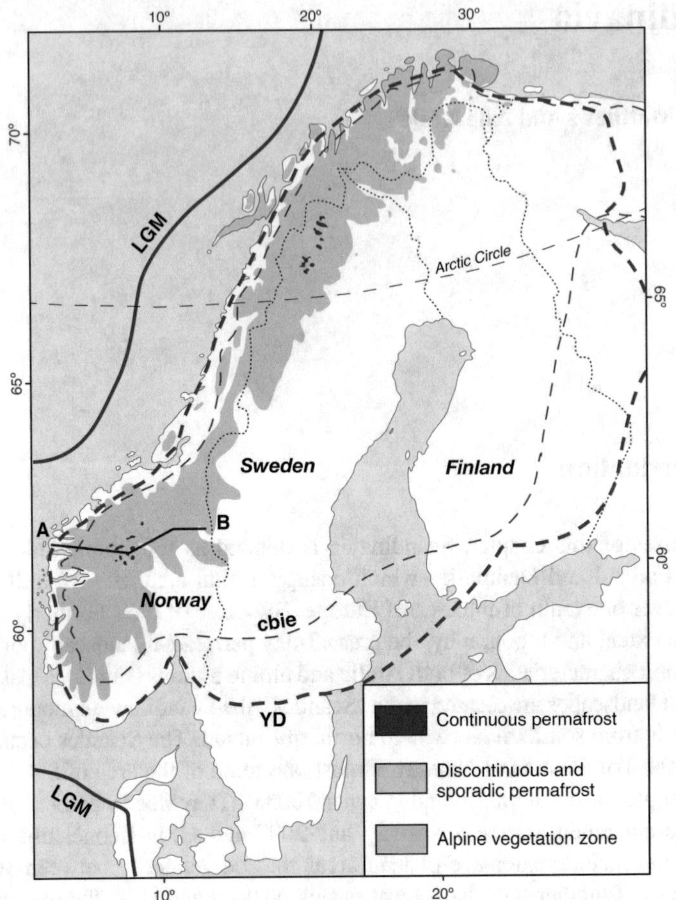

Fig. 1 Geographical background to Scandinavian periglacial landscapes: permafrost zonation (generalised from Gisnås et al. 2017); the alpine vegetation zone (based on Moen 1999; Heikkinen 2005), which approximates the zone of geomorphologically significant seasonal frost; limits of the Scandinavian Ice Sheet at the Last Glacial Maximum (LGM, 22–21 ka) and during the Younger Dryas (YD, ~12.7 ka), and the minimum cold-based ice extent (cbie) at the Last Glacial Maximum (all based on Strøeven et al. 2016). Continuous permafrost occupies 90–100% of the present landscape within the purple area; discontinuous and sporadic permafrost >10% within the pink area. Line (A–B) defines the topographic profile shown in Fig. 2

1.1 Geological, Climatic and Geo-Ecological Background

The mountains and plateaux consist mainly of metasedimentary and metavolcanic lithologies (including schist, quartzite and amphibolite) in Caledonian nappes, which overly and have incorporated slices and blocks of granite and gneiss of the crystalline basement rocks (Corner 2005a). These lithologies are relatively resistant to weathering and erosion by periglacial and other Earth surface processes. Exposed bedrock and patchy, generally thin, superficial covers of periglacial and glacigenic regolith are characteristic of the mountains and plateaux. Increasing thicknesses of this regolith occur at lower elevations and on lower slopes, with the addition of fluvial and lacustrine sediments on the floors of valleys and basins.

Mean annual air temperature (MAAT) in Scandinavia varies from <–4 °C on the highest mountains of the Scandes to >6 °C in the southern lowlands and around 0 °C at sea level in the far north (Tikkanen 2005; Gisnås et al. 2017). The corresponding ranges of mean January and July temperature are <–14 °C to >0 °C, and <6 °C to >16°, respectively. These climatic data reveal a very wide range of climatic regimes from permafrost, through seasonal frost to cool temperate climates, within which latitude, altitude and continentality gradients are important controls.

Other annual and seasonal climatic parameters have important effects in Scandinavian periglacial landscapes. Prevailing westerly wind systems bring mean annual precipitation totals of >2000 mm over wide areas of the western slopes of the Scandes. Orographic effects are extreme, with precipitation totals in parts of the mountains >4000 mm, reducing to <400 mm in the corresponding rain-shadow areas of both Lapland and interior southern Norway (Tikkanen 2005). Maximum snow depth is 2–5 m in the western parts of the mountains and normally <2 m over most areas east of the watershed, including many high plateau areas (Gisnås et al. 2017). Mean annual ground surface temperature (MAGT) tends to be ~1 °C lower than the MAAT at exposed sites and the thermal offset may be up to ~4 °C at sites with prolonged snow (Farbrot et al. 2011). Local snow distribution is the result of complex interactions between temperature, precipitation, wind, radiation, and local topography (Liston and Elder 2006), and affects many periglacial processes through microclimate and hydrology.

Geoecosystems are equally varied and are also largely determined on macro- to micro-scales by climate. Above and beyond the tree line, which normally involves mountain birch (*Betula pubescens*) and, in some regions, Scots pine (*Pinus sylvestris*), the alpine zone is commonly subdivided into high-, mid- and low-alpine belts on the basis of the plant communities (Moen 1999; Heikkinen 2005). In southern Scandinavia, permafrost is largely confined to the high- and mid-alpine belts, and the tree line (lower limit of the dwarf shrub-dominated low-alpine zone) can be considered as approximating to the lower limit of effective geomorphic activity associated with seasonal frost. On the northernmost coast, the tree line descends to sea level, and in the inland continental areas to the east of the northern Scandes, the tree line corresponds approximately with the lower limit of discontinuous permafrost.

1.2 Permafrost Distribution and Limits

In the context of Europe, Scandinavia is particularly important for the extent of periglacial environments in general and permafrost in particular. The estimated total land area with permafrost in Scandinavia is ~62,600 km^2, of which 2% has been classified as continuous, 20% as discontinuous, and 78% as sporadic (Gisnås et al. 2017).

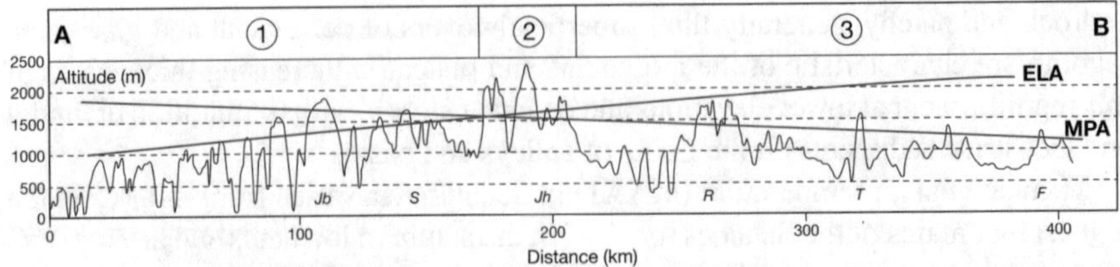

Fig. 2 A topographic profile across southern Norway defining three morphogenetic zones based on the lower altitudinal limit of (discontinuous) mountain permafrost (MPA, *red line*), and the modern glacial equilibrium-line altitude (ELA, *blue line*): (1) western zone with little or no permafrost, extensive seasonal frost and temperate glaciers (MPA > ELA); (2) central zone with extensive permafrost that co-exists in the landscape with polythermal glaciers and ice-cores moraines (MPA ≈ ELA); and (3) eastern zone where permafrost exists but glaciers are absent (ELA > MPA). The profile line (A–B), shown on Fig. 1, extends from the Norwegian west coast through the Jostedalsbreen ice-cap (Jb), Sognefjell (S), Jotunheimen (Jh), Rondane (R) and Tronfjell (T) to the Femund mountains (F) near the Swedish border (based on Etzelmüller et al. 2003; Etzelmüller and Hagen 2005)

The spatial distribution of permafrost is shown at the regional scale in Fig. 1. Altitudinal relationships are exemplified by a west-east transect across southern Norway at approximately latitude 62° N (Fig. 2) (Etzelmüller et al. 2003; Etzelmüller and Hagen 2005). This transect highlights the mountain permafrost altitude (MPA), which is equivalent to the lower altitudinal limit of discontinuous permafrost. The MPA declines from ~1750 m a.s.l. in the west to ~1350 m a.s.l. in the east, while the corresponding lower limits of sporadic permafrost occur at ~1450 m a.s.l. and ~1050 m a.s.l., respectively (Gisnås et al. 2017). Lower permafrost limits in the east reflect colder winters and decreasing snow depths associated with increasing continentality. In contrast, the equilibrium-line altitude (ELA) on the glaciers rises towards the east in response to warmer summers combined with less snow accumulation in winter.

In the Galdhøpiggen massif, Jotunheimen, in the centre of the west-east transect, several estimates place the lower altitudinal limit of discontinuous permafrost at ~1450 m a.s.l. (Ødegård et al. 1992; Isaksen et al. 2002; Farbrot et al. 2011; Lilleøren et al. 2012). There are also differences in permafrost limits between north-facing and south-facing aspects, which are most marked in relation to permafrost in steep bedrock slopes (Steiger et al. 2016; Gisnås et al. 2017). According to Hipp et al. (2014), the lower limit of discontinuous permafrost in the Galdhøpiggen massif occurs at 1500–1700 m a.s.l. in south-facing rock walls and 1200–1300 m a.s.l. in north-facing rock walls. Magnin et al. (2019) have demonstrated the occurrence of isolated, sporadic permafrost as low as 830 m a.s.l. in steep, north-facing bedrock slopes in southern Norway. However, the altitudinal limits of permafrost are not fixed. Based on borehole temperatures, Lilleøren et al. (2012) have demonstrated the scale of the altitudinal variations in permafrost limits during the Holocene. In the Galdhøpiggen massif, the lower limit of discontinuous permafrost was higher than today during the Holocene Thermal Maximum (~8.0 ka), when it was located at ~1600–1700 m a.s.l. (about 200 m higher than at present). Subsequently, a late-Holocene lowering of the limit culminated in the Little Ice Age (~AD 1750) with the aggradation of shallow permafrost possibly down to as low as ~1250 m a.s.l. (about 200 m lower than today).

In northern Scandinavia, where all permafrost limits are lower than in southern Norway, the lower limit of discontinuous permafrost decreases from ~1400 m a.s.l. in the west to generally ~400 m a.s.l. east of the Scandes (Gisnås et al. 2017). However, sporadic permafrost occurs in palsas close to sea level in northernmost Norway (Varanger peninsula) and down to ~50 m a.s.l. in Utsjoki, northernmost Finland (Seppälä 1997, 2011; Luoto and Seppälä 2002b; Miska Luoto, personal commumication). It may also occur close to sea level in steep north-facing bedrock slopes and rock glaciers (Magnin et al. 2019; Lilleoren et al. 2022). Holocene variations in permafrost limits in northern Scandinavia fluctuated within an altitudinal range of about ±100 m of today's limits (Lilleøren et al. 2012).

Permafrost thickness has been estimated as >300 m in the continuous permafrost zone at 1890 m a.s.l. in Jotunheimen, southern Norway, and at 1600 m a.s.l. in Tarfalaryggen, northern Sweden, respectively (Sollid et al. 2000; Isaksen et al. 2001; Etzelmüller et al. 2003). Active layer thicknesses of ~1.95–2.45 m have been recorded in boreholes through regolith in Jotunheimen (Harris et al 2009), ~3.0–5.0 m in rock walls in Norway (Hipp et al. 2014), and ~0.3–0.75 m in peat in northernmost Finland (Rönkkö and Seppälä 2003) where, in general, the range is ~0.5–4.0 m depending on soil type, snow depth and surface vegetation (Seppälä 2005a).

The lower regional-scale climatic limit of continuous permafrost is approximated by a MAAT of –4.0 °C, while the lower limit of sporadic permafrost can be set at –1.0 °C (King 1986; Ødegård et al. 1992; Etzelmüller and Hagen 2005). Although defining the lower climatic limit of seasonal frost is more arbitrary, a MAAT of 3.0 °C was suggested by Williams (1961) on the basis that this value is sufficient to achieve the >70 cm winter frost penetration necessary to counteract the amount of summer heat storage in the ground and to develop certain periglacial landforms (such as patterned ground and solifluction lobes). This climatic limit to the seasonal periglacial climate seems to have been widely accepted (cf. Ballantyne 2018; French 2018).

1.3 Periglacial Environments and Glacial History

Periglacial and glacial environments overlap in Scandinavia and cannot easily be separated, either as physical entities or conceptually (Etzelmüller and Hagen 2005; Berthling and Etzelmüller 2011; Miesen et al. 2021). Numerous temperate and polythermal glaciers are present in the landscape today, including the Jostedalsbreen ice cap, the largest ice body in continental Europe. During the Last (Weichselian) Glaciation, the Scandinavian Ice Sheet extended to the edge of the continental shelf and, by the end of the Late Glacial/Younger Dryas, ~11.7 ka, this ice sheet still covered most of the Scandinavian Peninsula (Hughes et al. 2016; Mangerud et al. 2016; Strøeven et al. 2016) (Fig. 1).

Throughout the Late Cenozoic Ice Age, all Scandinavian landscapes were intensely affected by numerous glacial-interglacial cycles. During the interglacials, periglacial environments were dominant. During the glacial episodes, cold-based ice sheets of varying size preserved much of the overridden landscape, while glacial erosion and deposition were most active in their marginal areas (Kleman 1994; Kleman and Hättestrand 1999; Hättestrand and Stroeven 2002; Kleman et al. 2008; Andersen et al. 2018, 2019). However, the ice sheets that occupied the Scandes for a large part of the late Pliocene and Pleistocene were smaller than those of the Last Glaciation and the other relatively recent glaciations. Such 'mountain glaciations', which are representative of 'average' glacial conditions (cf. Porter 1989), were present for an estimated 65% of the time during the last 1.88 Ma (Fredin 2002).

At the end of the Weichselian, the removal of the ice load led to uplift in response to glacio-isostatic rebound of the land mass. The land uplift was largest in the central part of Scandinavia where the relative ice thickness was largest. Uplift took place in two stages. The initial stage following deglaciation was almost immediate due to an elastic response of the crust as the ice load was reduced. Subsequently, uplift proceeded at an almost exponentially decreasing rate. The present relative land uplift rate in the northwestern part of the Gulf of Bothnia is ~9 mm/year. The total Holocene (<11.7 ka) land uplift relative to the present sea level in this area of the Gulf of Bothnia is estimated to have been ~250 m; in central southern Norway it was ~180 m; and in northern Sweden, northern Finland, and inner Finnmark, northern Norway, it was ~150 m (e.g. Vorren et al. 2006). Land uplift has shifted relative sea levels and the relative altitudinal limits of permafrost and other climate-dependant aspects of periglacial landscapes.

2 Historical Background to Research

2.1 *Early Twentieth Century Beginnings*

During the nineteenth century, there was little recognition of what are now known to be periglacial landforms and sediments. By the first decade of the twentieth century, however, specific landforms and sediments had been identified in Scandinavia and attributed to cold-climate processes. Lundqvist (1964) summarised influential early work from Sweden. Sernander (1905) and Andersson (1906) described different types of 'soil creep' (sorted and non-sorted patterned ground) and 'flow earth' (solifluction), while Svenonius (1909) discussed blockfield formation as being mainly formed by the disintegration of bedrock by frost, and Fries and Bergström (1910) were the first to report palsas.

A broader range of periglacial landforms was studied and theories of formation advanced during the early decades of the twentieth century. Högbom (1914, 1927), for example, classified patterned ground according to morphology, and recognised stripes, polygons, nets and circles. A Danish botanist's investigations in Greenland, carried out in the 1930s (Sørensen 1935), was important not only for his investigations into the effects of vegetation-patterned ground interaction but also for observations and measurements relating to micro-climate, soil texture, surface gradient, frost cracking, moisture-controlled density differences and 'convection currents' in the context of patterned ground, and for insights into the formation of sub-nival stone pavements involving nivation. A school of periglacial aeolian studies was established at Uppsala University in Sweden (Enquist 1916; Samuelsson 1926). At approximately the same time, coastal rock platforms in Norway were attributed to frost weathering (Vogt 1918), a process that was also seen as fundamentally important in the evolution of the Norwegian strandflat (Nansen 1922). As early as the 1930s, the effects of the freezing and thawing of sediments on roads and railways was studied in the laboratory by Beskow (1935). The English translation of his fundamental research on frost susceptibility is still widely cited (Beskow 1947).

In these early years, the development of periglacial geomorphology in Scandinavia was influenced by a number of different factors. First, Scandinavian scientists lived and worked in or near periglacial environments, which were seen as relevant. Second, explorer-scientists from Nansen (1922) to Sverdrup (1938) had taken a leading role in exploration of the Arctic, which included encounters with periglacial landscapes and active permafrost features that are not observable in Scandinavia today. Third, Scandinavian scientists were well connected with academics elsewhere in Europe and actively involved in the small but developing international periglacial research community. Following the 11th International Geological Congress held in Stockholm in 1910, the field excursion to Svalbard was particularly influential as it introduced scientists to unfamiliar permafrost environments and provided potential modern analogues for relict periglacial features elsewhere (French 2008). According to French and Karte (1988) the growth of periglacial geomorphology as an academic discipline began in 1949 with the establishment of the 'Commission on Periglacial Morphology' within the International Geographical Union. The first chairman of the Commission was the physical geographer and glaciologist, Professor H. W. Ahlmann of Stockholm University, later to become the Swedish Ambassador to Norway.

2.2 *Development in the Mid- to Late-Twentieth Century*

The periglacial zone had become an exemplar for climatic and climatogenic geomorphology, which developed and spread from Central Europe, especially Germany, France and Poland, during the 1940s to 1960s (André 2009). During this phase in the development of geomorphology, widespread belief in the dominance of frost action in periglacial landscapes was driven mainly by observation and mapping of landforms, combined with apparent associations with climate. Most conclusions about periglacial processes and landscape dynamics were based on inference and remained largely superficial. However, this gap in knowledge became a spur to understand better contemporary periglacial processes and climatic indicators.

André (2009) characterised the 1960s to 1980s as a period of 'periglacial fever', when there was unprecedented growth in laboratory experiment and quantitative process monitoring in the field. Research by Rapp (1960) in Kärkevagge, a valley in the alpine zone of of northern Sweden, was a milestone in this regard. For nine years, he monitored the frequency and magnitude of a wide spectrum of landscape processes using innovative field methods. Unexpectedly, his quantitative estimates showed generally low levels of activity and a minor role for frost weathering, solifluction and rapid mass movement, and the predominance of chemical weathering and fluvial transport as denudational processes. From the 1980s to the 1990s, field monitoring, combined with laboratory experiments on frost weathering, challenged what André (2009) termed the 'freeze-thaw dogma'—the traditional view that frost weathering is the dominant periglacial process—and raised awareness of the importance of azonal processes in periglacial landscapes.

Although monitoring and experiment continue to make an indispensable contribution to understanding periglacial landscapes, these approaches have weaknesses, two of which will be briefly mentioned here. First, the early laboratory studies, such as those at the Centre National de la Recherche Scientifique, Centre de Géomorphologie, Caen, France, reviewed by Lautridou and Ozouf (1982), which involved the disintegration of small rock particles under frequent freeze-thaw cycles, did not replicate field conditions well. The large-calibre debris accumulations that characterise many Scandinavian and other periglacial landscapes must be caused primarily by deep penetration of the annual freeze-thaw cycle into bedrock (cf. Matsuoka 2001). Later experiments, such as those of Murton et al. (2001, 2005), used sizable blocks of rock (450 × 300 × 300 mm) and were therefore closer simulations of conditions in the field. Linked to this was the lack of appreciation of the restricted areas of the landscape where frost shattering or frost wedging is important, namely areas of water availability. Thus, frost weathering can be slight on well-drained glacially-scoured rock outcrops, yet take place efficiently on adjacent cliffs subject to groundwater seepage. Second, field-monitoring studies in particular catchments over a small number of years are extremely problematic when attempting to scale-up in space and extrapolate in time. It is clear, therefore, that periglacial landscapes cannot be understood by reference to present processes alone.

Over time, Quaternary geology and other Quaternary sciences have provided the missing link. In Scandinavia, closer connections between periglacial geomorphology and the Quaternary sciences were being established in the closing decades of the twentieth century with a growing awareness of the effects and legacy of Pleistocene (Kleman 1992, 1994; Sollid and Sørbel 1994) and Holocene (Karlén 1988; Nesje and Kvamme 1991; Nesje et al. 1991; Matthews and Karlén 1992) environmental change. Although these insights originated from glacial geology and glacial geomorphology, they are equally applicable to the periglacial domain (as will be demonstrated later in this chapter).

2.3 Early Twenty-First Century Advances

Quaternary science diversified the techniques and approaches available for investigating periglacial landscapes. Foremost amongst these is a battery of numerical dating techniques. In combination with mapping and stratigraphic approaches, accurate and precise dating enables the reconstruction of landscapes of the past with high temporal resolution and also allows the separation of active from relict elements in the present landscape. Amongst the growing number of dating techniques available, Accelerator Mass Spectrometry radiocarbon dating, in situ cosmic-ray exposure-age dating, luminescence dating and Schmidt-hammer exposure-age dating have been found particularly useful in periglacial landscapes, where organic material is limited and bedrock outcrops and boulder deposits are abundant.

Other relatively new technical developments that are having a profound influence on research into periglacial landscapes in the twenty-first century include the use of geophysical survey, boreholes, satellite remote sensing such as light detection and ranging (LiDAR), airborne and terrestrial synthetic aperture radar interferometry (InSAR), computer algoritms for photogrammetry (structure from motion software), geographical information systems, and numerical and physical modelling. This is well illustrated in relation to the significant progress that has been made in understanding the nature, distribution and significance of permafrost in Scandinavia. Both French (2018) and Harris et al. (2018) define geocryology as the science of permafrost which, in respect of its core techniques, approaches and engineering aspects, may be considered as part of geophysics. Field studies using geophysical surveying of permafrost and ground temperature measurements began in both the northern and southern alpine areas in the 1980s (King 1984). Thirty boreholes were established in the International Polar Year, 2007–2009 (Christiansen et al. 2010). Data from boreholes were integrated with daily temperature and snow cover data in the CryoGRID1 model at 1 km^2 resolution, culminated in the first detailed and reliable permafrost map for Norway, Sweden and Finland (Gisnås et al. 2017).

Thus, today's studies of periglacial landscapes increasingly involve merging the research traditions of periglacial geomorphology, Quaternary science and geocryology (Harris and Murton 2005; Etzelmüller and Hagen 2005; French and Thorn 2006). Recent progress in dating periglacial features, obtaining chronological constraints, and positioning them in the palaeoclimatic context, have been particularly important. Like most fields of study focusing on the natural world, research programmes are being directed towards the climate change agenda in response to concerns over global warming (André 2009; Harris et al. 2009; Berthling et al. 2013; Etzelmüller et al. 2020). In Scandinavia where, due to polar amplification, atmospheric warming is likely to be above average, the importance of global climatic issues such as degrading permafrost, and its impacts on geohazards, carbon budgets, conservation and sustainability (e.g. Harris et al. 2009; Aalto et al. 2014; Karjalainen et al. 2019, 2020; Etzelmüller et al. 2020, 2021), are only likely to increase in the future.

3 Landforms, Processes and Age

Periglacial landscapes are commonly defined as the landforms and geomorphological processes characteristic of non-glacial cold climates (cf. Washburn 1979; Seppälä 2005a; Humlum 2008; Ballantyne 2018; French 2018; Murton 2021). The landforms bear an imprint of the processes of weathering, erosion and deposition associated with the freezing and thawing of ice in the ground (bedrock or regolith). However, landscapes in non-glacial cold-climates are also affected by river, lake and coastal ice, groundwater, snow and wind, all of which must be taken into accunt. Furthermore, periglacial and glacial landforms may be affected by transitional processes on different temporal scales, which leads to questioning the feasibility of strict differentiation.

Our aim is therefore to summarise and synthesise current knowledge of the periglacial landscapes of Scandinavia, pointing out their distinctive aspects in the context of Europe, and highlighting some unresolved research problems. Within this broad aim our approach is to focus on how and when the landforms evolved towards their present state. We examine plateau, steepland, low-gradient and glacier-foreland landscapes, the characteristic landforms of each having developed in response to periglacial processes under different topographic and other controls in the face of continuing environmental change. Owing to the scale of Scandinavia there is also the need to focus on exemplary and well-studied regions.

3.1 Plateau Landscapes

During Pleistocene interglacials, the high mountain and plateau landscapes of Scandinavia are likely to have experienced periglacial regimes similar to those of today. The dominant glaciation mode in the Scandes during glacials involved ice sheets and ice caps smaller than those of the Last Glacial Maximum with many plateau landscapes (along with the landscapes of the coasts and lowlands) subaerially exposed (not covered by LGM-like ice) and subjected to very cold periglacial climates (Porter 1989; Fredin 2002; Kleman et al. 2008). What are the implications of this history for the evolution of these periglacial landscapes? Recent debate has focused on the age and origin of blockfields and associated landforms on the plateaux, including whether they are wholly periglacial in origin, the role of frost weathering, whether they are in equilibrium with present environmental conditions, the extent of glacial erosion, and the possibility that these landforms could be pre-Pleistocene relicts.

3.1.1 Blockfields

Autochthonous (in situ) blockfields occur on low-angle plateau surfaces and mountain tops throughout the Scandes and are particularly extensive in present continuous and discontinuous permafrost zones (Rudberg 1977; Nesje et al. 1988; Juliussen and Humlum 2007; Goodfellow et al. 2014; Marr and Löffler 2018; Marr et al. 2018; Andersen et al. 2019). They seem typically to be composed of subangular to subrounded boulders at the surface (Fig. 3a), beneath which occur diamictons with more angular material, a sand or silt-sand matrix, and low clay content. The regolith seems normally to be up to several metres deep but excavations to bedrock have rarely been reported. A high proportion of the surface area of some blockfields is characterised by large sorted circles, transitioning to sorted stripes on sloping terrain (see Sect. 3.3.1).

Three main hypotheses for blockfield origin, development and survival have been proposed, suggesting they formed: (1) since deglaciation under periglacial climatic conditions similar to those existing today; (2) during Pleistocene periglacial climatic episodes and survived being overridden by one or more cold-based ice sheets; or (3) before the Pleistocene under a non-periglacial climatic regime and survived later episodes of glaciation (see, for example, Ballantyne 2010). The second hypothesis has been favoured by recent research and is exemplified by the results of 141 cosmic-ray exposure-age dates from regolith depth profiles, boulder erratics and bedrock outcrops combined with inverse modelling and sedimentological analysis of data from regolith profiles from the plateau landscape in Reinheimen National Park, central southern Norway (Andersen et al. 2019). Apparent ^{10}Be exposure ages for bedrock samples ranged from 78.0 $\pm$ 7.0 to 7.5 $\pm$ 0.7 ka with age clusters at ~11.0 and ~30.0 ka. The dates from the regolith profiles ranged between ~67.0 and 11.0 ka, and erratic boulders in the blockfield were dated to ~10.5 ka. The dates from erratics are fully consistent with deglaciation of the plateau in the early Holocene whereas the earlier dates from bedrock and regolith profiles demonstrate that parts of the blockfield fabric survived glaciation, including the Last Glacial Maximum (~22.0–21.0 ka). These ages demonstrate that blockfield formation in Reinheimen could not have begun post-Pleistocene, but appear not old enough to support a pre-Pleistocene origin.

Previous studies had suggested that the degree of chemical weathering indicated by the presence of secondary minerals, such as gibbsite and kaolinite, and high clay and fine silt contents in some blockfields might be the result of pre-Pleistocene weathering under a warm humid climate (e.g. Rea et al. 1996; André 2003; Paasche et al. 2006; Strømsøe and Paasche 2011; Olesen et al. 2012). However, Goodfellow (2012) and Goodfellow et al. (2014) have argued that low clay: silt ratios from blockfield matrix indicate the dominance of physical (frost) weathering, and that small amounts of secondary minerals result from chemical weathering in a cold climate under favourable hydrological conditions provided by a seasonally-wet active layer overlying permafrost. Furthermore, in situ cosmogenic nuclide concentrations measured in the regolith depth profiles by Andersen et al. (2019) indicate Reinheimen plateau summit erosion rates of ~6–8 m per Ma, increasing downslope. In contrast, the Last Glacial Maximum ice sheet was more erosive on the lower-lying plateaus towards its western margin (such as those surrounding the Sognefjord), where blockfields are

a
b
c
d
e
f
g
h

◀**Fig. 3** Selected periglacial landforms in Scandinavia. **a** Autochthonous blockfield with prominent quartzitic vein and scattered erratics, Melheimfjellet, Olden, western Norway (photo: Atle Nesje). **b** Highly weathered granulite tor, Paistunturi, northern Finland (photo: Jan Hjort). **c** Frost-shattered rock pinnacle ('The Blade'), Molladalen, Sunnmøre, western Norway (photo: Mons Rustøy). **d** Cryoplanation terrace, Svartkampan, Jotunheimen, southern Norway (photo: Richard Mourne). **e** Talus-derived rock glacier, Øyberget, Ottadalen, southern Norway (photo: Stefan Winkler). **f** Pronival rampart, Smørbotn, Romsdalsalpane, southern Norway (photo: Peter Wilson). **g** snow-avalanche boulder fan, Trollsteinkvelven, Jotunheimen, (photo: Jennifer Hill); **h** snow-avalanche impact crater, Meiadalen, Romsdalsalpane, western Norway (photo: John Matthews)

absent (Andersen et al. 2018). The evidence is therefore consistent with extremely slow lowering of the inland plateau surface by frost weathering where inefficient erosional processes dominated by frost creep, resulted in smooth, parabolic summits with occasional residual rock outcrops remaining as tors.

3.1.2 Tors and Rock Pinnacles

The widespread distribution of well-developed tors near the central area of glaciation in northern Sweden and northern Finland indicates that these areas have been protected from glacial erosion during all glacial cycles since ice-sheet inception in the late Cenozoic (Hättestrand and Stroeven (2002). Based on cosmic-ray exposure-age dating, these tors have yielded minimum ages which, when adjusted for ^{26}Al/^{10}Be ratios and the likely time shielded by ice sheets, suggests that this element of the landscape has survived as many as 14–16 episodes of glaciation (Fabel et al. 2002; Strøeven et al. 2002; Darmody et al. 2008). The highly weathered nature of the tors that rise above blockfields (Fig. 3b) can be attributed to relatively slow chemical weathering rates of resistant bedrock, the surface of which constitutes an exposed micro-environment with relatively low moisture availability (Darmody et al. 2008). The evidence seems to confirm the low weathering rates of exposed bedrock in a very old periglacial landscape. However, whereas resistant tors may survive multiple glaciations, the surrounding regolith of the blockfield elements of the landscape probably survived many fewer, perhaps only one or two. Sorted patterned ground that decorates the surface of many blockfields, and rare examples of blockstreams (Wilson et al. 2017), are the elements of these landscapes that are least likely to have survived glaciation, due to their greater sensitivity to warm-based glaciers during the transition from glacial to interglacial conditions.

At relatively low altitudes close to the west and north coasts of southern Norway, and in the Lofoten-Vesterålen archipelago of northern Norway, some tors and blockfields may not have been overridden by ice sheets during the Last Glacial Maximum and, by implication, earlier glaciations (Paasche et al. 2006; Nesje et al. 2007). Here, according to the 'Kleman model', warm-based ice streams descending steeply from the Scandes, eroded deep valleys and fjords, without overtopping the highest plateau surfaces and mountain summits (Nesje and Willans 1994; Kleman et al. 2008;

Strømsøe and Paasche 2011). This model is supported by an extensive, possibly ice-marginal trimline at ~250 m a.s.l. on the islands of Langøya and Hadseløya, which separates the 'glacial' landscapes from the 'periglacial' landscapes above (Paasche et al. 2006).

Perhaps the most distinctive landforms of periglacial landscapes throughout the coastal mountains of the Norwegian west coast are the jagged rock pinnacles that characterise the highest summits and ridges (Fig. 3c) and likely bear witness to uninterrupted frost weathering during glacial episodes (including the Last Glacial Maximum) when they appear to have been exposed as nunataks above the ice-sheet surface (Nesje et al. 1988). Such pinnacles are absent from inland mountains that were covered by the cold-based, central parts of ice sheets.

3.1.3 Cryoplanation and Nivation

Processes of cryoplanation have been proposed as a possible alternative to creep-dominated processes in the denudation of periglacial mountain and plateau landscapes (e.g. Demek 1969; Priesnitz 1988; Lauriol et al. 2006; Nyland et al. 2020). The most commonly cited evidence for the existence of cryoplanation is cryoplanation terraces, which are typically near-horizontal bedrock surfaces or benches backed by frost-weathered bedrock cliffs (Fig. 3d).

A recently proposed conceptual model of cryoplanation terrace formation in the alpine permafrost zone at Svartkampan, Jotunheimen, southern Norway (Matthews et al. 2019), develops the ideas of Boch and Krasnov (1943) on landscape evolution driven by frost weathering. Seasonal groundwater flow towards the freezing front in the cliff enhances frost weathering at the cliff-terrace junction, undermining and maintaining a near vertical cliff. Combined with the inefficient evacuation of regolith across the terrace tread by solifluction, this leads to parallel retreat of the cliff and terrace extension. Schmidt-hammer exposure-age dating of the bedrock cliffs and boulders embedded in the terrace surfaces at Svartkampen revealed Holocene ages ranging from modern to 8.9 ± 1.2 ka. Radiocarbon ages of up to ~5000 cal years BP were obtained from palaeosol material buried in the regolith at only 1.1–2.1 m from the cliff base, indicating *maximum* cliff recession rates of ~0.1 m per ka. Thus, cryoplanation at the site was shown to be a very slow process, incompatible with formation of the terraces entirely since deglaciation but compatible with their possible survival beneath cold-based ice sheets. It should be pointed out, however, that cryoplanation terraces do not appear to be common enough for cryoplanation to have had more than a minor influence on late-Pleistocene landscape evolution in the Scandinavian context.

Perhaps more common and often confused with cryoplanation is nivation, which comprises processes enhanced by the presence of late-lying snowbeds (Thorn 1988; Nyberg 1991; Hall 1998; Thorn and Hall 2002; Margold et al. 2011). It includes the enhancement of chemical weathering in water-saturated sediment and the transport of sediment by solifluction and meltwater flow from beneath a snowbed, but does not include frost weathering of bedrock. This is the main process involved in the formation of cryoplanation terraces but does not appear to be enhanced beneath snowbeds (see discussion in Matthews et al. 2019). Nivation can, however, produce nivation benches (or nivation hollows) in hillslopes covered in regolith (Christiansen 1998). Such benches may be fronted by solifluction lobes (see below) and/or stone pavements (surficial lag deposits resulting from the removal of the fine matrix sediment by slope wash). Little recent research has been carried out on nivation or nivation benches in Scandinavia (but see Rapp 1984; Rapp and Nyberg 1988). The only dated example from Scandinavia known to the authors is from Söderåsen, southern Sweden, where luminescence dates on aeolian sand on the surface of a nivation bench yielded Younger Dryas ages (Jonasson et al. 1997), which indicate a relict feature.

3.2 *Steepland Landscapes*

The term 'steepland' was introduced by Slaymaker (1995) to denote landscapes characterised by steep gradients, which profoundly influence mass movement processes. In this section, we focus on periglacial steeplands, the evolution of which in Scandinavia includes distinctive paraglacial and paraperiglacial aspects. Paraglacial processes are non-glacial processes directly conditioned by glaciation (Church and Ryder 1972; Ballantyne 2002; Slaymaker 2009) and include, for example, rock-slope failure in response to glacio-isostatic uplift or glacial de-buttressing. Paraperiglacial processes, which have been defined as non-permafrost processes directly conditioned by permafrost (Mercier 2008; Scarpozza 2016), include slope failures in both bedrock and regolith triggered by thawing permafrost. The concepts of periglaciation, glaciation, paraglaciation and paraperiglaciation are linked by the concept of cryoconditioning (Berthling and Etzelmüller 2011), important aspects of which include the thawing of ice associated with bedrock and sediments beneath cold-based glaciers, as well as the changing availability of meltwater to both groundwater and surface-water flows during and following deglaciation.

3.2.1 Rock-Slope Failures

Retreat of the Scandinavian Ice Sheet from its limits at the Last Glacial Maximum and in the Younger Dryas (Fig. 1) led to a large number of rock-slope failures in the steeplands of southern and northern Scandinavia, many of which have been dated using cosmic-ray exposure-age dating (e.g. Blikra and Christiansen 2014; Böhme et al. 2015; Hermanns et al. 2017; Oppikofer et al. 2017; Schleier et al. 2017; Hilger et al. 2018, 2021; Wilson et al. 2019; Curry 2021).

Several data sets have been compiled to examine the spatial distribution and/or temporal frequency of rock-slope failures in relation to the time elapsed since deglaciation in southern Norway. Hermanns et al. (2017) found nearly half of 22 accurately dated large rock avalanches occurred within 1.0 ka of deglaciation. The

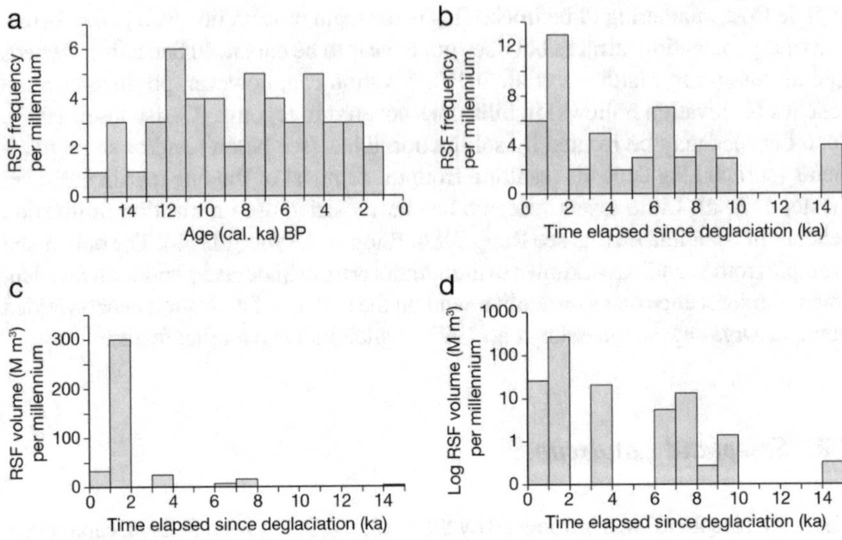

Fig. 4 Frequency and magnitude of moderate to large rock-slope failures (RSFs) in western Norway per millennium 16–0 ka: **a** age-frequency of RSFs; **b** frequency of RSFs related to time elapsed since local deglaciation; **c** volume of RSFs related to time elapsed since local glaciation; **d** volume of RSFs related to time elapsed since local deglaciation with volume on a logarithmic scale (based on Curry 2021)

larger data set of 49 moderate to large rock-slope failures compiled by Curry (2021) shows almost 40% of events occurred within 2.0 ka of deglaciation (Fig. 4). A similar, dominant activity peak is clear in even larger data sets of >100 events compiled by Longva et al. (2009), Böhme et al. (2015; Fig. 5a) and Bellwald et al. (2019), which include data derived from seismic stratigraphy of fjord sediments.

All these records give some indication of one or more secondary peaks or clusters in the frequency of events in the mid- to late Holocene, which may relate to century- to millennial-scale climatic variations. This is suggested in Fig. 5a by the small increase in the frequency of events from ~5.0–1.0 ka, which correlates with the fall in temperature and rise in precipitation during neoglaciation after the Holocene Thermal Maximum (Figs. 5j, k). Other temporal clusters of rock-slope failures have been tentatively identified, such as at ~5.0 ka (Hilger et al. 2018), and ~5.0–4.0 ka and ~8.0–7.0 ka (Curry 2021; Fig. 4), but the significance of such clusters is uncertain.

The concentration of rock-slope failures close in time to deglaciation points to several potential paraglacial trigger factors, including ice-sheet thinning, debuttressing of oversteepened glaciated valley slopes, and glacio-isostatic seismic shock. Those that occurred later in the Holocene may have involved progressive failure and/or transient triggering mechanisms. Historical rock-slope failures in Møre og Romsdal exhibit evidence of spatial clustering in the area of greatest current land-uplift rates, with the highest number of occurrences at a distance of 60–80 km from the Norwegian west coast (Henderson and Saintot 2011; Curry 2021), which suggests

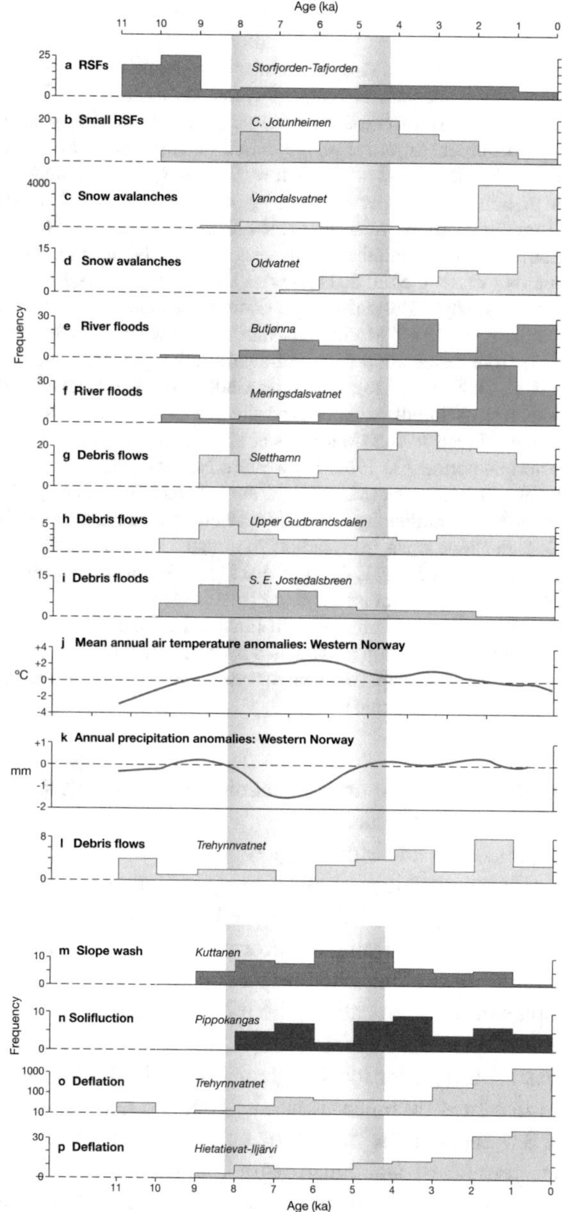

Age (ka)
Frequency
a RSFs
Storfjorden-Tafjorden
b Small RSFs
C. Jotunheimen
c Snow avalanches
Vanndalsvatnet
d Snow avalanches
Oldvatnet
e River floods
Butjønna
f River floods
Meringsdalsvatnet
g Debris flows
Sletthamn
h Debris flows
Upper Gudbrandsdalen
i Debris floods
S. E. Jostedalsbreen
j Mean annual air temperature anomalies: Western Norway
°C
k Annual precipitation anomalies: Western Norway
mm
l Debris flows
Trehynnvatnet
m Slope wash
Kuttanen
n Solifluction
Pippokangas
o Deflation
Trehynnvatnet
p Deflation
Hietatievat-Iljärvi

◂**Fig. 5** Holocene chronologies of the activity of landforms and geomorphic processes in the landscapes of southern (**a**–**i**) and northern (**l**–**p**) Scandinavia, in relation to climatic variations (**j**–**k**). All records except (**c**) and (**n**) depict frequency of events per millennium. Early-, mid- and late-Holocene subdivisions of the Holocene are shown according to Walker et al. (2018). **a** Rock-slope failures in the Storfjorden/Tafjorden region, based on cosmogenic exposure-age dating and fjord sediment stratigraphy (Böhme et al. 2015, n = 105). **b** Small rock-slope failures in central Jotunheimen, based on Schmidt-hammer exposure-age dating (Matthews et al. 2018, n = 92). **c**–**d** Snow avalanches from radiocarbon-dated lacustrine sediment stratigraphy: (**c**) Vanndalsvatnet, E Jostedalsbreen region, showing the number of particles > 1 mm (Nesje et al. 2007); (**d**) Oldvatnet, N Jostedalsbreen region (Vasskog et al. 2011, n = 49). **e**–**f** River floods from radiocarbon-dated lacustrine sediment stratigraphy: **e** Butjønna, upper Glomma catchment, Rondane region (Bøe et al. 2006; Nesje et al. 2007, n = 115); **f** Meringsdalsvatnet, E Jotunheimen (Støren et al. 2010, n = 108). **g**–**h** Debris flows from radiocarbon-dated colluvial stratigraphy: (**g**) Sletthamn, central Jotunheimen (Matthews et al. 2009, n = 123); (**h**) upper Gudbrandsdalen (Sletten and Blikra 2007, n = 64). **i** Debris floods from Schmidt-hammer exposure-age dating of surface boulder deposits on alluvial fans, SE Jostedalsbreen region (Matthews et al. 2020b, n = 47). **j**–**k** Climatic anomalies (smoothed) for the normal period AD 1961–91, western Norway (Mauri et al. 2015; Hilger et al. 2021): (**j**) mean annual air temperature (MAAT); (**k**) annual precipitation (AP). **l** Debris flows from radiocarbon-dated lacustrine stratigraphy at Trehynnvatnet, Vesterålen, northern Norway (Nielsen et al. 2016a, n = 35). **m** Slope wash from radiocarbon-dated colluvial stratigraphy at Kuttanen, Finnish Lapland (Matthews and Seppälä 2015, n = 131). **n** Solifluction from radiocarbon-dated lobe stratigraphy at Pippokangas, Finnish Lapland (Matthews et al. 2005, n = 46). **o** Deflation from radiocarbon-dated sand-dune stratigraphy in Finnish Lapland (Kotilainen 2004; Matthews and Seppälä 2014, n = 137). **p** Deflation as indicated by the number of sand grains >250 μm from radiocarbon-dated lacustrine stratigraphy at Trehynnvatnet, Vesterålen, northern Norway (Nielsen et al. 2016a, n = 35; note the logarithmic scale)

that long-lasting glacial-isostatic driven seismic paraglacial effects have remained important throughout the Holocene.

Many modern, historical and older large rock-slope failure events are associated with steep rock walls that contain permafrost today and/or are likely to have contained permafrost in the past (Magnin et al. 2019). Based on cosmic-ray exposure-age dating and permafrost modelling, the timing of deformation initiation and late-Pleistocene and Holocene sliding rates along slip surfaces were analysed for several actively deforming examples in northern and southern Norway (Hilger et al. 2021). At two sites (Oppstadhornet and Skjeringahaugane) without permafrost in the Holocene, deformation started shortly after local deglaciation at 16.0–14.0 and 11.0–10.0 ka, respectively. At three sites (Mannen, Revdalsfjellet and Gámanjunni), deformation started between 8.0 and 4.5 ka, during or at the end of the Holocene Thermal Maximum, when permafrost was most degraded. The evidence indicates that the presence of permafrost had a stabilising effect for several millennia after deglaciation and that sliding rates tended to decrease during the Holocene. In addition, modern measured sliding rates suggest recent acceleration at the three sites with permafrost, attributed to further permafrost degradation in response to recent climate warming. Hilger et al. (2021) concluded that permafrost degradation should be regarded as a first-order control on the stability and activity of steep rock slopes.

The importance of permafrost degradation has been further highlighted in an analysis of 92 small rock-slope failures in Jotunheimen ranging in volume from 12 to 2520 m^3, which have been dated by Schmidt-hammer exposure age dating (Matthews et al. 2018). These landforms comprise compact depositional fans of predominantly angular boulders beneath distinct erosional scars in bedrock cliffs. They were located at or close to the aspect-dependent lower limit of discontinuous permafrost in alpine rock faces as defined by Hipp et al. (2014). Age-frequency and probability-density analyses revealed low early-Holocene activity increasing to a dominant mode at 4.5 ± 1.4 ka (Fig. 5b). The increase in activity during and especially towards the end of the Holocene Thermal Maximum differs markedly from the paraglacial pattern exhibited by larger rock-slope failures. By combining this chronology with permafrost depth modelling, Matthews et al. (2018) revealed that peak activity was associated with minimum permafrost depth (maximum active-layer thickness) close to the end of the Holocene Thermal Maximum. This is essentially a paraperiglacial response. Earlier secondary activity phases seem to have represented earlier pulses of permafrost degradation. The continuous decline in activity after ~4.5 ka may signify the exhaustion of permafrost at depths shallow enough to host small rock-slope failures. A further reduction in activity would be expected in response to permafrost aggradation in the Little Ice Age (cf. Lilleøren et al. 2012).

3.2.2 Talus Slopes and Rockfalls

Several colluvial landforms develop at the foot of steep rock slopes due largely to the accumulation of rockfall debris. The most common of these landforms are talus slopes, which typically consist of sheets or cones of coarse, angular rock debris, the steep surface gradient of which (~33–39°) lies close to the angle of repose of the debris, at least in the straight upper parts of the slope. However, debris-mantled slopes, characterised by a thinner mantle of debris with a lower surface gradient reflecting that of the underlying terrain, are far more common than mature talus slopes in Scandinavia.

There are several reasons for the contrast with many other European steepland landscapes, where talus slopes are more abundant. First, many areas of the Scandes with permafrost or seasonal frost, where enhanced sediment supply from frost weathering would be expected, are characterised by plateau topography, which lacks the necessary steep bedrock slopes and is largely seismologically inactive. Second, much of the largely metamorphic bedrock in Scandinavia is resistant to weathering in general and to frost weathering in particular. The best developed talus slopes are associated with frost-susceptible, well-jointed rocks such as the quartzitic sparagmites of the Rondane massif, southern Norway (Sellier and Kerguillec 2021). Third, in most of the Scandinavian steeplands, insufficient time has been available for the development of mature talus slopes under the climatic regime of the Holocene; and only a few thousand years longer has been available in the peripheral areas deglaciated before the Younger Dryas. Fourth, although there is the possibility that inland talus slopes could have survived glaciation beneath cold-based ice sheets, the steepland areas towards the coast tend to have been scoured by ice streams during successive glaciations.

Rapp (1957, 1960) suggested, on the basis of his classic process-monitoring studies, that modern rockfall activity in the alpine valley of Kärkevagge, northern Sweden, is very limited, the talus slopes largely relict, and talus accumulation probably greatest shortly after deglaciation. This view of the evolution of talus slopes seems to be generally applicable in Scandinavia, even in areas most conducive to frost weathering, such as Rondane.

This paraglacial hypothesis for talus slope development in Scandinavia has been substantiated and the timing of talus accumulation clarified by stratigraphical studies. Blikra and Nemec (1998) dated colluvial sequences beneath coastal cliffs over a wide area of the outer coastal zone of Møre og Romsdal in southern Norway. They identified main stratigraphic units (A–C) separated by hiatuses, which represent a five-stage (1–5) evolution of what they termed 'rockfall-dominated colluvial-fan deltas' (Fig. 6). Unit B, consists of openwork rockfall boulders and cobbles (talus) alternating with low-viscosity debris-flow cobbles and pebbles (reworked talus). Within this talus unit, a frost-shattered palaeobeach facies lies at the height of a Younger Dryas shoreline, which effectively dates the main phase of talus accumulation and confirms a periglacial climate at sea level. Unit A (stage 1) consists of matrix-rich gravels, interpreted as high-viscosity debris flows (resedimented till), almost concurrent with local deglaciation, at a time of higher relative sea level and in a climate less conducive to frost weathering than unit B. Unit C represents renewed talus deposition, similar to unit B apart from interbedded humic palaeosol material, which was radiocarbon dated, and suggests variations in the rate of rockfall activity during the second half of the Holocene (see also Blikra and Nesje 1997). The hiatuses indicate stages without rockfall activity both before the Younger Dryas (during the Allerød interstadial) and during the early Holocene.

Unit B (Younger Dryas) dominates the stratigraphy and is significantly younger than unit A, which is a paraglacial feature. This suggests that rockfall-talus accumulation at this sea-level location was related to the climate of the Younger Dryas, rather than simply to paraglacial instability. Further support for climatic control of talus-slope development comes from late-Holocene variations in rockfall activity on talus slopes dated by lichenometry and further analysed by simulation modelling in Jotunheimen (McCarroll et al. 1998, 2001). These authors demonstrated that the rate of rockfall-talus accumulation in the eighteenth century, the coldest phase of the Little Ice Age, may have been at least five times the background late-Holocene rate.

3.2.3 Rock Glaciers

In their inventory based largely on aerial photographic interpretation, Lilleøren and Etzelmüller (2011) recognized 241 rock glaciers in Norway, 85% of which are in the north of the country. Barsch and Treter (1976) had previously identified a small number of rock glaciers at the foot of prominent talus slopes in Rondane, southern Norway, while Sollid and Sørbel (1992) recognised the relatively large number of

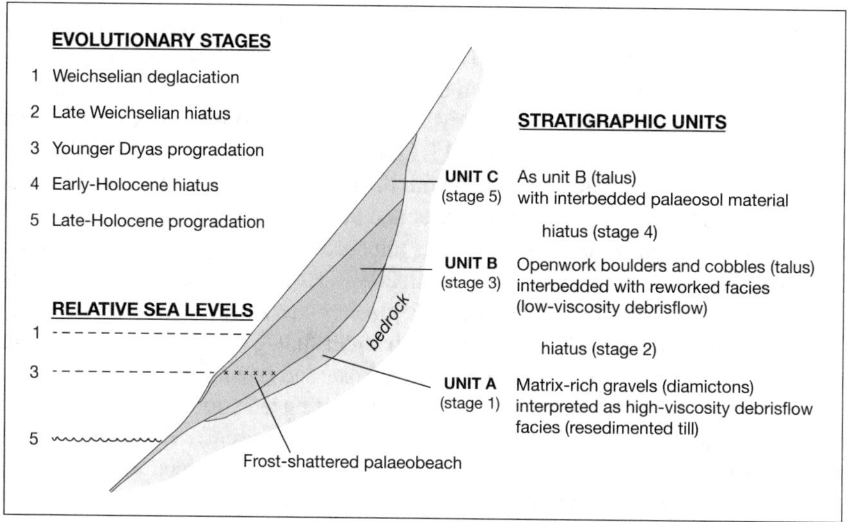

Fig. 6 Stratigraphic model of a 'rockfall-dominated colluvial-fan delta' in coastal Møre og Romsdal, southern Norway. Three stratigraphic units separated by hiatuses are assigned to five evolutionary stages and associated with three different relative sea levels. Note also the frost shattered palaeobeach in unit 2 at the Younger Dryas sea-level stand (based on Blikra and Nemec 1998)

rock glaciers in northern Norway outside Younger Dryas glacier limits, some of which occur close to sea level.

Lilleøren and Etzelmüller (2011) distinguished between talus- and moraine-derived rock glaciers, the former exhibiting evidence of permafrost creep at the foot of talus slopes (Fig. 3e), the latter derived from glacigenic sediments at locations below cirque glaciers. In northern Norway, their data included 188 talus-derived and 18 moraine-derived rock glaciers; in southern Norway, only 23 talus-derived and 12 moraine-derived rock glaciers were identified. The limited number of rock glaciers in southern Norway can be explained by similar reasoning to that proposed above for the sparsity of mature talus slopes—namely, insufficient sediment supply under a permafrost regime for development of such features beneath steep rock walls in a plateau landscapes. Lilleøren and Etzelmüller (2011) also attempted to distinguish between active and relict features but their 'intact' rock glaciers cannot be regarded as equivalent to 'active' ones. Morphological criteria such as steep frontal slopes, and deep ridges and furrows, are not reliable indicators of activity because of the potential for the persistence of such attributes in the landscape long after activity ceases. Active rock glaciers have been recently proven at Ádjet, Troms County, northern Norway by application of satellite InSAR measurements (Eriksen et al. 2018). Nevertheless, there can be little doubt, on the basis of the distribution and altitude of these landforms in non-permafrost zones, that most of the identified rock glaciers in both northern and southern Norway are relict.

There have been very few attempts to date rock glaciers in Scandinavia. Based on lacustrine sedimentary signals in Trollvatnet (Lyngen, northern Norway) a talus-derived rock glacier located at ~100 m a.s.l. formed following local deglaciation at >14.8 ka (Paasche et al. 2007). A steady sediment supply to the lake indicated that this rock glacier remained active until ~11.1 ka, becoming relict shortly after the end of the Younger Dryas, which coincided with the cessation of permafrost conditions. In southern Norway, both Schmidt-hammer and cosmic-ray exposure-age dating led to similar conclusions in relation to lobate, talus-derived rock glaciers at Øyberget, Ottadalen (Matthews et al. 2013; Linge et al. 2020). However, InSAR measurements have demonstrated that the Øyberget rock glaciers are active (Nesje et al. 2021) despite their location at ~530 m a.s.l. beneath south-facing rock walls and >1000 m below the lower limit of discontinuous permafrost, and having yielded dates up to 10.3 ± 1.3 and 11.2 ± 1.4 ka according to the two dating techniques, respectively. At this site, permafrost creep seems to have been initiated in the early Holocene as large quantities of paraglacial debris were released from the Øyberget rock wall. Sporadic permafrost developed and persisted in the blocky openwork sediments as a result of microclimatic undercooling mechanisms such as Balch ventilation and the chimney effect.

3.2.4 Ice-Cored Moraines

Another controversial issue related to rock glaciers in Scandinavia is their relationship to ice-cored moraines, which were investigated in detail for the first time in the 1950s (Østrem 1959, 1964, 1965). Ice-cored moraines are multiple-ridged, ramp-like structures that rise up to ~50 m above the surrounding terrain. They are most commonly located in the continuous or discontinuous permafrost zones of southern Norway where they are associated with small polythermal glaciers in relatively continental climatic regimes. Lilleøren and Etzelmüller (2011) recognised 26 ice-cored moraines in northern Norway and 40 in southern Norway, all of which were classified as intact/active permafrost indicators.

Barsch (1971, 1977) considered that ice-cored moraines *are* rock glaciers and interpreted the individual ridges as rock-glacier flow structures. Østrem (1971) argued that their position close to glacier fronts, their lack of movement, and their existence on level or near-level plateaux indicate glacial rather than mass-movement (permafrost creep) origins. Shakesby et al. (1987) and Matthews et al. (2014) supported Østrem's interpretation citing especially the lack of deformation induced by gravity. They considered that ice-cored moraine ridges are deposited and deform as a result of glacier-push processes during glacier advances. Hence, they used the term 'push-deformation moraine', which emphasises glacier push as the primary morphogenetic mechanism. However, the type of deformation exhibited by Scandinavian ice-cored moraines seems similar to rock-glacier creep and to be dependent on a permafrost environment (cf. Etzelmüller and Hagen 2005).

Matthews et al. (2014) applied Schmidt-hammer exposure-age dating to individual ridges within three ice-cored moraine complexes in Jotunheimen and Breheimen, southern Norway. At Gråsubreen, the ages of six ridges ranged from 3.9 ± 0.8 ka to modern and the outer ridges tended to be oldest; at Vesle Juvbreen both ridges yielded modern ages; and at Østre Tundradalskyrkjabreen, Breheimen, all four ridges dated to ~2 ka. These ages indicate significant activity during the late Holocene, which contrasts with the relict status of most rock glaciers in Scandinavia. The absence of older dates is explained by late-Holocene (including Little Ice Age) glacier advances, which reworked the moraine ridges, incorporating previously weathered boulders.

3.2.5 Pronival Ramparts

The defining characteristic of a pronival rampart is that it is formed at the distal margin of a snowbed. Some pronival ramparts occur at the foot of talus slopes but, as others abut directly against rock walls (Fig. 3f), the older term 'protalus rampart' is not appropriate (Shakesby 1997). Similarly, lack of generality leads to rejection of the idea that pronival ramparts are embryonic rock glaciers (cf. Haeberli 1985; Scapozza et al. 2011; Kääb 2013). Many active pronival ramparts investigated in northern and southern Scandinavia cannot be embryonic rock glaciers because they are developing in seasonal frost environments, not permafrost zones (e.g. Harris 1986; Ballantyne 1987; Shakesby et al. 1987, 1995, 1999; Matthews et al. 2017a).

Where sufficient boulders have enabled Schmidt-hammer exposure-age dating of pronival ramparts in southern Norway, both active and relict examples have been identified (Matthews et al. 2011, 2017a; Matthews and Wilson 2015; Marr et al. 2019; Wilson et al. 2020). Ages of 14.6 ± 1.1 to 8.7 ± 1.1 ka were obtained from relict examples located beyond ice-sheet and local-glacier limits during the Younger Dryas, while active ramparts (mostly inside the limits) yielded ages of 7.7 ± 1.1 ka to modern. The largest of these pronival ramparts seem to have developed under a permafrost regime during the Younger Dryas, or earlier when rockfall supply was enhanced by paraglacial slope instabilities. The ramparts that remained active after the Younger Dryas did so under the seasonal-frost regime of the Holocene with likely reduced sediment supply. These conclusions are consistent with four luminescence dates of 10.8 ± 1.0 to 10.4 ± 1.0 ka obtained by Jonasson et al. (1997) on aeolian sand in two possible pronival ramparts from southern Sweden. All the dating results are consistent with the supposed main process of pronival rampart formation, which involves the supranival transport of frost-weathered rockfall debris across steeply-sloping snowbed surfaces.

3.2.6 Snow-Avalanche Landforms

Snow avalanches undoubtedly contribute debris and snow to the development of some talus slopes, rock glaciers and pronival ramparts. Several small-scale landforms provide evidence of snow-avalanche activity on avalanche-affected slopes, including scattered perched angular boulders, boulder lines, gravel clumps, beaded gravel ridges, debris horns and debris shadows (Blikra and Nemec 1998; Owen et al.

2006a). Snow avalanches have also produced larger-scale landforms that are uniquely attributable to snow-avalanche processes, including snow-avalanche boulder fans (Fig. 3g) and snow-avalanche impact landforms (Fig. 3h), both of which are well represented in Scandinavia.

Snow-avalanche boulder fans were first recognised by Rapp (1959) who described distinctive avalanche 'roadbank tongues' in northern Sweden, which differ appreciably from talus slopes and other colluvial accumulations beneath steep bedrock slopes. Well-developed snow-avalanche boulder fans tend to be flat-topped, elongated, steep-sided, coarse-debris accumulations that extend at a low angle towards valley floors. The momentum of snow avalanches carries debris farther from the foot of the slope than is the case with talus slopes, producing strong concave-upwards curvature to the fan surface. Debris is transported from snow-avalanche headwall source areas by chutes or gullies eroded in bedrock.

In the Scandinavian examples investigated in detail by Matthews et al. (2020a) from two areas of Jotunheimen, southern Norway, the source areas lie in the high-alpine zone affected by permafrost, whereas the depositional fans are in the mid-alpine zone of seasonal frost. Schmidt-hammer exposure-age dating of boulders from the distal fringe of 11 fans yielded ages ranging from 7.4 ± 1.0 to 2.3 ± 0.7 ka. Geographical information systems-based morphometric analysis showed that the volume of rock material was generally less than the volume of the chutes. Although some unweathered boulders on the fan surface indicated continuing low-level activity by non-erosive, largely dry-snow avalanches, it was inferred that the fans were deposited entirely within the Holocene, and mainly within the early- to mid Holocene. The excess volume of the chutes was accounted for by enhanced subaerial erosion in the Younger Dryas (and possibly earlier) when they appear to have been located on frost-weathered nunataks protruding above the ice-sheet surface. The majority of the debris within the fans is therefore of paraglacial origin, probably supplemented by later additions from permafrost degradation associated with the Holocene Thermal Maximum.

Snow-avalanche impact landforms have been described from both northern (Corner 1980) and southern Scandinavia (Liestøl 1974; Hole 1981; Blikra et al. 1994; Matthews and McCarroll 1994; Owen et al. 2006a; Matthews et al. 2017b). In the steeplands of western Norway, erosional and depositional landforms of this type include valley-floor craters (Fig. 3h), lacustrine craters (subaquatic craters eroded in lake-floor sediments close to lake shores), river-bank ramparts (mounds and ridges of sedimentary material excavated from the channel bed and deposited on distal river banks). They occur in greater numbers in Vestlandet (western Norway) than anywhere else on Earth (Matthews and Owen 2021). Fifty-two valley-floor and lacustrine craters described by Matthews et al. (2017b) had an average diameter of 85 m (range 10–185 m) and were located at the bases of steep valley-sides with gradients of 28°–59°. It seems that the combination of exceptionally high snow volumes (reflected in mean annual snowfall amounts of > 4 m in the avalanche source areas), a gradient of at least ~15° for the lowest 200 m of the avalanche track, and a flat valley or lake floor, are all necessary for the development of such well-developed craters.

The composition of the depositional landforms associated with snow-avalanche impact reflects the composition of the sedimentary material excavated and ejected from the craters. Sufficient boulders were available on five snow-avalanche impact ramparts in Sprongdalen and Jostedalen to enable Schmidt-hammer exposure-age dating (Matthews et al. 2015). These yielded ages of 3.6 ± 0.5 to 0.6 ± 0.3 ka for the ramparts as a whole, whereas the distal fringes yielded ages up to 5.4 ± 1.0 ka, indicating that parts of the distal fringes survived burial by later impact events. The estimated maximum age of the surface boulders (defined as the age exceeded by the oldest 5%) was determined as up to ~9.0 ka. On the basis of this evidence alone, the evolution of the ramparts must have occurred throughout the Holocene as a result of frequent snow-avalanche events.

Comparative modelling by Matthews et al. (2017b) of the kinetic energy required to form meteorite and snow-avalanche impact craters with a diameter of 85 m indicates that the kinetic energy at impact of a single snow avalanche (~3.1×10^9 J) is two orders of magnitude less than that involved in excavating an 85-m diameter meterorite crater (~1.3×10^{11} J). This result is consistent with the number of snow-avalanche events that excavated the Vestlandet craters, estimated from meteorological data and dendrochronological and lacustrine proxy records in neighbouring valleys. These records suggest recurrence intervals for large snow-avalanche events of between 15 and 155 years (Vasskog et al. 2011; Decaulne et al. 2014).

However, such calculations take no account of variations in the frequency of snow-avalanche activity through time, for which there is evidence from stratigraphic studies. Based on terrestrial colluvial stratigraphy, Blikra and Nemec (1993, 1998), Blikra and Selvik (1998) reported relatively high frequency of snow-avalanche events during the Younger Dryas, contrasting with low activity levels in the early Holocene. A very late peak in activity (within the last two millennia) is clear in the lacustrine records of Nesje et al. (2007; Fig. 5c) and Vasskog et al. (2011; Fig. 5d). A similar pattern, with a dominant late-Holocene peak, seems also to characterise records of river floods (Bøe et al. 2006; Nesje et al. 2007; Støren et al. 2010, 2014; Engeland et al. 2020; Figs. 5e, f), which suggests possible links to winter snowfall, storminess and spring snowmelt. In addition, lichenometric dating and simulation modelling of lichen-size frequency distributions from 12 snow-avalanche ramparts (Matthews and McCarroll 1994) indicated variations in avalanche frequency over recent centuries, related to variations in snow climate.

3.2.7 Debris Flows and Debris Slides

Debris flows—a type of rapid mass movement of water-saturated unconsodidated sediment—are common in steeplands throughout Scandinavia today. Rubensdotter et al. (2021) recognise two major morphological types—open-slope debris flows and channel-dependent debris flows—together with sub-types of the latter. Several palaeoenvironmental reconstructions provide evidence of variations in the frequency of debris flows through time (Jonasson 1991; Blikra and Nemec 1993, 1998; Blikra and Nesje 1997; Matthews et al. 1997a, 2009; Sandvold et al. 2001; Sletten et al. 2003;

Sletten and Blikra 2007; Støren et al. 2008; Nielsen et al. 2016a). Larger-scale slope failures in unconsolidated material, such as debris slides (Sutinen et al. 2014; Ojala et al. 2019), have also been recognised in Scandinavia but dating is often difficult and trigger factors are likely to be different and more akin to those of rock-slope failures. Ojala et al. (2019) show that debris slides in northern Finland are located within 35 km of known postglacial faults and that the size of the slides decrease with distance from the faults, providing evidence of seismic triggering (see also Sutinen et al. 2014). Debris slides, like debris flows, are not exclusively periglacial landforms, although seasonal thawing or permafrost degradation may be involved as trigger factors. Here we focus on debris flows.

In southern Norway, detailed Holocene radiocarbon-dated terrestrial stratigraphic records are available from Sletthamn, Leirdalen, Jotunheimen (Matthews et al. 2009; Fig. 5g), and the upper Gudbransdalen region, farther east in southern Norway (Sletten and Blikra 2007; Fig. 5h). The sediments at Sletthamn represent distal debris-flow facies (cf. Matthews et al. 1999), and are illustrated in Fig. 7. The longer record, from Trehynnvatnet on Langøya in Vesterålen, northern Norway (Nielsen et al. 2016a; Fig. 5l), is based on turbulent subaqueous density currents triggered by debris flows. All three records show significant levels of debris-flow activity throughout the Holocene, with evidence of reduced activity in the mid-Holocene at ~7.0–6.0 ka, at the end of the Holocene Thermal Maximum (although millennial-scale variations are less clear in the Gudbrandsdalen record).

Debris flows are initiated by the failure of water-saturated sediments, which Matthews et al. (1999, 2009) attributed at Sletthamn primarily to the failure of till-mantled slopes following intense summer and autumn rainfall events. Till-mantled slopes are also involved at the Gudbrandsdalen sites whereas slopes around Trehynnvatnet involve both till-mantled and reworked-colluvial slopes. Rainfall events seem to be the main climatic trigger factor for modern debris flows in western Norway, rather than spring snowmelt, which is more important in eastern Norway (Sandersen 1997). However, all three records of debris-flow activity were probably affected by both summer rainstorms and environmental conditions in the spring, when both snowmelt and active-layer thaw are potential trigger factors. It can be speculated that such conditions were less likely during the Holocene Thermal Maximum (when debris-flow activity was at a minimum) than before or after.

3.2.8 Alluvial Fans and Debris Floods

Water flows, debris floods, debris flows and slush flows commonly produce fans or fan-like landforms on the lower valley-side slopes and valley floors of periglacial steeplands. The special environmental conditions affecting these landforms in periglacial landscapes today include the seasonal availability of meltwater from snow, ice and frozen ground. Differentiation of such fans is not straightforward and, indeed, composite fans that involve more than one of these transport processes may be the rule rather than the exception. Alluvial fans investigated in the Jotunheimen and SE Jostedalsbreen regions of southern Norway (Lewis and Birnie 2001; McEwen et al.

Fig. 7 **a** Modern debris-flow in Leirdalen, Jotunheimen. **b** Distal debris-flow sediments (*grey*) between peat layers (*brown*) in a valley-floor mire at Sletthamn, Leirdalen, excavated in 1996

2011, 2020; Matthews et al. 2020b) demonstrate a complex evolutionary history dominated by debris floods.

The subalpine fans in the SE Jostedalsbreen region (Matthews et al. 2020b) are located at 300–400 m a.s.l. and occupy areas of 0.16–0.51 km^2. The steep, rugged catchments are 1.17–3.44 m^2 in area, with a local relief of 1170–1380 m, and their present glacierized areas are between zero and 56%. The lower altitudinal limit of discontinuous permafrost currently lies at ~1600 m in the region (Fig. 2). Schmidt-hammer exposure-age dating applied to 47 distinctive boulder deposits (e.g. Fig. 8a) on the surface of four of the fans yielded ages ranging from 9.5 ± 0.8 to 2.0 ± 0.8 ka, with a peak age-frequency between ~9.0 and 8.0 ka (Fig. 5i). The main phase of fan aggradation occurred shortly after local deglaciation at ~9.7 ka. A secondary phase of renewed aggradation followed increasing glacierization of the catchments (Neoglaciation) after ~4.0 ka. The Holocene chronology of debris-flood events (Fig. 5i) differs appreciably from those of debris flows (Figs. 5g, h, l) and river floods (Figs. 5e, f).

a
b
c
d
e
f
g
h

◀**Fig. 8** Selected periglacial landforms in Scandinavia (*continued*). **a** Debris-flood boulder deposits on the Kupegjellet alluvial fan, SE Jostedalsbreen southern Norway (photo: Geraint Owen). **b** Sorted circles, Juvflye, Jotunheimen (photo: Jennifer Hill). **c** Sorted stripes, Juvflye, Jotunheimen (photo: Jennifer Hill). **d** Earth hummocks, Breidsæterdalen, Jotunheimen (photo; Stefan Winkler). **e** Palsa, Vaisjaggi, Utsjoki, northern Finland (photo: Olli Ruth). **f** Lacustrine rock platform, Bøverbrevatnet, Sognefjell, southern Norway (photo: John Matthews). **g** Strandflat, Helgeland coast, northern Norway (photo: www.visithelgeland.no); **h** boulder-cored frost boil, Styggedalsbreen glacier foreland, Jotunheimen (photo: Stefan Winkler)

The main aggradational phase on fans in the SE Jostedalsbreen region is attributed to debris floods (debris torrents or hyperconcentrated flows), which form a distinct flow type (intermediate between water flows and debris flows) characterised by debris concentrations of 40–70% by weight (cf. Costa 1984; Slaymaker 1988; Wilford et al. 2004; Pierson 2005; Ouellet and Germain 2014; De Haas et al. 2015; Heisser et al. 2015). Such high debris concentrations are explicable in terms of paraglacial conditions in the early Holocene, when till deposits on the extremely steep and unvegetated slopes of the catchments would have been first exposed to subaerial processes (cf. Curry 2000). Rainstorms, snow meltwater, melting glacier ice, thawing permafrost and/or thawing ground exposed from beneath cold-based glacier ice, in combination with the available sedimentary material, would have provided optimal environmental conditions for debris-flood generation. Such optimal conditions would have been attained with decreasing frequency through the mid Holocene and were never attained again, even during Neoglaciation in the late Holocene. By this time, available regolith on the upper slopes of the catchment is likely to have been reduced, if not exhausted, and tree cover would have extended over much of the lower catchment, adding to the stability of possible sediment sources there.

3.2.9 Slope Wash

Colluvial sand layers, typically of thickness 1–10 cm, separated by 1–5 cm thick charcoal layers in the bottom of seven kettle holes in an esker landscape at Kuttanen, Finnish Lapland, enabled Matthews and Seppälä (2015) to establish a unique Holocene chronology of slope-wash events (Fig. 5m).

Several qualities of these kettle holes enabled this reconstruction. First, the enclosed hollows provide efficient traps for sediment eroded from the slopes. Second, the well-sorted sand substrate is highly erodible. Third, the steepness of the side slopes of the kettle holes (maximum slope angles of up to 26° today) enabled slope-wash processes to operate throughout the Holocene. Fourth, the surrounding stands of fire-prone Scots pine (*Pinus sylvestris*) mixed with mountain birch (*Betula pubescens*) provided a source of charcoal for radiocarbon dating. Fifth, the microclimate in the kettle holes (cf. Rikkinen 1989; Tikkanen and Heikkila 1991) allowed cold-air drainage, frost damage and snow drifting to prevent the growth of trees on the lower slopes, enhancing instability on the slopes, and preserving the colluvial stratigraphy at the base.

Radiocarbon dating of 131 charcoal samples established the timing of forest fires that triggered colluvial events by destroying the vegetation cover, as first proposed by Seppälä (1981). Recovery of the forest then provided the fuel for the next fire, followed by re-activation of colluvial processes. Around 50 individual forest fires/colluvial events were identified by Matthews and Seppälä (2015), which represents an average fire frequency of 1 in ~200 years during the Holocene. The frequency of events increased rapidly between ~9.0 and 8.0 ka, attained its peak in the Holocene Thermal Maximum (when the pine forest reached its maximum density), and declined after ~4.0 ka (Fig. 5m). Thus, peak forest fire frequency and colluvial activity appear to have been associated with the warmest weather of the Holocene with its attendant droughts, convective storms and lightning strikes.

3.3 *Low-Gradient Landscapes: Inland and Coastal*

Where slope failure and rapid mass movement no longer dominate geomorphological activity in the landscape, a different set of processes and landforms becomes apparent. On low gradients, cryoturbation processes, such as frost heave, frost sorting, frost creep and gelifluction, take on a leading role. Characteristic landforms include, amongst others, patterned ground, solifluction lobes, permafrost mounds and stone pavements. In Scandinavia, these landforms and processes occur in the alpine zones with permafrost and seasonal frost, on plateau surfaces, in peatlands and dunefields, and along lacustrine and coastal shorelines.

3.3.1 Sorted Patterned Ground

Repeated freezing and thawing of regolith results in regular patterns at the ground surface, manifest either in terms of coarse and fine sediment (sorted patterned ground) or small-scale topographic variation (unsorted patterned ground). The difference between the two largely reflects whether the sediment involved is a diamicton that contains clasts in a fine, frost-susceptible matrix, or consists of fine sediment alone (Ballantyne 2018). On horizontal or near-horizontal land surfaces, circles and nets result from differential frost heave. Sorted forms are produced in response to the vertical and lateral frost sorting of different clast sizes within diamictons. Where the surface gradient exceeds a few degrees, circles are transformed by gelifluction into stripes.

Large-scale sorted circles (Fig. 8b) and sorted stripes (Fig. 8c) occur extensively in Scandinavia on the flattest parts of the high plateaux, commonly within the blockfield areas discussed previously. Although they are widespread within the permafrost zone, they do not appear to require permafrost for their formation, only a sufficiently deep active layer (Ballantyne 2018). Excellent examples have been described, mapped and analysed from the Juvflye plateau (Galdhøpiggen massif, Jotunheimen), on till surfaces within the permafrost zone at an altitude of ~1750–1950 m a.s.l., where sorted circles predominate on gradients of 0–3°, average of 2°, and sorted stripes on gradients of 3–17°, with an average of 7° (Ødegård et al. 1987, 1988; Winkler et al. 2021). Sorted patterned ground occupies 20–50% of the Juvflye plateau by area. Typically, the fine centres of circles have diameters of 3.0–5.0 m, which is also the average spacing between coarse stripes. At altitudes of >1950 m a.s.l., in areas dominated by in situ frost weathered regolith, they are typically smaller with diameters of 0.5–1.5 m (Ødegård et al. 1988).

Schmidt-hammer exposure-age dating of boulders in the coarse gutters of both sorted circles and sorted stripes at Juvflye (Winkler et al. 2016, 2020) indicate structural stability, at least since the early Holocene. Dates from sorted circles ranged from 8.2 ± 0.5 to 6.9 ± 0.5 ka (Winkler et al. 2016) and exhibited a significant decrease in age with increasing altitude between 1500 and 1950 m a.s.l. The age range from sorted stripes was similar, 8.0 ± 0.4 to 6.7 ± 0.4 ka (Winkler et al. 2020), but with little or no relationship to altitude. Areas of fine sediment between the coarse gutters commonly exhibit evidence of cryoturbation today, especially at relatively high-altitude sites, but this does not affect the boulders that are wedged together in the gutters. Cessation of major activity correlates with the onset of the Holocene Thermal Maximum at ~8.0 ka, declining moisture availability, and a rise in the lower altitudinal limit of discontinuous permafrost to ~1650–1700 m a.s.l. (Lilleøren et al. 2012) but there are several possible mechanistic explanations for the timing of stabilisation. Winkler et al. (2016) proposed three: (1) changes to moisture conditions and related thermodynamics in the active layer affecting frost sorting; (2) loss of fines and reduced frost susceptibility in the gutters; and (3) exhaustion of the supply of boulders from the fines-dominated areas.

The timing of the initiation of sorted patterned ground formation is more problematic. Although some areas of sorted patterned ground could have survived beneath cold-based ice sheets and could be as old as some blockfields (cf. Kleman et al. 2008; Andersen et al. 2019) others, including those at Juvflye, have formed in till and appear to have evolved over a relatively short period in deep, water-saturated active layers following deglaciation in the early Holocene. Rapid development of sorted circles on the Little Ice Age glacier foreland of Slettmarksbreen (Ballantyne and Matthews 1982) and their occurrence on ice-cored moraines at Gråsubreen (Matthews et al. 2014; Winkler et al. 2021), both in Jotunheimen, may provide modern analogues of this process. At Slettmarksbreen, sorted circles up to 3.5 m in diameter formed and stabilised within 50 years of glacier retreat from the site. There, rapid development occurred in water-saturated till immediately after deglacierisation, and stabilisation (indicated by vegetation development on the fine centres and lichen growth on boulders in the gutters) followed shortly after that.

In summary, therefore, we suggest that much of the large sorted patterned ground located in the present permafrost zone of Scandinavia evolved in the early Holocene (sometimes in older regolith of periglacial or glacial origin), is essentially relict, and may not have required permafrost for its formation.

3.3.2 Non-sorted Patterned Ground

Non-sorted patterned ground has been less intensively studied in Scandinavia than the sorted variety. Three types are considered here: (1) frost-crack polygons and sand-wedge polygons in areas of seasonal frost; (2) relict large-scale frost polygons, which traditionally were associated with thermal cracking and ice-wedge growth in formerly more extensive permafrost zones; and (3) active earth hummocks (thúfur) and peaty earth hummocks (pounus). Frost cracks, frost-crack polygons and sand-wedge polygons have been reported from northern Norway to southern Sweden (Svensson 1969; Aartolahti 1972; Seppälä 1982, 2004). Indeed, Svensson (1974) observed modern ground cracking in southern Sweden following the passage of extremely cold air. In the Hietatievat dunefield in northern Finland, Seppälä (1982, 2004) described irregular frost-crack polygons with sides 1–12 m in length, central open cracks 5–15 cm deep, and sand wedges <70 cm deep. The wedges form where the groundwater table is close to the surface in a deflation basin where the mean annual air temperature is –1.5 °C but the winter air temperature often drops to –40 °C.

Svensson (1964, 1988) recognised three lowland areas in Scandinavia with ice-wedge casts forming polygonal patterns: southwestern Jutland, Denmark (outside the limit of the Last Glacial Maximum shown in Fig. 1), Scania, southwestern Sweden, and the Varanger Peninsula, northern Norway (both outside the limit of the Younger Dryas Ice Sheet). Kolstrup (2004) investigated a site in Jutland (Tjæreborg) and assigned Saalian or early-Weichselian ages on the basis of thermoluminescence dates on both wedge infill and host sediments. Her dates ranged from 290 ± 20 to 133 ± 12 ka. In northern Norway, locations on raised shorelines of known age suggest ice wedges were last active under permafrost conditions during the Younger Dryas (see also Sollid et al. 1973). Early descriptions of similar forms inside the limit of the Younger Dryas Ice Sheet on valley floors in northern Sweden, where diameters of 10 to 40 m have been reported (Lundqvist 1964; Rapp and Clark 1971; Rapp 1982), are consistent with their survival beneath cold-based ice according to the 'Kleman model' (Kleman et al. 2008). Even in the areas of continuous permafrost, it appears that modern climatic conditions in Scandinavia have not been suitable for repeated thermal cracking, frost-wedge growth and active frost-polygon development, but this requires further investigation.

Earth hummocks (thúfur) are vegetated, dome-shaped mounds (Fig.8d), typically 0.5 m high and 1.0 m basal diameter, separated by troughs that are generally narrower than the mounds, and composed predominantly of minerogenic sediment (Schunke and Zoltai 1988; Grab 2005). In Scandinavia, they are active today in the low alpine zone throughout the Scandes (e.g. Lundqvist 1964). In northern Finland, similar peaty earth hummocks (pounus) are composed wholly or partly of peat with a minerogenic core and commonly contain permafrost (Luoto and Seppälä 2002a; Van Vliet-Lanoë and Seppälä 2002). Hummocks require moist, moderately well-drained sites to promote frost heave (Van Vliet Lanoë et al. 1993), while vegetation growing on mineral soil or peat has a role in preserving the mound shape. Examples of both types of hummocks have been dated with radiocarbon, and results have been interpreted as indicative of the onset of hummock development. Ellis (1983) obtained seven radiocarbon dates from two hummocks in Jotunheimen, southern Norway, which yielded ages of 4800 ± 220 to 3050 ± 80 cal year BP, whereas 11 ages obtained by Van Vliet-Lanoë and Seppälä (2002) from six pounus in Finnish Lapland were mostly <2.0 ka but ranging from modern to 4050 ± 140 (all dates ± 2σ). Both sites suggest that hummock formation began during late-Holocene climatic deterioration and that hummocks are active at present.

3.3.3 Solifluction Landforms

Solifluction lobes and sheets/terraces are equally if not more common than earth hummocks in the Scandinavian alpine zone, where they occur over a wide range of altitudes on permafrost and non-permafrost slopes (Lundqvist 1964; Williams 1957; Rudberg 1977; Ridefelt and Boelhouwers 2006; Harris et al. 2008; Hjort 2014). They are produced by the slow downslope flow of water-saturated regolith, mainly due to thaw consolidation of ice-rich sediments (gelifluction) and where flow is retarded by vegetation (turf-banked forms) and/or boulder accumulations (stone-banked forms). Alpine vegetation is particularly important in stabilising and maintaining the distinct shape of the step-like riser and tread components of the landform (Eichel et al. 2017), and contribute to the abundance and size of turf-banked solifluction lobes at relatively low altitudes in the zone of seasonal frost (Ulfstedt 1993; Ridefelt and Boelhouwers 2006). Tread dimensions are variable but risers are typically 0.5–1.0 m high and advance downslope at typical rates of 1.0–10.0 mm per year (Ridefelt et al. 2009; Ballantyne 2018). Ploughing blocks, which move downslope at faster rates than the surrounding material, are commonly found in areas of solifluction and are attributed to the thermal properties and weight of the block interacting with thaw consolidation processes. First recognised in Scandinavia by Sernander (1905), the largest known reported example with dimensions 4.9 × 5.0 × 5.6 m and an estimated weight of 36,400 kg, has a 42.5 m linear depression in its wake (Reid and Nesje 1988).

Frontal advance of solifluction lobes and concomitant burial of radiocarbon datable organic soil material have provided an approach to understanding the evolution of solifluction lobes, changing rates of solifluction, and relationships to Holocene climate. Scandinavian studies of this type have been attempted in northern Norway (Worsley and Harris 1974; Ellis 1979; Elliott and Worsley 1999, 2012), southern Norway (Matthews et al. 1986a; Nesje 1993; Nesje et al. 1989), northern Sweden (Rapp and Åkerman 1993) and Finnish Lapland (Kejonen 1979; Matthews et al. 2005). The general conclusion from these studies is that widespread solifluction lobe development occurred after the Holocene Thermal Maximum, between ~5.0 and 3.0 ka, and has continued until modern times with variations in rate. However, there are problems with this approach of which the apparent mean residence time of soil humus and the potential for erosion of the buried soil surface are of particular importance as they can lead to overestimates of time elapsed since burial by hundreds, if not thousands, of years (Matthews 1985, 1993).

Matthews et al. (2005) were able to overcome these two main potential errors by dating soil charcoal rather than soil humus, which was possible because of the special microclimate of their study site at Pippokangas in a 15-m deep kettle hole surrounded by mixed pine and birch forest. Stacked solifluction lobes with risers and treads up to 40 cm high and 8 m long occur on the <10° lower slopes of the kettle hole, where tree growth is prevented by frost damage and the accumulation of snow, as discussed above in relation to slope wash (cf. Rikkinen 1989; Tikkanen and Heikkila 1991). The Pippokangas study involved 46 radiocarbon dates from five solifluction lobes and identified phases of solifluction lobe inception at ~7.4–6.7, 4.2–3.4, 2.6–2.1 and 1.5–0.5 ka (Fig. 5n), which were attributed primarily to millennial-scale variations in effective moisture. However, solifluction activity has been continuous at the site since ~7.0 ka, and monitoring of solifluction rates in northern Sweden has revealed considerable complexity in relationships between solifluction and several climatic variables. Snow depth and both summer and winter temperatures seem to be effective via their influence on active-layer thickness (Åkerman and Johansson 2008; Ridefelt et al. 2009; Strand et al. 2020; see 'The future' below).

3.3.4 Palsas and Lithalsas

According to the classic Scandinavian definition, palsas (Fig. 8e) are peat-covered, dome-shaped perennial permafrost mounds up to 10 m high, which rise out of mires in discontinuous or sporadic permafrost zones (Seppälä 1986, 2011). However, their form may deviate considerably from a dome and include lower, irregular forms (so-called 'palsa plateaux') that have diameters of hundreds of metres (Sannel, 2020). Palsas may also be seen as one end of a continuum of landforms, from mounds formed wholly of peat, through peat-covered mounds with a mineral core, to lithalsas, which are mounds composed wholly of mineral sediments (Åhman 1976, 1977; Gurney 2001; Pissart 2002, 2013). Seppälä (1994, 1995b) demonstrated experimentally that the permafrost core of a palsa grows in areas of the mire where a shallow snow depth allows relatively deep frost penetration in winter. The palsa then survives the summer thaw because of the insulating properties of the peat cover. Palsas formed wholly of peat are uncommon and relatively small, whereas palsas with a mineral core grow larger because the frost susceptibility of the mineral substrate is much greater than that of the peat, promoting the development of ice lenses within the mound. Beyond the distribution of peatlands, lithalsas require cooler summers for preservation of the permafrost core and therefore tend to occur at higher altitudes and/or latitudes than palsas.

Palsas and lithalsas commonly occur in groups within which mounds of varying size represent different stages of development and decay, which is indicative of 'cyclic' development (Seppälä 1988; Matthews et al. 1997b). Permafrost degradation follows the growth of the mound, erosion of surface material, and exposure of the permafrost core to the atmosphere. This leads to thermokarst pools, of which those formed from lithalsas are more likely to be surrounded by rim-ridge ramparts and to survive as relicts in the landscape (Luoto and Seppälä 2002b, 2003).

The radiocarbon dating of palsas has been based on the ability to recognise changing ecological conditions on the mound as it emerges from the mire; specifically the transition from hygrophilous to xerophilous peat as indicated by pollen and macrofossil content. Although early attempts revealed dates from the frozen core in stratigraphic order, problems such as the removal of surface peat by deflation, and cyclic development were not sufficiently appreciated (Seppälä 2003, 2005a, b). In summarizing the dates obtained by Vorren and Vorren (1975), Vorren (1979) and others, Seppälä (1988, 2005a, b) thought that the palsas of Scandinavia are no older than ~3.0 ka and that most formed after 1.0 ka, some under the modern climatic regime. More recent research by Oksanen (2005, 2006) concluded that permafrost aggradation in the palsa mires of northern Finland and northern Pechoria (Russia) began no later than 2.5 and 3.0 ka, and possibly as early as ~4.0 and 5.0 ka, respectively.

3.3.5 Rivers, Lakes and Groundwater

Periglacial rivers, lakes and groundwater are influenced by distinctive hydrological regimes and processes linked to the presence of snow and ice in catchments, in channels and along shorelines (Beldring et al. 2008; Beylich and Laute 2021). Ice jams, flooding and erosion during the break-up of river ice (Lind et al. 2014), ice-push landforms around lakes (Alestalo and Häikiö 1979; Alestalo 1980; Worsley 2008), slushflows (Nyberg 1985, 1989; Clark and Seppälä 1988) and icings (naled or aufeis), and frost weathering associated with inland water bodies (Matthews et al. 1986b; Shakesby and Matthews 1987; McEwen and Matthews 1998; McEwen et al. 2002; Aarseth and Fosen, 2004a, b), are amongst the periglacial landforms and processes that have been described in Scandinavia.

Frost weathering of bedrock in the context of both river channels and lake shorelines provides insights into how effective this process can be in those places within periglacial landscapes where there is sufficient available water. In relation to four gorges on the Storutla, Jotunheimen, McEwen et al. (2002) estimated that frost weathering accounted for excavation of a minimum of 24–97% of gorge volume during the Holocene, at incision rates of 0.15–0.39 mm per year. These values for periglaciofluvial erosion of bedrock appear to exceed those of temperate fluvial systems unaffected by tectonic uplift. Deep penetration of seasonal frost into bedrock channel walls occurs in winter when water levels are low but there is still unfrozen water beneath river ice, especially in areas of the channel floor with relatively deep pools of water. Unfrozen water is drawn to the freezing front, enabling the growth of ice within cracks and inducing the frost weathering of bedrock close to the winter water level in the gorge. This model, which focuses on the annual freeze-thaw cycle, was first developed by Matthews et al. (1986b) to explain the rapid development of rock platforms around a short-lived moraine-dammed lake in Jotunheimen. During the existence of Bøverbrevatnet for between 75 and 125 years, platforms up to 5.3 m wide with backing cliffs up to 1.55 m high were produced (Fig. 8f). No other mechanism than frost weathering is capable of producing platforms in lacustrine environments where wave erosion is extremely limited and lake-ice activity is hardly capable of moving the large, angular (frost-shattered) blocks that remain largely in place as a fringing boulder pavement.

There are major questions about the importance of the fluvial component of denudation in periglacial landscapes where the traditional assumption was that frost weathering and mass movement processes dominate. The classical sediment budget approach developed in the Kärkevagge valley, northern Sweden (Rapp 1960) was instrumental in correcting this assumption. His monitoring of several different components of denudation showed that the solutional load of rivers (and hence chemical weathering) was the dominant process. Scaling-up such results is, however, difficult and later research has demonstrated the differences that exist within Scandinavian periglacial landscapes.

Dissolved load, suspended load and bedload have since been monitored in several catchments in northern (Latnjavagge, Swedish Lappland), middle (Homla, Trøndelag) and southern Scandinavia (Erdalen and Bødalen, western Norway (Beylich 2011; Beylich et al. 2004, 2005; Beylich and Laute 2012, 2018, 2021). This research has, in the main, corroborated Rapp's conclusions. In the partly glacierized western Norwegian oceanic catchments total fluvial denudation is 1.6–3 times that in the non-glacierized Homla catchment, where chemical denudation accounts for 77% of the total fluvial denudation, and is 3.4 times that of mechanical fluvial denudation (i.e. the sum of that attributed to suspended and bedload components combined). Corrected for atmospheric chemical inputs, the mean annual chemical denudation rate for the Homla catchment is 12.1 t km^{-2} a^{-1}, whereas the mean annual suspended sediment yield is 3.3 t km^{-2} a^{-1} and the mean annual bedload yield is 0.3 t km^{-2} a^{-1} (Beylich 2021). Comparative chemical denudation rates estimated for Kärkevagge (Darmody et al. 2000) and the neighbouring Latnjavagge (Beylich 2011) are 19.2 t km^{-2} a^{-1} and 4.9 t km^{-2} a^{-1}, respectively.

Most of the chemical load of rivers originates via groundwater from the regolith and reflects the effectiveness of chemical weathering in both sediments and soils (Darmody and Thorn 1997; Darmody et al. 2000; Thorn et al. 2007, 2011). Studies of differential weathering of minerals exposed to the atmosphere on glacially-scoured bedrock outcrops shed light on the effectiveness and rate of chemical weathering of bedrock throughout Scandinavian periglacial landscapes (André 2002; Nicholson, 2008, 2009; Owen et al. 2007; Matthews and Owen 2011). Based mainly on the negligible weathering of protruding quartzitic veins, average rates of rock-surface lowering at these Scandinavian periglacial sites appear to range from 0.2 to 4.8 mm per thousand years for resistant metamorphic lithologies. However, such estimates may also include the effects of granular disintegration by frost and biological weathering (McCarroll 1990; McCarroll and Viles 1995; Matthews and Owen 2011). On calcareous lithologies, which are not common in Scandinavia, rates of rock-surface lowering may be much higher, especially in hydrologically favourable localities, such as lake shorelines (Owen et al. 2006b).

3.3.6 Inland Aeolian Dunefields and loess

Wind is an important and somewhat neglected geomorphic agent of erosion, transport and deposition in periglacial landscapes. Open spaces and sparse vegetation in cold climates permit aeolian processes to act on the landscape both directly through abrasion, sand drift and deflation, and indirectly through snow-drift and frost-action effects (Seppälä 2004), to which must be added their role in drifting lake ice and, at the coast, sea ice (e.g. Alestalo and Häikiö 1975).

Erosive effects of wind abrasion in the form of ventifacts (faceted and grooved boulders) and polished boulders in southern Sweden and Denmark have been long recognised as indicators of wind blasting under former periglacial conditions (Milthers 1907; Svensson 1983; Schlyter 1995; Christiansen and Svensson 1998). The distribution and orientation of in situ wind-blasted rock surfaces in relation to ice-sheet limits, combined with associated luminescence dates, leads to the conclusion that formation was associated with easterly winds, mainly at or close to the Last (Weichselian) Glacial Maximum and during subsequent deglaciation. At these times, most probably in the period from 22 to 17 ka, strong easterly katabatic and zonal winds would have dominated in open, sparsely-vegetated landscapes (Christiansen and Svensson 1998).

Largely relict aeolian dunefields comprise a major element of the periglacial landscape across northern Scandinavia (Högbom 1923; Klemsdal 1969; Seppälä 1971, 1972; 1995a; Sollid et al. 1973; Matthews and Seppälä 2014). Luminescence dates up to 10.7 ± 1.2 ka (Kotilainen 2004) establish that mainly parabolic dunes, the arms of which are up to ~1500 m long, ~150 m wide and ~12 m high (Seppälä 1971) formed from the sand blown from eskers, valley trains and other glacigenic deposits shortly after retreat of the Younger Dryas Ice Sheet. The parabolic form of the dunes is the result of the arms being partly anchored by vegetation during their formation, and well-developed, undisturbed stratification (i.e. no niveo-aeolian features) within these dunes indicates that deposition occurred without snow in the summer months (Seppälä 2004).

The subsequent history of many inland dunefields has been investigated in some detail in northern Finland based on dune stratigraphy and the radiocarbon dating of buried palaeopodzols and charcoal layers (Seppälä 1995a, b; Käyhkö et al. 1999; Kotilainen 2004; Matthews and Seppälä 2014). In favourable sites for vegetation growth, initial stabilisation of the parabolic dunes probably occurred within a few hundred years of deglaciation (~10.9–10.2 ka) but at other sites dunes may have remained active for up to 2000 years. Based on analysis of a total of 137 radiocarbon dates on buried charcoal layers from dunes within the northern boreal forest ecotone (the subzone characterised as birch forest with pine stands at present), Matthews and Seppälä (2014) recognised at least 16 dune re-activation events (episodes of deflation) of more than local significance since ~8.3 ka. The available data indicate that dune re-activation was triggered by forest fires, which occurred throughout the Holocene but with increasing frequency in the late Holocene. They proposed a conceptual model summarising the complex geo-ecological interactions between climate and aeolian activity (deflation or stabilisation) modulated by the effects of fire and vegetation on fuel supply and successional recovery. The millennial-scale increase in deflation activity during the late Holocene (Fig. 5o) was attributed to relatively cool and moist climatic conditions, which increased the density of the tree cover in dune habitats and fuelled the fires, while centennial-scale successional recovery of the vegetation cover controlled the recurrence interval to the next fire.

A complementary approach to analysing the relationship between aeolian activity and Holocene climatic variation was adopted by Nielsen et al. (2016a, b), who analysed the input of sand grains to lacustrine sediments at two sites on Lofoten and Andøya on the west coast of northern Norway. At Latjønna, on Andøya (14.5 m a.s.l.), a record of the influx of medium to coarse sand (>250 μm) from 6.0 ka revealed a fluctuating increase in aeolian activity, which reached a distinct maximum at 1.7–0.7 ka. Nielsen et al. (2016b) considered falling sea level made sand available for deflation, and that increasing storminess was the main driver for increased niveo-aeolian transport of sand grains inland during winter. At Trehynnvatnet on Langøya (33 m a.s.l.) the longer record from 11.5 ka (Fig. 5p) showed generally low influx of sand before ~2.8 ka, after which the largest of several activity peaks occurred during the last 0.5 ka. Again, sand transport over snow by southwesterly winds during winter storms was invoked as the preferred process.

Aeolian silt deposits (loess) have seldom been reported from Scandinavia. However, Stevens et al. (2022) have confirmed that thin loess deposits occur at scattered locatities on relatively high ground in central and southern Sweden. Luminescence dating has here yielded mainly early- to mid-Holocene ages, which reflect early-Holocene origins from glacio-fluvial sediments followed by later sediment mixing during soil development and/or redeposition during later phases of landscape instability.

3.3.7 Coastal Rock Platforms and the Strandflat

Frost weathering has been proposed as the primary agent of cliff recession and platform extension in areas of restricted fetch within fjords along the coast of northern Norway for over a century. Vogt (1918) described rock platforms up to 12 m wide, close to high-tide level in Kvænangen fjord, on which angular fragmented bedrock showed little or no evidence of rounding by wave action. The Norwegian explorer, scientist and diplomat, Fridtjof Nansen described similar rock platforms up to 20 m wide close to his house on the Fornebo peninsula in inner Oslo fjord (Nansen 1922). More recently, Sollid et al. (1973) reported that the inner edge of 49 present-day coastal rock platforms in Finnmark occur at 1.4 ± 0.6 m above mean tide level, while 12 present-day coastal rock platforms on Hinnöya (Lofoten Islands) occur at 1.5 ± 0.5 m above mean tide level (Möller and Sollid 1972). At all these fjord locations, platform erosion depends on frost weathering at the cliff base associated with the winter formation of thick layers of shore-fast sea ice, possibly enhanced by brackish water in the fjords, fresh groundwater seepage from the cliffs, ice-foot development, and the subsequent removal of weathered material by storm waves and/or ice-rafting.

Frost weathering has been important also in the formation of relict rock platforms, which have been raised above present sea level by glacio-isostatic land uplift. This is particularly clear in the case of the main Younger Dryas shoreline in northern Norway, which is best developed in narrow fjords and sounds, but is absent from wave-exposed headlands (Andersen 1968; Sollid et al. 1973; Rasmussen 1981; Blikra and Longva 1985), and was eroded when a permafrost climatic regime existed down to sea level. Precise mechanisms of weathering and erosional processes associated with both active and relict coastal rock platforms are, however, poorly understood. Despite the moderating effect of saltwater, and a greater role for wave action at coastal locations, frost weathering on lake shores (e.g. Fig. 8f), as discussed above, may provide a model for the rapid formation of coastal rock platforms in periglacial environments (Matthews et al. 1986b; Dawson et al. 1987). Further investigations on modern periglacial coasts, such as that of Ødegård and Sollid (1993) in Svalbard, are required to clarify the disputed nature and effectiveness of the processes involved in cold-climate rock-platform erosion (cf. Dawson 1979; Byrne and Dionne 2002; Hansom et al. 2014).

On much longer timescales, frost weathering is one of the main processes responsible for the Norwegian strandflat (Fig. 8g). This is a horizontal to gently sloping, uneven bedrock coastal plain or foreland, with numerous low islands, skerries and submarine and subaerial rock platforms (Corner 2005b). The strandflat is present along much of the Norwegian coast but is regionally variable in extent and elevation: it is up to 60 km wide and ranges from 40 m below sea level to a maximum of 100 m a.s.l. After a long debate and many different theories concerning its age and origin (e.g. Reusch 1894; Nansen 1922; Klemsdal 1982; Larsen and Holtedahl 1985; Guilcher et al. 1986; Holtedahl 1998; Olesen et al. 2013; Fredin et al. 2017), the emerging picture is of a polygenetic landform that has evolved since pre-Quaternary times.

The latest major formative events involved processes of frost weathering, sea-ice erosion and wave abrasion, modified by glacial erosion, essentially as proposed by Nansen (1922). These events are encompassed by the theory, proposed by Porter (1989) that the age of the strandflat is primarily accounted for by average glacial conditions during the early Pleistocene (~2.6–0.9 Ma). During that interval, ice sheets were smaller than in the late Quaternary (cf. Kleman et al. 2008) and, crucially, would not have reached a coastal zone exposed to frost weathering during moderately low relative sea levels. Meanwhile, variable glacier extents and glacio-isostatic and glacio-eustatic changes in sea level would have affected the wide range of elevations and gradients exhibited by the strandflat today.

3.4 Glacier-Foreland Landscapes

Glacier forelands—the landscapes deglacierised since the Little Ice Age glacier maximum of the last few centuries, and continuing to be deglacierised today—provide an opportunity to investigate a range of periglacial landforms and processes.

The main advantage of these landscapes is that they can be viewed as microcosms or field laboratories, without the long and complex evolutionary history of the plateau, steepland and low-gradient landscapes already discussed. Furthermore, the short timescale for landform development can often be firmly established by invoking the chronosequence concept (cf. Matthews 1992; Ballantyne 2002; Heckmann and Morche 2019). In Scandinavia, several studies of this nature have provided insights into active (neo)paraglacial and periglacial processes following glacier retreat under present climatic regimes. In some cases, such studies may provide modern analogues for interpreting relict landforms, as already indicated in relation to the formation of frost-sorted patterned ground (*sorted circles*) and frost-weathered shorelines (*rock platforms*). Further selected examples are considered here.

Glacier retreat from steep Little Ice Age lateral moraines, and the ensuing modification of slope form by gully erosion and debris-flow activity, have been investigated in detail at Fåberstølsbreen and other outlet glaciers of the Jostedalsbreen ice cap (Ballantyne and Benn 1994, 1996; Ballantyne 1995; Curry and Ballantyne 1999; Curry 1999, 2000). Such neoparaglacial activity is initiated by rainstorms and snow meltwater. The main agent of erosion and sediment transport is debris flow with a minor role for snow avalanches, and colluvial fans are deposited at the foot of the moraine slope. Ballantyne and Benn (1994) recorded ~10 debris flows per year over a six-year observation period at Fåberstølsbreen. At this glacier foreland, and at neighbouring Bergsetbreen, Ballantyne (1995) showed that a stable neoparaglacial landsystem is produced within centuries, if not decades of deglaciation. This landsystem consists of gully-head areas eroded to bedrock, broad intersecting gullies in mid-slope position, coalescing debris-flow fans at the slope foot, and valley-floor deposits.

A small but perfectly formed rock glacier that developed from the outermost Little Ice Age lateral moraine of Bukkeholsbreen, Jotunheimen, exemplifies the rapid development and stabilisation of another steepland landform (Vere and Matthews 1985). The rock glacier is about 160 m long and 130 m wide, with an average surface slope of 10°, and several well-developed arcuate transverse ridges. According to lichenometric dating, it formed and stabilised within ~200 years of moraine deposition ~AD 1750. Formation of a rock glacier in this way indicates the importance of a sufficiently large debris supply, combined with the opportunity for incorporation of buried glacier ice or interstitial ice from overridden distal snowbeds.

On low-gradient parts of Norwegian glacier forelands, the processes of frost heave, frost sorting and slow mass movement (solifluction) are often in evidence. Boulder-cored frost boils (Fig. 8h) produced by the up-heaval of individual boulders in poorly drained areas of till were described by Harris and Matthews (1984) on the glacier forelands of Storbreen and Bøverbreen, Jotunheimen. Ballantyne and Matthews (1983) investigated the steady development of small-scale crack networks (probably desiccation cracks) into sorted polygons <0.5 m in diameter on terrain deglaciated from 0 to 35 years at Storbreen. Similar polygonal forms recognised at Styggedalsbreen by Matthews et al. (1998) and Haugland (2004, 2006), developed and stabilised close to the ice margin after ~60 years, on the flat tops of low, degraded annual moraines. The sorted polygons transitioned to sorted stripes on gradients of 13–17°, on the upper slopes of the degraded moraines. At the same location, Matthews et al. (1998) mapped boulder-cored frost boils, solifluction lobes with risers up to 30 cm high, ploughing blocks, and shallow accumulations of colluvial sand (possible slope wash and/or niveo-aeolian deposits), the sites of which were differentiated by moisture availability, gradient and exposure, as reflected in microtopography. Another study at Storbreen (Matthews and Vater 2015) demonstrated the rapid development of stone pavements within ~25 years of deglacierisation. Pavement formation was attributed to the combined action of pervection (downwashing of fine particles, particularly silt, from the surface layers), frost sorting (the raising of relatively large clasts towards the surface) and deflation (lateral removal of sand-sized particles by wind).

Biogeomorphological and geo-ecological approaches to glacier-foreland landscapes in Jotunheimen have improved understanding of the interaction of organisms with physico-chemical periglacial processes in landform, soil and vegetation development (Matthews 1992). Vegetation succession and soil development on sorted and non-sorted patterned ground signal rapid stabilisation of the landforms (Ballantyne and Matthews 1983; Matthews et al. 1998; Haugland and Beaty 2005; Haugland and Owen 2005; Haugland and Owen-Haugland 2008). Solifluction and ploughing boulders remain active on older terrain, where lobe development is promoted by the binding action of the vegetation cover (Matthews et al. 1998; Matthews 1999). Boulder-cored frost boils (Fig. 8h) also remain active but are unaffected by vegetation. Chemical weathering in sediments and soils is also enhanced as vegetation develops on glacier forelands (Darmody and Thorn 1997; Thorn et al. 2007, 2011); and on recently-deposited boulder surfaces, Matthews and Owen (2008) showed that endolithic lichens may enhance rock weathering rates by up to two orders of magnitude.

4 The Significance of Periglacial Dynamics in Scandinavia within the European context

This review concludes that Scandinavian periglacial landscapes have evolved over a long period of geological time; that only a minority of landforms are in equilibrium with present-day periglacial processes and climates; and that many elements of these landscapes are polygenetic and either relict or continuing to develop under changing Holocene and anthropocene environmental conditions.

4.1 Continual Environmental Change

It is perhaps surprising that so many elements of the periglacial landscapes of Scandinavia are so poorly adjusted to present environmental conditions, despite the widespread occurrence of permafrost and seasonal frost (Fig. 1). This is attributed to a number of factors, including the preservation of relatively old elements beneath cold-based ice sheets during the Pleistocene, paraglacial effects, and continuing environmental change during the Holocene when landforms may vary in their level of activity, become relict or be re-activated.

Possible pre-Quaternary elements are recognisable in the major topographic elements in today's periglacial landscapes, such as the inland plateaux and the coastal strandflat. During the Pleistocene, the cumulative effect of repeated glacial and interglacial episodes also left their mark on the distribution of specific landforms, such as blockfields. However, it is the environmental changes at the end of the Last Glaciation (the Weichselian Lateglacial) and the relatively small variations in climate of the Holocene that account for the active or relict status of many of the landforms and the extent of present activity. Many large rock-slope failures, rock glaciers, pronival ramparts, raised rock platforms and parabolic sand dunes provide examples of landforms that became relict at the Younger Dryas-Holocene transition.

Our survey of the current knowledge of the landforms, their formative processes, and especially their age as determined by the application of dating techniques (particularly radiocarbon dating, luminescance dating, cosmic-ray exposure-age dating and Schmidt-hammer exposure-age dating), has revealed the effectiveness of past environmental change in bringing about geomorphological change. In some cases, the same techniques have reduced our ignorance concerning which landforms and processes are active today and the extent of current activity.

4.2 *Landscape as Palimpsest*

In many cases it is now possible, using available dating techniques, to place at least approximate ages on relict periglacial landforms, such as blockfields, rock glaciers and sorted patterned ground. However, the allocation of active or relict status is often only a first approximation (cf. Hjort 2006), because recognising the extent of modern activity in the face of environmental change is problematic. Some stratigraphic contexts enable the reconstruction of the changing frequency of events since at least the early Holocene. At certain sites in Scandinavia, this has proved possible using radiocarbon dating; for example, in relation to records of debris flows, snow-avalanches and river floods in southern Norway; and solifluction, slope wash and aeolian activity in northern Finland (Fig. 5). Similarly, exposure-age dating techniques have established Holocene chronologies for coarse-debris deposits, as exemplified by large and small rock-slope failures, snow-avalanche impact ramparts, and debris floods on alluvial fans.

The periglacial landscapes of Scandinavia are therefore complex palimpsests in which active and relict forms exist in close proximity. The term 'palimpsest' has been used before in geomorphology (e.g. Kleman 1992) but not specifically in relation to periglacial landscapes. A palimpsest periglacial landscape is not simply a mosaic of active and relict landforms of specific ages. Although some landforms may be truly relict (formed and stabilised at specific times in the past), many others have more complex histories. Relict forms, such as parabolic dunes may be re-activated. Some may be relict only in part, as in the case of active sorted circles surrounded by relict gutters. Other landforms and processes exhibit varying activity levels and rate changes through time, as exemplified by the Holocene chronologies of Fig. 5.

A final, important implication of the palimpsest concept for understanding periglacial landscapes is the problem it poses for identifying a meaningful altitudinal zonation of landforms in relation to periglacial processes and environmental variables. Past attempts at defining altitudinal zones in Scandinavia, including those of Lundqvist (1964) and Rudberg (1977) in Sweden, Barsch and Treter (1976) and Kergillec (2015) in Rondane, southern Norway, Harris (1982) in Okstindan, northern Norway, Niessen et al. (1992) in northern Finland and, most recently, Winkler et al. (2021) in Jotunheimen, have all encountered this problem, which has yet to be resolved. Altitudinal variation in permafrost limits during the Holocene was of the order of 400 m in southern Scandinavia and 200 m in northern Scandinavia (Lilleøren et al. 2012), which emphasises the scale of the problem. Until active and relict elements of the landscape can be distinguished with confidence there seems little possibility of defining valid or useful altitudinal zonations of the periglacial landscape.

4.3 The Future

There are clear indications that future changes to the periglacial landscapes of Scandinavia are predictable. The dynamic nature of the altitudinal limits of permafrost during the Holocene gives reason to expect greater dynamism in the future. The present extent of permafrost is shrinking in response to current warming trends (Gisnås et al. 2017). Periglacial processes and landforms dependent upon both permafrost and seasonal frost, such as rock weathering, patterned ground, rock glaciers, rock-slope failures and solifluction, will be affected appreciably, if not catastrophically, by the projected future rise in mean annual air temperature (Harris et al. 2009; Aalto et al. 2017; Karjalainen et al. 2020).

The response of periglacial landforms and processes to climate change is, however, complex, as indicated by the following examples from Scandinavia. A good illustration of this complexity is the state of palsas in the sporadic permafrost zone. In northern Finland, small increases in temperature (1 °C) and precipitation (10%) are predicted to lead to considerable losses in areas suitable for palsa development (Fronzek et al. 2006, 2010). These authors tested models that predicted the total disappearance of palsas if mean annual temperature increases by 4 °C. The validity of such predictions is supported by observations and surveys of palsa mires from northern Finland (Luoto and Seppälä 2003) and northern Sweden (Zuidhoff and Kolstrup 2000; Sannel and Kuhry 2011; Sannel et al. 2016; Borge et al. 2017; Olvmo et al. 2020). The degradation of palsas, which can be a response to erosion of the peat cover on top of the aggrading permafrost core and/or melting of the ice core as thermal conditions change and/or increasing summer precipitation, is likely to be most marked near the limits of palsa distribution, as in southern Norway (Sollid and Sørbel 1998). However, local snow condition in particular may permit palsas to survive or new palsas to form even in a warming environment (Seppälä 1994, 2011; Nihlén 2003).

Current thawing of permafrost is also influencing both rapid and slow mass movement. Rock-slope failures tend to be stabilized by permafrost, as shown at the three rockwall sites in southern and northern Norway, where enhanced sliding rates were determined by Hilger et al. (2021) in response to permafrost degradation following the Holocene Thermal Maximum (see above). This implies rock-slope failures may become more frequent during accelerated warming, a suggestion validated by the failure of Veslemannen rockslide (part of the Mannen site investigated by Hilger) on September 5, 2019 (Kristensen et al. 2021). At Veslemannen, five years of monitoring revealed progressive rock-slope weakening as precursor movements, which coincided with precipitation events that increased seasonally from spring to autumn. The seasonal pattern implies thermal control on pore-water pressure, which Kristensen et al. (2021) attribute to the transition from permafrost to seasonal frost. This explanation is supported by thermal monitoring, geophysical surveying and numerical modelling at the Mannen site and the Gámanjunni site in northern Norway (Etzellmüller et al. 2021), and also by the observation by Blikra and Christiansen (2014) of ice-filled fractures at the rockslide site at Nordnesfjell in northern Norway.

Both aerial photography and satellite remote sensing (InSAR) have demonstrated recent velocity acceleration of the rock-glacier complex at Ádjet, northern Norway (Eriksen et al. 2018). Average annual horizontal velocity measured by aerial photography increased from ~0.5 m per year between 1954 and 1977 to ~3.6 m per year between 2006 and 2014. Average annual velocity measurements determined by InSAR increased from ~4.9 to 9.8 m per year between 2009 and 2016, while maximum velocities increased from ~12 to 69 m per year. Results of kinematic analysis are consistent with permafrost degradation and rock-glacier destabilisation following increases in mean annual air temperature of 1.8 °C and annual precipitation of 330 mm (55%) over the 62-year measurement period, leading to increased amounts of water reaching deeper layers within the rock glacier.

Active solifluction has been measured in the Abisko Mountains of northern Sweden (Kärkevagge in the west to Nissunvagge in the east) for over 50 years (Åkerman and Johansson 2008; Ridefelt et al. 2009; Strand et al. 2020). Various methods were used by different research workers to record movement rates. Ridefelt et al. (2009) compiled and analysed the available data in relation to the possible effects of climatic variables. Although variability was high, interpretable regional and temporal patterns were obtained. Generally higher movement rates were found in the western part of the region (Kärkevagge area) characterised by relatively low temperatures but relatively high precipitation and snowfall. Importantly, for data from the Kärkevagge area between 1979 and 2001, an increase in the rate of solifluction movement was detected, and statistically significant positive linear correlations of up to $r = 0.63$ and 0.56 ($p < 0.05$) were established between annual mean movement and annual mean air temperature and autumn (October to December) mean air temperature, respectively. Additional variables found to be significant in multiple regression analyses included summer (July to September) mean air temperature, annual precipitation, winter precipitation, melt start (first day with positive air temperatures) and freeze start (first day with negative air temperatures). The relationships involving freeze start and winter precipitation were negative. Overall, the results reported by Ridefelt et al. (2009) suggest an indirect causal relationship between warming air temperatures and solifluction rates involving an increase in active-layer thickness (affected by summer temperatures) and depth of frost penetration (affected by winter temperatures), moderated by snow depth and meltwater availability.

The complex interactions between periglacial landscapes and climate, some of which have been signalled in the selected examples above, will require a concerted research effort to understand fully likely future climate change impacts. There has already been a major effect on periglacial research agendas and this is likely to continue. In addition, observation, mapping, environmental reconstruction, stratigraphical investigation, dating, monitoring, experimentation, remote sensing, geographical information systems, and modelling, will all need to be embraced with imagination to improve what we already know about periglacial landscapes, past, present and future.

Acknowledgements This chapter is dedicated to Matti Seppälä, who sadly died on 24 November 2020; in appreciation of his immense contribution to periglacial geomorphology in Finland and internationally.

Several colleagues contributed photographs, including Jennifer Hill, Jan Hjort, Richard Mourne, Mons Rustøy, Olli Ruth, Peter Wilson and Stefan Winkler. Pål Ringkjøb Nielsen supplied the data for plotting Fig. 51. We are also grateful to Ola Fredin, Charles Harris, Miska Luoto, Richard Shakesby, Peter Wilson and Stefan Winkler for reviewing and commenting on the manuscript, and to Anna Ratcliffe who prepared the figures for publication. This paper represents Jotunheimen Research Expeditions, Contribution No. 225 (see http://jotunheimenresearch.wixsite.com/home).

References

Aalto J, Venäläinen A, Heikkinen RK, Luoto M (2014) Potential for extreme loss in high-latitude Earth surface processes due to climate change. Geophys Res Lett 41:3914–3924

Aalto J, Harrison S, Luoto M (2017) Statistical modelling predicts almost complete loss of major periglacial processes in Northern Europe by 2100. Nat Commun 8(1):1–8

Aarseth I, Fosen H (2004a) A Holocene lacustrine platform around Storavatnet, Osterøy, western Norway. Holocene 14:589–596

Aarseth I, Fosen H (2004b) Late Quaternary lacustrine cryoplanation of rock surfaces in and around Bergen, Norway. Norw J Geol 84:125–137

Aartolahti T (1972) Dyynien routahalkeamista ja routahalkeamapolygoneista (English summary: Frost cracks and frost polygons on dunes in Finland). Terra 84:124–131

Alestalo J (1980) Systems of ice movement on Lake Lappajärvi, Finland. Fennia 158:27–39

Alestalo J, Häikiö J (1975) Ice features and ice-thrust shore forms at Luodonselkä, Gulf of Bothnia, in winter 1972/73. Fennia 144:1–24

Åhman R (1976) The structure and morphology of minerogenic palsas in northern Norway. Biul Peryglac 26:25–31

Åhman R (1977) Palsar i Nordnorge. Meddelanden från Lunds Universitets Geografika Institution Avhandlingar 78:1–165

Åkerman HJ, Johansson M (2008) Thawing permafrost and thicker active layers in sub-arctic Sweden. Permafrost Periglac Process 19:279–292

Alestalo J, Häikiö J (1979) Forms created by thermal movement of lake ice in Finland in winter 1972–73. Fennia 157:51–92

Andersen BG (1968) Glcial geology of western Troms, north Norway. Nor Geol Unders 256:1–160

Andersen JL, Egholm DL, Knudsen MF, Linge H, Jansen JD, Pedersen VK, Nielsen SB, Tikhomirov D, Olsen J, Fabel D, Xu S (2018) Widespread erosion on high plateaus during recent glaciations in Scandinavia. Nat Commun 9:830. https://doi.org/10.1028/$41467-018-03280-2

Andersen JL, Egholm DL, Knudsen MF, Linge H, Jansen JD, Goodfellow BW, Pedersen VK, Tikhomirov D, Olsen J, Fredin O (2019) Pleistocene evolution of a Scandinavian plateau landscape. J Geophys Res Earth Surf 123. https://doi.org/10.1029/2018JF004670

Andersson JG (1906) Solifluction: a component of subaerial denudation. J Geol 14:91–112

André MF (2002) Rates of postglacial rock weathering on glacially scoured outcrops (Abisko-Riksgrånsen area, 68° N). Geogr Ann Ser (Phys Geogr) 64:139–150

André M-F (2003) Do periglacial landscapes evolve under periglacial conditions? Geomorphology 52:149–164

André, M-F (2009) From climate to global change geomorphology: contemporary shifts in periglacial geomorphology. In Knight J, Harrison S (eds) Periglacial and paraglacial processes and environments, vol 320. Geological Society, London, Special Publication, pp 5–28

Ballantyne CK (1987) Some observations on the morphology and sedimentology of two active protalus ramparts, Lyngen, northern Norway. Arct Alp Res 19:167–174

Ballantyne CK (1995) Paraglacial debris cone formation on recently deglaciated terrain. Holocene 5:25–33

Ballantyne CK (2002) Paraglacial geomorphology. Quat Sci Rev 21:1935–2017

Ballantyne CK (2010) A general model for autochthonous blockfield evolution. Permafrost Periglac Process 21:289–300

Ballantyne CK (2018) Periglacial geomorphology. Wiley-Blackwell, Chichester

Ballantyne CK, Benn DI (1994) Paraglacial slope adjustment and resedimentation following glacial retreat, Fåbergatølsbreen, Norway. Arct Alp Res 26:255–269

Ballantyne CK, Benn DI (1996) Paraglacial slope adjustment during recent deglaciation and its implications for slope evolution in formerly glaciated environments. In: Anderson MG, Brooks S (eds) Advances in hillslope processes, vol 2. Wiley, Chichester, pp 1173–1195

Ballantyne CK, Matthews JA (1982) The development of sorted circles on recently deglaciated terrain, Jotunheimen, Norway. Arct Alp Res 14:341–354

Ballantyne CK, Matthews JA (1983) Desiccation cracking and sorted polygon development, Jotunheimen, Norway. Arct Alp Res 15:339–349

Barsch D (1971) Rock glaciers and ice-cored moraines. Geogr Ann Ser (Phys Geogr) 53:203–206

Barsch D (1977) Nature and importance of mass wasting by rock glaciers in alpine permafrost environments. Earth Surf Proc Land 2:231–245

Barsch D, Treter U (1976) Zur Verbreitung von Periglazialphänomenen in Rondane/Norwegen. Geogr Ann Ser (Phys Geogr) 58:83–89

Beldring S, Engen-Skaugen T, Førland EJ, Roald LA (2008) Climate change impacts on hydrological processes in Norway based on two methods for transferring regional climate model results to meteorological station sites. Tellus (Dyn Meteorol Ocean) 60(3):439–450

Bellwald B, Hjelstuen BO, Sejrup HP, Stokowy T, Kuvås (2019) Holocene mass movments in west and mid-Norwegian fjords and lakes. Mari Geol 407:192–212

Berthling I, Etzelmüller B (2011) The concept of cryoconditioning in landscape evolution. Quat Res 75:378–384

Berthling I, Shomacker A, Benediktsson ÍÖ (2013) The glacial and periglacial research frontier: where from here? In: Giardino JR, Harbor JM (eds) Treatise on geomorphology, volume 8, glacial and periglacial geomorphology. Academic Press, San Diego, CA, pp 479–498

Beskow G (1935) Tjälbildningen och tjällyftningen med särskild hänsyn till vägar och järnvägar. Sveriges Geologiska Unersökning Avhandlingar och Uppsatser, Arsbok 26, Ser C 375:1–242
Beskow G (1947) Soil freezing and frost heaving with special applications to roads and railroads. Northwestern University Technological Institute: Evanston, IL. [English translation of Beskow (1935) by JO Osterberg.]
Beylich AA (2011) Mass transfers, sediment budgets and relief development in cold environments: results of long-term geomorphologic drainage basin stidies in Iceland, Swedish Lapland and Finnish Lapland. Zeitschrift Für Geomorphologie NF 55:145–174
Beylich AA, Laute K (2012) Spatial variations of surface water chemistry and chemical denudation in the Erdalen drainage basin, Nordfjord, western Norway. Geomorphology 167–168:77–90
Beylich AA, Laute K (2018) Morphoclimatic controls of contemporary chemical and mechanical denudation in a borel-oceanic drainage basin system in central Norway (Homla drainage basin, Trøndelag). Geogr Ann Ser (Phys Geogr) 100:116–139
Beylich AA, Laute K (2021) Fluvial processes and contemporary fluvial denudation in different mountain landscapes in western and central Norway. In: Beylich AA (ed) Landscapes and landforms of Norway. Springer, Berlin, pp 147–168
Beylich AA, Kolstrup E, Thysted T, Gintz D (2004) Water chemistry and its diversity in relation to local factors in the Latnajavagge drainage basin, arctic-oceanic Swedish Lapland. Geomorphology 58:125–143
Beylich AA, Molau U, Luthbom K, Gintz D (2005) Rates of chemical and mechanical fluvial denudation in an arctic-oceanic periglacial environment, Latnajavagge drainage basin, northernmost Swedish Lapland. Arct Antarct Alp Res 37:75–87
Blikra LH, Christiansen HH (2014) A field-based model of permafrost-controlled rockslide deformation in northern Norway. Geomorphology 208:34–49
Blikra LH, Nemec W (1993) Postglacial avalanche activity in western Norway: depositional facies sequences, chronostratigraphy and palaeoclimatic implications. In: Frenzel B, Matthews JA, Gläser B (eds) Solifluction and climatic variation in the Holocene. Gustav Fischer Verlag, Stuttgart, pp 143–162
Blikra LH, Nemec W (1998) Postglacial colluvium in western Norway: depositional processes, facies and palaeoenvironmental record. Sedimentology 45:909–959
Blikra LH, Nesje A (1997) Holocene avalanche activity in western Norway: chronostratigraphy and palaeoclimatic implications. In Matthews JA, Brunsden D, Frenzel B, Gläser B, Weiß (eds) Rapid mass movement as a source of climatic evidence for the Holocene. Gustav Fischer Verlag: Stuttgart, pp 299–312.
Blikra LH, Hole PA, Rye N (1994) Hurtige Massebevegelser og avsetningstyper i alpiner områder, Indre Nordfjord. Nor Geol Unders Skr 92:1–17
Blikra LH, Selvik SF (1998) Climatic signals recorded in snow avalanche-dominated colluvium in western Norway: depositional facies successions and pollen records. Holocene 8:631–658
Boch SG, Krasnov II (1943) O nagornykh terraskh i drevnikh poverkhnostyakh vyravnivaniya na Urale i svyazannykh s nimi problemakh. Vsesoyuznogo Geograficheskogo obshchestva Izvestiya 75:14–25. [Translated from Russian (1994) On altiplanation terraces and ancient surfaces of levelling in the Urals and associated problems. In Evans DJA (ed) Cold climate landforms. Wiley, Chichester, pp 177–204.]
Bøe AG, Dahl SO, Lie Ø, Nesje A (2006) Holocene river floods in the upper Glomma catchment, southern Norway: high resolution multiproxy record from lacustrine sediments. Holocene 16:445–455
Böhme M, Oppikofer T, Longva O, Jaboyedon M, Hermanns RL, Derron MH (2015) Analyses of past and present rock slope instabilities in a fjord valley: implications for hazard estimations. Geomorphology 248:464–474
Borge AF, Westermann S, Solheim I, Etzelmüller B (2017) Strong degradation of palsas and peat plateaus in northern Norway during the last 60 years. Cryosphere 11:1–16

Byrne M-L, Dionne J-C (2002) Typical aspects of cold regions shorelines. In Hewitt K, Byrne M-L, English M, Young G (eds) Landscapes of transition: landform assemblages and transformations in cold regions. Kluwer, Dordrecht, pp 141–158

Christiansen HH (1998) Nivation forms and processes in unconsolidated sediments, NE Greenland. Earth Surf Proc Land 23:751–760

Christiansen HH, Svensson H (1998) Windpolished boulders as indicators of a Late Weichselian wind regime in Denmark in relation to neighbouring areas. Permafrost Periglac Process 9:1–21

Christiansen HH, Etzelmüller B, Isakssen K, Juliussen H, Farbrot H, Humlum O, Johansson M, Ingeman-Nielsen T, Kristensen L, Hjort J, Holmlund P, Sannel ABK, Sidsgaard C, Åkerman HJ, Foged N, Blikra LH, Pernosky MA, Ødegård RS (2010) The thermal state of permafrost in the Nordic area during the international polar year 2007–2009. Permafrost Periglac Process 21:156–181

Church M, Ryder JM (1972) Paraglacial sedimentation: a consideration of fluvial processes conditioned by glaciation. Geol Soc Am Bull 83:3059–3072

Clark MJ, Seppälä M (1988) Slushflows in a subarctic environment, Kilpisjarvi, Finland. Arct Alp Res 20:97–105

Corner GD (1980) Avalanche impact landforms in Troms, North Norway. Geogr Ann Ser (Phys Geogr) 62:1–4

Corner GD (2005a) Scandes mountains. In: Seppälä M (ed) The physical geography of Fennoscandia. Wiley-Blackwell, Chichester, pp 229–254

Corner GD (2005b) Atlantic coasts and fjords. In: Seppälä M (ed) The physical geography of Fennoscandia. Wiley-Blackwell, Chichester, pp 203–228

Costa JE (1984) Physical geomorphology of debris flows. In: Costa JE, Fleisher PJ (eds) Developments and applications of geomorphology. Springer Verlag, Berlin, pp 268–317

Curry AM (1999) Paraglacial modification of slope form. Earth Surf Proc Land 24:1213–1228

Curry AM (2000) Observations on the distribution of paraglacial reworking of glacigenic drift in western Norway. Nor Geogr Tidsskr 54:139–147

Curry AM (2021) Paraglacial rock-slope failure following deglaciation in Western Norway. In Beylich AA (ed) Landscapes and landforms of Norway. Springer Nature: Cham, pp 97–130.

Curry AM, Ballantyne CK (1999) Paraglacial modification of glacigenic sediments. Geogr Ann Ser (Phys Geogr) 81:409–419

Darmody RG, Thorn CE (1997) Elevation, age, soil development, and chemical weathering at Storbreen, Jotunheimen. Geogr Ann Ser (Phys Geogr) 79:215–222

Darmody RG, Thorn CE, Harder RL, Schlyter JPL, Dixon JC (2000) Weathering implications of water chemistry in an arctic-alpine environment, northern Sweden. Geomorphology 34:89–100

Darmody RG, Thorn CE, Seppålå M, Campbell SW, Li YK, Harbor J (2008) Age and weathering status of granite tors in Arctic Finland (~68°N). Geomorphology 94:10–23

Dawson AG (1979) Polar and non-polar shore platform development. Department of Geography, Bedford College (University of London). Pap Geogr 6:1–28

Dawson AG, Matthews JA, Shakesby RA (1987) Rock platform erosion on periglacial shores: a modern analogue for Pleistocene rock platforms in Britain. In: Boardman J (ed) Periglacial processes and landforms in Britain and Ireland. Cambridge University Press, Cambridge, pp 173–182

De Haas T, Kleinhaus MG, Carbonneau PE, Rubensdottir L, Hauber E (2015) Morphology of fans in the high-arctic periglacial environment of Svalbard: controls and processes. Earth Sci Rev 146:163–182

Decaulne A, Eggertsson Ó, Laute K, Beylich AA (2014) A 100-year extreme snow-avalanche record based on tree-ring rsearch in upper Bødalen, inner Nordfjord, western Norway. Geomorphology 218:3–15

Demek J (1969) Cryoplanation terraces, their geographical distribution, genesis and development. Rozpravy Ĉeskoslovenské Akademie Věd, Rada Matematických a Prírodních Věd Rocnik 79(4):1–80

Eichel J, Draebling D, Klingbeil L, Wieland M, Wling C, Schmidtlein S, Kuhlmann H, Dikau R (2017) Solifluction meets vegetation: the role of biogeomorphic feedbacks for turf-banked solifluction lobe development. Earth Surf Proc Land 42:1623–1635

Elliott G, Worsley P (1999) The sedimentology, stratigraphy and ^{14}C dating of a turf-banked solifluction lobe: evidence for Holocene slope instability at Okstindan, northern Norway. J Quat Sci 14:175–188

Elliott G, Worsley P (2012) A solifluction lobe in Okstindan, north Norway and its paleoclimatic significance. In: Eberhardt E, Froese C, Turner K, Leroueil S (eds) Landslides and engineering slopes: protecting society through improved understanding. Taylor & Francis, London, pp 437–442

Ellis S (1979) Radiocarbon dating evidence for the initiation of solifluction ca. 5500 years B.P. at Okstindan, north Norway. Geogr Ann Ser (Phys Geogr) 61:29–33

Ellis S (1983) Stratigraphy and ^{14}C dating of two earth hummocks, Jotunheimen, south central Norway. Geogr Ann Ser (Phys Geogr) 65:279–287

Engeland K, Aano A, Steffensen I, Støren E, Paasche Ø (2020) New flood frequency estimates for the largest river in Norway based on the combination of short and long time series. Hydrol Earth Syst Sci 24:5595–5619

Enquist F (1916) Die Einfluss des Windes auf die Verteilung der Gletscher. Bull Geol Inst Univ Upps 14:1–108

Eriksen HØ, Rouyet L, Lauknes TR, Berthling I, Isaksen K, Hindberg H, Larsen Y, Corner GD (2018) Recent acceleration of a rock glacier complex, Ádjet, Norway, documented by 62 years of remote sensing observations. Geophys Res Lett 45(16):8314–8323

Etzelmüller B, Berthling I, Sollid JL (2003) Aspects and concepts on the geomorphological significance of Holocene permafrost in southern Norway. Geomorphology 52:87–104

Etzelmüller B, Hagen JO (2005) Glacier-permafrost interaction in Arctic and alpine mountain environments with examples from southern Norway and Svalbard. In Harris C, Murton JB (eds) Cryospheric systems: glaciers and permafrost, vol 242. Geological Society, London, Special Publication, pp 11–27

Etzelmüller B, Romstad B, Fjellanger J (2007) Automated regional classification of topography in Norway. Norw J Geol 87:167–180

Etzelmüller B, Guglielmin M, Hauck C, Hilbich C, Hoelzle M, Isaksen K, Noetzli J, Oliva M, Ramos M (2020) Twenty years of European mountain permafrost dynamics–the PACE legacy. Environ Res Lett 15: Article No. 104070.

Etzelmüller B, Czekirda J, Magnin F, Duvillard P-A, Malet E, Ravanel L, Aspaas A, Kristensen L, Skrede I, Majala D, Jacobs B, Leinauer J, Hauck C, Hilbich C, Böhme M, Hermanns R, Eriksen HØ, Krautblatter M, Westermann S (2021) Permafrost in monitored unstable rock slopes in Norway—new insights from rock wall temperature monitoring, geophysical surveying and numerical modelling. Earth Surf Dyn Discuss. https://doi.org/10.5194/esurf-2021-10

Fabel B, Stroeven AP, Harbor J, Kleman J, Elmore D, Fink D (2002) Landscape preservation under Fennoscandian ice sheets determined from *in situ* produced ^{10}Be and ^{26}Al. Earth Planet Sci Lett 201:397–406

Farbrot H, Hipp TF, Etzelmüller B, Isaksen K, Ødegård RS, Schuler TV, Humlum O (2011) Air and ground-temperature variations observed along elevation and continentality gradients in southern Norway. Permafrost Periglac Process 22:343–360

Fredin O (2002) Glacial incepton and Quaternary mountain glaciations in Fennoscandia. Quat Int 95–96:99–112

Fredin O, Viola G, Zwingmann H, Sørlie R, Brönner M, Lie J-E, Grandal EM, Müller A, Margreth A, Vogt C, Knies J (2017) The inheritance of a Mesozoic landscape in western Scandinavia. Nature Communications 8: Article No. 14879, pp 1–11.

French HM (2008) Periglacial processes and forms. In: Burt TP, Chorley RJ, Brunsden D, Cox NJ, Goudie AS (eds) The history of the study of landforms or the development of geomorphology, vol 4. The Geological Society, London, pp 622–676

French HM (2018) The periglacial environment. Wiley-Blackwell, Chichester

French HM, Karte J (1988) A periglacial overview. In: Clark MJ (ed) Advances in periglacial geomorphology. Wiley, Chichester, pp 463–473
French HM, Thorn CE (2006) The changing nature of periglacial geomorphology. Géomorphologie: Relief Process Environ 3:165–174
Fries TCE, Bergström E (1910) Några iakttagelser öfver palsar och deras förekomst i nordligaste Sverige. Geol Föreningen I Stock Förhandlingar 32:195–205
Fronzek S, Luoto M, Carter TR (2006) Potential effect of climate change on the distribution of palsa mires in subarctic Fennoscandia. Climate Res 32:1–12
Fronzek S, Carter TR, Räisänen J, Ruokolainen L, Luoto M (2010) Applying probabilistic projections of climate change with impact models: a case study for sub-arctic palsa mires in Fennoscandia. Clim Change 99:515–534
Gisnås K, Etzelmüller B, Lussana C, Hjort J, Sannel ABK, Isaksen K, Westermann S, Kuhry P, Christiansen H, Frampton A, Åkerman J (2017) Permafrost map for Norway, Sweden and Finland. Permafr Periglacial Process 28:359–378
Gjessing J (1967) Norway's palaeic surface. Nor Geogr Tidsskr 21:69–132
Goodfellow BW (2012) A granulometry and secondary mineral fingerprint of chemical weathering in periglacial landscapes and its application to blockfield origins. Quat Sci Rev 57:121–135
Goodfellow BW, Strøeven AP, Fabel D, Fredin O, Derron M-H, Bintanja R, Caffee MW (2014) Arctic-alpine blockfields in the northern Swedish Scandes: late Quaternary—not Neogene. Earth Surf Dyn 2:383–401
Grab S (2005) Aspects of the geomorphology, genesis and environmental significance of earth hummocks (thúfur, pounus): miniature cryogenic mounds. Prog Phys Geogr 29:139–155
Guilcher A, Bodéré J-C, Coudé A, Hansom JD, Moign A, Peulvast J-P (1986) Le problème des strandflats en cinq pays de hautes latitudes. Rev Géol Dynam Géog Phys 27:47–79
Gurney SD (2001) Aspects of the genesis, geomorphology and terminology of palsas: perennial cryogenic mounds. Prog Phys Geogr 25:249–260
Haeberli W (1985) Creep of mountain permafrost: internal structure and flow of alpine rock glaciers. Mitteilungen der Versuchsanstalt für Wasserbau, Hydrologie und Glaziologie 77:1–142
Hall K (1998) Nivation or cryoplanation: different terms, same features? Polar Geogr 22:1–16
Hansom JD, Forbes DL, Etienne S (2014) The rock coasts of polar and sub-polar regions. In Kennedy DM, Stephenson JJ, Naylor LA (eds) Rock coast geomorphology: a global synthesis, vol 40. Geological Society of London, Memoirs, pp 263–281
Harris C (1982) The distribution and altitudinal zonation of periglacial landforms, Okstindan, Norway. Zeitschrift Für Geomorphologie NF 26:283–304
Harris C (1986) Some observations concerning the morphology and sedimentology of a protalus rampart, Okstindan, Norway. Earth Surf Proc Land 11:673–676
Harris C, Matthews JA (1984) Some observations on boulder-cored frost boils. Geogr J 150:63–73
Harris C, Murton JB (2005) Interactions between glaciers and permafrost: an introduction. In Harris C, Murton JB (eds) Cryospheric systems: glaciers and permafrost, vol 242. Geological Society, London, Special Publication, pp 1–9
Harris C, Kern-Luetschg M, Smith F, Isaksen K (2008) Solifluction processes in an area of seasonal ground freezing, Dovrefjell, Norway. Permafrost Periglac Process 19:31–47
Harris C, Arenson LU, Christiansen HH, Etzelmüller B, Frauenfelder R, Gruber S, Haeberli W, Hauck C, Hoelzle M, Humlum O, Isaksen K, Kääb A, Kern-Luetschg MA, Lehning M, Matsuoka N, Murton JB, Noezli J, Phillips M, Ross N, Seppälä M, Springman SM, Vonder Mühll DV (2009) Permafrost and climate in Europe: monitoring and modelling thermal, geomorphological and geotechnical responses. Earth-Sci Rev 92:117–171
Harris SA, Brouchkov A, Guodong C (2018) Geocryology: characteristics and use of frozen ground and permafrost landforms. CRC Press-Balkema, Leiden
Hättestrand C, Stroeven AP (2002) A relict landscape in the centre of the Fennoscandian glaciation: geomorphological evidence of minimal Quaternary glacial erosion. Geomorphology 44:127–143
Haugland JE (2004) Formation of patterned ground and fine-scale soil development within two late Holocene glacial chronosequences: Jotunheimen, Norway. Geomorphology 61:287–301

Haugland JE (2006) Short-term periglacial processes, vegetation succession, and soil development within sorted patterned ground: Jotunheimen, Norway. Arct, Antarct AlpE Res 38:82–89
Haugland JE, Beaty SW (2005) Vegetation establishment, succession and microsite frost disturbance on glacier forelands within patterned ground chronosequences. J Biogeogr 32:145–153
Haugland JE, Owen BS (2005) Temporal and spatial variability of soil pH in patterned ground chronosequences: Jotunheimen, Norway. Phys Geogr 26:299–312
Haugland JE, Owen-Haugland BS (2008) Cryogenic disturbance and pedogenic lag effects as determined by profile developmental index: the Styggedalsbreen glacier chronosequence, Norway. Geomorphology 96:212–220
Heckmann T, Morche D (eds) (2019) Geomorphology of proglacial systems: landforms and sediment dynamics in recently deglaciated alpine landscapes. Springer Nature, Cham.
Heikkinen O (2005) Boreal forest and northern and upper timberlines. In: Seppålå M (ed) The physical geography of Fennoscandia. Wiley-Blackwell, Chichester, pp 185–200
Heisser M, Scheidl C, Eisl J, Spangl B, Hübl J (2015) Process type identification in torrential catchments in the eastern Swiss Alps. Geomorphology 232:239–247
Henderson IHC, Saintot (2011) Regional spatial variations in rockslide distribution from structural geology ranking: an example from Storfjorden, western Norway. In Jaboyedoff M (ed) Slope tectonics, vol 351. Geology Society of London, Special Publications, pp 59–70
Hermanns RL, Schleier M, Böhme M, Blikra LH, Gosse K, Ivy-Ochs S, Hilger P (2017) Rock-avalanche activity in W and S Norway peaks after the retreat of the Scandinavian ice sheet. In: Mikos M, Vilímek V, Yin Y, Sassa K (eds) Advancing culture of living with landslides, volume 5: Landslides in different environments. Springer, Heidelberg, pp 331–338
Hilger P, Hermanns RL, Gosse JC, Jacobs B, Etzelmüller B, Krautblatter M (2018) Multiple rock-slope failures from Mannen in Romsdal Valley, western Norway, revealed from Quaternary geological mapping and ^{10}Be exposure dating. Holocene 28:1841–1854
Hilger P, Hermanns RL, Czekirda J, Myhra KS, Gosse JC, Etzelmüller B (2021) Permafrost as a first order control on long-term rock-slope deformation in (Sub-) Arctic Norway. Quat Sci Rev 251:1–21. Article No. 106718
Hipp T, Etzelmüller B, Westermann S (2014) Permafrost in alpine rock faces from Jotunheimen and Hurrungane, southern Norway. Permafrost Periglac Process 25:1–13
Hjort J (2006) Environmental factors affecting the occurrence of periglacial landforms in Finnish Lapland: a numerical approach. Shaker Verlag, Aachen
Hjort J (2014) Which environmental factors determine recent cryoturbation and solifluction activity in a subarctic landscape? A comparison between active and inactive features. Permafrost Periglac Process 25:136–143
Högbom B (1914) Über die geologische Bedeutung des Frostes. Bull Geol Inst Univ Upps 12:251–389
Högbom I (1923) Ancient inland dunes of northern and middle Europe. Geogr Ann 5:113–242
Högbom B (1927) Beobachtungen aus Nord-Schweden über den Frost als geologischer Factor. Bull Geol Inst Univ Upps 20:1–38
Hole PA (1981) Groper danna av snøskred I Sunnylven og tilgrensande områder på Sunnmøre. Førbels resultat. Norsk Geografiske Tidsskrift 35:167–172
Holtedahl O (1960) Features of the geomorphology. In Holtedahl O (ed) Geology of Norway. Norges Geologiske Undersøkelse, Trondheim, pp 507–531
Holtedahl O (1998) The Norwegian strandflat—a geomorphological puzzle. Nor Geol Tidsskr 78:47–66
Hughes AL, Gyllencreutz R, Lohne ØS, Mangerud J, Svendsen JI (2016) The last Eurasian ice sheets—a chronological database and time-slice reconstruction, DATED-1. Boreas 45:1–45
Humlum O (2008) Alpine and polar periglacial processes: the current state of knowledge. In Kane DL, Hinkel KM (eds) Ninth International Conference on Permafrost, Fairbanks, Alaska, June 29–July 3 2008, pp 753–759
Isaksen K, Holmlund P, Sollid JL, Harris C (2001) Three deep alpine permafrost boreholes in Svalbard and Scandinavia. Permafrost Periglac Process 12:13–26

Isaksen K, Hauck C, Gudevang E, Ødegård RS, Sollid JL (2002) Mountain permafrost distribution in Dovrefjell and Jotunheimen, southern Norway, based on BTS and DC resistivity tomography data. Nor Geogr Tidsskr 56:122–136

Jonasson C (1991) Holocene slope processes of periglacial mountain areas in Scandinavia and Poland. Uppsala Universitet Naturgeografiska Institutionen Report 79:1–156

Jonasson C, Nyberg R, Rapp A (1997) Dating of rapid mass movements in Scandinavia: talus rockfalls, debris flows and slush avalanches. In Matthews JA, Brunsden D, Frenzel B, Gläser B, Weiß (eds) Rapid mass movement as a source of climatic evidence for the Holocene. Gustav Fischer Verlag: Stuttgart, pp 267–282

Juliussen H, Humlum O (2007) Preservation of blockfields beneath Pleistocene ice sheets on Sølen and Elgåhogna, central eastern Norway. Z Geomorphol Suppl 51:113–138

Kääb A (2013) Rock glaciers and protalus forms. In: Elias SA (ed) Encyclopedia of Quaternary Science, vol 3, 2nd edn. Elsevier, Amterdam, pp 535–541

Karjalainen O, Aalto J, Luoto M, Westermann S, Romanovsky VE, Nelson FE, Etzelmüller B, Hjort J (2019) Circunpolar permafrost maps and geohazard indices for near-future infrastructure risk assessment. Sci Data 6(1):1–6

Karjalainen O, Luoto M, Aalto J, Etzelmüller B, Grosse G, Jones BM, Lilleøren KS, Hjort J (2020) High potential for loss of permafrost landforms in a changing climate. Environ Res Lett 15(10):104065

Karlén W (1988) Scandinavian glacier and climatic fluctuations during the Holocene. Quat Sci Rev 7:199–208

Käyhkö JA, Worsley P, Pye K, Clarke ML (1999) A revised chronology for aeolian activity in subarctic Fennoscandia during the Holocene. Holocene 9:195–205

Kejonen A (1979) Vuotomaista Muotkatunterien alueella Pohjois-Lapissa. Publ Dep Quat Geol Univ Turku 40:1–43

Kergillec R (2015) Characteristics and altitudinal distribution of periglacial decay phenomena in the massif of Rondane, central Norway. Geogr Ann Ser (Phys Geogr) 97:299–315

King L (1984) Permafrost in Skandinavien Untersuchungsergebnisse aus Lappland, Jotunheimen und Dovre/Rondane. Heidelberger Geographische Arbeiten 76:1–174

King L (1986) Zonation and ecology of high mountain permafrost in Scandinavia. Geogr Ann Ser (Phys Geogr) 68:131–139

Kleman J (1992) The palimpsest glacial landscape in northwestern Sweden—late Weichselian deglaciation landforms and traces of older west-centred ice sheets. Geogr Ann Ser (Phys Geogr) 74:305–325

Kleman J (1994) Preservation of landforms under ice sheets and ice caps. Geomorphology 9:19–32

Kleman J, Hättestrand C (1999) Frozen-bed Fennoscandian and Laurentian ice sheets during the Last Glacial Maximum. Nature 402:63–66

Kleman J, Stroeven AP, Lundqvist J (2008) Patterns of Quaternary ice sheet erosion and deposition in Fennoscandia and a theoretical framework for explanation. Geomorphology 97:73–90

Klemsdal T (1969) Eolian forms in parts of Norway. Nor Geogr Tidsskr 23:49–66

Klemsdal T (1982) Coastal classification and the coast of Norway. Nor Geogr Tidsskr 36:129–152

Kolstrup E (2004) Stratigraphic and environmental implications of a large ice-wedge cast at Tjæreborg, Denmark. Permafrost Periglac Process 15:31–40

Kotilainen M (2004) Dune stratigraphy as an indicator of Holocene climatic change and human impact in northern Lapland, Finland. Ann Acad Sci Fenn: Geol-Geogr 166:1–156

Kristensen L, Czekirda J, Penna I, Etzelmüller B, Nicolet P, Pullarello JS, Blikra LH, Skrede I, Oldani S, Abellan A (2021) Movements, failure and climatic control of the Veslemannen rockslide, Western Norway. Landslides. https://doi.org/10.1007/s10346-020-01609-x

Lauriol B, Lamirande I, Lalonde AE (2006) The Giant Steps of Bug Creek, Richardson Mountains, N.W.T., Canada. Permafr Periglac Process 17:267–275

Lautridou JP, Ozouf JC (1982) Experimental frost shattering: 15 years of research in the Centre de Géomorphologie du CNRS. Prog Phys Geogr 6:215–232

Larsen E, Holtedahl H (1985) The Norwegian strandflat: a reconsideration of its age and origin. Nor Geol Tidsskr 65:247–254

Lewis SG, Birnie JF (2001) Little Ice Age alluvial fan development in Langedalen, western Norway. Geogr Ann Ser (Phys Geogr) 83:179–190

Lidmar-Bergström K, Ollier CD, Sulebak JR (2000) Landforms and uplift history of southern Norway. Global Planet Change 24:211–231

Liestøl O (1974) Avalanche plunge-pool effect. Norsk Polarinstitutt Arbok 1972:179–181

Lilleøren KS, Etzelmüller B (2011) A regional inventory of rock glaciers and ice-cored moraines in Norway. Geogr Ann Ser (Phys Geogr) 93:175–191

Lilleoren KS, Etzelmuller B, Rouyet L, Eiken T, Hilbich C (2022) Transitional rock glaciers at sea-level in Northern Norway. Earth Surface Dynamics 6

Lilleøren KS, Etzelmüller B, Schuler TV, Gisnås K, Humlum O (2012) The relative age of mountain permafrost—estimation of Holocene permafrost limits in Norway. Global Planet Change 92–93:209–223

Lind L, Nilsson C, Polvi LE, Weber C (2014) The role of ice dynamics in shaping vegetation in flowing waters. Biol Rev 89:791–804

Linge H, Nesje A, Matthews JA, Fabel D, Xu S (2020) Evidence for rapid paraglacial formation of rock glaciers in southern Norway from ^{10}Be surface-exposure dating. Quat Res 97:55–70

Liston GE, Elder K (2006) A distributed snow-evolution modelling system. J Hydrometeorol 7:1259–1276

Longva O, Blikra LH, Dehls JF (2009) Rock avalanches—distribution and frequencies in the inner part of Storfjorden, Møre og Romsdal County, Norway. NGU Rapport 2009.002. Geological Survey of Norway: Trondheim

Lundqvist J (1964) Patterned ground and related forest phenomena in Sweden. Sveriges Geologiska Unersökning Avhandlingar och Uppsatser Ser. C 583, pp 1–101

Luoto M, Seppälä M (2002a) Characteristics of earth hummocks (pounus) with and without permafrost in Finnish Lapland. Geogr Ann Ser (Phys Geogr) 84:127–136

Luoto M, Seppälä M (2002b) Modelling the distribution of palsas in Finnish Lapland with logistic regression and GIS. Permafrost Periglac Process 13:17–28

Luoto M, Seppålå M (2003) Thermokarst ponds as indicators of the former distribution of palsas in Finnish Lapland. Permafrost Periglac Process 14:19–27

Magnin F, Etzelmüller B, Westermann S, Isaksen K, Hilger P, Hermanns RL (2019) Permafrost distribution in steep rock slopes in Norway: measurements, statistical modelling and implications for geomorphic processes. Earth Surf Dyn 7:1019–1040

Mangerud J, Aarseth I, Hughes ALC, Lohne ØS, Skår K, Sønstegaard E, Svendsen JI (2016) A major re-growth of the Scandinavian Ice Sheet in western Norway during the Allerød-Younger Dryas. Quat Sci Rev 132:175–205

Margold M, Treml V, Petr L, Nyplová P (2011) Snowpatch hollows and pronival ramparts in the Krkonose Mountains, Czech Republic: distribution, morphology and chronology of formation. Geogr Ann Ser (Phys Geogr) 93:137–150

Marr P, Löffler G (2018) Establishing a multi-proxy approach to alpine blockfield evolution in south-central Norway. Acta Univ Carol Geogr 52:219–236

Marr P, Winkler S, Löffler G (2018) Investigations on blockfields and related landforms at Blåhø (southern Norway) using Schmidt-hammer exposure-age dating: palaeoclimatic and morphodynamic implications. Geogr Ann Ser (Phys Geogr) 100:285–306

Marr P, Winkler S, Löffler G (2019) Schmidt-hammer exposure-age dating (SHD) performed on periglacial and related landforms in Oppendskedalen, Geirangerfjellet, Norway: implications for mid- and late-Holocene climate variability. Holocene 29:97–109

Matsuoka N (2001) Microgelivation versus macrogelivation: towards bridging the gap between laboratory and field frost weathering. Permafrost Periglac Process 12:299–313

Matthews JA (1985) Radiocarbon dating of surface and buried soils: principles, problems and prospects. In Richards KS, Arnett RR, Ellis S (eds) Geomorphology and soils. George Allen & Unwin, London, pp 269–288

Matthews JA (1992) The ecology of recently-deglaciated terrain: a geoecological approach to glacier forelands and primary succession. Cambridge University Press, Cambridge

Matthews JA (1993) Radiocarbon dating of buried soils with particular reference to Holocene solifluction. In: Frenzel B, Matthews JA, Gläser B (eds) Solifluction and climatic variation in the Holocene. Gustav Fischer Verlag, Stuttgart, pp 309–324

Matthews JA (1999) Disturbance regimes and ecosystem response on recently-deglaciated substrates. In: Walker R (ed) Ecosystems of disturbed ground. Elsevier, Amsterdam, pp 17–37

Matthews JA, Karlén W (1992) Asynchronous neoglaciation and Holocene climatic change reconstructed from Norwegian glaciolacustrine sedimentary sequences. Geology 20:991–994

Matthews JA, McCarroll D (1994) Snow-avalanche impact landforms in Breheimen, southern Norway: origin, age and paleoclimatic implications. Arct Alp Res 26:103–115

Matthews JA, Owen G (2008) Endolithic lichens, rapid biological weathering and Schmidt hammer R-values on recently exposed rock surfaces: Storbreen glacier foreland, Jotunheimen, Norway. Geogr Ann Ser (Phys Geogr) 90:187–297

Matthews JA, Owen G (2011) Holocene chemical weathering, surface lowering and rock weakening rates from glacially-eroded bedrock surfaces in an alpine periglacial environment, Jotunheimen, southern Norway. Permafrost Periglac Process 22:279–290

Matthews JA, Owen G (2021) The snow-avalanche impact landforms of Vestlandet, southern Norway. In: Beylich AA (ed) Landscapes and landforms of Norway. Springer, Berlin, pp 131–145

Matthews JA, Seppälä, M (2014) Holocene environmental change in subarctic aeolian dunefields: the chronology of sand dune re-activation events in relation to forest fires, palaeosol development and climatic variations in Finnish Lapland. Holocene 24:149–164

Matthews JA, Seppälä M (2015) Holocene colluvial chronology in a sub-arctic esker landscape at Kuttanen, Finnish Lapland: kettle holes as geo-ecological archives of interactions amongst fire, vegetation, soil, climate and geomorphological instability. Boreas 44:343–367

Matthews JA, Vater AE (2015) Pioneer zone geo-ecological change: observations from a chronosequence on the Storbreen glacier foreland, Jotunheimen, southern Norway. Catena 135:219–230

Matthews JA, Wilson P (2015) Improved Schmidt-hammer exposure ages for active and relict pronival ramparts in southern Norway and their palaeoenvironmental implications. Geomorphology 246:7–21

Matthews JA, Harris C, Ballantyne CK (1986a) Studies on a gelifluction lobe, Jotunheimen, Norway: ^{14}C chronology, stratigraphy, sedimentology and palaeoenvironment. Geogr Ann Ser (Phys Geogr) 86:345–360

Matthews JA, Dawson AG, Shakesby, RA (1986b) Lake shoreline development, frost weathering and rock platform erosion in an alpine periglacial environment. Boreas 15: 33-50.

Matthews JA, Dahl SO, Berrisford MS, Dresser PQ, Dumayne-Peaty L (1997a) A preliminary history of Holocene colluvial (debris-flow) activity, Leirdalen, Jotunheimen, Norway. J Quat Sci 12:117–129

Matthews JA, Dahl SO, Berrisford MS, Nesje A (1997b) Cyclic development and thermokarstic degradation of palsas in the mid-alpine zone at Leirpullan, Dovrefjell, southern Norway. Permafr Periglac Process 8:107–122

Matthews JA, Shakesby RA, Berrisford MS, McEwen LJ (1998) Periglacial patterned ground on the Styggedalsbreen glacier foreland, Jotunheimen, southern Norway: micro-topographic, paraglacial and geoecological controls. Permafrost Periglac Process 9:147–166

Matthews JA, Shakesby RA, McEwen LJ, Berrisford MS, Owen G, Bevan P (1999) Alpine debris flows in Leirdalen, Jotunheimen, Norway, with particular reference to distal fans, intermediate-type deposits and flow types. Arct Antarct Alp Res 31:421–435

Matthews JA, Seppälä M, Dresser PQ (2005) Holocene solifluction, climate variation and fire in a subarctic landscape at Pippokangas, Finnish Lapland, based on radiocarbon-dated buried charcoal. J Quat Sci 20:533–548

Matthews JA, Dahl SO, Dresser PQ, Berrisford MS, Lie Ø, Nesje A, Owen G (2009) Radiocarbon chronology of Holocene colluvial (debris-flow) events at Sletthamn, Jotunheimen, southern

Norway: a window on the changing frequency of extreme climatic events and their landscape impact. Holocene 19:1107–1129

Matthews JA, Shakesby RA, Owen G, Vater AE (2011) Pronival rampart formation in relation to snow-avalanche activity and Schmidt-hammer exposure-age dating (SHD): three case studies from southern Norway. Geomorphology 130:280–288

Matthews JA, Nesje A, Linge H (2013) Relict talus-foot rock glaciers at Øyberget, upper Ottadalen, southern Norway. Permafrost Periglac Process 24:336–346

Matthews JA, Winkler S, Wilson P (2014) Age and origin of ice-cored moraines in Jotunheimen and Breheimen, southern Norway: insights from Schmidt-hammer exposure-age dating. Geogr Ann Ser (Phys Geogr) 96:531–548

Matthews JA, McEwen LJ, Owen G (2015) Schmidt-hammer exposure-age dating (SHD) of snow-avalanche impact ramparts in southern Norway: approaches, results and implications for landform age, dynamics and developmemt. Earth Surf Landfs Process 40:1705–1718

Matthews JA, Wilson P, Mourne RW (2017a) Landform transitions from pronival ramparts to moraines and rock glaciers: a case study from Smørbotn cirque, Romsdalsalpane, southern Norway. Geogr Ann Ser (Phys Geogr) 99:15–37

Matthews JA, Owen G, McEwen LJ, Shakesby RA, Hill JL, Vater AE, Ratcliffe AC (2017b) Snow-avalanche impact craters in southern Norway: their morphology and dynamics compared with small terrestrial meteorite craters. Geomorphology 296:11–30

Matthews JA, Winkler S, Wilson P, Tomkins MD, Dortch JM, Mourne RW, Hill JL, Owen G, Vater A (2018) Small rock-slope failures conditioned by Holocene permafrost degradation: a new approach and conceptual model based on Schmidt-hammer exposure-age dating, Jotunheimen, southern Norway. Boreas 47:1144–1169

Matthews JA, Wilson P, Winkler S, Mourne RW, Hill JL, Owen G, Hiemstra J, Hallang H, Geary AP (2019) Age and development of active cryoplanation terraces in the alpine permafrost zone at Svartkampan, Jotunheimen, southern Norway. Quat Res 92:641–664

Matthews JA, Haselberger S, Hill JL, Owen G, Winkler S, Hiemstra JF, Hallang H (2020a) Snow-avalanche boulder fans in Jotunheimen, southern Norway: Schmidt-hammer exposure-age dating, geomorphometrics, dynamics and evolution. Geogr Ann Ser (Phys Geogr) 102:118–140

Matthews JA, McEwen LJ, Owen G, Los SO (2020b) Holocene alluvial fan evolution, Schmidt-hammer exposure-age dating and paraglacial debris floods in the SE Jostedalsbreen region, southern Norway. Boreas 49:886–904

Mauri A, Davis BAS, Collins PM, Kaplan JO (2015) The climate of Europe during the Holocene: a gridded pollen-based reconstruction and its multi-proxy evaluation. Quat Sci Rev 112:109–127

McCarroll D (1990) Differential weathering of feldspar and pyroxene in an arctic-alpine environment. Earth Surf Proc Land 15:641–651

McCarroll D, Viles HA (1995) Rock weathering by the lichen Lecidea auriculata in an arctic-alpine environment. Earth Surf Proc Land 20:199–206

McCarroll D, Shakesby RA, Matthews JA (1998) Spatial and temporal patterns of late-Holocene rockfall activity on a Norwegian talus slope: a lichenometric and simulation modelling approach. Arct Alp Res 30:51–60

McCarroll D, Shakesby RA, Matthews JA (2001) Enhanced rockfall activity during the Little Ice Age: further lichenometric evidence from Norwegian talus. Permafrost Periglac Process 12:157–164

McEwen LJ, Matthews JA (1998) Channel form, bed material and sediment sources of the Sprongdøla, southern Norway: evidence for a distinct periglacio-fluvial system. Geogr Ann Ser (Phys Geogr) 80:17–36

McEwen LJ, Matthews JA, Shakesby RA, Berrisford MS (2002) Holocene gorge excavation linked to boulder fan formation and frost weathering in a Norwegian alpine periglaciofluvial system. Arct Antarct Alp Res 34:345–357

McEwen LJ, Owen G, Matthews JA, Hiemstra JF (2011) Late-Holocene development of a Norwegian alluvial fan affected by proximal glacier variations, distal undercutting and colluvial activity. Geomorphology 127:198–215

McEwen LJ, Matthews JA, Owen G (2020) Development of a Holocene glacier-fed composite alluvial fan based on surface exposure-age dating techniques: the Illåe fan, Jotunheimen, Norway. Geomorphology 363: Article No. 107200 (15 p).

Mercier D (2008) Paraglacial and paraperiglacial land systems: concepts, temporal scales and spatial distribution. Geomorphol: Relief Process Environ 14:223–233

Miesen F, Dahl SO, Schrott L (2021) Evidence of glacier-permafrost interactions associated with hydro-geomorphologicasl processes and landforms at Snøhetta, Dovrefjell, Norway. Geogr Ann Ser (Phys Geogr) 103:273–302

Milthers V (1907) Sandslebne stens form og danelse. Meddelelser fra Dansk Geologiske Forening 13:3–60.

Moen A (1999) National Atlas of Norway: Vegetation. Norwegian Mapping Authority, Hønefoss.

Möller JJ, Sollid JL (1972) Deglaciation chronology of Lofoten-Vesterålen-Ofoten, North Norway. Norske Geografisk Tidsskrift 26:101–133

Murton JB (2021) What and where are periglacial landscapes. Permafr Periglac Process 32:186–212

Murton JB, Coutard J-P, Lautridou J-P, Ozouf J-C, Robinson DA, Williams RBG (2001) Physical modelling of bedrock brecciation by ice segregation in permafrost. Permafrost Periglac Process 12:255–266

Murton JB, Peterson R, Ozouf J-C (2005) Bedrock fracture by ice segregation in cold regions. Science 314:1127–1129

Nansen F (1922) The strandflat and isostasy. Videnskapselskapet I Christiania Skrifter, Matematisk-Naturvitenskapelig klasse 2:1–313

Nesje A (1993) Neoglacial gelifluction in the Jostedalsbre region, western Norway: evidence from dated buried soils. In: Frenzel B, Matthews JA, Gläser B (eds) Solifluction and climatic variation in the Holocene. Gustav Fischer Verlag, Stuttgart, pp 37–47

Nesje A, Kvamme M (1991) Holocene glacier and climatic variations in western Norway: evidence for early-Holocene demise and multiple Neoglacial events. Geology 19:610–612

Nesje A, Willans IM (1994) Erosion of Sognefjord, Norway. Geomorphology 9:33–45

Nesje A, Dahl SO, Anda E, Rye N (1988) Blockfields in southern Norway. Significance for the Late Weichselian ice sheets. Nor Geol Tidsskr 68:149–169

Nesje A, Kvamme M, Rye N (1989) Neoglacial gelifluction in the Jostedalsbreen region, western Norway: evidence from dated buried palaeopodsols. Earth Surf Proc Land 14:259–270

Nesje A, Kvamme M, Rye N, Løvlie R (1991) Holocene glacial and climatic history of the Jostedalsbreen region, western Norway: evidence from lake sediments and terrestrial deposits. Quat Sci Rev 10:87–114

Nesje A, Bakke J, Dahl SO, Lie Ø, Bøe AG (2007) A continuous, high-resolution 8500-yr snow-avalanche record from western Norway. Holocene 17:269–277

Nesje A, Matthews JA, Linge H, Bredal M, Wilson P, Winkler S (2021) New evidence for active talus-foot rock glaciers at Øyberget, southern Norway, and their development during the Holocene. Holocene 31 (in press)

Nicholson DT (2008) Rock control on microweathering of bedrock surfaces in a periglacial environment. Geomorphology 101:655–665

Nicholson DT (2009) Holocene microweathering rates and processes on ice-eroded bedrock, Røldal area, Hardangervidda, southern Norway. In Knight J, Harrison S (eds) Periglacial and paraglacial processes and environments, vol 320. Geological Society, London, Special Publication, pp 29–49

Nielsen PR, Dahl SO, Jansen HL, Støren EWN (2016a) Holocene aeolian sedimentation and episodic mass-wasting events recorded in lacustrine sediments on Langøya in Vesterålen, northern Norway. Quat Sci Rev 148:146–162

Nielsen PR, Dahl SO, Jansen HL (2016b) Mid- to late-Holocene aeolian activity recorded in a coastal dunefield and lacustrine sediments on Andøya, northern Norway. Holocene 26:1486–1501

Niessen A, van Horssen P, Koster EA (1992) Altitudinal zonation of selected geomorphological phenomena in an alpine perglacial area (Abisko, northern Sweden). Geogr Ann Ser (Phys Geogr) 74:1835–2196

Nihlén T (2003) Palsas at Härjedalen, Sweden: 1910 and 1998 compared. Geogr Ann Ser (Phys Geogr) 82:39–44

Nyberg R (1985) Debris flows and slush avalanches in northern Swedish Lappland: distribution and geomorphological significance. Meddelanden Frå Lunds Universitets Geografiska Institution Avhandlingar 97:1–144

Nyberg R (1989) Observations of slushflows and their geomorphological effects in the Swedish mountain area. Geogr Ann Ser (Phys Geogr) 71:185–198

Nyberg R (1991) Geomorphic processes at snowpatch sites in the Abisko Mountains, northern Sweden. Zeitschrift für Geomorphologie N.F. 35: 321–343

Nyland K, Nelson FE, Figueiredo PM (2020) Cosmogenic ^{10}Be and ^{36}Cl geochronology of cryoplanation terraces in the Alaskan Yukon-Tanana upland. Quat Res 97:157–166

Ødegård RS, Sollid JL, Liestøl O (1987) Juvflya – Kvartærgeologi og geomorfologi M 1:10.000. Geografisk Institutt, Universitetet I Oslo, Oslo

Ødegård RS, Sollid JL, Liestøl O (1988) Periglacial forms related to terrain parameters in Jotunheimen, southern Norway. In: Senneset K (ed) 5th International Conference on Permafrost, Proceedings, vol 3. Tapir, Trondheim, pp 59–61

Ødegård RS, Sollid JL, Liestøl O (1992) Ground temperature measurements in mountain permafrost, Jotunheimen, southern Norway. Permafrost Periglac Process 3:231–234

Ødegård RS, Sollid JL (1993) Coastal cliff temperatures related to the potential for cryogenic weathering processes, western Spitsbergen, Svalbard. Polar Res 12:95–106

Ojala AEK, Mattila J, Markovaara-Koivisto M, Ruskeeniemi T, Palmu J-P, Sutinen R (2019) Distribution and morphology of landslides in northern Finland: an analysis of postglacial seismic activity. Geomorphology 326:190–121

Oksanen PO (2005) Development of palsa mires on the northern European continent in relation to Holocene climatic and environmental changes (Academic dissertation). Oulu University Press, Oulu, pp 1–50

Oksanen PO (2006) Holocene development of the Vaisjeäggi palsa mire, Finnish Lapland. Boreas 35:81–95

Olesen O, Bering D, Brönner M, Dalsegg E, Fabian K, Fredin O, Gellein J, Husteli B, Magnus C, Rønning JS, Solbakk T, Tønnesen JF, Øverland JA (2012) Tropical weathering in Norway. Norwegian Geological Survey Report 2012.005, 188 p

Olesen O, Kierulf HP, Brönner M, Dalsegg E, Fredin O, Solbakk T (2013) Deep weathering, neotectonics and strandflat formation in Nordland, northern Norway. Norw J Geol 93:189–213

Olvmo M, Holmer B, Thorsson S, Reese H, Lindberg F (2020) Sub-arctic palsa degradation and the role of climatic drivers in the largest coherent palsa mire complex in Sweden (Vissátvuopmi), 1955–2016. Scientific Reports Article No. 10: 8937 (10 p). https://doi.org/10.1038/s41598-020-65719-1

Oppikofer T, Saintot A, Hermanns RL, Böhme M, Scheiber T, Gosse J, Dreiås GM (2017) From incipient slope instability through slope deformation to catastrophic failure—different stages of failure development on the Ivasnasen and Vollan rock slopes (western Norway). Geomorphology 289:96–116

Østrem G (1959) Ice melting under a thin layer of moraine, and the existence of ice cores in moraine ridges. Geogr Ann 41:228–230

Østrem G (1964) Ice-cored moraines in Scandinavia. Geogr Ann 46:282–337

Østrem G (1965) Problems of dating ice-cored moraines. Geogr Ann 47:1–38

Østrem G (1971) Rock glaciers and ice-cored moraines: a reply to D. Barsch. Geogr Ann Ser (Phys Geogr) 53:207–213

Ouellet M-A, Germain D (2014) Hyperconcentrated flows on a forested alluvial fan of eastern Canada: geomorphic characteristics, return period and triggering scenarios. Earth Surf Proc Land 39:1876–1887

Owen G, Matthews JA, Shakesby RA, He X (2006a) Snow-avalanche impact landforms, deposits and effects at Urdvatnet, southern Norway: implications for avalanche style and process. Geogr Ann Ser (Phys Geogr) 88:295–307

Owen G, Matthews JA, Shakesby RA (2006b) Holocene chemical weathering on a calcitic lake shoreline in an alpine periglacial environment: Attgløyma, Sognefjell, southern Norway. Permafr Periglac Process 17:3–12

Owen G, Matthews JA, Albert PG (2007) Rates of Holocene chemical weathering, 'Little Ice Age' glacial erosion, and implications for Schmidt-hammer dating at a glacier-foreland boundary, Fåbergstølsbreen, southern Norway. Holocene 17:829–834

Paasche Ø, Strømsøe JR, Dahl SO, Linge H (2006) Weathering characteristics of arctic islands in northern Norway. Geomorphology 82:430–452

Paasche Ø, Dahl SO, Løvlie R, Bakke J, Nesje A (2007) Rockglacier activity during the Last Glacial-Interglacial transition and Holocene spring snowmelting. Quat Sci Rev 26:793–807

Pierson TC (2005) Hyperconcentrated flow—transitional process between water flow and debris flow. In: Jakob M, Hungr O (eds) Debris flows/avalanches. Geological Society of America, Boulder CO, pp 1–12

Pissart A (2002) Palsas, lithalsas and remnants of these periglacial mounds. A progress report. Prog Phys Geogr 26:605–621

Pissart A (2013) Palsas and lithalsas. In: Giardino JR, Harbor JM (eds) Treatise on geomorphology, volume 8, glacial and periglacial geomorphology. Academic Press, San Diego CA, pp 223–237

Porter SC (1989) Some geological implications of average Quaternary glacial conditions. Quat Res 32:245–261

Priesnitz K (1988) Cryoplanation. In: Clark MJ (ed) Advances in periglacial geomorphology. Wiley, Chichester, pp 49–67

Rapp A (1957) Studien über schutthalden in Lappland und auf Spitzbergen. Zeitschrift für Geomorphologie NF 1:179–200

Rapp A (1959) Avalanche boulder tongues in Lappland. Geogr Ann 41:34–48

Rapp A (1960) Recent development of mountain slopes in Kärkevagge and surroundings, northern Scandinavia. Geogr Ann 42:65–200

Rapp A (1982) Zonation of permafrost indicators in Swedish Lappland. Geogr Tidsskr/Dan J Geogr 82:37–38

Rapp A (1984) Nivation hollows and glacial cirques in Söderåsen, Scania, south Sweden. Geogr Ann Ser (Phys Geogr) 66:11–28

Rapp A, Åkerman HJ (1993) Slope processes and climate in the Abisko Mountains, northern Sweden. In: Frenzel B, Matthews JA, Gläser B (eds) Solifluction and climatic variation in the Holocene. Gustav Fischer Verlag, Stuttgart, pp 163–177

Rapp A, Clark M (1971) Large nonsorted polygons in Padjelanta National Park, Swedish Lappland. Geogr Ann Ser (Phys Geogr) 53:71–85

Rapp A, Nyberg R (1988) Mass movements, nivation processes and climatic fluctuations in northern Scandinavian mountains. Nor Geogr Tidsskr/Nor J Geogr 42:245–253

Rasmussen A (1981) The deglaciation of the coastal area NW of Svartisen, Northern Norway. Nor Geol Unders 369:1–31

Rea BR, Whalley WB, Rainey MM, Gordon JE (1996) Blockfields old or new? Evidence and implications for some plateaus in northern Norway. Geomorphology 15:109–121

Reid JR, Nesje A (1988) A giant ploughing block, Finse, southern Norway. Geogr Ann Ser (Phys Geogr) 70:27–33

Reusch H (1894) Strandfladen, et nyt træk i Norges geografi. Norges Geologiske Undersøgelse 14:1–14

Ridefelt H, Boelhouwers J (2006) Observations on regional variation in solifluction landform morphology and environment in the Abisko region, northern Sweden. Permafrost Periglac Process 17:253–266

Ridefelt H, Åkerman J, Beylich A, Boelhouwers J, Kolstrup E, Nyberg R (2009) 56 years of solifluction measurements in the Abisko mountains, northern Sweden—analysis of temporal and spatial variations of slow soil surface movements. Geogr Ann Ser (Phys Geogr) 91: 215–232

Rikkinen J (1989) Relations between topography, microclimates and vegetation in the Kalmari-Saarijarvi esker chain, central Finland. Fennia 167:87–150

Rönkkö M, Seppälä M (2003) Surface characteristics affecting active layer formation in palsas, Finnish Lapland. In: Phillips M, Springman SM, Arenson LU (eds) Permafrost: Proceedings of the Eighth International Conference on Permafrost. Swets and Zeitlinger, Lisse, pp 995–1000

Rubensdotter l, Sletten K, Sandøy G (2021) Morphological description of erosional and depositional landfoms formed by debris flow processes in mainland Norway. In Beylich AA (ed) Landscapes and landforms of Norway. Berlin: Springer, pp 225–240

Rudberg S (1977) Periglacial zonation in Scandinavia. Abhandlungen der Akademie der Wissenschaften in Göttingen Mathematisch-Physikalische Klasse 31:92–104

Samuelsson C (1926) Studien über die Wirkungen des Windes in den kalten und gemässigten Erdteilen. Bull Geol Inst Univ Upps 20:57–230

Sannel ABK (2020) Ground temperature and snow depth variability within a subarctic peat plateau landscape. Permafrost Periglac Process 31:255–263

Sannel ABK, Kuhry P (2011) Warming-induced destabilization of peat plateau/thermokarst lake complexes. J Geophys Res 116:GO3035

Sannel ABK, Hugelius G, Jansson P, Kuhry P (2016) Permafrost warming in a subarctic peatland—which meteorological controls are most important? Permafr Periglac Process 27:177–188

Sandersen F (1997) The influence of meteorological factors on the initiation of debris flows in Norway. In Matthews JA, Brunsden D, Frenzel B, Gläser B, Weiß (eds) Rapid mass movement as a source of climatic evidence for the Holocene. Gustav Fischer Verlag: Stuttgart, pp 321–332

Sandvold S, Lie Ø, Nesje A, Dahl SO (2001) Holocene glacial and colluvial activity in Leirungsdalen, eastern Jotunheimen, south-central Norway. Norw J Geol 81:25–40

Scapozza C, Lambiel C, Baron L, Marescot L, Reynard E (2011) Internal structure and permafrost distribution in two alpine periglacial talus slopes, Valais, Swiss Alps. Geomorphology 132:208–221

Scarpozza C (2016) Evidence of paraglacial and paraperiglacial crisis in Alpine sediment transfer since the last glaciation (Tincino, Switzerland). Quaternaire 27:139–155

Schleier M, Hermanns RL, Gosse JC, Oppikofer T, Rohn J, Tønnesen AF (2017) Subaqueous rock-avalanche deposits exposed by post-glacial isostatic rebound, Innfjorddalen, Western Norway. Geomorphology 289:117–133

Schlyter P (1995) Ventifacts as palaeo-wind indicators in southern Scandinavia. Permafrost Periglac Process 6:207–219

Schunke E, Zoltai SC (1988) Earth hummocks (thufur). In: Clark MJ (ed) Advances in periglacial geomorphology. Wiley, Chichester, pp 231–245

Sellier D, Kerguillec R (2021) Characterization of scree slopes in the Rondane mountains (south-central Norway). In Beylich AA (ed) Landscapes and landforms of Norway. Springer Nature, Cham, pp 203–223

Seppälä M (1971) Evolution of eolian relief of the Kaamasjoki—Kiellajoki river basin in Finnish Lapland. Fennia 104:1–88

Seppälä M (1972) Location, morphology and orientation of inland dunes in northern Sweden. Geogr Ann Ser (Phys Geogr) 54:85–104

Seppälä M (1981) Forest fires as activator of geomorphic processes in Kuttanen esker-dune region, northernmost Finland. Fennia 159:221–228

Seppälä M (1982) Present day periglacial phenomena in northern Finland. Biul Peryglac 29:231–243

Seppälä M (1986) The origin of palsas. Geogr Ann Ser (Phys Geogr) 68:141–147

Seppälä M (1988) Palsas and related forms. In: Clark MJ (ed) Advances in periglacial geomorphology. Wiley, Chichester, pp 247–278

Seppälä M (1994) Snow depth controls palsa growth. Permafrost Periglac Process 5:283–299

Seppälä M (1995a) Deflation and redeposition of sand dunes in Finnish Lapland. Quat Sci Rev 14:799–809

Seppälä M (1995b) How to make a palsa: a field experiment on formation of permafrost. Z Geomorphol Suppl 99:91–96

Seppälä M (1997) Distribution of permafrost in Finland. Bull Geol Soc Finl 69:87–96

Seppälä M (2003) Surface abrasion of palsas by wind action in Finnish Lapland. Geomorphology 52:141–148
Seppälä M (2004) Wind as a geomorphic agent in cold climates. Cambridge University Press, Cambridge
Seppälä M (2005a) Periglacial environment. In: Seppälä M (ed) The physical geography of Fennoscandia. Wiley-Blackwell, Chichester, pp 349–364
Seppälä M (2005b) Dating of palsas. In Ojala AEK (ed) Quaternary studies in the northern and Arctic regions of Finland. Geological Survey of Finland Special Paper 40, pp 79–84
Seppälä M (2011) Synthesis of studies of palsa formation underlining the importance of local environmental and physical characteristics. Quat Res 75:366–370
Sernander R (1905) Flytjord I svenska fjälltrakter. En botanisk-geologisk undersökning. Geol Föreningen I Stock Förhandlingar 27:41–84
Shakesby RA (1997) Pronival (protalus) ramparts: a review of forms, processes, diagnostic criteria and palaeoenvironmental implications. Prog Phys Geogr 21:394–418
Shakesby RA, Matthews JA (1987) Frost weathering and rock platform erosion on periglacial shorelines: a test of a hypothesis. Nor Geol Tidsskr 67:203
Shakesby RA, Dawson AG, Matthews JA (1987) Rock glaciers, protalus ramparts and related phenomena, Rondane, Norway: a continuum of large-scale talus-derived landforms. Boreas 16:305–317
Shakesby RA, Matthews JA, McCarroll D (1995) Pronival ("protalus") ramparts in the Romsdalsalpane, southern Norway: forms, terms, subnival processes, and alternative mechanisms of formation. Arct Alp Res 27:271–282
Shakesby RA, Matthews JA, McEwen LJ, Berrisford MS (1999) Snow-push processes in pronival (protalus) rampart formation: geomorphological evidence from Smørbotn, Romsdalsalpane, southern Norway. Geogr Ann Ser (Phys Geogr) 81:31–45
Slaymaker O (1988) The distinctive attributes of debris torrents. Hydrol Sci J 33:567–573
Slaymaker O (1995) Introduction. In: Slaymaker O (ed) Steepland geomorphology. Wiley, Chichester, pp 1–6
Slaymaker O (2009) Proglacial, periglacial or paraglacial? In Knight J, Harrison S (eds) Periglacial and paraglacial processes and environments, vol 320. Geological Society, London, Special Publication, pp 71–84
Sletten K, Blikra LH (2007) Holocene colluvial (debris-flow and water-flow) processes in eastern Norway: stratigraphy, chronology and palaeoenvironmental implications. J Quat Sci 22:619–635
Sletten K, Blikra LH, Ballantyne CK, Nesje A, Dahl SO (2003) Holocene debris flows recognized in a lacustrine sedimentary succession: sedimentology, chronostratigraphy and cause of triggering. Holocene 13:907–920
Sollid JL, Sørbel L (1992) Rock glaciers in Svalbard and Norway. Permafrost Periglac Process 3:215–222
Sollid JL, Sørbel L (1994) Distribution of glacial landforms in southern Norway in relation to the thermal regime of the last continental ice sheet. Geogr Ann Ser (Phys Geogr) 76:25–35
Sollid JL, Sørbel L (1998) Palsa bogs as a climatic indicator—examples ffom Dovrefjell, southern Norway. Ambio 27:287–291
Sollid JL, Andersen S, Hamre N, Kjeldsen O, Salvigsen O, Sturød S, Tveitå T, Wilhelmsen A (1973) Deglaciation of Finnmark, North Norway. Nor Geogr Tidsskr 27:233–325
Sollid JL, Holmlund P, Isaksen K, Harris C (2000) Deep permafrost boreholes in western Svalbard, northern Sweden and southern Norway. Nor Geogr Tidsskr 54:186–191
Sørensen T (1935) Bodenformen und Pflanzendecke in Nordostgrönland. Meddelelser om Gronland 93(4):1–69
Steiger C, Etzelmüller B, Westermann S, Myyhra KS (2016) Modelling the permafrost distribution in steep rockwalls in Norway. Norw J Geol 96:329–341
Stevens T, Sechl D, Tziavaras C, Schneider R, Banak A, Andreucci S, Hattestrand M, Pascucci V (2022) Age, formation and significance of loess deosits in central Sweden. Earth Surf Process Landf. https://doi.org/10.1002/esp.5456

Støren EN, Dahl SO, Lie Ø (2008) Separation of late-Holocene episodic paraglacial events and glacier fluctuations in eastern Jotunheimen, central southern Norway. Holocene 18:1179–1191

Støren EN, Dahl SO, Nesje A, Paasche Ø (2010) Identifying the sedimentary imprint of high-frequency Holocene river floods in lake sediments: development and application of a new method. Quat Sci Rev 29:3021–3033

Støren EN, Kolstad EW, Paasche Ø (2014) Linking past flood frequencies in Norway to regional atmospheric circulation anomalies. J Quat Sci 27:71–80

Strand SM, Christiansen HH, Johansson M, Åkerman J, Humlum O (2020) Active layer thickening and controls on interannual variability in the Nordic Arctic compared to the circum-Arctic. Permafrost Periglac Process 32:47–58

Strøeven AP, Fabel D, Hättestrand C, Harbor J (2002) A relict landscape in the centre of Fennoscandian glaciation: cosmogenic radionuclide evidence of tors preserved through multiple glacial cycles. Geomorphology 44:145–154

Strøeven AP, Hättestrand C, Kleman J, Heyman J, Fabel D, Fredin O, Goodfellow BW, Harbor JM, Jansen JD, Olsen L, Caffee MW, Fink D, Lundqvist J, Rosqvist GC, Strömberg B, Jansson KN (2016) Deglaciation of Fennoscandia. Quat Sci Rev 147:91–121

Strømsøe JR, Paasche Ø (2011) Weathering patterns in high-latitude regolith. J Geophys Res 116:F03021. https://doi.org/10.1029/2010JF001954

Sutinen R, Hyvönen E, Kukkonen I (2014) LiDAR detection of paleolandslides in the vicinity of the Suasselkä postglacial fault, Finnish Lapland. Int J Appl Earth Obs Geoinf 27:91–99

Svenonius FV (1909) Om scärf-eller blockhafven på våra högfjäll. Geol Föreningen I Stock Förhandlingar 32:169–181

Svensson H (1964) Tundra polygons. Photographic interpretation and field studies in North-Norwegian polygon areas. Norges Geologisk Undersøkelse Bulletin 223:298–327

Svensson H (1969) Open fissures in a polygonal net on the Norwegian Arctic coast. Biul Peryglac 19:389–398

Svensson H (1974) Distribution and chronology of relict polygon patterns on the Laholm Plain, the Swedish west coast. Geogr Ann Ser (Phys Geogr) 54:159–175

Svensson H (1983) Ventifacts as palaeowind indicators in a former periglacial area of southern Scandinavia. Proceedings of the Fourth International Conference on Permafrost, Fairbanks, Alaska, 18–22 July 1983, pp 1217–1220

Svensson H (1988) Ice-wedge casts and relict polygonal patterns in Scandinavia. J Quat Sci 3:57–68

Sverdrup HU (1938) Notes on erosion by drifting snow and transport of solid material by sea ice. Am J Sci 235:370–373

Thorn CE (1988) Nivation: a geomorphic chimera. In: Clark MJ (ed) Advances in periglacial geomorphology. Wiley, Chichester, pp 3–31

Thorn CE, Hall K (2002) Nivation and cryoplanation: the case for scrutiny and integration. Prog Phys Geogr 26:533–550

Thorn CE, Darmody RG, Campbell SW, Allen CE, Dixon JC (2007) Microvariability in the early stages of cobble weathering by microenvironment on a glacier foreland, Storbreen, Jotunheimen, Norway. Earth Surf Proc Land 32:2199–2211

Thorn CE, Darmody RG, Dixon JC (2011) Rethinking weathering and pedogenesis in alpine periglacial regions: some Scandinavian evidence. In Martini IP, French HM, Pérez Albert A (eds) Ice-marginal and periglacial processes and sediments, vol 354. Geological Society, London, Special Publication, pp 183–193

Tikkanen M (2005) Climate. In: Seppälä M (ed) The physical geography of Fennoscandia. Wiley-Blackwell, Chichester, pp 97–112

Tikkanen M, Heikkila R (1991) The influence of clear felling on temperature and vegetation in an esker area at Lammi, southern Finland. Fennia 169:1–24

Ulfstedt AC (1993) Solifluction in the Swedish mountains: distribution in relation to vegetation and snow cover. In: Frenzel B, Matthews JA, Gläser B (eds) Soliflucton and climatic variation in the Holocene. Gustav Fischer Verlag, Stuttgart, pp 217–223

Van Vliet-Lanoë B, Seppälä M (2002) Stratigraphy, age and formation of peaty earth hummocks (pounus), Finnish Lapland. Holocene 12:187–199

Van Vliet LB, Seppälä M, Käyhkö J (1993) Dune dynamics and cryoturbation features controlled by Holocene water level changes, Hietatievat, Finnish Lapland. Geol Mijnbouw 72:211–224

Vere DM, Matthews JA (1985) Rock glacier formation from a lateral moraine at Bukkeholsbreen, Jotunheimen, Norway: a sedimentological approach. Zeitschrift für Geomorphologie NF 28:397–415

Vasskog K, Nesje A, Støren EN, Waldmann N, Chapron E, Ariztegui D (2011) A Holocene record of snow-avalanche and flood activity reconstructed from a lacustrine sedimentary sequence at Oldevatnet, western Norway. Holocene 21:597–614

Vogt T (1918) Om recente og gamle strandlinjer I fast fjell. Norsk Geologiske Tidsskrift 4:107–127

Vorren KD (1979) Recent palsa datings, a brief survey. Nor Geogr Tidsskr 33:217–219

Vorren KD, Vorren B (1975) The problem of dating a palsa, Two attempts involving pollen diagrams, determination of moss subfossils, and C14-datings. Astarte 8:73–81

Vorren TO, Mangerud J, Blikra LH, Nesje A, Sveian H (2006) Landet trer fram. In: Ramberg IB, Bryhni I, Nøttvedt A (eds) (2006) Landet blir til: Norges geologi (Chapter 16, 532–555). Norsk Geologisk Forening (NGF): Trondheim. 608 pp

Walker MJC, Head MJ, Berkelhammer M, Björck S, Cheng H, Cwynar L, Fisher D, Gkinis V, Long A, Lowe JJ, Newnham RJ, Rasmussen SO, Weiss H (2018) Formal ratification of the subdivision of the Holocene Series/Epoch (Quaternary System/Period): two new Global Boundary Stratotype Sections and Points (GSSPs) and three new stages/subseries. Episodes 41:213–223

Washburn AL (1979) Geocryology: a survey of periglacial processes and environments. Arnold, London

Wilford DJ, Sakals ME, Innes JL, Sidle RC, Bergerud WA (2004) Recognition of debris flow, debris flood and flood hazard through watershed morphometrics. Landslides 1:61–66

Williams PJ (1957) Some investigations into solifluction features in Norway. Geogr J 23:42–58

Williams PJ (1961) Climatic factors controlling the distribution of certain frozen ground phenomena. Geogr Ann 43:339–347

Wilson P, Matthews JA, Mourne RW (2017) Relict blockstreams at Insteheia, Valldalen-Tafjorden, southern Norway: their nature and Schmidt-hammer exposure age. Permafrost Periglac Process 28:286–297

Wilson P, Linge H, Matthews JA, Mourne RW, Olsen J (2019) Comparative numerical surface exposure-age dating (^{10}Be and Schmidt hammer) of an early-Holocene rock avalanche at Alstadfjellet, Valldalen, southern Norway. Geogr Ann Ser (Phys Geogr) 101:293–309

Wilson P, Matthews JA, Mourne RW, Linge H, Olsen J (2020) Interpretation, age and significance of a relict paraglacial and periglacial boulder-dominated landform assemblage in Alnesdalen, Romsdalsalpane, southern Norway. Geomorphology 369: Article No. 107362 (16 p).

Winkler S, Matthews JA, Mourne RW, Wilson P (2016) Schmidt-hammer exposure ages from periglacial patterned ground (sorted circles) in Jotunheimen, Norway, and their interpretive problems. Geogr Ann Ser (Phys Geogr) 98:265–285

Winkler S, Matthews JA, Haselberger S, Hill JL, Mourne RW, Owen G, Wilson P (2020) Schmidt-hammer exposure-age dating (SHD) of sorted stripes on Juvflye, Jotunheimen (central southern Norway): morphodynamic and palaeoclimatic implications. Geomorphology 353: Article No. 107014 (19 p).

Winkler S, Donner A, Tintrup gen Suntrup A (2021) Periglacial landforms in Jotunheimen, central southern Norway, and their altitudinal distribution. In Beylich AA (ed) Landscapes and landforms of Norway. Berlin: Springer, pp 169-202

Worsley P (2008) Some observations on lake ice-push features, Grasvatn, northern Scandinavia. Nor Geogr Tidsskr 29:10–19

Worsley P, Harris C (1974) Evidence for Neoglacial solifluction at Okstindan, north Norway. Arctic 27:128–144

Zuidhoff FZ, Kolstrup E (2000) Changes in palsa distribution in relation to climate change in Laivadalen, northern Sweden, especially 1960–1997. Permafr Periglac Process 11:55–69

Iceland

José M. Fernández-Fernández, Bernd Etzelmüller, Costanza Morino, and Þorsteinn Sæmundsson

1 Geographical Framework

Isolated in the North Atlantic Ocean and on the Mid-Atlantic Ridge, Iceland extends along 103,546 km^2, between the parallels 63° 23' and 66° 32' N and meridians 13° 30' and 24° 32' W, thus being the second largest island in Europe and the third largest in the Atlantic Ocean (European Environmental Agency, 2015). The closest surrounding lands are Greenland (290 km) to the west, the Faroe Islands (420 km) to the east, the Jan Mayen Island (550 km) to the north, Scotland (800 km) to the southeast, and Norway (970 km) to the east. The island is characterized by mountainous terrain, with an average elevation of 510 m a.s.l. (~25% of the surface is below 200 m a.s.l.), culminating in the Hvannadalshnjúkur Peak (64° 0′ 51" N, 16° 40′ 37" W, 2110 m) in the south-east. The Icelandic coastline varies in shape at different regions, being smoothest at the southern lowlands, while it gets sinuously and rough at western, northern, and eastern basaltic fjords (Einarsson 1994) (Fig. 1).

Iceland is a young island, whose origin is found at the Miocene and whose existence and landscapes are mostly explained by the volcanic activity, manifested in frequent eruptions and earthquakes (Einarsson 1991a; Larsen 1998; Guðmundsson et al. 2008; Jakobsdóttir 2008). As the only emerging point of the Mid-Atlantic Ridge,

J. M. Fernández-Fernández (✉)
Department of Geography, Universidad Complutense de Madrid, Madrid, Spain
e-mail: josemariafernandez@ucm.es

B. Etzelmüller
Department of Geosciences, University of Oslo, Oslo, Norway

C. Morino
CNRS-Laboratoire EDYTEM - Université Savoie Mont Blanc, Chambéry, France

Þ. Sæmundsson
Faculty of Life- and Environmental Sciences, University of Iceland, Reykjavík, Iceland

Institute of Earth Sciences, University of Iceland, Reykjavík, Iceland

M. Oliva et al. (eds.), *Periglacial Landscapes of Europe*,
https://doi.org/10.1007/978-3-031-14895-8_15

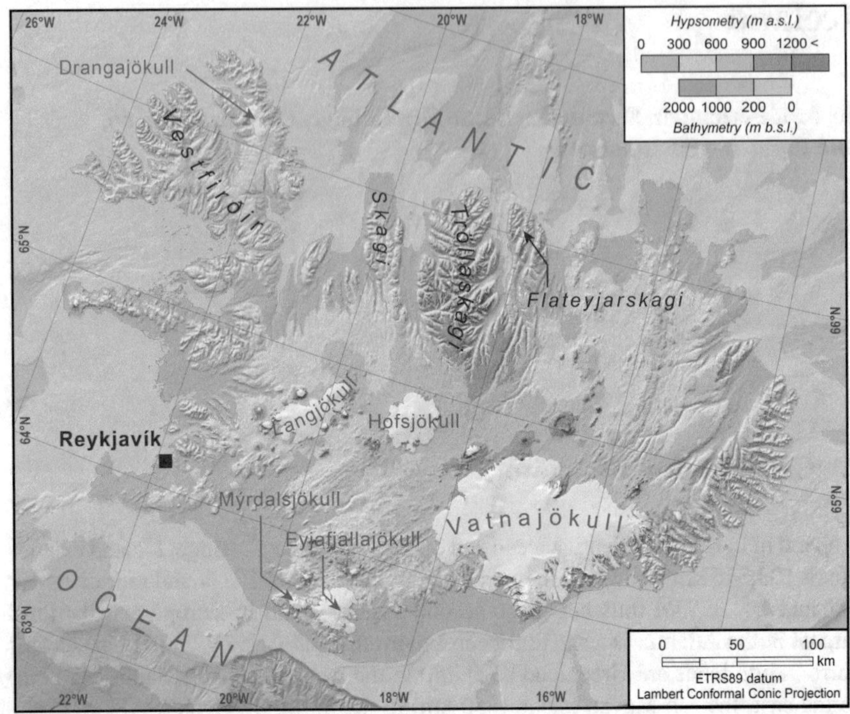

Fig. 1 Location map of the main Icelandic areas mentioned throughout the manuscript

its origin is associated to the interaction of the asthenospheric flow with a deep-seated mantle plume (Wolfe et al. 1997; Shen et al. 1998) of ascending magma (*hotspot*) at the tectonic plate boundaries, close to the northwestern tip of the Vatnajökull ice cap (Einarsson 2008; Thordarson and Höskuldsson 2002). Such boundaries are extended through two main ridges to the north and the south (i.e. Kolbeinseyjarhryggur and Reykjaneshryggur) as submarine segments of the Mid-Atlantic Ridge. In surface, they are manifested as narrow belts of active faulting and volcanism from west to east: Snæfellsnes Volcanic Belt, Reykjanes Volcanic Belt, Western Volcanic Zone, Mid-Iceland Belt, Tjörnes Fracture Zone and North Volcanic Zone (Thordarson and Hoskuldsson 2002; Arnalds 2015).

The Icelandic bedrock is composed of basaltic lavas arranged in three main geological formations from the oldest to the youngest (Sæmundsson 1980; Johannesson and Sæmundsson 1989):

i. Tertiary Basalt Formation (16-3 Ma; Miocene-Pliocene age) extends mainly at the eastern-north-eastern zone, the central northern peninsulas and at the north-western quadrant, and is composed of sub-horizontal jointed basaltic lava flows. Thickness of lava strata range from 2 to 30 m, and are usually alternated with lithified sedimentary horizons, corresponding to paleosols, the so-called "red interbed layers", ranging from a few centimeters to tenths of meters (Thordarson and Hoskuldsson 2002).

ii. Grey Basalt Formation (3-0.7 Ma; Pliocene-Pleistocene) surrounds both sides of the active rift zones and covers the area between the West and East Volcanic Zones and the Snæfellsnes and Skagi peninsulas. These series are composed of subglacial basalt lava flows—pillow lavas, breccias and hyaloclastites—interlayered with subaerial lavas originated during the interglacials. They have a reduced thickness compared to the Tertiary series because of the higher erosion (Thordarson and Hoskuldsson 2002).

iii. Móberg Formation (<0.7 Ma; Late Pleistocene) is located mainly in the active volcanic zones. It is constituted by hyaloclastic volcanic ridges, table mountains originated in subglacial eruptions, interglacial lavas later sculpted by glaciers, and the lava fields—active volcanic zone—originated in postglacial times.

Throughout the Quaternary glacial cycles until 10 ka, the bedrock and associated seafloor (platform) of the island has been considerably reshaped by the so-called Icelandic Ice Sheet (IIS, hereafter), which expanded up to 50–100 km beyond the current coastline (Norðdahl and Pétursson 2005; Norðdahl et al. 2008). Subsequently, when glaciers disappeared, periglacial dynamics reshaped the landscape. In post-glacial times, explosive volcanic events produced a number of tephra layers of which the more ryolitic ones—especially those spread by the Hekla volcano—are the most easily identifiable in soil profiles given their characteristic white colour, and often used in tephrochronological dating (e.g. Thorarinsson 1981a; Dugmore and Buckland 1991).

The Icelandic terrain is the result of volcanic, glacial, fluvial and marine erosive and depositional processes. The core of the island, i.e. the volcanic active zone, is characterized by the existence of hyaloclastic ridges, extensive lava fields and table mountains with high-gradient slopes. In the northwestern, central north and eastern regions, glacial erosion has carved U-shaped fjords and valleys, deeply incised in the highland plateaus (Thordarson and Hoskuldsson 2002; Einarsson 1994). Valleys are characterized by steep slopes, with vertical cliffs in the higher sectors and glacial, fluvial, colluvial and talus deposits in the lower parts. The landscape is different in southern Iceland, bordered on either side by the West and East Volcanic Zones, and characterised by contrasting topography, with flat-lying lowlands sourrounded by mountains up to 1000–1500 m a.s.l. Here is where the most extensive glaciers are located (Björnsson and Pálsson 2008); large *sandur* plains originated after glacial retreat are the most outstanding elements of the landscape.

Iceland's location determines a boundary position between the main large-scale atmospheric and oceanic circulation systems of the polar and temperate zones of the Northern Hemisphere. Therefore, cold and dry north-easterly winds coming from the Polar high-pressure cells converge at this latitude with the warm and moist south-westerly ones emitted by the subtropical high-pressure cells (Einarsson 1984; Arnalds 2015), generating at ~60° N a *Polar front* along which successive subpolar low-pressure cells originate (Bjerknes and Solberg 1922). The most constant of these is the so-called 'Icelandic low', whose influence is propagated far beyond Iceland, towards western Europe. On the other hand, the oceanic circulation around Iceland reproduces contrasts, analogous to the atmospheric ones, with the interaction of sea currents with opposite properties, coming from the north and the south: the cold Eastern Iceland (a ramification of the East Greenland one) and the warm Irminger's (a ramification of the Gulf stream), which contributes to a mild climate (Einarsson 1984; Aranlds 2015).

Iceland's location and the large-scale circulation systems determine a temperate climate with cool summers and mild winters (Einarsson 1984). Driven by the relief distribution, temperature shows a great spatial variability throughout the island. The mean annual air temperature (MAAT) along the coasts ranges from 0 to 4 ºC and can even decrease to −4 ºC in the highlands, where permafrost can be found. Higher temperatures (4.5 ºC and even above 5 ºC) prevail in the southern zone, while they can decrease to 3–4 °C at inner western Iceland, and even to 2 ºC at inland and highlands of the northern regions (Einarsson 1984; Arnalds 2015; Icelandic Meteorological Office 2021).

Precipitation distribution is driven both by prevailing moist winds (easterly and southerly) and orography, decreasing from the south to the north. The highest annual values are found in the higher glaciated areas of the south and southeast, with more than 2000 mm, and even 4000–5000 and 7000 mm in the Vatnajökull and Mýrdalsjökull ice-caps areas (Björnsson and Pálsson 2008), creating a rain shadow to the south-easterly winds. Consequently, precipitation decreases to 1000–1600 mm and 700–1000 mm at the coastal and inland areas of southwestern and western Iceland, and to 400–600 mm or less in the northern regions, and especially north of Vatnajökull (Einarsson 1984; Ólafsson et al. 2007; Arnalds 2015; Icelandic Meteorological Office 2021). Snow represents in northern Iceland more than a half of the winter precipitation, whereas in the southern part decreases to only 5–10% (Arnalds et al. 2001a, b), with its depth and duration directly controlled by the relief, decreasing from the southern and south-eastern glaciated areas (>300 days) to the lowlands (<100 days) (Dietz et al. 2012). Both the topographic and the precipitation patterns control the distribution of glaciers, which cover 11% of the Icelandic surface—in many cases, active volcanoes, being specially clustered at the central, southern, and south-eastern highlands, where the largest ice caps are located (Björnsson and Pálsson 2008).

The sequence of Pliocene–Pleistocene subaerial and subglacial volcanic materials in Iceland provides evidence of the occurrence of numerous glacial cycles, during which the IIS covered most of the island and even the surrounding shelves (Eiriksson 2008). During the Last Glacial Maximum (LGM), the IIS covered the whole island with a maximum thickness of 2000 m (Hubbard et al. 2006), and reached the north shelf edge (40 km away from the current shoreline) by 24.4 cal ka BP (Pétursson et al. 2015). The subsequent evolution of the IIS is summarized in several publications (Kaldal and Víkingssson 1990; Norðdahl and Pétursson 2005; Norðdahl et al. 2008; Geirsdóttir et al. 2009; Geirsdóttir 2011; Pétursson et al. 2015).

During the Bølling/Allerød (B/A) interstadial, the sea level rise triggered rapid collapse and catastrophic retreat from the shelf (Norðdahl and Pétursson 2005) until it only covered the central and southeastern regions of the island by 15 ka (Andrews et al. 2000; Norðdahl and Einarsson 2001; Geirsdóttir et al. 2009; Brynjólfsson et al. 2015). Following the Bølling interstadial, the IIS advanced during Older Dryas (Ingólfsson et al. 1997). However, the atmospheric warming during the Allerød interstadial triggered again the retreat of the IIS (Rundgren 1995, 1999; Rundgren and Ingólfsson 1999). After that, a new readvance of the glaciers occurred during the Younger Dryas (YD, hereafter), reaching the current coastline and entering the most important fjords, which in many cases generated the obturation of lateral tributary valleys and formed lakes (Geirsdóttir et al. 2002; Norðdahl and Pétursson 2005), although many elevated areas remained ice-free (Norðdahl and Einarsson 2001). By these times, glacial advances occurred in the cirques of Tröllaskagi in the north (Fernández-Fernández et al. 2020; Tanarro et al. 2021). Following the retreat at the end of the YD, the IIS readvanced at 11.2 cal ka BP (Early Preboreal; Ingólfsson et al. 1997; Norðdahl and Einarsson 2001; Norðdahl and Pétursson 2005; Sæmundsson 1995).

After that, the IIS retreated again so rapidly (Kaldal and Víkingssson 1990; Andrews et al. 2000; Norðdahl and Einarsson 2001; Geirsdóttir et al. 2002, 2009; Larsen et al. 2012; Andrés et al. 2019) that by 10.3 cal ka BP it was rapidly retreating across the highlands, with the central and northern regions of Iceland already ice-free (Stötter et al. 1999; Caseldine et al. 2003). The existence of subaerial lava flows shows that the IIS had definitively disappeared and the highlands were ice-free by 8.6 cal ka BP. By this time, glacial extent was fairly similar to the present (Geirsdóttir et al. 2009; Hjartarson and Ingólfsson 1988; Kaldal and Víkingsson 1990). Then, during the Holocene Thermal Maximum (HTM; 8–6 cal ka BP), most of the remaining ice caps disappeared in response to prevailing high temperatures, up to 2–3 °C higher than the reference period 1961–1990 (Caseldine et al. 2003, 2006; Flowers et al. 2008; Geirsdóttir et al. 2009, 2013). Glaciers readvanced again following 6 cal ka BP during the Neoglaciation (Gudmundsson 1997a; Stötter et al. 1999; Geirsdóttir et al. 2019; Fernández-Fernández et al. 2019), with the ice caps of central Iceland reaching their maximum Holocene extent during the Little Ice Age (LIA, hereafter) alongside the minimum temperatures of the Holocene (Stötter et al. 1999; Kirkbride and Dugmore 2001, 2006). Following the LIA, glacier retreat has prevailed until recent times (see e.g. Fernández-Fernández et al. 2017, 2019).

2 The Development of Periglacial Resarch in Iceland

Research on periglacial processes and landforms in Iceland started with the first explorations at the end of the 19th and the onset of the 20th century. They consisted of thorough descriptions of some of the characteristic periglacial landforms such as patterned ground and palsas (e.g. Thoroddsen 1899, 1905–1906, 1911, 1913, 1913–1915, 1914; Hawkes 1924; Jónsson 1909; Gruner 1912; Hawkes 1924; Hannesson 1927). The studies continued with some cases of local detailed geomorphological maps and some attempts to map the spatial distribution of the periglacial phenomena throughout Iceland (e.g. Thorarinsson 1951; Bout et al. 1955).

First research on the Icelandic permafrost started in the 1950s (Thorarinsson 1951; Bout et al. 1955), and continued in the following decades with studies on permafrost-related landforms (i.e. palsas) and mapping of permafrost conditions in the central highlands (Thorarinsson 1964; Friedman et al. 1971; Schunke 1974; Stingl and Herrmann 1976; Priesnitz and Schunke 1983; Hirakawa 1986; Thórhallsdóttir 1994, 1996, 1997; Sæmundsson et al. 2012). Since the 1970s, the periglacial research advanced a step forward with the implementation of the first measurements of landforms, quantification of processes and dating techniques (Friedman et al. 1971; Bergmann 1973; Webb 1972; Schunke 1973, 1974, 1975, 1977; Thorarinsson 1981b; Hirakawa 1989; Schunke and Zoltai 1988; Van Vliet-Lanoë et al. 1998). By this time the investigation of rock glaciers started, with discussions on their activity, age and origin, supported on a number of techniques such as geomorphological mapping, lichenometric and Schmidt hammer dating and displacement monitoring based on geodetic measurements (e.g. Martin and Whalley 1987; Martin et al. 1991, 1994; Hamilton and Whalley 1995; Whalley et al. 1995a). Following the catastrophic snow avalanche and debris flow events occurred in 1995, systematic research on such events and further planning in Iceland started and continued during the next decade (e.g. Bell and Glade 2004; Bessason et al. 2007; Decaulne 2001, 2004, 2005, 2007; Decaulne and Sæmundsson 1996, 2003, 2004, 2006a, 2006b, 2007a, 2007b, 2008, 2010; Conway et al. 2010; Decaulne et al. 2005, 2012; Sæmundsson and Decaulne 2003; Sæmundsson and Pétursson 2018; Sæmundsson et al. 2018; Morino et al. 2019a).

The installation of boreholes in central and eastern Iceland to record ground surface temperatures and the modelling of the current and past permafrost established a more quantitative basis of permafrost distribution in Iceland. In addition, investigation on seasonal frost features such as solifluction landforms (e.g. Kirkbride and Dugmore 2005; Veit et al. 2012) and earth hummocks (Van Vliet-Lanoë et al. 1998; Grab 2005) advanced. Studies also included inventories of permafrost related-landforms (rock glaciers and stable ice-cored moraines) and climatological data (Etzelmüller et al. 2007, 2020; Farbrot et al. 2007a, 2007b; Kellerer-Pirklbauer et al. 2007; Lilleøren et al. 2013; Kneisel et al. 2007; Kneisel 2010; Czekirda et al. 2019; Petersen 2021). The 21st century also saw the implementation of new techniques for the study of permafrost and periglacial processes such as georadar and electric resistivity (e.g. Kneisel et al. 2007; Kneisel 2010), photogrammetry and remote sensing (Wangensteen et al. 2006; Kellerer-Pirklbauer et al. 2007; Lilleøren et al. 2013; Campos et al. 2019; Tanarro et al. 2019, 2021). In the last decade, research in geochronology greatly advanced by different dating techniques (radiocarbon, tephrochronology, Schmidt hammer, cosmic-ray exposure dating; CRE) on rock slope failures (e.g. Cossart et al. 2014; Coquin et al. 2015, 2016; Decaulne et al. 2016) and rock glaciers (Fernández-Fernández et al. 2020; Palacios et al. 2021; Tanarro et al. 2021). However, chronological investigation still needs further work, as the timing of relict landforms and the chronological constraints of past periglacial processes are still very limited.

Most published studies concerning permafrost in Iceland focus on geomorphological features indicative of permafrost, such as palsas (Priesnitz and Schunke

1978; Thórhallsdóttir 1996; Sæmundsson et al. 2012) or frost-related landforms like patterned ground (e.g. Thorarinsson 1951; Stingl and Herrmann 1976) and ice-wedges (e.g. Schunke 1974). A great part of these studies was carried out in the central highlands of Iceland. Priesnitz and Schunke (1978) explicitly limited the permafrost distribution in Iceland to this low-relief highland area and to poorly drained and organic-rich soils, with palsas located in topographic depressions. They estimated that permafrost was mainly present between 460 and 720 m a.s.l. and covered an area of 180 km^2. The official International Permafrost Association (IPA) map classifies the extensive palsa-rich areas in the central part of Iceland as falling within the zones of sporadic discontinuous permafrost and isolated patches of permafrost (Brown et al. 1995), which has been recently confirmed (Obu et al. 2019).

Widespread mountain permafrost in the sense of Haeberli et al. (1993) has not been mapped in Iceland. Glacially derived rock glaciers and ice-cored moraines were reported from the Tröllaskagi area in northern Iceland (e.g. Whalley et al. 1995b), but were not connected to the overall distribution of mountain permafrost. However, Whalley and Martin (1994) suggested that, based on temperature lapse rate calculations from climate stations situated along the coast, permafrost might exist on high-elevation plateaus above 1200 m a.s.l., leaving the presently active rock-glaciers (e.g. Wangensteen et al. 2006) below this permafrost limit.

Direct permafrost observations have been made in volcanic mountains such as the Askja caldera (Helgason 2000) and the Hekla stratovolcano ridge (Kellerer-Pirklbauer et al. 2006), where permafrost was shown by the perennial survival of snow layers underneath tephra from the volcano eruptions in 1875 and 2000. Persistent ice has been found in several lava caves and tubes, such as in upper Borgarfjörður in Eastern Iceland, in the caves Víðgelmir and Surtshellir (Sigurdur Sveinn Jónsson, pers. comm.). Excavations for a large hydro-power dam built as part of the Kárahnjúkar Hydroelectric Project in eastern Iceland revealed several sections of frozen gravelly sediments at elevations of 550 to 650 m a.s.l. At around the same elevation, road cuts in debris revealed frozen material between Jökuldalur and Mývatn in northern Iceland and in Boöðarsdalur between Vopnafjörður and Hérað (Konráð Vilhjálmsson, Arnarfell contractors, pers. comm. 2003). At the northern and north-eastern glacier margins and forefields around the glacier outlets of Mýrdalsjökull (e.g. Kruger 1996; e.g. Van der Meer et al. 1999), Hofsjökull, Langjökull (Bennett et al. 2003) and Vatnajökull, sub-freezing temperatures and frozen sediments are present. More recently Van Vliet-Lanoë et al. (1998, p. 349) speculated that "mountain permafrost seems common above 800 m and continuous above 1000 m to the centre of the island".

3 Distribution and Chronology of Periglacial Landforms

When discussing the relationship between permafrost and climate in the North Atlantic, Iceland is considered the nexus between Scandinavia and Greenland (Farbrot et al. 2007a). First direct observations revealed the existence of permafrost based on the preservation of snow layers underneath tephra layers (Askja caldera Helgason 2000; and the Hekla volcano, Kellerer-Pirklbauer et al. 2006), the existence of ice in lava caves and tubes in eastern Iceland, frozen sediments in eastern and northern Iceland (Etzelmüller et al. 2007, p. 186) and subfreezing temperatures and frozen sediments in the vicinity of the outlets of the Mýrdalsjökull (Kruger 1996; Van der Meer et al. 1999), Hofsjökull, Langjökull (Bennett et al. 2003) and Vatnajökull glaciers.

The maritime climate prevailing in Iceland determines low thermal amplitude and greatly limits the number of freezing and thawing degree days. Initial hypothesis pointed to the likely occurrence of mountain permafrost above 900 m a.s.l. in northern and eastern Iceland (Farbrot et al. 2007b). Measurements done by Etzelmüller et al. (2007) showed mean ground surface temperatures between −2.2 and 3.7 ºC throughout northern and eastern Iceland, with the 0 ºC-isotherm located at ~950 m a.s.l. and sub-zero temperatures above 800 m a.s.l. The authors concluded that the lower permafrost limit was located in northern and central Iceland at 800–850 and 950–1000 m a.s.l., respectively, following a southwards increasing trend with a similar pattern to precipitation and glaciated area (Fig. 2). Farbrot et al. (2007b) showed that mountain permafrost is present at sites with MAAT lower than −2 ºC.

Data from shallow boreholes in northeastern Iceland (Farbrot et al. 2007b; Etzelmüller et al. 2007) revealed permafrost thicknesses from less than 10 m to over than 30 m, and active layer thicknesses between 2 and 6 m. To the south of the Skagafjörður fjord, Kneisel et al. (2007) and Kneisel (2010) detected shallow permafrost (8–10 m thick) and a thinner active layer in palsa sites of the central highlands, ranging from 0.35 to 0.81 m. It is "warm" permafrost, with temperatures between −2 and 0 ºC (Farbrot et al. 2007a, b; Etzelmüller et al. 2007; Czekirda et al. 2019), and whose active layer is currently increasing in thickness in response to the higher summer temperatures recorded over the recent years (Kneisel et al. 2007; Sæmundsson et al. 2012).

Regional permafrost modelling by Czekirda et al. (2019) and Etzelmüller et al. (2007) showed that the mountain permafrost zone covers around 6900 and 8000 km^2 of the island (7–8%), respectively. New results show that the largest area with continuous permafrost is in Tröllaskagi—coldest and deepest permafrost (Czekirda et al. 2019)—where also other markers such as the time-variable distribution of perennial snow patches (Tussetschläger et al. 2020) have been used as permafrost indicators. Other secondary areas exist to the NE of Vatnajökull and the Askja volcano (Etzelmüller et al. 2007). Sporadic permafrost is also found around the palsa and peat areas of the plateaus (e.g. Thórhallsdóttir 1996; Sæmundsson et al. 2012). Permafrost distribution is mainly controlled by wind and snow, especially between 700 and 1000 m a.s.l. Prevailing strong south-easterly winds blow the snow from flat areas and south-easterly slopes, which remain snow-free while northerly slopes receive important accumulation. This determines a faster cooling of the snow-blown slopes in winter and a snow-driven ground cooling in the northerly slopes (Etzelmüller et al. 2007).

Recent modelling has reconstructed the spatio-temporal variation of permafrost in Iceland since the LGM (Etzelmüller et al. 2020). Its distribution was controlled

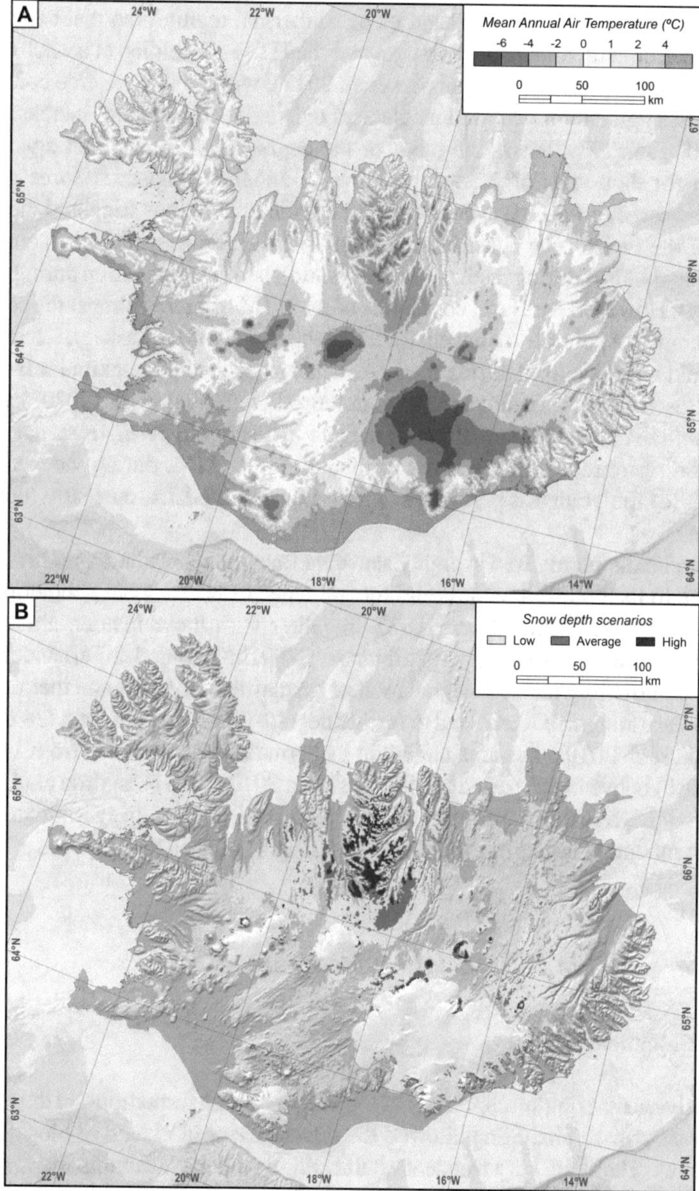

Fig. 2 (A) Map of the Mean Annual Air Temperature (MAAT) (B) Permafrost distribution in Iceland according to different snow depth scenarios (*Low snow* corresponds to permafrost reproduced in a 50% precipitation scenario; *Average snow* corresponds to permafrost reproduced in 50–100% scenarios; and *High snow* corresponds to permafrost reproduced in all precipitation scenarios). For more information, the reader is referred to the original publication in Czekirda et al. (2019)

mainly by the extent of the IIS—and its basal thermal regime—so that between the LGM, a period with total land coverage, and the HTM (Caseldine et al. 2003, 2006; Geirsdóttir et al. 2009, 2013; Renssen et al. 2012), with mostly ice-free conditions, permafrost aggradation occurred as glaciers receded, and its extent increased from 0 to 20,000 km^2. Following the onset of the deglaciation, permafrost aggradation occurred for thousands of years in the first mountainous coastal ice-free areas of the northwest, north and east. Subsequent climatic fluctuations triggered the degradation of the Pleistocene coastal permafrost—while it developed in the central and eastern highlands—during the B/A (warm-humid), and its aggradation during the YD (cold/dry). However, the warm conditions of the HTM triggered almost the complete disappearance of permafrost, except for the highest areas of Tröllaskagi. Later expansion would have occurred during the LIA, when it reached the maximum Holocene extent. Based on their findings from modelling, Etzelmüller et al. (2020) hypothesized on three classes of permafrost in Iceland: (a) very old permafrost, not thawed during the Holocene; (b) permafrost formed after the HTM, during the Neoglaciation; and (c) the youngest permafrost formed during the LIA, currently located in the palsa areas.

The Icelandic permafrost is highly sensitive to climate oscillations. This is especially due to its modest thickness and the influence of the regional geothermal heat flux (Flóvenz and Sæmundsson 1993). Together with these factors, atmospheric warming of the last decades (e.g. Einarsson 1991b; Fernández-Fernández et al. 2017) is contributing to a fast degradation of permafrost, in extent and thermal state. Projected warming will likely lead to further degradation during the next few decades (Czekirda et al. 2019). The area underlain by permafrost has reduced from 11 to 7% (i.e. −40%) between the periods 1980–1989 and 2010–2016 (Czekirda et al. 2019), which may have major implications for slope instability (e.g. triggering landslides at frozen mountain slopes; Sæmundsson et al. 2018; Morino et al. 2019a, 2021).

3.1 Large-Scale Periglacial Landscapes

3.1.1 Blockfields

The highly-weathering susceptibility of the Tertiary basalt formation and the frequent occurrence of frost shattering allowed the development of extensive blockfields in the plateaus. These are very frequent on flat and snow-blown summit areas at 1200–1400 m a.s.l. (MAAT −1.9 °C; Farbrot et al. 2007a) in the highlands of Tröllaskagi (Fig. 3), where they evolved to patterned ground with polygons up to 1.5 m. Some examples have been mapped bordering the cirques of the Hóladalur (Tanarro et al. 2018; Fig. 4) and Héðinsdalur valleys (Rodríguez-Mena et al. 2021) cirques, in the central zone of the peninsula.

Blockfield (Northern Iceland)
Costanza Morino

Aeolian ventifacts (Iceland)
Bernd Etzelmüller

Fig. 3 Pictures of large-scale periglacial landscapes

3.1.2 Inland and Lowland Aeolian Fields

Given its position in the North Atlantic and next to the 'Icelandic low', Iceland is a windy territory, with frequent high-velocity winds (Einarsson 1984; Ólafsson et al. 2007). This, together with the high number of freeze–thaw cycles and the aridity of soils, contributes to an intense aeolian activity (see e.g. Olafsson 2004) and a great extension of sandy deserts (>20,000 km^2; 21% of the country's surface) along coastal sand-fields and the highlands. Sandy deserts in Iceland are classified into three morphological classes (Arnalds et al. 1997, 2001a, b; Arnalds 2015): (i) *sand-fields and pumice*, bare, active and the most unstable, in glacial-margin areas; (ii) *sandy lavas*, where large amounts of sand have been deposited on Holocene lava fields; and (iii) *sandy lag gravel*, formed on lag gravel surfaces (*melur*, in Icelandic) such as till.

The texture of the Icelandic aeolian deposits ranges between 0.2 and 1 mm in active aeolian areas (Sigurdardottir 1992; Arnalds et al. 2001a, b). They include ash and physically weathered materials, composed mainly by dark basaltic volcanic glass (see Jonsson 1995; Arnalds et al. 2001a, b). The sources of the Icelandic aeolian sediments are volcanic eruptions, glaciers (Bullard 2013) and glacial rivers through periodic floods (daily or after rain events) or geothermal meltwater events (*jökulhlaups*) (Palsson and Zophoniasson 1992; Maizels 1997; Arnalds et al. 1997, 2001a, b, 2016; Old et al. 2005; Russell et al. 2005; Arnalds 2015).

As silty and sandy particles are deposited, wind erosion sorts the materials so that the finer ones (dust) are dispersed while the larger ones are left behind. The best expression of this process can be observed at the fluvial ice-marginal areas (Arnalds et al. 2016), which are periodically flooded (i.e. recharged with silty and coarse sediments) and where dry katabatic winds, channelled between mountains, transport materials downwind forming a *river of sand* or *sand-path* (see Gisladottir et al. 2005).

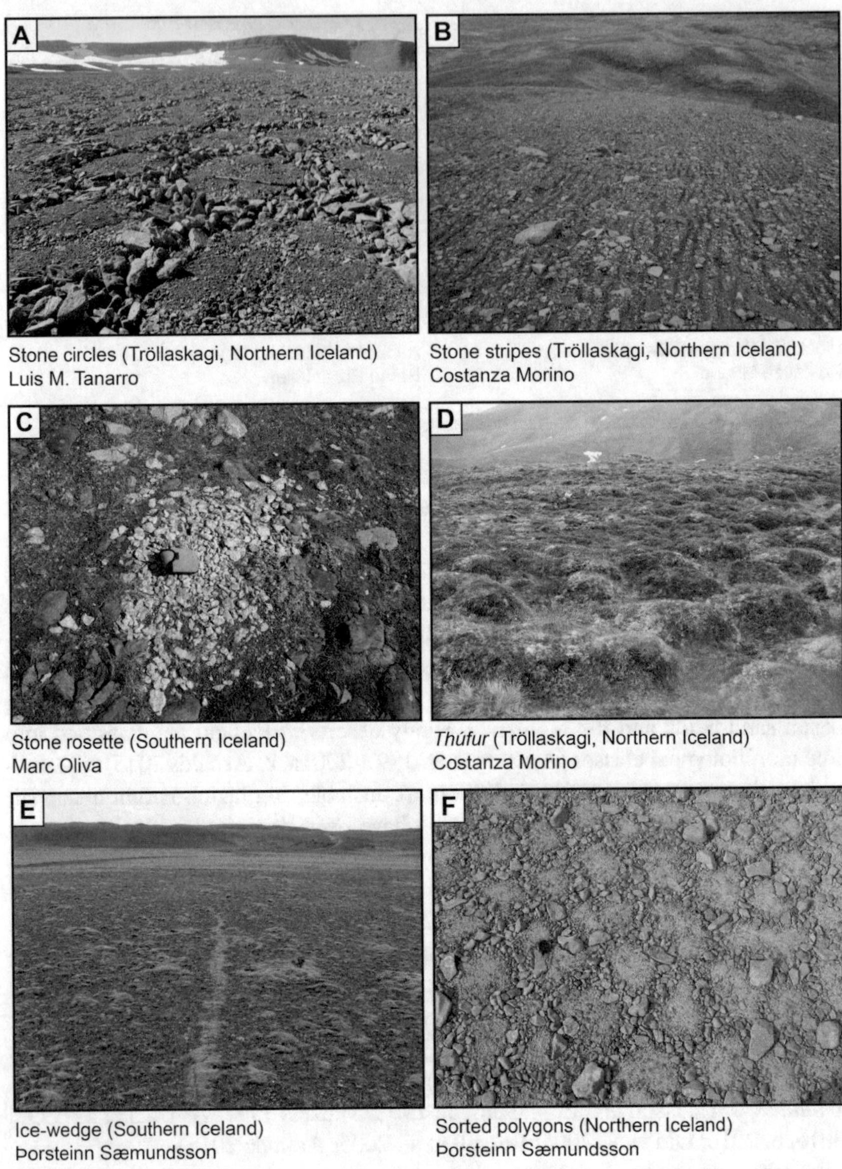

Fig. 4 Pictures of seasonally frozen ground features

At the same time, dust is produced in very large quantities in some places (*plume areas*), such as fluvial and glacial outwash plains (see Arnalds 2010b), where the snow cover is thin and thaws in spring, when the peak of sediment supply in rivers occurs (Gislason et al. 1997). Dust redistribution occurs mainly during *dust events* or *dust storms*. Near to 50% of them (in southern Iceland) occur during winter below the freezing point, allowing a mixture of snow and dust, which reduces snow and ice reflectance, thus enhancing their melting (Dagsson-Waldhauserova et al. 2015). Gisladottir et al. (2005) and Prospero et al. (2012), amongst others, hypothesize that the glacial shrinking triggered by the global warming will likely increase dust emissions.

As aeolian materials are transported, three main morphogenetic processes occur, namely: denudation, abrasion and sedimentation, which generate a spectrum of landforms:

i. *Denudation* consist on the lifting and transport of rock particles. Cryoturbation contributes to its effectiveness through damaging vegetation, promoting needle ice occurrence, and delivering sand particles. Aeolian *deflation* often generates *stone* or *gravel pavements* which are very common in the sandy lag gravel deserts of the extensive areas of the moraines and sandur surfaces of the central parts of the island. Some examples are the Sprengisandur sandur plain (770 m a.s.l.) (Priesnitz and Schunke 1983); the Askja region in north-eastern parts, (Mountney and Russell 2004); and the so-called 'Main Palagonite Terrace' in the Sólheimasandur sandur plain in the south (see Mountney and Russell 2006 and references therein), formed following the deflation of the finer materials (Arnalds et al. 2001a, b). Some of the most outstanding landforms derived from aeolian deflation are *vegetation cliffs*, with heights ranging from 0.3 to 2.5 m. *Deflation cliffs* are also common in those areas covered by fines in the highlands and lowlands (Priesnitz and Schunke 1983).
ii. *Erosion* and *abrasion* produce the wastage of rock surfaces, soil or vegetation due to the impacts of the particles transported by wind. As a result, it produces smooth and polished-like surfaces. In case of loose debris (pebbles, cobbles, boulders), it generates *facets* separated by sharp angles, thus being these debris called *ventifacts* (Priesnitz and Schunke 1983). They often appear in armoured surface lags (Fig. 3), demonstrating the efficiency of aeolian transport (Mountney and Russell 2004) and being indicative of recent prevailing wind directions (e.g. Baratoux et al. 2011). On outcrops, wind erosion produces *wind-sculpted rocks*, conditioned also by the differential hardness of the strata so that the less resistant ones are the firstly eroded and *caves* and concavities are generated while the hardest are lest eroded and form *ledges* and *overhangs*. *Micro-yardangs* also appear in the highlands (e.g. the Lambahraun lava field); they consist on aeolian scour-remnants or jagged ridges of centimetric scale, formed on wet sand in deflation areas, which then disappear as the sand dries (Baratoux et al. 2011). Aeolian erosion can easily be observed in the Late Pleistocene móberg formation (Einarsson 1994) and in soils, which remain in some cases after sand encroachment as wind-eroded banks or soil scarpments (*rofabarð*). Erosion rates in these features range from < 1 to 41 cm yr^{-1} (Arnalds and Ragnarsson 1994; Fridriksson 1995; Fridriksson and Gudbergsson 1995; Arnalds 2000; Dugmore et al. 2009). Aeolian abrasion has been quantified in a very few locations, such as the Elborgir volcanic centre (the southwest of Langjökull), where laves cordées have been affected by rates up to 12 $\mu m\ yr^{-1}$ (Baratoux et al. 2011).

iii. *Sedimentation* occurs where wind finds an obstacle, or it simply slows down. The finer materials are blown from there to large distances, contributing to form a surface loessic horizon at larger distances (Boulton and Dent 1974). In the gravel lag areas, sand accumulates at the leeward side of the protruding rocks. Subsequently, winter frost-heave processes uplift of coarse clasts (see Ashwell 1966), covering the sand layers, and as a result, surfaces become gravelly and irregular, and sand keeps accumulating (rates of 0.1–1 mm yr^{-1}; Arnalds 2010b) as it drifts on the surface. In general, *dunes* are one of the most characteristic aeolian landforms, although they are rare in areas under humid boreal or sub-arctic climates due to the limitation of the aeolian sediment transport by soil moisture (Mountney and Russell 2006). Dunes usually develop on soils composed of silt, fines and coarse sand (Johannesson 1960), and cover near to 2000 km^2 inland (Runólffsson 1978). Repeated sediment supply (*jökulhlaups*), wind regime, type and coverage of vegetation, water table level, glacial retreat and freezing of the sand surface (ponds, pore ice sublimation) are amongst the factors that control the development, distribution, and preservation of dunes (Mountney and Russell 2004, 2009). The most extensive dune fields are located in general in the sandur plains of southern Iceland, e.g. Sólheimasandur (Mountney and Russell 2006), Skógarsandur, Mýrdalssandur (Einarsson 1994) and Skeiðarársandur (Bogacki 1970; Hine and Boothroyd 1978; Mountney and Russell 2009); however, some other dune fields occur at the western and northern coastal areas (Doody 2005). Dune fields can develop in a few decades following the end of episodic flood events as Mountney and Russell (2006, 2009) showed for Sólheimasandur and Skeiðarársandur. In such areas, important aeolian dune systems developed, with ellipsoidal dune individuals up to 3–8 m high and 8–12 m long. Such fields are usually vegetated at different degree, by lime grass (*Leymus arenarius*, *Leymus mollis*) and marram grass (*Ammophila*) amongst other species, which contribute to trap and accumulate sediment, stabilize and form isolated and non-migratory dune individual (*nebkhas*). These features are usually characterized by steep flanks (30–50°) lacking vegetation, which originate *vegetation cliffs*. Dune individuals are usually separated by low relief areas (*interdune* corridors or depressions) up to 5–10 m wide and several hundred meters long (Mountney and Russell 2009). Other features that have been reported from this area are *climbing dunes* (e.g. 24 m high near the Ingólfshöfði mountain; King 1956) and *transverse dunes* (up to 2 m high, in Skeiðarársandur; Hine and Boothroyd 1978; Mountney and Russell 2009). In addition to southern Iceland, this type of dunes—up to 1–4 m high—can be also found in northern (Kvensodull; Arnalds et al. 2001a, b) and north-western Iceland, in the tephra deposits of the highlands (Einarsson 1994).

Some other minor aeolian sedimentation features are *wind ripples* and *megaripples*, also characteristic of sandy sheet areas. Large-scale and sinuous-crested ripples and megaripples originated by secondary winds have been reported from Sólheimasandur, Skeiðarársandur and the Askja region amongst other areas, occurring in dune flanks, wet and damp interdune corridors (*adhesion ripples*) and at the lee of pebbles. Such features show wavelengths up to 1.5–3 m, heigths up to 0.1–0.3 m and spacings of 0.2–6 m, increasing downwind (Mountney and Russell 2004, 2006).

3.2 Seasonally Frozen Ground Features

3.2.1 Patterned Ground Features

Ballantyne (2007) uses this term referring to "terrain that exhibits regular or irregular surface patterning, most commonly in the form of circles, polygons, irregular networks, or stripes", which "are characteristic of, but not necessarily confined to, mantle subject to intensive frost action" (Washburn 1956). According to Washburn (1979), deep freezing and moisture content are decisive for the development of these landforms.

Hawkes (1924), Troll (1944), and Thorarinsson (1951) conducted the first descriptive studies on this topic in Iceland. Thorarinsson (1951) reported the occurrence of patterned ground features throughout the island both in vegetated and bare areas, but especially in the former, to the detriment of coastal *sandurs* and extensive ground moraine areas inland. The alternation of fine and coarse debris at ground surface are the main characteristic of the *sorted patterned ground*. However, in other occasions the alternation alludes to small topographic features (i.e. microrelief; e.g. mounds-depressions or ridges-furrows) or vegetated and unvegetated surfaces, which are identified as *non-sorted patterned ground* (Ballantyne 2007).

a. Sorted patterned ground

Sorted patterned ground usually develops on barely vegetated or even completely bare ground, where abundant clasts are deposited or embedded in fine sediment, resulting in 'cells' or 'bands' depending on the surface slope gradient (Ballantyne 2007). Schunke (1975) and Priesnitz and Schunke (1983) suggested that sorted patterned ground is widespread throughout Reykjavík. This group of landforms includes stone polygons, stone stripes, debris islands, debris stripes and stone rosettes, whose sizes (diameter) range from 0.1–0.5 m for the small features to more than 0.5 and even 1–3 m for the larger ones (Priesnitz and Schunke 1983).

Sorted polygons are more evident in areas with low aeolian sedimentation (Arnalds 2015). Early descriptions from Iceland are found in Thoroddsen (1914), amongst others. Thorarinsson (1951) stated that polygons can easily form following soil erosion and showed that their diameter increases with elevation from 60 cm to even 10 m from low-level coastal areas to the inland and basalt plateaus at 600–1000 m a.s.l. Similar conclusions were reached by Priesnitz and Schunke (1983) and Feuillet et al. (2012). The latter explained the positive correlation between altitude and diameter/clast size to a more effective snow deflation at the plateau, and the subsequent greater frost severity and freezing depth. The largest active polygons at the plateau may be associated with permafrost, at least discontinuous (see Etzelmüller et al. 2007; Farbrot et al. 2007b). The topography above 1000 m a.s.l. seems not to be adequate for the development of polygons. Thorarinsson (1964) and Friedman et al. (1971) proposed > 200–300 m a.s.l. as the upper limit for these landforms. However, Priesnitz and Schunke (1983), suggested that the lower altitudinal boundary of polygons is 650–700 in the south, and 340 m a.s.l. in the north.

Sorted nets up to 1.2 m diameter (cells of 10–30 m) have been observed on recent terminal, flute and ground moraines of the Fláajökull (southeast Iceland) (Dąbski 2005; Dąbski and Gryglewicz 1998) and Breiðarmerkurjökull outlet glaciers (Czerwiński 1973), in Hrútafjörður fjord (northern Iceland) and the Búlandstindur mountain (eastern Iceland), where Thorarinsson (1953) described 'anchored' stone and 'floating' gravel polygons with diameters up to 2–3 m and 30–80 cm, respectively. Where slope gradient increases above 2–3°, sorted nets and polygons evolve to *sorted stripes / striated soils* up to decametric scale (Fig. 4) (Bout et al. 1955; Priesnitz and Schunke 1983; Müller et al. 1986; Dąbski 2005), which can be found even on solifluction treads (Douglas and Harrison 1996).

The occurrence of past frost activity originating these landforms is seldom documented. Thorarinsson (1964) attributed big subsoil ('fossil') polygons in coastal plains and valley bottoms to past colder periods. Feuillet et al. (2012) hypothesized that some large relict polygons located on raised beaches in Skagafjörður might have been formed in colder Holocene climate conditions. Conversely, Schunke (1974) suggested that frost-crack macropolygons (15–35 m in diameter) at the surroundings of Hofsökull and Langjökull (western central Iceland) were formed between the 16th and 19th centuries (LIA), although they were reactivated during the 1960s cold spell (see Einarsson 1984, 1991b).

b. Non-sorted patterned ground

In other cases, seasonal frost produces irregular networks of *frost-crack polygons*, indicative of hard continental frost climate (Schunke 1974). They usually form in detritic materials, and where cracking occurs regularly because of a stable frost climate (Friedman et al. 1971). Their existence was first reported by Thorarinsson (1954, 1964), Bout et al. (1955) and Thorsteinsson (1956) around Hofsjökull, Tungnafellsjökull and the vicinity of Eyjafjallajökull and Mýrdalsjökull (200–260 m a.s.l.). The origin of these landforms has been attributed to an intensification of the frost action in cold periods and the impact of katabatic winds at the glacier margins (Schunke 1979; Priesnitz and Schunke 1983; Krüger 1994). In fact, polygons up to 1.8 m in diameter and cracks up to 0.3 m deep have been reported from a moraine on the Mælifellssandur field in the southern highlands.

According to Priesnitz and Schunke (1983), cracks observed in Iceland do not contain ice, and usually appear as fissures up to 1–5 cm wide and 0.6–0.8 m deep. Friedman et al. (1971) identified analogue large-scale tetragons, pentagons, and irregular polygons (20–30 m to 100 m diameter) in several locations, namely Hekla, Vatnajökull, Hofsjökull and Langjökull, between 300 and 800 m a.s.l. Such polygons were associated to the existence of frost cracks and furrows from 1.5 mm to 30–45 cm wide and 20–25 cm deep. The authors ruled out the relation of these polygons to *ice-wedge polygons*; they rather pointed out to an increase of the frost-crack occurrence during the 1960s. Conversely, Thorarinsson (1954) identified some very regular patterns of squares (10–15 m in cellsize) in a flat area at 750 m a.s.l. in Sprengisandur, between Hofsjökull and Langjökull, in central Iceland. Although he recognised that the area was probably not under permafrost regime, he was 'inclined to interpret these patterns as nets of *ice-wedges*', which might have formed in extremely cold periods, such as during the end of the LIA, around the 1880s.

3.2.2 Thúfur

Thúfur consist of miniature cryogenic and minerogenic vegetated mounds originated on periglacial environments resulting from the lateral-horizontal cryostatic pressure derived from deep seasonal freezing, differences in penetration of the freezing front and the moisture movement towards it during the freezing period (Schunke 1977; Grab 2005; Van Vliet-Lanoë et al. 1998). Thus, they are synonym of earth hummock, frost hummocks, earth grass mounds or soil hummocks amongst others. Its meaning and classification as a periglacial landform have been debated in the literature (see Thorarinsson 1951, p. 148; Sveinbjarnardóttir et al. 1982; Van Vliet-Lanoë et al. 1998; Grab 2005, p. 140).

This landform was firstly described at the beginning of the twentieth century (Jónsson 1909; Gruner 1912; Thoroddsen 1913, 1914; Hawkes 1924; amongst others), as 'one of the most common microtopographical features of the Icelandic landscape' (Webb 1972) if not the most one (Priesnitz and Schunke 1983). It is well known by farmers, as agriculture and livestock tend to ease the growth and regrowth of thúfur (Thoroddsen 1914; Hawkes 1924; Webb 1972; Sveinbjarnardóttir et al. 1982; Arnalds 2015), but when undisturbed they contribute to erosion (Ólafsdóttir and Guðmundsson 2002). Amongst the most detailed studies of this landform, stand out those carried out by Mölholm-Hansen (1930), Thorarinsson (1951), Preusser (1976), Schunke (1977), Gerrard (1992), and Van Vliet-Lanoë et al. (1998).

Thúfur form in most of the vegetated surfaces in Iceland, especially in areas of gentle slope within a MAAT range of 3–6 °C (Grab 2005; Arnalds 2015) and their development is favoured by the wetness of the Icelandic climate (Van Vliet-Lanoë et al. 1998). The ground thermal regimes under which these landforms appear has been also subject of debate. Schunke and Zoltai (1988) restricted the term to those found in areas with seasonal frost, while e.g. Grab (2005) included also permafrost areas. However, Priesnitz and Schunke (1983) indicated that this feature is not conditioned by a special frost regime and not indicative of permafrost. At Icelandic scale, these landforms appear both in the lowlands and highlands (Priesnitz and Schunke 1983), but especially at the north-eastern, northern and western zones, in areas with fine soil textures and higher degree of soil development, whereas they are less present in those areas with tephra layers, a thick snow cover or at birch forests (Thorarinsson 1951; Arnalds 2015). Thúfur occurs in clay-rich soils (mean silt/clay fraction of

12–60% in Iceland; Schunke and Zoltai 1988; Gerrard 1992; Thorarinsson 1951; Arnalds 2015), wind-blown sediment from volcanic loess, openwork till or glacial substratum (Van Vliet-Lanoë et al. 1998).

This landform grows in numerous groups, with spacing of 10–15 cm (Schunke 1977), and have flat to convex shapes (Van Vliet-Lanoë et al. 1998) (Fig. 4). Their diameter typically ranges from 50 to 200 cm, and they have heights from less than 10 cm to even more than 50 cm and 1 m (Thorodssen 1914; Sveinbjarnardóttir et al. 1982; Gerrard 1992). The tallest form in the transition areas between wetlands and heathlands, and where the water table is at 20–60 cm depth (Johannesson 1960; Arnalds 2015). However, with a high-water content in the soil, predominance of allophane as clay materials, and animal grazing, they can also occur even with the water table being at higher depth (Arnalds 2010a, 2015), although an excessive moisture may prevent from its development (Sveinbjarnardóttir et al. 1982, p. 86).

Webb (1972) found that thúfur resulted from a cyclical process and that the larger features—sometimes flat-topped (Preusser 1976; Van Vliet-Lanoë et al. 1998)—were located at higher elevations due to a more intense frost action (see e.g. Thorarinsson 1951). Nevertheless, its growth in altitude is only possible until the presence of grasses diminishes enough to impede hummock development (Schunke 1977). In addition, it has been suggested that they may be polygenetic, with a cumulative building process resulting from both frost heave and aeolian accretion events (Van Vliet-Lanoë et al. 1998). Its life cycle was suggested to last for about 25 years (Webb 1972), although they may redevelop within 10–20 years after tilling (Gruner 1912; Van Vliet-Lanoë et al. 1998), which makes them distinguishable from other hummock-like landforms. From a geochronological point of view, they are linked to the climate deterioration following the end of the HTM, with four phases of development at around 4.5 ka, after 2.8–2.6 ka BP, prior to 1104 CE and during two stages of the LIA (Van Vliet-Lanoë et al. 1998; Webb 1972).

3.3 Landforms with Active Ice Core and Permafrost Indicators

3.3.1 The Icelandic Flá: Palsas, Lithalsas and Thermokarst Mounds

Although most of the Icelandic inland plateaus are bare (e.g. lava fields or ground moraines amongst others), there are some densely vegetated areas in the form of bogs, which are affected by permafrost and are indicated with the Icelandic word *flá*. The first thorough studies of these areas were carried out by Thoroddsen (1899, 1905–1906, 1911, 1913, 1913–1915, 1914), Hannesson (1927) and Steindórsson (1945). Hannesson (1927) characterized this landscape as a sub-arctic tundra. Those bogs are characterized by an undulating topography where numerous depressions in the form of lakes, pools and ponds alternate with large hummocks (Fig. 5), and were defined originally as the only permafrost area of Iceland in the official IPA map (Brown et al. 1995).

In this landscape, hummocks are usually called *rústs* (Thorarinsson 1951) or *rústir* (Saemundsson et al. 2012) in Icelandic, and most commonly with the Lapp-derived word *palsas* (see Friedman et al. 1971). Their structure consists of a cover

Lithalsa (Central Northern Iceland)
Þorsteinn Sæmundsson

Thermokarst relief (Central Norhtern Iceland)
Þorsteinn Sæmundsson

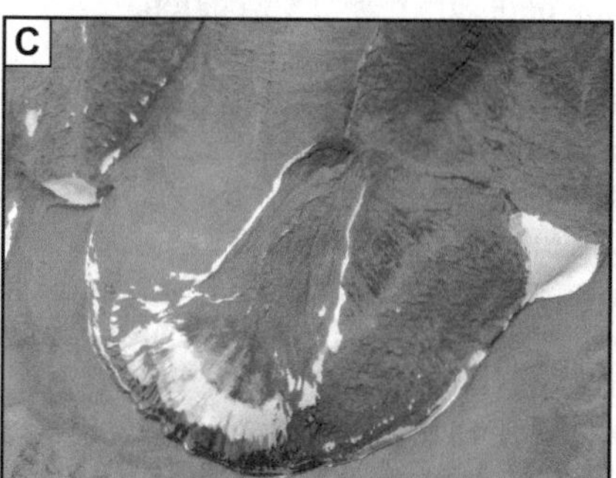

Rock glacier (Nautárdalur, Tröllaskagi)
Landmælingar Íslands

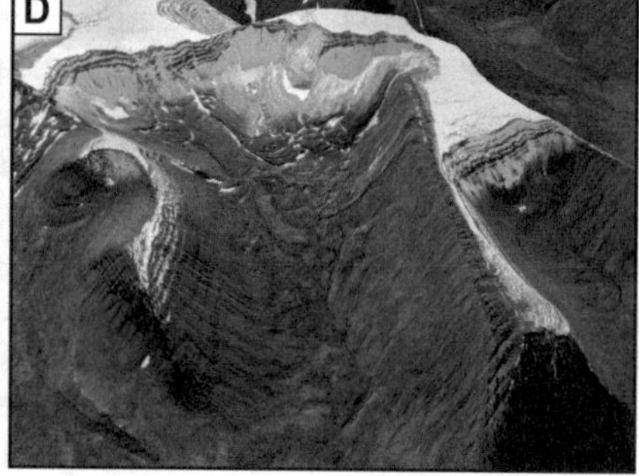

Rock glaciers (Fremri-Grjótárdalur, Tröllaskagi)
Google Earth®

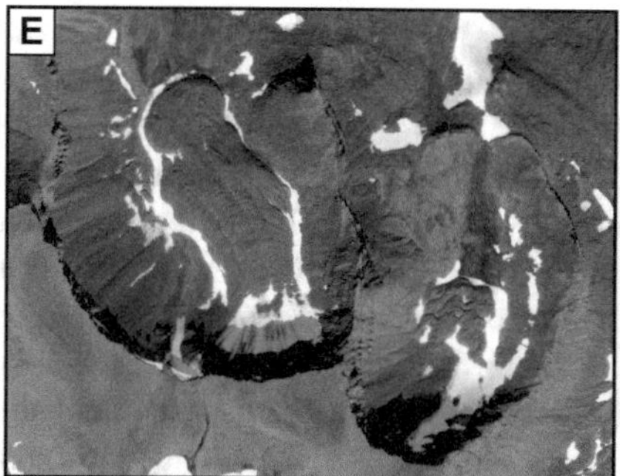

Rock glaciers (Hofsdalur, Tröllaskagi)
Landmælingar Íslands

Ice-cored moraine (Iceland)
Bernd Etzelmüller

Fig. 5 Pictures of landforms with active ice core and permafrost indicators of peat (15–25 cm thick), and underlying core of ice cement and segregated ice on sand layers resting on the substratum (e.g. ground moraine, sandur) regardless their stage of development. Icelandic palsas may be both organic (frozen peat core) and mineral (frozen mineral soil), being classified as *lithalsas* in the latter case. According to field measurements, their active layer thickness ranges between 40 and 80 cm, with permafrost exceeding 5 m depth (Sæmundsson et al. 2012). The frozen surface has been usually found at 0.5-2 m depth, identified as a hardly penetrable layer (Friedman et al. 1971; Hirakawa 1986). Thus, these landforms are a good indicator of (discontinuous) permafrost. The areas between palsas tend to thaw in summer, which evidences the collapse stage and partial decay of permafrost, resulting in the development of a *thermokarst* topography (Hirakawa 1986). These humps are usually level, with flat tops (*plateau-based landforms*; cf. Hirakawa 1986) free of snow but with abundant alpine vegetation. Usually, hummocks lack stones and their (perpendicular) sides show abundant fissures. Palsas are variable in shape, from circular to oval, and in size, as they can be between 0.5 and 3 m high—increasing with elevation—, 8–15 m long and 4–12 m wide (Thorarinsson 1951; Hirakawa 1986; Sæmundsson et al. 2012; Emmert and Kneisel 2021) (Fig. 5). Their flanks change in slope angle depending on their stage of evolution and are steeper towards more mature stages. Interestingly, the slopes facing pools are free of vegetation.

The existence of palsas is conditioned to a dense peat and vegetation cover, which insulates the palsa core form summer melting (Seppälä 1986, 1988, 1995). At their top, palsas show horizontal and dry (i.e. snow blown) surfaces, limiting a thin snow cover, which precludes their existence in the vicinity of slopes with a thicker snow cover (Thorarinsson 1951; Hirakawa 1986; Emmert and Kneisel 2021). Thórhallsdóttir (1996) highlighted the existence of water near the surface, and Saemundsson et al. (2012) pointed to the crucial effect of periodic deposition of tephras, which enhance the insulating effect of the ice core (Kellerer-Pirklbauer et al. 2007).

These landforms are found in the inland plateaus, to the north of the largest ice caps Langjökull, Hofsjökull and Vatnajökull, where a continental climate, a thin snow cover and large amounts of aeolian dust and tephra deposition prevail (Thorarinsson 1951; Saæundsson et al. 2012). Other important palsa areas are found also in eastern Iceland, covering about 93 km^2 (Ottósson et al. 2016). Hannesson (1927) was the first to map the regional distribution of flá. He found it in the inland plateaus of Langjökull (NW-N; 400–550 m a.s.l.), Hofsjökull (NW- N-S; 550–720 m a.s.l.), the terrain between Hofsjökull and Vatnajökull (750–800 m a.s.l.), the Möðrudalur valley (SE-E; 460–560 m a.s.l.) and the Mt. Snæfell (NW–SE; 600–700 m a.s.l.), which were then studied by later researchers (Friedman et al. 1971; Schunke 1975; Hirakawa 1986). The lower altitudinal limit is marked by the occurrence of peat thawing during the summer, probably close to the 0 ºC isotherm (Thorarinsson 1951). The best studied palsa sites are in the highlands of central Iceland, namely Orravatnsrústir (Hirakawa 1986; Sæmundsson et al. 2012; Emmert and Kneisel 2021) and Þjórsárvel (Thórhallsdóttir, 1994, 1996), where the greatest concentration exists. Other marginal secondary palsa areas are found in Blanda (northern Iceland), Jökuldalsheiði and Fljótsdalsheiði (eastern Iceland) (Sæmundsson et al. 2012).

Flá is a dynamic landscape with constant and balanced development and destruction of palsas (see e.g. Kristinsson and Sigurðardóttir 2002; Saundsson et al. 2012; Emmert and Kneisel 2021). Once they reach a mature stage, they produce a morphology of hollows with ring-shaped borders. Such depressions show various shapes, sizes (5–50 m wide) and depths (0.5–3 m), and are filled with water or peat (Hirakawa 1986). Upheaving increases palsa's exposure to wind abrasion (see Emmert and Kneisel 2021) and water collection, and triggers bending and cracking of the surrounding peat surface, resulting in a diminished insulating protection of the frozen core (see Friedman et al. 1971). Abrupt climate fluctuations—warming, drying and desiccation—can also lead to the degradation and disappearance of many of them (Seppälä 1986, 1988, 1995; Sæmundsson et al. 2012). Conversely, local environmental factors (hydrology) can trigger their resume after they have demised (cf. Thórhallsdóttir 1994). Thorarinsson (1951) informed on the degradation of palsas during the warm decades of 1930s and 1940s (see Einarsson 1991b), in good agreement with the disappearance by 1950s of those from the 1920s, reported by Friedman et al. (1971). Icelandic palsas are highly sensitive to the current atmospheric warming, as field observations and aerial photo imagery from Orravatnsrústir show that palsa degradation is in progress, with a decreasing size (11–31% from 1960 to 2004; 15–20% from 2001 to 2010) and increasing thickness of the active layer (Sæmundsson et al. 2012; Emmert and Kneisel 2021), similarly to other places of Iceland (Bergman 1973; Arnalds 2010a). Nevertheless, new palsas are still forming (Sæmundsson et al. 2012).

The origin of these landforms as permafrost degrades still remains unclear, although literature informs on the development of new palsas to the north of Langjökull and Hofsjökull by the 1960s-1970s. Hirakawa (1986) suggested that the thermokarst morphology of the palsa landscape might have developed during the LIA or even earlier, during the Medieval Warm Period (MWP). However, it is well known that the development of the ground ice in the Orravatnsrústir area started after 4.5 ka BP (Hirakawa 1986; Sæmundsson et al. 2012; Emmert and Kneisel 2021), in good agreement with the Neoglaciation cooling trend throughout Iceland (Gudmundsson 1997a; Geirsdóttir et al. 2019).

3.3.2 Rock Glaciers

Rock glaciers are perennially frozen unconsolidated material that moves mainly by downslope creep under permafrost conditions (Barsch 1996; Haeberli 2000; Haeberli et al. 2006; Berthling 2011; Lilleøren et al. 2013). The highest abundance of Icelandic rock glaciers clusters almost exclusively in the alpine cirques of Tröllaskagi (Lilleøren et al. 2013; Andrés et al. 2016). This spatial exclusiveness is explained by: (i) the high efficiency of frost shattering in rock walls; (ii) the high susceptibility of basalt lava flows to weathering (Beylich 2000); and (iii) the abundant rock slope failures, with high volumes of debris supplied. These landforms usually developed at the foot of steep north facing slopes (Lilleøren et al. 2013), due to reduced solar incoming radiation, snow overaccumulation from snow blowing from the plateaus by southerly winds and high rock wall retreat rates (0.4–1.2 mm yr^{-1}) linked to frost shattering (Farbrot et al. 2007a).

The first studies on rock glaciers in Tröllaskagi started between the 1980s and 1990s, with the description and interpretation of the main landforms, measurement of their surface velocities and a tentative dating based on lichenometry (e.g. Martin and Whalley 1987; Martin et al. 1991; Whalley and Martin 1994; Hamilton and Whalley 1995; Whalley et al. 1995a, 1995b). Following these works, a systematic inventory of rock glaciers was performed (Guðmundsson 2000; Lilleøren et al. 2013), their features and structures have been mapped in high detail (Tanarro et al. 2018; Rodríguez-Mena et al. 2021). In addition, their dynamics have been analysed (e.g. Wangensteen et al. 2016; Tanarro et al. 2019, 2021; Campos et al. 2019) and their first absolute ages have been determined through CRE dating (Andrés et al. 2016; Fernández-Fernández et al. 2019; Palacios et al. 2021; Tanarro et al. 2021).

Lilleøren et al. (2013) identified up to 178 rock glaciers in Tröllaskagi, which were classified in several categories according to their: (i) origin, distinguishing between *talus-derived* and *moraine derived*, and (ii) activity, differentiating between *intact* and *relict* rock glaciers, depending on the existence or inexistence of internal ice and surface creep, respectively. A total of 118 intact rock glaciers were identified between 370 and 1160 m a.s.l. (average ~960 m). Relict rock glaciers accounted for 60, between 130 and 950 m a.s.l. (average ~570 m); such low altitudes, close to sea level, indicate a lower limit of permafrost in the past and ice-free conditions. Etzelmüller et al. (2020) suggested that permafrost conditions might have lasted for millennia to develop coastal rock glaciers between the onset of the deglaciation and the HTM. Intact rock glaciers clustered in southern and central Tröllaskagi, while the relict ones distributed more evenly. The most numerous rock glaciers in Iceland are those connected to and derived from glacial activity, i.e. *moraine-derived rock glaciers* (Lilleøren et al. 2013). Etzelmüller et al. (2020) found a good fit between the active rock glacier distribution and the area modelled as Holocene permafrost.

Lilleøren et al. (2013) showed velocities of active intact rock glaciers in the range of 0.2–0.5 m yr^{-1} and maximum of >3 m yr^{-1} for a 46-day period in 2007 based on remote-sensing analyses. First surface displacement measurements in Tröllaskagi were carried out in the Nautárdalur valley by Martin and Whalley (1987) and Whalley et al. (1995a, 1995b), who measured velocities from less than 0.05 to 0.31 m yr^{-1} over a 17-year period (by means of theodolite) and reconstructed past (assumed LIA age) velocities one order of magnitude greater. Supported on cross-correlation matching orthophotos (1985–1994), Wangensteen et al. (2006) monitored boulder mobility of a rock glacier complex in the Fremri-Grjótárdalur and Hóladalur valleys (central Tröllaskagi) and reported displacement rates of <0.74 m yr^{-1} and in the range of 0.11–0.39 m yr^{-1} for active and relict rock glacier generations, respectively. The same rock glaciers were later studied by Kellerer-Pirklbauer et al. (2007) with the same methodology, obtaining similar displacement rates on the range of 0.06–0.74 m yr^{-1}. However, more recently Tanarro et al. (2019) obtained much lower values based on 'point-to-point' boulder tracking over multitemporal aerial photographs (1980–1994), with maximum surface velocities of 0.29 m yr^{-1} and a surface subsidence of 0.5 m. Campos et al. (2019) obtained for the same period similar mean boulder displacements of 0.30 m yr^{-1}, and a mean surface subsidence of 0.63 m on the Júllogil rock glacier. Even lower displacement rates (around ~0.04 and ~0.14 m yr^{-1}) and subsidence have been recently measured in the Hofsdalur valley in Tröllaskagi (Tanarro et al. 2021). All in all, these authors concluded that the main dynamics of such active landforms is subsidence linked to permafrost degradation.

Concerning rock glacier origin, the first studies in Nautárdalur suggested that the rock glacier (Fig. 5) resulted from the abundant debris supply triggered by frequent rockfalls, which finally mantled the surface of a small cirque glacier (Martin and Whalley 1987; Martin et al. 1991; Whalley and Martin 1994; Hamilton and Whalley 1995), neglecting the periglacial nature of the rock glacier. Whalley and Martin (1994) suggested that permafrost was only possible at the surrounding snow-blown plateau areas at 1250–1300 m a.s.l., even during the last 200 years. However, empirical modelling by Etzelmüller et al. (2007) indicated probable permafrost above 850–950 m a.s.l, very close to the elevation of the Fremri-Grjótárdalur rock glaciers, which makes them valuable permafrost indicators. Recent publications considered that some of the rock glaciers of central Tröllaskagi derive from the degradation of glaciers (e.g. Fernández-Fernández et al. 2020; Rodríguez-Mena et al. 2021; Palacios et al. 2021; Tanarro et al. 2021). However, this issue is still under a heated debate, as rock glaciers have been identified and interpreted in coastal areas as resulting from rockslides in permafrost environments through secondary creep and cumulative long-term deformation by means of viscous flow (Cossart et al. 2017; Etzelmüller et al. 2020).

In the first systematic studies of the Icelandic rock glaciers (Martin et al. 1994; Hamilton and Whalley 1995), it was suggested that the Nautárdalur rock glacier was 200-year-old (i.e. formed in during the LIA), based on headwall recession rates and lichenometric dating. From streamline interpolations of surface velocity fields and surface relative dating (Schmidt hammer), Wangensteen et al. (2006) proposed that the oldest landforms of Fremri-Grjótárdalur rock glacier complex (Fig. 5) formed at ~5–4.5 ka, while the younger lobes reactivated at 1.5–1 ka. Later, Kellerer-Pirklbauer et al. (2007) proposed older ages, contemporary with the GH-8.2 ka event and 3.2–3.0 ka BP for both rock glacier generations. However, more recently, ^{36}Cl CRE dating revealed that the origin of the rock glacier complex was older than previously suggested, with the relict lobes being formed at ~11–9 ka (Early Holocene) and the higher active ones after 10 ka, being stagnant since 7–3 ka (Mid-Late Holocene) (Fernández-Fernández et al. 2020). Similar results have been recently obtained by Tanarro et al. (2021), with rock glacier development after 9.2 ± 0.7 ka and stagnation since 2.8 ± 0.4 ka. The last area with available ^{36}Cl CRE data is the Héðinsdalur valley, where the development of an active rock glacier and its immediate stagnation was achieved by 6.3 ± 0.8 ka (Palacios et al. 2021). Recently, Etzelmüller et al. (2020) considered that the distribution of active rock glaciers is in good agreement with the area modelled as Holocene permafrost, while the authors suggest that relict rock glaciers of the coastal areas (northernmost coastline and the Vestfirðir area) may be older than the HTM and may have formed under long enough cold ice-free conditions during the IIS deglaciation.

3.3.3 Ice-Cored Moraines

Ice-cored moraines can be considered a permafrost indicator as long as they appear as stable, steer-fronted and oversized in front of small glaciers (Østrem 1964; Etzelmüller and Hagen 2005) (Fig. 5). Nevertheless, not all of them are permafrost indicators, such as those observed in a partially stable state in the vicinity of the large outlet glaciers of Vatnajökull, Öræfajökull and Mýrdalsjökull, whose existence is explained by a thick insulating debris-cover rather than permafrost conditions. Lilleøren et al. (2013) pointed to the abundancy of ice-cored moraines (87) in the interior of Tröllaskagi: they are located adjacent to active cirque glaciers and coexist with moraine-derived rock glaciers at similar elevations, and with preference of the north-east component. Such landforms exist in Tröllaskagi under a probable permafrost environment (Etzelmüller et al. 2007; Lilleøren et al. 2013).

3.4 Mass Movement Landforms and Periglacial Slope Evolution

3.4.1 Solifluction Landforms

Solifluction is a very common process in the vegetated slopes of Iceland, and even near to sea level (Veit et al. 2012). Priesnitz and Schunke (1983) restricted the occurrence of this process to the highlands, with the lower limits at 450–500 and 250–300 m to the south and the north, respectively. It results from the combination of the freeze–thaw cycles, gravity forces, and the resulting periglacial processes (frost creep, frost heave) and thaw consolidation, which contribute to a slow downward movement of the surficial soil layers during thaw (gelifluction) (Washburn 1979; Harris et al. 1993, 1997; Harris and Davies 2000). Hirakawa (1989) concluded that past solifluidal movement was not derived by shearing, but rolling-over phenomena. The analysis of solifluction facies by Kirkbride and Dugmore (2005) revealed that vertical accretion of soil occurs on vegetated surfaces, while perturbation by frost action is observed on unvegetated surfaces, which together with downslope motion contribute to the attenuation of the vertical profile upslope and folding and overturning (i.e. thickening) downslope.

In Iceland, solifluction appears as amorphous gelifluction sheets, gelifluction terraces and gelifluction lobes (Fig. 6) (Priesnitz and Schunke 1983). As a whole, this process originates terraces and lobes, locally known as *stallar / paldrar* and *jarðsilstungur* (Arnalds 2015). Coarse sediment or turf are commonly found at the front of the lobes—slowing down the movement, becoming thus *stone-banked* or *turf-banked lobes.* In such cases, *stone stripes* can also occur in some sloping treads, as indicators of active frost sorting (Douglas and Harrison 1996). Lobes are greatly vegetated, even including over than 20 plant species (Kirbride and Dugmore, 2005).

These authors report riser heights and tread lengths up to 1.5 and 10–30 m, respectively, at 750–800 m a.s.l. at the Mt. Snæfell (central eastern Iceland). However, Veit et al. (2012) discovered greater sizes, with solifluction lobes up to 2 m high (risers) and even exceeding 60 m long, at 700–750 m a.s.l. in northern Iceland, evidencing deep soil frost and probably permafrost occurrence. These features often include a superimposed subordinate scale, with treads and risers of ~1 and ~0.1 m respectively.

Hirakawa (1989) suggested that this process may have been operating in Iceland since at least 7 ka BP, although later, Kirkbride and Dugmore (2005) hypothesized on a likely age overestimation for such date and suggested that this solifluction activity in eastern Iceland might have started at 2.5 ka BP under a cooler and wetter climate. Overall, it seems that solifluction activity roughly coincided with the main Holocene glacier advance stages (see e.g. Gudmundsson 1997a; Stötter et al. 1999 amongst others). Veit et al. (2012) identified three main phases of solifluction development and aquatic erosion/sedimentation, driven by climate fluctuations in a number of locations, namely the Flateyjardalsheiði heathland, the Mt. Hlíðarfjall, the Krafla volcano, the Vaðlaheiði and Öxnaðalsheiði heathlands. (i) YD (sometime after ~12.2 ka BP); (ii) onset of the Neoglaciation (~3 ka BP); and (iii) onset of the LIA (~1300 CE). After the LIA, the activity of these features would have decreased significantly. The authors suggested that slopes were stable from the early Holocene until 3 ka BP (i.e. HTM), and that solifluction activity around the onset of the LIA might have resulted not only from the climate cooling, but because of human activity (i.e., loss of vegetation cover).

Past climate fluctuations necessarily coincided in variations of solifluidal activity. Hirakawa (1989) found at several sites of central and western Iceland (100–850 m a.s.l.) enhanced downslope rates between 4 and 2 ka BP (Neoglacial times; up to 2.4 and 2.6 mm yr^{-1}) compared to the last 900 yr (1.4 mm yr^{-1}) and the period prior to 4–4.5 ka BP (i.e. HTM, 0.64 m yr^{-1}). In the same way, Kirkbride and Dugmore (2005) report on active solifluction at 2.5 ka BP, followed by plant-induced slope stabilization during the MWP, and reactivation of the pre-MWP lobes during the LIA as the climate and vegetal cover deteriorated. At present, surface displacements up to 11 mm yr^{-1}, decreasing to zero at > 24 cm depth have been recorded in southeast Iceland (Douglas and Harrison 1996).

3.4.2 Ice Needles

It is one of the most ecologically impacting processes in Iceland, occurring in barren surfaces of fine-rich soils, especially andosols due to their lack of cohesion and high vulnerability to erosion. These features form especially during frost-nights following rain events (spring and fall, without a heavy snow cover), when the soil moisture is high. According to field experimentation in the Agricultural University of Hvanneyri, over 40 freeze–thaw cycles conducting to needle-ice development was reported within a period of two months (Madsen 2013). The occurrence of several consecutive frost events leads to a stratified appearance and extraordinary lengths

Fig. 6 Pictures of landforms linked to mass movements and periglacial slope evolution

(Arnalds 2015, p. 124). Its occurrence is widespread in Iceland, although documentary evidence with measurements is only available from a number of locations such at the Mt. Akrafjall (near to Reykjavík), the Hraunhreppur area (western Iceland), the Mt. Laugafell (central Iceland) (Lawler 1988) and at valley slopes in Tröllaskagi (Fig. 6). Such cases consist on polycyclic ice needles up to 50–60 mm long, growing both on unvegetated NE slopes and horizontal silty or sandy surfaces, and raising up stones. This process involves several negative effects: impedes vegetal colonization, enhances surface runoff, and increases fluvial sediment loads (Arnalds 2015).

3.4.3 Talus Slopes and Associated Phenomena

Temperature oscillations around the freezing point, thermal gradient and build-up of the rocks determine an efficient frost shattering, being the bedrock of the upper slopes an efficient debris supplier. The high susceptibility of the Tertiary basalts to weathering (Beylich 2000) through frost shattering makes talus slopes a common feature in the Icelandic valleys. In addition, the maritime climate determines a temperate and isothermal snow at high altitudes and high sub-surface temperature gradients where snow is close to permafrost, thus enhancing frost shattering in rock walls and resulting in high headwall retreat rates (Farbrot et al. 2007a).

Moreover, since the onset of the last deglaciation, gravitational processes activated in the slopes (e.g. landslides, snow avalanches, debris flows, rock falls and rockslides) have generated extensive deposits, in the form of *talus slopes* and *talus cones* in the mountains of north-western, northern and eastern Iceland (Fig. 6), with a high instability throughout the Holocene. Decaulne and Sæmundsson (2006a) studied the morphometry of talus in a number of locations in north-western Iceland and measured slope angles ranging from 19 to 25°. Talus do not resemble simple profiles (i.e. concave, convex, rectilinear), but rather complex, dominated by concave-shaped sectors combined with convex and rectilinear sections. The authors suggest that limited distal concavities and proximal convexities show that debris redistribution is predominant to supply.

Given the similarity of fjord and tributary valleys in central north and eastern Iceland, similar characteristics are also expectable in these areas. In Tröllaskagi, talus slopes are very common in the valleys and deglaciated cirques, where they are reworked by other processes, such as debris-flows (Fig. 6). In this area, talus slopes usually connect the upper rock walls with glaciers and rock glaciers, located in the lower sectors. A higher frequency of *rockfall* events and higher volumes of debris supply in a paraglacial and periglacial environment may lead to the subsequent evolution from debris-free to debris-covered glaciers and rock glaciers, as it has been observed in this area (Andrés et al. 2016; Tanarro et al. 2018, 2019, 2021; Campos et al. 2019; Fernández-Fernández et al. 2020; Rodríguez-Mena et al. 2021; Palacios et al. 2021).

3.4.4 Debris Flows

Debris flows consist of fast-moving gravitational flows mobilizing poorly sorted rock and soil, mixed—and saturated—with water (Morino et al. 2019a). Fjords constitute environments especially prone to its occurrence due to the steep slopes, with high vertical developments from 400 to 900 m, especially in the north-western (e.g. Gudmundsson 1997b), northern and eastern zones of Iceland (Fig. 6). Densities up to 37 debris flow per km^2 have been reported from north-western Iceland (Decaulne and Sæmundsson, 2006a). Shallow regoliths, extensive covers of loose material (e.g. colluvium and till) and weathered basaltic cliffs constitute the sources of debris (Decaulne and Sæmundsson 2007a). Morino et al. (2019a) recognize the role of slope failures and of the "fire-hose effect" as the main triggering processes, with discrete zones of accumulated debris at high angles (<35º) along the chutes/channels and alternating zones of fill and scour. A thorough analysis of the areas, scenarios and sediment requirements in Bíldudalur (north-western Iceland) is found in Glade (2005).

These rapid events are triggered by sudden supplies of water after heavy precipitation or rapid snowmelt that infiltrate in the sediments reaching the impermeable basalt bedrock (e.g. Decaulne et al. 2005), although these circumstances vary both in space and time throughout Iceland (Sæmundsson et al. 2003; Decaulne and Sæmundsson 2007a). In the north-western zone, debris flows are known to occur in autumn and late spring with rain on snow and long episodes of rain. In northern Iceland, they occur in late spring and late summer, after snowmelt (Decaulne et al. 2005). In eastern Iceland, they tend to cluster in summer and autumn, being associated just to rain (Sæmundsson and Decaulne 2007).

Together with slope avalanches, debris flows contribute actively to the development of talus cones and talus slopes (Fig. 6) (Blikra and Sæmundsson 1998; Decaulne 2001). They mobilize huge amounts of materials (up to 3000 m^3), including boulders, blocks, gravel and matrix, stored in the topographic benches of the upper sectors of the slopes, contributing to denudation rates up to less than 10 mm yr^{-1} (Decaulne and Sæmundsson, 2003; Decaulne et al. 2005; Decaulne and Sæmundsson 2006a). Usually, they generate landforms easily identifiable (Decaulne and Sæmundsson, 2003, 2004) and follow an upper lower erosion–deposition pattern (Morino et al. 2019a; Conway et al. 2010). Decaulne et al. (2005) differentiated three sectors common for Icelandic debris flows: (i) an upper *dissection area* up to 5 m deep and 50–70 m long, without deposits; (ii) an intermediate sector 250–370 m long with lateral *levées* constraining a channel up to 2 m deep; and (iii) an *accumulation area* or *impact area* with a slope angle < 5º, limited erosion and lateral spreading. Often, different generations may overlap each other (Decaulne et al. 2005).

3.4.5 Snow Avalanches and Slush Flows

Decaulne and Sæmundsson (2006a) differentiate three types of snow avalanches in north-western Iceland, namely: *dry slab* avalanches, *wet slab* and *slush avalanche*, constituted by water-saturated snow. In general, snow avalanche areas agree well with the extent of the Tertiary Basalts area (Sæmundsson et al. 2003). The highest avalanche risk is concentrated in the western, northern and eastern zones of Iceland, close to the coast, where steep mountain slopes (up to 600–900 and > 1300 m a.s.l.), connect with flat plateaus from which snow is blown (Björnsson 1980). This setting is especially evident in north-western Iceland, while mountains constitute narrow ridges in the eastern zone.

Since the Early Holocene, snow avalanches contribute actively to slope denudation, the redistribution of materials and the production of extensive talus slopes and cones. Decaulne (2001), Decaulne and Sæmundsson (2006a, 2007b) studied the morphometries of snow-avalanche deposits. Boulders are transported from the upper part of the track and deposited at the foot of the slope as diffuse and fresh unsorted accumulations (>0.5 m), parallel-oriented and disconnected from the foot of the talus (Decaulne and Sæmundsson 2010). Accumulation landforms range from *perched* and *balanced boulders* to *avalanche tongues* (Decaulne 2007). The latter have been surveyed in north-western Iceland, and are characterized by concave profiles, steeper in the upper parts (37–56º) and gentler in the lower Sect. (24–34º). Level *fans*, and elevated and flat *roadbanks* have also been reported in the lower areas (Decaulne 2001). Impact marks such as ploughing depressions, *scouring marks* or shifted boulders may appear following snowmelt after the snow avalanche (see Sæmundsson et al. 2008).

Slush flows in Iceland (*krapaflóð*) have been classified as avalanches, debris flow or simply floods, and therefore, there is not abundant specific data on this kind of events (Tómasson and Hestnes 2000). Some of the most prone areas are in north-western Iceland, such as the village of Bíldudalur and the Vesturbyggð municipality, amongst others. Decaulne and Sæmundsson (2006b) linked meltwater of two *slush-flow* releases (years 1997 and 1998) to sustained thaw, triggered by southerly winds and warm/rainy weather. The collapse of hanging snow cornices, ice barriers in lakes upslope and water ponding in gullies are amongst the main intervening mechanisms. The erosive impact of such slush releases is limited to channel scouring, although they can transfer up to 30,000 m^3 (see Tómasson and Hestnes 2000; Decaulne and Sæmundsson, 2006b), with deposition of a chaotic assemblage of blocks on a cone.

3.4.6 Landslides

Kjartansson (1967) used the Icelandic word *hlaup* to englobe rock falls, landslides and avalanches amongst other rapid mass movements. They constitute a real threaten to the society as a whole (Jóhannesson and Arnalds 2000; Sæmundsson and Petursson 2018), which justify the study of these processes. Research on these processes in Iceland started in the second third of the last century (Líndal 1936; Þorarinsson 1954; Jónsson 1957, 1976; Sæmundsson 1973; Whalley et al. 1983; Jónsson and Pétursson 1992; Bentley and Dugmore 1998; Coquin et al. 2015; Peras et al. 2016), being the works by Jónsson (1957), Jóhannesson and Arnalds (2000), Pétursson and

Sæmundsson (2008) and Feuillet et al. (2014) some of the most exhaustive local and regional landslide inventories.

Landslides in Iceland are widespread (Fig. 6), particularly in the weathered Tertiary basaltic area (Sæmundsson and Pétursson 2000). This bedrock is easily erodible by phenomena like frost shattering, creating a constant supply of new debris on the mountain slopes, which recent glaciations have steepened, forming the perfect conditions for the occurrence of rapid mass movements (Pétursson et al. 2010). The main triggering factors promoting landslides in Iceland are heavy precipitation, snowmelt, permafrost thaw and earthquakes (Decaulne et al. 2005; Decaulne and Sæmundsson, 2007a; Sæmundsson et al. 2003, 2014, 2018). In particular, atmospheric temperatures are increasingly contributing to the degradation of permafrost, which is among the hypothesized causes of the unusual higher frequency in landslide occurrence in Iceland over the last decade (Sæmundsson et al. 2018; Morino et al. 2021; Czekirda et al. 2019). Two examples of complex landslides occurred recently on the Mt. Móafellshyrna (central north Iceland) and on Árnesfjall mountain (in northwest Iceland) (Sæmundsson et al. 2018; Morino et al. 2021). Both involved ice-cemented talus and were associated to ground ice thawing. Such failures originated from the detachment of blocks of ice-cemented talus, composed of layers of clay to boulders and evolved into two different phases: fluid-dominated (debris flow, by incorporating oversaturated material), and solid-dominated (coarse debris arranged in ridges and channels). Morino et al. (2021) link the enhanced mobility of the aforementioned failures to the availability of a liquid component originated mainly from the thaw of ground ice. Sæmundsson et al. (2018) and Morino et al. (2021) agree on the permafrost degradation as the decisive trigger for the Móafellshyrna failure, based on: (i) the source located at the lower limit of the (discontinuous) permafrost (Etzelmüller et al. 2007; Czekirda et al. 2019); (ii) sudden temperature increase under a general warming trend; (iii) the original source material (talus cone) slid as a single cohesive mass, due to basal lubrication (due to ground-ice thaw). The authors pointed also to other factors such as heavy precipitation in the preceding days and seismic activity in the case of Mófellshyrna failure.

In the deposits associated with such complex landslides, some peculiar conical mounds of loose talus, i.e. *molards*, have been observed. They derive from ice-rich blocks of loose deposits (soil and rock) mobilized by landslides (Morino et al. 2019b). However, their final and distinctive conical shape (Fig. 6) result from the degradation (thawing) of the cementing ice in the new emplacement. Molards up to 27 m long and 4 m high were surveyed in the abovementioned landslides (Morino et al. 2021). They can occur both isolated and in densely packed elongated individuals, depending on whether they come from a rock-fall-like process or from a rotational movement of the frozen talus, respectively (Morino et al. 2019b). Sæmundsson et al. (2018) estimate the content of ground ice in these features in a 15–20%, thus they constitute a valuable indicator of the presence and ongoing degradation of permafrost. Future atmospheric warming will exacerbate the permafrost degradation, and thus the frequency of permafrost-induced landslides (Czekirda et al. 2019).

The monitoring of active landslides is one of the priority issues of risk management in Iceland to avoid damages in infrastructures, through the University of Iceland,

Icelandic Road Authority and Icelandic Meteorological Office programs. One of the recently surveyed areas is Almenningar, between Skagafjörður and the city of Siglufjörður, where three landslides are monitored (Sæmundsson et al. 2005, 2007), namely Hraun, Þúfnavellir and Tjarnardalir. A displacement rate of 0.7 m yr^{-1} was measured for the last one (Sæmundsson et al. 2005, 2007; Sæmundsson and Petursson 2018). Wangensteen et al. (2006) monitored the displacement of the Tjarnardalur landslide, which they classified as a translational slide, whose movement significantly increases in wet periods such as the spring and autumn months, strongly connected to snowmelt and heavy rainfall, suggesting that its movement is due to the ground water (Sæmundsson et al. 2005, 2007; Sæmundsson and Petursson 2018). Remote sensing is being applied also to monitor landslide displacements in northern and eastern Iceland (e.g. Jónsson 2007). According to their study, displacements have been recorded yearly in all the landslide masses.

Large landslides have also occurred on glaciers (e.g. Friedl et al. 2018), resulting in exceptionally long runout distances. Different examples have been reported in Iceland in the last decades. Kjartansson (1967) reported a rockslide from the Steinsholtsdalur valley (to the north of Eyjafjallajökull) after days of intense and heavy rain and snow melt. The landslide covered part of the glacier, entrained ice and reached a nearby proglacial lake (Steinsholtslón), deriving a glacial lake outburst flood (GLOF) down the walley. In other cases, rockslides and rock avalanches falling over glaciers have reduced their melting and retreating rate (Sigurðsson and Williams 1991; Decaulne et al. 2010; Sæmundsson et al. 2011; Ben-Yehoshua et al. 2022 and Lacroix et al. 2022).

3.4.7 Paraglacial Rock Avalanches

Paraglacial rock avalanches (with paraglacial referring to non-glacial earth-surface processes, sediment accumulations, landforms, land systems and landscapes that are directly conditioned by glaciation and deglaciation; Ballantyne 2002) have shaped the landscape of Iceland. These processes are particularly intense in the Tertiary Basalts formation in north-western, northern and eastern Iceland (see e.g. Jónsson 1957; Sæmundsson 1973; Whalley et al. 1983), with the Skagafjördur area densely populated by these landforms (Pétursson and Sæmundsson, 2008; Cossart et al. 2014). Icelandic climate and geology (steep slopes, sub-horizontal strata, sub-vertical fracture networks) generate the ideal conditions for *rock-slope failures* in north-western, central north and eastern Iceland (e.g. Sæmundsson et al. 2003). The retreat of rock walls and the continued accumulation of talus cones at their foot due to rock avalanches caused the gradual erosion of hillslopes and valley widening (Beylich 2000).

Paraglacial rock avalanches in Iceland have been reported and analysed by different authors (e.g., Coquin et al. 2015, 2016; Decaulne et al. 2016; Feuillet et al. 2014; Jónsson et al. 2004; Jónsson 1957; Líndal 1936; Mercier et al. 2017; Peras et al. 2016; Thorarinsson et al. 1959). After Jónsson's (1957) theory of Icelandic landsliding, it is widely accepted that rock-avalanche activity was due to to post-glacial rebound (Cossart et al. 2014) and was especially intense durngduring and following the Late Weichselian deglaciation and the disappearance of the IIS, especially during the Pleistocene-Holocene transition, as many studies have shown (Jónsson 1957, 1976; Hafliðason 1982; Jónsson et al. 2004; Van Vliet-Lanoë 2005; Decaulne et al. 20102016; Sigurgeirsson and Hjartarson 2011; Mercier et al. 20132017; Feuillet et al. 2014; Cossart et al. 2014; Coquin et al. 2015, 2016; Peras et al. 2016). The millennia that followed the deglaciation were favourable for rock avalanches, as it has been demonstrated in several locations, namely: the Höfðahólar landslide in north Iceland (10.2–8 cal ka BP; Cossart et al. 2014); the Sjávarhólar landslide in the Kjalarnes area (southwest Iceland; ca. 10 cal ka BP; Sigurgeirsson and Hjartarson 2011); the Vatnsdalshólar landslide (northern Iceland; 9.6–6.1 ^{14}C ka BP; Jónsson et al. 2004) and in the Loðmundarfjörður fjord (eastern Iceland; after 2 ka; Hjartarson 1997). However, there has been also a controversial debate about some of these large debris bodies, which showed clear creep movement and features related to rock-glacier development (Guðmundsson 2000), which was later discussed in Etzelmüller et al. (2020). These dissussions have been questioned and rejected by Sæmundsson et al. (2005, 2007), who hypothesized that these large debris bodies are old landslides dated to the beginning of Holocene, as defended by several authors (e.g. Jónsson 1957, 1976; Hafliðason 1982 amongst others).

3.4.8 Cryoplanation and Nivation Landforms

Such phenomena occur especially at higher elevations of the Central Highlands. In such areas, snow patches are perennial or last until mid-summer, depending on the elevation, being the lower limit of the perennial snow patches ~800 m a.s.l. in the south, and ~500 m in the north. The great majority of nivation and cryoplanation features are located in basaltic areas; in hyaloclastite areas, the most common features are the *nivation ravines*. Other common nivation features are *nivation hollows* and *benches* (Priesnitz and Schunke 1983).

4 The Significance of Periglacial Dynamics in Iceland Within the European Context

In the middle of the North Atlantic and close to the Arctic Polar Circle, Icelandic periglacial landscapes show a wide altitudinal spectrum (i.e., large areas in the periglacial zone): from the large areas of the central plateau (i.e. highlands), exceeding 800 m and even 1200–1400 m, down to sea level. This is possible due to the high thermal (frost) and weathering susceptibility of the ubiquitous volcanic rocks and sediments, which counteracts the relatively mild climate and determines a high activity of the periglacial processes. However, the location of Iceland in the interaction area of a hotspot and the mid-oceanic ridge implies a high geothermal flux, which determines a highly dynamic ('warm') permafrost, very thin and shallow, properties that exacerbate its climatic sensitivity. The modelled shifts of permafrost within the last decades provide evidence of that. According to models (there are no direct observations), permafrost occupies less than 10% of the island today, being Tröllaskagi the most extensive area of continuous permafrost. In this sense, the most visible evidence of the distribution of permafrost throughout Iceland today is essentially restricted to palsa sites at the highlands and the active rock glaciers in Tröllaskagi.

Nevertheless, periglacial dynamics in Iceland are not exclusively related to permafrost but also seasonal frost, through frost shattering, freeze–thaw, cryoturbation, and mass movement processes. Current periglacial activity is observed in solifluction lobes and terraces, thúfur, palsas, small-scale patterned ground (sorted polygons/nets, stripes), slope talus, younger generations of rock glaciers, slush, avalanche, debris-flow and landlside deposits, amongst others. Relict landforms also occur, especially at lower elevations in the valleys and close to the coastal areas in north-western and northern Iceland, e.g., rock glaciers of Early Holocene age and large sorted polygons on raised beaches. Moreover, the windy climate, the high number of freeze–thaw cycles, the aridity of soils, glacier shrinking, and the extension of sandy areas enhance the efficiency of aeolian activity, generating a range of erosive, accumulative, and deflationary landforms.

Together with heat flux associated to the volcanic nature of Iceland, the warming trend of the last decades has led to the progressive thinning of permafrost and the thickening of the active layer. Consequently, the Icelandic permafrost is highly vulnerable to further degradation if the current climatic trend continues. One of the most evident signs of permafrost degradation is the thawing of ground ice, which is affecting slope stability and triggering the detachment of blocks of ice-cemented talus, which can lead to the occurrence of complex landslides (involving different processes such as rock falls, debris slides, debris flows). As a result of these processes, *molards* form, being valuable indicators of permafrost degradation. This decaying trend is observed also in the active rock glaciers of northern Iceland whose main current dynamics are surface subsidence and development of collapse depressions driven by the melting of the inner ice core. Projected atmospheric warming will likely lead to the degradation of permafrost and the increase of natural hazards related to thaw processes.

Since the end of the LIA, the interference of the periglacial dynamics and human activity has received more attention given the increase of risk and damage reports. The late 19th century coincided with economical and social changes leading to a population growth, especially in the fjords. Consequently, population spread on the inhabited areas such as slopes, resulting in a superimposition of human activities, infrastructures, and natural processes, and hence an increased risk. This is reflected in an augmented number of reports on avalanches and debris flows throughout the 20th century, especially around the decades of the 1980s and 1990s. In fact, snow avalanches are known to be the largest mass movement threat, responsible of many accidents and economic losses in Iceland since the Norse settlement (e.g. catastrophic snow-avalanche events in 1974 and 1995). This has led to the urgent need of mitigate and prevent future events through new studies and building defence structures in western, northern and eastern Iceland.

References

Ágústsson K, Pétursson HG (2013) Skriður og grjóthrun. In: Sólnes, J. (Ed.), Náttúruvá á Íslandi, eldgos og jarðskjálftar. Háskólaútgáfan, pp 639–645 (in Icelandic)

Andrés N, Palacios D, Sæmundsson Þ, Brynjólfsson S, Fernández-Fernández JM (2019) The rapid deglaciation of the Skagafjörður fjord, northern Iceland. Boreas 48:92–106. https://doi.org/10.1111/bor.12341

Andrés N, Tanarro LM, Fernández JM, Palacios D (2016) The origin of glacial alpine landscape in Tröllaskagi Peninsula (North Iceland). Cuadernos de Investigación Geográfica 42, 341–368. https://doi.org/10.18172/cig.2935

Andrews JT, Harðardóttir J, Helgadóttir G, Jennings AE, Geirsdóttir Á, Sveinbjörnsdóttir ÁE, Schoolfield S, Kristjánsdóttir GB, Smith LM, Thors K, Syvitski J (2000) The N and W Iceland Shelf: Insights into Last Glacial Maximum ice extent and deglaciation based on acoustic stratigraphy and basal radiocarbon AMS dates. Quatern Sci Rev 19:619–631

Arnalds O (2000) The Icelandic 'rofabard' soil erosion features. Earth Surf Proc Land 25:17–28

Arnalds O (2010a) Cryoturbation, frost and soil (Kulferli, frost og mold). Agricultural University of Iceland Publication no 26 (in Icelandic).

Arnalds O (2010b) Dust sources and deposition of aeolian materials in Iceland. Icelandic Agric. Sci. 23:3–21

Arnalds O (2015) The Soils of Iceland. Springer, Dordrecht, p 183

Arnalds O, Dagsson-Waldhauserova P, Olafsson H (2016) The Icelandic volcanic aeolian environment: Processes and impacts—A review. Aeol Res 20:176–195. https://doi.org/10.1016/j.aeolia.2016.01.004

Arnalds O, Gisladottir FO, Sigurjonsson H (2001a) Sandy deserts of Iceland: An overview. J Arid Environ 47:359–371. https://doi.org/10.1006/jare.2000.0680

Arnalds O, Kimble J (2001) Andisols of Deserts in Iceland. Soil Sci Soc Am J 65:1778–1786

Arnalds O, Ragnarsson O (1994) The Djupholar rofabards. SCS Yearbook 5:39–44 (in Icelandic)

Arnalds O, Thorarinsdottir EF, Metusalemsson S, Jonsson A, Gretarsson E, Arnorsson A (1997) Soil Erosion in Iceland. Reykjavik: Soil Conservation Service and Agricultural Research Institute. English translation 2000, orginally published in Icelandic. 157 pp

Arnalds P, Jónasson K, Sigurdsson S (2004) Avalanche hazard zoning in Iceland based on individual risk. Ann Glaciol 38:285–290. https://doi.org/10.3189/172756404781814816

Ashwell IY (1966) Glacial Control of Wind and of Soil Erosion in Iceland. Ann Assoc Am Geogr 56:529–540

Ballantyne CK (2002) Paraglacial geomorphology. Quat Sci Rev 21:1935–2017. https://doi.org/10.1016/S0277-3791(02)00005-7

Ballantyne CK (2007) Periglacial landforms | Patterned Ground, in: Encyclopedia of Quaternary Science. Elsevier, Dordrecht, pp. 2182–2191. https://doi.org/10.1016/B0-44-452747-8/00107-1

Baratoux D, Mangold N, Arnalds O, Bardintzeff JM, Platevoët B, Grégoire M, Pinet P (2011) Volcanic sands of Iceland—diverse origins of aeolian sand deposits revealed at Dyngjusandur and Lambahraun. Earth Surf Proc Land 36:1789–1808. https://doi.org/10.1002/esp.2201

Barsch D (1996) Rockglaciers, Springer Series in Physical Environment. Springer Berlin Heidelberg, Berlin, Heidelberg https://doi.org/10.1007/978-3-642-80093-1

Bell R, Glade T (2004) Quantitative risk analysis for landslides-Examples from Bíldudalur, NW-Iceland. Nat Hazard 4(1):117–131

Bennett MR, Waller RI, Midgley NG, Huddart D, Gonzalez S, Cook SJ, Tomio A (2003) Subglacial defor- mation at sub-freezing temperatures? Evidence from Hagafellsjökull-Eystri, Iceland. Quatern Sci Rev 22:915–923

Bentley MJ, Dugmore A (1998) Landslides and the rate of glacial trough formation in Iceland. Quaternary Proceedings 6:11–15

Ben-Yehoshua D, Sæmundsson Þ, Helgason JK, Belart JMC, Sigurðsson JV, Erlingsson S (2022) Paraglacial exposure and collapse of glacial sediment: The 2013 landslide onto Svínafellsjökull, southeast Iceland. Earth Surface Processes and Landforms 47: 2612–2627. https://doi.org/10.1002/esp.5398

Bergmann B (1973) Um rústir á Húnvetnskum heiðum (Notes on palsas on the north-west highlands). Náttúrufræðingurinn 42:190–198 (In Icelandic)

Berthling I (2011) Beyond confusion: Rock glaciers as cryo-conditioned landforms. Geomorphology 131:98–106. https://doi.org/10.1016/j.geomorph.2011.05.002

Bessason B, Eiríksson G, Thórarinsson Ó, Thórarinsson A, Einarsson S (2007) Automatic detection of avalanches and debris flows by seismic methods. J Glaciol 53(182):461–472

Beylich AA (2000) Geomorphology, Sediment Budget, and Relief Development in Austdalur, Austfirðir, East Iceland. Arct Antarct Alp Res 32:466–477

Bjerknes J, Solberg H (1922) Life cycle of cyclones and the polar front theory of atmospheric circulation. Geophys. Publ. 3:1–18

Bjornsson H (2002) Veður í aðdraganda snjóflóðahrina á norðanverðum Vestfjörðum [Weather preceding avalanche cycles in the North-western Peninsula of Iceland], Icelandic Meteorological Office, G02019, 75 p. http://www.vedur.is/snjoflod/haettumat/index.htm

Björnsson H (1980) Avalanche activity in Iceland, climatic conditions, and terrain features. J Glaciol 26:13–23. https://doi.org/10.3189/s0022143000201007

Björnsson H, Pálsson F (2008) Icelandic glaciers. Jökull 58:365–386

Blikra LH, Sæmundsson Th (1998) The potential of sedimentology and stratigraphy in avalanche-hazard research. In: Hestnes E (ed) 25 years of snow avalanche research. NGI Publication 203, pp 60–64

Bogacki M (1970) Eolian processes on the forefield of the Skeiðarárjökull (Iceland). Bull. Acad. Pol. Sci. Ser. Sci. Geol. Geogr. 18:279–287

Boulton GS, Dent DL (1974) The nature and rate of postdepositional changes in recently deposited till from south-east Iceland. Geogr Ann 56A:121–134

Bout P, Corbel J, Derruau M, Garavel L, Péguy CP (1955) Géomorphologie et Glaciologie en Islande Centrale. Norois 8:461–574

Brynjólfsson S, Schomacker A, Ingólfsson Ó, Keiding JK (2015) Cosmogenic 36Cl exposure ages reveal a 9.3 ka BP glacier advance and the Late Weichselian-Early Holocene glacial history of the Drangajökull region, northwest Iceland. Quatern Sci Rev 126:140–157

Brown J, Ferrians OJ, Jr, Heginbottom JA, Melnikov ES (1995) Circum-Arctic Map of Permafrost and Gound-ice Conditions. IPA and USGS: Reston, Virginia.

Bullard JE (2013) Contemporary glacigenic inputs to the dust cycle. Earth Surf Proc Land 38:71–89. https://doi.org/10.1002/esp.3315

Campos N, Tanarro LM, Palacios D, Zamorano JJ (2019) Slow dynamics in debris-covered and rock glaciers in Hofsdalur, Tröllaskagi Peninsula (northern Iceland). Geomorphology 342:61–77. https://doi.org/10.1016/j.geomorph.2019.06.005

Caseldine C, Geirsdottir A, Langdon P (2003) Efstadalsvatn—a multiproxy study of a Holocene lacustrine sequence from NW Iceland. J Paleolimnol 30:55–73

Caseldine C, Langdon P, Holmes NJQSR (2006) Early Holocene climate variability and the timing and extent of the Holocene thermal maximum (HTM) in northern Iceland. Quat Sci Rev 25:2314–2331

Conway SJ, Decaulne A, Balme MR, Murray JB, Towner MC (2010) A new approach to estimating hazard posed by debris flows in the Westfjords of Iceland. Geomorphology 114(4):556–572

Coquin J, Mercier D, Bourgeois O, Cossart E, Decaulne A (2015) Gravitational spreading of mountain ridges coeval with Late Weichselian deglaciation: Impact on glacial landscapes in Tröllaskagi, northern Iceland. Quatern Sci Rev 107:197–213. https://doi.org/10.1016/j.quascirev.2014.10.023

Coquin J, Mercier D, Bourgeois O, Feuillet T, Decaulne A (2016) Is gravitational spreading a precursor for the Stífluhólar landslide (Skagafjörður, Northern Iceland)? Géomorphologie: relief. Processus. Environment 22(1):9–24

Cossart E, Mercier D, Coquin J, Decaulne A, Feuillet T, Jónsson HP, Sæmundsson Þ (2017) Denudation rates during a postglacial sequence in Northern Iceland: example of Laxárdalur valley in the Skagafjörður area. Geogr Ann Ser B 99(3):240–261

Cossart E, Mercier D, Decaulne A, Feuillet T, Jónsson HP, Sæmundsson T (2014) Impacts of post-glacial rebound on landslide spatial distribution at a regional scale in northern Iceland (Skagafjörður). Earth Surf Proc Land 39:336–350. https://doi.org/10.1002/esp.3450

Czekirda J, Westermann S, Etzelmüller B, Jóhannesson T (2019) Transient Modelling of Permafrost Distribution in Iceland. Front Earth Sci 7:1–23. https://doi.org/10.3389/feart.2019.00130

Czerwiński J (1973) Niektóre elementy mikroreliefu na przedpolu Breidamerkurjökull (Islandia) i zagadnienie tzw. 'fluted moraine'. Czasopismo Geograficzne 44(2), 305–314

Dąbski M, Gryglewicz E (1998) Selected forms of frost sorting in the marginal zone of Fláajökull (Iceland). Biul Peryglac 37:19–34

Dąbski M (2005) Small-scale sorted nets on glacial till, Fláajökull (southeast Iceland) and Elisbreen (northwest Spitsbergen). Permafrost Periglac Process 16:305–310. https://doi.org/10.1002/ppp.527

Dagsson-Waldhauserova P, Arnalds O, Olafsson H, Hladil J, Skala R, Navratil T, Chadimova L, Meinander O (2015) Snow-Dust Storm: Unique case study from Iceland, March 6–7, 2013. Aeol Res 16:69–74. https://doi.org/10.1016/j.aeolia.2014.11.001

Decaulne A (2001) Dynamique des versants et risques naturels dans les fjords d'Islande du nord-ouest: l'impact géomorphologique et humain des avalanches et des debris flows. Department of Geography, University Blaise Pascal, Clermont-Ferrand France, PhD, p 391

Decaulne A (2004) Combining geomorphological, historical and lichenometrical data for assessment of risk due to present-day slope processes, a case study from the Icelandic Westfjords, in: Risk Analysis 4, edited by: Brebbia, C. A., WIT Press: Southampton, 177–186

Decaulne A (2005) Slope processes and related risk appearance within the Icelandic Westfjords during the twentieth century. Natural Hazards and Earth System Science 5:309–318. https://doi.org/10.5194/nhess-5-309-2005

Decaulne A (2007) Snow-avalanche and debris-flow hazards in the fjords of north-western Iceland, mitigation and prevention. Nat Hazards 41:81–98. https://doi.org/10.1007/s11069-006-9025-x

Decaulne A, Sæmundsson T (2010) Distribution and frequency of snow-avalanche debris transfer in the distal part of colluvial cones in central north iceland. Geografiska Annaler, Series A: Physical Geography 92, 177–187. https://doi.org/10.1111/j.1468-0459.2010.00388.x

Decaulne A, Eggertsson O, Sæmundsson P (2012) A first dendrogeomorphologic approach of snow avalanche magnitude-frequency in Northern Iceland. Geomorphology 167–168:35–44. https://doi.org/10.1016/j.geomorph.2011.11.017

Decaulne A, Sæmundsson P (2006) Meteorological conditions during slush-flow release and their geomorphological impact in northwestern Iceland: A case study from the Bíldudalur valley. Geogr Ann Ser B 88:187–197. https://doi.org/10.1111/j.1468-0459.2006.00294.x

Decaulne A, Sæmundsson T (2006a) Geomorphic evidence for present-day snow-avalanche and debris-flow impact in the Icelandic Westfjords. Geomorphology 80:80–93. https://doi.org/10.1016/J.GEOMORPH.2005.09.007

Decaulne A., Sæmundsson P (2006b) Meteorological conditions during slush-flow release and their geomorphological impact in northwestern Iceland: A case study from the Bíldudalur valley. Geografiska Annaler, Series A: Physical Geography 88:187–197. https://doi.org/10.1111/j.1468-0459.2006.00294.x

Decaulne A, Saemundsson T (2007b) The role of geomorphological evidence for snow-avalanche hazard and mitigation research in Northen Icelandic fjords, in: Schaefer, V.R., Schuster, R.L.,

Turner, A.K. (Eds.), First North America Landslide Conference. AEG Publication No. 23. Vali, Colorado, pp. 583–592

Decaulne A, Sæmundsson T (2010) Distribution and frequency of snow-avalanche debris transfer in the distal part of colluvial cones in central north iceland. Geogr Ann Ser B 92:177–187. https://doi.org/10.1111/j.1468-0459.2010.00388.x

Decaulne A, Saemundsson Th (2003) Debris-flow characteristics in the Gleidarhjalli area, North-western Iceland. In: Rickenman D, Chen CI (eds) Debris-flow hazards mitigation: mechanics, prediction, and assessment. Mill Press, Rotterdam vol 2, pp 1107–1118

Decaulne A, Saemundsson Th (2004) The 10–12th June, 1999 snowmelt triggered debris flows in the Gleiðarhjalli area, north- western Iceland. In: Beylich, A.A., Saemundsson, Th., Decaulne, A., Sandberg, O. (Eds.), First Science Meeting of the ESF Network SEDIFLUX, Extended Abstracts of Science Meeting Contributions, NNV-2004–03, Saudarkrokur, Iceland, pp. 74–75

Decaulne A, Saemundsson Þ (2007b) Spatial and temporal diversity for debris-flow meteorological control in subarctic oceanic periglacial environments in Iceland. Earth Surf Proc Land 34:613–628. https://doi.org/10.1002/esp

Decaulne A, Sæmundsson Þ (2008) Dendrogeomorphology as a tool to unravel snow-avalanche activity: Preliminary results from the Fnjóskadalur test site, Northern Iceland. Norsk Geografisk Tidsskrift—Norwegian Journal of Geography 62:55–65. https://doi.org/10.1080/00291950802094742

Decaulne A, Sæmundsson Þ, Jónsson HP (2008) Extreme runout distance of snow-avalanche transported boulders linked to hazard assessment; some case studies in Northwestern and Northern Iceland, in: International Symposium on Mitigative Measures against Snow Avalanches. Egilsstaðir, Iceland, pp. 131–136

Decaulne A, Sæmundsson Þ, Pétursson HG, Jónsson HP, Sigurðsson IA (2010) A large rock avalanche onto Morsarjökull glacier, south-east Iceland. Its implications for ice-surface evolution and glacier dynamics. Iceland in the Central Northern Atlantic: hotspot, sea currents and climate change, Plouzane: France 6–8

Decaulne A, Th, Saemundsson, Petursson O (2005) Debris flows triggered by rapid snowmelt in the Gleidarhjalli area, northwestern Iceland. Geogr Ann 87A:487–500

Decaulne A, Cossart E, Mercier D, Feuillet T, Coquin J, Jónsson HP (2016) An early Holocene age for the Vatn landslide (Skagafjörður, central northern Iceland): Insights into the role of postglacial landsliding on slope development. Holocene 26:1304–1318. https://doi.org/10.1177/0959683616638432

Decaulne, A., Saemundsson, T., 1996. The role of geomorphological evidence for snow-avalanche hazard and mitigation research in Northen Icelandic fjords, in: Schaefer, V.R., Schuster, R.L., Turner, A.K. (Eds.), First North America Landslide Conference. AEG Publication No. 23. Vali, Colorado, pp. 583–592.

Dietz AJ, Wohner C, Kuenzer C (2012) European Snow Cover Characteristics between 2000 and 2011 Derived from Improved MODIS Daily Snow Cover Products. Remote Sensing 4:2432–2454. https://doi.org/10.3390/rs4082432

Doody JP (2005) Sand dune inventory of Europe. In: Herrier J-L, Mees J, Salman A, Seys H, Van Nieuwenhuyse H, Dobbelaere I (eds) Proceedings Dunes and Estuaries International Conference on Nature Restoration Practices in European Coastal Habitats. pp 45–54

Douglas TD, Harrison S (1996) Turf-Banked Terraces in Oraefi, Southeast Iceland: Morphometry, Rates of Movement, and Environmental Controls. Arct Alp Res 28:228. https://doi.org/10.2307/1551764

Dugmore A., Buckland P (1991) Tephrochronology and late holocene soil erosion in South Iceland. In: Maizels, J.K., Caseldine, C. (eds) Environmental change in Iceland: past and present. Glaciology and Quaternary Geology, vol 7. Springer, Dordrecht

Dugmore AJ, Gísladóttir G, Simpson IA, Newton A (2009) Conceptual models of 1200 years of Icelandic soil erosion reconstructed using tephrochronology. Journal of the North Atlantic 2:1–18

Eddudóttir SD, Erlendsson E, Tinganelli L, Gísladóttir G (2016) Climate change and human impact in a sensitive ecosystem: the Holocene environment of the Northwest Icelandic highland margin. Boreas 45:715–728

Egilsson JG (1996) Two Destructive Avalanches in Iceland, in: Proceedings of the 1996 International Snow Science Workshop, Banff, Canada. pp 264–267

Einarsson MÁ (1984) Climate of Iceland. In: van Loon H (ed) World Survey of Climatology: 15: Climates of the Oceans. Elsevier, Amsterdam, pp 673–697

Einarsson MA (1991a) Temperature conditions in Iceland 1901–1990. Jökull 41:1–20

Einarsson P (1991b) Earthquakes and present-day tectonism in Iceland. Tectonophysics 189:261–279. https://doi.org/10.1016/0040-1951(91)90501-I

Einarsson P (1994) Geology of Iceland. Rocks and landscape. Second Edition. Mál og Menning, Iceland. 309 p

Einarsson P (2008) Plate boundaries, rifts and transforms in Iceland. Jökull 58:35–58

Eiríksson J (2008) Glaciation events in the Pliocene-Pleistocene volcanic succession of Iceland. Jökull 58:315–329

Emmert A, Kneisel C (2021) Internal structure and palsa development at Orravatnsrústir Palsa Site (Central Iceland), investigated by means of integrated resistivity and ground-penetrating radar methods. Permafrost Periglac Process 32:503–519. https://doi.org/10.1002/ppp.2106

Etzelmüller B, Farbrot H, Guðmundsson Á, Humlum O, Tveito OE, Björnsson H (2007) The regional distribution of mountain permafrost in Iceland. Permafrost Periglac Process 18:185–199. https://doi.org/10.1002/ppp.583

Etzelmüller B, Hagen JO (2005) Glacier-permafrost interaction in Arctic and alpine mountain environments with examples from southern Norway and Svalbard. Geological Society, London, Special Publications 242:11–27. https://doi.org/10.1144/GSL.SP.2005.242.01.02

Etzelmüller B, Patton H, Schomacker A, Czekirda J, Girod L, Hubbard A, Lilleøren KS, Westermann S (2020) Icelandic permafrost dynamics since the Last Glacial Maximum – model results and geomorphological implications. Quatern Sci Rev 233:106236. https://doi.org/10.1016/j.quascirev.2020.106236

European Environment Agency (2015) Country profile—Distinguishing factors (Iceland). Available online at: https://www.eea.europa.eu/soer/2010/countries/is/country-introduction-iceland (last access, 06/09/2021).

Eyþórsson J, Sigtryggsson H (1971) The climate and weather of Iceland. Vald. Pedersen Bogtrykkeri, Copenhagen, 62 pp

Farbrot H, Etzelmüller B, Gudmundsson A, Humlum O, Kellerer-Pirklbauer A, Eiken T, Wangensteen B (2007a) Rock glaciers and permafrost in Trollaskagi, northern Iceland. Zeitschrift Fur Geomorphologie 51:1–16

Farbrot H, Etzelmüller B, Schuler TV, Gudmundsson A, Eiken T, Humlum O, Björnsson H (2007b) Thermal characteristics and impact of climate change on mountain permafrost in Iceland. J. Geophys. Res. 112

Fernández-Fernández JM, Andrés N, Sæmundsson Þ, Brynjólfsson S, Palacios D (2017) High sensitivity of North Iceland (Tröllaskagi) debris-free glaciers to climatic change from the 'Little Ice Age' to the present. The Holocene 27:1187–1200. https://doi.org/10.1177/0959683616683262

Fernández-Fernández JM, Palacios D, Andrés N, Schimmelpfennig I, Tanarro LM, Brynjólfsson S, López-Acevedo FJ, Sæmundsson Þ, Team ASTER (2020) Constraints on the timing of debris-covered and rock glaciers: an exploratory case study in the Hólar area, northern Iceland. Geomorphology 361:107196. https://doi.org/10.1016/j.geomorph.2020.107196

Fernández-Fernández JM, Palacios D, Andrés N, Schimmelpfennig I, Brynjólfsson S, Sancho LG, Zamorano JJ, Heiðmarsson S, Sæmundsson Þ (2019) A multi-proxy approach to Late Holocene fluctuations of Tungnahryggsjökull glaciers in the Tröllaskagi peninsula (northern Iceland). Sci Total Environ 664:499–517. https://doi.org/10.1016/j.scitotenv.2019.01.364

Feuillet T, Mercier D, Decaulne A, Cossart E (2012) Classification of sorted patterned ground areas based on their environmental characteristics (Skagafjörður, Northern Iceland). Geomorphology 139–140:577–587. https://doi.org/10.1016/j.geomorph.2011.12.022

Feuillet T, Coquin J, Mercier D, Cossart E, Decaulne A, Jónsson HP, Sæmundsson T (2014) Focusing on the spatial non-stationarity of landslide predisposing factors in northern Iceland: Do paraglacial factors vary over space? Prog Phys Geogr 38:354–377. https://doi.org/10.1177/0309133314528944

Flóvenz OG, Sæmundsson K (1993) Heat flow and geothermal processes in Iceland. Tectonophysics 225:123–138

Flowers GE, Björnsson H, Geirsdóttir Á, Miller GH, Black JL, Clarke GK (2008) Holocene climate conditions and glacier variation in central Iceland from physical modelling and empirical evidence. Quat Sci Rev 27:797–813

Fridriksson S (1995) Alarming rate of soil erosion of some Icelandic soils. Environ Conserv 22:167

Fridriksson S, Gudbergsson G (1995) Rate of vegetation retreat at rofabards. Freyr, 224–231 (The Icelandic Farmers Society, Reykjavik in Icelandic).

Friedl B, Hölbling D, Tiede D, Dittrich J, Robson BA, Sæmundsson T, Pedersen GBM (2018) Delineation of rock avalanche deposits on glaciers from different remote sensing data. European Geosciences Union General Assembly. Geophysical Research Abstracts 13689

Friedman JD, Johansson CE, Oskarsson N, Svensson H, Thorarinsson S, Williams RS (1971) Observations on Icelandic Polygon Surfaces and Palsa Areas. Photo Interpretation and Field Studies. Geogr Ann Ser B 53:115–145. https://doi.org/10.1080/04353676.1971.11879841

Geirsdóttir A (2011) Pliocene and pleistocene glaciations of Iceland: A brief overview of the glacial history. In: Ehlers J, Gibbard PL, Hughes PD (eds), Quaternary Glaciations – Extent and Chronology. Amsterdam, ISBN 978–0–444–53447–7, pp. 199–210

Geirsdóttir Á, Miller GH, Andrews JT, Harning DJ, Anderson LS, Thordarson T (2019) The onset of Neoglaciation in Iceland and the 4.2 ka event. Clim Past Discuss 15, 25–40.

Geirsdóttir Á, Miller GH, Axford Y, Ólafsdóttir S (2009) Holocene and latest Pleistocene climate and glacier fluctuations in Iceland. Quat Sci Rev 28:2107–2118

Geirsdóttir Á, Andrews JT, Ólafsdóttir S, Helgadóttir G, Harðardóttir J (2002) A 36 Ky record of iceberg rafting and sedimentation from north-west Iceland. Polar Res 21:291–298

Geirsdóttir A, Miller GH, Larsen DJ, Ólafsdóttir S (2013) Abrupt Holocene climate transitions in the northern North Atlantic region recorded by synchronized lacustrine records in Iceland. Quatern Sci Rev 70:48–62

Gerrard AJ (1992) The nature and geomorphological relationships of earth hummocks (thufur) in Iceland. Zeitschrift Für Geomorphologie Supplement Bd 86:173–182

Gisladottir FO, Arnalds O, Gisladottir G (2005) The effect of landscape and retreating glaciers on wind erosion in south Iceland. L. Degrad. Dev. 16:177–187. https://doi.org/10.1002/ldr.645

Gislason SR, Olafsson JO, Snorrason A (1997) Dissolved constituents, suspended concentration and discharge of rivers in Southern Iceland. The Database of the Science Institute, the Marine Institute and the National Energy Authority of Iceland. Science Institute Progress Report RH-25–97

Glade T (2005) Linking debris-flow hazard assessments with geomorphology. Geomorphology 66:189–213. https://doi.org/10.1016/j.geomorph.2004.09.023

Grab S (2005) Aspects of the geomorphology, genesis and environmental significance of earth hummocks (thúfur, pounus): Miniature cryogenic mounds. Prog Phys Geogr 29:139–155

Gruner M (1912) Die Bodenkultur Islands. Archiev für Biontologie 3, 213 pp

Guðmundsson Á (2000) Frerafjöll, urðarbingir á Tröllaskaga. MSc thesis, University of Iceland

Gudmundsson AT (1997a) A lively neighbour—occasionally bad tempered. Atlantica 2:16–22

Gudmundsson HJ (1997b) A review of the Holocene environmental history of Iceland. Quatern Sci Rev 16:81–92

Guðmundsson MT, Larsen G, Hoskuldsson A, Gylfason AG (2008) Volcanic hazards in Iceland. Jökull 58:251–268

Gylfadóttir SS (2003) Spatial interpolation of Icelandic monthly mean temperature data, report. Icelandic Meteorol. Off., Reykjavik. 27 pp

Haeberli W (2000) Modern Research Perspectives Relating to Permafrost Creep and Rock Glaciers: A Discussion. Permafrost Periglac Process 11:290–293

Haeberli W, Cheng G, Gorbunov AP, Harris SA (1993) Mountain permafrost and climatic change. Permafrost Periglac Process 4:165–174

Haeberli W, Hallet B, Arenson L, Elconin R, Humlum O, Kääb A, Kaufmann V, Ladanyi B, Matsuoka N, Springman S, Mühll DV (2006) Permafrost creep and rock glacier dynamics. Permafr. Periglac. Process. 17, 189–214. https://doi.org/ https://doi.org/10.1002/ppp.561

Hafliðason H (1982) Jarðfræðiskýrsla vegna jarðsigs á Almenningum við Siglufjörð. Unnið fyrir: Vegagerð ríkisins, 15 bls.(in Icelandc)

Halldórsson P (1984) Skagafjarðarskjálftinn 1963. Veðurstofa Íslands, Reykjavík (in Icelandic)

Hamilton SJ, Whalley WB (1995) Preliminary results from the lichenometric study of the Nautardalur rock glacier, Trollaskagi, Northern Iceland. Geomorphology 12:123–132

Hannesson P (1927) Frá obygdum I. Rjettur, XII. Akureyri

Haraldsdóttir SH (2004) Snow, snowdrift and avalanche hazard in a windy climate. University of Iceland, Department of Physics, p 47

Haraldsdóttir SH, Olafsson H, Durand Y, Giraud G, Meindol L (2004) A system for prediction of avalanche hazard in the windy climate of Iceland. Ann Glaciol 38(1):319–324

Harris C, Davies MCR (2000) Gelifluction: observations from large-scale laboratory simulations. Arct Alp Res 32:202–207

Harris C, Davies MCR, Coutard J-P (1997) Rates and processes of periglacial solifluction: an experimental approach. Earth Surf Proc Land 22:849–868

Harris C, Gallop M, Coutard J-P (1993) Physical modelling of gelifluction and frost creep: some results of a large-scale laboratory experiment. Earth Surf Proc Land 18:383–398

Hawkes L (1924) Frost Action in Superficial Deposits, Iceland. Geol Mag 61:509–513. https://doi.org/10.1017/S0016756800102018

Helgason, J., 2000. Ground ice in Iceland; possible analogs for equatorial Mars. Second International Conference on Mars Polar Science and Exploration, 21–25 August, 2000, Reykjavik, Iceland. Lunar and Planetary Institute, Houston, TX, USA.

Hine AC, Boothroyd JC (1978) Morphology, processes and recent sedimentary history of a glacial-outwash plain shoreline, southern Iceland. J Sediment Petrol 48:901–920

Hirakawa K (1986) Development of palsa bog in central highland, Iceland. Geographical Reports of Tokyo Metropolitan University 21:111–122

Hirakawa K (1989) Downslope movement of solifluction lobes in Iceland: a tephrostratigraphic approach. Geographical Reports of Tokyo Metro- Politan University 24:15–30

Hjartarson Á, Ingólfsson Ó (1988) Preboreal glaciation of Southern Iceland. Jökull 38:1–16

Hubbard A, Sugden J, Dugmore A, Norðdahl H, Pétursson HG (2006) A modelling insight into the Icelandic Late Glacial Maximum ice sheet. Quatern Sci Rev 25:2283–2296

Hjartarsson Á (1997) Loðmundarskriður. Náttúrufræðingurinn 67:97–103 (in Icelandic)

Icelandic Meteorological Office, 2021. Climate of Iceland. Available online at: https://en.vedur.is/climatology/iceland (last access: 07/09/2021).

Ingólfsson Ó, Björck S, Hafliðason H, Rundgren M (1997) Glacial and climatic events in Iceland reflecting regional North Atlantic climatic shifts during the Pleistocene-Holocene transition. Quatern Sci Rev 16:1135–1144

Jakobsdóttir SS (2008) Seismicity in Iceland: 1994–2007. Jökull 58:75–100

Jensen EH (2000) Úttekt á jarðfræðilegum hættum eftir jarðskjálftana 17. og 21. Júní 2000. Veðurstofa Íslands, Reykjavík (in Icelandic)

Johannesson B (1960) Soils of Iceland. University Research Institute, Dept. of Agriculture, reports series B, no 13.

Johannesson H, Sæmundsson K (1989) Geological map of Iceland, 1:500000, Bedrock geology. Museum of Natural History and the Icelandic Geodetic Survey, Reykjavik

Jóhannesson T (2001) Run-up of two avalanches on the deflecting dams at Flateyri, Northwestern Iceland. Ann Glaciol 32:350–354. https://doi.org/10.3189/172756401781819382

Jóhannesson T, Arnalds Þ (2000) Accidents and economic damage due to snow avalanches and landslides in Iceland. Jökull 81–94

Johannesson T, Jonsson T (1996) Weather in Vestfirðir before and during several avalanche cycles in the period 1949 to 1995. Icelandic Meteorological Office, VI-G98O13. http://www.vedur.is/snjoflod/haettumat/index.htm

Jónsson H (1909) Thufur. Freyr 6(13–15):69–70

Jónsson HB, Norðdahl H, Pétursson HG (2004) Myndaði Berghlaup Vatnsdalshóla? Náttúrufræðingurinn 72(3–4):129–138 (in Icelandic)

Jonsson J (1995) Sandur-sandar. Observations. Research Institute Neðri Ás Report 52, Hveragerði, Iceland. (In Icelandic)

Jónsson Ó (1957) Skriðuföll og snjóflóð, I bindi. Bókaútgáfan Norðri hf, Akureyri (in Icelandic)

Jónsson Ó (1976) Berghlaup. Ræktunarfélag Norðurlands, Akureyri 622 pp. (in Icelandic)

Jónsson Ó, Pétursson HG (1992) Skriðuföll og snjóflóð, II bindi. Skriðuannáll (2. útgáfa). Skjaldborg, Reykjavík (in Icelandic)

Jónsson Ó, Rist S (1972) Snjóflóð og snjóflóðahætta á Islandi. Jökull 21:24–44

Jónsson S (2007) A survey of active landslide movement in east Iceland from satellite radar interferometry. Veðurstofa Íslands. Report 07004 VÍ-VS-03, 85 pp

Kaldal I, Víkingsssson S (1990) Early Holocene Deglaciation in Central Iceland. Jökull 40:51–66

Kellerer-Pirklbauer A, Farbrot H, Etzelmuller B (2007) Permafrost aggradations caused by tephra accumulation over snow-covered surfaces: examples from the Hekla-2000 eruption in Iceland. Permafrost Periglac Process 18:269–284

Kellerer-Pirklbauer A, Farbrot H, Etzelmüller B (2006) The potential of volcanic eruptions for permafrost aggradation: Hekla volcano in Iceland. Geophys Res Abstr 8:09982

King CAM (1956) The Coast of South-East Iceland near Ingólfshöfði. Geogr J 122:241–246

Kirkbride MP, Dugmore AJ (2001) Timing and significance of mid-Holocene glacier advances in northern and central Iceland. J Quat Sci 16:145–153

Kirkbride MP, Dugmore AJ (2005) Late Holocene solifluction history reconstructed using tephrochronology. Geol Soc Spec Pub 242:145–155. https://doi.org/10.1144/GSL.SP.2005.242.01.13

Kirkbride MP, Dugmore AJ (2006) Responses of mountain lee caps in central Iceland to Holocene climate change. Quatern Sci Rev 25:1692–1707

Kjartansson G (1967) Steinsholtshlaup, Central-South Iceland on January 15th, 1967. Jökull 17:249–262

Kneisel C (2010) The nature and dynamics of frozen ground in alpine and subarctic periglacial environments. The Holocene 20(3):423–445

Kneisel C, Sæmundsson Þ, Beylich AA (2007) Reconnaissance Surveys of Contemporary Permafrost Environments in Central Iceland Using Geoelectrical Methods: Implications for Permafrost Degradation and Sediment Fluxes. Geografiska Annaler. Series a, Physical Geography 89:41–50

Kristinsson H, Sigurðardóttir R (2002) Freðmæyrarústir á áhrifasvæði Norðlingaölduveitu. Breytingar á 30 ára tímabili. (Palsas near the Norðlingaalda water diversion: 30 yr changes). The Icelandic Institute for Natural History, NI-02002, Reykjavík, Iceland. (26 pp. In Icelandic)

Krüger J (1994) Sorted Polygons on Recently Deglaciated Terrain in the Highland of Mælifellssandur, South Iceland. Geogr Ann Ser B 76:49–55. https://doi.org/10.1080/04353676.1994.11880405

Kruger J (1996) Moraine ridges formed from subglacial frozen-on sediment slabs and their differentiation from push moraines. Boreas 25:57–63

Lacroix P, Belart JMC, Berthier E, Sæmundsson Þ, Jónsdóttir K (2022) Mechanisms of landslide destabilization induced by glacier-retreat on Tungnakvíslarjökull area, Iceland. Geophys Res Lett 49:1–11. https://doi.org/10.1029/2022GL098302

Larsen G (1998) Eight centuries of periodic volcanism at the center of the Iceland hotspot revealed by glacier tephrostratigraphy. Geology 26:943–946. https://doi.org/10.1130/0091-7613(1998)026%3c0943:ECOPVA%3e2.3.CO;2

Larsen D, Miller GH, Geirsdottir A, Ólafsdóttir S (2012) Non-linear Holocene climate evolution in the North Atlantic: a high-resolution, multi-proxy record of glacier activity and environmental change from Hvítárvatn, central Iceland. Quatern Sci Rev 39:14–25

Lawler DM (1988) Environmental limits of needle ice: a global survey. Arct Alp Res 20:137–159. https://doi.org/10.2307/1551494

Liebricht H (1983) Das frostklima Islands seit dem Beginn der Instrumentenbeobachtung. Bamberger Geographische Schriften, 5

Lilleøren KS, Etzelmüller B, Gärtner-Roer I, Kääb A, Westermann S, Gudmundsson Á (2013) The Distribution, Thermal Characteristics and Dynamics of Permafrost in Tröllaskagi, Northern Iceland, as Inferred from the Distribution of Rock Glaciers and Ice-Cored Moraines. Permafrost Periglac Process 24:322–335. https://doi.org/10.1002/ppp.1792

Líndal J (1936) Hvernig Eru Vatnsdalshólar Til Orðnir. Náttúrufræðingurinn 6:65–75

Madsen KS (2013) Needle ice in Icelandic Andosols: a field experiment in Hvanneyri, Iceland. Unpublished BS thesis, University of Copenhagen and the Agricultural University of Iceland

Maizels J (1997) Jökulhlaup deposits in proglacial areas. Quat Sci Rev 16:793–819

Martin HE, Whalley WB (1987) A glacier ice- cored rock glacier, Tröllaskagi, Iceland. Jökull 37:49–56

Martin HE, Whalley WB, Caseldine C (1991) Glacier fluctuations and rock glaciers in Tröllaskagi, northern Iceland, with special reference to 1946–1986. In: Maizels JK, Caseldine C (eds) Environmental Change in Iceland: Past and Present. Kluwer Academic Publisher, Netherlands, pp 255–265

Martin HE, Whalley B, Orr J, Caseldine C (1994) Dating and interpretation of rock gla- ciers using lichenometry, south Tröllaskagi. North Iceland. Münchener Geogr. Arb. 12:205–224

Mercier D, Cossart E, Decaulne A, Feuillet T, Jónsson HP, Sæmundsson Þ (2013) The Höfðahólar rock avalanche (sturzström): chronological constraint of paraglacial landsliding on an Icelandic hillslope. The Holocene 23(3):432–446

Mercier D, Coquin J, Feuillet T, Decaulne A, Cossart E, Jónsson HP, Sæmundsson Þ (2017) Are Icelandic rock-slope failures paraglacial? Age evaluation of seventeen rock-slope failures in the Skagafjörður area, based on geomorphological stacking, radiocarbon dating and tephrochronology. Geomorphology 296:45–58. https://doi.org/10.1016/j.geomorph.2017.08.011

Mölholm-Hansem H (1930) Studies on the vegetation of Iceland. The Botany of Iceland, Vol. IV., 1

Morino C, Conway SJ, Balme MR, Helgason JK, Sæmundsson Þ, Jordan C, Hillier J, Argles T (2021) The impact of ground-ice thaw on landslide geomorphology and dynamics: two case studies in northern Iceland. Landslides 18:2785–2812. https://doi.org/10.1007/s10346-021-01661-1

Morino C, Conway SJ, Balme MR, Hillier J, Jordan C, Sæmundsson Þ, Argles T (2019a) Debris-flow release processes investigated through the analysis of multi-temporal LiDAR datasets in north-western Iceland. Earth Surf Proc Land 44(1):144–159. https://doi.org/10.1002/esp.4488

Morino C, Conway SJ, Sæmundsson Þ, Helgason JK, Hillier J, Butcher FE, Balme MR, Jordan C, Argles T (2019b) Molards as an indicator of permafrost degradation and landslide processes. Earth Planet Sci Lett 516:136–147

Mountney NP, Russell AJ (2004) Sedimentology of cold-climate aeolian sandsheet deposits in the Askja region of northeast Iceland. Sed Geol 166:223–244. https://doi.org/10.1016/j.sedgeo.2003.12.007

Mountney NP, Russell AJ (2006) Coastal aeolian dune development, Sólheimasandur, southern Iceland. Sed Geol 192:167–181. https://doi.org/10.1016/j.sedgeo.2006.04.004

Mountney NP, Russell AJ (2009) Aeolian dune-field development in a water table-controlled system: Skeiðarársandur, Southern Iceland. Sedimentology 56:2107–2131. https://doi.org/10.1111/j.1365-3091.2009.01072.x

Müller HN, Stütter J, Schubert A, Betzler A (1986) Glacial and periglacial investigations in skidadalur, trollaskagi, northern iceland. Polar Geogr Geol 10:1–18. https://doi.org/10.1080/10889378609377267

Norðdahl H, Pétursson HG (2005) Relative sea level changes in Iceland. New aspects of the Weichselian deglaciation of Iceland. In: C. Caseldine, A. Russel, J. Harðardottir, Ó. Knudsen (eds.), Iceland—modern processes and past environments, Elsevier, Amsterdam, pp 25–78

Norðdahl H, Ingólfsson Ó, Pétursson HG, Hallsdóttir M (2008) Late Weichselian and Holocene environmental history of Iceland. Jökull 58:343–364

Norðdahl H, Th, Einarsson (2001) Concurrent changes of relative sea-level and glacier extent at the Weichselian – Holocene boundary in Berufjörður, Eastern Iceland. Quatern Sci Rev 20:1607–1622

Obu J, Westermann S, Bartsch A, Berdnikov N, Christiansen HH, Dashtseren A, Delaloye R, Elberling B, Etzelmüller B, Kholodov A, Khomutov A (2019) Northern Hemisphere permafrost map based on TTOP modelling for 2000–2016 at 1 km2 scale. Earth Sci Rev 193:299–316

Ólafsdóttir R, Guðmundsson HJ (2002) Holocene land degradation and climatic change in northeastern Iceland. The Holocene 12:159–167. https://doi.org/10.1191/0959683602hl531rp

Ólafsson H (1998) Veður fyrir snjóflóðahrinur í Neskaupstað 1974–1995. Icel. Met. Office, rep. VÍ-G98015- ÚR12

Olafsson H (2004) The dust storm weather Oktober 5th, 2004 (Sandfoksveðrið 5. október 2004). Náttúrufræðingurinn 72, 93–95. In Icelandic, English summary

Ólafsson H, Furger M, Brümmer B (2007) The weather and climate of Iceland. Meteorol Z 16:5–8. https://doi.org/10.1127/0941-2948/2007/0185

Old GH, Lawler DM, Snorrason A (2005) Discharge and suspended sediment dynamics during two jökulhlaups in the Skaftá River, Iceland. Earth Surf Proc Land 30:1441–1460

Østrem G (1964) Ice-cored moraines in Scandinavia. Geogr Ann 46:282–337

Ottósson JG, Sveinsdóttir A, Harðardóttir M (eds) (2016) Vistgerðir á Íslandi, Fjölrit Náttúrufræðistofnunar 54. Garðabær: Náttúrufræðistofnun Íslands (Icelandic Institute of Natural History). In Icelandic with English summary

Palacios D, Rodríguez-Mena M, Fernández-Fernández JM, Schimmelpfennig I, Tanarro LM, Zamorano JJ, Andrés N, Úbeda J, Sæmundsson Þ, Brynjólfsson S, Oliva M, Team ASTER (2021) Reversible glacial-periglacial transition in response to climate changes and paraglacial dynamics: A case study from Héðinsdalsjökull (northern Iceland). Geomorphology 388:107787. https://doi.org/10.1016/j.geomorph.2021.107787

Palsson S, Zophoniasson S (1992) The Skaftárhlaup, 1991, sediments and suspended material. Reykjavík: Orkustofnun, OS-92014/VOD-02. ISBN 9979–827–07–6. In Icelandic. 26 pp.

Peras A, Decaulne A, Cossart E, Coquin J, Mercier D (2016) Distributionand spatial analysis of rockslides failures in the Icelandic Westfjords: first results. Géomorphologie 22(1):25–35. https://doi.org/10.4000/geomorphologie.11303

Petersen GN (2021) Trends in soil temperature in the Icelandic highlands from 1977 to 2019. Int J Climatol 1–12. https://doi.org/10.1002/joc.7366

Pétursson HG, Sæmundsson þ (2008) Skriðuföll í Skagafirði, In : Sæmundsson þ, Decaulne, A, Jónsson, HP (Eds.), Skagfirsk náttúra 2008, Náttúrustofa Norðurlands vestra NNV, Sauðárkrókur, 02:25–30

Pétursson HG, Sæmundsson Þ, Jónsson HP, Brynjólfsson S (2010) Landslides in Iceland, a short review. In Proceedings of the International Workshop on Earthquakes in North Iceland, Húsavík, North Iceland; 44–45

Pétursson HG, Norðdahl H, Ingólfsson O (2015) Late Weischelian history of relative sea level changes in Iceland during a collapse and subsequent retreat of marine based ice sheet. Cuadernos De Investigación Geográfica 41(2):261–277

Preusser H (1976) The Landscape ofIceland: Types and Regions. PhD Thesis. University of Saarland. Junk, The Hague, 363 pp

Priesnit K, Schunke E (1978) An approach to the ecology of permafrost in Central Iceland. Third International Conference on Permafrost, 1978 Edmonton, Canada. National Research Council of Canada, Ottawa, 474–479

Prospero JM, Bullard JE, Hodgkins R (2012) High-latitude dust over the North Atlantic: Inputs from Icelandic proglacial dust storms. Science 335(80):1078–1082. https://doi.org/10.1126/science.1217447

Priesnitz K, Schunke E (1983) The significance of periglacial phenomena in Iceland. Polarforschung 53:9–19

Renssen H, Seppä H, Crosta X, Goosse H, Roche DM (2012) Global characterization of the Holocene Thermal Maximum. Quatern Sci Rev 48:7–19. https://doi.org/10.1016/j.quascirev.2012.05.022

Rodríguez-Mena M, Fernández-Fernández JM, Tanarro LM, Zamorano JJ, Palacios D (2021) Héðinsdalsjökull, northern Iceland: geomorphology recording the recent complex evolution of a glacier. J Maps 17:301–313. https://doi.org/10.1080/17445647.2021.1920056

Rundgren M (1995) Biostratigraphic evidence of the Allerød-Younger Dryas-Preboreal oscillation in Northern Iceland. Quatern Res 44:405–416

Rundgren M (1999) A summary of the environmental history of the Skagi Peninsula, northern Iceland, 11,300–7800 BP. Jökull 47:1–19

Rundgren M, Ingólfsson Ó (1999) Plant survival in Iceland during periods of glaciations. J Biogeogr 26:387–396

Runólffsson S (1978) Soil conservation in Iceland. In: Holdgate MW, Woodman MJ (eds) The Breakdown and Restoration of Ecosystems. Plenum Press, New York, pp 231–240

Russell AJ, Roberts MJ, Fay H, Marren PM, Cassidy MJ, Tweed FS, Harris T (2005) Icelandic jökulhlaup impacts: Implications for ice-sheet hydrology, sediment transfer and geomorphology. Geomorphology 75:33–64

Sæmundsson K (1980) Outlines of the geology of Iceland. Jökull 29:7–28

Sæmundsson K (1973) Straumrákaðar Klappir í Kringum Ásbyrgi. Náttúrufræðingurinn 43:52–60

Sæmundsson T, Decaulne A (2003) Meteorological triggering factors and threshold conditions for shallow landsliding and debris-flow activity in Iceland. In: Schaefer VR, Schuster RL, Turner AK (eds) First North America Landslide Conference – Vail, Colorado, AEG Publication No. 23. pp. 1475–1485

Sæmundsson T, Petursson H (2018) Causes and triggering factors for large scale displacements in the Almenningar landslide area, in central North Iceland, in: European Geosciences Union General Assembly. Geophysical Research Abstracts. pp. 6482–1

Sæmundsson Þ (1995) Deglaciation and shoreline displacement in Vopna¬fjördur, Northeast Iceland. LUNDQUA Thesis, 33. Department of Quaternary Geology, University of Lund, p 106

Saemundsson T, Arnalds O, Kneisel C, Jonsson HP, Decaulne A (2012) The Orravatnsrustir palsa site in Central Iceland-Palsas in an aeolian sedimentation environment. Geomorphology 167–168:13–20. https://doi.org/10.1016/j.geomorph.2012.03.014

Sæmundsson Þ, Decaulne A (2007) Meteorological triggering factors and threshold conditions for shallow landslides and debris-flow activity in Iceland. In: Schaefer VR, Schuster RL, Turner AK (eds) First North American Landslide Conference, Vail Colorado, AEG Publication No. 23, 1475–1485

Sæmundsson Þ, Decaulne A, Jónsson HP (2008) Sediment transport associated with snow avalanche activity and its implication for natural hazard management in Iceland, in: International Symposium on Mitigative Measures against Snow Avalanches. Egilsstaðir, Iceland, pp 137–142

Sæmundsson Þ, Morino C, Helgason JK, Conway SJ, Pétursson HG (2018) The triggering factors of the Móafellshyrna debris slide in northern Iceland: Intense precipitation, earthquake activity and thawing of mountain permafrost. Sci Total Environ 621:1163–1175. https://doi.org/10.1016/J.SCITOTENV.2017.10.111

Sæmundsson Þ, Pétursson HG (2000) The Sölvadalur debris-slide. Local Authorities Confronting Disasters and Emergencies, Reykjavk, Iceland. Abstract book 26

Sæmundsson Þ, Pétursson HG, Jónsson HB, Jónsson HP (2005) Kortlagning á sigi á Siglufjarðavegi um Almenninga, Lokaskýrsla 2004, NNV-2005- 003. Náttúrustofa Norðurlands Vestra: Sauðárkrókur, 45 pp

Sæmundsson Þ, Pétursson HG, Kneisel C, Beylich A (2007) Monitoring of the Tjarnardalir landslide, in central North Iceland. In: Schaefer, V.R., Schuster, R.L., Turner, A.K. (Eds.): First North American Landslide Conference, Vail Colorado, AEG Publication No. 23, 1029–1040

Sæmundsson Þ, Sigurðsson IA, Pétursson HG, Jónsson HP, Decaulne A, Roberts MJ, Jensen EH (2011) Bergflóðið sem féll á Morsárjökul 20. mars 2007 – hverjar hafa afleiðingar þess orðið? Náttúrufræðingurinn 81:131–141

Saemundsson, Th, Petursson HG, Decaulne A (2003) Triggering factors for rapid mass movements in Iceland. In: Rickenman D, Chen CI (eds) Debris-Flow Hazards Mitigation: Mechanics, Prediction, and Assessment. Millpress, Rotterdam, pp 167–178

Sæmundsson Þ, Helgason JK, Pétursson HP (2014) The melting ofmountain perma- frost and the Móafellshyrna debris slide in Northern Iceland. 31st Nordic Geological Winter Meeting. 8–10 January 2014. Lund University

Schunke E (1973) Palsen und Kryokarst in Zenrtal-Island. Nachr. Akad. Wiss. Göttingen, Math.-Phys. K1. 2, 65–102

Schunke E (1974) Frostspaltenmakropolygone im westlichen Zentral-Island-ihre klimatischen und edaphischen Bedingungen. Eiszeitalter u. Gegenwart 25:157–165

Schunke E (1975) Die Periglazialerscheinungen Islands in Abhängigkeit von Klirna und Substrat. Abh. Akad. Wissensch. Göttingen, Mathem.-Phys. Kl., Folge 3, 30:1–273.

Schunke E (1977) Zur Genese der Thufur islands und Ost-Grönlands (On the Genesis of Thufurs in Iceland and East Greenland). Erdkunde 31:279–287

Schunke E (1979) Aktuelle thermische Klimaveränderungen am Polarrand der Ökumene Europas – Ausmaß, Ursachen und Auswirkungen. Erdkunde 33:282–291

Schunke E, Zoltai SC (1988) Earth hummocks (thufur). In: Clark MJ (ed) Advances in periglacial geomorphology. John Wiley, New York, pp 231–245

Seppälä M (1986) The origin of palsas. Geogr Ann A68:141–147

Seppälä M (1988) Palsas and related forms. In: Clark MJ (ed) Advances in Periglacial Geomorphology. Wiley, New York, pp 247–297

Seppälä M (1995) How to make a palsa: a field experiment on permafrost formation. Zeitschrift Fur Geomorphologie, Supplement 99:91–96

Shen Y, Solomon SC, Bjarnason IT, Wolfe CJ (1998) Seismic evidence for a lower-mantle origin of the Iceland Plume. Nature 395:62–65

Sigbjarnarson G (1967) The changing level of Hagavatn and glacial recession in this century. Jökull 17:263–279

Sigbjarnasson G (1983) The Quaternary alpine glaciation and marine erosion in Iceland. Jökull 33:87–98

Sigurdardottir R (1992) Eolian processes in the Myvatn area. B.S. thesis. Reykjavik: University of Iceland. In Icelandic. 100 pp

Sigurðsson O, Williams RS (1991) Rockslides on the terminus of "Jökulsárgilsjökull", southern Iceland." Geografiska Annaler, Series A 73 A, 129–140. https://doi.org/10.1080/04353676.1991.11880338

Sigurgeirsson MÁ, Hjartarson Á (2011) Gjóskulög og fjörumór á berghlaupi við Sjávarhóla á Kjalarnesi Náttúrufræðingurinn 81 (3–4), 123–130 (in Icelandic)

Steindórsson S (1945) Studies on the Vegetation of the Central Highland of Iceland. The Botany of Iceland, Vol. III, 4. Copenhagen

Stingl H, Herrmann R (1976) Untersuchungen zum Strukturbodenproblem auf Island; Gelaendebeobachtungen und statistische Auswertung. Z Geomorphol 20:205–226

Stötter J, Wastl M, Caseldine C, Häberle T (1999) Holocene palaeoclimatic reconstruction in northern Iceland: Approaches and results. Quat Sci Rev 18:457–474

Sveinbjarnardóttir G, Buckland PC, Gerrard AJ (1982) Landscape change in eyjafjallasveit, southern iceland. Nor Geogr Tidsskr 36:75–88

Tanarro LM, Palacios D, Andrés N, Fernández-Fernández JM, Zamorano JJ, Sæmundsson Þ, Brynjólfsson S (2019) Unchanged surface morphology in debris-covered glaciers and rock glaciers in Tröllaskagi peninsula (northern Iceland). Sci Total Environ 648:218–235. https://doi.org/10.1016/j.scitotenv.2018.07.460

Tanarro LM, Palacios D, Fernández-Fernández JM, Andrés N, Oliva M, Rodríguez-Mena M, Schimmelpfennig I, Brynjólfsson S, Sæmundsson Þ, Zamorano JJ, Úbeda J, Aumaître G, Bourlès D, Keddadouche K (2021) Origins of the divergent evolution of mountain glaciers during deglaciation: Hofsdalur cirques. Northern Iceland. Quaternary Science Reviews 273:107248. https://doi.org/10.1016/j.quascirev.2021.107248

Tanarro LM, Palacios D, Zamorano JJ, Andrés N (2018) Proposal for geomorphological mapping of debris-covered and rock glaciers and its application to Tröllaskagi Peninsula (Northern Iceland). J Maps 14:692–703. https://doi.org/10.1080/17445647.2018.1539417

Thorarinsson S (1981a) The application of tephrochronology in Iceland. In: Self S, Sparks RSJ (eds) Tephra Studies. Springer, Dordrecht, pp 9–134

Thorarinsson S (1937) Das Dalvik-Beben in Nord-island, 2. juni 1934. Geogr Ann 19:232–277

Thorarinsson S (1951) Notes on patterned ground in Iceland, with particular reference to the Icelandic 'flás.' Geogr Ann 33:144–156

Thorarinsson S (1953) Anchored stone polygons at low levels within the Iceland basalt regions. Jökull 3:38–39

Þorarinsson S (1954) Séð frá Þjóðvegi [Seen from the national road]. Náttúrufræðingurinn 24:7–15 (in Icelandic)

Thorarinsson S (1954) Fleygsprungnanet á Sprengisandi. Jökull 4:38–39

Thorarinsson S (1964) Additional notes on patterned ground in Iceland with a particular reference to icewedge polygons. Biul Peryglac 14:327–336

Thorarinsson S (1981b) Solifluction terraces on south slopes of Pétursey. Jökull 31:73–74

Thorarinsson S, Einarsson T, Kjartansson G (1959) On the geology and geomorphology of Iceland. Geogr Ann 41(2/3):135–169

Thordarson T, Hoskuldsson A (2002) Iceland. Classic Geology in Europe 3 (2nd edition). Terra Publishing, 224 pp

Thórhallsdóttir TE (1998) Flowering phenology in the central highland of Iceland and implications for climatic warming in the Arctic. Oecologia 114:43–49

Thórhallsdóttir ThE (1994) Effects of changes in groundwater level on palsas in Central Iceland. Geogr Ann A76:161–167

Thórhallsdóttir ThE (1996) Seasonal and annual dynamics of frozen ground in the central highland of Iceland. Arct Alp Res 28:237–243

Thórhallsdóttir ThE (1997) Tundra ecosystems of Iceland. In: Wiegolaski FE (ed) Polar and Alpine Tundra. Elsevier, Amsterdam, pp 85–96

Thoroddsen Th (1899) Uppi á heidum. Ferdaskýrsla 1898. Andvari. 24. Reykjavík

Thoroddsen Th (1905–1906) Island. Grundriss der Geographie und Geologie. Peterm. Geogr. Mitt. Erg.-H. 152 u 153. Gotha

Thoroddsen Th (1911) Lýsing Íslands II. Conpenhagen

Thoroddsen Th (1913) Polygonböden and 'thufur' auf Island. Petermanns Geogr Mitt 59:253–255

Thoroddsen Th (1913–1915) Ferdabók I-IV. Copenhagen

Thoroddsen Th (1914) An account of the Physical Geography of Iceland with special reference to the plant life. In "The Botany of Iceland", pp. 187–343

Thorsteinsson S (1956) Frostsprungur. Jökull 6:37

Tómasson GG, Hestnes E (2000) Slushflow hazard and mitigation in Vesturbyggd, northwest Iceland. Nord Hydrol 31:399–410. https://doi.org/10.2166/nh.2000.0024

Troll C (1944) Strukturboden, Solifluktion und Frostklimate der Erde. Geol. Rundschau 34. H. 7 8 (Klimaheft). Stuttgart

Tussetschläger H, Brynjólfsson S, Brynjólfsson S, Nagler T, Sailer R, Stötter J, Wuite J (2020) Perennial snow patch detection based on remote sensing data on Tröllaskagi Peninsula, northern Iceland. Jökull 69:103–128

Tveito OE, Førland EJ, Heino R, Hannssen-Bauer I, Alexandersson H, Dahlstrøm B, Drebs A, Kern-Hansen C, Vaarby Laursen E, Westman Y (2000) Nordic temperature maps. Report No.09/00, 28 pp

Van der Meer JJM, Kjær KH, Krüger J (1999) Subglacial water-escape structures and till structures, Sléttjökull, Iceland. J Quat Sci 14:191–205

Van Vliet-Lanoë B, Van Bourgeois O, Dauteuil O (1998) Thufur formation in Northern Iceland and its relation to Holocene climate change 365

Van Vliet-Lanoë B (2005) La planéte des glaces: histoire et environnements de notre ère glaciaire. Vuibert, Paris

Veit H, Marti T, Winiger L (2012) Environmental changes in Northern Iceland since the Younger Dryas inferred from periglacial slope deposits. Holocene 22:325–335. https://doi.org/10.1177/0959683611423695

Wangensteen B, Guðmundsson Á, Eiken T, Kääb A, Farbrot H, Etzelmüller B (2006) Surface displacements and surface age estimates for creeping slope landforms in Northern and Eastern Iceland using digital photogrammetry. Geomorphology 80:59–79

Washburn AL (1956) Classification of patterned ground and review of suggested origins. Bulletin of the Geological Society of America 67:823–865

Washburn AL (1979) Geocryology. a survey of periglacial processes and environments. London, Edward Arnold. 406 pp

Webb R (1972) Vegetation cover on Icelandic thúfur. Acta Botanica Islandica

Whalley WB, Douglas GR, Jonnson A (1983) The magnitude and frequency of large rockslides in Iceland in the postglacial. Geogr Ann 65A:99–110

Whalley WB, Hamilton SJ, Palmer CF, Gordon JE, Martin HE (1995a) The dynamics of rock glaciers: data from Tröllaskagi, north Iceland. In: Slaymaker O (ed) Steepland geomorphology. John Wiley & Sons Ltd., Chichester, pp 129–146

Whalley WB, Martin HE (1994) Rock glaciers in Tröllaskagi: their origin and climatic significance. In: Stötter J, Wilhelm F (eds) Environmental change in Iceland. Münchener Geographische Abhandlungen, Germany, pp 289–308

Whalley WB, Palmer CF, Hamilton SJ, Martin HE (1995b) An Assessment of Rock Glacier Sliding Using Seventeen Years of Velocity Data: Nautárdalur Rock Glacier, North Iceland. Arct Alp Res 27:345–351

Wolfe CJ, Bjarnason IT, VanDecar JC, Solomon SC (1997) Seismic structure of the Iceland mantle plume. Nature 385:245–247

Conclusions

The Periglaciation of Europe

Marc Oliva, José M. Fernández-Fernández, and Daniel Nývlt

1 Geographical Distribution of Periglacial Phenomena in Europe

The vast surface of the European continent (~10 million km^2) extending from subtropical (36 ºN) to polar latitudes (71 ºN) displays a complex relief configuration including numerous peninsulas, islands, archipelagos, capes, bays and high mountain ranges exceeding 2,500–3,000 m. In addition, mainland Europe constitutes a large peninsula surrounded by the relatively warm Mediterranean Sea and the cooler Atlantic Ocean in its southern and western-northern flanks, respectively. This unique geographical setting determines the occurrence of a wide spectrum of climate regimes, which influence the nature of the geomorphological processes affecting the Earth surface processes in this continent.

Taking into account French's (2007) broad definition of the periglacial environment as the non-glacierised areas where the mean annual air temperatures (MAAT) oscillate between -2 °C and $+3$ °C, currently in Europe periglacial regions are only located in mountain ranges, northern Scandinavia and the high-altitude islands (i.e. Iceland, Jan Mayen, Svalbard) (Fig. 1). Following north–south temperature gradients, there is a strong latitudinal influence of the periglacial domain: whereas the lower elevation of the periglacial belt in the Sierra Nevada is located at ~2,650 m, in the Alps it is located at ~1,950 m and in northern Scandinavia periglacial processes

M. Oliva (✉)
Department of Geography, Universitat de Barcelona, Barcelona, Spain
e-mail: marcoliva@ub.edu

J. M. Fernández-Fernández
Department of Geography, Universidad Complutense de Madrid, Madrid, Spain
e-mail: josemariafernandez@ucm.es

D. Nývlt
Department of Geography, Masaryk University, Brno, Czechia
e-mail: daniel.nyvlt@sci.muni.cz

M. Oliva et al. (eds.), *Periglacial Landscapes of Europe*,
https://doi.org/10.1007/978-3-031-14895-8_16

are currently active down to sea level. In addition, in some of the high-latitude islands, such as Svalbard, periglacial processes are also widespread in glacier-free areas.

Therefore, the present-day active periglacial domain in southern and central Europe is restricted to upper parts of the highest mountain ranges although periglacial evidence is also found at lower altitudes, and even in the flat lowlands, and in mid-low altitude massifs (Oliva et al. 2018b). Actually, deep permafrost at a depth of several hundred metres from the surface has been also detected in Poland (see chapter "The North European Plain"). In these regions, we find a combination of active features with others inherited from previous cold-climate stages, namely during Quaternary glacial periods. In northern Europe, where vast regions were under large ice sheets during the Last Glacial Cycle (i.e. Scandinavia, Iceland, Svalbard, large part of Great Britain, and northern part of North European Plain) active periglacial processes can coexist in the coldest non-glaciated areas together with periglacial phenomena formed during the deglaciation of current glacier-free environments (Fig. 2).

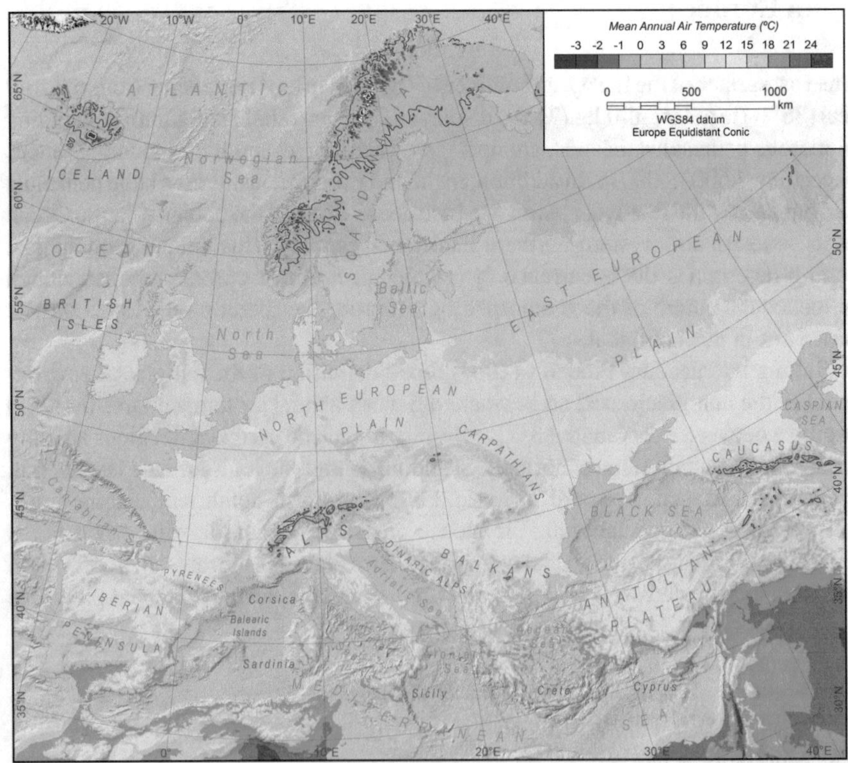

Fig. 1 Map of Europe showing present-day MAAT. The white and black contour line encircles the areas included between −3 and 2 ºC following the general definition of French (2007) for the periglacial environment. The line represents the transects shown in Fig. 2

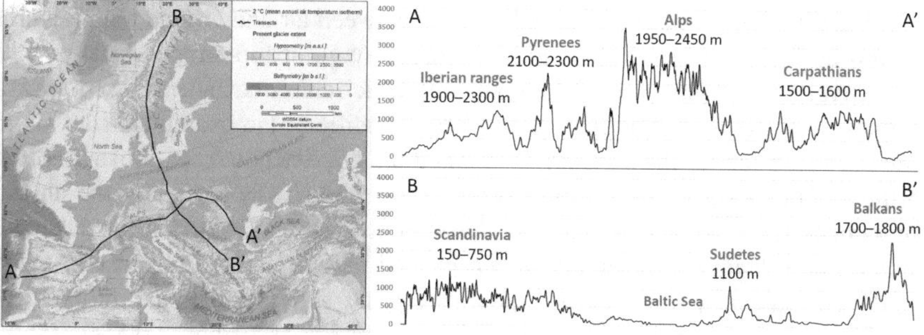

Fig. 2 North-South (A-A') and West-East (B-B') transects showing the lower limit of periglacial processes and permafrost belt in European mountains

In this section, we will integrate knowledge presented separately in the different chapters in order to provide a more general overview on European periglacial landscapes. We will first focus both on present-day landforms, their current activity, ground thermal dynamics and factors controlling their movement. Subsequently, we will examine the distribution and types of relict landforms formed in the past, which can be used to infer Late Quaternary climate conditions. Finally, we will also describe how, in some cases, these periglacial features have been altered by human activities.

2 Ground Thermal Regime of Present-Day Periglacial Processes

Present-day periglacial processes in Europe are generally driven by seasonal frost dynamics, as permafrost conditions are only recorded in the highest mountain ranges as well as in the lowland regions of northern Scandinavia and high-latitude archipelagos, such as Svalbard. Consequently, periglacial landscapes include, thus, cold non-glacial landforms generated by both seasonal frost and permafrost ground thermal regimes.

Seasonal frost in Europe is favoured by the moderately low temperatures recorded in winter throughout the continent, particularly across the mountains, which promoted the freezing of the first decimetres to several metres of the ground. Negative temperatures during the cold season of the year are widespread across the continent and diurnal one-sided freezing can result in the ephemeral formation of some frost action features (i.e. frost heaving, needle ice). However, periglacial environments are those where processes related to ground freezing are continuous during a certain time of the year and the consequences of freeze-thaw oscillations result in identifiable features across the landscape (French 2007). For example, even if the average winter air temperature of Stockholm and Warsaw are approximate −1 °C—their mean annual air temperatures (MAAT)s are ~7 and 9 °C, respectively—these climate conditions are not able to significantly modify the landscape.

Normally, there are substantial differences between air and ground temperatures as well as factors driven by the local lithology (e.g., spatio-temporal patterns of snow cover, topographical conditions, albedo), which can contribute to the cooling or warming of the ground regardless atmospheric conditions. For example, in the Sierra Nevada, south Iberia, the seasonal frost environment extends from above 2,500 m to the summit level (3,300–3,400 m) where the MAATs is 0 ºC, but the annual soil temperatures average is 2–2.5 ºC (Oliva et al. 2016a). Generally, the calendar, duration and depth of seasonally-frozen ground in Europe is highly variable and largely controlled by the snow precipitation regime, topographical and lithological conditions. In latitudes >65 ºN and in high mountains exceeding 2,000–2,500 m, the frozen ground can reach depths >1–1.5 m and persist between 6 and 9 months, and even longer in sheltered environments (chapters "The Iberian Peninsula"–"The Mediterranean Islands").

Snow cover constitutes a key element controlling periglacial activity in mid-latitude mountain environments (Zhang 2005). Its duration and thickness have a direct effect on ground temperatures as well as on soil moisture conditions. Generally, a thicker snow cover isolates the ground from air temperature oscillations and determines rather stable ground temperatures at negative values close to ~0 ºC, favouring also a delayed thawing of the ground; by contrast, years with a thinner snow cover favour a deeper seasonal frost layer with colder ground temperatures, although these values tend to increase rapidly as we approach the warm season of the year (Oliva et al. 2014a). These contrasted patterns are even more pronounced in southern Europe, where Mediterranean climates determine low precipitation—albeit with high inter-annual and intraanual variability—together with relatively higher temperatures and high insolation. Here, the role of snow is crucial for ground temperatures and thus for periglacial processes, such as it happens in the semiarid mountain range of the Sierra Nevada, southern Spain (Fig. 3) (Oliva et al. 2014b; Gómez-Ortiz et al. 2019).

Snow melting provides water during late spring and summer, when the ground subsurface can still be frozen. This moisture supply is important for the vegetation cover, which in turn affect the activity of periglacial slope dynamics, such as cryoturbation or solifluction processes (Ulfstedt 1993). The degree of vegetation cover and moisture availability are key factors controlling soil development in the periglacial belt of mid-latitude mountains, particularly in southern Europe. Well-developed soils favour geomorphic stability, whereas areas where soil processes are weak enhance periglacial activity, such as it usually happens in the upper periglacial belt (Oliva et al. 2009). Other factors such as the lithology and topography play also a decisive role on the ground thermal regime. Certain lithologies, such as volcanic coarse-grained soils, have a major influence on the freezing and thawing of the ground depending on their degree of saturation (Ishikawa et al. 2016). This is clearly seen in Iceland, where seasonal frost prevails up to 800–900 m shifting to permafrost conditions at higher elevations (Etzelmüller et al. 2007). The texture of the sediments is also a key element as shown by rock glaciers, where the openwork coarse blocks favour the presence of cold air trapped within inter-clast voids and thus the existence (and preservation) of permafrost beneath (Jones et al. 2019). Besides, mineralogical composition of sediments or regolith also significantly affects thermal conductivity, which is highly

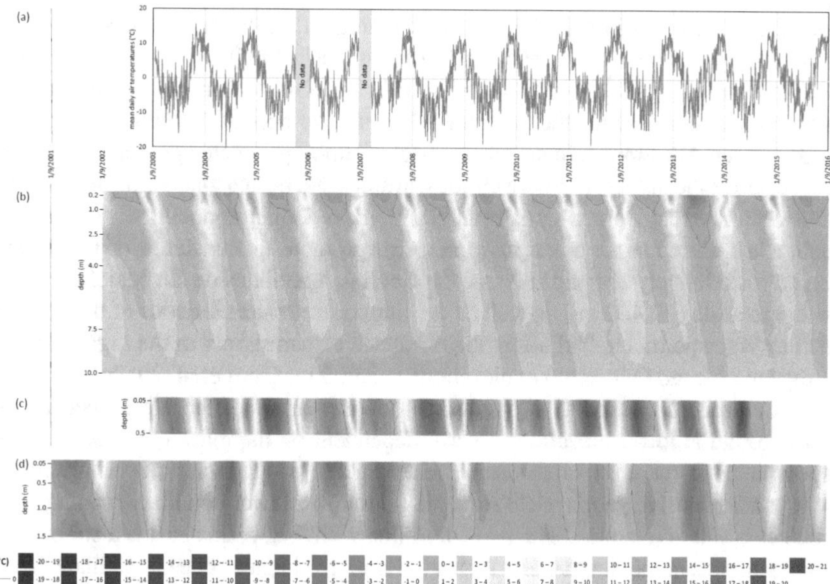

Fig. 3 Ground thermal regime in several periglacial landforms in the Sierra Nevada: sorted circle at the Machos summit plateau at 3300 m (c), rock glacier of the Veleta cirque floor at 3100 m (d), and bedrock temperatures at the Veleta Peak at 3380 m (b) and evolution of air temperatures at the Veleta peak at 3398 m. Figure published in Gómez-Ortiz et al. (2019). The ground generally freezes in autumn and thaws in late spring. The high interannual variability of the seasonal frost thickness (ranging between 0.6 to 2 m) is determined by the presence/absence, thickness and duration of snow in the ground

variable for siliciclastic and carbonaceous material (Hrbáček et al. 2017). Topographical conditions create a wide spectrum of microclimate regimes within the mountain systems, determining substantially different patterns of radiation in northern and southern slopes that have a critical importance for ground thermal regime (Gruber and Haeberli 2007). In Europe, as in the rest of the Northern Hemisphere, this is reflected in colder conditions in north-exposed slopes than in those facing south. As a result, for a similar altitude in the Alps, Pyrenees or Scandinavian Alps, we may have permafrost in the northern slope and a deep and long seasonal frost regime in the southern side (Fig. 2). In addition, the aspect has major implications for periglacial dynamics as it conditions, among others, the intensity of frost shattering, persistence of long-lying snow-patches and the vegetation cover.

Nowadays, the altitude of the lower limit of alpine permafrost in Europe generally increases towards the east and from north to south (Fig. 2). In the mountain ranges surrounding the Mediterranean Sea, permafrost is rare <2500 m, turns discontinuous between 2,500 and 2,800/3,000 m and becomes continuous in ice-free areas above this elevation. In the Pyrenees, the lower boundary of permafrost is set at 2,650 m, and its occurrence is possible above this level and probable >2,900 m, always lower in northern aspects (Serrano et al. 2019a). This is slightly higher than in the Southern

Alps (>2,400 m; Bodin et al. 2009), Rila Mountain (>2,350 m; Dobiński 2005) but lower than in the Mount Olympus (>2,700 m; Dobiński 2005), or central Anatolia and NE Turkey (>2,800–3,400 m; Gorbunov 2012). In more southern (and higher) mountains exceeding 3,000 m, such as Sierra Nevada, no permafrost belt is detected—neither in the Atlas range, in northern Africa—although isolated permafrost patches are located in the highest northern cirques, in areas that hosted glaciers during the LIA (Oliva et al. 2016a; Gómez-Ortiz et al. 2019). Indeed, microtopoclimatic conditions, such as a reduced snow cover in winter, can promote the existence of permafrost patches at relatively low altitudes of the Central Apennines on La Majella and M. Velino massifs (~2,400 m), as well as the lithological characteristics of the ground, such as volcanic rocks, that allow the existence of permafrost in the highest active European volcano (Mt. Etna) at elevations (>2,900 m; Oliva et al. 2018a, b).

In Central Europe north of 45 ºN, as MAAT decrease, the alpine permafrost belt is found at lower elevations always dependent on the relief configuration and aspect. This is the case of the Alps, where active rock glaciers, borehole monitoring and extensive geophysical surveying point to a lower boundary ranging from 2,200 to 2,600 m (chapter "The Alps"). In line with this, permafrost is also found at lower altitudes in the South Carpathians (>1,900 m; chapter "The Carpathians"), the Eastern Carpathians (>1,800 m; chapter "The Carpathians"), or the Tatra Mountains in Western Carpathians (>1,930 m; Dobiński 2005). In northern Europe, permafrost is only found in Scandinavia and Iceland, or High Arctic Archipelagos such as Svalbard. In the Scandinavia case, permafrost appears between 1750 and 1050 m (Gisnås et al. 2017), most of which is either discontinuous or sporadic, being continuous only about the 2%. However, in northern and central Iceland, it can be found at lower elevations, down to 800 m (Etzelmüller et al. 2007; Czekirda et al. 2019).

Permafrost temperatures in Europe range between −1 to −5 ºC (Fig. 4). At 2,900 m in the Pyrenees, permafrost temperatures are of ca. −1 ºC (Serrano et al. 2019a), whereas in the Alps they range between −2 and −3 ºC at 3,300–3,400 m, and between 0 and −1.5 ºC at 2,600–2,950 m (Bodin et al. 2015; PERMOS 2021). In Scandinavia, permafrost temperatures ate close to 0 ºC on bedrock at elevations below 1,000 m, as well as in the northern part of the peninsula at altitudes of ~300–500 m; in the mountains, however, at 1,500–1,800 m, permafrost temperatures are of ~−2.5 ºC (Christiansen et al. 2010). In Svalbard, cold permafrost is widespread, with values of ~−1.2 to −5.1 ºC (Christiansen et al. 2019). The active layer is highly variable and depends on the nature of the surface. Bedrock sites tend to show much deeper active layer thicknesses than permafrost environments composed of unconsolidated sediments (e.g. active rock glaciers). In recent years, boreholes have provided a wide range of active layer thicknesses, oscillating from 2.8 to 11 m in the Swiss Alps (PERMOS 2021) to 0.6–4.6 m in Svalbard (Christiansen et al. 2019). In all areas, permafrost temperatures record a warming trend, which oscillated between 1.5 °C on Svalbard and 0.4 °C in the Alps between the late 1990s and 2018, accompanied by a deepening of the active layer, which thickened between ~10% and 200% (Etzelmüller et al. 2020).

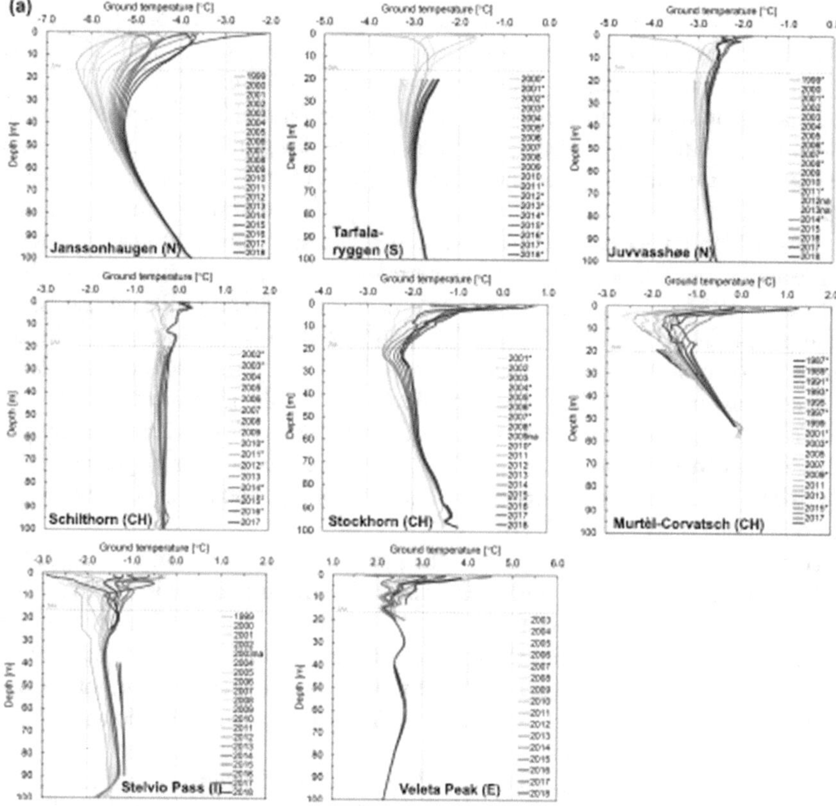

Fig. 4 Mean annual ground temperature profiles from 1987 to 2018 at different sites along a transect from Sierra Nevada in Spain (37 °N) to Svalbard (78 °N) established during the PACE (Permafrost and Climate in Europe) project (figure published in Etzelmüller et al. 2020). The figure shows how ground temperatures have warmed across all the permafrost sites over the last decades

3 Present-Day Periglacial Processes

Present-day activity of periglacial landforms in Europe is conditioned by the distribution of seasonal frost and permafrost conditions. In the coldest regions where permafrost is widespread, periglacial processes are generally active and most associated phenomena show a dynamic pattern under current climate regimes. This is the case of ice-free regions in high-latitude regions (e.g. Svalbard, northern Scandinavia and northern Iceland) as well as in the highest mountain ranges (e.g. Alps, Pyrenees, Anatolia mountains) that include a broad spectrum of active periglacial landforms and processes, including widespread thermal-contraction-crack polygons (e.g. ice and sand wedges), permafrost creeping landforms (e.g. rock glaciers, protalus lobes) and other hillslope phenomena (e.g. solifluction lobes, debris flows), cryoturbation features (e.g. sorted circles, stone stripes), or thermokarst phenomena (e.g. retrogressive-thaw-slumps, lakes), amongst others. However, some permafrost-related features (i.e. pingos) may undergo a process of collapse by hydrostatic and topographic adjustment as part of their natural evolution (French 2007).

However, in seasonal frost environments, active periglacial landforms coexist with others that are inactive or weakly active as present-day climate conditions are not favourable for maintaining periglacial activity. This may be the case of periglacial features that formed under colder climates prevailing during Quaternary glacial periods, which favoured the expansion of permafrost to lower latitudes/altitudes. For example, most rock glaciers in Mediterranean mountains are currently inactive and developed following Late Glacial cold phases (i.e. Bølling–Allerød and Early Holocene; Palacios et al. 2017a, b, 2022; Andrés et al. 2018; Oliva et al. 2018a, b). These colder conditions also favoured the decline of the periglacial belt associated with seasonal frost conditions to the mid-low slopes of the high massifs as well as the occurrence of seasonal frost in moderately high massifs that are currently not affected by cryogenic processes. Stratified scree deposits or talus slopes currently covered by vegetation show evidence of this fact in many mountains in southern and central Europe (Oliva et al. 2018b).

Hence, whereas in southern Europe the periglacial domain is only found in the highest mountains where no settlements and little infrastructure exists (Oliva et al. 2016b), in the high latitudes of Europe a large number of socio-economic activities take place in active periglacial terrain. Here, a deep comprehension of periglacial processes is needed for a proper assessment of landscape management, infrastructures potentially affected by ground ice dynamics as well as reduce the potential risks of societies to natural hazards.

Together with ground thermal regime, the prevailing lithology in periglacial landscapes constitutes also key factor controlling also their degree of activity (French 2007). Whereas some processes are widespread and show a greater development in certain lithologies, such as solifluction dynamics in metamorphic and limestone massifs, others are critically constrained by the bedrock type, such as rock glaciers, which are almost absent in limestone massifs, such as the Picos de Europa in Iberia, or in the Alps. In addition, topography also determines contrasting degrees of activity within the regional area, particularly in mountain areas. Whereas northern slopes tend to show greater periglacial dynamics than southern sites in non-glaciated massifs of southern and central Europe (Fig. 2). In glaciated environments in northern Europe and south-facing slopes of high-mountain regions can be affected by intense periglacial dynamics. However, this pattern is highly influenced by microclimatic conditions, the presence of late-lying snow patches, past environmental conditions, etc.

3.1 Southern Europe

In southern Europe, the active periglacial domain resembles the glaciated domain of the last glacial cycle: The elevation difference between the lower Last Glacial

Maximum (LGM) periglacial zone and present-day periglacial processes is about 800–1000 m (Oliva et al. 2016b). At latitudes below ~45 ºN in the European continent, periglacial processes and landforms are mostly concentrated in the main mountain ranges, with minor evidence in mid- and low-mountainous regions as well as in interior plateaus and basins. Many periglacial landforms and deposits are located in environments with relatively high modern MAATs, and are thus inactive or very weakly active. This is the case of solifluction landforms in the Iberian mountains, which are generally relict and inherited from past colder and wetter periods and only show activity above 2,500 m in the Pyrenees to 3,000 m in the Sierra Nevada, where MAAT are <2 ºC (chapter "The Iberian Peninsula"). A wide range of active periglacial phenomena are found across the periglacial belt driven by seasonal frost, such as features associated with cryoturbation (e.g. patterned ground, garlands, earth hummocks), rock weathering (e.g. blockfields, tors), talus slopes and related processes (talus cones, debris flows, or relict rock glaciers), periglacial mass movements (solifluction landforms, landslides) and others that are not so common and associated with certain lithologies (e.g. ice caves) (Fig. 5). In addition, there are also several active morphologies within the marginal areas underlined by permafrost; rock glaciers and protalus lobes are amongst the most abundant, although other features such as block streams and frost mounds have been also reported from high mountain ranges of Southern Europe (chapters "The Iberian Peninsula"–"The Mediterranean Islands").

Monitoring of active periglacial features is, however, very limited and has been mostly done in some of the highest Iberian mountains. This is the case of solifluction, which has been shown to be inactive below 2,800 m in the Sierra Nevada, and only the landforms located at ~3,000 m near late-lying snow patches reported annual displacement rates <1 cm yr^{-1} (Oliva et al. 2009, 2014b). Other studies in the Pyrenees and the Picos de Europa showed higher movement rates of solifluction features up to 1.8–3.6 cm yr^{-1} (chapter "The Iberian Peninsula"). Several rock glaciers have been monitored in the Pyrenees since the early 1990s showing highly variable annual horizontal displacement rates between ~10 and 40 cm yr^{-1} (Chueca and Julián 2011; Serrano et al. 2019a), with lower rates of 4–11 cm yr^{-1} in the case of an active protalus lobe in the Maladeta massif (González-García et al. 2017). Some rock glaciers showed evidence of undergoing a rapid degradation of underlying buried ice and permafrost that resulted in higher vertical than horizontal displacements (4 vs 1.3 m during 2001–2016; Gómez-Ortiz et al. 2019). The recent evolution of periglacial talus cones has also been surveyed at elevations of 1900–2200 m in the Cantabrian Mountains, where very active dynamics resulted in surface changes up to 0.5–1 m yr^{-1} (Serrano et al. 2019b).

3.2 Central Europe

In Central Europe, the present active periglacial domain resembles the highest mountain ranges, i.e. the Alps, Carpathians, or the highest Variscan ones such as

Fig. 5 Examples of active periglacial features in southern Europe

Karkonosze, Bohemian Forest, or Hrubý Jeseník Mts. A significant longitudinal gradient connected with the continentality of the climate could be seen for mountain ranges along ~50 °N. The lowermost limit of active periglacial processes in the Sudetes lies around 1400 m (chapter "The Central European Variscan Ranges"), however this altitude rises to 1500–1600 m in the Western and Eastern Carpathians further to the east (chapter "The Carpathians"). Significantly higher is the location of the current periglacial realm in the Alps: for the Northern-Eastern Alps it starts at altitudes above 1950 m, but in Southern-Western Alps this altitude lies at least at 2350 m (chapter "The Alps"). The areas of active periglacial processes in Central Europe were not always glaciated during the last glacial period, as these mountain ranges were mostly covered by valley glaciers during the Pleistocene cold periods and periglacial domain existed also on summits of younger Alpine belts of the Carpathians, or flat unglaciated surfaces of older Variscan ranges. In such location periglacial landforms survived till present and the best evolved periglacial landscape could be found there. This was different for much higher European Alps, where 55% of the area was glaciated during the LGM with the largest glacier coverage in Northern-Western Alps (chapter "The Alps") and non-glaciated summit areas were, therefore, spatially restricted. While large areas of the Alps and Carpathians have negative MAAT with the 0 °C isotherm being located at ~2200–2600 m and ~1760–1885 m in the Alps and Carpathians, respectively, the summit areas of Central European Variscan ranges show at present only positive MAAT, although the MAATwas negative at the highest parts of the Karkonosze Mts. even during the LIA (Sobik et al. 2019). The large difference of the altitude of the 0 °C isotherm throughout the Alps reflects rather the north–south elevation gradient with higher altitudes generally for Southern Alps. This is also supported by diverse precititation amounts, being more than double in Northern Alps compared to Southern Alps.

The Alps and Carpathians represent a classical area of European periglacial landforms research as a very wide spectrum of different active periglacial features induced by frost weathering, cryoturbation, regelation sorting, periglacial slope movement, cryoplanation, nivation, or aeolian and thermokarst processes can be found there (chapters "The Alps", "The Carpathians"). Rather different features induced by periglacial processes could be found in carbonate and igneous rocks of both main Central European mountain ranges, where active periglacial processes occur. On the contrary, the only characteristic periglacial landforms in significantly lower Central European Variscan ranges, which are currently considered as a marginal periglacial domain, are regelation-related sorting features (chapter "The Central European Variscan Ranges").

Permafrost monitoring has one of the longest European tradition in the Alps, which is since 1990s transnational and collaborative. However, long term monitoring of rock glaciers started in Switzerland and Austria already in 1920s (Chaix 1923; Finsterwalder, 1928) and many other monitoring activities emerged in 1950s–1980s (e.g.; Pillewizer 1957; Vietoris 1972; Francou and Reynaud 1992). Most of the rock glaciers in the Alps move annually in decimetres to metres (chapter "The Alps"), but some rock glaciers show an accelerating trendwith extraordinary velocities of up to 100 m a^{-1}, probably connected with rapid degradation of permafrost (Delaloye et al. 2010; Eriksen et al. 2018). The slower current periglacial processes are monitored at different places of the Alps and the Carpathians. Present-day solifluction movement rates are 1–20 cm a^{-1} in the Alps (Matsuoka 2001; Veit 2002; Kellerer-Pirklbauer 2018) and from <1 cm a^{-1} to >15 cm a^{-1} in the Carpathians (Urdea 2000; Rączkowska

2007; Onaca et al. 2017b) showing highly variable displacement rates depending on microtopographic and microclimatological conditions (Fig. 6).

Fig. 6 Examples of active periglacial features in central Europe

3.3 Northern Europe

North European regions were extensively glaciated and shaped by extensive cold-based ice sheets during successive glacial cycles. Thus, active and relict landforms often intercalate with those originated by glaciers, and in some cases, they even survived the passage of the ice. Currently, MAAT are below the freezing point at the highest peaks of the Scandes (<−6 °C) and the Icelandic highlands (−4 °C). However, in Great Britain, the MAAT does not reach the negative values even at the highest Scottish summits (1–3 °C). The zonation of the periglacial landforms is problematic in these two areas as active and inactive landforms coexist. Nowadays, periglacial activity is limited to landforms and processes associated to mass movements (solifluction, complex landslides, rock slope failures and debris flows linked to permafrost degradation, and avalanches, with permafrost degradation as a major trigger), cryoturbations (earth hummocks / thúfur, small-scale patterned ground), talus slopes (talus cones, debris flows) and ice-cored formations (rock glaciers, palsas and ice-cored moraines).

Monitoring of the active periglacial features is restricted to a reduced number of landforms. Solifluction lobes have been monitored in the Great Britain, Scandinavia and Iceland. In the first case, studies were carried out in the highest areas of northern Scotland (900 m) for over than 30 years with average rates between 7.8 and 10.6 mm a^{-1} (Ballantyne 2013). In Scandinavia, solifluction lobes have been studied in a number of locations, and the measurement of their displacement velocities revealed a clear relationship with the atmospheric warming (Åkerman and Johansson 2008; Ridefelt et al. 2009; Strand et al. 2020). Douglas and Harrison (1996) measured displacement rates of solifluction lobes in south-eastern Iceland reaching up to 11 mm a^{-1}. Active landslides have also been monitored in Scandinavia, revealing a clear acceleration of their sliding, linked to the degradation of permafrost (Hilger et al. 2021). Rates up to 0.57 and 0.7 m a^{-1} were measured in active landslides of northern Iceland (Wangensteen et al. 2006; Sæmundsson and Petursson 2018). Active rock glaciers are showing evidence of degradation in northern Iceland, with surface subsidence and an almost static state (velocities between 0.29 and 0.74 m a^{-1}; Wangensteen et al. 2006; Tanarro et al. 2019). In Scandinavia, however, the recent trend of rock glaciers was characterised by an acceleration of their displacement, linked to the permafrost degradation (Eriksen et al. 2018) (Fig. 7).

4 The Legacy of the Past

Quaternary climate variability determined large-scale variations of the cold-climate geomorphological processes affecting the European continent. The surface extent of glacial and periglacial dynamics expanded to lower latitudes and elevations during colder (glacial) periods and migrated to the higher latitudes and the upper parts of the mountain ranges during warmer (interglacial) periods. Colder climate conditions

Fig. 7 Examples of active periglacial features in northern Europe

prevailing during glacial periods promoted the inception of the European Ice Sheet Complex that reached latitudes 48 ºN (Palacios et al. 2022) as well as a large expansion of glaciers in the currently still glaciated mountain ranges and the formation of numerous glaciers in currently ice-free massifs (Woodward 2009; Allard et al. 2021). By contrast, warmer conditions during interglacial phases—such as the current one, the Holocene—determined the final melting or substantial retreat of glaciers as well as the shift of permafrost and periglacial processes to higher elevations and northerly latitudes (Oliva et al. 2018b) (Fig. 8).

The timing of the different glaciations and the surface extent occupied by the ice masses across the European continent has been recently summarised in a book that complements the present one focusing on the periglacial dynamics (Palacios et al. 2022). In fact, research on past glacial processes in Europe has received much more attention from scholars than studies on the impact that past periglacial processes had on the current landscape of the continent. Indeed, the timing of development of periglacial landforms is still little known and has been mostly inferred using chronostratigraphic approaches and relative dating techniques.

Reconstructions of past periglacial dynamics have been done using different methods depending on the nature of periglacial deposits. Radiocarbon dating has been the most widespread method to infer the past activity of periglacial features

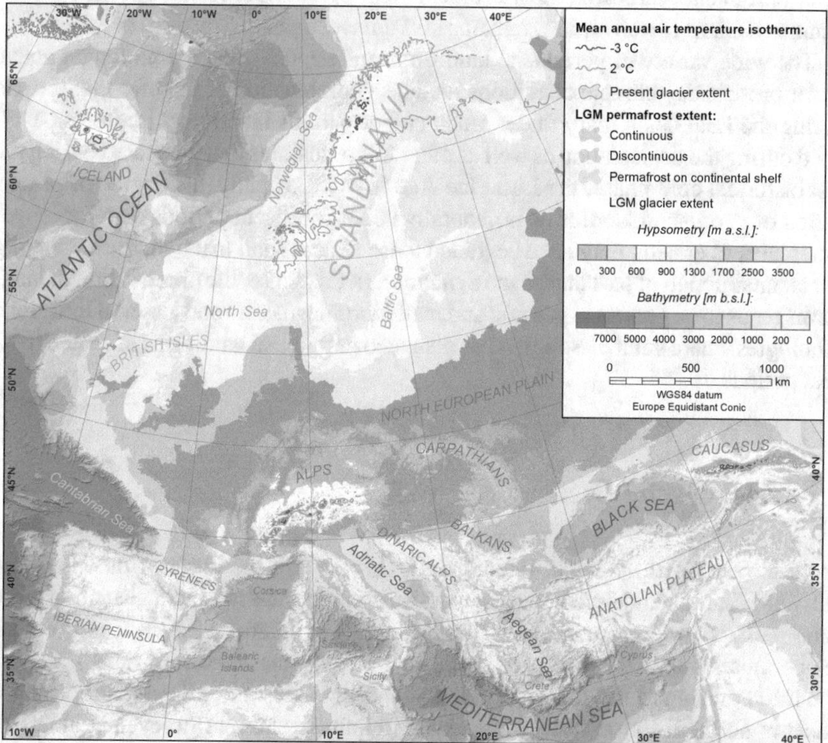

Fig. 8 Map of Europe showing the distribution of continuous, discontinuous permafrost, permafrost on continental shelf and glaciers during the LGM (Lindgren et al. 2016) versus current Mean annual air temperatures (MAAT) isotherms of 2 °C (defining current periglacial realm) and −3 °C (defining current continuous permafrost) and current glacier extent

as it allowed dating organic remnants trapped by periglacial processes within sedimentary sequences; it has been used to date stratified screes (Ruiz-Fernández et al. 2019), solifluction beds (Oliva et al. 2011), ice-wedges (Opel et al. 2018) or pingos (Yoshikawa and Nakamura 1996) or aeolian loess deposits (Oliva et al. 2014c). More recently, research studies on periglacial deposits promoted the cross-comparison of different (absolute) dating techniques, such as radiocarbon and luminescence dating (Palstra et al. 2021), as well as the implementation of cosmic-ray exposure (CRE) dating on some periglacial features whose chronology was only relative. This is the case of CRE on relict rock glaciers that are reporting ages of development following their deglaciation, and revealing thus a paraglacial origin in many features from both south (Palacios et al. 2017b; Andrés et al. 2018) and north Europe (Andrés et al. 2019; Fernández-Fernández et al. 2020). In addition, relative dating techniques, such as Schmidt-hammer exposure-age dating, have been recently used to establish absolute ages for nival landforms, such as the formation of pronival ramparts by frequent snow-avalanche (Matthews and Wilson 2015). However, the age of development of some periglacial landforms, such as blockfields or features associated pattern ground features such as sorted circles, remains still uninvestigated.

The wide variety of periglacial landforms and deposits that are no longer active under present-day climate conditions include a combination of features that formed during the Late Quaternary under much colder periods than today to others developed during the deglaciation as well as other minor features that generated during the last historical cold phase, the Little Ice Age (LIA). Logically, the degree of preservation of periglacial landforms is generally better for recent periods than for longer timescales as erosion may have degraded these features and make more challenging the reconstruction of past landscape dynamics. In this sense, the preservation of landforms tends to be greater in granitic and metamorphic massifs and lesser in limestone lithologies where karst dissolution has reshaped periglacial landforms inherited from past periods.

4.1 Last Glacial Cycle

From the end of the Eemian Interglacial (~114 ka) till the onset of the Holocene (11.7 ka), Europe underwent the last massive cooling and associated glaciation, with some periods of warmer conditions (interstadials) and others with significantly colder temperatures (stadials) (chapter "Periglacial Landforms Across Europe"). Colder conditions favoured the formation of mountain glaciers and a large European Ice Sheet in northern Europe advancing South to latitudes 50 ºN (Stroeven et al. 2016; Batchelor et al. 2019), together with an expansion of winter sea ice to latitudes surrounding the southern fringe of the European Ice Sheet (Denton et al. 2010). These conditions favoured significantly different oceanographic and atmospheric circulation patterns than present-day, which in turn affected spatial and temporal precipitation regimes. The combination of both temperatures and moisture regimes during the Last Glacial Cycle favoured the expansion of cold-climate geomorphological processes, including glacial and periglacial dynamics.

Southern Europe was not extensively glaciated and glaciers remained confined in mountain regions as small ice fields and alpine glaciers descending through the valleys. The depression of the Equilibrium Line Altitude (ELA) in Mediterranean mountains during the LGM—compared to present-day values—ranged between 800 and 1,200 m, which suggests a minimum temperature difference of ca. 6–10 ºC (Oliva et al. 2018b). The ELAs in glaciated mountains in southern Europe increased towards the East: whereas in NW Iberia the ELA was located <1,000 m, it increased to >1,500–2,000 m in the western and central Mediterranean Basin (Hughes et al. 2006), and reached elevations of 2,500–3,000 m in mountains of Turkey and Lebanon (Messerli 1967).

The northern fringe of the Mediterranean regions included areas with permafrost occurrence during the Last Glacial Cycle, such as central France and the northern Balkan region (Brown et al. 2001; Vandenberghe et al. 2014). The cold and arid conditions prevailing in these permafrost regions during the coldest phases of the last glaciation, such as the LGM (chapter "Quaternary Climate Variability and Periglacial Dynamics"), favoured a scarce vegetation cover and an intense aeolian activity, which explains the thick loess deposits existing in central-northern Europe (Zeeden et al. 2018; Chapter 11) and northern Italy (Cremaschi et al. 2015). In some internal depressions across the Mediterranean region, evidence of the existence of ice wedges and cryoturbations features has been associated with deep seasonal frost conditions or even isolated patches of sporadic permafrost, such as in the Spanish Meseta (chapter "The Iberian Peninsula"), or in the Pannonian basin (chapter "The Italian Peninsula"). Loess deposits including cryoturbation evidence may reveal also the existence of permafrost conditions in areas surrounding the valley glaciers in the southern Alps, depressions at the foot of the northern Apennines and even at the Po floodplain (Cremaschi et al. 2015).

At higher elevations in the mountain regions of southern Europe, below the glaciated domain was the periglacial belt conditioned by the occurrence of permafrost at moderately high elevations and seasonal frost at the lower parts of the mountains high-altitude plateaus (e.g. at 600–1,200 m). Due to topographical conditions, permafrost was also widespread in non-glaciated environments above the snowline. This is confirmed by the existence of permafrost-related landforms that are inactive under current climate regime (Fig. 9), such as metre-sized stone circles developed in relatively flat summit surfaces where the intense wind did not favour snow and ice accumulation (Oliva et al. 2018b). Present-day analogues of these patterned ground features exist in polar regions, where permafrost is widespread with MAAT ~−6 ºC (French 2007). It is thus likely to associate the formation of these present-day inactive landforms to the occurrence of permafrost. At the summit level, in palaeonunataks protruding the former glaciers, permafrost must have been widespread. In these areas, under very intense periglacial conditions, large block fields and tors also developed in favourable lithological settings. Other features typically associated with permafrost landscapes are those developed in glacial-free slopes, such as rock glaciers, which also formed during the Last Glacial Cycle at a broad range of elevations (~1,000 to

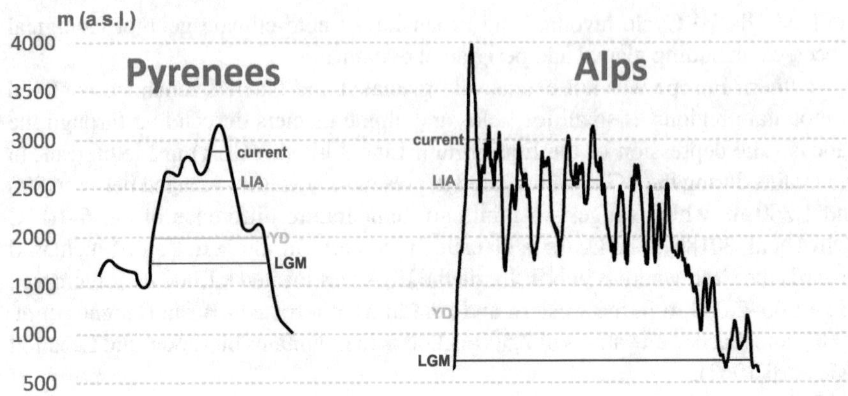

Fig. 9 Altitudinal changes of permafrost distribution during different periods of the last climate cycle in the Pyrenees and Alps

2,500 m; Fig. 9), or block streams—some several km long and hundreds of metres wide—distributed between 700 and 1800 m of altitude.

At lower elevations, seasonal frost was widespread and affected the low-mid mountain slopes <1,000–1,500 m in most mountain ranges, including those close to the Mediterranean coast. In Iberia, some stratified periglacial deposits have been dated of pre-LGM age in NW mountains, whereas others coincided with the LGM in the Central Pyrenees (chapter "The Iberian Peninsula"). In these areas, the existence of stratified slope deposits (or *grèzes litées*) has been used to reconstruct the occurrence of periglacial conditions during the Last Glacial Cycle (Fig. 10). In addition, the lowlands in more continental areas (e.g. at 300–800 m) were also affected by intense seasonal frost dynamics, such as in Central Iberia (chapter "The Iberian Peninsula"), or the Balkan region (chapter "The Balkans (Without Carpathians)"), although geomorphic evidence of that period in these areas is scarce and mostly documented based on the existence of stratified screes.

The North European Plain building the northern part of Central Europe was repeatedly covered by the southern margin of the European Ice Sheet (Batchelor et al. 2019; Palacios et al. 2022), with the most extensive advances reaching the northern forelands of Central European Variscan Ranges and Northern Carpathians during the Elsterian, MIS 12 (Nývlt et al. 2011). However, the largest advance took place during the Donian (MIS 16) and Saalian (MIS 6) glaciations in the Eastern European Plain, where the ice sheet front crossed by extensive frontal lobes the latitude of 50°N and advanced to Southern Russia and Ukraine, respectively (Ehlers et al. 2011). However, the European Ice Sheet was of a limited extent during the LGM (MIS 2) with its maximum extent occurring during the Leszno Phase, dated at 20.7 $\pm$ 0.8 ka (Tylmann et al. 2019; Marks et al. 2022) near the current state boundary of Germany and Poland and vast areas of Central Europe were glacier-free apart from mountain glaciers. The most extensive mountain glaciation in Central Europe covered during the LGM the Alps, where ~55% of the mountain system was glacierised (Ivy-Ochs

Blockfields (Iberian NW ranges)
Augusto Pérez-Alberti

Landslide (Western Carpathians, Czechia)
Tomáš Pánek

Sand wedge (Tvarozna, Czechia)
Petr Škrdla

Stone sorted-circles (Sierra Nevada)
Marc Oliva

Periglacial involutions (Kent, England)
Julian Murton

Talus slopes (Tramuntana massif, Mallorca)
Francesco X. Roig Munar

Fig. 10 Examples of periglacial features formed during the Last Glacial Cycle in Europe

et al. 2022; Chapter 8). Mountain valley glaciers were also present in Carpathians (Urdea et al. 2022; Zasadni et al. 2022), or in the highest parts of Central European Variscan Ranges (Mentlík et al. 2013; Engel et al. 2014; Hofmann et al. 2020). The vast glacier-free corridor between the European Ice Sheet in the north and the Alps ice cap in the south (with only minor mountain glaciers) represented a periglacial zone with negative MAAT and highly dynamic periglacial processes during the LGM. The area was underlain by a continuous permafrost (Vandenberghe et al. 2014; Lindgren et al. 2016) and numerous periglacial features indicative of its presence (e.g. ice wedge pseudomorphs, relic sand wedges). Due to gradual long-term cooling leading to peak (pleni)glacial conditions, the permafrost thickness in low-lying areas of Central Europe could reach 200–250 m, according to cryogenic deformations of the rocks detected in deep boreholes and model calculations (Šafanda and Kubík 1992; Czudek 2005). This was also the reason why the North European Plain and adjoining mountain ranges became a classical area of the study of past periglacial features since the early twentieth century, as is well described in chapters "The Central European Variscan Ranges"–"The North European Plain". The periglacial features are mostly connected with their origin during Pleistocene cold phases and vast majority of them would stem from the Last Glacial.

The MAAT dropped by 6 to 12 °C during the LGM compared to current interglacial temperatures (Heyman et al. 2013; Pačes et al. 2017). In the non-mountain areas, MAAT ranged between −7 and −2 °C (Šafanda and Rajver 2001). Peak pleniglacial conditions also represented the driest periods of the Quaternary climatic cycles, as mean annual precipitation values in Central Europe declined by 50–60% and were estimated to be from 250 to 300 mm in lower lying areas (Hatté and Guiot 2005). Low temperatures and decreased precipitation amounts led to scarce vegetation cover, which persisted only in low-lying areas and especially along the south-facing slopes of rivers (Feurdean et al. 2014; Magyari et al. 2014). These climatic conditions favoured periglacial dynamics not only in mountain ranges of Central Europe, such as the Alps, and Carpathians, but also in lowlands of the North European Plain. Detailed descriptions of Last Glacial periglacial landforms and their time frame of formation is given in chapters "The Alps"–"The North European Plain". For example, hillslope activity and accumulation of periglacial slope deposits during the LGM and slightly after was proven in some of the Central European Variscan ranges (Mt. Ślęża with ages from 17.1 ± 1.1 to 19.5 ± 1.3 ka; Waroszewski et al. 2020). However, even pre-LGM periglacial landsliding activity (MIS 3; 48 to 56 cal. ka BP) has been described and radiocarbon dated in the Carpathians (e.g.; Pánek et al. 2014; Jankovská et al. 2018). However, it should also be noted that pre-Last Glacial periglacial features could often be found in Mid-Pleistocene aeolian, glacial or fluvial deposits preserved outside the southern maximum extent of the European Ice Sheet during the LGM. Accordingly, pre-glacial tors were recently found buried by Elsterian glacial deposits at the southernmost margin of European Ice Sheet (Hanáček et al. 2021).

North European regions were extensively glaciated by European Ice Sheet (Batchelor et al. 2019; Palacios et al. 2022). In Iceland, the Pliocene–Pleistocene succession of volcanic rocks evidence the occurrence of many glaciations covering great part of the Island and the surrounding shelves. During the LGM, the Icelandic Ice Sheet (IIS) covered the entire island, reaching the shelf edge at 40 km off the coastline (chapter "Iceland"). In Great Britain and Scandinavia, the successive cold-based ice sheets allowed the preservation of periglacial landforms originated prior to the last glacial cycle. This was the case of the blockfields and tors formed in the plateau areas before the passing of the Late Devensian ice sheet (>30 ka) (Ballantyne and Stone

2015) in Great Britain. They coexisted with the formation of patterned ground during a number of episodes (Bateman et al. 2014; West 2015). During the Pleistocene, different climatic stages alternated, and during the cold ones, aeolian activity was especially intense, with the deposition of periglacial aeolian deposits such as loess and coversands (chapter "Great Britain and Ireland"). The influence of the air ice masses during the cold stages of the Pleistocene led also to significant changes of the ground thermal regime in the ice-free areas, from continuous/discontinuous permafrost to seasonally frozen ground. In these circumstances, is likely that permafrost developed under the ice sheets and at their vicinity, as deformed sediments wedges, polygonal structures (eastern England), remnants of frost mounds, thermokarst depressions and involutions have suggested (Waller et al. 2011). Conversely, in the flat lowland areas of the British Isles periglacial weathering occurred, together with river incision, formation of abundant dry valleys, fluvio-colluvial deposits and river terraces (chapter "Great Britain and Ireland").

In Scandinavia, the situation was relatively similar. Ice sheets and smaller ice caps than those of the LGM covered successively the plateau areas, with some glacier-free parts likely being subaerially exposed to periglacial conditions during interglacials. Block fields formed under that morphoclimatic domain and were successively overridden by successive ice sheets whose cold-based regime allowed them to survive even to the LGM glaciation (chapter "Scandinavia"). The same occurred with tors, which benefited from shielding by the cold-based ice sheet and survived multiple glacial cycles, alike to what it must have occurred to more fragile sorted patterned ground and block streams appearing in the blockfield areas (Wilson et al. 2017). In some lowland areas, periglacial activity occurred in the form of ice-wedge casts (~290–130 ka) forming polygonal patterns that remained active until the YD (chapter "Scandinavia").

4.2 *Last Deglaciation*

The deglaciation process of European Ice Sheet that started by 20.7 ka (Tylmann et al. 2019) promoted major environmental changes in Europe (Clark et al. 2009). Its shrinking towards the high latitudes was parallel to the deglaciation of most Northern Hemisphere mountain environments. Warmer climate conditions favoured also shifts in the intensity and spatial extent of periglacial processes in Europe, which became progressively confined to the highest mountain systems. Indeed, as glaciers retreated, periglacial dynamics occupied the recently glacier-free terrain with very active processes during the paraglacial phase (Ballantyne 2008).

However, the deglaciation was not a gradual process, but included an alternation of colder and warmer phases (chapter "Periglacial Landforms Across Europe"). During these major periods, the glacial evolution in European mountains followed a much more homogenous pattern than during the Last Glacial Cycle. Recent studies show interruptions during the long-term glacial retreat and even some phases of glacial re-expansion that took place in full synchrony within the Mediterranean region and with the mountainous systems of central Europe, as well as with the European Ice Sheet. The onset of a rapid and massive retreat of this ice sheet occurred at 20.7 ka (Tylmann et al. 2019) and accelerated at 18 ka as it also occurred in most European mountains (Palacios et al. 2022). In the Alps, for example, some of the largest glaciers in the range had already lost ca. 80% of their mass by 18 ka (Ivy-Ochs 2015), whereas the Pyrenees could have been fully deglaciated at that time, as it was probably the case of most south European mountain ranges (Palacios et al. 2017a).

Within the last deglaciation, the first glacial advance of mountain glaciers and of the European Ice Sheet occurred during the Oldest Dryas (17.5–14.6 ka; Palacios et al. 2022). Glacial expansion was driven by a significant reduction of the temperature of the North Atlantic Ocean that favoured a decrease in the intensity of the Atlantic Meridional Overturning Circulation, with extreme seasonality of cold winters and mild summers (Denton et al. 2005). Glaciers advanced to elevations of only 400 m above the LGM moraines in the Alps (Ivy-Ochs 2015), but also advanced to positions close to the LGM limits in several mountains in the Iberian Peninsula (Oliva et al. 2019); substantial glacial advances were also recorded in other European mountains, such as in the Tatra Mountains, Apennines, Balkans and Anatolian mountains (see details in Palacios et al. 2022). In the Alps, MAATs have been reconstructed to be 0 °C lower than present-day, and precipitation was one-third less (Ivy-Ochs 2015). This colder climate must have favoured the expansion of periglacial conditions, including permafrost, in mountain areas. Some fossil periglacial phenomena in European mountains may have formed during this period, including blockfields, patterned ground features or rock glaciers (Fig. 11; Oliva et al. 2018a, b). However, there is a lack of absolute ages for periglacial landforms formed during this period.

Similarly, in Central Europe the development of the periglacial realm is linked to the retreat of European Ice Sheet, which had completely withdrawn from North European Plain after 16.5 ka (the age of Gardno deglaciation phase in Southern Baltic Sea sector; Tylmann and Uścinowicz 2022). The deglaciation of the highest Central European ranges (the Alps and Carpathians, as well as Central European Variscan ranges) took also place just after the local LGM, as the maximum advance of mountain glaciers took place mostly by 22 ka (Ivy-Ochs et al. 2022; Zasadni et al. 2022). Relict periglacial landforms are widespread in the Alps and Carpathians, as they are related either to glacier-free conditions and hence periglacial weathering during the LGM above the highest glacierised areas, or to periglacial landscape modification of deglaciated areas in the paraglacial phase during the Late Pleniglacial, Late Glacial and Holocene periods (chapters "The Alps", "The Carpathians"). However, the currently available geochronological data are by far sufficient to address the timing of the periglacial landforms formation especially for the time period preceding the Younger Dryas, i.e. 18–13 ka. Besides, numerous periglacial features experienced multiple stage development during the colder phases of the Late Glacial, the Early Holocene and may have been activated also during the LIA in higher mountainous areas of Central Europe.

In the Alps, the relationship between glacier recession and formation of permafrost-related landforms was analysed by Kellerer-Pirklbauer and Kaufmann (2018). They also proposed a mass-balance approach of rock glacier evolution, as

Fig. 11 Examples of periglacial features formed during the last deglaciation in Europe

most of them evolved significantly during the paraglacial phase after glacier retreat. Most of the currently relict rock glaciers of the Alps were formed shortly after, or even simultaneously with the pre-Bølling retreat of glaciers and stabilised around 15.7 ka (Steinemann et al. 2020). The beginning of the paraglacial slope activity in the high mountains of Central Europe is connected with the early deglaciation phase (starting from Older Dryas) as evidenced by CRE dating of deep-seated rock-slope failures (sackungen) in Tatra Mts., the Western Carpathians by Pánek et al. (2017). In the Carpathians, the geochronological data on Older Dryas rock glacier activity are missing. However, Urdea (1998) was the first to claim that the formation of rock glaciers in the Southern Carpathians took place already during the Older Dryas. Besides, periglacial hillslope accumulations of Oldest Dryas and Older Dryas ages are described from Central European Variscan ranges not only based on their stratigraphic position in relation to LGM loess accumulation and Laacher See tephra unit, but also based on OSL dating (Mt. Ślęża with ages from 17.1 ± 1.1 to 19.5 ± 1.3 ka; Waroszewski et al. 2020). In the lower-lying North European Plain, frost wedges and fissures are common features proving the presence of permafrost during the cold stages of the Late Glacial (Kolstrup 2004; Buylaert et al. 2009). During the Late Pleniglacial and colder phases of the Late Glacial, aeolian processes were facilitated by periglacial climate leading to the deposition of cover sands in the North European Plain and the formation of the European Sand Belt stretching from Belgium and the Netherlands in the west, across Germany, southern Denmark, and Poland into Russia in the east (Zeeberg 1998; Kasse 2002; Koster 2009, Chapter 11). Different wind directions in Central Europe were connected with disappearance of northern katabatic winds flowing from the centre of European Ice Sheet and their replacement by westerlies during the Late Pleniglacial to Late Glacial periods (Holuša et al. 2022).

The retreat of the local ice sheets in Great Britain and Scandinavia after the end of the LGM was followed by a similar response. The unloading of the steep walls, glacio-isostatic rebound, frost wedging and the degradation of permafrost triggered the operation of the paraglacial processes in the form of rock slope failures of different types (rockfalls, large-scale slope deformation, catastrophic events, etc.). The connection of these events with the deglaciation is evident given that they took place just after the onset of the deglaciation; in fact, in Great Britain (chapter "Great Britain and Ireland") and Scandinavia (chapter "Scandinavia"), rock slope failures clustered during the early stages of the deglaciation; lags up to 1.6–2.0 ka were reported from Great Britain and Scandinavia.

In parallel, the retreat of the European Ice Sheet was followed by periglacial conditions prevailing in the glacier-free areas. In fact, an intense accumulation of aeolian deposits took place in Great Britain, with most of the loess deposition occurring between 25 and 10 ka (Stevens et al. 2020) and coversand being deposited especially between 26 and 15 ka (chapter "Great Britain and Ireland"). In Scandinavia, aeolian activity during the LGM was also inferred from ventifacts, and luminescence dating of ice-sheet limits (chapter "Scandinavia"). The Late Glacial glacier-free areas of Great Britain experienced periglacial conditions that determined the formation of sorted patterned ground features above former permafrost in frost-susceptible regolith (earth hummocks) and the occurrence of mass movements producing terraces and lobes of debris, especially in granite mountains, extending their influence beyond the slopes reaching valley floors in some upland areas. During the transition to the warm Bølling-Allerød interstadial, very few rock slope failures occurred in Great Britain, which evidences that permafrost degradation was not the main driver at this time. By this time, following local deglaciation, some (talus-derived) rock glaciers developed at low elevations in Scandinavia, even remaining active until ~11.1 ka when permafrost conditions ceased (chapter "Scandinavia").

In Iceland, the marine-based character of the IIS determined its abrupt collapse eustatic sea level rise associated to the Bølling-Allerød interstadial, rapidly retreating from the shelf (Pétursson et al. 2015; Chapter 14). The retreat of the IIS determined that the outermost areas of north-western, north-eastern and eastern Iceland were subaerially exposed, enabling permafrost aggradation. However, the degradation of this early permafrost occurred rapidly due to the warm conditions of the Bølling-Allerød period (Etzelmüller et al. 2020). The response of the slopes of north-western, northern and eastern Iceland to the deglaciation was similar to that observed in Great Britain and Scandinavia, with the occurrence of numerous rock slope failures, concentrated especially within the first millennia of the deglaciation process, around ~10 ka, when the uplift rates were the highest. In addition to the rock slope failures, there is also evidence of solifluction processes starting following ~12.2 ka BP in a number of locations throughout Iceland (chapter "Iceland").

Subsequently, at the transition between the Oldest Dryas and the onset of the Bølling–Allerød (14.6–12.9 ka) there was a substantial temperature increase that promoted a massive glacial retreat across European mountains, including the European Ice Sheet (see details in Palacios et al. 2022). Indeed, the use of CRE dating on relict rock glaciers provides a better comprehension of the timing of their formation and stabilisation, showing evidence that most features stabilised soon after formation, although their upper parts remained active during several millennia. This is the case of many rock glaciers in Iberian Mountains, which formed during the paraglacial phase at the onset of the Bølling–Allerød at ~15–14.5 ka, and must be linked to a glacigenic origin (Palacios et al. 2017a; Andrés et al. 2018). As temperatures increased during this period, permafrost conditions moved to higher altitudes and those rock glaciers became relict. A similar process for rapid paraglacial formation of rock glaciers has been also described in other mountain regions, including the Alps (Hippolyte et al. 2009), southern Scandinavia (Linge 2020) and the British Isles (Ballantyne et al. 2009) (see chapter "The Alps, Great Britain and Ireland and Scandinavia". At lower elevations, stratified screes in valleys and slopes have been associated with this phase (García-Ruiz et al. 2001).

Temperatures dropped again during the Younger Dryas (12.9–11.7 ka), which favoured the last major advance of the European Ice Sheet as well as of most mountain glaciers in Europe. In moderately high mountain ranges, such as the Pyrenees, Sierra Nevada or the Apennines, glaciers rarely exceeded the limits of the cirques as this period was characterised by cold, but arid conditions in southern Europe (Oliva et al. 2019). In other higher ranges, such as in the Alps - where MAAT were 3.5–5 ºC lower than present and precipitation was up to 30% less (Kerschner and Ivy-Ochs 2008)—glaciers had a larger expansion and the largest units reoccupied the valley floors (see details in Palacios et al. 2022). Similarly to what occurred during the latest stages of the Oldest Dryas, the end of the Younger Dryas promoted the rapid shrinking of mountain glaciers that enhanced the reactivation of paraglacial processes in the cirque walls, which in turn favoured the development of rock glaciers and protalus lobes (Oliva et al. 2018b). CRE ages from inactive rock glaciers distributed across several Iberian mountain ranges showed evidence that their fronts became relict immediately after their formation, although their roots remained active until the Mid-Holocene, when the higher temperatures favoured the melting of the last permafrost patches located beneath them (Palacios et al. 2015; García-Ruiz et al. 2016; Oliva et al. 2016b; Andrés et al. 2018). Although many rock glaciers in the Mediterranean mountains, such as in the southern Alps (Colucci et al. 2016a), could have also developed during this stage and be thus associated to a paraglacial origin, further research needs to confirm this hypothesis. The Younger Dryas promoted a widespread down-valley expansion of periglacial activity in southern Europe. This very cold climate regime expanded the permafrost domain to lower elevations and must have favoured the development of block streams in the highest mountain ranges and reactivated patterned ground features across the snow-drifted plateaus (Oliva et al. 2016b).

Consequently, the cold phases interrupting the long-term deglaciation process (Oldest and Younger Dryas) left both glacial and periglacial evidence in the highest mountain ranges in southern Europe, mainly with the formation of permafrost-related features derived from paraglacial dynamics. In fact, the presence of these inactive landforms under present-day climate conditions is indicative of the minimum altitude for the past occurrence of permafrost during those phases.

Numerous relict periglacial landforms found in the Alps and Carpathians were active during the Younger Dryas. However, they were often reactivated and does not constitute newly developed landforms of Younger Dryas age. Many nowadays relict rock glaciers in the Alps were formed between Heinrich Stadial 1 and the Younger Dryas and stabilised during the Early Holocene (Charton et al. 2021). Their past presence was supported by modelling of their Younger Dryas distribution for example in Hohe Tauern (Avian and Kellerer-Pirklbauer 2012), or in Err-Julier area in Swiss Alps (Frauenfelder et al. 2001). In the Carpathians, present data on the formation of rock glaciers indicate that most of their relict forms were active during the Younger Dryas (12.9–11.5 ka BP; Kotarba 1991–1992; Kłapyta 2013; Onaca et al. 2017a; Uxa and Mida 2017; Popescu et al. 2017). In the Tatras and other parts of Carpathians, the stabilisation of the rock glaciers took place with the prominent Early Holocene warming associated with the Preboreal period (11.8–10.0 ka; Zasadni et al. 2020). Younger Dryas cooling also enhanced hillslope activity, as shown by OSL ages for the upper periglacial cover beds in the Taunus Mts and Ślęża Mt. spanning 8.7 ± 1.0 to 12.5 ± 1.4 ka (Hülle et al. 2009) and 12.0 ± 1.0 to 14.2 ± 1.1 ka (Waroszewski et al. 2020), respectively. Furthermore, ice loading of the lithosphere induced by the proximity of the European Ice Sheet margin during the LGM reactivated the Sudetic Marginal Fault (Štěpančíková et al. 2010, 2022), which led to hillslope processes such as gelifluction (Fig. 9) and enhanced hillslope activity continued till the beginning of the Holocene (10.9 ± 0.1 cal. ka BP; Štěpančíková et al. 2010). The Younger Dryas also represents the final phase of the formation of aeolian landforms (Kasse 1999, 2002; Schirmer 1999), even though aeolian activity might have continued till the beginning of the Holocene at some places of North European Plain according to recent OSL dates (Moska et al. 2022).

4.3 Holocene

Marine sedimentary records, Greenland ice cores and a large number of terrestrial proxies showed evidence of a significant temperature increase in the North Atlantic region at the onset of the Holocene (chapter "Quaternary Climate Variability and Periglacial Dynamics"). The European continent has recorded substantial climate changes during the Holocene, with temperature oscillations of the order of ca. ±2 °C that have caused significant disruptions in human civilizations (Mayewski et al. 2004). Temperature and moisture changes have thus affected type and intensity of cold-climate geomorphological processes prevailing in Europe (Oliva et al. 2018b). The sharp temperature increase recorded during the Early Holocene promoted an accelerated shrinking of mountain glaciers, which caused their final disappearance in many massifs, as well a rapid retreat and the complete melting of the European Ice Sheet during the Early Holocene (Palacios et al. 2022). The recently deglaciated environments were gradually replaced by periglacial processes that also migrate to higher elevations. Most of the periglacial landforms that were active during the Younger Dryas became progressively inactive, and only reactivated during the coldest Holocene phases (e.g. LIA).

In general, during the Holocene periglacial processes in Southern and Central Europe have mostly been restricted to the high mountain areas, where glaciers only existed during the coldest phases (Oliva et al. 2018b). As occurred in the Alps (chapter "The Alps"), the Apennines (chapter "Periglacial Landforms Across Europe") or Iberian mountains (chapter "The Iberian Peninsula"), the cold-climate conditions during the Holocene promoted the formation and/or expansion of small glaciers in the highest mountains and decrease of periglacial processes to lower elevations whereas warmer phases favoured a lower periglacial activity, which also migrated to higher altitudes and favoured the spread of vegetation cover through the periglacial belt. In northern Europe, periglacial activity was limited to the summits and slopes of the highest mountains and cols of the British Islands (chapter "Great Britain and Ireland") and to the new deglaciated terrain after the retreat of the local ice sheets in Scandinavia (chapter "Scandinavia") and Iceland (chapter "Iceland").

In most mid-altitude mountain ranges in southern Europe—with maximum elevations around 2000 m in the central and western Mediterranean and 2500–3000 m in the eastern fringe of the basin—the periglacial belt disappeared during the Holocene and was significantly reduced in the highest mountains. During the warmest periods, such as the Holocene Thermal Maximum (HTM; Renssen et al. 2012) and the Medieval Warm Period, only the mountain regions exceeding 2500–3000 m included a periglacial environment dominated by seasonal frost conditions, with permafrost only confined to the highest sectors of the Maritime Alps, as well as in the highest summits in Anatolia. However, geochronological evidence of Holocene periglacial processes in southern Europe is mostly related to rock glaciers and solifluction lobes, as well as other landforms and deposits, for which few ages are available (i.e., talus screes) (Fig. 11).

Actually, permafrost-derived landforms were probably active in many mountain ranges during the Early Holocene in the highest massifs in southern Europe (Oliva et al. 2018b). Indeed, in the central Taurus Mountains (Çiner et al. 2017) and the Pyrenees (Serrano et al. 2010) there is evidence of the formation of some still active rock glaciers during the HTM at ~6 ka. Similarly, in the southern Alps, some currently active rock glaciers showed activity during Holocene warm phases (chapter "The Alps"), with evidence for their formation during dry periods: 8 ka, 3.7–3.2 ka, and possibly 1.2 ka cal BP (Dramis et al. 2003). In some limestone massifs, below the periglacial environment, some ice caves containing perennial accumulations of ice started accumulating ice during at the onset of the Neoglacial period at 5.5–5 ka (Serrano et al. 2018a); the frozen ground in contact with the ice inside the caves has been also described as permafrost (Colucci et al. 2016b); thus, permafrost may have aggraded since then in these areas, persisting until today despite the MAAT well above 0 °C due to specific microtopographic conditions (very low solar radiation, existence of stable temperature inversion, or chimney effects) (Oliva et al. 2018b). Other periglacial phenomena associated with seasonal frost, such as solifluction lobes distributed across the present-day periglacial belt of Sierra Nevada at elevations between 2500 and 3000 m, were also active during this warm phase until 5 ka cal BP. These features include sedimentological evidence of solifluction activity during Mid and Late Holocene cold and wet phases, namely at 5–4 ka, 3.6–3.4 ka, 3–2.8 ka, 2.5–2.3 ka, 1.8–1.6 ka, 0.85–0.7 ka, and 0.4–0.15 ka cal BP (Oliva et al. 2009). The intensification of arid conditions in southern Iberia by 4.2 ka cal BP resulted in a long-term increase in the elevation of active solifluction processes in this semiarid massif (Oliva et al. 2011). As for solifluction, the alternation between cold and warm phases during the Holocene accompanied also by changing precipitation regimes must have also influenced the intensity of periglacial processes and the spatial distribution of periglacial processes. Stratified scree deposits at mid-altitude regions of the Picos de Europa revealed a lithological and topographical control on their formation regardless the prevailing climatic conditions of the Early-Mid Holocene (Ruiz-Fernández et al. 2019) (Fig. 12).

As in Sierra Nevada, the Alps (Veit 1988), or the Apennines (Giraudi 2005), warmer periods favoured soil formation and colder phases enhanced solifluction processes in seasonal frost environments, with variable intensities depending on the moisture regime. In fact, a similar timing for solifluction chronology has been inferred in these three mountain regions. Many of the debris-mantled slopes and blockstreams in the Alps and Carpthians were formed during the Pleistocene with only minor reshaping during the Holocene, but their reactivation only occurred in highest location during the Holocene's coldest period, such as the LIA (Kotarba 1995; Puţan 2015; Şerban et al. 2019). This was accompanied by enhanced physical weathering and rock fall activity (Kotarba and Pech 2002).

With the onset of the Holocene, associated with permafrost degradation combined with the establishment of groundwater circulation, numerous paraglacial landslides, slope failures and rock avalanches in the Alps and Carpathians originated (Margielewski 2006; Krautblatter et al. 2010, 2012; Pánek et al. 2014, 2016). Similarly, Early Holocene hillslope activity was also described from Variscan Ranges

Ice-wedge polygonal terrain (Svalbard) Marc Oliva

Landslide (Malenik Ridge, Western Carpathians) Ivo Baroň

Solifluction deposits (Sierra Nevada) Marc Oliva

Stratified screes (NW Iberian ranges) Augusto Pérez-Alberti

Thufur (north Iceland) Costanza Morino

Pingo (Svalbard) Marc Oliva

Fig. 12 Examples of periglacial features formed during the Holocene in Europe

of the Sudetes Mts. (e.g.; Štěpančíková et al. 2010). At some higher places of the Alps and Carpathians, rock glacier stabilisation did not take place earlier than the Early Holocene (Vespreameanu Stroe et al. 2012; Zasadni et al. 2020; Charton et al. 2021) and their activity increased during Holocene colder periods, especially the LIA (Urdea 1998; Avian and Kellerer-Pirklbauer 2012). This reactivation only occurred in the highest massifs of the mountains. Howver, in some Carpathian ranges (such as Tatras) no rock glacial activity was detected during the LIA (Kędzia 2014). Special attention was paid on the modelling of mountain permafrost distribution and the distribution of rock glaciers and permafrost in European Alps in the past and today showing its larger distribution during Holocene cold periods, such as the LIA (e.g.; Avian and Kellerer-Pirklbauer 2012; Kellerer-Pirklbauer et al. 2012).

In Great Britain and Scandinavia, climatic conditions characterised by relatively warm summers prevailed during great part of the Early Holocene. Under these circumstances, Holocene frost weathering activity was restricted to granular disaggregation and flaking, especially in granitic and sandstone lithologies. In parallel, processes conducting to patterned ground sorting (circles, nets, stone stripes) were especially active in frost-susceptible soils in the mountains of the British Islands (chapter "Great Britain and Ireland") and in till with water-saturated active layers, after the Early Holocene deglaciation in Scandinavia (chapter "Scandinavia"). In steep slopes inside the limits of the ice during the Loch Lomond Stadial/YD ice sheet, rock slope failure activity continued throughout the Holocene. Accordingly, there is evidence of rock slope deformation, permafrost creeping and rock glacier formation in Scandinavia starting at 11–10 ka in some places. In Iceland, the cirque glaciers of northern Iceland (Tröllaskagi Peninsula), retreated accordingly with the trend of the IIS outlets after culminating a main glacial advance at ~11 ka, triggering paraglacial processes at the steep slopes of the cirques and valleys. Consequently, cirque glaciers evolved to debris-covered glaciers and rock glaciers, which rapidly stabilised by 10–9 ka. Subsequently, new generations of rock glaciers formed, some of which are currently active albeit they remain static since 6.3 ka (chapter "Iceland").

In Scandinavia, YD climate and the high debris supply linked to the paraglacial conditions prevailing at the Early Holocene allowed the development of pronival ramparts—that remained active throughout the Holocene—and the aggradation of fans, attaining a later peak around 9–8 ka. Climate was also prone for the occurrence of colluvial events (slope wash) and the formation of sorted circles and sorted stripes, dated at ~8–6 ka. After the retreat of the YD ice sheet, aeolian activity was intense in the new ice-free areas, with the formation and stabilisation of parabolic dunes at ~10.9–10.2 ka due to sand blasting from glacial deposits. However, some dunes were reactivated successively during the Holocene, especially after ~8.3 ka (chapter "Scandinavia"). Aeolian activity started also in Iceland as the IIS retreated at the highlands by ~10 ka, and subsequently, with the glacier disappearance during the Holocene Thermal Maximum (HTM). Consequently, glaciers left extensive ice-free surfaces covered by till. Such surfaces and the glacial rivers became important sources of aeolian sediments that contributed to the construction of sand-fields in the glacial margin areas and the accumulation of sand in subaerial Holocene lava flows and originating different types of dunes, especially in southern Iceland. On the other hand, aeolian activity contributed to the generation of deflation surfaces such as stone or gravel pavements (sandy lag gravel) and abrasion features such as sculpted rocks, micro-yardangs and ventifacts (chapter "Iceland").

Throughout the Holocene, rockfall activity continued, with major events of rockslides and rock avalanches during the Middle and Late Holocene, peaking at ~6–4 ka and ~1.7 ka in Great Britain (chapter "Great Britain and Ireland"), an at 8.0–7.0, 5.4–4.0 and 3.0 ka in Scandinavia (chapter "Scandinavia"). In Great Britain, landslides and debris flows were also amongst the dominant processes reshaping mountain slopes (chapter "Great Britain and Ireland"). Warm climate conditions prevailing during the HTM, enhanced rock slope activity in Scandinavia, especially at the end of such period, due to permafrost degradation and its resulted minimum depth. The rise of the lower altitudinal level of discontinuous permafrost (1650–1700 m) led to the cessation of sone circles / stripes activity, with stabilization ages between ~8 and ~7 ka. The climatic setting of this period, especially at the end (~7.0–6.0 ka), contributed to a decrease of debris flow activity as snow melt and the degradation of the permafrost active layer were more limited and less likely to occur. Conversely, significant avalanche activity occurred during the Mid- and Late Holocene, with the deposition of numerous fans (boulders dated at 7.4–2.3 ka, possibly linked to paraglacial processes and permafrost degradation during the HTM) and the formation of snow-avalanche impact ramparts (younger than 3.6 ka). However, after the HTM, a decline in rock fall activity was evident due to the exhaustion of permafrost, not enough for triggering large rock-slope failures (chapter "Scandinavia").

In Great Britain, the climate underwent a significant cooling trend during the last 5 ka, where MAAT fluctuated in the range of ±2 °C. Consequently, during, the Late Holocene cold climate conditions in Great Britain supported solifluction processes (above 600 m) and the movement of ploughing boulders, active since at least the Mid-Holocene. The same was found in Scandinavia, with solifluction lobe activity and development starting during the Mid-Holocene (phases at ~7.4–6.7 ka) and continuing after the HTM during the Late Holocene (4.2–3.4 ka, 2.6–2.1 ka and 1.5–0.5 ka) driven by millennial changes of moisture supply (chapter "Scandinavia"). Enhanced solifluction activity was also reported from several sites in central and western Iceland between 4 and 2 ka (chapter "Iceland"). Moreover, cooling trend starting after the HTM led to the formation of earth hummocks, some of which have been dated at between ~4.8 and ~3.1 ka in southern Norway and younger than ~4.1 ka in Finnish Lapland. Earth hummocks also formed in Iceland following the HTM, with different phases of formation at ~4.5 cal ka BP, or after 2.8–2.6 cal ka BP and prior to 1104 CE (chapter "Iceland"). The prevailing climate might have been cold enough to allow permafrost aggradation in the palsas of northern Finland and northern Pechoria by 3.0–2.5 ka, possibly starting between 5.0 and 4.0 ka (chapter "Scandinavia"). The cooling trend that followed the HTM and characterised the Neoglaciation would have led to the formation of the ground ice of the Icelandic palsas (Orravatnsrústir), starting at 4.5 ka BP (chapter "Iceland").

However, cryogenic processes were not the only ones that characterised the Mid- and Late Holocene periods. Evidence of increased aeolian activity since 6.0 ka was found in north-western Norway, culminating by around 1.7–0.7 ka. Strong winds and intense aeolian activity have been inferred also in Great Britain on the basis of features such as ventifacts, wind-patterned ground, wind stripes, gravelly/bouldery surfaces and vegetated aeolian sand deposits, accumulated in north-western Ireland especially during the Middle and Late Holocene (chapter "Iceland").

4.4 Little Ice Age

From approximately 1300 to 1850 CE, the climate of Europe underwent the coldest most recent period, that is known as the LIA (Oliva et al. 2018a). During these centuries, which have been used as a reference for Holocene cold phases, geomorphological processes in Europe reacted accordingly to the lower temperatures: (i) glaciers that survived the Medieval Warm Period expanded to their largest volume over the last several millennia and several other small ice masses formed in the highest massifs (Bradley and Jones 1992), and (ii) the spatial domain of periglacial dynamics expanded down-valleys and cryogenic processes reappeared in some areas where seasonal frost activity was reduced during the preceding warm period and is no longer active at present.

In Europe, average temperatures during the coldest stages of the LIA recorded during the Minimum Maunder (1675–1715 CE) were ~0.5–1 °C colder than today, with changing precipitation patterns (Grove 2004; Luterbacher et al. 2006, 2016; Mann et al. 2009). In some regions of southern Europe, such as the Iberian and Balkan peninsulas, the difference of temperatures between the LIA and the present could have been even larger. Based on glacial and periglacial geomorphic evidence, temperatures during the last cold stages of the LIA were ~1 ºC lower compared to present-day values, up to ~2 ºC with respect to the Minimum Maunder (Oliva et al. 2018a).

Colder conditions favoured more extensive long-lasting and perennial snow-patches than today in the high mountains in southern Europe. In turn, nival processes were more intense, with increased avalanched activity (García-Hernández et al. 2017) that favoured the formation of (well-preserved) protalus ramparts at the foot of slopes as well as more active and widespread solifluction processes (Oliva et al. 2009). Colder conditions must have also favoured the reactivation of cryoturbation processes, including pattern ground phenomena in the highest sectors of the mountains (Fig. 13). Permafrost-related features, such as rock glaciers and protalus lobes formed and/or reactivated in the highest slopes of the main massifs, at ~2350 m in the Balkan region (chapter "The Balkans (Without Carpathians)"), ~2400–2500 m in the southern Alps (chapter "The Alps") and ~2560 m in the Pyrenees (chapter "The Iberian Peninsula"). In addition, ice caves accumulated more ice during this period in several high karstic mountains in southern Europe, in areas located several hundreds of metres below the concurrently glaciated environments, even at ca. 1000 m in the Dinaric Alps (Kern et al. 2018).

As temperatures started rising during the second half of the twentieth century, glaciers start to retreat and many disappeared (Hughes 2014, 2018), paraglacial activity increased (Serrano et al. 2018b), periglacial processes migrate to higher elevations and alpine permafrost environments in mid-latitude high mountain started to degrade and reduce in extent (Oliva et al. 2016c). All these processes were parallel to substantial geological changes in the upper sectors of the highest ranges (Pauli et al. 2012). Pattern ground developed in the forelands of glaciated environments in the Pyrenees (Feuillet and Mercier 2012) as well as frost mounds that also formed near

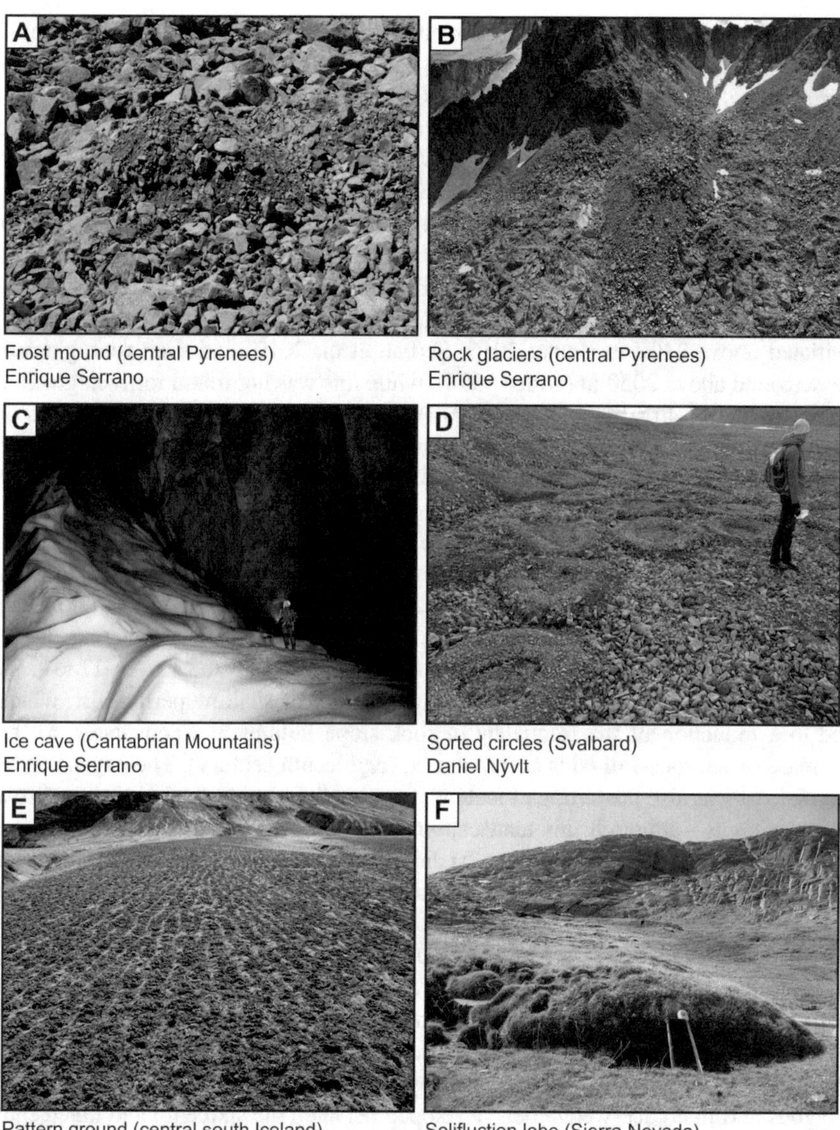

Fig. 13 Examples of periglacial features formed during the Little Ice Age in Europe

the glacier margins on unconsolidated sediments (Serrano et al. 2018b). Proglacial areas and Little Ice Age moraines were also reshaped by debris flows, solifluction processes and alluvial dynamics, with the formation of several new lakes and lagoons (Serrano et al. 2018b). Shrinking glaciers favoured also rock avalanches and landslides that fed the talus slopes within the cirques, where, in some cases, even promoted the development of incipient rock glaciers and protalus lobes that tend to become relict as temperatures keep increasing and buried ice masses and isolated patches of permafrost degrade (Gómez-Ortiz et al. 2014, 2019).

Glaciers in European Alps reached their LIA maximum extent in the fourteenth, seventeenth and nineteenth century AD with most of them reaching their maxima around 1850/1860 AD (Ivy-Ochs et al. 2009). Advances of local glaciers has also effect on periglacial dynamics such as physical weathering, debris flows and other hillslope processes, which account also for non-glaciated Central European mountain ranges of Central Europe during the LIA, such as Carpathians (Kotarba 1995; Kotarba and Pech 2002). In the Southern Carpathians, the formation of blockstreams was re-initiated above 2100 m (Puţan 2015; Şerban et al. 2019) and even rock glaciers re-activated above 2050 m (Urdea 1998), while this was not found for rock glaciers in the Western Carpathians (Kędzia 2014). Nevertheless, the climatic conditions of the LIA were not sufficient to change significantly the periglacial dynamics in the lower Variscan ranges of Central Europe, such as the Sudetes Mts. (chapter "The Central European Variscan Ranges").

The LIA was the coldest period of the last 10 ka in Great Britain, with climate conditions that maintained perennial snow patch in some Scottish mountains. Its enhanced storminess brought intense wind erosion and deposition at the plateau areas (chapter "Great Britain and Ireland"). In Scandinavia, the lowering of the lower limit of discontinuous permafrost culminated during the LIA (~1750 CE), reaching ~1250 m a.s.l., leading to the aggradation of shallow permafrost, which led to a reduction of the frequency of rock slope failures and conversely, to the increase of the rock-fall talus accumulation (eighteenth century). The expansion of permafrost was also prevailing in Iceland, forming the youngest existing permafrost in palsa areas—although this landscape might have developed earlier, during the Medieval Warm Period (MWP)—so far and reaching the maximum Holocene extent (Etzelmüller et al. 2020) (chapter "Iceland").

Climate of the LIA was also favourable for glacial advances—which in some cases reworked ice-cored moraines—and snow avalanches, some of which were dated to 0.6 ka in Scandinavia (chapter "Scandinavia"). In Iceland, the largest ice caps and smaller glaciers advanced, reaching their maximum Holocene extent and reworking older Neoglacial moraines in northern Iceland (Fernández-Fernández et al. 2019). In the surroundings of some ice caps at central Iceland, frost-crack macropolygons and likely ice-wedge nets developed between the sixteenth and nineteenth centuries, in the coldest decades of the LIA (chapter "Iceland"). Meanwhile, in most of the low-gradient vegetated surfaces, thúfur formed in accordance with the coldest and wettest periods. In other vegetated slopes, the loss of vegetation cover linked to the climate deterioration triggered solifluidal processes, with the reactivation of some lobes formed before the MWP. Conversely, the potential formation of some rock glaciers in the glacial cirques of northern Iceland (Tröllaskagi) has also been attributed to the LIA (chapter "Iceland").

Once the climate conditions of the LIA ended, glacial retreat from the LIA positions started. Subsequently, patterned ground such as stone circles rapidly developed in some recently deglaciated forelands, and gully erosion and debris-flow activity initiated in the steep slopes of the lateral moraines, contributing to their degradation (chapters "Scandinavia" and "Iceland"). Even there is evidence of rock glacier formation from moraines thank to increased debris supply and the incorporation of ice (interstitial ice or buried glacier ice) in Scandinavia (chapter "Scandinavia").

5 The Interaction Between Periglacial Landscapes and Human Activities

The landscapes of Europe have been continuously transformed by human activities: from the local impacts in their surrounding environments carried out by hunter-gatherers since the last glacial cycle to the major changes imposed by present-day industrialised societies (Pinto-Correia et al. 2018). Obviously, these transformations followed different spatial and temporal patterns with variables impacts, but have almost affected the entire continent. Very few areas remain virtually intact and periglacial landscapes are not an exception as they have been also massively disturbed by human activities. This is the case of inherited periglacial landscapes in Central and Northern Europe above latitude 50 ºN, where most of the infrastructures and settlements networks have been established on past permafrost terrain although current MAAT are well above 0 ºC (Fig. 1). In these areas, periglacial landforms are almost no longer visible at the surface although periglacial deposits are preserved below and geotechnically condition the building processes, foundation of houses, construction of roads, etc. (Fig. 14).

Actually, inactive periglacial landforms are widespread across Europe, particularly in mountain regions where they tend to be better preserved because of generally lower anthropogenic pressure. However, present-day periglacial landscapes in many mountain areas have been also largely transformed by the construction of ski complexes, particularly in gently sloping glacial valleys and valley slopes currently affected by periglacial conditions. The initial conditioning of the ski slopes also involved the removal of (glacial erratic) blocks, dismantling of moraines, the use of glacial lagoons to produce artificial snow, soil disturbance, and other impacts across the entire ski area associated with the construction of the resorts, ropeway transport systems, high-altitude restaurants, huts, etc. As a result, this tourism massification induced a significant alteration of the natural ecosystem dynamics of the periglacial belt, together with severe impacts in some periglacial features abundant in these areas, such as solifluction landforms and even some larger scale landforms, such as rock glaciers (Fig. 14). In addition, snow grooming in ski resorts alters also the thickness and persistence of snow in the ground, which in turn affects the natural regime of ground temperatures and frost action dynamics.

Periglacial processes are still active in high mountain environments and in Northern Europe, particularly in high areas in Scandinavia and Iceland. Here, rapid mass movements (debris flows, mud flows, rock falls, etc.) within the periglacial domain frequently affect the road network, mountain tracks and trails and even

Ski resort, Pas de la Casa (central Pyrenees)
Marc Oliva

Road cutting a rock glacier (Swiss Alps)
José M. Fernández-Fernández

Snow-avalanche defense structures (Swiss Alps)
José M. Fernández-Fernández

Upslope treeline migration (central Pyrenees)
Marc Oliva

Solifluction lobes in Sierra Nevada (S Iberia)
Marc Oliva

Rock falls in permafrost terrain (Mt. Blanc massif)
Ludovic Ravanel

Fig. 14 Examples of transformations of periglacial landscapes by human activities: (**A**) Ski resort built occupying an ancient glacial cirque currently affected by periglacial conditions, Pas de la Casa (central Pyrenees), (**B**) Rock glacier cut by the construction of the road in the Albula (Swiss Alps); (**C**) Defence structures against snow avalanches and debris flows (Swiss Alps); (**D**) Upslope migration of the treeline in the lower periglacial belt (central Pyrenees); (**E**) Solifluction lobes generated by the expansion of wetlands created by the spreading of irrigation channels in the Sierra Nevada (South Iberia); (**F**) Rock fall triggered by permafrost degradation at the foot of the Aiguille des Grands Montets (3285 m; Mt. Blanc massif, French Alps); almost affecting ski facilities some settlements and mountain huts. They are frequent in spring when snowmelt favours slope processes, but they also occur in summer, particularly at high altitudes where permafrost degradation may also trigger slope failures. Slow mass wasting processes, namely solifluction, also affect widespread gentle and steep slopes in these regions, and some infrastructures such as ski resort facilities or electricity lines, must be constructed considering these processes. This is also valid for mountain regions in Central Europe, particularly the Alps and Carpathians. The Alps are under high anthropic pressure with tens of millions of inhabitants settling across the Alpine region, and therefore susceptible of being affected by periglacial processes. However, several Alpine countries have implemented effective and extensive monitoring systems in order to control potential natural hazards associated with permafrost thawing (Gruber and Haeberli 2007; Deline et al. 2011). However, rock falls during summer heat waves are becoming more widespread and a threatening risk for climbers and visitors of the highest massifs (Bodin et al. 2015; Magnin et al. 2017), and scientists are pushing stakeholders to implement new strategies to reduce their impacts on infrastructure in high mountain areas (Duvillard et al. 2021). Although to a lesser extent, mountain regions in Southern Europe are also affected by periglacial slope processes, which can damage ski facilities and huts (Serrano et al. 2018b). At the highest areas at ca. 3,000 m, some mountains still include permafrost, which is also undergoing a rapid degradation and favouring rock falls that can also affect climbers and mountaineers (Rico et al. 2021). Although not strictly periglacial, snow avalanches in mountain regions constitute also major hazards triggered in the periglacial belt; ski resorts and mountain villages are frequently affected by these processes and local administrations implement a wide range of measures to mitigate the risk (Fig. 14). Similarly, the alpine channels affected by snow avalanches can be also frequently used by debris flows in late spring and summer mobilising unconsolidated sediments from the periglacial zone to the valley floor, and may constitute also a potential hazard for mountain communities.

The strong relationship between human practices and geomorphic processes in some mountain ranges has also favoured an increase/decrease of the intensity of periglacial dynamics. Widespread grazing activities in steep hillslopes in the lower periglacial belt where soils are shallow and poorly developed have favoured shallow landslides and the formation of *terracettes*. By contrast, in other areas such as in in the Pyrenees, the accelerated abandonment of grazing activities have promoted a decrease of surface runoff and soil erosion (García-Ruiz et al. 2021). In Iceland, livestock—especially horse and cattle—accelerates the formation of *thúfur* (or earth hummocks; Schunke and Zoltai 1988) due to the frequent grazing in winter, when water saturation of the soil occurs: the animals put the feet in the depressions between the hummocks, and given the low soil cohesion, a gradual upwards push of the hummocks occurs (Arnalds 2015). Despite efforts on levelling terrain to remove hummocks, or other agricultural activities such as tilling and ploughing, these landforms may redevelop relatively fast, being a major problem for Icelandic farmers (Webb 1972). In parallel, the unvegetated periglacial belt is reducing as there has been an upslope migration of treeline over the last decades in response to both climate warming and lower human pressure (Camarero et al. 2015). Human-induced deforestation since the Viking colonisation of Iceland (Church et al. 2007) lead to extensive deflation and damages to soils and vegetation cover (Brewington et al. 2015). Similarly, most of the Central European mountain ranges were deforested during the seventeenth-twentieth century (Cebecauer and Hofierka 2008; Kapustová et al. 2018) especially for the production of timber and charcoal, however, the deforestation started already during the Late Mediaeval period in some mountain ranges (Speranza et al. 2000). Deforestation lead to an increase of humidity in local soils and enhanced also periglacial processes in the summit areas of the mountains. The most intensive processes of soil erosion coincided with the early phases of the LIA in Central European mountains (Klimek et al. 2006). In the Iberian Central Mountain System, the increasing occurrence of wildfires have triggered an enhanced activity of debris flows (Sequeira et al. 2019), together with grazing, vegetation degradation and increases of frost processes (Vieira et al. 2003). In the Sierra Nevada, the bypassing the waters in glacial cirques and high valleys to transport water to the surrounding lowlands favoured the expansion of wetlands and promoted the formation of solifluction lobes in gentle slopes (Oliva 2009).

Despite these impacts, a long-term perspective allows us to be optimistic about the future preservation of the European periglacial landscapes. Since the creation of the first National Parks more than a century ago, many countries have preserved vast regions including periglacial landscapes. Currently, there are several environmental protection approaches (e.g. National Parks, Natural Parks, Geoparks, Geosites, Sites of Community Importance etc.) to protect glacial and periglacial landscapes, particularly in European mountain regions being visited by an increasing number of tourists (Debarbieux et al. 2014). In addition, over the last decades there has been an increase of the awareness of the uniqueness of the natural heritage of the periglacial landscapes. These measures for their protection must be expanded and improved, so that these features can be enjoyed and further studied by future generations.

Acknowledgements Marc Oliva is supported by the Ramón y Cajal Program (RYC-2015-17597) and the Research Group ANTALP (Antarctic, Arctic, Alpine Environments; 2017-SGR-1102) funded by the Government of Catalonia through the AGAUR agency. This chapter complements the research topics examined in the NEOGREEN project (PID2020-113798GB-C31) and the NUNANTAR project (02/SAICT/2017 - 32002) of the Fundação para a Ciência e a Tecnologia, Portugal. We appreciate the help of Jakub Holuša in drawing of some of the figures.

References

Åkerman HJ, Johansson M (2008) Thawing permafrost and thicker active layers in sub-arctic Sweden. Permafrost and Periglacial Processes 19:279–292

Allard JL, Hughes PD, Woodward JC (2021) Heinrich Stadial aridity forced Mediterranean-wide glacier retreat in the last cold stage. Nat Geosci 14:197–205

Andrés N, Gómez-Ortiz A, Fernández-Fernández JM et al (2018) Timing of deglaciation and rock glacier origin in the southeastern Pyrenees: a review and new data. Boreas 47:1050–1071

Andrés N, Palacios D, Saemundsson Þ et al (2019) The rapid deglaciation of the Skagafjörður fjord, northern Iceland. Boreas 48:92–106
Arnalds O (2015) The Soils of Iceland. Springer, Dordrecht
Avian M, Kellerer-Pirklbauer A (2012) Modelling of potential permafrost distribution during the Younger Dryas, the Little Ice Age and at present in the Reisseck Mountains, Hohe Tauern Range. Austria. Austrian J Earth Sci 105(2):140–153
Ballantyne CK (2008) After the ice: Holocene geomorphic activity in the Scottish Highlands. Scot Geogr J 124(1):8–52
Ballantyne CK (2013) A 35-year record of solifluction in a maritime periglacial environment. Permafrost Periglac Process 24:56–66
Ballantyne CK, Schnabel C, Xu S (2009) Exposure dating and reinterpretation of coarse debris accumulations ('rock glaciers') in the Cairngorm Mountains, Scotland. J Quat Sci 24:19–31
Ballantyne CK, Stone JO (2015) Trimlines, blockfields and the vertical extent of the last ice sheet in southern Ireland. Boreas 44:277–287
Batchelor CL, Margold M, Krapp M et al (2019) The configuration of Northern Hemisphere ice sheets through the Quaternary. Nat Commun 10:1–10
Bateman MD, Hitchens S, Murton JB et al. (2014) The evolution of periglacial patterned ground of central East Anglia, UK. J Quat Sci 29:301–317
Bodin X, Schoeneich P, Deline P et al (2015) Mountain permafrost and associated geomorphological processes: recent changes in the French Alps. Rev géographie Alp 103–2
Bodin X, Thibert E, Fabre D et al (2009) Two decades of responses (1986–2006) to climate by the Laurichard rock glacier, French Alps. Permafr Periglac Process 20:331–344
Bradley RS, Jones PD (1992) Climate since 1500 A.D. Routeledge, London
Brewington S, Hicks M, Edwald A et al (2015) Islands of change vs. islands of disaster: managing pigs and birds in the Anthropocene of the North Atlantic. Holocene 25:1676–1684
Brown J, Ferrians OJ, Heginbottom JA, Melnikov ES (2001) Circum-Arctic map of permafrost and ground-ice conditions, version 2. National Snow and Ice Data Center, Boulder, Colorado (USA)
Buylaert JP, Ghysels G, Murray AS et al (2009) Optical dating of relict sand wedges and composite-wedge pseudomorphs in Flanders, Belgium. Boreas 38:160–175
Camarero JJ, García-Ruiz JM, Sangüesa-Barreda G et al (2015) Recent and intense dynamics in a formerly static Pyrenean Treeline. Arctic, Antarct Alp Res 47:773–783
Cebecauer T, Hofierka J (2008) The consequences of land-cover changes on soil erosion distribution in Slovakia. Geomorphology 98:187–198
Chaix A (1923) Les coulées de blocs du Parc National Suisse d'Engadine (Note préliminaire). Le Globe 62:1–35
Charton J, Verfaillie D, Jomelli V, Francou B (2021) Early Holocene rock glacier stabilisation at col du Lautaret (French Alps): palaeoclimatic implications. Geomorphology 394:107962. https://doi.org/10.1016/j.geomorph.2021.107962
Christiansen HH, Etzelmüller B, Isaksen K et al (2010) The thermal state of permafrost in the nordic area during the international polar year 2007–2009. Permafr Periglac Process 21:156–181
Christiansen HH, Gilbert GL, Demidov N et al (2019) Permafrost temperatures and active layer thickness in Svalbard during 2017/2018 (PermaSval). State Environ Sci Svalbard 2018:236–249
Chueca J, Julián A (2011) Besiberris glacigenic rock glacier (Central Pyrenees, Spain): mapping surface horizontal and vertical movement (1993–2003). Cuad Investig Geogr 37:7–24
Church MJ, Dugmore AJ, Mairs KA et al (2007) Charcoal production during the norse and early medieval periods in Eyjafjallahreppur, southern Iceland. Radiocarbon 49:659–672
Çiner A, Sarıkaya MA, Yıldırım C (2017) Misleading old age on a young landform? The dilemma of cosmogenic inheritance in surface exposure dating: Moraines vs. rock glaciers. Quat Geochronol 42:76–88
Clark PU, Dyke AS, Shakun JD et al (2009) The Last Glacial Maximum. Science (80-) 325:710–714
Colucci RR, Boccali C, Žebre M, Guglielmin M (2016a) Rock glaciers, protalus ramparts and pronival ramparts in the south-eastern Alps. Geomorphology 269:112–121

Colucci RR, Fontana D, Forte E et al (2016b) Response of ice caves to weather extremes in the southeastern Alps, Europe. Geomorphology 261:1–11

Cremaschi M, Zerboni A, Nicosia C et al (2015) Age, soil-forming processes, and archaeology of the loess deposits at the Apennine margin of the Po plain (northern Italy): New insights from the Ghiardo area. Quat Int 376:173–188

Czudek T (2005) Vývoj reliéfu krajiny České republiky v kvartéru. Moravské zemské muzeum, Brno (in Czech with German Summary)

Czekirda J, Westermann S, Etzelmüller B, Jóhannesson T (2019) Transient Modelling of Permafrost Distribution in Iceland. Frontiers in Earth Science 7, 1–23. https://doi.org/10.3389/feart.2019.00130

Debarbieux B, Oiry Varacca M, Rudaz G et al (2014) Tourism in mountain regions: hopes, fears and realities. Sustainable Mountain Development Series

Delaloye R, Lambiel C, Gärtner-Roer I (2010) Overview of rock glacier kinematics research in the Swiss Alps. Seasonal rhythm, interannual variations and trends over several decades. Geogr Helvetica 65/2:135–145

Deline P, Chiarle M, Curtaz M et al (2011) Chapter 3: Rockfalls. In Schoeneich P et al (eds): Hazards related to permafrost and to permafrost degradation. PermaNET project, state-of-the-art report 6.2. pp. 67–105

Denton GH, Alley RB, Comer GC, Broecker WS (2005) The role of seasonality in abrupt climate change. Quat Sci Rev 24:1159–1182

Denton GH, Anderson RF, Toggweiler JR et al (2010) The last glacial termination. Science (80-) 328:1652–1656

Dobiński W (2005) Permafrost of the Carpathian and Balkan Mountains, eastern and southeastern Europe. Permafr Periglac Process 16:395–398

Douglas TD, Harrison S (1996) Turf-Banked Terraces in Oraefi, Southeast Iceland: Morphometry, Rates of Movement, and Environmental Controls. Arctic and Alpine Research 28:228. https://doi.org/10.2307/1551764

Dramis F, Giraudi C, Guglielmin M (2003) Rock glacier distribution and paleoclimate in Italy. In: Philips M, Springman SM, Arenson LU (eds) Permafrost. Taylor & Francis, London, pp 199–204

Duvillard PA, Ravanel L, Schoeneich P et al (2021) Qualitative risk assessment and strategies for infrastructure on permafrost in the French Alps. Cold Reg Sci Technol 189:103311

Ehlers J, Gibbard PL, Hughes PD (2011) Quaternary glaciations-extent and chronology: a closer look. Elsevier, Amsterdam

Engel Z, Braucher R, Traczyk A, Leanni L (2014) Be-10 exposure age chronology of the last glaciation in the Krkonose Mountains, Central Europe. Geomorphology 206:107–121

Eriksen HØ, Rouyet L, Laukness TR et al (2018) Recent Acceleration of a Rock Glacier Complex, Ádjet, Norway, documented by 62 years of remote sensing observations. Geophys Res Lett 45:8314–8323. https://doi.org/10.1029/2018GL077605

Etzelmüller B, Farbrot H, Guomundsson Á et al (2007) The regional distribution of mountain permafrost in Iceland. Permafr Periglac Process 18:185–199

Etzelmüller B, Guglielmin M, Hauck C et al (2020) Twenty years of European mountain permafrost dynamics-the PACE legacy. Environ Res Lett 15

Fernández-Fernández JM, Palacios D, Andrés N et al (2020) Constraints on the timing of debris-covered and rock glaciers: an exploratory case study in the Hólar area, northern Iceland. Geomorphology 361:107196

Fernández-Fernández JM, Palacios D, Andrés N et al (2019) A multi-proxy approach to Late Holocene fluctuations of Tungnahryggsjökull glaciers in the Tröllaskagi peninsula (northern Iceland). Science of the Total Environment 664:499–517. https://doi.org/10.1016/j.scitotenv.2019.01.364

Feuillet T, Mercier D (2012) Post-little ice age patterned ground development on two pyrenean proglacial areas: from deglaciation to periglaciation. Geogr Ann Ser A Phys Geogr 94:363–376

Feurdean A, Persoiu A, Tantau I et al (2014) Climate variability and associated vegetation response throughout Central and Eastern Europe (CEE) between 60 and 8 ka. Quatern Sci Rev 106:206–224

Finsterwalder S (1928) Begleitworte zur Karte des Gepatschferners. Z Gletscherkunde 16:20–41

Francou B, Reynaud L (1992) 10 year surficial velocities on a rock glacier (Laurichard, French Alps). Permafr Periglac Process 3:209–213. https://doi.org/10.1002/ppp.3430030306

Frauenfelder R, Haeberli W, Hoelzle M, Maisch M (2001) Using relict rockglaciers in GIS-based modelling to reconstruct Younger Dryas permafrost distribution patterns in the Err-Julier area, Swiss Alps. Norsk Geogr Tidsskr 55:195–202

French HM (2007) The Periglacial Environment. Wiley

García-Hernández C, Ruiz-Fernández J, Sánchez-Posada C et al (2017) Reforestation and land use change as drivers for a decrease of avalanche damage in mid-latitude mountains (NW Spain). Glob Planet Change 153:35–50

García-Ruiz JM, Arnáez J, Sanjuán Y et al (2021) Landscape changes and land degradation in the subalpine belt of the Central Spanish Pyrenees. J Arid Environ 186:104396

García-Ruiz JM, Palacios D, González-Sampériz P et al (2016) Mountain glacier evolution in the Iberian Peninsula during the Younger Dryas. Quat Sci Rev 138:16–30

García-Ruiz JM, Valero-Garcés B, González-Sampériz P et al (2001) Stratified scree in the Central Spanish Pyrenees: palaeoenvironmental implications. Permafr Periglac Process 12:233–242

Giraudi C (2005) Late-Holocene alluvial events in the Central Apennines, Italy. Holocene 15(5):768–773

Gisnås K, Etzelmüller B, Lussana C et al (2017) Permafrost map for Norway, Sweden and Finland. Permafrost and Periglacial Processes 28:359–378

Gómez-Ortiz A, Oliva M, Salvador-Franch F et al (2014) Degradation of buried ice and permafrost in the Veleta cirque (Sierra Nevada, Spain) from 2006 to 2013 as a response to recent climate trends. Solid Earth 5:979–993

Gómez-Ortiz A, Oliva M, Salvador-Franch F et al (2019) Monitoring permafrost and periglacial processes in Sierra Nevada (Spain) from 2001 to 2016. Permafr Periglac Process. 30(4):278–291. https://doi.org/10.1002/PPP.2002

González-García M, Serrano E, Sanjosé JJ, González-Trueba JJ (2017) Surface dynamic of a protalus lobe in the temperate high mountain. Western Maladeta. Pyrenees. Catena 149:689–700

Gorbunov AP (2012) Rock glaciers, kurums, glaciers and permafrost in the mountains of Turkey. Earth Cryosph 16:3–8

Grove J (2004) Little ice ages: ancient and modern. Routledge, London

Gruber S, Haeberli W (2007) Permafrost in steep bedrock slopes and its temperatures-related destabilization following climate change. J Geophys Res Earth Surf 112:1–10

Hanáček M, Nývlt D, Jennings SJA (2021) Thermal basal regime of the Elsterian ice-sheet marginal zone in a hilly mountain foreland, Rychleby Mts., Eastern Sudetes. Boreas 50:582–605

Hatté C, Guiot J (2005) Palaeoprecipitation reconstruction by inverse modelling using the isotopic signal of loess organic matter: application to the Nussloch loess sequence (Rhine Valley, Germany). Clim Dynam 25:315–327

Heyman BM, Heyman J, Fickert T, Harbor JM (2013) Paleo-climate of the central European uplands during the last glacial maximum based on glacier mass-balance modeling. Quat Res 79:49–54

Hilger P, Hermanns RL, Czekirda J et al (2021) Permafrost as a first order control on long-term rock-slope deformation in (Sub-) Arctic Norway. Quaternary Science Reviews 251:106718, pp 1–21

Hippolyte JC, Bourlès D, Braucher R et al (2009) Cosmogenic 10Be dating of a sackung and its faulted rock glaciers, in the Alps of Savoy (France). Geomorphology 108:312–320

Hofmann FM, Rauscher F, McCreary W, Bischoff J-P, Preusser F (2020) Revisiting Late Pleistocene glacier dynamics north-west of the Feldberg, southern Black Forest, Germany. E&G Quat Sci J 69:61–87

Holuša J, Nývlt D, Woronko B, Matějka M, Stuchlík R (2022) Environmental factors controlling the Last Glacial multi-phase development of the Moravian Sahara dune field, Lower Moravian Basin, Central Europe. Geomorphology, 413:108355

Hrbáček F, Nývlt D, Láska K (2017) Active layer thermal dynamics at two lithologically different sites on James Ross Island, Eastern Antarctic Peninsula. Catena 149:592–602

Hughes P (2014) Little ice age glaciers in the Mediterranean mountains. Geophys Res Abstr 16:2014
Hughes PD (2018) Little ice age glaciers and climate in the Mediterranean mountains: a new analysis. Geogr Res Lett 44:15–45
Hughes DP, Woodward JC, Gibbard PL et al (2006) Quaternary glacial history of the Mediterranean mountains. Prog Phys Geogr 30:334–364
Hülle D, Hilgers A, Kühn P, Radtke U (2009) The potential of optically stimulated luminescence for dating periglacial slope deposits—a case study from the Taunus area, Germany. Geomorphology 109:66–78
Ishikawa T, Tokoro T, Miura S (2016) Influence of freeze–thaw action on hydraulic behavior of unsaturated volcanic coarse-grained soils. Soils Found 56:790–804
Ivy-Ochs S (2015) Variaciones glaciares en los Alpes europeos al final de la última glaciación. Cuad Investig Geogr 41:295–315
Ivy-Ochs S, Kerschner H, Maisch M et al (2009) Latest Pleistocene and Holocene glacier variations in the European Alps. Quaternary Sci Rev 28:2137–2149. https://doi.org/10.1016/j.quascirev.2009.03.009
Ivy-Ochs S, Monegato G, Reitner JM (2022) The Alps: glacial landforms from the Last Glacial Maximum. In: Palacios D, Hughes PD, García-Ruiz JM, Andrés N (eds) (2022) European Glacial Landscapes. Elsevier, pp 449–460
Jankovská V, Baroň I, Nývlt D, Krejčí O, Krejčí V (2018) Last Glacial to Holocene vegetation succession recorded in polyphase slope-failure deposits on the Malenik Ridge, Outer Western Carpathians. Quat Int 470:38–52
Jones DB, Harrison S, Anderson K, Whalley WB (2019) Rock glaciers and mountain hydrology: a review. Earth-Science Rev 193:66–90
Kapustová V, Panek T, Hradecky J et al (2018) Peat bog and alluvial deposits reveal land degradation during 16th- and 17th-century colonisation of the Western Carpathians (Czech Republic). Land Degrad Dev 29:894–906
Kasse C (1999) Late Pleniglacial and Late Glacial aeolian phases in The Netherlands. GeoArchaeoRhein 3:61–82
Kasse C (2002) Sandy aeolian deposits and their relation to climate during the Last Glacial Maximum and Lateglacial in northwest and central Europe. Prog Phys Geogr 26:507–532
Kędzia S (2014) Are there any active rock glaciers in the Tatra Mountains? Stud Geomorphol Carpatho Balc 48:5–16. https://doi.org/10.1515/sgcb-2015-0001
Kellerer-Pirklbauer A (2018) Solifluction rates and environmental controls at local and regional scales in central Austria. Norsk Geogr Tidsskr 72:37–56. https://doi.org/10.1080/00291951.2017.1399164
Kellerer-Pirklbauer A, Kaufmann V (2018) Deglaciation and its impact on permafrost and rock glacier evolution: new insight from two adjacent cirques in Austria. Sci Total Environ 621:1397–1414. https://doi.org/10.1016/j.scitotenv.2017.10.087
Kellerer-Pirklbauer A, Lieb GK, Avian M, Carrivick J (2012) Climate change and rock fall events in high mountain areas: numerous and extensive rock falls in 2007 at Mittlerer Burgstall, Central Austria. Geogr Ann A 94:59–78. https://doi.org/10.1111/j.1468-0459.2011.00449.x
Kern Z, Bočić N, Sipos G (2018) Radiocarbon-dated vegetal remains from the cave ice deposits of velebit mountain, Croatia. Radiocarbon 60:1391–1402
Kerschner H, Ivy-Ochs S (2008) Palaeoclimate from glaciers: examples from the Eastern Alps during the Alpine Lateglacial and early Holocene. Glob Planet Change 60:58–71
Kłapyta P (2013) Application of Schmidt hammer relative age dating to Late Pleistocene moraines and rock glaciers in the Western Tatra Mountains, Slovakia. Catena 111:104–121. https://doi.org/10.1016/j.catena.2013.07.004
Klimek K, Lanczont M, Nogaj-Chachaj J (2006) Historical deforestation as a cause of alluviation in small valleys, Subcarpathian loess plateau, Poland. Regional Environ Change 6:52–61
Kolstrup E (2004) Stratigraphic and environmental implications of a large ice-wedge cast at Tjæreborg, Denmark. Permafrost Periglac Process 15:31–40

Koster EA (2009) The "European Aeolian Sand Belt": geoconservation of drift sand landscapes. Geoheritage 1:93–110

Kotarba A (1991–1992) Reliktowe lodowce gruzowe jako element deglacjacji Tatr Wysokich. Stud Geomorphol Carpatho Balc 25–26:133–150 (in Polish)

Kotarba A (1995) Rapid mass wasting over the last 500 years in the High Tatra Mountains. Quest Geogr Spec Issue 4:177–183

Kotarba A, Pech P (2002) The recent evolution of talus slopes in the High Tatra Mountains (with the Pańszczyca valley as example). Stud Geomorphol Carpatho Balc 36:69–76

Krautblatter M, Huggel C, Deline P, Hasler A (2012) Research perspectives on unstable high-alpine bedrock permafrost: measurement, modelling and process understanding. Permafr Periglac Process 23(1):80–88. https://doi.org/10.1002/ppp.740

Krautblatter M, Verleysdonk S, Flores-Orozco A, Kemna A (2010) Temperature-calibrated imaging of seasonal changes in permafrost rock walls by quantitative electrical resistivity tomography (Zugspitze, German/Austrian Alps). J Geophys Res-Earth 115:F2. https://doi.org/10.1029/2008JF001209

Lindgren A, Hugelius G, Kuhry P, Christensen TR, Vandenberghe J (2016) GIS-based maps and area estimates of Northern Hemisphere permafrost extent during the last glacial maximum. Permafr Periglac Process 27:6–16

Linge H, Nesje A, Matthews JA, Fabel D, Xu S (2020) Evidence for rapid paraglacial formation of rock glaciers in southern Norway from 10Be surface-exposure dating. Quat Res 97:55–70

Linge H (2020) Evidence for rapid paraglacial formation of rock glaciers in southern Norway from 10 Be surface-exposure dating. https://doi.org/10.1017/qua.2020.10

Luterbacher J, Werner JP, Smerdon JE et al (2016) European summer temperatures since Roman times. Environ Res Lett 11:024001

Luterbacher J, Xoplaki E, Casty C et al (2006) Chapter 1 Mediterranean climate variability over the last centuries: A review. In: 148. pp 27–148

Magnin F, Josnin JY, Ravanel L et al (2017) Modelling rock wall permafrost degradation in the Mont Blanc massif from the LIA to the end of the 21st century. Cryosphere 11:1813–1834

Magyari EOK, Kunes P, Jakab G et al (2014) Late Pleniglacial vegetationn in eastern-central Europe: are there modern analogues in Siberia? Quat Sci Rev 95:60–79

Mann ME, Zhang Z, Rutherford S et al (2009) Global signatures and dynamical origins of the Little Ice Age and Medieval Climate Anomaly. Science (80-) 326:1256–1260

Margielewski W (2006) Records of the Late Glacial-Holocene palaeoenvironmental changes in landslide forms and deposits in the Beskid Makowski and Beskid Wyspowy Mts. Area (Polish Outer Carpathians). Folia Quat 76:1–149

Marks L, Bitinas A, Blaszkiewicz M et al (2022) Northern Central Europe: glacial landforms from the Last Glacial Maximum. In: Palacios D, Hughes PD, García-Ruiz JM, Andrés N (eds) (2022) European Glacial Landscapes. Elsevier, pp 381–388

Matsuoka N (2001) Solifluction rates, processes and landforms: a global review. Earth Sci Rev 55:107–134

Matthews JA, Wilson P (2015) Improved Schmidt-hammer exposure ages for active and relict pronival ramparts in southern Norway, and their palaeoenvironmental implications. Geomorphology 246:7–21

Mayewski PA, Rohling EE, Stager JC et al (2004) Holocene climate variability. Quat Res 62:243–255

Mentlik P, Engel Z, Braucher R, Leanni L, Aster Team (2013) Chronology of the Late Weichselian glaciation in the Bohemian Forest in Central Europe. Quat Sci Rev 65:120–128

Messerli B (1967) Die eiszeitliche und die gegenwartige Vergletscherung in Mittelmeerraum. Geogr Helv 22:105–228

Moska P, Sokołowski RJ, Jary Z et al (2022) Stratigraphy of the Late Glacial and Holocene aeolian series in different sedimentary zones related to the Last Glacial Maximum in Poland. Quat Int 630:65–83. https://doi.org/10.1016/j.quaint.2021.04.004

Nývlt D, Engel Z, Tyráček J (2011) Pleistocene Glaciations of Czechia. In: Ehlers J, Gibbard PL, Hughes PD (eds) Quaternary glaciations—extent and chronology, A closer look. Developments in Quaternary Science 15:37–46, Elsevier

Oliva M (2009) Holocene alpine environments in Sierra Nevada (southern Spain). University of Barcelona

Oliva M, Gómez Ortiz A, Salvador F et al (2014a) Long-term soil temperature dynamics in the Sierra Nevada, Spain. Geoderma 235–236:170–181

Oliva M, Gómez-Ortiz A, Salvador-Franch F et al (2016a) Inexistence of permafrost at the top of the Veleta peak (Sierra Nevada, Spain). Sci Total Environ 550:484–494

Oliva M, Ortiz AG, Franch FS, Catarineu MS (2014b) Present-day solifluction processes in the semi-arid range of Sierra Nevada (Spain). Arctic, Antarct Alp Res 46:365–370

Oliva M, Palacios D, Fernández-Fernández JM et al (2019) Late Quaternary glacial phases in the Iberian Peninsula. Earth-Science Rev 192:564–600

Oliva M, Ruiz-Fernández J, Barriendos M et al (2018a) The Little Ice Age in Iberian mountains. Earth-Science Rev 177:175–208

Oliva M, Schulte L, Gómez-Ortiz A (2009) Morphometry and Late Holocene activity of solifluction landforms in the Sierra Nevada, southern Spain. Permafr Periglac Process 20:369–382

Oliva M, Schulte L, Ortiz AG (2011) The role of aridification in constraining the elevation range of Holocene solifluction processes and associated landforms in the periglacial belt of the Sierra Nevada (southern Spain). Earth Surf Process Landforms 36:1279–1291

Oliva M, Serrano E, Gómez-Ortiz A et al (2016b) Spatial and temporal variability of periglaciation of the Iberian Peninsula. Quat Sci Rev 137:176–199

Oliva M, Vieira G, Pina P et al (2014c) Sedimentological characteristics of ice-wedge polygon terrain in adventdalen (Svalbard) environmental and climatic implications for the late Holocene. Solid Earth 5:901–914

Oliva M, Žebre M, Guglielmin M et al (2018b) Permafrost conditions in the Mediterranean region since the Last Glaciation. Earth-Science Rev 185:397–436

Onaca A, Ardelean F, Urdea P, Magori B (2017a) Southern Carpathian rock glaciers: inventory, distribution and environmental controlling factors. Geomorphology 293:391–404. https://doi.org/10.1016/j.geomorph.2016.03.032

Onaca A, Urdea P, Ardelean AC, Șerban R, Ardelean F (2017b) Present-day periglacial processes in the Alpine Zone. In: Rădoane M, Vespremeanu-Stroe A(eds) Landform dynamics and evolution in Romania. Springer Geography, pp 147–176. https://doi.org/10.1007/978-3-319-32589-7_7

Opel T, Meyer H, Wetterich S et al (2018) Ice wedges as archives of winter paleoclimate: a review. Permafr Periglac Process 29:199–209

Pačes T, Dobrovolný P, Holeček J, Nývlt D, Rukavičková L (2017) Future water-rock interaction in deep repository of spent nuclear fuel. Proc Earth Planet Sci 17:100–103

Palacios D, de Andrés N, Gómez-Ortiz A, García-Ruiz JM (2017a) Evidence of glacial activity during the Oldest Dryas in the mountains of Spain. Geol Soc Spec Publ 433:87–110

Palacios D, de Andrés N, López-Moreno JI, García-Ruiz JM (2015) Late Pleistocene deglaciation in the upper Gállego Valley, central Pyrenees. Quat Res (United States) 83:397–414

Palacios D, García-Ruiz JM, Andrés N et al (2017b) Deglaciation in the central Pyrenees during the Pleistocene-Holocene transition: timing and geomorphological significance. Quat Sci Rev 162:111–127

Palacios D, Hughes PD, García-Ruiz JM, Andrés N (eds) (2022) European glacial landscapes. Elsevier

Palstra SWL, Wallinga J, Viveen W et al (2021) Cross-comparison of last glacial radiocarbon and OSL ages using periglacial fan deposits. Quat Geochronol 61:101128

Pánek T, Engel Z, Mentlík P et al (2016) Cosmogenic age constraints on post-LGM catastrophic rock slope failures in the Tatra Mountains (Western Carpathians). Catena 138:52–67. https://doi.org/10.1016/j.catena.2015.11.005

Pánek T, Hartvich F, Jankovská V et al (2014) Large Late Pleistocene landslides from the marginal slope of the Flysch Carpathians. Landslides 11:981–992

Pánek T, Mentlík P, Engel Z, Braucher R, Zondervan A, ASTER Team (2017) Late Quaternary sackungen in the highest mountains of the Carpathians. Quat Sci Rev 159:47–62
Pauli H, Gottfried M, Dullinger S et al (2012) Recent plant diversity changes on Europe's mountain summits. Science (80-) 336:353–355
PERMOS (2021) Swiss Permafrost Bulletin 2019/2020
Pétursson HG, Norðdahl H, Ingólfsson O (2015) Late Weischelian history of relative sea level changes in Iceland during a collapse and subsequent retreat of marine based ice sheet. Cuadernos de Investigación Geográfica 41 (2):261–277
Pillewizer W (1957) Untersuchungen an Blockströmen der Ötztaler Alpen. Geomorph Abhandl Geograph Inst FU Berlin (Otto-Maull-Festschrift) 5:37–50
Pinto-Correia T, Primdahl J, Pedroli B (2018) European landscapes in transition, Implications for policy and practice. Studies in landscape ecology, Cambridge University Press, Cambridge
Popescu R, Urdea P, Vespremeanu-Stroe A (2017) Deglaciation history of high massifs from the Romanian Carpathains: towards an Integrated View. In Rădoane M, Vespremeanu-Stroe A (eds) Landform dynamics and evolution in Romania. Springer Geography, pp 87–116. https://doi.org/10.1007/978-3-319-32589-7_7
Puţan R (2015) Analiza unor procese și forme periglaciare din bazinul superior al văii Capra din Munții Făgăraș. Unpublised PhD Thesis, Universitatea de Vest din Timişoara, 36 p (in Romanian)
Rączkowska Z (2007) Współczesna rzeźba peryglacjalna wysokich gór Europy. Pr Geogr Inst Geogr Przestrz Zagospod PAN 212:1–252 (in Polish)
Renssen H, Seppä H, Crosta X, Goosse H, Roche DM (2012) Global characterization of the Holocene Thermal Maximum. Quaternary Science Reviews 48:7–19. https://doi.org/10.1016/j.quascirev.2012.05.022
Rico I, Magnin F, López Moreno JI et al (2021) First evidence of rock wall permafrost in the Pyrenees (Vignemale peak, 3,298 m a.s.l., 42°46′16″N/0°08′33″W). Permafr Periglac Process 32(4):673–680
Ridefelt H, Åkerman J, Beylich A et al (2009) 56 years of solifluction measurements in the Abisko mountains, northern Sweden–analysis of temporal and spatial variations of slow soil surface movements. Geografiska Annaler Series A Physical Geography 91:215–232
Ruiz-Fernández J, García-Hernández C, del Fernández SC (2019) Holocene stratified scree of praón (Picos de Europa, cantabrian mountains) = Los derrubios estratificados holocenos de praón (Picos de Europa, montañas cantábricas). Cad Do Lab Xeol Laxe 41:23–46
Sæmundsson T, Petursson H (2018) Causes and triggering factors for large scale displacements in the Almenningar landslide area, in central North Iceland, in: European Geosciences Union General Assembly. Geophysical Research Abstracts: 6482–1.
Šafanda J, Kubík J (1992) Evidence of ground surface temperature changes from two boreholes in the Bohemian Massif. Palaeogeogr Palaeoclimatol Palaeoecol 98:199–208
Šafanda J, Rajver D (2001) Signature of the last ice age in the present subsurface temperatures in the Czech Republic and Slovenia. Global Planet Change 29:241–258
Schunke E, Zoltai SC (1988) Earth hummocks (thufur). In: Clark MJ (ed) Advances in periglacial geomorphology. Wiley, Chichester, pp 231–245
Sequeira CR, Rego FC, Montiel-Molina C, Morgan P (2019) Half-century changes in lulc and fire in two iberian inner mountain areas. Fire 2:1–22
Şerban RD, Onaca A, Şerban M, Urdea P (2019) Block stream characteristics in Southern Carpathians (Romania). CATENA 178:20–31. https://doi.org/10.1016/j.catena.2019.03.003
Serrano E, de Sanjosé JJ, González-Trueba JJ (2010) Rock glacier dynamics in marginal periglacial environments. Earth Surf Process Landforms 35:1302–1314
Serrano E, Gómez-Lende M, Belmonte Á et al (2018a) Ice Caves in Spain. Elsevier Inc.
Serrano E, Lende MG, Ignacio J et al (2019a) Periglacial environments and frozen ground in the central Pyrenean high mountain area: ground thermal regime and distribution of landforms and processes. Permafr Periglac Process 30:292–309
Serrano E, Oliva M, González-García M et al (2018b) Post-little ice age paraglacial processes and landforms in the high Iberian mountains: a review. L Degrad Dev 29:4186–4208

Serrano E, Sanjosé JJ, Gómez-Gutiérrez Á, Gómez-Lende M (2019b) Surface movement and cascade processes on debris cones in temperate high mountain (Picos de Europa, northern Spain). Sci Total Environ 649:1323–1337

Schirmer W (1999) Dune phases and fossil soils in the European sand belt. In: Schirmer W (ed) Dunes and fossil soils. – GeoArchaeoRhein 3:11–42

Sobik M, Błaś M, Migała K, Godek M, Nasiółkowski T (2019) Klimat. In: Knapik R, Migoń P, Raj A (eds) Przyroda Karkonoskiego Parku Narodowego. Karkonoski Park Narodowy, Jelenia Góra, pp 147–186.

Stevens T, Sechi D, Bradák B et al. (2020) Abrupt last glacial dust fall over southeast England associated with dynamics of the British-Irish ice sheet. Quat Sci Rev 250:106641

Strand SM, Christiansen HH, Johansson M, Åkerman J, Humlum O (2020) Active layer thickening and controls on interannual variability in the Nordic Arctic compared to the circum-Arctic. Permafrost and Periglacial Processes 32:47–58

Speranza A, Hanke J, van Geel B, Fanta J (2000) Late-Holocene human impact and peat development in the Černá Hora bog, Krkonoše Mountains, Czech Republic. Holocene 10:575–585

Steinemann O, Reitner JM, Ivy-Ochs S, Christl M, Synal HA (2020) Tracking rockglacier evolution in the Eastern Alps from the Lateglacial to the early Holocene. Quatern Sci Rev 241:106424. https://doi.org/10.1016/j.quascirev.2020.106424

Štěpančíková P, Hók J, Nývlt D et al (2010) Active tectonics research using trenching technique on the south-eastern section of the Sudetic Marginal Fault (NE Bohemian Massif, central Europe). Tectonophysics 485:269–282

Štěpančíková P, Rockwell TK, Stemberk J et al (2022) Acceleration of Late Pleistocene activity of a Central European fault driven by ice loading. Earth Planet Sci Lett 591:117596

Stroeven AP, Hättestrand C, Kleman J et al (2016) Deglaciation of Fennoscandia. Quat Sci Rev 147:91–121

Tanarro LM, Palacios D, Andrés N et al (2019) Unchanged surface morphology in debris-covered glaciers and rock glaciers in Tröllaskagi peninsula (northern Iceland). Science of the Total Environment 648:218–235. https://doi.org/10.1016/j.scitotenv.2018.07.460

Tylmann K, Rinterknecht VR, Wozniak PP et al (2019) The Local Last Glacial Maximum of the southern Scandinavian Ice Sheet front: cosmogenic nuclide dating of erratics in northern Poland. Quat Sci Rev 219:36–46

Tylmann K, Uścinowicz S (2022) Timing of the last deglaciation phase in the southern Baltic area inferred from Bayesian age modeling. Quat Sci Rev, 287:107563

Ulfstedt AC (1993) Solifluction in the Swedish mountains: distribution in relation to vegetation and snow cover. In: Frenzel B, Matthews JA, Gläser B (eds) Solifluction and climatic variation in the Holocene. Gustav Fischer Verlag, Stuttgart, pp 217–224

Urdea P (1998) Rock glaciers and permafrost reconstruction in the Southern Carpathians Mountains, Romania. In: Permafrost—seventh international conference proceedings, Yellowknife, Canada. Collection Nordicana, 57, Univ. Laval, pp 1063–1069

Urdea P (2000) Munţii Retezat. Studiu geomorfologic. Edit, Academiei Române, Bucureşti (in Romanian)

Urdea P, Ardelean F, Ardelean M, Onaca A (2022) The Romanian Carpathians: glacial landforms from the Last Glacial Maximum (29–19 ka). In: Palacios D, Hughes PD, García-Ruiz JM, Andrés N (eds) (2022) European Glacial Landscapes. Elsevier, pp 441–448

Uxa T, Mida P (2017) Rock glaciers in the Western and High Tatra Mountains, Western Carpathians. J Maps 13(2):844–857. https://doi.org/10.1080/17445647.2017.1378136

Vandenberghe J, French HM, Gorbunov A et al (2014) The Last Permafrost Maximum (LPM) map of the Northern Hemisphere: permafrost extent and mean annual air temperatures, 25–17ka BP. Boreas 43:652–666

Veit H (1988) Fluviale und solifluidale morphodynamik des spät- und postglazials in einem zentralalpinen flusseinzugsgebiet (Südliche Hohe Tauern, Osttirol). Bayreuther Geowissenschaftliche Arbeiten 13:1–167

Veit H (2002) Die Alpen - Geoökologie und Landschaftsentwicklung. Verlag Eugen Ulmer, Stuttgart

Vespremeanu-Stroe A, Urdea P, Popescu R, Vasile M (2012) Rock glacier activity in the Retezat Mountains, Southern Carpathians, Romania. Permaf Permafr Periglac Process 23:127–137. https://doi.org/10.1002/ppp.1736

Vieira G, Mora C, Ramos M, Benett M (2003) Ground temperature regimes and geomorphological implications in a Mediterranean mountain (Serra da Estrela, Portugal). Geomorphology 52:57–72

Vietoris L (1972) Über die Blockgletscher des Äußeren Hochebenkars. Z Gletscherk Glazialgeol 8:169–188

Waller RI, Phillips E, Murton JB et al. (2011) Sand intraclasts as evidence of subglacial deformation of Middle Pleistocene permafrost, north Norfolk, UK. Quat Sci Rev 30:3481–3500

Wangensteen B, Guðmundsson Á, Eiken T et al (2006) Surface displacements and surface age estimates for creeping slope landforms in Northern and Eastern Iceland using digital photogrammetry. Geomorphology 80:59–79

Waroszewski J, Sprafke T, Kabala C et al (2020) Chronostratigraphy of silt-dominated Pleistocene periglacial slope deposits on Mt. Ślęża (SW, Poland): palaeoenvironmental and pedogenic significance. Catena 190:104549

Webb R (1972) Vegetation cover on Icelandic thúfur. Acta Bot Islandica 1:51–60

West RG (2015) Evolution of a Breckland Landscape: Chalkland under a Cold Climate in the Area of Beachamwell, Norfolk. Suffolk Naturalists's Society, Ipswich

Wilson P, Matthews JA, Mourne RW (2017) Relict blockstreams at Insteheia, Valldalen-Tafjorden, southern Norway: Their nature and Schmidt-hammer exposure age. Permafrost and Periglacial Processes 28:286–297

Woodward JJ (2009) The physical geography of the Mediterranean lands. Ecosyst World 11:663

Yoshikawa K, Nakamura T (1996) Pingo growth ages in the delta area, Adventdalen, Spitsbergen. Polar Rec (gr Brit) 32:347–352

Zasadni J, Kłapyta P, Broś E et al (2020) Latest Pleistocene glacier advances and post-Younger Dryas rock glacier stabilization in the Mt. Kriváň group, High Tatra Mountains, Slovakia. Geomorphology 358. https://doi.org/10.1016/j.geomorph.2020.107093

Zasadni J, Klapyta P, Makos M (2022) The Tatra Mountains: glacial landforms from the Last Glacial Maximum. In: Palacios D, Hughes PD, García-Ruiz JM, Andrés N (eds) (2022) European Glacial Landscapes. Elsevier, pp 435–440

Zeeberg J (1998) The European sand belt in eastern Europe—and comparison of Late Glacial dune orientation with GCM simulation results. Boreas 27:127–139

Zeeden C, Hambach U, Veres D et al (2018) Millennial scale climate oscillations recorded in the Lower Danube loess over the last glacial period. Palaeogeogr Palaeoclimatol Palaeoecol 509:164–181

Zhang T (2005) Influence of the seasonal snow cover on the ground thermal regime: an overview. Rev Geophys 43:RG4002

Correction to: Periglacial Landscapes of Europe

Marc Oliva, Daniel Nývlt, and José M. Fernández-Fernández

Correction to:
M. Oliva et al. (eds.), *Periglacial Landscapes of Europe*, https://doi.org/10.1007/978-3-031-14895-8

The original version of the book was inadvertently published with missing letter 'a' in section heading "Climtic Frmework of Europe During the Last Glacil Cycle" in Chapter 2, which has now been corrected to "Climatic Framework of Europe During the Last Glacial Cycle", and the following belated correction in Chapter 6 has been incorporated: The number of the funding "P1-001", has been changed as "P1-0419" in Page "107". The book has been updated with the changes.

The updated original versions of these chapters can be found at
https://doi.org/10.1007/978-3-031-14895-8_2
https://doi.org/10.1007/978-3-031-14895-8_6

M. Oliva et al. (eds.), *Periglacial Landscapes of Europe*,
https://doi.org/10.1007/978-3-031-14895-8_17